Handbook of the Thermodynamics of Organic Compounds

Richard M. Stephenson
Professor of Chemical Engineering
University of Connecticut
Storrs, Connecticut

Stanislaw Malanowski
Institute of Physical Chemistry
Polish Academy of Science
Warsaw, Poland

SECTION ON
Vapor-Liquid Critical Constants of Fluids
D. Ambrose
Department of Chemistry
University College London
London, England

Elsevier

New York • Amsterdam • London

Elsevier Science Publishing Co., Inc.
52 Vanderbilt Avenue, New York, New York 10017

Distributors outside the United States and Canada:

Elsevier Applied Science Publishers Ltd.
Crown House, Linton Road, Barking, Essex IG11 8JU, England

Library of Congress Cataloging in Publication Data

Stephenson, Richard Montgomery, 1917–
    Handbook of the thermodynamics of organic compounds.

    Bibliography: p.
    1. Thermodynamics—Tables.   2. Organic coumpounds—
Tables.   3. Organic compounds—Tables.   I. Malanowski,
Stanislaw.   II. Ambrose, D. (Douglas)   III. Title.
QD504.S735   1987   547.1′369   87-10042
ISBN 0-444-01240-0

Current printing (last digit)
10 9 8 7 6 5 4 3 2 1

Manufactured in the United States of America

Handbook of the Thermodynamics of Organic Compounds

Contents

Foreword

This book brings together data from Czechoslovakia on vapor pressures, data from England on critical properties, and data from America on physical properties of organic and organometallic compounds to provide a basic reference book for engineers and scientists involved with research and design in the chemical and petroleum industries.

We would like to acknowledge Jaroslav Dykyj, Milan Repas, and Josef Svoboda of Czechoslovakia for providing the material on Antoine constants and Douglas Ambrose of the University of London for providing the material on critical properties. Stanislaw Malanowski pointed out and made available the sources of data from Eastern Europe. Richard Stephenson translated and correlated the data in tabular form.

We would like to thank Dr. Matej Andras of the Slovenska Literarna Agentura for granting permission to use the data from Czechoslovakia and Dr. Marjan Bace of Elsevier Science Publishing Co., Inc., who encouraged preparation of this manuscript and handled the publishing arrangements. Particular thanks go to Mary Stephenson for typing the entire camera-ready copy.

Richard M. Stephenson
University of Connecticut
Storrs, Connecticut

Stanislaw Malanowski
Institute of Physical Chemistry
Warsaw, Poland

Introduction

All scientific and engineering calculations are dependent on the availability of thermodynamic and physical property data for the materials or systems in question. This dependency is particularly true in engineering design, which relies almost exclusively on computers for accurate data to produce meaningful final designs.

Because industry cannot afford the time nor money to make experimental measurements of quantities needed for design calculations most designers are limited to standard reference books. For this reason, we have compiled basic design information for a large number of organic and organometallic compounds. Much of this information is not readily available elsewhere.

We have arbitrarily divided into two general classes: properties of organic compounds, which includes all compounds made up of carbon, hydrogen, oxygen, nitrogen, sulfur, phosphorus, and the halogens; properties of organometallic compounds, which includes all compounds containing any other chemical element such as silicon, boron, and aluminum.

For each compound listed in the table we give, where available, the following basic data:

1. The name of the compound, including important synonyms.

2. The *Chemical Abstracts* entry number, which identifies the particular compound.

3. Toxicity and hazard warnings.

4. The melting point (*T*m) at one atmosphere pressure, given in kelvins. In some cases we give the freezing point (*T*f), which is usually substantially lower than the melting point.

5. The boiling point (*T*b) at one atmosphere pressure, given in kelvins.

6. The liquid molar volume (*V*m) calculated from the density. It is given in cubic meters per kilomole (m^3/kmol) at a particular temperature given in kelvins.

7. The critical temperature (*T*c) is the temperature above which a gas cannot be liquified. It is given in kelvins.

8. The critical pressure (*P*c) is the minimum pressure required for liquefaction at the critical temperature. It is given in kilopascals (kPa).

9. The critical volume (*V*c) is the volume occupied by one kilomole of the substance at the critical temperature and pressure. It is given in cubic meters per kilomole (m^3/kmol).

The Antoine constants, A, B, and C, are the constants for the Antoine equation

$$\log_{10} P = A - \frac{B}{C + T}$$

or

$$T = \frac{B}{A - \log_{10} P} - C$$

where *P* is the saturated vapor pressure in kilopascals (kPa) and *T* is the temperature in kelvins. In most cases the Antoine constants are for the liquid (liq), but in some cases the constants are for the solid (sol). The listed deviation (dev) gives the estimated average difference in degrees Kelvin between the experimental values of temperature and values calculated using the given values of A, B, and C. The range is the temperature range in kelvins over which the given values of A, B, and C are valid.

SOURCES OF DATA

The best and most comprehensive source of physical property data for organics is the *Dictionary of Organic Compounds* edited by J. Buckingham.[2] His recently published *Dictionary of Organometallic Compounds*[3] is the best source for the organometallics. The *CRC Handbook of Data on Organic Compounds*[9] is a useful source, although it is little different from the *CRC Handbook of Chemistry and Physics*. Much of the data comes from Beilstein. The *Merck Index*[7] is particularly good for medicinal chemicals. *Kirk-Othmer*[6] is useful for industrial chemicals, and the *Aldrich Catalog*[1] is a good source for shipping and handling requirements. The DIPPR[4]·project is extremely comprehensive, but it includes only some two hundred chemicals of greatest industrial interest. Sax[8] is by far the best source for toxicity and handling precautions.

Vapor pressure data are taken from the extremely comprehensive study by Dykyj, Repas, and Svoboda[5] published in Czechoslovakia in two volumes. This important survey of some five thousand organic compounds is practically unknown outside of eastern Europe. It is an extremely important source for vapor pressure data, which are of great practical interest to the chemical and petroleum industries.

Following the data on organic and organometallic compounds is a listing of vapor-liquid critical constants of fluids. The fluids are listed by formula.

REFERENCES

1. *Aldrich Catalog of Fine Chemicals*, Aldrich Chemical Co., Inc., Milwaukee, Wisconsin, 1986–1987.

2. Buckingham, J., editor, *Dictionary of Organic Compounds*, 5th Ed. (plus supplements), Chapman and Hall, New York, 1982.

3. Buckingham, J. editor, *Dictionary of Organometallic Compounds*, Chapman and Hall, New York, 1984.

4. Design Institute for Physical Property Data (DIPPR), *Data Compilation Tables of Properties of Pure Compounds*, American Institute of Chemical Engineers, New York, 1985.

5. Dykyj, Jaroslav, Repas, Milan, and Svoboda, Josef, *Tlak Nasytenej Pary Organickych Zlucenin*, Vydavatelstvo Slovenskej Akademie Vied, Bratislava, Czechoslovakia. Volume I, 1979; Volume II, 1984.

6. *Kirk-Othmer Encyclopedia of Chemical Technology*, 3rd Ed., John Wiley and Sons, Inc., New York, 1978.

7. *The Merck Index*, 10th Ed., Merck and Sons, Inc., Rahway, New Jersey, 1983.

8. Sax, N. Irving, *Dangerous Properties of Industrial Materials*, Van Nostrand Reinhold Co., New York, 1984.

9. Weast, Robert, and Astle, Melvin, editors, *CRC Handbook of Data on Organic Compounds*, CRC Press, Inc., Boca Raton, Florida, 1985.

Handbook of the Thermodynamics of Organic Compounds

Properties of Organic Compounds

CBrClF$_2$ MW 165.36

Bromochlorodifluoromethane, CA 353-59-3: Used as refrigerant and fire ex-
tinguisher. Tm = 112.15; Tb = 269.15; Vm = 0.0894 at 298.15; Tc = 426.88;
Antoine I (liq) A = 5.95024, B = 930.716, C = -33.26, dev = 0.1, range =
194 to 287; Antoine II (liq) A = 6.1309, B = 1020.262, C = -21.885, dev =
0.1, range = 268 to 324; Antoine III (liq) A = 6.34598, B = 1157.096, C =
-1.885, dev = 0.1, range = 321 to 403; Antoine IV (liq) A = 7.12694, B =
1829.935, C = 96.199, dev = 0.1, range = 403 to 427.

CBrCl$_3$ MW 198.27

Bromotrichloromethane, CA 75-62-7: Highly toxic, causes liver damage;
soluble ethanol, ether. Tm = 294.15; Tb = 377.85; Vm = 0.0985 at 293.15;
Antoine (liq) A = 5.99115, B = 1294.08, C = -53.15, dev = 0.1 to 1.0,
range = 294 to 443.

CBrFO   MW 126.91

Carbonic bromide fluoride, CA 753-56-0: Tm = 153.15; Tb = 255.65 to
263.15; Antoine (liq) A = 6.7434, B = 1196.7, C = 0, dev = 5.0, range =
197 to 256.

CBrF$_3$ MW 148.91

Bromotrifluoromethane, CA 75-63-8: Toxic; irritant; fire-extinguishing
agent, refrigerant; soluble chloroform. Tm = 105.15; Tb = 215.35; Vm =
0.0942 at 293.15; Tc = 340.15; Pc = 3960; Vc = 0.200; Antoine I (liq) A =
5.91830, B = 738.40, C = -26.64, dev = 0.1, range = 160 to 267; Antoine II
(liq) A = 6.41474, B = 983.612, C = 8.63, dev = 0.1, range = 276 to 340.

CBrN   MW 105.92

Cyanogen bromide, CA 506-68-3: Highly irritant; toxic; impure material
decomposes and may explode; soluble water, ethanol, ether. Tm = 325.15;
Tb = 334.15 to 335.15; Antoine (sol) A = 9.71635, B = 2697.49, C = 18.61,
dev = 1.0, range = 273 to 318.

CBrN$_3$O$_6$ MW 229.93

Bromotrinitromethane, CA 560-95-2: Irritant; can explode if heated. Tm =
290.15 to 291.15; Vm = 0.1132 at 293.15; Antoine (liq) A = 7.7901, B =
2496.32, C = 0, dev = 1.0, range = 318 to 335.

CBr$_2$F$_2$ MW 209.82

Dibromodifluoromethane, CA 75-61-6: Moderately irritant; fire-
extinguishing agent, refrigerant; soluble ethanol, ether, acetone, benzene.
Tm = 163.15; Tb = 297.65; Vm = 0.0916 at 293.15; Tc = 471.30; Antoine (liq)
A = 6.28693, B = 1185.98, C = -18.91, dev = 0.1, range = 247 to 297.

1

CBr$_3$F   MW 270.72

Tribromofluoromethane, CA 353-54-8: Irritant; soluble ethanol. Tm =
199.55; Tb = 381.11; Vm = 0.0981 at 293.15; Antoine (liq) A = 6.73669, B =
1793.439, C = -0.298, dev = 1.0, range = 315 to 380.

CBr$_4$ MW 331.63

Carbon tetrabromide, CA 558-13-4: Toxic; soluble ethanol, ether, chloroform; insoluble water. Tm = 363.25; Tb = 462.65, decomposes; Vm = 0.112 at
372.65; Antoine I (sol) A = 8.5116, B = 2841.4, C = 0, dev = 1.0 to 5.0,
range = 294 to 319; Antoine II (sol) A = 7.6919, B = 2578.9, C = 0, dev =
1.0 to 5.0, range = 320 to 329; Antoine III (liq) A = 5.30743, B = 1097.81,
C = -130.113, dev = 1.0, range = 369 to 463.

CClFO   MW 82.46

Carbonic chloride fluoride, CA 353-49-1: Toxic. Tm = 125.15; Tb = 228.15;
Antoine (liq) A = 7.2298, B = 1187.2, C = 0, dev = 1.0, range = 165 to 211.

CClF$_2$NO   MW 115.47

Difluorocarbamoyl chloride, CA 16847-30-6: Irritant. Tb = 268.15; Antoine
(liq) A = 7.03661, B = 1347.345, C = 0, dev = 1.0, range = 189 to 234.

CClF$_3$ MW 104.46

Chlorotrifluoromethane, CA 75-72-9: Narcotic in high concentrations;
refrigerant, fire-extinguishing agent. Tm = 92.15; Tb = 191.15; Vm =
0.0805 at 243.15; Tc = 302.05; Pc = 3870; Vc = 0.181; Antoine I (liq) A =
6.03488, B = 692.39, C = -19.81, dev = 0.1, range = 145 to 192; Antoine II
(liq) A = 5.99404, B = 681.375, C = -20.784, dev = 0.1, range = 133 to 185;
Antoine III (liq) A = 6.01518, B = 694.106, C = -18.568, dev = 0.1, range =
184 to 246; Antoine IV (liq) A = 6.38143, B = 863.583, C = 6.651, dev =
0.1, range = 243 to 271; Antoine V (liq) A = 7.52662, B = 1630.607, C =
112.164, dev = 0.1, range = 268 to 302.

CClF$_3$O   MW 120.46

Trifluoromethyl hypochlorite, CA 22082-78-6: Antoine I (liq) A = 6.538,
B = 1025, C = 0, dev = 5.0, range = 142 to 219; Antoine II (liq) A =
7.00026, B = 1159.205, C = 4.684, dev = 1.0 to 5.0, range = 160 to 226.

CClF$_3$O$_2$ MW 136.46

Peroxyhypochlorous acid, trifluoromethyl ester, CA 32755-26-3: Tm =
141.15; Tb = 251.15; Antoine (liq) A = 6.867, B = 1221, C = 0, dev = 5.0,
range = 163 to 296.

CClF$_3$O$_3$S   MW 184.52

Fluorosulfuric acid, chlorodifluoromethyl ester, CA 6069-31-4: Antoine
(liq) A = 5.36772, B = 745.576, C = -86.423, dev = 5.0, range = 228
to 310.

CClF$_3$O$_4$ MW 168.46

Perchloric acid, trifluoromethyl ester, CA 52003-45-9: Antoine (liq) A = 6.6077, B = 1301, C = 0, dev = 1.0, range not specified.

CClF$_3$S   MW 136.52

Methanesulfenylchloride, trifluoro, CA 421-17-0: Highly toxic. Tb = 272.45; Antoine (liq) A = 5.25548, B = 639.89, C = -75.65, dev = 1.0, range = 247 to 272.

CClF$_4$N   MW 137.46

Difluoro(difluorochloromethyl) amine, CA 13880-71-2: Irritant. Tm = 177.15; Antoine (liq) A = 6.83952, B = 1388.771, C = 0, dev = 1.0, range = 209 to 277.

CClF$_4$NO$_2$S   MW 201.52

Chloro(trifluoromethyl) sulfamoyl fluoride, CA 19419-95-5: Antoine (liq) A = 6.901, B = 1503, C = 0, dev = 5.0, range = 253 to 288.

CClF$_4$NO$_{12}$S$_4$ MW 457.70

Fluorosulfuric acid, bis[(fluorosulfonyl)oxy]amino chloromethylene ester, CA 53684-03-0: Antoine (liq) A = 7.955, B = 2520, C = 0, dev = 1.0 to 5.0, range not specified.

CClF$_7$S   MW 212.51

Tetrafluorochloro(trifluoromethyl) sulfur, CA 25030-42-6: Antoine (liq) A = 6.615, B = 1352, C = 0, dev = 5.0, range not specified.

CClN   MW 61.47

Cyanogen chloride, CA 506-77-4: Highly irritant; toxic; lachrymator; similar to HCN in effects; soluble water, ethanol, ether: Tf = 267.15; Tb = 286.95; Vm = 0.0518 at 293.15; Antoine (sol) A = 7.15381, B = 1232.34, C = -43.887, dev = 1.0, range = 196 to 259.

CCl$_2$FNO   MW 131.92

Dichlorocarbamic fluoride, CA 32751-02-3: Irritant; decomposes on standing. Tb = 344.15; Antoine (liq) A = 8.185, B = 2125.7, C = 0, dev = 1.0 to 5.0, range not specified.

CCl$_2$F$_2$ MW 120.91

Dichlorodifluoromethane, CA 75-71-8: Nonflammable refrigerant; narcotic in high concentrations; can react violently with aluminum or magnesium; insoluble water; soluble ethanol, ether, acetone. Tm = 118.15; Tb = 243.35; Vm = 0.0922 at 298.15; Tc = 385.15; Pc = 4110; Vc = 0.217; Antoine I (liq) A = 5.94677, B = 839.6, C = -30.311, dev = 0.1, range = 173 to *(continues)*

CCl_2F_2 *(continued)*

244; Antoine II (liq) A = 6.0058, B = 860.828, C = -28.11, dev = 0.1, range = 173 to 240; Antoine III (liq) A = 5.92289, B = 826.707, C = -32.274, dev = 0.1, range = 236 to 285; Antoine IV (liq) A = 6.30541, B = 1035.857, C = -1.496, dev = 0.1, range = 282 to 345; Antoine V (liq) A = 7.51271, B = 2016.711, C = 132.578, dev = 0.1, range = 341 to 385.

CCl_2F_3N MW 153.92

Methyl amine, *N,N*-dichloro-1,1,1-trifluoro, CA 13880-73-4: Irritant. Tb = 288.05; Antoine (liq) A = 6.65638, B = 1347.371, C = 0, dev = 1.0 to 5.0, range = 226 to 291.

CCl_2F_3NS MW 185.98

(Trifluoromethyl)imidosulfurous dichloride, CA 10564-47-3: Irritant. Tb = 362.15 to 363.15; Antoine (liq) A = 8.27066, B = 2724.4, C = 72.44, dev = 1.0, range = 283 to 362.

CCl_2F_3PS MW 202.95

Dichloro(trifluoromethylthio) phosphine, CA 18799-78-5: Irritant; toxic. Tb = 371.15; Antoine (liq) A = 6.4699, B = 1655, C = 0, dev = 5.0, range = 293 to 363.

CCl_2O MW 98.92

Phosgene, carbonyl chloride, CA 75-44-5: Irritant; highly toxic, used as military gas in first world war; important chemical raw material and intermediate; slightly soluble water; soluble benzene, acetic acid, chloroform. Tm = 145.31; Tb = 280.63; Vm = 0.0713 at 293.15; Tc = 455.15; Pc = 5680; Vc = 0.190; Antoine I (liq) A = 6.06819, B = 986.45, C = -37.88, dev = 0.1, range = 240 to 281; Antoine II (liq) A = 6.81263, B = 1428.299, C = 16.439, dev = 1.0, range = 280 to 341; Antoine III (liq) A = 6.37426, B = 1144.238, C = -19.373, dev = 1.0, range = 338 to 410; Antoine IV (liq) A = 6.58798, B = 1303.455, C = 4.738, dev = 1.0, range = 406 to 455.

CCl_3F MW 137.37

Trichlorofluoromethane, CA 75-69-4: Nonflammable refrigerant; narcotic at high concentrations; insoluble water; soluble ethanol, ether. Tm = 162.15; Tb = 296.97; Vm = 0.0931 at 298.15; Tc = 471.15; Pc = 4410; Vc = 0.247; Antoine I (liq) A = 5.99210, B = 1032.23, C = -37.85, dev = 0.1, range = 213 to 249; Antoine II (liq) A = 5.99652, B = 1034.048, C = -37.672, dev = 0.1, range = 213 to 301; Antoine III (liq) A = 6.03083, B = 1053.874, C = -34.955, dev = 0.1 to 1.0, range = 295 to 363; Antoine IV (liq) A = 6.36472, B = 1285.088, C = -0.653, dev = 0.1 to 1.0, range = 357 to 429; Antoine V (liq) A = 7.75501, B = 2744.806, C = 196.225, dev = 1.0, range = 424 to 468.

CCl_3F_2N MW 170.37

Difluoro(trichloromethyl) amine, CA 24708-52-9: Irritant. Tm = 252.15; Antoine (liq) A = 7.31983, B = 1745.398, C = 0, dev = 1.0 to 5.0, range = 252 to 325.

CCl_3F_2P MW 187.34

Difluoro(trichloromethyl) phosphine, CA 1112-03-4: Irritant. Tm = 288.95 to 289.55; Tb = 346.25; Antoine I (sol) A = 7.6679, B = 1920, C = 0, dev = 5.0, range = 264 to 283; Antoine II (liq) A = 6.9019, B = 1699, C = 0, dev = 5.0, range = 289 to 313.

CCl_3NO MW 148.38

Trichloronitrosomethane, CA 3711-49-7: Irritant; decomposes at room temperature. Tb = 330.15 to 331.15, decomposes; Vm = 0.1016 at 293.15; Antoine (liq) A = 7.085, B = 1690, C = 0, dev = 5.0, range = 253 to 333.

CCl_3NO_2 MW 164.38

Trichloronitromethane, chloropicrin, CA 76-06-2: Strong irritant; highly toxic; explosive; lachrymator; stable to water; can detonate by shock; fumigation insecticide; insoluble water; soluble benzene, ether, carbon disulfide. Tf = 204.15; Tb = 385.15; Vm = 0.0972 at 273.15; Antoine I (liq) A = 6.15825, B 1369.7, C = -55.15, dev = 0.1, range = 301 to 449; Antoine II (liq) A = 7.40016, B = 2054.3, C = 0, dev = 1.0, range = 273 to 333.

CCl_4 MW 153.82

Carbon tetrachloride, CA 56-23-5: Highly toxic by inhalation and skin contact; reacts violently with powdered aluminum; can decompose to phosgene on strong heating; nonflammable solvent used for dry cleaning, as grain fumigant, and organic raw material; soluble ethanol, ether; insoluble water. Tm = 251.95, highest melting of three solid forms; Tb = 349.87; Vm = 0.09425 at 273.15; Tc = 556.35; Pc = 4600; Vc = 0.276; Antoine I (sol) A = 8.214, B = 2027, C = 0, dev = 1.0 to 5.0, range = 208 to 225; Antoine II (sol) A = 7.946, B = 1975.3, C = 0, dev = 1.0 to 5.0, range = 226 to 248; Antoine III (liq) A = 5.99114, B = 1202.9, C = -48.01, dev = 0.1, range = 262 to 349; Antoine IV (liq) A = 5.97092, B = 1195.903, C = -48.217, dev = 0.1 to 1.0, range = 349 to 416; Antoine V (liq) A = 6.22882, B = 1392.458, C = -19.19, dev = 1.0, range = 412 to 497; Antoine VI (liq) A = 6.36976, B = 1439.651, C = -25.734, dev = 1.0, range = 494 to 555.

CFIO   MW 173.91

Carbonyl fluoride iodide: Tm = 183.15; Antoine (liq) A = 6.6133, B = 1366.6, C = 0, dev = 5.0, range = 230 to 292.

CFN   MW 45.02

Cyanogen fluoride, CA 1495-50-7: Tm = 201.15; Antoine I (sol) A = 8.9229, B = 1508, C = 0, dev = 1.0, range = 147 to 191; Antoine II (liq) A = 7.1549, B = 1169.2, C = 0, dev = 1.0, range = 201 to 227.

$CFNO_3S$ MW 125.07

Sulfuryl fluoride isocyanate, CA 1495-51-8: Irritant; toxic. Antoine (liq) A = 7.68377, B = 1905.365, C = 0, dev = 1.0, range = 294 to 335.

$CFNO_6S_2$ MW 205.13

Pyrosulfuryl fluoride isocyanate, CA 27931-74-4: Irritant; toxic. Antoine (liq) A = 7.2599, B = 2134, C = 0, dev = 1.0, range = 330 to 405.

CFN_3O_6 MW 169.03

Fluorotrinitromethane, CA 1840-42-2: Irritant; lachrymator; oxidant for liquid rocket propellants. Tm = 244.15; Tb = 356.15 to 357.15; Vm = 0.1063 at 293.15; Antoine (liq) A = 6.9436, B = 1785.71, C = 0, dev = 1.0, range = 274 to 358.

CF_2NO_2P MW 126.99

Difluorophosphoryl isocyanate, CA 1495-54-1: Irritant; toxic. Antoine (liq) A = 7.52, B = 1880, C = 0, dev = 5.0, range from melting point to 341.

CF_2N_2 MW 78.02

Difluorocyanamide, CA 7127-18-6: Tm = 253.55; Antoine (sol) A = 7.095, B = 1075, C = 0, dev = 5.0, range = 179 to 198.

CF_2N_2OS MW 126.08

Cyanoimidosulfuryl fluoride, CA 19073-57-5: Tm = 244.65; Antoine (liq) A = 7.485, B = 1945, C = 0, dev = 5.0, range = 262 to 354.

$CF_2N_2O_4$ MW 142.02

Difluorodinitromethane, CA 1185-11-1: Oxidant for liquid rocket propellants. Tb = 307.15; Vm = 0.0910 at 283.15; Antoine (liq) A = 8.995, B = 2163.9, C = 0, dev = 1.0 to 5.0, range = 283 to 310.

CF_2N_2S MW 110.08

N-Cyano-S,S-difluorosulfilimine, CA 14453-41-9: Tm = 240.65; Antoine (liq) A = 8.11, B = 2302, C = 0, dev = 5.0, range = 271 to 320.

CF_2O MW 66.01

Carbonyl fluoride, CA 353-50-4: Irritant; highly toxic by inhalation; hygroscopic; easily decomposed by water. Tm = 159.15; Tb = 190.05; Vm = 0.0580 at 159.15; Antoine I (sol) A = 6.65565, B = 717.5, C = -32.95, dev = 5.0, range = 130 to 159; Antoine II (liq) A = 6.06499, B = 591.84, C = -42.77, dev = 1.0, range = 159 to 189.

CF_2O_4S MW 146.07

Fluoroformyl fluorosulfate: Antoine (liq) A = 6.61678, B = 1511.3, C = 8.45, dev = 1.0, range = 250 to 296.

CF$_2$S   MW 82.07

  Thiocarbonyl fluoride, CA 420-32-6: Tm = 109.65; Tb = 219.15; Antoine I
  (liq) A = 6.765, B = 1002, C = 0, dev = 5.0, range = 133 to 211; Antoine II
  (liq) A = 6.3069, B = 908, C = 0, dev = 1.0 to 5.0, range = 178 to 211.

CF$_3$I   MW 195.91

  Iodotrifluoromethane, CA 2314-97-8: Fire-extinguishing agent. Tb =
  250.65; Vm = 0.0830 at 241.15; Antoine (liq) A = 6.6914, B = 1174.29, C = 0,
  dev not specified, range = 188 to 296.

CF$_3$NO   MW 99.01

  (Difluoroamino) carbonyl fluoride, CA 2368-32-3: Highly toxic; strong
  oxidizing agent. Tm = 120.95; Tb = 221.15; Antoine (liq) A = 7.1089, B =
  1129, C = 0, dev = 1.0, range = 143 to 217.

  Trifluoronitrosomethane, CA 334-99-6: Synthetic rubber intermediate.
  Tm = 76.55; Tb = 189.15; Antoine (liq) A = 6.799, B = 895.86, C = 0, dev =
  1.0 to 5.0, range = 141 to 174.

CF$_3$NOS   MW 131.07

  S,S-Difluoro-N-(fluoroformyl)-sulfilimine, CA 3855-41-2: Highly toxic.
  Tm = 178.15; Tb = 327.15; Antoine (liq) A = 8.0655, B = 1950, C = 0, dev =
  1.0, range = 220 to 323.

  (N-Sulfinyl)-trifluoromethylamine, CA 10564-49-5: Tb = 291.15; Antoine
  (liq) A = 6.858, B = 1413, C = 0, dev = 1.0, range = 239 to 289.

  Thionitrous acid, S-(trifluoromethyl) ester, CA 24892-54-4: Decomposes
  rapidly above 243; Tm = below 157.15; Tb = 270.15; Antoine (liq) A = 7.005,
  B = 1350, C = 0, dev = 5.0, range not specified.

CF$_3$NO$_2$ MW 115.01

  Trifluoronitromethane, fluoropicrin, CA 335-02-4: Tb = 253.15; Antoine
  (liq) A = 6.666, B = 1128.3, C = 0, dev = 1.0, range = 238 to 243.

CF$_3$NO$_4$ MW 147.01

  (Trifluoromethyl) peroxynitrate, CA 50311-48-3: Antoine (liq) A = 6.74837,
  B = 1297.361, C = 0, dev = 1.0 to 5.0, range = 193 to 247.

CF$_3$NO$_6$S$_2$ MW 243.13

  N-(Fluoroformyl)-N,O-bis-(fluorosulfonyl) hydroxylamine, CA 19252-48-3:
  Antoine (liq) A = 6.7944, B = 1895, C = 0, dev = 1.0, range = 325 to 392.

CF$_4$ MW 88.00

  Carbon tetrafluoride, CA 75-73-0: Narcotic in high concentrations; can
  react violently with aluminum powder; low temperature   *(continues)*

CF_4 *(continued)*

refrigerant; gaseous insulator; slightly soluble water; soluble organic solvents. $Tm = 89.55$; $Tb = 145.09$; $Vm = 0.0546$ at 143.15; $Tc = 227.55$; $Pc = 3740$; $Vc = 0.140$; Antoine I (sol, *beta*) $A = 8.9729$, $B = 876.5$, $C = 0$, $dev = 1.0$, range = 70 to 76; Antoine II (sol, *alpha*) $A = 7.6289$, $B = 770.0$, $C = 0$, $dev = 1.0$, range = 76 to 90; Antoine III (liq) $A = 5.9617436$, $B = 511.69474$, $C = -15.7745$, $dev = 0.02$, range = 90 to 146; Antoine IV (liq) $A = 5.96254$, $B = 513.129$, $C = -15.474$, $dev = 0.02$ to 0.1, range = 89 to 163; Antoine V (liq) $A = 6.23758$, $B = 599.591$, $C = -3.252$, $dev = 0.1$, range = 160 to 197; Antoine VI (liq) $A = 6.99759$, $B = 936.128$, $C = 45.844$, $dev = 0.1$, range = 195 to 227.

CF_4N_2O MW 132.02

Fluoro(trifluoromethyl) diimidoxide, CA 815-10-1: Antoine (liq) $A = 5.70258$, $B = 744.07$, $C = -64.25$, $dev = 1.0$, range = 233 to 267.

$CF_4N_2O_3S_2$ MW 228.14

Carbonylbis(imidosulfuryl fluoride), CA 25523-80-2: $Tm = 249.65$; Antoine (liq) $A = 7.5069$, $B = 2159$, $C = 0$, $dev = 1.0$, range = 316 to 331.

CF_4O MW 104.00

Hypofluorous acid, trifluoromethyl ester, CA 373-91-1: Strong oxidizing agent. $Tm =$ below 58.15; $Tb = 178.15$; Antoine (liq) $A = 6.19992$, $B = 689.23$, $C = -13.78$, $dev = 1.0$, range = 153 to 194.

CF_4OS MW 136.06

Trifluoromethane sulfinyl fluoride, CA 812-12-4: Antoine (liq) $A = 5.56188$, $B = 804.05$, $C = -44.8$, $dev = 1.0$ to 5.0, range = 204 to 271.

CF_4O_2 MW 120.00

Hydroperoxyfluoric acid, trifluoromethyl ester, CA 34511-13-2: $Tm =$ below 77.15; Antoine (liq) $A = 6.16405$, $B = 749.708$, $C = -23.301$, $dev = 1.0$, range = 156 to 203.

CF_4O_2S MW 152.06

Trifluoromethane sulfonyl fluoride, CA 335-05-7: $Tb = 251.45$; Antoine (liq) $A = 6.861$, $B = 1221$, $C = 0$, $dev = 1.0$, range = 226 to 249.

$CF_4O_5S_2$ MW 232.12

Anhydride fluorosulfonic acid and trifluoromethane sulfonic acid, CA 21595-44-8: Antoine (liq) $A = 7.043$, $B = 1721$, $C = 0$, $dev = 1.0$, range = 308 to 338.

CF$_5$N   MW 121.01

Pentafluoro methylamine, CA 335-01-3: Tm = 141.15; Tb = 195.15; Antoine I
(sol) A = 7.075, B = 970, C = 0, dev = 5.0, range = 128 to 141; Antoine II
(liq) A = 6.5989, B = 901, C = 0, dev = 1.0, range = 151 to 198.

CF$_5$NO   MW 137.01

Pentafluoro methoxyamine, CA 4217-93-0: Tm = 93.15 to 113.15; Tb = 213.15;
Antoine (liq) A = 6.4999, B = 968.6, C = 0, dev = 1.0, range = 167 to 210.

CF$_5$OPS   MW 186.04

Difluorothiophosphoric acid, S-(trifluoromethyl) ester, CA 52752-66-6:
Antoine (liq) A = 6.0269, B = 1207.4, C = 0, dev = 1.0 to 5.0, range not
specified.

CF$_5$O$_3$P   MW 185.98

Difluoroperoxyphosphoric acid, trifluoromethyl ester, CA 39125-42-3: Tm =
184.55; Antoine (liq) A = 7.8019, B = 1672.8, C = 0, dev = 1.0 to 5.0,
range = 241 to 280.

CF$_6$N$_2$S$_2$ MW 218.13

Difluoromethane bis(S,S-difluorosulfilimine), CA 17686-45-2: Tm = 188.15;
Antoine (liq) A = 7.206, B = 1880, C = 0, dev = 5.0, range = 230 to 313.

CF$_6$PS   MW 189.04

Difluoro(trifluoromethylthio) phosphine, CA 52752-65-5: Antoine (liq) A =
6.2109, B = 1164.2, C = 0, dev = 1.0 to 5.0, range not specified.

CF$_8$OS   MW 212.06

Pentafluoro(trifluoromethoxy) sulfur, CA 1873-23-0: Tm = 130.15; Antoine
(liq) A = 6.16109, B = 949.88, C = -33.622, dev = 1.0, range = 217 to 262.

CF$_8$S   MW 196.06

(Pentafluorothio)methane, trifluoro, CA 373-80-8: Tm = 186.25; Tb =
253.15; Antoine (liq) A = 5.96611, B = 849.09, C = -38.36, dev = 1.0,
range = 205 to 262.

CF$_9$NOS   MW 245.06

Tetrafluoro(difluoroamino)(trifluoromethoxy) sulfur: Antoine (liq) A =
6.60149, B = 1294.535, C = -19.011, dev = 1.0, range = 257 to 298.

CF$_{10}$O$_5$S$_2$ MW 346.11

Carbonodicarboxato-decafluoro disulfur, CA 60672-59-5: Tm = 227.65;
Antoine (liq) A = 7.2147, B = 1988.4, C = 0, dev = 1.0 to 5.0, range not
specified.

CHBrF$_2$ MW 130.92

Bromodifluoromethane, CA 1511-62-2: Soluble ethanol. Tb = 258.65; Vm =
0.0844 at 289.15; Antoine (liq) A = 6.870, B = 1255, C = 0, dev = 1.0,
range = 194 to 259.

CHBr$_3$ MW 252.73

Tribromomethane, bromoform, CA 75-25-2: Highly toxic vapor; irritant;
lachrymator; slightly soluble water; soluble ethanol, ether, acetone,
benzene. Tm = 280.85; Tb = 422.65; Vm = 0.0874 at 293.15; Antoine (liq) A
= 6.20911, B = 1544.81, C = -54.77, dev = 0.1, range = 320 to 412.

CHClF$_2$ MW 86.47

Chlorodifluoromethane, CA 75-45-6: Refrigerant; can react exothermically
with aluminum powder, narcotic in high concentrations; soluble ethanol,
acetone, chloroform. Tm = 113.15; Tb = 232.4; Vm = 0.0724 at 298.15; Tc =
369.15; Pc = 4970; Vc = 0.165; Antoine I (liq) A = 6.33292, B = 919.834,
C = -19.718, dev = 0.1, range = 170 to 233; Antoine II (liq) A = 6.19138,
B = 863.436, C = -26.04, dev = 0.1 to 1.0, range = 230 to 275; Antoine III
(liq) A = 6.35713, B = 950.38, C = -13.474, dev = 0.1 to 1.0, range = 275
to 327; Antoine IV (liq) A = 7.13064, B = 1490.048, C = 64.627, dev = 1.0,
range = 324 to 366.

CHCl$_2$F$_2$ MW 102.92

Dichlorofluoromethane, CA 75-43-4: Moderately toxic; refrigerant; bac-
teriocide; insoluble water; soluble ethanol, ether, chloroform. Tm =
138.15; Tb = 282.07; Vm = 0.0760 at 298.15; Tc = 451.65; Pc = 5170; Vc =
0.197; Antoine I (liq) A = 6.22023, B = 1052.833, C = -32.317, dev = 0.1,
range = 225 to 282; Antoine II (liq) A = 6.02210, B = 957.338, C = -43.675,
dev = 0.1, range = 279 to 344; Antoine III (liq) A = 6.35759, B = 1156.802,
C = -15.644, dev = 0.1, range = 341 to 399; Antoine IV (liq) A = 7.66239,
B = 2394.666, C = 155.01, dev = 0.1 to 1.0, range = 397 to 450.

CHCl$_2$FO$_3$S   MW 182.98

Fluorosulfuric acid, dichloromethyl ester, CA 42016-50-2: Irritant. Tm =
207.15; Antoine (liq) A = 7.125, B = 1890, C = 0, dev not specified,
range = 275 to 293.

CHCl$_3$ MW 119.38

Trichloromethane, chloroform, CA 67-66-3: Toxic vapor; eye irritant; suspected carcinogen; can explode with powdered aluminum or magnesium; can explode with acetone if base is present; nonflammable solvent; important
chemical intermediate; slowly decomposes in air and light; slightly soluble
water; soluble ethanol, ether, benzene. Tm = 209.95; Tb =   *(continues)*

CHCl$_3$ *(continued)*

344.45; Vm = 0.0806 at 298.15; Tc = 536.55; Pc = 5450; Vc = 0.239; Antoine
I (liq) A = 6.07853, B = 1170.42, C = -46.98, dev = 0.1 to 1.0, range = 262
to 334; Antoine II (liq) A = 5.38327, B = 948.979, C = -61.73, dev = 1.0,
range = 227 to 269; Antoine III (liq) A = 6.11152, B = 1173.606, C = -48.54,
dev = 1.0, range = 333 to 416; Antoine IV (liq) A = 7.89882, B = 2879.244,
C = 161.978, dev = 1.0, range = 410 to 481; Antoine V (liq) A = 4.58922,
B = 181.802, C = -325.374, dev = 1.0, range = 479 to 523.

CHFI$_2$ MW 285.83

Diiodofluoromethane, CA 1493-01-2: Soluble ethanol, ether. Tm = 238.65;
Tb = 373.45; Vm = 0.0894 at 296.15; Antoine (liq) A = 6.355, B = 1720,
C = 0, dev = 1.0 to 5.0, range = 299 to 332.

CHFN$_2$O$_4$ MW 124.03

Fluorodinitromethane, CA 7182-87-8: Antoine (liq) A = 7.775, B = 2278.1,
C = 0, dev = 5.0, range = 298 to 338.

CHFO   MW 48.02

Formyl fluoride, CA 1493-02-3: Strong irritant. Tm = 131.15; Tb = 244.15;
Vm = 0.0404 at 243.15; Antoine (liq) A = 7.1989, B = 1273.4, C = 0, dev =
1.0, range = 178 to 235.

CHF$_2$I   MW 177.92

Difluoroiodomethane, CA 1493-03-4: Tm = 151.15; Tb = 294.75; Vm = 0.0549
at 254.15; Antoine (liq) A = 6.6108, B = 1358.3, C = 0, dev = 1.0, range =
227 to 287.

CHF$_3$ MW 70.01

Trifluoromethane, fluoroform, CA 75-46-7: Low temperature refrigerant;
slightly soluble water; soluble ethanol, acetone, benzene. Tm = 117.95;
Tb = 190.99; Vm = 0.0461 at 173.15; Tc = 298.85; Pc = 4830; Vc = 0.133;
Antoine I (sol) A = 10.20097, B = 1338.7, C = 0, dev = 1.0, range = 59 to
118; Antoine II (liq) A = 6.28009, B = 724.13, C = -21.6, dev = 1.0,
range = 138 to 190; Antoine III (liq) A = 6.64071, B = 892.116, C = 2.04,
dev = 1.0, range = 198 to 298.

CHF$_3$O$_2$ MW 102.01

Trifluoromethyl hydroperoxide, CA 16156-36-8: Tf = 198.15 to 199.15; Tb =
284.45; Vm = 0.0699 at 293.15; Antoine (liq) A = 7.6817, B = 1614.5, C = 0,
dev = 1.0, range = 248 to 285.

CHF$_3$O$_3$S   MW 150.07

Methanesulfonic acid, trifluoro, CA 1493-13-6: Corrosive; irritant; very
hygroscopic; very strong acid; soluble water, ether. Tm = 307.15; Tb =
435.15; Antoine (liq) A - 7.73233, B = 2492.152, C = 0, dev = 5.0, range =
354 to 435.

CHF$_3$S   MW 102.07

  Trifluoromethanethiol, CA 1493-15-8: Irritant. Tm = 116.03; Tb = 236.45;
  Antoine (liq) A = 6.25881, B = 922.28, C = -18.303, dev = 0.1, range = 167
  to 236.

CHF$_7$S   MW 178.07

  (Difluoromethyl) sulfur pentafluoride: Tm = 186.15; Antoine (liq) A =
  6.01731, B = 932.73, C = -45.76, dev = 1.0, range = 221 to 293.

CHI$_3$ MW 393.73

  Triiodomethane, iodoform, CA 75-47-8: Toxic; powerful antiseptic; nonflammable; decomposes slowly in light; almost insoluble water; soluble acetone,
  ether, benzene, chloroform. Tm = 392.15; Tb = 491.15; Antoine (sol) A =
  8.2566, B = 3650, C = 0, dev = 1.0, range = 308 to 365.

CHN   MW 27.03

  Hydrogen cyanide, hydrocyanic acid, CA 74-90-8: Very toxic; absorbed
  through skin; flammable, flash point 255; may become unstable and explode
  if stored for some time; used as fumigant; important chemical intermediate;
  soluble water, ethanol, ether. Tm = 259.91; Tb = 298.85; Vm = 0.0393 at
  293.15; Tc = 456.65; Pc = 5400; Vc = 0.139; Antoine I (sol) A = 9.55866,
  B = 2451.86, C = 36.049, dev = 0.1, range = 236 to 259; Antoine II (liq) A
  = 6.54538, B = 1271.284, C = -18.778, dev = 0.02, range = 259 to 299;
  Antoine III (liq) A = 7.13596, B = 1631.43, C = 18.953, dev = 1.0, range =
  298 to 457.

CHNO   MW 43.02

  Cyanic acid, CA 420-05-3: Powerful irritant; vesicant; polymerizes;
  decomposes with water; soluble ether, benzene, chloroform. Tm = 187.15;
  Tb = 296.15; Vm = 0.0377 at 293.15; Antoine (liq) A = 6.66616, B = 1239.94,
  C = -30.404, dev = 0.1, range = 233 to 268.

CHNS   MW 59.09

  Thiocyanic acid, CA 463-56-9: Toxic; decomposes on standing; soluble water,
  ethanol, ether. Tm = 278.15; Tb = 419.15 to 424.15, decomposes; Antoine
  (liq) A = 1.4076, B = 49.280, C = -123.42, dev = 1.0, range = 278 to 396.

CHN$_3$O$_6$ MW 151.04

  Trinitromethane, nitroform, CA 517-25-9: Irritant; decomposes over 298;
  can explode on heating; soluble water, ethanol, ether, acetone. Tm =
  287.45, decomposes; Vm = 0.1027 at 298.15; Antoine (liq) A = 5.3909, B =
  1702, C = 0, dev = 1.0, range = 290 to 317.

CH$_2$BrCl   MW 129.38

  Bromochloromethane, methylene bromochloride, CA 74-97-5: Moderately toxic
  vapor; irritant; nonflammable, used in fire extinguishers;   *(continues)*

CH$_2$BrCl *(continued)*

slightly soluble water; soluble organic solvents. Tf = 185.15; Tb = 341.25;
Vm = 0.0650 at 283.15; Antoine (liq) A = 5.51146, B = 891.345, C = -86.965,
dev = 0.1, range = 226 to 341.

CH$_2$Br$_2$ MW 173.83

Dibromomethane, methylene bromide, CA 74-95-3: Moderately toxic; nonflammable; slightly soluble water; soluble ethanol, ether, acetone. Tm =
220.45; Tb = 370.05; Vm = 0.0701 at 298.15; Antoine (liq) A = 6.1874, B =
1327.8, C = -52.57, dev = 0.1 to 1.0, range = 290 to 409.

CH$_2$ClF   MW 68.48

Chlorofluoromethane, methylene chlorofluoride, CA 593-70-4: Soluble
chloroform. Tm = 140.15; Tb = 264.05; Vm = 0.0539 at 293.15; Antoine (liq)
A = 6.6208, B = 1219.13, C = 0, dev = 1.0 to 5.0, range = 140 to 264.

CH$_2$Cl$_2$ MW 84.93

Dichloromethane, methylene chloride, CA 75-09-2: Moderately toxic; eye
irritant; expected carcinogen at high doses; nonflammable under normal cir-
cumstances; low-boiling solvent; slightly soluble water; soluble ethanol,
ether. Tf = 176.45; Tb = 312.95; Vm = 0.0639 at 288.15; Tc = 518.15; Pc =
6171; Vc = 0.180, Antoine I (liq) A = 6.18649, B = 1126.53, C = -43.46,
dev = 0.1, range = 264 to 312; Antoine II (liq) A = 6.88926, B = 1545.323,
C = 3.375, dev = 1.0, range = 311 to 383; Antoine III (liq) A = 5.87285,
B = 861.817, C = -94.102, dev = 1.0, range = 379 to 455; Antoine IV (liq)
A = 5.20540, B = 449.586, C = -193.701, dev = 1.0, range = 450 to 510.

CH$_2$F$_2$ MW 52.02

Difluoromethane, methylene fluoride, CA 75-10-5: Tm = 137.15; Tb = 221.55;
Vm = 0.0434 at 223.15; Tc = 351.6; Antoine I (liq) A = 6.02873, B = 738.88,
C = -37.8, dev = 1.0, range = 191 to 222; Antoine II (liq) A = 6.37612,
B = 864.292, C = -23.728, dev = 1.0, range = 191 to 258; Antoine III (liq)
A = 6.66467, B = 1013.473, C = -3.229, dev = 1.0, range = 256 to 321;
Antoine IV (liq) A = 8.07302, B = 2088.572, C = 133.275, dev = 1.0, range =
316 to 351.

CH$_2$F$_3$NOS   MW 133.09

Methanesulfinamide, 1,1,1-trifluoro, CA 30957-48-3: Antoine (liq) A =
6.625, B = 1943, C = 0, dev = 1.0 to 5.0, range not specified.

CH$_2$F$_3$NS   MW 117.09

1,1,1-Trifluoromethanesulfenamide, CA 1512-33-0: Tm = 184.15; Tb = 319.65;
Antoine (liq) A = 7.585, B = 1783, C = 0, dev = 1.0 to 5.0, range = 218
to 291.

CH_2I_2 MW 267.84

Diiodomethane, methylene iodide, CA 75-11-6: Flammable, flash point over
385; slightly soluble water; soluble ethanol, ether, benzene, light hydrocarbons. Tm = 278.15; Tb = 453.15, decomposes; Vm = 0.0806 at 293.15;
Antoine (liq) A = 6.06736, B = 1567.8, C = -69.15, dev = 5.0, range = 356
to 505.

CH_2N_2 MW 42.04

Cyanamide, CA 420-04-2: Irritant; highly toxic; flammable, flash point 414;
can explode above 323 in contact with water, acids, or alkalis; polymerizes
at 395; soluble water, ethanol, ether, benzene, chloroform. Tm = 319.15;
Antoine (liq) A = 8.915, B = 3580, C = 0, dev = 5.0, range not specified.

CH_2O MW 30.03

Formaldehyde, CA 50-00-0: Highly toxic and irritant; carcinogen; polymerizes readily; flammable, flash point 358 for 37 percent solution; important
intermediate and polymer raw material; soluble water, most organic solvents.
Tm = 155.15; Tb = 254.15; Vm = 0.0368 at 253.15; Tc = 410.35 to 414.35;
Pc = 6637 to 6784; Antoine I (liq) A = 6.5475, B = 1062.4, C = -19.92,
dev = 1.0, range = 184 to 251; Antoine II (liq) A = 6.4306, B = 1013.206,
C = -24.883, dev = 1.0, range = 163 to 251.

CH_2O_2 MW 46.03

Formic acid, CA 64-18-6: Corrosive; highly toxic; causes severe burns;
esters may be toxic and/or flammable; flammable, flash point 342; important
intermediate, strong reducing agent; soluble water, ether, ethanol,
acetone. Tm = 281.55; Tb = 373.85; Vm = 0.0377 at 293.15; Antoine I (sol)
A = 11.611, B = 3160, C = 0, dev = 1.0, range = 268 to 281; Antoine II
(liq) A = 6.5028, B = 1563.28, C = -26.09, dev = 0.1, range = 283 to 384.

CH_3Br MW 94.94

Methyl bromide, bromomethane, CA 74-83-9: Irritant; poisonous vapor;
causes burns; nonflammable under most conditions; refrigerant; fumigant;
solvent; slightly soluble water; soluble ethanol, ether, chloroform, benzene. Tm = 179.45; Tb = 276.71; Vm = 0.0549 at 273.15; Tc = 467.15,
calculated. Antoine (liq) A = 6.08455, B = 986.59, C = -34.83, dev = 0.1,
range = 201 to 296.

CH_3Cl MW 50.49

Methyl chloride, chloromethane, CA 74-87-3: Toxic by inhalation; mild
narcotic; suspected carcinogen; may ignite or explode on contact with
aluminum or magnesium; extremely flammable, flash point 228; refrigerant;
important chemical intermediate; moderately soluble water; soluble organic
solvents. Tm = 175.45; Tb = 249.42; Vm = 0.0549 at 293.15; Tc = 416.25;
Pc = 6679.2; Vc = 0.143; Antoine I (liq) A = 6.11935, B = 902.451, C =
-29.55, dev = 0.1, range = 180 to 266; Antoine II (liq) A = 6.04835, B =
869.887, C = -33.773, dev = 0.1, range = 247 to 310; Antoine III (liq) A =
6.94638, B = 1448.913, C = 47.996, dev = 0.1, range = 308 to 373; Antoine
IV (liq) A = 6.94002, B = 1447.601, C = 48.385, dev = 0.1, range = 368
to 416.

CH_3Cl_2P MW 116.91

Dichloromethyl phosphine, CA 676-83-5: Irritant; air-sensitive liquid.
Tm = 206.15; Tb = 354.15 to 355.15; Vm = 0.0897 at 293.15; Antoine (liq) A
= 7.32735, B = 1857, C = 0, dev = 1.0, range = 229 to 297.

CH_3F MW 34.03

Methyl fluoride, fluoromethane, CA 593-53-3: Narcotic in high concentrations; flammable; moderately soluble water; soluble ethanol, ether, benzene,
chloroform. Tm = 131.35; Tb = 194.95; Vm = 0.0388 at 195.15; Tc = 318.05;
Pc = 5877; Antoine I (liq) A = 6.22251, B = 740.218, C = -19.26, dev = 0.1,
range = 141 to 208; Antoine II (liq) A = 6.32764, B = 784.457, C = -13.248,
dev = 0.1, range = 205 to 242; Antoine III (liq) A = 6.59639, B = 913.481,
C = 5.008, dev = 0.1, range = 240 to 288.

CH_3F_2N MW 67.04

N,N-Difluoro methylamine, CA 753-58-2: Tm = 158.35; Antoine (liq) A =
5.97577, B = 871.05, C = -37.78, dev = 1.0, range = 203 to 257.

CH_3F_2NS MW 99.10

Imidosulfurous difluoride, methyl, CA 758-20-3: Tm = 150.15; Antoine (liq)
A = 7.1777, B = 1498.46, C = 0, dev = 1.0, range = 194 to 258.

CH_3F_2OPS MW 132.06

Difluorothiophosphoric acid, S-methyl ester, CA 25237-37-0: Antoine (liq)
A = 5.90079, B = 1107.494, C = -43.936, dev = 1.0, range = 236 to 298.

CH_3F_2P MW 84.01

Difluoromethyl phosphine, CA 753-59-3: Antoine (liq) A = 6.955, B = 1220,
C = 0, dev = 5.0, range = 174 to 236.

$CH_3F_2PS_2$ MW 148.13

Difluorodithiophosphoric acid, methyl ester, CA 21348-13-0: Antoine (liq)
A = 7.28734, B = 1857.075, C = -12.216, dev = 1.0, range = 253 to 298.

$CH_3F_4NP_2S_2$ MW 231.10

N,N-Bis(difluorothiophosphoral) methylamine, CA 25741-62-2: Antoine (liq)
A = 7.11581, B = 2092.119, C = 4.887, dev = 5.0, range = 273 to 325.

CH_3I MW 141.94

Methyl iodide, iodomethane, CA 74-88-4: Corrosive; toxic by inhalation and
skin absorption; suspected carcinogen; nonflammable under most conditions;
slightly soluble water; soluble ethanol, ether, acetone, carbon tetrachloride. Tf = 207.05; Tb = 315.65; Vm = 0.0623 at 293.15; Tc = 528;
Antoine I (liq) A = 6.10541, B = 1142.67, C = -36.87, dev = 0.1, *(continues)*

CH_3I *(continued)*

range = 228 to 337; Antoine II (liq) A = 6.17197, B = 1240.17, C = -17.887, dev = 1.0 to 5.0, range = 315 to 502.

CH_3NO MW 45.04

Formamide, CA 75-12-7: Irritant; expected teratogen; flammable, flash point 428; chemical intermediate; soluble water, methanol, ethanol, acetone; hygroscopic. Tf = 275.7; Tb = 483.15, decomposes; Vm = 0.0397 at 293.15; Antoine I (liq) A = 9.40789, B = 3397.688, C = -12.866, dev = 1.0, range = 293 to 377; Antoine II (liq) A = 5.15582, B = 907.24, C = -200.64, dev = 1.0, range = 415 to 466.

CH_3NOS MW 77.10

Methanamine, *N*-sulfinyl, CA 4291-05-8: Antoine (liq) A = 7.04991, B = 1659.955, C = 0, dev = 1.0, range = 252 to 277.

CH_3NO_2 MW 61.04

Methyl nitrite, CA 624-91-9: Highly toxic; soluble ethanol, ether. Tb = 261.15; Vm = 0.0616 at 288.15; Antoine (liq) A = 6.58183, B = 1181.36, C = 2.85, dev = 1.0, range = 218 to 273.

Nitromethane, CA 75-52-5: Toxic; flammable, flash point 318; potential explosive; polar solvent; slightly soluble water; soluble ethanol, ether, acetone. Tf = 244.6; Tb = 374.35; Vm = 0.0539 at 298.15; Tc = 588.15; Pc = 6309; Antoine I (liq) A = 6.40194, B = 1444.38, C = -45.786, dev = 0.1, range = 328 to 410; Antoine II (liq) A = 12.8267, B = 3905.39, C = -13.15, dev = 1.0, range = 405 to 476.

CH_3NO_3 MW 77.04

Methyl nitrate, CA 598-58-3: Highly irritant; flammable; may explode on heating; slightly soluble water; soluble ethanol, ether. Tm = 190.25; Tb = 338.15, explodes; Vm = 0.0642 at 298.15; Antoine (liq) A = 6.62964, B = 1392.69, C = -35.995, dev = 0.1, range = 273 to 303.

CH_4 MW 16.04

Methane, CA 74-82-8: Extremely flammable; flash point 85; forms explosive mixtures with air; chief constituent of natural gas; important fuel and chemical raw material; moderately soluble water; soluble methanol, most organic solvents. Tm = 90.7; Tb = 111; Vm = 0.0344 at 129.15; Tc = 190.6; Pc = 4600; Vc = 0.100; Antoine I (sol) A = 6.8199, B = 532.2, C = 1.852; dev = 1.0, range = 53 to 91; Antoine II (liq) A = 5.82051, B = 405.42, C = -5.373, dev = 0.1, range = 90 to 120; Antoine III (liq) A = 5.88277, B = 421.82, C = -2.824, dev = 0.1, range = 115 to 149; Antoine IV (liq) A = 6.49246, B = 620.151, C = 28.44, dev = 0.1, range = 148 to 189.

CH_4F_2NPS MW 131.08

Difluorothiophosphoric acid, *N*-methylamide, CA 31411-30-0: Antoine (liq) A = 9.46303, B = 3721.225, C = 100.455, dev = 1.0, range = 273 to 325.

CH_4N_2 MW 44.06

Methyl diazene, CA 26981-93-1: Can explode on heating. Tm = 151.15;
Antoine (liq) A = 7.23568, B = 1436.885, C = 0, dev = 5.0, range = 195
to 236.

CH_4N_2O MW 60.06

Urea, CA 57-13-6: Moderately toxic; fertilizer; polymer raw material; very
soluble water; soluble methanol, ethanol, pyridine. Tm = 408.15, decomposes; Antoine (sol) A = 9.565, B = 4579, C = 0, dev = 5.0, range = 345
to 368.

$CH_4N_4O_2$ MW 104.07

Nitroguanidine, CA 556-88-7: Highly toxic; explosive; almost insoluble
water; slightly soluble methanol, ethanol; soluble concentrated acids.
Tm = approximately 503.15, decomposes; Antoine (sol) A = 12.10, B = 7452,
C = 0, dev = 5.0, range = 402 to 473.

CH_4O MW 32.04

Methanol, methyl alcohol, CA 67-56-1: Moderately toxic vapor, can damage
eyes; highly flammable, flash point 284; can react violently with many
materials; important solvent, chemical raw material; soluble water, most
organic solvents. Tm = 175.47; Tb = 337.85; Vm = 0.0406 at 293.15; Tc =
512.58; Pc = 8096; Vc = 0.118; Antoine I (liq) A = 7.4182, B = 1710.2, C =
-22.25, dev = 1.0, range = 175 to 273; Antoine II (liq) A = 7.25164, B =
1608.39, C = -31.07, dev = 0.02, range = 274 to 337; Antoine III (liq) A =
7.09498, B = 1521.23, C = -39.18, dev = 1.0, range = 338 to 487; Antoine IV
(liq) A = 5.86277, B = 1105.884, C = -64.272, dev = 1.0, range = 188 to 228;
Antoine V (liq) A = 7.44355, B = 1712.316, C = -22.61, dev = 0.1 to 1.0,
range = 224 to 290; Antoine VI (liq) A = 7.26415, B = 1615.59, C = -30.437,
dev = 0.1, range = 285 to 345; Antoine VII (liq) A = 7.14736, B = 1544.804,
C = -37.235, dev = 0.1, range = 335 to 376; Antoine VIII (liq) A = 7.27446,
B = 1641.542, C = -25.789, dev = 0.1 to 1.0, range = 373 to 458; Antoine IX
(liq) A = 8.18215, B = 2546.019, C = 83.019, dev = 1.0, range = 453 to 513.

CH_4O_2 MW 48.04

Methyl hydroperoxide, CA 3031-73-0: Explosive; soluble water, ethanol,
ether, benzene. Tm = approximately 198.15; Vm = 0.0241 at 288.15; Antoine
(liq) A = 7.505, B = 1972, C = 0, dev = 5.0, range = 253 to 313.

CH_4O_3S MW 96.10

Methanesulfonic acid, CA 75-75-2: Corrosive; highly irritant; suspected
carcinogen; corrodes iron, brass, copper, lead; flammable, flash point over
385; soluble water, ethanol, ether. Tm = 293.15; Antoine (liq) A = 8.896,
B = 3861, C = 0, dev = 5.0, range = 395 to 440.

CH_4S MW 48.10

Methyl mercaptan, methanethiol, CA 74-93-1: Moderately toxic vapor; extremely flammable, flash point 255; chemical intermediate;   *(continues)*

CH_4S *(continued)*

soluble water, ethanol, ether.   Tf = 150.18; Tb = 279.05; Vm = 0.0553 at
293.15; Tc = 469.95; Pc = 7235; Antoine I (liq) A = 6.19283, B = 1031.216,
C = -32.816, dev = 0.02, range = 221 to 283; Antoine II (liq) A = 6.19219,
B = 1030.938, C = -32.845, dev = 0.02, range = 222 to 279; Antoine III
(liq) A = 6.13669, B = 1006.199, C = -35.529, dev = 1.0, range = 267 to
359; Antoine IV (liq) A = 6.53487, B = 1278.361, C = 5.318, dev = 1.0,
range = 345 to 424; Antoine V (liq) A = 8.49935, B = 3497.599, C = 283.722,
dev = 1.0 to 5.0, range = 414 to 470.

CH_5N MW 31.06

Methylamine, CA 74-89-5:   Irritant; flammable, flash point 273; refrigerant,
chemical intermediate; very soluble water; soluble ethanol, ether, acetone.
Tm = 179.65; Tb = 266.85; Vm = 0.0444 at 262.35; Tc = 430.05; Pc = 7458;
Antoine I (liq) A = 6.6218, B = 1079.15, C = -32.92, dev = 0.1, range = 223
to 273; Antoine II (liq) A = 6.76954, B = 1174.666, C = -20.186, dev = 1.0,
range = 263 to 329; Antoine III (liq) A = 6.32072, B = 936.232, C =
-50.047, dev = 1.0, range = 319 to 381; Antoine IV (liq) A = 8.61285, B =
3135.822, C = 231.226, dev = 1.0, range = 373 to 430.

CH_5NO MW 47.06

N-Methylhydroxylamine, CA 593-77-1:   Gradually decomposes on standing; very
soluble water; soluble methanol, ethanol; hygroscopic.   Tm = 311.65; Tb =
388.15; Antoine I (sol) A = 9.844, B = 2958.4, C = 0, dev = 1.0, range =
273 to 308; Antoine II (liq) A = 8.6949, B = 2597, C = 0, dev = 1.0,
range = 293 to 338.

O-Methylhydroxylamine, CA 67-62-9:   Highly irritant; toxic; soluble water,
ethanol, ether.   Tm = 186.75; Tb = 321.25; Antoine (liq) A = 6.39483, B =
1178.62, C = -52.75, dev = 1.0, range = 228 to 322.

CH_6ClN MW 67.52

Methylamine, hydrochloride: Tm = 498.15; Antoine (liq) A = 12.0909, B =
5981.4, C = 0, dev = 1.0, range = 518 to 593.

CH_6N_2 MW 46.07

Methylhydrazine, CA 60-34-4:   Highly toxic; carcinogen; flammable, flash
point 265; may ignite spontaneously in air; used in rocket fuels; soluble
water, ethanol, ether, light hydrocarbons.   Tm = 220.75; Tb = 360.65; Vm =
0.0527 at 298.15; Tc = 585.15; Pc = 8240; Vc = 0.159; Antoine (liq) A =
5.98766, B = 1122.94, C = -80.821, dev = 0.1, range = 274 to 299.

CIN   MW 152.92

Cyanogen iodide, CA 506-78-5:   Very toxic when absorbed; soluble hot water,
ethanol, ether.   Antoine I (sol) A = 9.585, B = 3128.6, C = 0, dev = 1.0,
range = 337 to 419; Antoine II (liq) A = 6.895, B = 2090.1, C = 0, dev =
1.0, range = 419 to 426.

CN_4O_8 MW 196.03

Tetranitromethane, CA 509-14-8: Highly toxic; explosive; flammable, flash point above 385; used as oxidizer for rocket propellants; slightly soluble water; soluble ethanol, ether. Tm = 286.95; Tb = 398.85; Vm = 0.1208 at 298.15; Antoine I (sol) A = 8.5331, B = 2478.6, C = 0, dev = 1.0, range = 255 to 286; Antoine II (liq) A = 6.82194, B = 1739.5, C = -36.2, dev = 1.0, range = 286 to 373.

CO   MW 28.01

Carbon monoxide, CA 630-08-0: Highly toxic by inhalation; extremely flammable; forms highly explosive mixtures with air; important fuel, reducing agent, chemical raw material; slightly soluble water; soluble copper solutions. Tm = 68.09; Tb = 81.65; Tc = 132.9; Pc = 3496; Vc = 0.0931; Antoine I (sol) A = 13.83128, B = 1555.83, C = 55.915, dev = 0.1, range = 54 to 61; Antoine II (liq) A = 5.3651, B = 230.274, C = -13.15, dev = 1.0, range = 68 to 108.

COS   MW 60.07

Carbonyl sulfide, carbon oxysulfide, CA 463-58-1: Irritant; highly toxic by inhalation; moderately soluble water; soluble ethanol, toluene. Tm = 134.35; Tb = 222.95; Vm = 0.505 at 220.15; Tc = 378.15; Pc = 5946; Vc = 0.138; Antoine (liq) A = 5.91043, B = 764.67, C = -27.525, dev = 1.0, range = 140 to 224.

CO_2 MW 44.01

Carbon dioxide, CA 124-38-9: Solid causes skin burns; carbonation of beverages, fire protection, frozen foods as dry ice; slightly soluble water; soluble alkali solutions, ethanolamine solutions. Tm = 216.15 at 5 atmospheres pressure; Tb = 194.95, sublimes; Tc = 304.25; Pc = 734; Vc = 0.0942; Antoine I (sol) A = 8.5853, B = 1242.55, C = -6.84, dev = 1.0, range = 83 to 153; Antoine II (sol) A = 8.61502, B = 1232.15, C = -8.243, dev = 0.02 to 0.1, range = 153 to 198; Antoine III (sol) A = 10.12622, B = 1861.87 C = 34.619, dev = 0.02, range = 198 to 216; Antoine IV (liq) A = 6.46212, B = 748.283, C = -16.904, dev = 0.02 to 0.1, range = 216 to 273; Antoine V (liq) A = 7.76331, B = 1566.08, C = 97.87, dev = 0.1, range = 273 to 304.

CS_2 MW 76.13

Carbon disulfide, CA 75-15-0: Highly toxic by inhalation; extremely flammable, flash point 243; ignition temperature 363; vapors may ignite from contact with steam line or an ordinary lamp bulb; important solvent, fumigant, chemical intermediate; slightly soluble water; soluble ethanol, ether, benzene. Tm = 161.11; Tb = 319.4; Vm = 0.0589 at 273.15; Tc = 546.15; Pc = 7700; Vc = 0.201; Antoine I (liq) A = 6.03694, B = 1153.5, C = -33.22, dev = 0.1, range = 256 to 319; Antoine II (liq) A = 6.07588, B = 1174.112, C = -30.896, dev = 0.1, range = 260 to 353; Antoine III (liq) A = 6.19814, B = 1231.307, C = -26.024, dev = 1.0, range = 338 to 408; Antoine IV (liq) A = 6.80466, B = 1728.903, C = 43.404, dev = 1.0, range = 388 to 497; Antoine V (liq) A = 7.58592, B = 2639.181, C = 165.312, dev = 1.0 to 5.0, range = 490 to 533.

$C_2BrCl_2F_3O_4$ MW 295.82

Perchloric acid, 1,2,2-trifluoro-1-chloro-2-bromoethyl ester, CA 38217-36-6:
Antoine (liq) A = 7.72172, B = 2217.896, C = 0, dev not specified, range =
273 to 294.

C_2BrCl_3O MW 226.28

Trichloroacetyl bromide, CA 34069-94-8: Irritant; soluble ether, benzene.
Tb = 416.15; Vm = 0.1191 at 288.15; Antoine (liq) A = 6.31204, B = 1613.22,
C = -41.513, dev = 1.0, range = 265 to 416.

C_2BrCl_5 MW 281.19

Bromopentachloroethane, CA 79504-02-2: Tm = approximately 463.15; Antoine
(sol) A = 6.455, B = 2320, C = 0, dev = 5.0, range = 383 to 433.

C_2BrF_3 MW 160.92

Bromotrifluoroethylene, CA 598-73-2: Highly toxic; flammable; polymerizes
on standing. Tb = 270.15; Vm = 0.0865 at 298.15; Antoine (liq) A = 6.8476,
B = 1308.04, C = 0, dev = 1.0, range = 260 to 340.

C_2BrF_3O MW 176.92

Trifluoroacetyl bromide, CA 354-31-4: Irritant; reacts with water. Tm =
137.15; Antoine (liq) A = 5.995, B = 1069, C = 0, dev = 5.0, range not
specified.

C_2BrF_9S MW 306.97

Pentafluoro(1-bromo-1,2,2,2-tetrafluoroethyl) sulfur, CA 63011-81-4:
Antoine (liq) A = 6.8569, B = 1603, C = 0, dev = 1.0 to 5.0, range = 294
to 330.

$C_2Br_2ClF_3$ MW 276.28

2-Chloro-1,2-dibromo-1,1,2-trifluoroethane, CA 354-51-8: Tm = approximately 323.15; Antoine (liq) A = 6.4843, B = 1639, C = 0, dev = 1.0,
range = 343 to 428.

$C_2Br_2Cl_4$ MW 325.64

1,2-Dibromotetrachloroethane, CA 630-25-1: Fungicide, fire-proofing agent.
Tm = 493.15 to 495.15, decomposes; Antoine (sol) A = 7.305, B = 2740, C = 0,
dev = 5.0, range = 383 to 453.

$C_2Br_2F_4$ MW 259.82

1,2-Dibromotetrafluoroethane, CA 124-73-2: Tm = 163.15, Tb = 320.45; Vm =
0.1201 at 298.15; Antoine I (liq) A = 6.893, B = 1567.0, C = 0, dev = 1.0,
range = 246 to 295; Antoine II (liq) A = 6.10296, B = 1176.556, C =
-33.003, dev = 1.0, range = 283 to 357; Antoine III (liq) A =   *(continues)*

$C_2Br_2F_4$ *(continued)*

6.22735, B = 1276.659, C = -17.094, dev = 1.0, range = 354 to 443; Antoine IV (liq) A = 8.69578, B = 4230.98, C = 331.076, dev = 1.0, range = 440 to 488.

C_2Br_4 MW 343.64

Tetrabromoethylene, CA 79-28-7: Soluble ethanol, ether, acetone, benzene. Tm = 329.65; Tb = 499.15 to 500.15; Antoine (sol) A = 8.5643, B = 2309.6, C = 0, dev = 1.0, range = 221 to 310.

C_2ClFN_2 MW 106.49

Acetonitrile, chloro(fluoroimino), *cis*, CA 30915-40-3: Antoine (liq) A = 6.63272, B = 1358.264, C = -25.452, dev = 1.0, range = 254 to 320.

Acetonitrile, chloro(fluoroimino), *trans*, CA 30915-39-0: Antoine (liq) A = 6.54215, B = 1278.807, C = -36.658, dev = 1.0, range = 257 to 320.

$C_2ClF_2NO_2$ MW 143.48

Carbamic fluoride, chloro(fluorocarbonyl), CA 42016-33-1: Antoine (liq) A = 7.0949, B = 1913, C = 0, dev = 1.0 to 5.0, range not specified.

C_2ClF_3 MW 116.47

Chlorotrifluoroethylene, CA 79-38-9: Moderately toxic; extremely flammable, flash point 246; reacts violently with many substances; refrigerant, polymer raw material; soluble benzene. Tm = 114.95; Tb = 246.95; Vm = 0.0844 at 273.15; Antoine I (liq) A = 6.03432, B = 853.33, C = -32.98, dev = 0.1, range = 206 to 262; Antoine II (liq) A = 6.87902, B = 1392.82, C = 46.55, dev = 0.1, range = 298 to 379.

C_2ClF_3O MW 135.47

Trifluoroacetyl chloride, CA 354-32-5: Irritant; reacts with water. Tm = 127.15; Tb = 255.15; Antoine (liq) A = 6.195, B = 1028, C = 0, dev = 5.0, range not specified.

$C_2ClF_3O_2$ MW 148.47

Chloroformic acid, trifluoromethyl ester, CA 23213-83-4: Antoine (liq) A = 5.62844, B = 849.321, C = -45.927, dev = 1.0, range = 195 to 273.

$C_2ClF_3O_4S$ MW 212.53

Anhydride difluorochloroacetic acid and fluorosulfuric acid: Antoine (liq) A = 6.42722, B = 1293.436, C = -59.036, dev = 1.0, range = 265 to 352.

C_2ClF_4NO MW 165.47

Carbamic fluoride, chloro(trifluoromethyl), CA 42016-31-9: Antoine (liq) A = 6.9669, B = 1540, C = 0, dev = 1.0 to 5.0, range not specified.

$C_2ClF_4NO_4S$ MW 245.53

Carbamic acid, chloro(trifluoromethyl) anhydride with fluorosulfuric acid,
CA 42016-34-2: Antoine (liq) A = 5.7749, B = 1500, C = 0, dev = 1.0 to
5.0, range not specified.

C_2ClF_5 MW 154.47

Chloropentafluoroethane, CA 76-15-3: Nonflammable gas; insoluble water;
soluble ethanol, ether. Tm = 167.15; Tb = 234.05; Vm = 0.1197 at 298.15;
Tc = 353.15; Pc = 3120; Vc = 0.259; Antoine I (liq) A = 5.99579, B =
816.306, C = -29.463, dev = 0.02, range = 176 to 235; Antoine II (liq) A =
6.01929, B = 828.337, C = -27.658, dev = 0.02, range = 234 to 265; Antoine
III (liq) A = 6.2600, B = 947.562, C = -11.015, dev = 0.1, range = 262 to
317; Antoine IV (liq) A = 6.73898, B = 1256.751, C = 34.474, dev = 0.1,
range = 312 to 353.

C_2ClF_5O MW 170.47

Hypochlorous acid, pentafluoroethyl ester, CA 22675-67-8: Antoine (liq)
A = 5.87822, B = 855.496, C = -44.038, dev = 1.0, range = 193 to 248.

C_2ClF_5OS MW 202.53

Ethanesulfinyl chloride, pentafluoro, CA 39937-08-1: Antoine (liq) A =
7.0989, B = 1707.4, C = 0, dev = 1.0 to 5.0, range = 273 to 338.

$C_2ClF_5O_4$ MW 218.46

Perchloric acid, pentafluoroethyl ester, CA 53011-52-2: Antoine (liq) A =
6.7605, B = 1430.8, C = 0, dev = 1.0, range not specified.

$C_2ClF_5O_6S_2$ MW 314.58

Ethane, 1,2,2-trifluoro-1-chloro-1,2-bis(fluorosulfate): Antoine (liq) A =
4.25414, B = 490.934, C = -186.551, dev = 1.0 to 5.0, range = 308 to 406.

C_2ClF_6NOS MW 235.53

Imidosulfuryl fluoride, 2-chloro-1,1,2,2-tetrafluoroethyl, CA 34495-79-9:
Antoine (liq) A = 7.5349, B = 1815, C = 0, dev = 1.0, range not specified.

C_2ClF_9S MW 262.52

Pentafluoro(1,1,2,2-tetrafluoro-2-chloroethyl) sulfur, CA 646-63-9: Tm =
160.15; Antoine (liq) A = 6.6189, B = 1479, C = 0, dev = 1.0 to 5.0, range
not specified.

Tetrafluorochloro(pentafluoroethyl) sulfur, CA 42769-85-7: Antoine (liq)
A = 6.625, B = 1497, C = 0, dev = 5.0, range not specified.

$C_2ClF_{10}NS$ MW 295.53

Sulfur, (N-chloro-1,1,1-trifluoromethanaminato)tetrafluoro(trifluoromethyl), CA 56868-57-6: Antoine (liq) A = 7.135, B = 1785, C = 0, dev = 1.0 to 5.0, range not specified.

$C_2Cl_2F_2$ MW 132.92

1,2-Dichloro-1,2-difluoroethylene, CA 598-88-9: Can react with water or steam. Tm = 142.65; Tb = 294.25; Vm = 0.0889 at 273.15; Antoine (liq) A = 6.09133, B = 1012.69, C = -46.152, dev = 1.0, range = 191 to 294.

$C_2Cl_2F_2N_2$ MW 160.94

Acetonitrile, dichloro(difluoroamino), CA 30913-21-4: Antoine (liq) A = 8.77118, B = 3177.191, C = 128.005, dev = 1.0 to 5.0, range = 238 to 341.

$C_2Cl_2F_2O$ MW 148.92

Fluorodichloroacetyl fluoride, CA 354-18-7: Antoine (liq) A = 5.91713, B = 1297.37, C = 17.459, dev = 1.0, range = 208 to 273.

$C_2Cl_2F_3NO$ MW 181.93

Acetamide, N,N-dichloro-2,2,2-trifluoro, CA 32751-03-4: Antoine (liq) A = 8.015, B = 2136.1, C = 0, dev = 1.0 to 5.0, range not specified.

$C_2Cl_2F_3NOS$ MW 213.99

S,S-Dichloro-N-(trifluoroacetyl) sulfilimine, CA 24433-67-8: Tm = 227.65; Antoine (liq) A = 7.6579, B = 2309, C = 0, dev = 1.0 to 5.0, range = 306 to 333.

$C_2Cl_2F_3NO_2S$ MW 229.99

Carbonimidic dichloride, (trifluoromethyl)sulfonyl, CA 51587-33-8: Antoine (liq) A = 7.02514, B = 1878.75, C = -31.617, dev = 1.0 to 5.0, range = 312 to 405.

$C_2Cl_2F_4$ MW 170.92

1,1-Dichlorotetrafluoroethane, CA 374-07-2: Narcotic at high concentrations; can react violently with hot aluminum; soluble ethanol, ether, benzene. Tm = 179.15; Tb = 276.75; Vm = 0.1175 at 298.15; Tc = 418.75; Pc = 3290; Antoine (liq) A = 6.4494, B = 1227, C = 0, dev = 1.0 to 5.0, range = 231 to 373.

1,2-Dichlorotetrafluoroethane, CA 76-14-2: Narcotic at high concentrations; can react violently with hot aluminum; soluble ethanol, ether. Tm = 179.15; Tb = 276.92, Vm = 0.1174 at 298.15; Tc = 418.85; Pc = 3260; Vc = 0.293; Antoine I (liq) A = 6.02676, B = 962.61, C = -37.3, dev = 1.0, range = 210 to 277; Antoine II (liq) A = 6.07469, B = 986.4, C = -34.233, dev = 1.0 to 5.0, range = 277 to 391.

$C_2Cl_2F_4O_4$ MW 234.92

Perchloric acid, 1,1,2,2-tetrafluoro-2-chloroethyl ester, CA 38126-28-2:
Tm = below 195.15; Antoine (liq) A = 5.75529, B = 1088.506, C = -55.729,
dev = 1.0, range = 249 to 294.

$C_2Cl_2F_5NS$ MW 235.99

S,S-Dichloro-N-(pentafluoroethyl) sulfilimine, CA 10564-48-4: Tm = 163.15;
Antoine (liq) A = 7.1559, B = 1956, C = 0, dev = 1.0 to 5.0, range = 297
to 375.

$C_2Cl_2F_7NS$ MW 273.98

Sulfur, (carbonimidic dichloridato)tetrafluoro(trifluoromethyl),
CA 56868-53-2: Antoine (liq) A = 6.145, B = 1527, C = 0, dev = 1.0 to 5.0,
range not specified.

$C_2Cl_3F_3$ MW 187.38

1,1,1-Trichlorotrifluoroethane, CA 354-58-5: May react violently with hot
aluminum; nonflammable refrigerant, blowing agent; soluble ethanol, ether,
chloroform. Tm = 287.35; Tb = 318.95; Vm = 0.1186 at 293.15; Antoine (liq)
A = 5.1537, B = 702.35, C = -95.72, dev = 1.0, range = 286 to 310.

1,1,2-Trichlorotrifluoroethane, CA 76-13-1: May react violently with hot
aluminum; soluble ethanol, ether, benzene. Tm = 238.15; Tb = 320.72; Vm =
0.1197 at 298.15; Tc = 487.25; Pc = 3410, Vc = 0.325; Antoine I (liq) A =
5.91657, B = 1094.37, C = -39.61, dev = 0.1, range = 273 to 319; Antoine II
(liq) A = 6.01641, B = 1115.812, C = -42.515, dev = 1.0, range = 238 to 364;
Antoine III (liq) A = 6.53093, B = 1500.489, C = 12.469, dev = 1.0, range =
360 to 473; Antoine IV (liq) A = 5.95163, B = 1082.588, C = -46.427, dev =
0.02 to 0.1, range = 297 to 317.

$C_2Cl_3F_3O_4$ MW 251.37

Perchloric acid, 1,2,2-trifluoro-1,2-dichloroethyl ester, CA 38126-27-1:
Tm = below 195.15; Antoine (liq) A = 5.26945, B = 1405.233, C = 0, dev not
specified, range = 273 to 296.

C_2Cl_3N MW 144.39

Trichloro acetonitrile, CA 545-06-2: Strong irritant; toxic; lachrymator;
can react with water to produce toxic fumes. Tm = 231.15; Tb = 358.15;
Vm = 0.1003 at 298.15; Antoine (liq) A = 6.1379, B = 1275, C = -50.35,
dev = 1.0, range = 289 to 357.

C_2Cl_4 MW 165.83

Tetrachloroethylene, perchloroethylene, CA 127-18-4: Irritant; moderately
toxic; suspected carcinogen; nonflammable solvent used for dry cleaning and
degreasing; insoluble water; soluble organic solvents. Tm = 254.15; Tb =
394.35; Vm = 0.1022 at 293.15; Tc = 620.25; Pc = 9740; Antoine (liq) A =
6.1017, B = 1386.9, C = -55.63, dev = 0.1, range = 310 to 393.

$C_2Cl_4F_2$ MW 203.83

1,1-Difluorotetrachloroethane, CA 76-11-9: Nonflammable refrigerant,
blowing agent; soluble ethanol, ether, chloroform. Tm = 311.15 to 313.15;
Tb = 364.15; Vm = 0.1235 at 293.15.

1,2-Difluorotetrachloroethane, CA 76-12-0: Nonflammable refrigerant,
blowing agent; soluble ethanol, ether, chloroform. Tm = 299.15; Tb =
365.95; Vm = 0.1247 at 303.15; Tc = 551.15; Pc = 3440; Vc = 0.370; Antoine
I (sol) A = 6.50788, B = 1526.24, C = -28.93, dev = 1.0, range = 235 to
293; Antoine II (liq) A = 6.67780, B = 1567.8, C = -29.58, dev = 1.0,
range = 301 to 365; Antoine III (liq) A = 5.97313, B = 1239.993, C =
-53.468, dev = 0.02, range = 312 to 362.

$C_2Cl_4F_2O_4$ MW 267.83

Perchloric acid, 1,2-difluoro-1,1,2-trichloroethyl ester, CA 38126-29-3:
Tm = below 195.15; Antoine (liq) A = 5.20493, B = 1578.533, C = 0, dev not
specified, range = 273 to 294.

$C_2Cl_4F_4N_2$ MW 269.84

1,2-Ethanediamine, N,N,N',N'-tetrachloro-1,1,2,2-tetrafluoro, CA 35695-53-5:
Antoine (liq) A = 7.265, B = 2248.5, C = 0, dev = 1.0 to 5.0, range not
specified.

$C_2Cl_4F_6OS$ MW 327.88

Pentafluoro(2-fluoro-1,1,2,2-tetrachloroethoxy) sulfur: Antoine (liq) A =
5.71413, B = 1229.937, C = -84.723, dev = 1.0, range = 314 to 418.

C_2Cl_4O MW 181.83

Tetrachloroethylene oxide, CA 16650-10-5: Tm = 214.35; Antoine (liq) A =
5.39815, B = 996.798, C = -90.485, dev = 1.0, range = 308 to 348.

Trichloroacetyl chloride, CA 76-02-8: Corrosive; highly toxic; soluble
ether. Tm = 216.23; Tb = 391.15; Vm = 0.1122 at 293.15; Antoine (liq) A =
6.10932, B = 1386.68, C = -53.447, dev = 1.0, range = 305 to 393.

C_2Cl_6 MW 236.74

Hexachloroethane, perchloroethane, CA 67-72-1: Irritant; very toxic; suspected carcinogen; insoluble water; soluble organic solvents. Tm = 458.15,
sublimes; Tb = 459.15; Antoine I (sol, rhombic) A = 9.015, B = 3077, C = 0,
dev = 1.0, range = 317 to 345; Antoine II (sol, cubic) A = 7.856, B = 2677,
C = 0, dev = 1.0, range = 345 to 460; Antoine III (liq) A = 6.6075, B =
2103.6, C = 0, dev = 1.0, range = 460 to 513.

C_2FNO_2 MW 89.03

Fluorocarbonyl isocyanate, CA 15435-14-0: Toxic; reacts with water. Tm =
174.15; Antoine (liq) A = 7.82, B = 1750, C = 0, dev = 5.0, range = 228
to 264.

$C_2F_2N_2O$ MW 106.03

Carbonocyanidic amide, difluoro, CA 32837-63-1: Tm = 172.15; Antoine (liq)
A = 7.455, B = 1544, C = 0, dev = 1.0 to 5.0, range not specified.

$C_2F_2N_2O_2$ MW 122.03

Carbonisocyanatidic amide, difluoro, CA 32837-64-2: Tm = 174.15; Antoine
(liq) A = 7.385, B = 1762, C = 0, dev = 1.0 to 5.0, range not specified.

$C_2F_2N_4O_8$ MW 246.04

1,2-Difluoro-1,1,2,2-tetranitroethane, CA 20165-39-3: Antoine (liq) A =
10.625, B = 3280.4, C = 0, dev = 1.0 to 5.0, range = 297 to 323.

$C_2F_2O_2$ MW 94.02

Oxalyl fluoride, CA 359-40-0: Tm = 260.73; Tb = 273.15 to 275.15; Antoine
I (sol) A = 8.45156, B = 1340.93, C = 59.27, dev = 0.1, range = 234 to 260;
Antoine II (liq) A = 4.93276, B = 409.554, C = -130.21, dev = 0.1, range =
264 to 272.

$C_2F_2O_4$ MW 126.02

Bis(fluorocarbonyl) peroxide, CA 692-74-0: Tm = 230.65; Antoine I (sol)
A = 9.180, B = 1994, C = 0, dev = 5.0, range not specified; Antoine II
(liq) A = 7.389, B = 1555.9, C = 0, dev = 1.0, range not specified.

C_2F_3N MW 95.02

Trifluoroacetonitrile, CA 353-85-5: Tm = 128.71; Tb = 209.15; Antoine I
(liq) A = 6.26708, B = 778.252, C = -22.837, dev = 1.0, range = 151 to 206;
Antoine II (liq) A = 5.53607, B = 558.489, C = -47.505, dev = 1.0, range =
141 to 203; Antoine III (liq) A = 6.31783, B = 794.126, C = -21.301, dev =
1.0, range = 197 to 241; Antoine IV (liq) A = 6.46987, B = 903.958, C =
-1.135, dev = 1.0, range = 236 to 282; Antoine V (liq) A = 7.66859, B =
1720.073, C = 107.272, dev = 1.0, range = 272 to 311.

C_2F_3NO MW 111.02

Trifluoromethyl isocyanate, CA 460-49-1: Irritant; toxic; reacts with
water. Antoine (liq) A = 6.963, B = 1176, C = 0, dev = 1.0 to 5.0, range =
195 to 228.

Trifluoro nitrosoethylene, CA 2713-04-4: Antoine (liq) A = 7.3764, B =
1340, C = 0, dev = 1.0, range = 247 to 250.

C_2F_3NOS MW 143.08

Methanesulfenyl isocyanate, trifluoro, CA 691-03-2: Irritant; reacts with
water. Tm = 176.15; Antoine (liq) A = 6.8659, B = 1458, C = 0, dev = 1.0
to 5.0, range = 231 to 293.
                                                                        (continues)

C_2F_3NOS *(continued)*

Methanesulfinyl cyanide, trifluoro, CA 61951-27-7: Antoine (liq) A = 7.925, B = 2090, C = 0, dev = 1.0 to 5.0, range not specified.

$C_2F_3NO_2S$ MW 159.08

Acetamide, 2,2,2-trifluoro-*N*-sulfinyl, CA 26454-68-2: Tm = 219.15; Antoine (liq) A = 7.46314, B = 1902.727, C = 0, dev = 5.0, range = 267 to 302.

$C_2F_3NO_2S_2$ MW 191.14

Methanesulfonyl isothiocyanate, trifluoro, CA 51587-30-5: Antoine (liq) A = 6.76780, B = 1655.074, C = -37.606, dev = 1.0 to 5.0, range = 297 to 385.

$C_2F_3NO_3S$ MW 175.08

Methanesulfonyl isocyanate, trifluoro, CA 30227-06-6: Antoine (liq) A = 6.84387, B = 1511.124, C = -33.004, dev = 1.0 to 5.0, range = 275 to 345.

C_2F_3NS MW 127.08

Thiocyanic acid, trifluoromethyl ester, CA 690-24-4: Tm = 203.15; Antoine (liq) A = 7.5199, B = 1704, C = 0, dev = 1.0 to 5.0, range = 226 to 294.

$C_2F_3N_3O_6$ MW 219.03

1,1,2-Trifluoro-1,2,2-trinitroethane, CA 20165-38-2: Antoine (liq) A = 9.525, B = 3016.3, C = 0, dev = 1.0 to 5.0, range = 313 to 353.

C_2F_4 MW 100.02

Tetrafluoroethylene, CA 116-14-3: Explodes under pressure; forms explosive peroxides; monomer for Teflon; very exothermic polymerization. Tm = 130.65; Tb = 196.85; Vm = 0.0658 at 197.15; Tc = 306.45; Pc = 392000; Vc = 0.1724; Antoine I (liq) A = 6.91436, B = 972.692, C = 0, dev = 1.0, range = 142 to 208; Antoine II (liq) A = 6.4595, B = 875.14, C = 0, dev = 1.0, range = 197 to 273; Antoine III (liq) A = 6.4291, B = 866.84, C = 0, dev = 1.0, range = 273 to 306.

$C_2F_4N_2$ MW 128.03

Tetrafluoroaminoacetic acid, nitrile, CA 5131-88-4: Antoine (liq) A = 6.6461, B = 1018.9, C = -21.35, dev = 5.0, range = 193 to 238.

$C_2F_4N_2O_2S$ MW 192.09

Diimidosulfuryl fluoride, bis (fluorocarbonyl), CA 63697-48-3: Tm = 278.15; Antoine (liq) A = 7.725, B = 2178, C = 0, dev = 1.0 to 5.0, range not specified.

$C_2F_4N_2O_3$ MW 176.03

1,1,2,2-Tetrafluoro-1-nitro-2-nitrosoethane, CA 679-08-3: Antoine (liq)
A = 7.0579, B = 1503, C = 0, dev = 1.0, range = 233 to 293.

$C_2F_4N_2O_4$ MW 192.03

1,1,2,2,-Tetrafluoro-1,2-dinitroethane, CA 356-16-1: Soluble acetone.
Tm = 231.65; Tb = 331.15 to 332.15; Vm = 0.1198 at 298.15; Antoine (liq)
A = 6.52062, B = 1324.2, C = -39.73, dev = 1.0, range = 259 to 333.

$C_2F_4N_2O_6S_2$ MW 288.15

Hydrazine, 1,2-bis(fluoroformyl)-1,2-bis(fluorosulfonyl), CA 19252-50-7:
Antoine (liq) A = 8.77183, B = 2599.064, C = 0, dev = 1.0 to 5.0, range =
273 to 296.

C_2F_4O MW 116.02

Trifluoroacetyl fluoride, CA 354-34-7: Tm = 113.69; Antoine (liq) A =
6.23413, B = 770.919, C = -31.76, dev = 0.1, range = 161 to 215.

$C_2F_4O_2S$ MW 164.07

(Trifluoroethylene) sulfonyl fluoride, CA 684-10-6: Antoine (liq) A =
2.4029, B = 1412, C = 0, dev = 1.0 to 5.0, range = 270 to 313.

$C_2F_4O_3$ MW 148.01

Fluoroperoxyformic acid, trifluoromethyl ester, CA 16118-40-4: Tm =
145.15; Antoine (liq) A = 7.5341, B = 1423.8, C = 0, dev = 1.0, range = 194
to 249.

$C_2F_4O_4S$ MW 196.07

Trifluoroacetyl fluorosulfate, CA 5762-53-8: Antoine (liq) A = 6.43081,
B = 1184.3, C = -51.73, dev = 1.0, range = 262 to 321.

$C_2F_4S_2$ MW 164.14

2,2,4,4-Tetrafluoro-1,3-dithietane, CA 1717-50-6: Antoine (liq) A =
6.7649, B = 1524, C = 0, dev = 1.0 to 5.0, range not specified.

C_2F_5I MW 245.92

Iodopentafluoroethane, CA 354-64-3: Tb = 286.15; Vm = 0.1187 at 301.15;
Antoine (liq) A = 5.8339, B = 1087.4, C = 0, dev = 1.0, range = 248 to 283.

C_2F_5NO MW 149.02

Pentafluoroacetamide, CA 32822-49-4: Antoine (liq) A = 6.925, B = 1240,
C = 0, dev = 1.0 to 5.0, range not specified.   *(continues)*

C_2F_5NO *(continued)*

Pentafluoronitrosoethane, CA 354-72-3: Tb = 231.15; Antoine (liq) A = 6.8149, B = 1094, C = 0, dev = 1.0, range = 193 to 227.

C_2F_5NOS MW 181.08

Difluorocarbamothioic acid, *S*-(trifluoromethyl) ester, CA 32837-66-4: Tm = 154.15; Antoine (liq) A = 5.845, B = 1213, C = 0, dev = 1.0 to 5.0, range not specified.

S,S-Difluoro-*N*-(trifluoroacetyl) sulfilimine, CA 24433-65-6: Tm = 207.15; Antoine (liq) A = 7.3899, B = 1798, C = 0, dev = 1.0 to 5.0, range = 240 to 282.

Methanesulfinimidyl fluoride, 1,1,1-trifluoro-*N*-(fluoroformyl), CA 28103-61-9: Antoine (liq) A = 7.8449, B = 2033, C = 0, dev = 1.0, range = 276 to 323.

1,1,2,2,2-Pentafluoro-*N*-sulfinyl ethylamine, CA 10564-50-8: Tm = 188.65; Antoine (liq) A = 6.7679, B = 1516, C = 0, dev = 1.0, range = 245 to 303.

$C_2F_5NO_4S$ MW 229.08

(Fluorosulfonyl)(trifluoromethoxy) carbamoyl fluoride, CA 19252-49-4: Antoine (liq) A = 6.3557, B = 1582.3, C = 0, dev = 1.0, range not specified.

$C_2F_5N_3O_3$ MW 209.03

Diimide, fluoro(1,1,2,2-tetrafluoro-2-nitroethyl)-2-oxide, CA 755-68-0: Tm = 222.35; Antoine (liq) A = 5.97425, B = 1115.5, C = -67.84, dev = 1.0, range = 257 to 350.

C_2F_6 MW 138.01

Hexafluoroethane, perfluoroethane, CA 76-16-4: Tm = 172.55; Tb = 194.95; Vm = 0.0863 at 193.15; Tc = 292.85; Pc = 2990; Vc = 0.224; Antoine (liq) A = 6.641, B = 903.219, C = 0, dev = 1.0, range = 172 to 200.

C_2F_6IN MW 278.92

N-Iodo-bis(trifluoromethyl) amine, CA 5764-87-4: Tm = 189.15; Antoine (liq) A = 6.525, B = 1490, C = 0, dev = 5.0, range = 261 to 318.

$C_2F_6N_2$ MW 166.03

Hexafluoroazomethane, CA 372-63-4: Tm = 135.15; Tb = 241.55; Antoine (liq) A = 6.9569, B = 1196, C = 0, dev = 1.0, range = 205 to 242.

$C_2F_6N_2O$ MW 182.03

Hexafluoroazoxymethane, CA 371-56-2: Antoine (liq) A = 7.07162, B = 1418.61, C = 0, dev = 0.1, range = 274 to 281.

$C_2F_6N_2O_2$ MW 198.02

Methanamine, 1,1,1-trifluoro-*N*-(nitrosooxy)-*N*-(trifluoromethyl),
CA 359-75-1: Antoine (liq) A = 6.8409, B = 1399, C = 0, dev = 1.0 to 5.0,
range = 245 to 285.

N-Nitroso-*O*,*N*-(bis(trifluoromethyl) hydroxylamine: Antoine (liq) A =
6.7009, B = 1329.1, C = 0, dev = 1.0, range = 272 to 283.

C_2F_6OS MW 186.07

Bis(trifluoromethyl) sulfoxide, CA 30341-37-8: Tm = 221.15; Antoine (liq)
A = 6.717, B = 1456, C = 0, dev = 5.0, range = 248 to 303.

Pentafluoroethyl sulfinyl fluoride, CA 20621-31-2: Antoine (liq) A =
6.78405, B = 1356.1, C = -10.19, dev = 1.0, range = 234 to 293.

$C_2F_6OS_2$ MW 218.13

Methanesulfinothioic acid, trifluoro, *S*-(trifluoromethyl) ester,
CA 63548-94-7: Tm = 135.15; Antoine (liq) A = 6.6711, B = 1605.98, C = 0,
dev = 1.0, range not specified.

$C_2F_6O_3$ MW 186.01

Bis(trifluoromethyl) trioxide, CA 1718-18-9: Tm = 135.15; Antoine (liq) A
= 7.84062, B = 1712.427, C = 37.406, dev = 1.0, range = 193 to 248.

$C_2F_6O_3S$ MW 218.07

Trifluoromethanesulfonic acid, trifluoromethyl ester, CA 3582-05-6: Tm =
164.95; Antoine (liq) A = 5.68419, B = 842.252, C = -65.496, dev = 1.0,
range = 238 to 294.

$C_2F_6O_5S$ MW 250.07

Peroxysulfuric acid, bis(trifluoromethyl) ester, CA 41765-14-4: Antoine
(liq) A = 7.0196, B = 1601.134, C = 0, dev = 1.0, range not specified.

$C_2F_6O_6S_2$ MW 298.13

Tetrafluoroethylene glycol, bis(fluorosulfate), CA 1479-53-4: Tm = 245.35;
Antoine (liq) A = 5.2494, B = 820.64, C = -123.71, dev = 1.0, range = 295
to 378.

C_2F_6S MW 170.07

Bis(trifluoromethyl) sulfide, CA 371-78-8: Tb = 250.95; Antoine (liq) A =
6.941, B = 1239.1, C = 0, dev = 1.0 to 5.0, range not specified.

$C_2F_6S_2$ MW 202.13

Bis(trifluoromethyl) disulfide, CA 372-64-5: Tm = over 274.15; Tb =
307.75; Antoine (liq) A = 6.890, B = 1503.1, C = 0, dev = 1.0 to 5.0,
range not specified.

C_2F_7N MW 171.02

Perfluorodimethylamine, CA 359-62-6: Tb = 236.15; Antoine I (liq) A =
6.1249, B = 972.7, C = 0, dev = 1.0, range = 203 to 233; Antoine II (liq)
A = 6.7779, B = 1118, C = 0, dev = 1.0 to 5.0, range = 199 to 230.

Perfluoroethylamine, CA 354-80-3: Tb = 238.15; Antoine (liq) A = 6.560,
B = 1088, C = 0, dev = 1.0, range = 171 to 236.

C_2F_7NOS MW 219.08

Imidosulfuryl fluoride, pentafluoroethyl, CA 59617-28-6: Antoine (liq) A =
7.405, B = 1604, C = 0, dev = 1.0 to 5.0, range not specified.

Sulfoxime, S-fluoro-N,S-bis(trifluoromethyl), CA 59665-14-4: Antoine (liq)
A = 6.015, B = 1195, C = 0, dev = 1.0 to 5.0, range not specified.

$C_2F_7NO_2S$ MW 235.08

Hydroxylamine, O-(fluorosulfonyl)-N,N-bis(trifluoromethyl), CA 21950-99-2:
Antoine (liq) A = 6.605, B = 1479, C = 0, dev = 1.0 to 5.0, range not
specified.

$C_2F_7NO_3S$ MW 251.08

Fluorosulfuric acid, 1,1,2,2-tetrafluoro-2-(difluoroamino)ethyl ester,
CA 4188-34-5: Antoine (liq) A = 7.000, B = 1625, C = 0, dev = 1.0, range =
276 to 326.

$C_2F_7NO_{12}S_4$ MW 491.25

Fluorosulfuric acid, 1-bis[(fluorosulfonyl)oxy]amino-2,2,2-trifluoroethyl-
idene ester, CA 53684-02-9: Antoine (liq) A = 8.135, B = 2560, C = 0,
dev = 1.0 to 5.0, range not specified.

C_2F_8OS MW 224.07

Difluorooxo-bis(trifluoromethyl) sulfur, CA 33716-15-3: Antoine (liq) A =
6.85227, B = 1771.004, C = 65.762, dev = 1.0 to 5.0, range = 239 to 299.

$C_2F_8O_3S$ MW 256.07

Pentafluoro(trifluoroethaneperoxoato) sulfur, CA 60672-61-9: Antoine (liq)
A = 6.5996, B = 1465.3, C = 0, dev = 1.0, range not specified.

C_2F_8S MW 208.07

Difluorobis(trifluoromethyl) sulfur, CA 30341-38-9: Antoine (liq) A =
7.125, B = 1507, C = 0, dev not specified, range not specified.

Pentafluoro(trifluorovinyl) sulfur, CA 1186-51-2: Antoine (liq) A = 6.485,
B = 1310, C = 0, dev not specified, range not specified.

C_2F_9NS MW 241.07

Sulfur, trifluoro[1,1,1-trifluoromethanaminato-2-(trifluoromethyl)],
CA 56868-56-5: Antoine (liq) A = 7.135, B = 1572, C = 0, dev = 1.0 to 5.0,
range not specified.

$C_2F_{10}OS$ MW 262.07

Pentafluoro(pentafluoroethoxy) sulfur: Antoine (liq) A = 5.98633, B =
938.914, C = -52.27, dev = 1.0 to 5.0, range = 245 to 287.

$C_2F_{10}O_2S$ MW 278.06

Tetrafluorobis(trifluoromethoxy) sulfur, CA 2004-38-8: Tm = 143.15;
Antoine (liq) A = 6.03015, B = 1014.515, C = -50.323, dev = 1.0, range =
246 to 302.

$C_2F_{10}O_3S$ MW 294.06

(Trifluoromethoxy)-[(trifluoromethyl)dioxy] sulfur tetrafluoride,
CA 41938-43-6: Antoine (liq) A = 7.1703, B = 1637.72, C = 0, dev = 1.0,
range not specified.

$C_2F_{10}O_3S_2$ MW 326.12

Pentafluoro[1,2,2,2-tetrafluoro-1-[(fluorosulfonyl)oxy]ethyl] sulfur,
CA 68010-32-2: Antoine (liq) A = 7.105, B = 1819, C = 0, dev = 1.0 to 5.0,
range not specified.

$C_2F_{10}S$ MW 246.07

trans-Tetrafluorobis(trifluoromethyl) sulfur, CA 42179-02-2: Antoine (liq)
A = 6.1519, B = 1217.6, C = 0, dev = 1.0, range = 233 to 293.

$C_2F_{11}NS$ MW 279.07

[Bis(trifluoromethyl)amino] sulfur pentafluoride, CA 13888-13-6: Antoine
(liq) A = 6.995, B = 1530, C = 0, dev = 5.0, range = 233 to 306.

Sulfur, tetrafluoro(*N*,1,1,1-tetrafluoromethanaminato)(trifluoromethyl),
CA 59665-16-6: Antoine (liq) A = 6.975, B = 1576, C = 0, dev = 1.0 to 5.0,
range not specified.

$C_2F_{12}S_2$ MW 316.12

1,1,1,1,2,2,3,3,3,3,4,4-Dodecafluoro-1,1,1,1,3,3,3,3-octahydro-1,3-
dithietane, CA 42060-66-2: Antoine (liq) A = 7.415, B = 1857, C = 0, dev =
1.0 to 5.0, range not specified.

$C_2F_{14}S_2$ MW 354.12

Octafluorobis(trifluoromethyl) disulfur, CA 1580-11-6: Antoine (liq) A =
6.945, B = 1792, C = 0, dev not specified, range not specified.

C_2HBr MW 104.93

Bromoacetylene, CA 593-61-3: Unstable; spontaneously flammable; may explode in contact with air; polymerizes in light; soluble water, ether, dilute nitric acid. Tb = 277.85; Antoine (liq) A = 6.16013, B = 1015.83, C = -32.86, dev = 0.1 to 1.0, range = 214 to 273.

$C_2HBrClF_3$ MW 197.38

2-Bromo-2-chloro-1,1,1-trifluoroethane, Halothane, CA 151-67-7: Anesthetic; expected teratogen; sensitive to light; slightly soluble water; soluble most organic solvents. Tb = 323.35; Vm = 0.1061 at 293.15; Antoine I (liq) A = 5.9762, B = 1082.495, C = -50.71, dev = 0.1, range = 222 to 329; Antoine II (liq) A = 5.66153, B = 915.806, C = -72.844, dev = 0.1, range = 298 to 323.

$C_2HBr_2FO_2$ MW 235.84

Fluorodibromoacetic acid. Tm = 246.65; Antoine (liq) A = 7.22386, B = 2029.9, C = -81.99, dev = 5.0, range = 403 to 468.

C_2HBr_3O MW 280.74

Tribromoacetaldehyde, bromal, CA 115-17-3: Irritant; once used as hypnotic; habit-forming; flammable, flash point 338; forms water-bromal hydrate; soluble water, ethanol, ether, acetone. Tb = 447.15, decomposes; Vm = 0.1053 at 298.15; Antoine (liq) A = 6.96352, B = 2088.8, C = -25.921, dev = 1.0, range = 291 to 447.

C_2HCl MW 60.48

Chloroacetylene, CA 593-63-5: Highly explosive; spontaneously flammable in air; very poisonous; forms explosive salts; soluble water. Tm = 147.15; Tb = 241.15 to 243.15; Antoine (liq) A = 5.68112, B = 719.713, C = -45.52, dev = 0.1, range = 204 to 238.

C_2HClF_2 MW 98.48

1-Chloro-2,2-difluoroethylene, CA 359-10-4: Tm = 134.65; Tb = 255.45; Antoine (liq) A = 6.93502, B = 1254.974, C = 0, dev = 1.0, range not specified.

C_2HClF_4 MW 136.48

1-Chloro-1,1,2,2-tetrafluoroethane, CA 354-25-6: Tm = 156.15; Tb = 262.95; Vm = 0.0990 at 293.15.

C_2HClF_8S MW 244.53

(1,1,2-Trifluoro-2-chloroethyl) sulfur pentafluoride, CA 22756-13-4: Antoine (liq) A = 6.801, B = 1580, C = 0, dev = 1.0 to 5.0, range = 279 to 323.

C_2HCl_3 MW 131.39

Trichloroethylene, CA 79-01-6: Irritant; moderately toxic by inhalation
and skin absorption; suspected carcinogen; nonflammable under normal use;
industrial degreasing solvent; insoluble water; soluble most organic solvents. Tm = 186.05; Tb = 359.85; Vm = 0.0897 at 293.15; Tc = 544.15; Pc =
5020; Vc = 0.256; Antoine (liq) A = 6.15298, B = 1315, C = -43.15, dev =
0.1 to 1.0, range = 280 to 428.

$C_2HCl_3F_2O_3S$ MW 249.44

Fluorosulfuric acid, 2-fluoro-1,1,2-trichloroethyl ester, CA 42087-88-7:
Antoine (liq) A = 6.555, B = 1911, C = 0, dev not specified, range = 317
to 353.

C_2HCl_3O MW 147.39

Dichloroacetyl chloride, CA 79-36-7: Causes severe burns; corrosive;
lachrymator; flammable, flash point 339; decomposed by water; soluble
ether. Tb = 381.15 to 384.15; Vm = 0.0962 at 289.15.

Trichloroacetaldehyde, chloral, CA 75-87-6: Irritant; toxic; hydrate is
the well-known "knockout drops"; organic intermediate; soluble water,
ethanol, ether. Tm = 215.15 to 216.15; Tb = 371.15; Vm = 0.0956 at 273.15;
Antoine (liq) A = 6.39135, B = 1503.93, C = -28.034, dev = 1.0, range = 235
to 371.

$C_2HCl_3O_2$ MW 163.39

Trichloroacetic acid, CA 76-03-9: Highly irritant; highly toxic and carcinogenic; causes severe burns; flammable, flash point over 385; strong
acid; soluble water, ethanol, ether. Tm = 331.15; Tb = 470.65; Vm = 0.1003
at 334.15; Antoine (liq) A = 6.43547, B = 1618.97, C = -105.268, dev = 0.1,
range = 326 to 473.

C_2HCl_4FS MW 217.90

(Dichloromethyl)(fluorodichloromethyl) sulfide: Antoine (liq) A = 7.71867,
B = 2427.19, C = 0, dev = 5.0, range = 322 to 352.

C_2HCl_5 MW 202.29

Pentachloroethane, CA 76-01-7: Highly toxic by inhalation and skin absorption, narcotic; nonflammable liquid with chloroform odor; insoluble water;
soluble ethanol, ether. Tf = 244.15; Tb = 435.1; Vm = 0.1206 at 293.15;
Antoine (liq) A = 6.69025, B = 1982.65, C = -10.577. dev = 1.0, range = 274
to 434.

$C_2HF_3N_2$ MW 110.04

Ethane, 2-diazo-1,1,1-trifluoro, CA 371-67-5: Soluble water, ether. Tb =
286.15; Antoine (liq) A = 7.0759, B = 1442, C = 0, dev = 1.0 to 5.0, range
not specified.

$C_2HF_3O_2$ MW 114.02

Trifluoroacetic acid, CA 76-05-1: Moderately toxic; corrosive; causes
severe burns; strong acid used in synthesis; soluble water, ethanol, ether,
acetone, benzene. Tf = 257.79; Tb = 344.95; Vm = 0.0768 at 298.15; Antoine
(liq) A = 6.59971, B = 1408.69, C = -38.289, dev = 0.1 to 1.0, range = 285
to 345.

C_2HF_5 MW 120.02

Pentafluoroethane, CA 354-33-6: Tm = 170.15; Tb = 224.65.

C_2HF_6N MW 153.03

Bis(trifluoromethyl) amine, CA 371-77-7: Tm = 143.15; Tb = 266.45; Antoine
(liq) A = 7.021, B = 1335, C = 0, dev = 5.0, range = 207 to 265.

C_2HF_6NOS MW 201.09

Sulfoxime, S,S-bis(trifluoromethyl), CA 34556-22-4: Antoine (liq) A =
7.304, B = 1830.4, C = 0, dev = 1.0 to 5.0, range not specified.

C_2HF_6OP MW 185.99

Phosphinous acid, bis(trifluoromethyl) ester, CA 359-65-9: Tm = 252.05;
Antoine I (sol) A = 8.9384, B = 1998.3, C = -22.75, dev = 1.0, range = 233
to 251; Antoine II (liq) A = 5.95542, B = 1025.8, C = -73.07, dev = 1.0,
range = 256 to 288.

C_2HF_6OPS MW 218.05

Bis(trifluoromethyl) thiophosphinic acid, CA 35814-49-4: Antoine (liq) A =
7.1782, B = 2001, C = 0, dev = 1.0, range = 283 to 324.

C_2HF_6PS MW 202.05

Thiophosphinous acid, bis(trifluoromethyl) ester, CA 1486-19-7: Tm =
173.65; Antoine (liq) A = 6.98036, B - 1561.96, C = -12.18, dev = 1.0,
range = 217 to 280.

$C_2HF_6PS_2$ MW 234.11

Phosphinodithioic acid, bis(trifluoromethyl) ester, CA 18799-75-2: Tm =
287.15; Antoine I (sol) A = 8.0279, B = 2188, C = 0, dev = 1.0, range = 273
to 287; Antoine II (liq) A = 5.89755, B = 1225.71, C = -63.57, dev = 0.1 to
1.0, range = 291 to 315.

C_2HF_7S MW 190.08

Pentafluoro(2,2-difluoroethenyl) sulfur, CA 58636-78-5: Antoine (liq) A =
6.815, B = 1447.6, C = 0, dev not specified, range not specified.

C_2HF_9S MW 228.08

Pentafluoro(1,2,2,2-tetrafluoroethyl) sulfur, CA 63011-80-3: Antoine (liq)
A = 6.885, B = 1463, C = 0, dev not specified, range not specified.

$C_2HF_{10}NS$ MW 261.08

Sulfur, tetrafluoro(1,1,1-trifluoromethanaminato)(trifluoromethyl),
CA 56868-58-7: Antoine (liq) A = 6.045, B = 1352, C = 0, dev = 1.0 to 5.0,
range not specified.

C_2H_2 MW 26.04

Acetylene, CA 74-86-2: Extremely flammable; may explode on heating; forms
many explosive compounds; polymerizes to explosive low polymers; flash
point 256; slightly soluble water; moderately soluble ethanol, benzene,
acetic acid; very soluble acetone. Tm = 192.04; Tb = 189.15, sublimes;
Vm = 0.0420 at 192.04; Tc = 308.35; Pc = 6190; Vc = 0.1130; Antoine I (sol)
A = 8.11522, B = 1162.05, C = 0.854, dev = 0.1, range = 150 to 192; Antoine
II (sol) = 8.6028, B = 1231.7, C = 0.28, dev = 1.0, range = 98 to 145;
Antoine III (liq) A = 6.27098, B - 726.768, C = -18.008, dev = 0.1, range =
192 to 308; Antoine IV (liq) A = 6.22829, B = 710.803, C = -20.109, dev =
0.02, range = 192 to 225; Antoine V (liq) A = 6.53573, B = 849.10, C =
-0.164, dev = 0.1, range = 225 to 259; Antoine VI (liq) A = 7.10920, B =
1185.808, C = 48.683, dev = 1.0, range = 258 to 308.

$C_2H_2Br_2$ MW 185.85

1,2-Dibromoethylene, *cis*, CA 590-11-4: Nonflammable; insoluble water;
soluble organic solvents. Tm = 220.15; Tb = 385.65; Vm = 0.0813 at 290.15;
Antoine (liq) A = 7.525, B = 2120, C = 0, dev = 5.0, range = 299 to 351.

1,2-Dibromoethylene, *trans*, CA 590-12-5: Nonflammable; insoluble water;
soluble organic solvents. Tm = 266.65; Tb = 381.15; Vm = 0.0820 at 290.65;
Antoine (liq) A = 6.835, B = 1840, C = 0, dev = 5.0, range = 277 to 344.

$C_2H_2Br_2Cl_2$ MW 256.75

1,2-Dibromo-1,1-dichloroethane, CA 75-81-0: Soluble organic solvents.
Tm = 339.95; Tb = 449.15 to 451.15; Antoine (liq) A = 6.16268, B = 1593.35,
C = -68.15, dev = 0.1, range = 354 to 519.

1,2-Dibromo-1,2-dichloroethane, CA 683-68-1: Soluble organic solvents.
Tm = 247.15; Tb = 468.15; Vm = 0.1126 at 298.15; Antoine (liq) A = 7.5729,
B = 2722.5, C = 22.26, dev = 1.0, range = 320 to 379.

$C_2H_2Br_4$ MW 345.65

1,1,1,2-Tetrabromoethane, CA 630-16-0: Irritant; decomposes above 385;
nonflammable under normal conditions; soluble organic compounds. Tm =
257.15; Vm = 0.1213 at 298.15; Antoine (liq) A = 8.16136, B = 2757.31, C =
-25.355, dev = 1.0, range = 331 to 473.

(continues)

$C_2H_2Br_4$ *(continued)*

1,1,2,2-Tetrabromoethane, CA 79-27-6: Irritant; very potent mutagen;
highly toxic vapor; nonflammable under normal conditions; insoluble water;
soluble ethanol, ether, chloroform. Tm = 273.25; Tb = 516.65; Vm = 0.1165
at 293.15; Antoine (liq) A = 6.50284, B = 1963, C = -80.15, dev = 1.0,
range = 413 to 573.

C_2H_2ClFO MW 96.49

Chloroacetyl fluoride, CA 359-14-8: Antoine (liq) A = 7.7279, B = 1982.5,
C = 0, dev = 1.0, range = 273 to 333.

Fluoroacetyl chloride, CA 359-06-8: Antoine (liq) A = 7.6203, B = 1916.6,
C = 0, dev = 1.0, range = 273 to 333.

$C_2H_2ClF_3$ MW 118.49

1,1,1-Trifluorochloroethane, CA 75-88-7: Tm = 172.15; Tb = 279.25; Vm =
0.0855 at 273.15.

$C_2H_2ClF_3O_2S$ MW 182.54

2,2,2-Trifluoroethanesulfonic acid, chloride, CA 1648-99-3: Antoine (liq)
A = 7.045, B = 1881, C = 0, dev not specified, range not specified.

$C_2H_2ClF_7S$ MW 226.54

Pentafluoro(2,2-difluoro-2-chloroethyl) sulfur, CA 68010-35-5: Antoine
(liq) A = 7.005, B = 1719.9, C = 0, dev not specified, range not specified.

$C_2H_2Cl_2$ MW 96.94

1,1-Dichloroethylene, vinylidene chloride, CA 75-35-4: Moderately toxic by
inhalation; expected carcinogen; rapidly absorbs oxygen to form explosive
peroxides; polymerizes; extremely flammable, flash point 245; polymer
intermediate; reacts with alcohols and halides; insoluble water; soluble
most organic solvents. Tf = 150.59; Tb = 304.71; Vm = 0.0796 at 293.15;
Tc = 493.95; Pc = 5210; Vc = 0.218; Antoine (liq) A = 6.10046, B = 1100.67,
C = -35.9, dev = 0.1, range = 244 to 306.

1,2-Dichloroethylene, *cis*, CA 156-59-2: Suspected carcinogen; gradually
decomposed by air and light; moisture-sensitive; flammable, flash point 279;
degreasing solvent; soluble organic solvents. Tm = 191.68; Tb = 333.35;
Vm = 0.0751 at 288.15; Antoine I (liq) A = 6.30025, B = 1293.95, C = -32.41,
dev = 1.0, range = 273 to 334; Antoine II (liq) A = 6.22178, B = 1271.55,
C = -30.557, dev = 1.0, range = 332 to 495.

1,2-Dichloroethylene, *trans*, CA 156-60-5: Suspected carcinogen; moderately
toxic; gradually decomposed by air and light; flammable, flash point 277;
degreasing solvent; soluble organic solvents. Tm = 223.71; Tb = 320.85;
Vm = 0.0766 at 288.15; Antoine I (liq) A = 5.93307, B = 1059.93, C = -50.83,
dev = 1.0, range = 263 to 323; Antoine II (liq) A = 6.27465, B = 1226.69,
C = -33.653, dev = 1.0, range = 321 to 473.

$C_2H_2Cl_2F_2$ MW 134.94

1,2-Dichloro-1,1-difluoroethane, CA 1649-08-7: Tm = 171.95; Tb = 319.95;
Vm = 0.0953 at 293.15; Antoine (liq) A = 6.5782, B = 1453.91, C = 0, dev =
1.0 to 5.0, range = 323 to 493.

$C_2H_2Cl_2O$ MW 112.94

Chloroacetyl chloride, CA 79-04-9: Highly irritant; lachrymator; causes
severe burns; corrosive; nonflammable; decomposed by water; soluble organic
solvents. Tm = approximately 251.15; Tb = 381.15 to 383.15; Vm = 0.0755 at
273.15; Tc = 581, calculated; Pc = 5110, calculated; Vc = 0.245, calculated;
Antoine (liq) A = 6.25901, B = 1331.94, C = -65.974, dev = 1.0, range = 253
to 379.

Dichloroacetaldehyde, CA 79-02-7: Highly irritant; polymerizes on standing;
soluble ethanol. Tb = 363.15 to 364.15.

$C_2H_2Cl_2O_2$ MW 128.94

Dichloroacetic acid, CA 79-43-6: Highly irritant; corrosive; toxic; causes
severe burns; flammable, flash point over 385; strong acid; soluble water,
ethanol, ether, acetone. Tm = 286.65; Tb = 467.15; Vm = 0.0825 at 293.15;
Antoine (liq) A = 7.47122, B = 2385.6, C = -31.197, dev = 1.0, range = 317
to 468.

$C_2H_2Cl_4$ MW 167.85

1,1,1,2-Tetrachloroethane, CA 630-20-6: Highly toxic; suspected carcinogen;
nonflammable; insoluble water; soluble most organic solvents. Tm = 204.45;
Tb = 403.65; Vm = 0.1089 at 298.15; Antoine (liq) A = 6.1005, B = 1410.7,
C = -59.15, dev = 1.0, range = 316 to 447.

1,1,2,2-Tetrachloroethane, CA 79-34-5: Highly toxic by inhalation and skin
contact; suspected carcinogen; nonflammable solvent; slightly soluble water;
soluble most organic solvents. Tm = 230.65; Tb = 419.45; Vm = 0.1056 at
298.15; Tc = 661.15; Pc = 3990; Antoine I (liq) A = 6.1073, B = 1465.1, C =
-62.15, dev = 1.0, range = 328 to 464; Antoine II (liq) A = 5.98913, B =
1365.692, C = -76.476, dev = 0.1 to 1.0, range = 377 to 419.

$C_2H_2Cl_4S$ MW 199.91

Bis(dichloromethyl) sulfide: Antoine (liq) A = 7.373, B = 2484.5, C = 0,
dev = 1.0, range = 355 to 462.

C_2H_2FN MW 59.04

Acetonitrile, fluoro, CA 503-20-8: Highly toxic; flammable, flash point
259. Tb = 352.15 to 353.15; Vm = 0.556 at 293.15; Antoine (liq) A =
7.7211, B = 1992.8, C = 0, dev = 1.0, range = 273 to 333.

$C_2H_2F_2$ MW 64.03

1,1-Difluoroethylene, vinylidene fluoride, CA 75-38-7: Flammable; polymerizes; slightly soluble water; soluble ethanol, ether;   *(continues)*

$C_2H_2F_2$ *(continued)*

Tf = 129.15; Tb = 187.45; Tc = 303.25; Pc = 4433; Vc = 0.1535; Antoine
(liq) A = 6.3015, B = 805, C = 0, dev = 1.0 to 5.0, range not specified.

$C_2H_2F_3NO$ MW 113.04

Acetamide, 2,2,2-trifluoro, CA 354-38-1: Tm = 343.15 to 348.15; Tb =
435.15 to 436.15; Antoine (sol) A = 7.47964, B = 1660.074, C = -112.457,
dev = 1.0, range = 287 to 329.

$C_2H_2F_4$ MW 102.03

1,1,1,2-Tetrafluoroethane, CA 811-97-2: Soluble ether. Tm = 172.15; Tb =
246.65.

$C_2H_2F_4O_2S$ MW 166.09

Fluorosulfurous acid, 2,2,2-trifluoroethyl ester, CA 75988-14-6: Antoine
(liq) A = 7.305, B = 1755, C = 0, dev not specified, range not specified.

$C_2H_2F_6P_2$ MW 201.98

1,2-Bis(trifluoromethyl) diphosphine, CA 462-57-7: Antoine (liq) A =
6.2761, B = 1285.028, C = -40.384, dev = 1.0, range = 233 to 292.

$C_2H_2F_8S$ MW 210.08

Pentafluoro(2,2,2-trifluoroethyl) sulfur, CA 65227-29-4: Antoine (liq) A =
6.895, B = 1530.8, C = 0, dev not specified, range not specified.

Tetrafluoro(fluoromethyl)(trifluoromethyl) sulfur, CA 56919-36-9: Antoine
(liq) A = 6.325, B = 1425, C = 0, dev = 1.0 to 5.0, range not specified.

$C_2H_2I_2$ MW 279.85

1,2-Diiodoethylene, *cis*, CA 590-26-1: Soluble ether, chloroform. Tm =
259.15; Tb = 461.15, decomposes; Vm = 0.0914 at 293.15; Antoine (liq) A =
7.305, B = 2430, C = 0, dev = 5.0, range = 302 to 425.

1,2-Diiodoethylene, *trans*, CA 590-27-2: Tm = 351.15; Tb = 465.15, decomposes; Antoine I (sol) A = 4.985, B = 2130, C = 0, dev = 5.0, range = 265
to 293; Antoine II (liq) A = 6.745, B = 2210, C = 0, dev = 5.0, range = 350
to 403.

C_2H_2O MW 42.04

Ketene, CA 463-51-4: Highly toxic vapor; polymerizes readily; flammable;
important intermediate; decomposes in water, ethanol; soluble ether,
acetone. Tm = 122.15; Tb = 223.35; Vm = 0.0530 at 223.35; Tc = 370, calculated; Pc = 5810, calculated; Vc = 0.144, calculated; Antoine (liq) A =
5.80297, B = 711.14, C = -36.39, dev = 1.0, range = 159 to 224.

$C_2H_2O_2$ MW 58.04

Glyoxal, CA 107-22-2: Irritant; reactive dialdehyde; polymerizes on standing; reacts violently with water and other materials. Tm = 288.15; Tb =
324.15; Vm = 0.0450 at 293.15.

$C_2H_2O_4$ MW 90.04

Oxalic acid, CA 144-62-7: Highly toxic by oral or skin absorption; salts
are highly toxic; hygroscopic; very soluble water; moderately soluble
ethanol. Tm = 462.65, decomposes; Antoine I (sol, *alpha*) A = 12.295, B =
5130, C = 0, dev = 5.0, range = 310 to 325; Antoine II (sol, *alpha*) A =
14.42973, B = 6971.431, C = 61.606, dev = 1.0, range = 303 to 328; Antoine
III (sol, *beta*) A = 11.695, B = 4875, C = 0, dev = 5.0, range = 310 to 325.

C_2H_3Br MW 106.95

Vinyl bromide, CA 593-60-2: Toxic; suspected carcinogen; extremely flammable, flash point 265; polymerizes easily in light; insoluble water; soluble ethanol, ether. Tf = 135.35; Tb = 288.95; Vm = 0.0718 at 293.15;
Antoine (liq) A = 6.03266, B = 1014, C = -37.15; dev = 1.0, range = 224
to 319.

C_2H_3BrO MW 122.95

Acetyl bromide, CA 506-96-7: Corrosive; highly toxic; causes burns;
lachrymator; reacts violently with water or ethanol; soluble ether, benzene, chloroform. Tm = 177.15; Tb = 349.15, decomposes; Vm = 0.0740 at
289.15; Antoine (liq) A = 4.3953, B = 573.3, C = -118.15, dev = 1.0,
range = 289 to 334.

$C_2H_3BrO_2$ MW 138.95

Bromoacetic acid, CA 79-08-3: Corrosive; irritant and highly toxic;
lachrymator; flammable, flash point over 385; hygroscopic; soluble water,
ethanol, ether, acetone. Tm = 323.15; Tb = 481.15; Vm = 0.0719 at 323.15;
Antoine (liq) A = 7.7104, B = 2621.03, C = -21.75, dev = 1.0, range = 327
to 481.

$C_2H_3Br_3$ MW 266.76

1,1,2-Tribromoethane, CA 78-74-0: Soluble ethanol, ether, benzene, chloroform. Tm = 247.15; Tb = 462.15; Vm = 0.1018 at 292.15; Antoine (liq) A =
6.48876, B = 1757, C = -70.15, dev = 1.0, range = 368 to 511.

C_2H_3Cl MW 62.50

Vinyl chloride, CA 75-01-4: Irritant; potent carcinogen; extremely flammable, flash point 195; polymerizes readily; may form unstable peroxides;
polymer raw material; slightly soluble water; soluble ethanol, ether, most
organic solvents. Tm = 119.35; Tb = 259.75; Vm = 0.0685 at 293.15; Tc =
429.75; Pc = 5600; Vc = 0.169; Antoine I (liq) A = 5.98598, B = 892.757,
C = -35.051, dev = 1.0, range = 208 to 260; Antoine II (liq) A = 5.20198,
B = 556.26, C = 187.765, dev = 1.0, range = 259 to 328.

$C_2H_3ClF_2$ MW 100.50

1-Chloro-1,1-difluoroethane, CA 75-68-3: Refrigerant; insoluble water; soluble most organic solvents. Tm = 142.15; Tb = 263.35; Vm = 0.0903 at 298.15; Tc = 410.25; Pc = 4120; Antoine (liq) A = 6.5025, B = 1184.5, C = 0, dev = 1.0, range = 248 to 390.

$C_2H_3ClF_3N$ MW 133.50

Ethanamine, *N*-chloro-*N*,1,1-trifluoro, CA 16276-45-2: Tm = 161.55; Antoine (liq) A = 8.39485, B = 2343.795, C = 57.507, dev = 1.0 to 5.0, range = 220 to 294.

$C_2H_3ClF_3P$ MW 150.47

Chloromethyl(trifluoromethyl) phosphine, CA 4669-76-5: Antoine (liq) A = 6.08028, B = 1138.235, C = -44.461, dev = 1.0, range = 236 to 294.

$C_2H_3ClF_6OS$ MW 224.55

Pentafluoro(2-fluoro-2-chloroethoxy) sulfur: Antoine (liq) A = 6.36179, B = 1372.372, C = -47.592, dev = 1.0, range = 285 to 363.

C_2H_3ClO MW 78.50

Acetyl chloride, CA 75-36-5: Highly toxic and irritant; corrosive; causes burns; highly flammable, flash point 277; reacts violently with water and ethanol; soluble ether, benzene, chloroform. Tm = 160.29; Tb = 324.95; Vm = 0.0710 at 293.15; Tc = 508, calculated; Pc = 5740, calculated; Vc = 0.196, calculated; Antoine (liq) A = 5.96568, B = 1062.86, C = -55.531, dev = 1.0 to 5.0, range = 267 to 324.

Chloroacetaldehyde, CA 107-20-0: Highly irritant; polymerizes on standing. Tb = 358.15.

$C_2H_3ClO_2$ MW 94.50

Chloroacetic acid, CA 79-11-8: Corrosive; highly irritant; causes severe burns; flammable, flash point 399; used as herbicide; very soluble water; soluble most organic solvents. Tm = 336.15, *alpha* form, highest melting of three solid forms. Tb = 462.50; Vm = 0.0686 at 336.15; Tc = 686, calculated; Pc = 5780, calculated; Vc = 0.221, calculated; Antoine I (liq) A = 6.69087, B = 1733.96, C = -92.154, dev = 1.0, range = 336 to 463; Antoine II (liq) A = 6.67975, B = 1727.293, C = -92.742, dev = 0.1 to 1.0, range = 377 to 464.

Methyl chloroformate, CA 79-22-1: Toxic vapor, highly irritant; highly flammable, flash point 285; slightly soluble water, with decomposition; soluble ethanol, ether, benzene, chloroform. Tb = 344.15; Vm = 0.0756 at 293.15.

$C_2H_3Cl_3$ MW 133.40

1,1,1-Trichloroethane, methyl chloroform, CA 71-55-6: Irritant; moderately toxic; suspected carcinogen; nonflammable cleaning and   *(continues)*

42

$C_2H_3Cl_3$ *(continued)*

degreasing solvent; insoluble water; soluble most organic solvents. Tm = 240.65; Tb = 347.25; Vm = 0.0996 at 293.15; Antoine I (liq) A = 5.98755, B = 1182.527, C = -50.256, dev = 0.02, range = 295 to 372; Antoine II (liq) A = 6.00452, B = 1193.604, C = -48.743, dev = 0.1, range = 349 to 408; Antoine III (liq) A = 6.36873, B = 1474.394, C = -8.08, dev = 1.0, range = 399 to 487; Antoine IV (liq) A = 7.3598, B = 2563.458, C = 142.906, dev = 1.0 to 5.0, range = 479 to 545.

1,1,2-Trichloroethane, CA 79-00-5: Irritant; moderately toxic; suspected carcinogen; nonflammable industrial solvent; insoluble water; soluble most organic solvents. Tm = 237.65; Tb = 386.15 to 387.15; Vm = 0.0926 at 293.15; Antoine I (liq) A = 6.09017, B = 1351, C = -56.15, dev = 0.1, range = 302 to 428; Antoine II (liq) A = 6.13875, B = 1351.685, C = -59.953, dev = 0.1, range = 316 to 384.

$C_2H_3Cl_3O_2$ MW 165.40

Chloral hydrate, CA 302-17-0: Irritant; moderately toxic; powerful hypnotic and sedative; addictive drug; soluble water, ethanol, ether, chloroform. Tm = 326.15; Tb = 369.45, decomposes; Antoine I (sol) A = 9.6406, B = 2865.22, C = 8.579, dev = 1.0, range = 263 to 320; Antoine II (liq) A = 9.02003, B = 2590.78, C = 0, dev = 1.0, range = 325 to 370.

C_2H_3F MW 46.04

Vinyl fluoride, CA 75-02-5: Polymerizes readily; flammable; slightly soluble water; soluble ethanol, acetone. Tf = 112.65; Tb = 200.95; Tc = 327.85; Pc = 5100; Vc = 0.1439; Antoine (liq) A = 6.07797, B = 780.10, C = -9.633, dev = 1.0, range = 124 to 201.

C_2H_3FO MW 62.04

Acetyl fluoride, CA 557-99-3: Corrosive; lachrymator; flammable; decomposed by water; moderately soluble ethanol, ether, acetone. Tm = 189.15; Tb = 293.15 to 294.15; Vm = 0.0601 at 273.15; Antoine (liq) A = 4.578, B = 748.1, C = 0, dev = 5.0, range = 195 to 281.

$C_2H_3FO_2$ MW 78.04

Fluoroacetic acid, CA 144-49-0: Highly toxic; flammable; used as rat poison; soluble water, ethanol. Tm = 306.15; Tb = 438.15; Vm = 0.0570 at 309.15; Antoine (liq) A = 8.25875, B = 2775.1, C = 2.46, dev = 1.0, range = 293 to 443.

$C_2H_3F_3$ MW 84.04

1,1,1-Trifluoroethane, CA 420-46-2: Refrigerant; insoluble water; soluble ether, chloroform. Tm = 161.85; Tb = 225.75; Vm = 0.0715 at 323.15; Tc = 346.25; Pc = 3760; Vc = 0.194; Antoine (liq) A = 6.5048, B = 1014.7, C = 0, dev = 1.0 to 5.0, range not specified.

$C_2H_3F_3N_2$ MW 112.05

Azomethane, 1,1,1-trifluoro, CA 690-21-1: Tb = 275.75; Antoine (liq) A =
7.0009, B = 1377, C = 0, dev = 1.0, range = 240 to 273.

$C_2H_3F_3O$ MW 100.04

Methyl(trifluoromethyl) ether, CA 421-14-7: Tm = 176.95; Antoine (liq) A =
7.8678, B = 1760.81, C = 0, dev = 1.0 to 5.0, range not specified.

2,2,2-Trifluoroethanol, CA 75-89-8: Moderately toxic; flammable, flash
point 306; soluble most organic solvents. Tm = 228.15; Tb = 346.75; Vm =
0.06815 at 293.15; Antoine I (liq) A = 8.776, B = 2325, C = 0, dev = 1.0 to
5.0, range = 298 to 328; Antoine II (liq) A = 5.9656, B = 952.466, C =
-106.563, dev = 1.0, range = 276 to 302.

$C_2H_3F_3O_2S$ MW 148.10

Trifluoromethanesulfonic acid, methyl ester, CA 333-27-7: Antoine (liq)
A = 6.795, B = 1656, C = 0, dev = 1.0 to 5.0, range not specified.

$C_2H_3F_5O_3S$ MW 202.10

Sulfur, (ethaneperoxoato)pentafluoro, CA 60672-60-8: Tm = 217.55; Antoine
(liq) A = 7.0303, B = 1895, C = 0, dev = 1.0 to 5.0, range not specified.

C_2H_3IO MW 169.95

Acetyl iodide, CA 507-02-8: Irritant; highly toxic; flammable; slowly decomposes in air; decomposed by water or ethanol; soluble ether, benzene.
Tb = 381.15; Antoine (liq) A = 3.00167, B = 285.36, C = -177.53, dev = 1.0,
range = 276 to 302.

C_2H_3N MW 41.05

Acetonitrile, CA 75-05-8: Toxic; lachrymator; flammable, flash point 279;
good solvent, used in extractive distillation; soluble water, most organic
solvents. Tf = 227.45; Tb = 354.75; Vm = 0.0524 at 293.15; Tc = 545.5;
Pc = 4833; Vc = 0.173; Antoine (liq) A = 6.34522, B = 1388.446, C = -34.856,
dev = 0.02 to 0.1, range = 314 to 355.

C_2H_3NO MW 57.05

Methyl isocyanate, CA 624-83-9: Irritant; lachrymator; highly toxic; flammable, flash point below 266; intermediate for insecticides and synthetic
rubber; reacts with water, soluble organic solvents. Tb = 316.15 to
318.15; Vm = 0.0594 at 293.15; Antoine I (liq) A = 7.0308, B = 1562, C = 0,
dev = 5.0, range = 265 to 308; Antoine II (liq) A = 5.6419, B = 865.515,
C = -74.094, dev = 1.0, range = 253 to 310.

$C_2H_3NO_3$ MW 89.05

Oxalic acid, monoamide, CA 471-47-6: Slightly soluble water; insoluble
ethanol, ether. Tm = 483.15, decomposes; Antoine (sol) A = 11.7049, B =
5639, C = 0, dev = 1.0 to 5.0, range = 348 to 364.

$C_2H_3NO_5$ MW 121.05

Peroxide, acetyl nitro, CA 2278-22-0: Violently explosive when condensed as a pure liquid; important air pollutant. Tm = 224.65; Antoine (liq) A = 6.66449, B = 1808.47, C = 0, dev = 5.0, range = 277 to 330.

C_2H_3NS MW 73.11

Methyl isothiocyanate, CA 556-61-6: Highly irritant; highly toxic; flammable, flash point 305; slightly soluble water; very soluble ethanol, ether. Tm = 309.15; Tb = 392.15; Vm = 0.0684 at 310.15; Antoine I (sol) A = 5.883, B = 1570.53, C = -5.76, dev = 1.0, range = 238 to 303; Antoine II (liq) A = 6.59162, B = 1696.42, C = -22.23, dev = 1.0, range = 309 to 392.

Methylthiocyanate, CA 556-64-9: Highly toxic; lachrymator; flammable, flash point 311; slightly soluble water; soluble ethanol, ether. Tf = 222.15; Tb = 403.65; Vm = 0.0679 at 288.15; Antoine (liq) A = 6.39459, B = 1634.79, C = -33.65, dev = 1.0, range = 259 to 406.

C_2H_4 MW 28.05

Ethylene, CA 74-85-1: Extremely flammable gas, explosive reaction with many materials; most important chemical raw material; slightly soluble water; very soluble ethanol, ether, acetone. Tm = 104.15; Tb = 169.44; Vm = 0.0496 at 169.15; Tc = 282.35; Pc = 5042; Vc = 0.1310; Antoine I (sol) A = 8.1413, B = 955.76, C = 0, dev = 1.0, range = 79 to 104; Antoine II (liq) A = 5.87246, B = 585.0, C = -18.15, dev = 0.1, range = 120 to 182; Antoine III (liq) A = 6.2375, B = 716.494, C = 0, dev = 0.1 to 1.0, range = 170 to 273; Antoine IV (liq) A = 5.88692, B = 588.485, C = -17.786, dev = 0.1, range = 120 to 170; Antoine V (liq) A = 5.97705, B = 617.674, C = -13.878, dev = 0.1, range = 169 to 211; Antoine VI (liq) A = 6.27462, B = 742.522, C = 5.408, dev = 0.1, range = 209 to 254; Antoine VII (liq) A = 7.38145, B = 1426.696, C = 105.443, dev = 0.1, range = 252 to 282.

C_2H_4BrCl MW 143.41

1-Bromo-1-chloroethane, CA 593-96-4: Irritant. Tm = 289.75; Tb = 353.65 to 354.65; Vm = 0.0859 at 289.15; Antoine I (sol) A = 8.9397, B = 2326.5, C = 0, dev = 1.0, range = 237 to 289; Antoine II (liq) A = 6.8695, B = 1730.6, C = 0, dev = 1.0, range = 290 to 356.

1-Bromo-2-chloroethane, CA 107-04-0: Irritant; toxic; nonflammable; soluble ethanol, ether. Tf = 256.45; Tb = 380.15; Vm = 0.0824 at 293.15; Antoine (liq) A = 6.34539, B = 1469.72, C = -41.103, dev = 1.0, range = 244 to 380.

$C_2H_4Br_2$ MW 187.86

1,1-Dibromoethane, ethylidene bromide, CA 557-91-5: Irritant; slightly soluble water; soluble most organic solvents. Tm = 210.15; Tb = 381.15 to 383.15; Vm = 0.0912 at 293.15; Antoine (liq) A = 6.3303, B = 1412, C = -54.65, dev = 1.0, range = 301 to 421.

1,2-Dibromoethane, ethylene dibromide, CA 106-93-4: Irritant; toxic by inhalation and skin contact; suspected carcinogen; nonflammable; slightly soluble water; soluble most organic solvents. Tm = 283.15; Tb = 404.85; Vm = 0.08625 at 293.15; Tc = 650.15; Pc = 5477; Vc = 0.262;   *(continues)*

$C_2H_4Br_2$ *(continued)*

Antoine I (sol) A = 10.03, B = 2863, C = 0, dev = 1.0, range = 228 to 248; Antoine II (sol) A = 9.009, B = 2606.5, C = 0, dev = 1.0, range = 251 to 281; Antoine III (liq) A = 7.501, B = 2181.1, C = 0, dev = 1.0, range = 283 to 317; Antoine IV (liq) A = 6.18735, B = 1469.7, C = -53.15, dev = 0.1 to 1.0, range = 316 to 488; Antoine V (liq) A = 8.16841, B = 3200, C = 117.25, dev = 1.0, range = 404 to 578.

C_2H_4ClF MW 82.50

1-Chloro-2-fluoroethane, CA 762-50-5: Tm = below 223; Tb = 330.15; Vm = 0.0702 at 296.15; Antoine (liq) A = 7.33341, B = 1788.6, C = 9.73, dev = 1.0, range = 288 to 327.

$C_2H_4ClN_3$ MW 105.53

1-Chloro-2-azidoethane, CA 53422-48-3: Antoine (liq) A = 7.8361, B = 2287.3, C = 0, dev = 1.0 to 5.0, range = 273 to 333.

$C_2H_4Cl_2$ MW 98.96

1,1-Dichloroethane, ethylidene chloride, CA 75-34-3: Moderately toxic; expected teratogen; highly flammable, flash point 267; slightly soluble water; soluble most organic solvents. Tm = 176.45; Tb = 330.45; Vm = 0.0842 at 293.15; Tc = 534.65; Pc = 5060; Antoine I (liq) A = 6.1102, B = 1171.42, C = -45.03, dev = 0.1, range = 258 to 365; Antoine II (liq) A = 6.14443, B = 1216.12, C = -36.579, dev = 1.0, range = 323 to 535; Antoine III (liq) A = 6.22839, B = 1288.092, C = -24.381, dev = 1.0 to 5.0, range = 363 to 535.

1,2-Dichloroethane, ethylene dichloride, CA 107-06-2: Toxic; skin irritant; expected carcinogen; highly flammable, flash point 286; intermediate for vinyl chloride; slightly soluble water; soluble most organic solvents. Tm = 237.85; Tb = 356.85; Vm = 0.0790 at 293.15; Tc = 563.15; Pc = 5360; Vc = 0.225; Antoine I (liq) A = 6.07712, B = 1247.8, C = -50.15, dev = 0.1, range = 279 to 434; Antoine II (liq) A = 6.53278, B = 1599.07, C = -3.303, dev = 1.0, range = 356 to 558; Antoine III (liq) A = 6.07287, B = 1245.488, C = -50.392, dev = 0.1 to 1.0, range = 279 to 374; Antoine IV (liq) A = 6.40918, B = 1505.414, C = -14.406, dev = 5.0, range = 368 to 524; Antoine V (liq) A = 7.5198, B = 2129.577, C = 0, dev = 5.0, range = 523 to 561.

$C_2H_4Cl_2O$ MW 114.96

Bis(chloromethyl) ether, CA 542-88-1: Strong irritant; very toxic; carcinogen; decomposed by water; soluble ethanol, ether. Tf = 231.65; Tb = 377.15 to 378.15; Vm = 0.0874 at 293.15.

$C_2H_4Cl_2S$ MW 131.02

Bis(chloromethyl) sulfide, CA 3592-44-7: Decomposed by hot water. Tm = 219.15; Tb = 429.65; Vm = 0.0931 at 293.15; Antoine (liq) A = 7.4896, B = 2358.2, C = 0, dev = 1.0, range = 320 to 430.

$C_2H_4FNO_2$ MW 93.06

2-Fluoroethyl nitrate, CA 4528-33-0: Antoine (liq) A = 7.955, B = 2000, C = 0, dev = 5.0, range = 273 to 333.

$C_2H_4F_2$ MW 66.05

1,1-Difluoroethane, ethylidene fluoride, CA 75-37-6: Tm = 156.15; Tb = 247.35; Vm = 0.0646 at 243.15; Tc = 386.65; Pc = 4500; Vc = 0.181; Antoine I (liq) A = 6.155, B = 910, C = -29.15, dev = 1.0, range = 193 to 275; Antoine II (liq) A = 6.6034, B = 1140.95, C = 0, dev = 1.0, range = 250 to 386.

$C_2H_4F_3N$ MW 99.06

2,2,2-Trifluoroethyl amine, CA 753-90-2: Tb = 309.15; Vm = 0.0796 at 298.15; Antoine (liq) A = 7.062, B = 1568, C = 0, dev = 5.0, range not specified.

$C_2H_4F_3NS$ MW 131.12

Methanesulfenamide, 1,1,1-trifluoro-N-methyl, CA 62067-12-3: Antoine (liq) A = 7.485, B = 1754, C = 0, dev = 1.0 to 5.0, range = 223 to 294.

$C_2H_4F_3OP$ MW 132.02

(Trifluoromethyl)phosphinous acid, methyl ester, CA 6395-71-7: Antoine (liq) A = 6.29667, B = 1174.801, C = -36.238, dev = 1.0, range = 194 to 291.

$C_2H_4F_6OS$ MW 190.10

Pentafluoro(2-fluoroethoxy) sulfur: Antoine (liq) A = 5.75548, B = 1023.869, C = -89.385, dev = 1.0, range = 290 to 364.

$C_2H_4I_2$ MW 281.86

1,2-Diiodoethane, ethylene diiodide, CA 624-73-7: Irritant; decomposes in light; soluble most organic solvents. Tm = 354.15 to 355.15; Tb = 473.15; decomposes; Antoine (liq) A = 6.11324, B = 1647.1, C = -72.15, dev = 5.0, range = 371 to 526.

$C_2H_4N_2O_2$ MW 88.07

1,2-Diformyl hydrazine, CA 628-36-4: Soluble water; slightly soluble ethanol. Tm = 435.15; Antoine (sol) A = 10.655, B = 5264, C = 0, dev = 5.0, range = 370 to 403.

Oxalic acid, diamide, oxamide, CA 471-46-5: Irritant; slightly soluble ethanol, hot water; decomposes above 593. Antoine (sol) A = 11.6929, B = 5893, C = 0, dev = 1.0 to 5.0, range = 353 to 370.

$C_2H_4N_2O_4$ MW 120.06

1,1-Dinitroethane, CA 600-40-8: Slightly soluble water; soluble ethanol,
ether. Tb = 458.15 to 459.15; Vm = 0.0889 at 296.65; Antoine (liq) A =
7.8073, B = 2663.8, C = 0, dev = 1.0, range = 303 to 363.

$C_2H_4N_2O_6$ MW 152.06

Ethylene glycol, dinitrate, CA 628-96-6: Toxic; explodes when heated; soluble ethanol, ether. Tm = 250.85; Tb = 470.15 to 473.15, explodes; Vm =
0.1019 at 293.15; Antoine I (liq) A = 15.0271, B = 10208.5, C = 314.85,
dev = 5.0, range = 343 to 465; Antoine II (liq) A = 4.9662, B = 1170.6, C =
-129.55, dev = 5.0, range = 283 to 353; Antoine III (liq) A = 6.84945, B =
2125.093, C = -58.0, dev = 1.0 to 5.0, range = 240 to 298.

$C_2H_4N_2S_2$ MW 120.19

Dithiooxamide, rubeanic acid, CA 79-40-3: Irritant; toxic; decomposes
above about 443; slightly soluble water; soluble ethanol. Antoine (sol)
A = 11.8379, B = 5515, C = 0, dev = 1.0 to 5.0, range = 359 to 379.

$C_2H_4N_4$ MW 84.08

Dicyandiamide, CA 461-58-5: Moderately soluble water, ethanol; decomposed
by water over 353. Tm = 481.15; Tb, decomposes.

C_2H_4O MW 44.05

Acetaldehyde, CA 75-07-0: Irritant; moderately toxic; undergoes exothermic
polymerization in presence of alkali; extremely flammable, flash point 235;
important chemical intermediate; soluble water, organic solvents. Tm =
149.65; Tb = 293.31; Vm = 0.0566 at 293.15; Tc = 461.0; Pc = 5550, calculated; Vc = 0.157, calculated; Antoine I (liq) A = 6.1410, B = 1034.5, C =
-43.15, dev = 0.1, range = 272 to 294; Antoine II (liq) A = 6.03292, B =
1012.828, C = -41.823, dev = 1.0 to 5.0, range = 293 to 377.

Ethylene oxide, epoxyethane, CA 75-21-8: Highly irritant; toxic; possible
carcinogen; extremely flammable, flash point 244; important intermediate
for ethylene glycol and other derivatives; soluble water, ethanol, ether.
Tm = 162.15; Tb = 283.55; Vm = 0.0499 at 283.15; Tc = 468.95; Pc = 7190;
Vc = 0.140; Antoine I (liq) A = 6.3859, B = 1115.1, C = -29.015, dev = 0.1,
range = 238 to 285; Antoine II (liq) A = 6.31809, B = 1083.31, C = -32.409,
dev = 0.1, range = 223 to 285; Antoine III (liq) A = 6.45597, B = 1170.93,
C = -20.498, dev = 1.0, range = 283 to 385.

C_2H_4OS MW 76.11

Thioacetic acid, CA 507-09-5: Corrosive; irritant; flammable, flash point
296; soluble water, ethanol, ether. Tm = below 256.15; Tb = 366.15; Vm =
0.0713 at 293.15; Antoine (liq) A = 7.1056, B = 1836, C = 0, dev = 5.0,
range = 307 to 360.

$C_2H_4O_2$ MW 60.05

Acetic acid, CA 64-19-7: Corrosive; causes severe burns; flammable, flash
point 312; important solvent, chemical intermediate.   *(continues)*

48

$C_2H_4O_2$ *(continued)*

Tm = 289.79; Tb = 391.02; Vm = 0.0572 at 293.15; Tc = 594.75; Pc = 5786; Vc = 0.171; Antoine I (sol) A = 7.672, B = 2177, C = 0, dev = 1.0, range = 238 to 283; Antoine II (sol) A = 9.9268, B = 2847, C = 0, dev = 1.0 to 5.0, range = 243 to 289; Antoine III (liq) A = 6.68206, B = 1642.54, C = -39.764, dev = 0.1, range = 289 to 392; Antoine IV (liq) A = 7.39226, B = 2258.22, C = 27.82, dev = 1.0, range = 391 to 550; Antoine V (liq) A = 6.5729, B = 1572.32, C = -46.777, dev = 0.1, range = 290 to 396; Antoine VI (liq) A = 6.82561, B = 1748.572, C = -28.259, dev = 0.1, range = 391 to 447; Antoine VII (liq) A = 7.22638, B = 2101.805, C = 12.244, dev = 1.0, range = 437 to 535; Antoine VIII (liq) A = 8.44129, B = 3628.209, C = 182.674, dev = 1.0, range = 525 to 593.

Methyl formate, CA 107-31-3: Irritant; moderately toxic; extremely flammable, flash point 254; moderately soluble water, methanol; soluble ethanol. Tm = 174.15; Tb = 304.65; Vm = 0.0608 at 288.15; Tc = 487.2; Pc = 5998; Vc = 0.172; Antoine I (liq) A = 6.225963, B = 1088.955, C = -46.675, dev = 0.1, range = 279 to 305; Antoine II (liq) A = 6.39684, B = 1196.323, C = -32.629, dev = 1.0 to 5.0, range = 305 to 443.

$C_2H_4O_2S$ MW 92.11

Mercaptoacetic acid, thioglycolic acid, CA 68-11-1: Corrosive; toxic; the ammonium salt is highly toxic; flammable, flash point over 385; oxidizes in air; soluble water, ethanol, ether, benzene, chloroform. Tf = 256.65; Vm = 0.0695 at 293.15; Antoine (liq) A = 9.08435, B = 3743.87, C = 42.978, dev = 1.0, range = 333 to 427.

$C_2H_4O_3$ MW 76.05

Ethylene ozonide: Antoine (liq) A = 7.294, B = 1816.5, C = 0, dev = 5.0, range = 273 to 289.

Glycolic acid, hydroxyacetic acid, CA 79-14-1: Irritant; loses water when vaporized; soluble water, ethanol, methanol, acetone. Tm = 352.15 to 353.15; Tb = 373.15, decomposes; Antoine (liq) A = 9.2596, B = 2708.09, C = 0, dev = 1.0, range = 350 to 375.

Peroxyacetic acid, peracetic acid, CA 79-21-0: Highly corrosive to skin; explodes violently above 383; flammable; powerful oxidizing agent, disinfectant; stable in dilute solutions; soluble water, ethanol, ether. Tf = 273.25; Vm = 0.0620 at 288.15; Antoine (liq) A = 8.036, B = 2311, C = 0, dev = 1.0 to 5.0, range = 273 to 383.

C_2H_4S MW 60.11

Thiirane, ethylene sulfide, CA 420-12-2: Highly irritant; suspected carcinogen; flammable, flash point 283; rapidly polymerizes on standing. Tm = 164.15; Tb = 328.15 to 329.15; Vm = 0.0580 at 293.15; Antoine (liq) A = 6.15386, B = 1189.396, C = -41.337, dev = 0.1, range = 291 to 361.

C_2H_5Br MW 108.97

Ethyl bromide, bromoethane, CA 74-96-4: Irritant; toxic; narcotic; nonflammable under normal conditions; slightly soluble water; *(continues)*

C_2H_5Br *(continued)*

soluble most organic solvents. Tm = 153.85; Tb = 311.55; Vm = 0.0746 at
293.15; Antoine I (liq) A = 6.04485, B = 1090.81, C = -41.44, dev = 0.1,
range = 225 to 333; Antoine II (liq) A = 6.66835, B = 1511.96, C = 12.999,
dev = 1.0, range = 334 to 504; Antoine III (liq) A = 6.77490, B = 1602.705,
C = 25.282, dev = 1.0, range = 326 to 454; Antoine IV (liq) A = 6.99873,
B = 1619.697, C = 0, dev = 1.0 to 5.0, range = 452 to 503.

C_2H_5Cl MW 64.51

Ethyl chloride, chloroethane, CA 75-00-3: Mild irritant; moderately toxic;
extremely flammable, flash point 223; important solvent, refrigerant;
slightly soluble water; soluble ethanol, ether. Tm = 134.85; Tb = 285.55;
Vm = 0.0719 at 293.15; Tc = 459.75; Pc = 5270; Vc = 0.200; Antoine I (liq)
A = 6.07404, B = 1012.771, C = -36.48, dev = 0.1, range = 207 to 305;
Antoine II (liq) A = 6.14258, B = 1053.998, C = -30.686, dev = 1.0, range =
285 to 344; Antoine III (liq) A = 6.4495, B = 1248.788, C = -3.798, dev =
1.0, range = 334 to 413; Antoine IV (liq) A = 6.70739, B = 1465.147, C =
29.696, dev = 1.0, range = 403 to 460.

C_2H_5ClO MW 80.51

2-Chloroethanol, ethylene chlorohydrin, CA 107-07-3: Highly toxic by inhalation and skin absorption; narcotic; moisture-sensitive; flammable,
flash point 333; soluble water, ethanol. Tm = 210.55; Tb = 401.85; Vm =
0.0668 at 293.15; Antoine (liq) A = 7.6496, B = 2262.5, C = 0, dev = 1.0,
range = 328 to 401.

Methyl(chloromethyl) ether, CA 107-30-2: Very toxic; carcinogen; flammable,
flash point 293. Tm = 169.65; Tb = 332.65; Vm = 0.0752 at 298.15; Antoine
(liq) A = 6.259, B = 1240, C = -43.15; dev = 5.0, range = 290 to 332.

$C_2H_5ClO_2S$ MW 128.57

Ethane sulfonyl chloride, CA 594-44-5: Corrosive; lachrymator; flammable,
flash point over 385; soluble ether, methylene dichloride. Tb = 444.15;
Vm = 0.0947 at 295.65; Antoine (liq) A = 9.293, B = 2943.7, C = 0, dev =
1.0 to 5.0, range = 233 to 263.

$C_2H_5Cl_2P$ MW 130.94

Dichloroethyl phosphine, CA 1498-40-4: Irritant; toxic. Antoine (liq) A =
5.464, B = 1029, C = -87.75, dev = 1.0 to 5.0, range = 313 to 385.

C_2H_5F MW 48.06

Ethyl fluoride, fluoroethane, CA 353-36-6: Flammable; slightly soluble
water; soluble ethanol, ether. Tf = 129.95; Tb = 235.45; Vm = 0.0589 at
235.45; Antoine I (liq) A = 6.10343, B = 854.211, C = -26.99, dev = 0.1,
range = 173 to 251; Antoine II (liq) A = 6.15533, B = 867.112, C = -26.466,
dev = 1.0, range = 235 to 280; Antoine III (liq) A = 6.64052, B = 1136.166,
C = 10.91, dev = 1.0, range = 275 to 353; Antoine IV (liq) A = 7.27489, B =
1635.913, C = 82.541, dev = 1.0, range = 343 to 375.

C_2H_5FO MW 64.06

2-Fluoroethanol, ethylene fluorohydrin, CA 371-62-0: Very toxic; flammable, flash point 304; soluble water, ethanol, ether, acetone. Tm = 246.7; Tb = 376.7; Vm = 0.0580 at 293.15; Antoine (liq) A = 8.185, B = 2304, C = 0, dev = 5.0, range = 273 to 333.

$C_2H_5FO_3S$ MW 128.12

Ethyl fluorosulfonate, CA 371-69-7: Highly toxic; highly irritant; flammable, flash point 305. Tb = 385.15 to 386.15; Antoine (liq) A = 7.2129, B = 2010.7, C = 0, dev = 1.0, range = 273 to 385.

$C_2H_5F_2N$ MW 81.06

N,N-Difluoroethylamine, CA 758-18-9: Tm = 122.85; Antoine (liq) A = 6.04079, B = 1009.53, C = -39.78, dev = 1.0, range = 241 to 259.

$C_2H_5F_3NP$ MW 131.04

Methyl(trifluoromethyl) phosphinic acid, amide, CA 4669-74-3: Antoine (liq) A = 6.36104, B = 1311.913, C = -48.465, dev = 1.0, range = 238 to 294.

C_2H_5I MW 155.97

Ethyl iodide, iodoethane, CA 75-03-6: Moderately toxic and irritant vapor; corrosive; moisture-sensitive; sensitive to light; nonflammable under normal conditions; slightly soluble water; soluble most organic solvents. Tm = 164.65; Tb = 345.35; Vm = 0.0801 at 288.15; Tc = 561, calculated; Pc = 5990, calculated; Vc = 0.238, calculated; Antoine (liq) A = 5.95686, B = 1175.709, C = -47.89, dev = 0.1, range = 249 to 369.

C_2H_5N MW 43.07

Aziridine, ethyleneimine, CA 151-56-4: Highly toxic; carcinogen; reacts violently with many materials; polymerizes readily; flammable, flash point 262; soluble water, ethanol, ether. Tf = 199.19; Tb = 329.87; Vm = 0.0518 at 297.15; Antoine (liq) A = 6.12442, B = 1081.631, C = -66.206, dev = 0.1, range = 274 to 303.

C_2H_5NO MW 59.07

Acetaldehyde, oxime, CA 107-29-9: Flammable, flash point 311; soluble water, ethanol, ether. Tm = 319.65; Tb = 387.65; Vm = 0.0612 at 293.15; Antoine (liq) A = 7.43647, B = 1893.47, C = -39.543, dev = 1.0, range = 288 to 388.

Acetamide, CA 60-35-5: Mild irritant; possible carcinogen; excellent dipolar solvent and chemical intermediate; very soluble water, ethanol, chloroform. Tm = 354.15; Tb = 494.35; Vm = 0.0592 at 358.15; Antoine I (sol) A = 10.9717, B = 4050.1, C = 0, dev = 1.0, range = 298 to 349; Antoine II (liq) A = 6.97079, B = 1998.3, C = -89.32, dev = 1.0, range = 381 to 492.

(continues)

C_2H_5NO *(continued)*

N-Methylformamide, CA 123-39-7: Irritant; expected teratogen; flammable, flash point 372; soluble water, ethanol, acetone, most organic solvents. Tm = 267.75; Tb = 453.15 to 458.15; Vm = 0.0585 at 292.15; Antoine I (liq) A = 6.8372, B = 2010.2, C = -56.54, dev = 5.0, range = 369 to 473; Antoine II (liq) A = 7.69505, B = 2569.296, C = -16.089, dev = 1.0, range = 310 to 391.

$C_2H_5NO_2$ MW 75.07

Aminoacetic acid, glycine, CA 56-40-6: Soluble water; slightly soluble ethanol. Tm = 535.15, decomposes; Antoine I (sol) A = 13.595, B = 7120, C = 0, dev = 5.0, range = 423 to 473; Antoine II (sol) A = 13.62467, B = 7131.296, C = 0, dev = 1.0 to 5.0, range = 403 to 429.

Ethyl nitrite, CA 109-95-5: Toxic; mutagenic; decomposes on standing; flammable, flash point 238; may explode from heating or shock; slightly soluble water, with decomposition; soluble ethanol, ether. Tb = 290.15; Vm = 0.0834 at 288.15; Antoine (liq) A = 6.625, B = 1340, C = 0, dev = 5.0, range = 252 to 276.

Methyl carbamate, CA 598-55-0: Very soluble water, ethanol. Tm = 327.15; Tb = 450.15; Vm = 0.0661 at 329.15; Antoine I (sol) A = 11.0909, B = 3883, C = 0, dev = 1.0 to 5.0, range = 287 to 305; Antoine II (liq) A = 8.39114, B = 3272.016, C = 59.43, dev = 1.0 to 5.0, range = 333 to 388.

Nitroethane, CA 79-24-3: Irritant; moderately toxic; flammable, flash point 301; used as propellant and in explosives; slightly soluble hot water; soluble methanol, ethanol, ether, chloroform. Tf = 183.63; Tb = 387.22; Vm = 0.0718 at 298.15; Tc = 661.15; Antoine (liq) A = 6.300057, B = 1435.402, C = -52.966, dev = 0.02 to 0.1, range = 324 to 388.

$C_2H_5NO_3$ MW 91.07

Ethyl nitrate, CA 625-58-1: Moderately toxic; flammable, flash point 283; may explode from heating or shock; soluble water, ethanol, ether. Tm = 161.15; Tb = 360.65 to 360.85; Vm = 0.0907 at 298.15; Antoine (liq) A = 6.2699, B = 1329, C = -49.15, dev = 1.0, range = 273 to 361.

$C_2H_5N_3$ MW 71.08

Azidoethane, CA 871-31-8: May explode on heating. Tb = 323.15; Antoine I (liq) A = 7.1384, B = 1644.4, C = 0, dev = 1.0, range = 296 to 320; Antoine II (liq) A = 6.91503, B = 1623.317, C = 10.285, dev = 1.0 to 5.0, range = 253 to 298.

$C_2H_5N_3O_2$ MW 103.08

Bis(nitrosomethyl) amine: Antoine (liq) A = 6.84257, B = 1962.99, C = -20.492, dev = 1.0, range = 276 to 426.

C_2H_6 MW 30.07

Ethane, CA 74-84-0: Extremely flammable, flash point 138; important fuel, feedstock for ethylene production; slightly soluble water;   *(continues)*

C_2H_6 *(continued)*

soluble ethanol. Tm = 90.4; Tb = 184.55; Vm = 0.0526 at 173.15; Tc = 305.3; Pc = 4870; Vc = 0.147; Antoine I (liq) A = 5.95942, B = 663.7, C = -16.68, dev = 0.02, range = 154 to 185; Antoine II (liq) A = 6.75274, B = 822.53, C = -6.37, dev = 1.0, range = 95 to 129; Antoine III (liq) A = 6.00479, B = 684.003, C = -13.505, dev = 0.02, range = 185 to 229; Antoine IV (liq) A = 6.29311, B = 815.056, C = 6.4, dev = 0.1, range = 228 to 274; Antoine V (liq) A = 7.38133, B = 1536.358, C = 110.577, dev = 0.1, range = 273 to 305.

$C_2H_6BrF_4NS$ MW 232.03

Sulfur, bromotetrafluoro(*N*-methylmethanaminato), CA 63324-17-4: Antoine (liq) A = 7.335, B = 1983, C = 0, dev = 1.0 to 5.0, range not specified.

$C_2H_6ClF_4NS$ MW 187.58

Sulfur, chlorotetrafluoro(*N*-methylmethanaminato), CA 63324-16-3: Antoine (liq) A = 7.225, B = 1874, C = 0, dev = 1.0 to 5.0, range not specified.

C_2H_6ClP MW 96.50

Chlorodimethyl phosphine, CA 811-62-1: Irritant; flammable; ignites in air. Tm = 271.15; Tb = 346.15 to 347.15; Vm = 0.0786 at 289.15; Antoine I (sol) A = 11.31603, B = 2899.976, C = 0, dev = 1.0, range = 233 to 268; Antoine II (liq) A = 9.76377, B = 3710.161, C = 134.73, dev = 1.0, range = 273 to 306.

C_2H_6FN MW 63.07

Fluorodimethyl amine: Antoine (liq) A = 7.19484, B = 1562.74, C = 0, dev = 1.0, range = 249 to 273.

$C_2H_6FO_3P$ MW 128.04

Dimethylfluorophosphate, CA 5954-50-7: Antoine (liq) A = 7.550, B = 2319.5, C = 0, dev = 1.0 to 5.0, range = 273 to 333.

$C_2H_6F_3NS$ MW 133.13

(Dimethylamino) sulfur trifluoride, CA 3880-03-3: Corrosive; moisture-sensitive. Tm = 194.45; Tb = 390.65; Antoine (liq) A = 7.4185, B = 2115.3, C = 0, dev = 1.0, range = 296 to 327.

$C_2H_6N_2$ MW 58.08

Azomethane, dimethyl diimide, CA 503-28-6: Explodes readily; soluble water, most organic solvents. Tf = 195.15; Tb = 275.15; Antoine I (liq) A = 7.052, B = 1380, C = 0, dev = 5.0, range = 195 to 273; Antoine II (liq) A = 6.8435, B = 1324.757, C = 0, dev = 1.0 to 5.0, range = 209 to 236.

Methylammonium cyanide: Tm = 223.25; Antoine (liq) A = 9.86573, B = 2560.061, C = 0, dev = 1.0, range = 250 to 295.

$C_2H_6N_2O$ MW 74.08

Dimethyl amine, *N*-nitroso, CA 62-75-9: Highly toxic; potent carcinogen; flammable, flash point 334; soluble water, ethanol, ether. Tb = 426.15; Vm = 0.0737 at 283.15; Antoine (liq) A = 7.10632, B = 2159.476, C = 0, dev = 1.0, range = 309 to 423.

$C_2H_6N_2O_2$ MW 90.08

N-Nitrodimethyl amine, CA 4164-28-7: Potent carcinogen; soluble water, ethanol, ether, acetone. Tm = 331.15; Tb = 460.15; Vm = 0.0812 at 345.15; Antoine (sol) A = 10.925, B = 3650, C = 0, dev = 5.0, range = 314 to 324.

$C_2H_6N_2S$ MW 90.14

Sulfur diimide, dimethyl, CA 13849-02-0: Tm = 181.35; Antoine (liq) A = 7.20374, B = 1941.7, C = 0, dev = 1.0, range = 248 to 298.

C_2H_6O MW 46.07

Dimethyl ether, CA 115-10-6: Extremely flammable, flash point 231; narcotic; moderately soluble water; soluble most organic solvents. Tm = 134.65; Tb = 249.55; Vm = 0.0631 at 249.55; Tc = 400.1; Pc = 5370; Vc = 0.170; Antoine I (liq) A = 6.44136, B = 1025.56, C = -17.1, dev = 0.1, range = 183 to 265; Antoine II (liq) A = 6.30358, B = 982.46, C = -20.894, dev = 1.0, range not specified; Antoine III (liq) A = 6.36322, B = 995.747, C = -19.864, dev = 0.1 to 1.0, range = 180 to 249; Antoine IV (liq) A = 6.09534, B = 880.813, C = -33.007, dev = 1.0, range = 241 to 303; Antoine V (liq) A = 6.28318, B = 987.484, C = -16.813, dev = 1.0, range = 293 to 360; Antoine VI (liq) A = 7.48877, B = 1971.127, C = 122.787, dev = 1.0, range = 349 to 400.

Ethanol, ethyl alcohol, CA 64-17-5: Highly flammable, flash point 286; reacts violently with oxidizing agents; solvent, antiseptic, important chemical intermediate; hygroscopic; soluble water, most organic solvents. Tm = 159.05; Tb = 351.47; Vm = 0.0584 at 293.15; Tc = 516.25; Pc = 6383; Vc = 0.167; Antoine I (liq) A = 7.15946, B = 1547.464, C = -51.177, dev = 0.02, range = 320 to 359; Antoine II (liq) A = 7.23347, B = 1591.28, C = -47.056, dev = 0.02 to 0.1, range = 292 to 367; Antoine III (liq) A = 8.9391, B = 2381.5, C = 0, dev = 1.0 to 5.0, range = 210 to 271; Antoine IV (liq) A = 8.5224, B = 2299, C = 0, dev = 1.0 to 5.0, range = 193 to 223; Antoine V (liq) A = 7.16386, B = 1550.006, C = -50.941, dev = 0.02, range = 320 to 359; Antoine VI (liq) A = 7.27664, B = 1615.127, C = -45.012, dev = 0.1, range = 292 to 353; Antoine VII (liq) A = 6.95131, B = 1423.668, C = -63.568, dev = 0.1, range = 349 to 374; Antoine VIII (liq) A = 6.84806, B = 1358.124, C = -71.034, dev = 0.1, range = 370 to 464; Antoine IX (liq) A = 7.64893, B = 2073.007, C = 22.965, dev = 1.0, range = 459 to 514.

C_2H_6OS MW 78.13

Dimethyl sulfoxide, CA 67-68-5: Irritant; readily penetrates skin; flammable, flash point 368; reacts violently or explosively with many materials; very hygroscopic; powerful solvent; soluble water, most organic solvents. Tm = 291.7; Tb = 462.15; Vm = 0.0710 at 293.15; Antoine (liq) A = 6.72167, B = 1962.05, C = -47.258, dev = 1.0, range = 305 to 464.

(continues)

C_2H_6OS *(continued)*

2-Mercaptoethanol, CA 60-24-2: Highly toxic; stench; flammable, flash point 347; soluble water, most organic solvents. Tb = 430.15 to 431.15, decomposes; Vm = 0.0701 at 293.15; Antoine (liq) A = 6.54372, B = 1616.5, C = -74.82, dev = 1.0, range = 293 to 440.

$C_2H_6O_2$ MW 62.07

Ethylene glycol, 1,2-ethanediol, CA 107-21-1: Irritant; moderately toxic; flammable, flash point 384; high-boiling solvent; antifreeze; chemical and polymer intermediate; soluble water, methanol, ethanol, acetone. Tm = 260.15; Tb = 470.75; Vm = 0.0560 at 293.15; Tc = 645, calculated; Pc = 7530, calculated; Vc = 0.191, calculated; Antoine I (liq) A = 6.98465, B = 1928.08, C = -83.45, dev = 1.0, range = 323 to 473; Antoine II (liq) A = 6.86912, B = 1817.439, C = -95.859, dev = 1.0, range = 363 to 418.

Ethyl hydroperoxide, CA 3031-74-1: Flammable; explosive when heated; soluble water, ethanol, ether, benzene. Tm = 173.15; Tb = 366.15 to 370.15; Vm = 0.0665 at 293.15; Antoine (liq) A = 7.959, B = 2228, C = 0, dev = 5.0, range = 253 to 363.

$C_2H_6O_2S$ MW 94.13

Dimethyl sulfone, CA 67-71-0: High-temperature solvent; soluble water, methanol, ethanol, acetone. Tm = 382.15; Tb = 511.15; Vm = 0.0804 at 383.15; Antoine (liq) A = 6.749, B = 2241, C = -50.15, dev = 1.0 to 5.0, range = 387 to 523.

$C_2H_6O_4$ MW 94.07

Bis(hydroxymethyl) peroxide, CA 17088-73-2: Antoine (sol) A = 14.548, B = 4910, C = 0, dev = 5.0, range = 288 to 413.

$C_2H_6O_4S$ MW 126.13

Dimethyl sulfate, CA 77-78-1: Highly toxic; easily absorbed through skin; suspected carcinogen; flammable, flash point 356; hydrolyzes above 290; moderately soluble water; soluble ether, acetone, aromatics. Tm = 241.35; Tb = 461.95, decomposes; Vm = 0.0950 at 293.15; Antoine (liq) A = 7.28235, B = 2437.54, C = 0, dev = 5.0, range = 340 to 470.

C_2H_6S MW 62.13

Dimethyl sulfide, CA 75-18-3: Toxic; stench; highly flammable, flash point 239; slightly soluble water; soluble ethanol, ether. Tf = 174.88; Tb = 310.65 to 311.15; Vm = 0.0732 at 293.15; Tc = 503.04; Pc = 5530; Vc = 0.201; Antoine I (liq) A = 6.07043, B = 1088.851, C = -42.594, dev = 0.02, range = 268 to 319; Antoine II (liq) A = 6.13402, B = 1124.998, C = -37.961, dev = 0.1 to 1.0, range = 307 to 379; Antoine III (liq) A = 6.42655, B = 1334.329, C = -7.456, dev = 1.0, range = 372 to 453; Antoine IV (liq) A = 7.36327, B = 2293.043, C = 130.243, dev = 1.0 to 5.0, range = 447 to 503.

Ethyl mercaptan, ethanethiol, CA 75-08-1: Toxic; stench; highly flammable, flash point below 255; odorant for natural gas; chemical intermediate; slightly soluble water; soluble ethanol, ether. Tm = 152.15; *(continues)*

C_2H_6S *(continued)*

Tb = 308.15; Vm = 0.0740 at 293.15; Tc = 498.65; Pc = 5492; Antoine I (liq)
A = 6.07243, B = 1081.984, C = -42.085, dev = 0.02, range = 273 to 340;
Antoine II (liq) A = 6.08254, B = 1086.982, C = -41.517, dev = 0.02, range
= 273 to 313; Antoine III (liq) A = 6.10279, B = 1099.374, C = -39.807,
dev = 0.1, range = 303 to 375; Antoine IV (liq) A = 6.42565, B = 1328.598,
C = -6.231, dev = 1.0, range = 365 to 448; Antoine V (liq) A = 7.84948,
B = 2874.377, C = 200.657, dev = 1.0 to 5.0, range = 442 to 499.

$C_2H_6S_2$ MW 94.19

Dimethyl disulfide, CA 624-92-0: Irritant; stench; flammable, flash point
297; soluble ethanol, ether. Tm = 175.15; Tb = 383.15; Vm = 0.0884 at
293.15; Antoine (liq) A = 6.10018, B = 1345.066, C = -54.389, dev = 0.02 to
0.1, range = 297 to 402.

C_2H_7N MW 45.08

Dimethyl amine, CA 124-40-3: Irritant; extremely flammable; accelerator
for vulcanization of rubber; very soluble water; soluble ethanol, ether.
Tm = 180.95; Tb = 280.05; Vm = 0.0663 at 273.15; Tc = 437.65; Pc = 5309;
Vc = 0.187; Antoine I (liq) A = 6.29301, B = 993.586, C = -48.12, dev =
0.1, range = 201 to 280; Antoine II (liq) A = 6.20646, B = 965.728, C =
-50.151, dev = 1.0, range = 277 to 360; Antoine III (liq) A = 7.81489, B =
2369.425, C = 141.433, dev = 1.0, range = 358 to 438.

Ethyl amine, CA 75-04-7: Irritant; highly toxic; corrosive; highly flammable, flash point 228; chemical intermediate; soluble water, ethanol,
ether. Tm = 192.15; Tb = 289.15; Vm = 0.0654 at 288.15; Tc = 458.15; Pc =
5624; Vc = 0.182; Antoine I (liq) A = 6.57462, B = 1167.57, C = -34.18,
dev = 0.1, range = 213 to 297; Antoine II (liq) A = 6.43082, B = 1140.62,
C = -32.133, dev = 1.0, range = 290 to 449; Antoine III (liq) A = 6.21526,
B = 1009.66, C = -49.804, dev = 1.0, range = 291 to 387; Antoine IV (liq)
A = 6.48782, B = 1176.995, C = -26.674, dev = 1.0, range = 377 to 456.

C_2H_7NO MW 61.08

N,N-Dimethylhydroxyl amine, CA 5725-96-2: Irritant; flammable. Tm =
290.75; Antoine (liq) A = 6.90076, B = 1530.34, C = -60.69, dev = 1.0,
range = 290 to 363.

N,O-Dimethylhydroxyl amine, CA 1117-97-1: Irritant; flammable. Tm =
176.15; Tb = 315.35 to 315.75; Antoine (liq) A = 6.51247, B = 1238.32, C =
-40.66, dev = 1.0, range = 228 to 316.

Ethanolamine, 2-aminoethanol, CA 141-43-5: Irritant; corrosive; toxic;
flammable, flash point 358; solvent for acid gases; chemical intermediate;
hygroscopic; soluble water, methanol, ethanol, acetone. Tm = 283.65; Tb =
444.15; Vm = 0.0604 at 298.15; Tc = 638, calculated; Pc = 6870, calculated;
Vc = 0.225, calculated; Antoine (liq) A = 6.8629, B = 1732.11, C = -86.6,
dev = 1.0, range = 310 to 444.

$C_2H_7O_3P$ MW 110.05

Dimethyl phosphite, CA 868-85-9: Irritant; flammable, flash point 302;
moisture-sensitive; soluble ethanol, pyridine.   *(continues)*

56

$C_2H_7O_3P$ *(continued)*

Tb = 443.15 to 444.15; Vm = 0.0924 at 298.15; Antoine (liq) A = 6.5589, B = 2022.6, C = 0, dev = 1.0, range = 346 to 456.

C_2H_8ClN MW 81.54

Dimethyl ammonium chloride, dimethyl amine, hydrochloride, CA 506-59-2: Soluble water, ethanol, chloroform. Tm = 444.15; Antoine I (liq) A = 10.5159, B = 4994.8, C = 0, dev = 1.0, range = 429 to 533; Antoine II (liq) A = 15.2429, B = 7519, C = 0, dev = 1.0, range = 533 to 569.

Ethyl ammonium chloride, ethyl amine, hydrochloride, CA 557-66-4: Soluble water, ethanol, chloroform, acetone. Tm = 382.15 to 383.15; Tb = 588.15, decomposes; Antoine (liq) A = 4.1609, B = 1794.4, C = 0, dev = 1.0, range = 382 to 480.

$C_2H_8N_2$ MW 60.10

1,1-Dimethylhydrazine, CA 57-14-7: Irritant; corrosive; toxic; expected carcinogen; highly flammable, flash point 258; rocket propellant; very hygroscopic; soluble water, ethanol, ether. Tf = 215.95; Tb = 336.15; Vm = 0.0760 at 295.15; Tc = 523.15; Pc = 5420; Vc = 0.261; Antoine (liq) A = 6.50791, B = 1294.48, C = -48.615, dev = 0.1, range = 250 to 293.

1,2-Dimethylhydrazine, CA 540-73-8: Highly irritant; expected carcinogen; flammable, flash point below 296; hygroscopic; soluble water, ethanol, ether, light hydrocarbons. Tm = 264.24; Tb = 354.15; Vm = 0.0727 at 293.15; Antoine (liq) A = 4.90635, B = 688.31, C = -123.316, dev = 0.1, range = 274 to 298.

Ethylenediamine, CA 107-15-3: Irritant; corrosive; toxic; allergen; flammable, flash point 307; reacts violently with many materials; solvent, chemical intermediate; soluble water, ethanol. Tm = 283.95; Tb = 390.15; Vm = 0.0669 at 298.15; Antoine I (sol) A = 25.26281, B = 14058.005, C = 268.979, dev = 0.1 to 1.0, range = 242 to 278; Antoine II (liq) A = 6.62961, B = 1541.1, C = -57.16, dev = 1.0, range = 303 to 391; Antoine III (liq) A = 6.43416, B = 1424.931, C = -68.336, dev = 1.0, range = 284 to 419.

C_2N_2 MW 52.04

Cyanogen, CA 460-19-5: Toxic vapor; irritant; extremely flammable; soluble water, ethanol, ether. Tm = 239.15; Tb = 252.15; Vm = 0.05455 at 252.15; Antoine I (sol) A = 8.55529, B = 1574.008, C = -9.962, dev = 0.02, range = 202 to 239; Antoine II (liq) A = 6.01852, B = 808.004, C = -50.655, dev = 0.02, range = 240 to 253.

$C_2N_6O_{12}$ MW 300.06

Hexanitroethane, CA 918-37-6: May explode by heating or shock. Tm = 423.15; Antoine (sol) A = 4.1493, B = 1590, C = 0, dev = 5.0, range = 293 to 343.

$C_3BrClF_6O_4$ MW 329.38

Perchloric acid, 1,1,2,3,3,3-hexafluoro-2-bromopropyl ester, CA 38126-26-0:
Tm = below 195.15; Antoine (liq) A = 7.25954, B = 1991.137, C = 0, dev not
specified, range = 273 to 293.

C_3BrF_5O MW 226.93

Propanoyl fluoride, 2-bromo-2,3,3,3-tetrafluoro, CA 6129-62-0: Tm = below
195.15; Antoine (liq) A = 5.90062, B = 943.971, C = -60.107, dev = 1.0,
range = 224 to 282.

C_3BrF_6NO MW 259.93

N,N-Bis(trifluoromethyl) carbamoyl bromide: Antoine (liq) A = 6.9009, B =
1603, C = 0, dev = 1.0 to 5.0, range = 233 to 293.

$C_3BrF_{10}NS$ MW 351.99

Sulfur, bromotrifluoro(1,1,1,2,3,3,3-heptafluoro-2-propanaminato),
CA 62977-73-5: Antoine (liq) A = 6.685, B = 1884, C = 0, dev = 1.0 to 5.0,
range not specified.

$C_3Br_2F_6O$ MW 325.83

Trifluoromethyl(1,2-dibromo-1,2,2-trifluoroethyl) ether, CA 2356-57-2:
Antoine (liq) A = 7.309, B = 1806, C = 0, dev = 5.0, range = 299 to 335.

$C_3Br_3F_6NO$ MW 419.74

Dimethylamine, 1,1,1,1',1',1'-hexafluoro-N-(tribromomethoxy), CA 29528-78-7:
Antoine (liq) A = 4.845, B = 1510, C = 0, dev = 1.0 to 5.0, range = 297
to 338.

$C_3ClF_4NO_2$ MW 193.49

Carbamic fluoride, chloro(trifluoroacetyl), CA 42016-32-0: Antoine (liq)
A = 7.6249, B = 2083, C = 0, dev = 1.0 to 5.0, range not specified.

C_3ClF_5O MW 182.48

Chloropentafluoroacetone, CA 79-53-8: Irritant. Tm = 140.15; Antoine
(liq) A = 5.95123, B = 916.15, C = -48.72, dev = 0.1, range = 232 to 303.

2-Chloro-2,3,3,3-tetrafluoropropionyl fluoride, CA 28627-00-1: Tm = below
195.15; Antoine (liq) A = 6.4059, B = 1248, C = 0, dev = 1.0, range = 195
to 273.

C_3ClF_6NOS MW 247.54

Ethanamine, 1-chloro-2,2,2-trifluoro-N-sulfinyl-1-(trifluoromethyl),
CA 39095-51-7: Antoine (liq) A = 7.385, B = 1954, C = 0, dev = 1.0 to 5.0,
range not specified.

58

$C_3ClF_6NO_2$ MW 231.48

O-(Chloroformyl)-N,N-bis(trifluoromethyl) hydroxylamine, CA 15496-01-2:
Antoine (liq) A = 7.8212, B = 1804.5, C = 0, dev = 5.0, range = 227 to 286.

C_3ClF_6NS MW 231.54

Amidosulfenyl chloride, [2,2,2-trifluoro-1-(trifluoromethyl)ethylidene],
CA 38005-18-4: Antoine (liq) A = 7.295, B = 1960, C = 0, dev = 1.0 to 5.0,
range not specified.

Ethanimidoyl chloride, 2,2,2-trifluoro-N-[(trifluoromethyl)thio],
CA 62067-05-4: Antoine (liq) A = 6.965, B = 1676.9, C = 0, dev = 1.0 to
5.0, range not specified.

C_3ClF_7O MW 220.47

Heptafluoroisopropyl hypochlorite, CA 22675-68-9: Antoine I (liq) A =
6.7398, B = 1395.6, C = 0, dev = 5.0, range = 196 to 287; Antoine II (liq)
A = 5.59636, B = 966.204, C = -24.913, dev = 5.0, range = 194 to 273.

C_3ClF_8N MW 237.48

Ethylamine, N-chloro-N,1,2,2,2-pentafluoro-1-(trifluoromethyl),
CA 33757-13-0: Antoine (liq) A = 6.83371, B = 1503.003, C = 0, dev = 1.0
to 5.0, range = 240 to 311.

$C_3ClF_{10}NS$ MW 307.54

Sulfur, chlorotrifluoro(1,1,1,2,3,3,3-heptafluoro-2-propanaminato),
CA 62977-71-3: Antoine (liq) A = 6.465, B = 1747, C = 0, dev = 1.0 to 5.0,
range not specified.

Sulfur, tetrafluoro(2,2,2-trifluoroethanimidoyl chloridato)(trifluoromethyl),
CA 56868-52-1: Antoine (liq) A = 6.655, B = 1664, C = 0, dev = 1.0 to 5.0,
range not specified.

$C_3ClF_{11}S$ MW 312.53

Chlorotetrafluoro(heptafluoropropyl) sulfur, CA 42769-86-8: Antoine (liq)
A = 7.055, B = 1768, C = 0, dev = 1.0 to 5.0, range not specified.

$C_3ClF_{12}NS$ MW 345.53

Sulfur, tetrafluoro(N-chloro-1,1,2,2,2-pentafluoroethanaminato)(trifluoromethyl), CA 56868-59-8: Antoine (liq) A = 7.145, B = 1882, C = 0, dev =
1.0 to 5.0, range not specified.

$C_3Cl_2F_5N$ MW 215.94

Ethylidenimine, 2,2-difluoro-1,2-dichloro-N-(trifluoromethyl): Antoine
(liq) A = 6.955, B = 1629, C = 0, dev = 1.0 to 5.0, range = 283 to 318.

$C_3Cl_2F_6O$ MW 236.93

Hypochlorous acid, 2-chloro-1,1,2,3,3,3-hexafluoropropyl ester,
CA 22675-69-0: Tm = below 195.15; Antoine (liq) A = 6.04355, B = 1546.951,
C = 0, dev not specified, range = 273 to 293.

$C_3Cl_2F_7N$ MW 253.93

Ethylamine, N,N-dichloro-1,2,2,2-tetrafluoro-1-(trifluoromethyl),
CA 32751-04-5: Antoine (liq) A = 6.97194, B = 1706.105, C = 0, dev = 1.0,
range = 299 to 344.

$C_3Cl_2F_7NS$ MW 285.99

Sulfilimine, S,S-dichloro-N-[tetrafluoro-1-(trifluoromethyl)ethyl],
CA 26454-66-0: Tm = 197.15; Antoine (liq) A = 7.26193, B = 2055.268, C = 0,
dev = 1.0, range = 313 to 347.

$C_3Cl_3F_5O$ MW 253.38

Chlorodifluoromethyl 2,2-dichloro-1,1,2-trifluoroethyl ether, CA 37136-24-6:
Antoine (liq) A = 6.00476, B = 1174.797, C = -56.905, dev = 0.02, range =
302 to 350.

$C_3Cl_4F_4$ MW 253.84

1,3,3,3-Tetrachlorotetrafluoropropane, CA 2268-46-4: Antoine (liq) A =
6.8115, B = 1856.9, C = 0, dev = 1.0 to 5.0, range not specified.

$C_3Cl_5F_3O$ MW 286.29

Trichloromethyl 2,2-dichloro-1,1,2-trifluoroethyl ether, CA 428-73-9:
Antoine (liq) A = 6.02059, B = 1421.247, C = -69.914, dev = 0.02, range =
341 to 423.

C_3Cl_6 MW 248.75

Hexachloropropylene, CA 1888-71-7: Irritant; highly toxic by inhalation;
nonflammable; insoluble water; soluble most organic solvents. Tm = 216.06;
Tb = 482.15 to 483.15; Vm = 0.1409 at 293.15; Antoine (liq) A = 6.04819,
B = 1649.33, C = -79.28, dev = 0.1, range = 382 to 540.

$C_3F_3N_2P$ MW 152.02

Dicyano(trifluoromethyl) phosphine, CA 58310-46-6: Irritant. Antoine
(liq) A = 6.66608, B = 1490.55, C = -63.73, dev = 1.0, range = 291 to 334.

$C_3F_3N_3$ MW 135.05

2,4,6-Trifluoro-1,3,5-triazine, CA 675-14-9: Tm = 235.15; Tb = 345.55;
Vm = 0.0844 at 293.15; Antoine (liq) A = 6.56268, B = 1320.9, C = -56.04,
dev = 1.0, range = 277 to 344.

60

C_3F_4 MW 112.03

Tetrafluoropropyne, CA 20174-11-2: Slowly polymerizes at room temperature. Tb = 223.15; Antoine (liq) A = 6.405, B = 980, C = 0, dev = 1.0 to 5.0, range = 179 to 218.

C_3F_4O MW 128.03

Trifluoroacryloyl fluoride, CA 667-49-2: Antoine (liq) A = 6.895, B = 1462, C = 0, dev = 1.0 to 5.0, range not specified.

C_3F_5N MW 145.03

2-H-Azirine, 2,2-difluoro-3-(trifluoromethyl), CA 3291-42-7: Antoine (liq) A = 6.955, B = 1254, C = 0, dev = 1.0 to 5.0, range = 193 to 249.

2-H-Azirine, 2,3-difluoro-2-(trifluoromethyl), CA 3291-41-6: Antoine (liq) A = 6.925, B = 1268, C = 0, dev = 1.0 to 5.0, range = 193 to 298.

$C_3F_5NO_3$ MW 193.03

2-Propanone, 1,1,1,3,3-pentafluoro-3-nitro, CA 388-00-4: Antoine (liq) A = 7.0179, B = 1532, C = 0, dev = 1.0 to 5.0, range = 284 to 303.

C_3F_6 MW 150.02

Hexafluoropropylene, CA 116-15-4: Moderately toxic; insoluble water; soluble most organic solvents. Tm = 116.95; Tb = 243.75; Vm = 0.0948 at 233.15; Tc = 358.15; Pc = 3254; Vc = 0.250; Antoine (liq) A = 6.51264, B = 1028.15, C = -14.71, dev = 0.1, range = 232 to 293.

$C_3F_6N_2OS$ MW 226.10

Sulfoxime, N-cyano-S,S-bis(trifluoromethyl), CA 34556-28-0: Antoine (liq) A = 6.215, B = 1605.7, C = 0, dev = 1.0 to 5.0, range not specified.

C_3F_6O MW 166.02

Hexafluoroacetone, CA 684-16-2: Highly irritant. Tm = 147.7; Tb = 245.87; Vm = 0.1260 at 298.15; Pc = 2834; Antoine (liq) A = 5.84836, B = 748.94, C = -50.886, dev = 0.1, range = 193 to 246.

Trifluoromethyl trifluorovinyl ether, CA 1187-93-5: Antoine (liq) A = 6.836, B = 1194, C = 0, dev = 5.0, range = 208 to 241.

$C_3F_6O_2$ MW 182.02

Pentafluoropropionyl hypofluorite, CA 5930-63-2: Tm = 171.15; Antoine (liq) A = 6.8870, B = 1345.3, C = 0, dev = 5.0, range = 214 to 248.

$C_3F_6O_3$ MW 198.02

Trifluoroperoxyacetic acid, trifluoromethyl ester, CA 30321-53-0: Antoine (liq) A = 7.6913, B = 1603.4, C = 0, dev not specified, range not specified.

$C_3F_6O_4$ MW 214.02

Carbonoperoxoic acid, bis(trifluoromethyl) ester, CA 41864-59-9: Tm = 192.15; Antoine (liq) A = 7.33298, B = 1588.98, C = 0, dev = 1.0 to 5.0, range not specified.

$C_3F_6O_4S$ MW 246.08

Anhydride of pentafluoropropionic acid and fluorosulfuric acid, CA 51689-98-6: Antoine (liq) A = 5.85971, B = 969.967, C = -82.948, dev = 1.0 to 5.0, range = 252 to 335.

$C_3F_6O_5$ MW 230.02

Diperoxycarbonic acid, bis(trifluoromethyl) ester, CA 16156-35-7: Tm = 188.15; Antoine (liq) A = 7.3294, B = 1672.4, C = 0, dev not specified, range not specified.

C_3F_7N MW 183.03

Ethylamine, N-(difluoromethylene), CA 428-71-7: Antoine (liq) A = 7.028, B = 1341, C = 0, dev = 5.0, range not specified.

Ethylideneimine, tetrafluoro-1-(trifluoromethyl), CA 2802-70-2: Antoine (liq) A = 6.346, B = 1119, C = 0, dev = 5.0, range not specified.

C_3F_7NO MW 199.03

Perfluoro(N-methylethylideneamine): Tm = 122.15; Antoine (liq) A = 6.950, B = 1303, C = 0, dev = 5.0, range not specified.

Propionamide, heptafluoro, CA 32822-50-7: Antoine (liq) A = 7.085, B = 1418, C = 0, dev = 1.0 to 5.0, range not specified.

C_3F_7NOS MW 231.09

2-Propanamine, 1,1,1,2,3,3,3-heptafluoro-N-sulfinyl, CA 26454-67-1: Tm = 165.15; Antoine (liq) A = 7.3341, B = 1781.095, C = 0, dev =·1.0, range = 252 to 280.

$C_3F_7NO_2$ MW 215.03

O-(Fluoroformyl)-N,N-bis(trifluoromethyl) hydroxylamine, CA 15496-00-1: Antoine (liq) A = 5.575, B = 1494, C = 0, dev = 5.0, range = 194 to 273.

Perfluoro-1-nitropropane, CA 423-33-6: Tb = 298.15; Antoine (liq) A = 6.9899, B = 1491, C = 0, dev = 5.0, range = 247 to 296.

C_3F_7NS MW 215.09

Ethanimidoyl fluoride, 2,2,2-trifluoro-N-[(trifluoromethyl)thio], CA 62067-06-5: Antoine (liq) A = 6.815, B = 1464.9, C = 0, dev = 1.0 to 5.0, range not specified.

C_3F_8 MW 188.02

Octafluoropropane, perfluoropropane, CA 76-19-7:   Tf = 90.15; Tb = 237.15;
Antoine (liq) A = 6.44883, B = 983.09, C = -15.15, dev = 1.0, range = 195
to 237.

$C_3F_8N_2O_2$ MW 248.03

Methanamine, N-[(difluoroamino)carbonyl]oxy-1,1,1-trifluoro-N-(trifluoromethyl), CA 32837-67-5:   Tm = 170.15; Antoine (liq) A = 7.315, B = 1645,
C = 0, dev = 1.0 to 5.0, range not specified.

C_3F_8OS MW 236.08

(Trifluoromethyl)(pentafluoroethyl) sulfoxide, CA 33622-17-2:   Antoine (liq)
A = 6.785, B = 1483, C = 0, dev = 1.0 to 5.0, range not specified.

$C_3F_8O_5$ MW 268.02

Difluoro[(trifluoromethyl)dioxy][(trifluoromethyl)trioxy] methane,
CA 29291-73-4:   Antoine (liq) A = 6.266, B = 1440, C = 0, dev = 5.0, range
not specified.

C_3F_8S MW 220.08

(Trifluoromethyl)(pentafluoroethyl) sulfide, CA 33547-10-3:   Antoine (liq)
A = 6.405, B = 1228, C = 0, dev = 1.0 to 5.0, range not specified.

C_3F_9N MW 221.03

Tris(trifluoromethyl) amine, CA 432-03-1:   Tm = 158.45; Antoine (liq) A =
6.735, B = 1250, C = 0, dev = 5.0, range = 193 to 263.

C_3F_9NO MW 237.02

Methanamine, 1,1,1-trifluoro-N-(trifluoromethoxy)-N-(trifluoromethyl),
CA 671-63-6:   Antoine (liq) A = 7.1565, B = 1410, C = 0, dev = 1.0, range =
226 to 268.

C_3F_9NOS MW 269.08

Imidosulfuryl fluoride, [1,2,2,2-tetrafluoro-1-(trifluoromethyl)ethyl],
CA 59617-29-7:   Antoine (liq) A = 6.745, B = 1497, C = 0, dev = 1.0 to 5.0,
range not specified.

$C_3F_9NOS_2$ MW 301.14

Sulfoxime, S,S-bis(trifluoromethyl)-N-[(trifluoromethyl)thio],
CA 34556-26-8:   Antoine (liq) A = 6.525, B = 1627.7, C = 0, dev = 1.0 to
5.0, range not specified.

$C_3F_9NO_2S_2$ MW 317.14

Sulfoxime, S,S-bis(trifluoromethyl)-N-[(trifluoromethyl)sulfinyl],
CA 34556-27-9: Antoine (liq) A = 7.015, B = 1944.3, C = 0, dev = 1.0 to
5.0, range not specified.

$C_3F_9NO_2S_3$ MW 349.20

Methanesulfonamide, 1,1,1-trifluoro-N,N-bis[(trifluoromethyl)thio],
CA 29749-02-8: Tm = 281.15; Antoine (liq) A = 6.90338, B = 1799.356, C =
-33.296, dev = 5.0, range = 288 to 403.

$C_3F_9N_3O$ MW 265.04

Nitrosotris(trifluoromethyl) hydrazine, CA 10405-30-8: Antoine (liq) A =
7.0449, B = 1540, C = 0, dev = 1.0, range = 279 to 300.

$C_3F_9N_3O_2$ MW 281.04

Nitrotris(trifluoromethyl) hydrazine, CA 10405-31-9: Antoine (liq) A =
7.0299, B = 1650, C = 0, dev = 1.0, range = 293 to 321.

C_3F_9P MW 237.99

Tris(trifluoromethyl) phosphine, CA 432-04-2: Irritant; spontaneously flammable. Tm = 161.15; Tb = 290.45; Antoine (liq) A = 6.4479, B = 1289.6,
C = 0, dev = 1.0 to 5.0, range = 248 to 285.

$C_3F_9PS_2$ MW 302.11

(Trifluoromethyl)dithiophosphite acid, bis(trifluoromethyl) ester,
CA 36121-49-0: Tm = 179.65; Antoine (liq) A = 7.5279, B = 1980, C = 0,
dev = 1.0, range = 273 to 296.

$C_3F_{10}OS$ MW 274.08

Difluorooxo(trifluoromethyl)(pentafluoroethyl) sulfur, CA 33564-24-8:
Antoine (liq) A = 6.95821, B = 1598.811, C = 0, dev = 1.0 to 5.0, range =
291 to 324.

$C_3F_{10}O_3S$ MW 306.08

Sulfur, pentafluoro(pentafluoropropaneperoxoato), CA 60672-62-0: Tm =
212.15; Antoine (liq) A = 7.4035, B = 1798.4, C = 0, dev = 1.0 to 5.0,
range not specified.

$C_3F_{10}S$ MW 258.08

Difluoro(trifluoromethyl)(pentafluoroethyl) sulfur, CA 31222-06-7: Antoine
(liq) A = 6.815, B = 1525, C = 0, dev = 1.0 to 5.0, range not specified.

Pentafluoro[2,2-difluoro-1-(trifluoromethyl)ethenyl] sulfur, CA 68010-33-3:
Antoine (liq) A = 6.885, B = 1561.5, C = 0, dev = 1.0 to 5.0, range not
specified.

$C_3F_{11}NO_3S_2$ MW 371.14

Sulfur, trifluoro(fluorosulfato)(1,1,1,2,3,3,3-heptafluoro-2-propanaminato),
CA 65844-08-8: Antoine (liq) A = 6.275, B = 1695, C = 0, dev = 5.0, range
not specified.

$C_3F_{12}O_3S_2$ MW 376.13

Sulfur, pentafluoro[2,2,2-trifluoro-1-(fluorosulfonyl)oxy-1-(trifluoromethyl)ethyl], CA 68010-30-0: Antoine (liq) A = 7.145, B = 1942.5, C = 0,
dev = 1.0 to 5.0, range not specified.

$C_3F_{13}NS$ MW 329.08

Sulfur, tetrafluoro(N,1,1,2,2,2-hexafluoroethanaminato)(trifluoromethyl),
CA 59665-17-7: Antoine (liq) A = 6.935, B = 1652, C = 0, dev = 1.0 to 5.0,
range not specified.

$C_3HClF_6O_2S$ MW 250.54

Chlorosulfurous acid, 2,2,2-trifluoro-1-(trifluoromethyl)ethyl ester,
CA 57169-81-0. Antoine (liq) A = 7.185, B = 1918, C = 0, dev = 1.0 to 5.0,
range not specified.

$C_3HClF_{10}S$ MW 294.54

Pentafluoro[(1-chlorodifluoromethyl)-2,2,2-trifluoroethyl] sulfur,
CA 68010-36-6: Antoine (liq) A = 6.635, B = 1628.5, C = 0, dev = 1.0 to
5.0, range not specified.

C_3HCl_7 MW 285.21

1,1,1,2,2,3,3-Heptachloropropane, CA 594-89-8: Irritant; nonflammable;
soluble chloroform. Tm = 302.55; Tb = 520.15 to 521.15; Vm = 0.1580 at
307.15; Antoine (liq) A = 5.235, B = 1820, C = 0, dev = 5.0, range = 413
to 473.

C_3HF_3 MW 94.04

3,3,3-Trifluoropropyne, CA 661-54-1: Polymerizes readily; liable to explode. Tb = 224.85; Antoine (liq) A = 7.0057, B = 1124, C = 0, dev = 1.0,
range = 138 to 213.

C_3HF_6N MW 165.04

Aziridine, 2,2,3-trifluoro-3-trifluoromethyl, CA 3291-64-3: Antoine (liq)
A = 7.215, B = 1576, C = 0, dev = 1.0 to 5.0, range = 268 to 298.

$C_3HF_7O_2S$ MW 234.09

Fluorosulfurous acid, 2,2,2-trifluoro-1-(trifluoromethyl)ethyl ester,
CA 52225-56-6: Antoine (liq) A = 7.335, B = 1766, C = 0, dev = 1.0 to 5.0,
range not specified.

C_3HF_8NOS MW 251.09

Sulfoxime, *S*-(pentafluoroethyl)-*S*-(trifluoromethyl), CA 34556-23-5:
Antoine (liq) A = 7.285, B = 1891.2, C = 0, dev = 1.0 to 5.0, range not
specified.

$C_3HF_{11}S$ MW 278.08

Pentafluoro[2,2,2-trifluoro-1-(trifluoromethyl)ethyl] sulfur, CA 68010-34-4:
Antoine (liq) A = 6.835, B = 1571.5, C = 0, dev = 1.0 to 5.0, range not
specified.

$C_3HF_{12}NS$ MW 311.09

Sulfur, tetrafluoro(1,1,2,2,2-pentafluoroethanaminato)(trifluoromethyl),
CA 56868-60-1: Antoine (liq) A = 6.025, B = 1401, C = 0, dev = 5.0, range
not specified.

C_3HN MW 51.05

Propargylonitrile, cyanoacetylene, CA 1070-71-9: Decomposes rapidly; decomposed by air and light; soluble ethanol. Tm = 278.15; Tb = 315.65; Vm =
0.0625 at 290.15; Antoine I (sol) A = 9.305, B = 2210, C = 0, dev = 5.0,
range = 247 to 279; Antoine II (liq) A = 6.665, B = 1470, C = 0, dev = 5.0,
range = 279 to 315.

$C_3H_2ClF_5O$ MW 184.49

Enflurane, 1-chloro-1,2,2-trifluoro-2-(difluoromethoxy) ethane,
CA 13838-16-9: Anesthetic; soluble most organic solvents including fats
and oils. Tb = 329.65; Vm = 0.1216 at 298.15; Antoine (liq) A = 6.1133,
B = 1107.839, C = -60.087, dev = 1.0, range = 290 to 329.

Isoflurane, 2-chloro-1,1,1-trifluoro-2-(difluoromethoxy) ethane,
CA 26675-46-7: Anesthetic; soluble most organic solvents including fats
and oils. Tb = 321.65; Vm = 0.1272 at 298.15; Antoine (liq) A = 6.08923,
B = 1077.738, C = -58.097, dev = 1.0, range = 283 to 312.

$C_3H_2Cl_2F_4$ MW 184.95

3,3-Dichloro-1,1,1,3-tetrafluoropropane, CA 64712-27-2: Antoine (liq) A =
5.99811, B = 1105.633, C = -56.85, dev = 0.1, range = 297 to 333.

$C_3H_2Cl_3F_3$ MW 201.40

1,1,1-Trichloro-3,3,3-trifluoropropane, CA 7125-84-0: Tm = 231.45; Antoine
(liq) A = 6.03629, B = 1247.555, C = -58.809, dev = 0.02, range = 320
to 365.

$C_3H_2Cl_4$ MW 179.86

1,1,2,3-Tetrachloropropylene, CA 10436-39-2: Tb = 440.15 to 440.65; Vm =
0.1160 at 293.15; Antoine (liq) A = 7.4518, B = 2495, C = 20.13, dev = 1.0,
range = 347 to 416.

C_3H_2FNOS MW 119.11

Fluoroacetyl isothiocyanate, CA 459-71-2: Antoine (liq) A = 8.1502, B = 2576.8, C = 0, dev = 1.0, range = 273 to 353.

$C_3H_2F_6N_2S$ MW 212.11

Ethanimidamide, 2,2,2-trifluoro-N-[(trifluoromethyl)thio], CA 62067-09-8: Antoine (liq) A = 7.145, B = 2080.3, C = 0, dev = 1.0 to 5.0, range = 322 to 390.

Sulfoxylic diamide, [2,2,2-trifluoro-1-(trifluoromethyl)ethylidene], CA 38005-20-8: Antoine (liq) A = 7.295, B = 2052, C = 0, dev = 1.0 to 5.0, range not specified.

$C_3H_2F_6O$ MW 168.04

1,1,1,3,3,3-Hexafluoro-2-propanol, CA 920-66-1: Irritant; nonflammable solvent. Tm = 269.15; Tb = 330.15 to 331.15; Vm = 0.1151 at 293.65; Antoine (liq) A = 5.8082, B = 811.709, C = -117.063, dev = 1.0, range = 294 to 331.

$C_3H_2F_6O_2S$ MW 216.10

Trifluoromethanesulfonic acid, 2,2,2-trifluoroethyl ester, CA 30957-44-9: Antoine (liq) A = 7.325, B = 1931, C = 0, dev = 1.0 to 5.0, range not specified.

$C_3H_2F_8N_2S$ MW 250.11

S,S-Difluoro-N-[1-amino-2,2,2-trifluoro-1-(trifluoromethyl)ethyl] sulfil-imine, CA 24433-66-7: Tm = 211.65; Antoine (liq) A = 7.4609, B = 2023, C = 0, dev = 1.0 to 5.0, range = 295 to 313.

$C_3H_2F_{14}S_2$ MW 368.15

Octafluoromethylene bis(trifluoromethyl) disulfur, CA 56919-35-8: Antoine (liq) A = 6.365, B = 1635, C = 0, dev = 5.0, range not specified.

$C_3H_2N_2$ MW 66.06

Malononitrile, CA 109-77-3: Highly toxic by inhalation and skin absorption; may polymerize violently on heating or with alkalis; flammable, flash point 385; soluble water, ethanol, ether, benzene. Tm = 303.15 to 303.65; Tb = 491.15 to 492.15; Vm = 0.0629 at 307.35; Antoine (sol) A = 12.2009, B = 4139, C = 0, dev = 1.0, range = 255 to 282.

$C_3H_2O_3$ MW 86.05

Vinylene carbonate, CA 872-36-6: Toxic; expected carcinogen; flammable, flash point 345. Tm = 295.15; Tb = 435.15; Vm = 0.0637 at 298.15; Antoine (liq) A = 1.6759, B = 97.865, C = -256.81, dev = 5.0, range = 308 to 350.

C_3H_3Cl MW 74.51

1-Chloropropyne, CA 7747-84-4:   Antoine (liq) A = 6.865, B = 1480, C = 0, dev = 1.0, range = 200 to 289.

$C_3H_3Cl_2F_3$ MW 166.96

1,1-Dichloro-3,3,3-trifluoropropane, CA 460-69-5:   Tm = 179.95; Antoine (liq) A = 6.07353, B = 1168.397, C = -58.263, dev = 0.02, range = 301 to 342.

$C_3H_3Cl_5$ MW 216.32

1,1,2,2,3-Pentachloropropane, CA 16714-68-4:   Tb = 464.15 to 464.65; Vm = 0.1325 at 298.15; Antoine (liq) A = 6.0311, B = 1624.4, C = -68.43, dev = 1.0, range = 365 to 447.

$C_3H_3F_3$ MW 96.05

3,3,3-Trifluoro-1-propene, CA 677-21-4:   Antoine (liq) A = 6.13437, B = 859.095, C = -40.082, dev = 1.0, range = 283 to 363.

$C_3H_3F_3O_3$ MW 144.05

Peroxyacetic acid, trifluoromethyl ester, CA 33017-08-2:   Tm = 191.15; Antoine (liq) A = 7.0412, B = 1698.8, C = 0, dev = 1.0, range not specified.

$C_3H_3F_4I$ MW 241.95

1,1,1,2-Tetrafluoro-3-iodopropane, CA 1737-76-4:   Antoine (liq) A = 6.1699, B = 1486, C = 0, dev = 1.0 to 5.0, range = 295 to 356.

1,1,1,3-Tetrafluoro-3-iodopropane, CA 460-74-2:   Antoine (liq) A = 6.5669, B = 1630, C = 0, dev = 1.0 to 5.0, range = 301 to 356.

$C_3H_3F_5$ MW 134.05

1,1,1,2,2-Pentafluoropropane, CA 1814-88-6:   Antoine (liq) A = 5.88038, B = 802.883, C = -48.338, dev = 0.1, range = 232 to 283.

$C_3H_3F_5O$ MW 150.05

2,2,3,3,3-Pentafluoro-1-propanol, CA 422-05-9:   Irritant; nonflammable. Tm = 258.15; Tb = 353.15; Vm = 0.0997 at 293.15; Antoine (liq) A = 5.8163, B = 965.375, C = -106.105, dev = 1.0, range = 273 to 297.

$C_3H_3F_6NOS$ MW 215.11

Sulfoximine, N-methyl-S,S-bis(trifluoromethyl), CA 34556-25-7:   Antoine (liq) A = 6.745, B = 1602.5, C = 0, dev = 1.0 to 5.0, range not specified.

$C_3H_3F_6NOS_2$ MW 247.17

1,1,1,1',1',1'-Hexafluoro-N-methyldimethanesulfenamide: Antoine (liq) A =
7.155, B = 2275, C = 0, dev = 5.0, range not specified.

$C_3H_3F_6NS$ MW 199.11

N,N-Bis(trifluoromethyl)methanesulfenamide: Antoine (liq) A = 7.025, B =
1625, C = 0, dev = 1.0 to 5.0, range = 269 to 309.

$C_3H_3F_6O_2P$ MW 216.02

Bis(trifluoromethyl)phosphinic acid, methyl ester, CA 25439-11-6: Antoine
(liq) A = 7.76551, B = 2114.998, C = 0, dev = 1.0, range = 258 to 313.

$C_3H_3F_6PS_2$ MW 248.14

Bis(trifluoromethyl) dithiophosphinic acid, methyl ester, CA 18799-79-6:
Tm = 255.95; Antoine (liq) A = 7.43103, B = 2168.9, C = 0, dev = 1.0,
range = 273 to 344.

$C_3H_3F_7N_2S$ MW 232.12

Sulfur, fluoro(methanaminato)(trifluoromethyl)(trifluoromethanaminato),
CA 59665-15-5: Antoine (liq) A = 6.475, B = 1510, C = 0, dev = 5.0, range
not specified.

C_3H_3N MW 53.06

Acrylonitrile, CA 107-13-1: Toxic; extremely carcinogenic; flammable,
flash point 273; moderately soluble water; soluble most organic solvents;
polymer raw material. Tm = 189.6; Tb = 350.45; Vm = 0.0658 at 293.15; Tc =
535.0; Pc = 4480; Vc = 0.212; Antoine I (liq) A = 6.12021, B = 1288.9, C =
-38.74, dev = 1.0, range = 257 to 352; Antoine II (liq) A = 6.4811, B =
1518.381, C = -12.003, dev = 1.0, range = 283 to 343.

C_3H_3NO MW 69.06

Oxazole, CA 288-42-6: Irritant; flammable, flash point 291. Tm = 186.15
to 189.15, Tb = 342.15 to 343.15; Vm = 0.0658 at 293.15; Antoine (liq) A =
6.3082, B = 1254.643, C = -51.106, dev = 0.02 to 0.1, range = 293 to 344.

$C_3H_3NO_2$ MW 85.06

Cyanoformic acid, methyl ester, CA 17640-15-2: Irritant. Tb = 374.15;
Antoine (liq) A = 7.5682, B = 2053.6, C = 0, dev = 1.0, range = 273 to 333.

C_3H_3NS MW 85.12

Thiazole, CA 288-47-1: Irritant; foul odor; flammable, flash point 295;
slightly soluble water; soluble ethanol, ether, acetone. Tb = 390.15 to
391.65; Vm = 0.0709 at 290.15; Antoine I (liq) A = 6.26425,   *(continues)*

C_3H_3NS *(continued)*

B = 1424.137, C = -56.978, dev = 0.02, range = 335 to 392; Antoine II (liq)
A = 6.26602, B = 1424.8, C = -56.956, dev = 0.02, range = 333 to 393.

C_3H_4 MW 40.07

Allene, propadiene, CA 463-49-0: May explode under pressure; readily isomerizes to methyl acetylene on heating; soluble benzene, light hydrocarbons.
Tm = 136.87; Tb = 238.65; Vm = 0.0601 at 238.65; Tc = 393.15; Pc = 5470;
Vc = 0.162; Antoine I (liq) A = 4.8386, B = 458.06, C = -77.08, dev = 1.0,
range = 193 to 246; Antoine II (liq) A = 6.8739, B = 1114.5, C = -7.08,
dev = 5.0, range = 136 to 274.

Propyne, methyl acetylene, CA 74-99-7: Anesthetic vapors; extremely flammable; polymerizes; forms metallic compounds; moderately soluble water;
soluble ethanol, ether, chloroform. Tm = 168.15; Tb = 249.95; Antoine I
(liq) A = 5.90975, B = 803.73, C = -44.07, dev = 1.0, range = 183 to 257;
Antoine II (liq) A = 6.81779, B = 1321.342, C = 27.993, dev = 1.0, range =
257 to 402; Antoine III (liq) A = 5.53364, B = 615.795, C = -75.396, dev =
1.0, range = 249 to 306; Antoine IV (liq) A = 6.89118, B = 1379.717, C =
36.577, dev = 1.0, range = 303 to 361; Antoine V (liq) A = 7.76698, B =
2194.145, C = 143.849, dev = 1.0, range = 359 to 402.

$C_3H_4Br_2$ MW 199.87

2,3-Dibromopropylene, CA 513-31-5: Irritant; lachrymator; flammable, flash
point 354; soluble ether, acetone, chloroform. Tb = 412.15 to 413.15,
slight decomposition; Vm = 0.0982 at 298.15; Antoine (liq) A = 6.50162, B =
1696.13, C = -37.177, dev = 1.0, range = 267 to 415.

$C_3H_4Br_4$ MW 359.68

1,2,2,3-Tetrabromopropane, CA 54268-02-9: Tm = 283.15 to 284.15; Vm =
0.1331 at 293.15; Antoine (liq) A = 6.50635, B = 1989.3, C = -81.15, dev =
5.0, range = 418 to 580.

$C_3H_4ClFO_2$ MW 126.51

Carbonochloridic acid, 2-fluoroethyl ester, CA 462-27-1: Antoine (liq) A =
8.126, B = 2435, C = 0, dev = 5.0, range = 273 to 333.

$C_3H_4ClF_3$ MW 132.51

1-Chloro-3,3,3-trifluoropropane, CA 460-35-5: Tm = 166.65; Tb = 318.25;
Vm = 0.1000 at 293.15; Antoine (liq) A = 6.00255, B = 1058.165, C = -54.116,
dev = 0.1, range = 297 to 315.

$C_3H_4ClF_3O_2S$ MW 196.57

Trifluoromethanesulfinic acid, 2-chloroethyl ester, CA 61915-99-9: Antoine
(liq) A = 6.925, B = 2117, C = 0, dev = 5.0, range = 320 to 403.

$C_3H_4Cl_2$ MW 110.97

1,2-Dichloropropylene, mixed *cis* and *trans* isomers, CA 563-54-2: Vm = 0.0939 at 293.15; Antoine (liq) A = 7.9358, B = 2366.9, C = 35.55, dev = 1.0, range = 307 to 349.

$C_3H_4Cl_2F_2O$ MW 164.97

Methoxyflurane, 2,2-dichloro-1,1-difluoro-1-methoxy ethane, CA 76-38-0: Tm = 238.15; Tb = 378.15; Vm = 0.1157 at 293.15; Antoine (liq) A = 6.20709, B = 1336.58, C = -59.67, dev = 1.0, range = 279 to 378.

$C_3H_4Cl_2O$ MW 126.97

1,1-Dichloroacetone, CA 513-88-2: Irritant; lachrymator; flammable, flash point 297; slightly soluble water; soluble ethanol, ether. Tb = 393.15; Vm = 0.0973 at 291.15; Antoine (liq) A = 5.9228, B = 1355.5, C = -45.61, dev = 1.0, range = 292 to 392.

1,3-Dichloroacetone, CA 534-07-6: Moderate irritant; corrosive; lachry- mator; vesicant; flammable, flash point 362; soluble water, ethanol, ether. Tm = 318.15; Tb = 446.15; Vm = 0.0918 at 319.15; Antoine (liq) A = 6.900, B = 1937.4, C = -49.1, dev = 1.0, range = 348 to 445.

$C_3H_4Cl_2O_2$ MW 142.97

Methyl dichloroacetate, CA 116-54-1: Highly irritant; corrosive; lachry- mator; flammable, flash point 353; soluble ethanol. Tm = 221.25; Tb = 416.15 to 417.15; Vm = 0.1038 at 293.15; Antoine (liq) A = 6.48562, B = 1589.44, C = -61.15, dev = 0.1, range = 331 to 481.

$C_3H_4Cl_4$ MW 181.88

1,1,1,2-Tetrachloropropane, CA 812-03-3: Soluble ethanol, ether, chloro- form. Tm = 208.15; Tb = 425.15 to 426.15; Vm = 0.1235 at 293.15; Antoine (liq) A = 6.10562, B = 1476, C = -63.15, dev = 5.0, range = 331 to 469.

1,1,1,3-Tetrachloropropane, CA 1070-78-6: Soluble ethanol, ether, benzene. Tb = 432.15; Vm = 0.1258 at 298.15; Antoine (liq) A = 3.95632, B = 511.493, C = -184.592, dev = 5.0, range = 300 to 377.

1,2,2,3-Tetrachloropropane, CA 13116-53-5: Soluble ethanol, ether, chloro- form. Tb = 438.65; Vm = 0.1213 at 291.15; Antoine (liq) A = 6.4378, B = 1760.1, C = -40.55, dev = 1.0, range = 346 to 415.

$C_3H_4F_2O_2$ MW 110.06

Methyl difluoroacetate, CA 433-53-4: Tb = 358.35; Antoine (liq) A = 8.2547, B = 2187.9, C = 0, dev = 1.0, range = 273 to 333.

$C_3H_4F_4O$ MW 132.06

2,2,3,3-Tetrafluoro-1-propanol, CA 76-37-9: Soluble acetone, ethanol, chloroform. Tm = 258.15; Tb = 382.15 to 383.15; Vm = 0.0889 at 293.15; Antoine (liq) A = 6.5244, B = 1333.9, C = -85.55, dev = 1.0, range = 303 to 380.

C_3H_4O MW 56.06

Acrolein, CA 107-02-8: Highly toxic; irritant; lachrymator; unstable;
undergoes exothermic polymerization, particularly in presence of alkali;
highly flammable, flash point 247; very soluble water; soluble ethanol,
ether, acetone. Tm = 185.15; Tb = 325.65; Vm = 0.0667 at 293.15; Tc =
527.15, calculated; Pc = 5080, calculated; Vc = 0.189, calculated; Antoine
I (liq) A = 6.19181, B = 1204.95, C = -37.8, dev = 1.0, range = 208 to 326;
Antoine II (liq) A = 6.2878, B = 1231.003, C = -38.405, dev = 0.1 to 1.0,
range = 250 to 306.

Propargyl alcohol, CA 107-19-7: Irritating to skin and mucous membranes;
polymerized by heat or alkali; flammable, flash point 309; soluble water,
most organic solvents. Tm = 221.15; Tb = 387.15; Vm = 0.0591 at 293.15;
Antoine (liq) A = 7.66721, B = 2191.846, C = 0, dev = 1.0, range = 293
to 387.

$C_3H_4O_2$ MW 72.06

Acrylic acid, CA 79-10-7: Irritant; corrosive; toxic; causes burns;
expected teratogen; flammable, flash point 323; polymerizes in presence of
oxygen; important polymer raw material; soluble water, ethanol, ether,
acetone. Tm = 286.65; Tb = 414.15; Vm = 0.0689 at 298.15; Tc = 615, calculated; Pc = 5660, calculated; Vc = 0.208, calculated; Antoine (liq) A =
6.93296, B = 1827.9, C = -43.15, dev = 1.0, range = 341 to 414.

beta-Propiolactone, CA 57-57-8: Highly irritant and toxic; suspected carcinogen; slowly decomposes with moisture; flammable, flash point 343; Tm =
239.75; Tb = 435.15, decomposes; Vm = 0.0629 at 293.15; Antoine (liq) A =
7.601, B = 2422, C = 0, dev = 5.0, range = 324 to 435.

$C_3H_4O_3$.   MW 88.06

Ethylene carbonate, CA 96-49-1: Flammable, flash point 433; soluble
ethanol. Tm = 311.65 to 312.15; Tb = 521.15; Vm = 0.0666 at 311.15;
Antoine I (sol) A = 9.55107, B = 3588.3, C = 0, dev = 1.0, range = 273 to
297; Antoine II (liq) A = 5.88959, B = 1602.71, C = -111.737, dev = 1.0,
range = 381 to 437.

Pyruvic acid, 2-oxopropionic acid, CA 127-17-3: Flammable, flash point
356; impure samples decompose on standing; soluble water, ethanol, ether,
acetone. Tm = 286.75; Tb = 438.15, decomposes; Vm = 0.0718 at 293.15;
Antoine (liq) A = 7.3295, B = 2195.01, C = -25.94, dev = 1.0, range = 204
to 438.

$C_3H_4O_4$ MW 104.06

Malonic acid, CA 141-82-2: Strong irritant; soluble water, methanol,
ethanol, ether, pyridine. Tm = 408.75, decomposes; Antoine (sol) A =
9.176, B = 3799, C = 0, dev = 1.0, range = 291 to 320.

$C_3H_4S_3$ MW 136.26

Ethylene trithiocarbonate, CA 822-38-8: Irritant; stench; flammable, flash
point 436; soluble methanol. Tm = 327.65 to 328.65; Tb = 580.15; Antoine
(sol) A = 10.66469, B = 4288.942, C = 0, dev = 1.0, range = 294 to 304.

C_3H_5Br MW 120.98

Allyl bromide, CA 106-95-6: Irritant; lachrymator; highly toxic; flammable, flash point 272; used as fumigant; slightly soluble water; soluble ethanol, ether, chloroform. Tm = 153.75; Tb = 343.15 to 344.15; Vm = 0.0865 at 293.15; Antoine (liq) A = 6.43143, B = 1397.08, C = -27.486, dev = 0.1, range = 297 to 338.

1-Bromopropylene, *cis*, CA 590-14-7: Irritant; flammable, flash point 277; soluble ether, acetone, chloroform. Tm = 160.15; Tb = 331.65; Vm = 0.0846 at 293.15; Antoine (liq) A = 6.06317, B = 1159.6, C = -45.15, dev = 1.0, range = 257 to 366.

1-Bromopropylene, *trans*, CA 590-14-7: Irritant; flammable, flash point 277; soluble ether, acetone, chloroform. Tm = 196.65; Tb = 336.15; Antoine (liq) A = 6.06702, B = 1178.6, C = -46.15, dev = 1.0, range = 262 to 372.

$C_3H_5Br_3$ MW 280.78

1,2,3-Tribromopropane, CA 96-11-7: Insoluble water; soluble ethanol, ether, chloroform. Tm = 298.65; Tb = 493.15; Vm = 0.1155 at 296.15; Antoine (liq) A = 6.22024, B = 1779.2, C = -73.15, dev = 1.0, range = 390 to 595.

C_3H_5Cl MW 76.53

Allyl chloride, 3-chloropropylene, CA 107-05-1: Highly toxic and irritant vapor; lachrymator; can cause liver damage; highly flammable, flash point 241; intermediate in manufacture of glycerol and epichlorohydrin; slightly soluble water; soluble most organic solvents. Tf = 138.65; Tb = 318.11; Vm = 0.0816 at 293.15; Tc = 513.85; Antoine (liq) A = 6.05409, B = 1114.9, C = -42.72, dev = 1.0, range = 203 to 318.

1-Chloropropylene, *cis*, CA 590-21-6: Extremely flammable, flash point below 267; soluble ether, acetone, benzene, chloroform. Tf = 138.35; Tb = 305.95; Vm = 0.0819 at 293.15; Antoine (liq) A = 6.04968, B = 1074.9, C = -40.15, dev = 1.0, range = 237 to 338.

1-Chloropropylene, *trans*, CA 16136-85-9: Extremely flammable, flash point below 267; soluble ether, acetone, benzene, chloroform. Tf = 174.15; Tb = 310.55; Vm = 0.0819 at 293.15; Antoine (liq) A = 6.05105, B = 1089.8, C = -41.15, dev = 1.0, range = 241 to 343.

2-Chloropropylene, CA 557-98-2: Highly toxic; suspected carcinogen; lachrymator; extremely flammable, flash point below 253; refrigerant; soluble ether, acetone, benzene, chloroform. Tm = 134.55; Tb = 295.8; Vm = 0.0840 at 293.15; Antoine (liq) A = 6.04314, B = 1040.2, C = -38.15, dev = 1.0, range = 229 to 327.

C_3H_5ClO MW 92.52

Chloroacetone, CA 78-95-5: Strong irritant; corrosive; highly toxic by inhalation and skin contact; suspected carcinogen; lachrymator; can polymerize explosively on storage; flammable, flash point 300; very soluble water; soluble ethanol, ether, chloroform. Tm = 228.65; Tb = 392.15; Vm = 0.0796 at 289.15; Tc = 610, calculated; Pc = 4900, calculated; Vc = 0.233, calculated; Antoine (liq) A = 7.6166, B = 2276.8, C = 14.13, dev = 1.0, range = 316 to 392.

(continues)

C_3H_5ClO *(continued)*

Epichlorohydrin, mixed isomers, CA 106-89-8: Toxic by inhalation and skin contact; irritant; can cause kidney damage; flammable, flash point 314; reacts violently with some materials; moderately soluble water; soluble most organic solvents. Tm = 215.95; Tb = 389.26; Vm = 0.0783 at 293.15; Antoine (liq) A = 6.5958, B = 1587.9, C = -43.15, dev = 1.0, range = 328 to 388.

$C_3H_5ClO_2$ MW 108.52

Ethyl chloroformate, CA 541-41-3: Irritant; highly toxic; flammable, flash point 289; slowly hydrolyzed by water; almost insoluble water; soluble ethanol, ether, benzene, chloroform.

Methyl chloroacetate, CA 96-34-4: Irritant; corrosive; lachrymator; flammable, flash point 291; soluble ethanol, ether, benzene, acetone. Tm = 241.05; Tb = 403.15; Vm = 0.0877 at 293.15; Antoine (liq) A = 6.13997, B = 1324.16, C = -82.41, dev = 1.0, range = 318 to 403.

$C_3H_5Cl_3$ MW 147.43

1,1,1-Trichloropropane, CA 7789-89-1: Moderately toxic; soluble ethanol, ether, chloroform; Antoine (liq) A = 6.35846, B = 1503.91, C = -35.884, dev = 1.0, range = 244 to 382.

1,1,3-Trichloropropane, CA 20395-25-9: Soluble ethanol, ether, chloroform. Tm = 214.15; Tb = 419.15 to 421.15; Vm = 0.1084 at 288.15; Antoine (liq) A = 6.10554, B = 1461.8, C = -62.15, dev = 0.1, range = 328 to 464.

1,2,3-Trichloropropane, CA 96-18-4: Irritant; toxic; flammable, flash point 355; soluble ethanol, ether, chloroform; Tm = 258.45; Tb = 431.15; Vm = 0.1038 at 288.15; Antoine (liq) A = 6.11206, B = 1502.3, C = -64.15, dev = 0.1, range = 337 to 477.

C_3H_5FO MW 76.07

Epifluorohydrin, 1,2-epoxy-3-fluoropropane, CA 503-09-3: Corrosive; lachrymator; flammable, flash point 277; Tb = 357.15 to 358.15; Vm = 0.0711 at 293.15; Antoine (liq) A = 7.9187, B = 2084.6, C = 0, dev = 1.0, range = 273 to 333.

$C_3H_5FO_2$ MW 92.07

Methyl fluoroacetate, CA 453-18-9: Flammable; soluble water; slightly soluble lighter hydrocarbons. Tm = 238.15; Tb = 377.65; Vm = 0.0793 at 288.15; Antoine (liq) A = 7.9109, B = 2229.4, C = 0, dev = 1.0, range = 273 to 333.

$C_3H_5F_3O$ MW 114.07

1,1,1-Trifluoro-2-propanol, CA 374-01-6: Soluble ethanol, ether, acetone, benzene. Tm = 221.15; Tb = 349.85; Vm = 0.0891 at 288.15; Antoine (liq) A = 6.8491, B = 1334.238, C = -73.51, dev = 0.1, range = 292 to 333.

$C_3H_5F_3OS$ MW 146.13

[(Trifluoromethyl)sulfinyl] ethane, CA 56919-38-1: Antoine (liq) A = 4.865, B = 1198, C = 0, dev = 5.0, range not specified.

$C_3H_5F_3O_2S$ MW 162.13

(Trifluoromethyl)methanesulfinic acid, ethyl ester, CA 30957-43-8: Antoine (liq) A = 7.125, B = 1893, C = 0, dev = 5.0, range not specified.

$C_3H_5F_3S_2$ MW 178.19

Ethyl(trifluoromethyl) disulfide, CA 691-05-4: Irritant; stench. Antoine (liq) A = 6.975, B = 1766, C = 0, dev = 5.0, range = 253 to 303.

$C_3H_5F_7S$ MW 222.12

Ethyltetrafluoro(trifluoromethyl) sulfur, CA 56919-37-0: Irritant. Antoine (liq) A = 6.745, B = 1630, C = 0, dev = 5.0, range not specified.

C_3H_5N MW 55.08

Propionitrile, ethyl cyanide, CA 107-12-0: Highly toxic by inhalation and skin absorption; flammable, flash point 275; moderately soluble water; soluble ethanol, ether, acetone. Tm = 169.65; Tb = 370.15; Vm = 0.0705 at 293.15; Tc = 564.35; Pc = 4177; Vc = 0.2295; Antoine (liq) A = 6.32058, B = 1429.07, C = -39.3, dev = 1.0, range = 288 to 371.

C_3H_5NO MW 71.08

Acrylamide, CA 79-06-1: Irritant; highly toxic; polymerizes on heating; polymer raw material; soluble water, methanol, ethanol, ether, acetone. Tm = 357.65; Antoine I (liq) A = 7.395, B = 3213, C = 0, dev = 5.0, range = 357 to 413; Antoine II (liq) A = 10.31055, B = 3994.667, C = 0, dev = 1.0 to 5.0, range = 373 to 413.

3-Hydroxypropionitrile, 2-cyanoethanol, CA 109-78-4: Irritant; highly toxic; flammable, flash point over 385; soluble water, ethanol. Tm = 227.15; Tb = 501.15; Vm = 0.0683 at 298.15; Antoine (liq) A = 8.72002, B = 3654.42, C = 49.871, dev = 1.0, range = 331 to 494.

2-Propenal, oxime, CA 5314-33-0: Antoine (liq) A = 7.20632, B = 2203.14, C = 0, dev = 5.0, range = 303 to 381.

$C_3H_5NO_2$ MW 87.08

1-Nitropropylene, CA 3156-70-5: Irritant; lachrymator; flammable; soluble ethanol, acetone, chloroform. Vm = 0.0817 at 293.15; Antoine (liq) A = 7.5841, B = 2306.3, C = 0, dev = 1.0, range = 273 to 333.

2-Nitropropylene, CA 4749-28-4: Irritant; lachrymator; flammable; soluble ether, acetone, chloroform. Vm = 0.818 at 293.15; Antoine (liq) A = 7.0521, B = 1993.1, C = 0, dev = 1.0, range = 273 to 333.

C_3H_5NS MW 87.14

Ethyl isothiocyanate, CA 542-85-8: Highly irritant; highly toxic; used as
war gas; flammable, flash point 305; insoluble water; soluble ethanol,
ether. Tm = 267.25; Tb = 404.15 to 405.15; Vm = 0.0869 at 291.15; Antoine
(liq) A = 7.04615, B = 2000.43, C = -7.234, dev = 1.0, range = 283 to 404.

Ethyl thiocyanate, CA 542-90-5: Toxic; flammable; insoluble water; soluble
ethanol, ether. Tm = 187.65; Tb = 419.15; Vm = 0.0854 at 289.15; Antoine
(liq) A = 6.258, B = 1439.9, C = -78.36, dev = 1.0, range = 358 to 422.

$C_3H_5N_3O_9$ MW 227.09

Glycerol trinitrate, nitroglycerine, CA 55-63-0: Heart stimulant; produces
headaches if inhaled; decomposes above about 323; explosive by heating or
shock; slightly soluble water; soluble ethanol, ether, acetone, benzene.
Tm = 286.25 for higher-melting solid form; Vm = 0.1419 at 288.15; Antoine
(liq) = 2.6576, B = 712.8, C = -195.85, dev = 5.0, range = 293 to 373.

C_3H_6 MW 42.08

Cyclopropane, CA 75-19-4: Extremely flammable gas; used as inhalation
anesthetic; moderately soluble water; very soluble ethanol, ether, benzene,
light hydrocarbons. Tm = 146.15 to 147.15; Tb = 239.15; Vm = 0.0584 at
194.15; Tc = 397.75; Pc = 5487; Vc = 0.171; Antoine I (liq) A = 5.81229,
B = 803.26, C = -28.66, dev = 1.0, range = 188 to 239; Antoine II (liq) A =
6.40545, B = 1068.315, C = 3.164, dev = 0.1 to 1.0, range = 239 to 298;
Antoine III (liq) A = 6.37093, B = 1055.413, C = 2.448, dev = 0.1, range =
297 to 359; Antoine IV (liq) A = 7.74179, B = 2280.626, C = 172.518, dev =
0.1, range = 358 to 398.

Propylene, propene, CA 115-07-1: Extremely flammable, flash point 165;
forms an ozonide which is explosive at room temperature; moderately soluble
ethanol, acetic acid; absorbed by sulfuric acid and by mercuric solutions.
Tm = 87.95; Tb = 225.35; Vm = 0.0810 at 293.15; Tc = 365.25; Pc = 4613;
Vc = 0.181; Antoine I (liq) A = 7.6059, B = 1244, C = 5.24, dev = 1.0,
range = 104 to 161; Antoine II (liq) A = 5.9445, B = 785.0, C = -26.15,
dev = 0.1, range = 161 to 242; Antoine III (liq) A = 6.341585, B = 977.553,
C = 0, dev = 0.1 to 1.0, range = 297 to 363; Antoine IV (liq) A = 5.99046,
B = 805.281, C = -23.377, dev = 0.1, range = 228 to 271; Antoine V (liq)
A = 6.27253, B = 948.998, C = -2.474, dev = 0.1, range = 270 to 327;
Antoine VI (liq) A = 7.22022, B = 1661.046, C = 102.004, dev = 0.1 to 1.0,
range = 325 to 363.

C_3H_6BrCl MW 157.44

1-Bromo-3-chloropropane, CA 109-70-6: Irritant; nonflammable; soluble
ethanol, ether, chloroform. Tm = 214.25; Tb = 416.55; Vm = 0.0991 at
298.15; Antoine (liq) A = 6.15705, B = 1475.23. C = -61.15, dev = 0.1,
range = 326 to 488.

C_3H_6BrNO MW 151.99

2-Bromo-2-nitrosopropane, CA 7119-91-7: Antoine (liq) A = 7.19112, B =
1704.51, C = -27.437, dev = 1.0 to 5.0, range = 239 to 356.

$C_3H_6Br_2$ MW 201.89

1,1-Dibromopropane, CA 598-17-4: Slightly soluble water; soluble ethanol, ether, chloroform. Tm = 218.15; Tb = 406.15 to 407.15; Vm = 0.1030 at 298.15; Antoine (liq) A = 6.339, B = 1504, C = -59.55, dev = 1.0, range = 322 to 449.

1,2-Dibromopropane, propylene dibromide, CA 78-75-1: Mutagen; nonflammable; slightly soluble water; soluble ethanol, ether, chloroform; Tm = 217.95; Tb = 414.75; Vm = 0.1044 at 293.15; Antoine I (liq) A = 6.47365, B = 1572.7, C = -61.15, dev = 1.0, range = 329 to 456; Antoine II (liq) A = 6.00898, B = 1409.6, C = -62.856, dev = 0.1, range = 312 to 403.

1,3-Dibromopropane, CA 109-64-8: Decomposes on prolonged heating; slightly soluble water; soluble ethanol, ether. Tf = 238.95; Tb = 438.15; Vm = 0.1025 at 290.15; Antoine I (liq) A = 6.48273, B = 1675.8, C = -66.15, dev = 1.0, range = 351 to 487; Antoine II (liq) A = 6.12286, B = 1539.764, C = -65.852, dev = 0.02 to 0.1, range = 307 to 437.

$C_3H_6Br_2O$ MW 217.89

2,3-Dibromo-1-propanol, CA 96-13-9: Soluble ethanol, ether, acetone, benzene. Tb = 492.15, decomposes; Vm = 0.1051 at 293.15; Antoine (liq) A = 7.22794, B = 2374.02, C = -37.583, dev = 1.0, range = 330 to 492.

C_3H_6ClFO MW 112.53

1-Chloro-3-fluoro-2-propanol, CA 453-11-2: Antoine (liq) A = 8.8538, B = 2801.8, C = 0, dev = 1.0, range = 273 to 333.

$C_3H_6Cl_2$ MW 112.99

1,1-Dichloropropane, CA 78-99-9: Moderately toxic by inhalation and skin absorption; flammable, flash point 294; slightly soluble water; soluble ethanol, ether, benzene, chloroform. Tb = 361.45; Vm = 0.0989 at 283.15; Antoine (liq) A = 6.107, B = 1273, C = -50.85, dev = 1.0, range = 282 to 399.

1,2-Dichloropropane, CA 78-87-5: Irritant; moderately toxic; flammable, flash point 289; slightly soluble water; soluble ethanol, ether, benzene, chloroform. Tm = 172.75; Tb = 370.15 to 371.15; Vm = 0.0969 at 287.15; Antoine (liq) A = 6.08324, B = 1292.64, C = -52.52, dev = 0.1, range = 239 to 373.

1,3-Dichloropropane, CA 142-28-9: Irritant; moderately toxic; flammable, flash point 305; slightly soluble water; soluble ethanol, ether, benzene. Tm = 173.65; Tb = 393.55; Vm = 0.0951 at 293.15; Antoine (liq) A = 6.09676, B = 1376.2, C = -57.15, dev = 1.0, range = 307 to 435.

2,2-Dichloropropane, CA 594-20-7: Irritant; moderately toxic; flammable, flash point 246; slightly soluble water; soluble ethanol, ether, benzene, chloroform. Tm = 239.35; Tb = 343.65; Vm = 0.1031 at 293.15; Antoine (liq) A = 6.0731, B = 1201.1, C = -47.15, dev = 1.0, range = 267 to 378.

$C_3H_6Cl_2O$ MW 128.99

2,3-Dichloro-1-propanol, CA 616-23-9: Highly toxic; flammable, flash point 364; moderately soluble water; soluble ethanol, ether, acetone, benzene. Tb = 455.15; Vm = 0.0943 at 284.15; Antoine (liq) A = 7.64152, B = 2532.7, C = 0, dev = 1.0 to 5.0, range = 384 to 419.

1,3-Dichloro-2-propanol, CA 96-23-1: Irritant; toxic by inhalation and skin absorption; flammable, flash point 358; solvent for nitrocellulose and resins; moderately soluble water; soluble ethanol, ether, acetone, benzene. Tm = 269.15; Tb = 448.15; Vm = 0.0955 at 290.15; Antoine (liq) A = 7.50741, B = 2374.57, C = -15.87, dev = 1.0, range = 301 to 448.

$C_3H_6F_2$ MW 80.08

1,1-Difluoropropane, CA 430-61-5: Antoine (liq) A = 6.164, B = 1023, C = -35.15, dev = 1.0, range = 219 to 311.

2,2-Difluoropropane, CA 420-45-1: Tm = 168.35; Tb = 272.75; Vm = 0.0870 at 293.15; Antoine (liq) A = 6.21977, B = 1035, C = -27.15, dev = 1.0, range = 211 to 302.

$C_3H_6F_3N$ MW 113.08

N-Methyl-2,2,2-trifluoroethylamine, CA 2730-67-8: Antoine (liq) A = 7.300, B = 1694, C = 0, dev = 5.0, range not specified.

$C_3H_6F_3NOS$ MW 161.14

Methanesulfinamide, 1,1,1-trifluoro-N,N-dimethyl, CA 30957-45-0: Antoine (liq) A = 6.855, B = 1882, C = 0, dev = 1.0 to 5.0, range not specified.

$C_3H_6F_3NS$ MW 145.14

Dimethyl(trifluoromethylthio) amine, CA 62067-13-4: Antoine (liq) A = 6.03096, B = 1130.84, C = -47.58, dev not specified, range = 273 to 329.

$C_3H_6F_3O_2P$ MW 162.05

(Trifluoromethyl)phosphonic acid, dimethyl ester, CA 684-56-0: Antoine (liq) A = 6.81801, B = 1633.618, C = -21.545, dev = 0.1 to 1.0, range = 237 to 318.

$C_3H_6N_2$ MW 70.09

Dimethylcyanamide, CA 1467-79-4: Toxic; moisture-sensitive; flammable, flash point 331; soluble ethanol, ether, acetone. Tb = 435.15 to 436.15; Vm = 0.0808 at 293.15; Antoine (liq) A = 6.9339, B = 2141.4, C = 0, dev = 1.0, range not specified.

$C_3H_6N_2O_4$ MW 134.09

1,1-Dinitropropane, CA 601-76-3: Tm = 231.15; Tb = 457.15; Vm = 0.1063 at 298.15; Antoine (liq) A = 7.1074, B = 2025.3, C = -61.36, dev = 1.0, range = 323 to 383. *(continues)*

78

$C_3H_6N_2O_4$ *(continued)*

2,2-Dinitropropane, CA 595-49-3: Insoluble water; soluble ethanol, ether. Tm = 327.15; Tb = 458.15; Vm = 0.1031 at 298.15; Antoine (liq) A = 9.2264, B = 4153.3, C = 117.21, dev = 1.0, range = 363 to 553.

$C_3H_6N_2O_6$ MW 166.09

1,2-Propanediol, dinitrate, CA 6423-43-4: Tm = below 253.15; Antoine (liq) A = 9.3057, B = 3335.1, C = 0, dev = 1.0, range = 288 to 328.

1,3-Propanediol, dinitrate, CA 3457-90-7: Tm = 243.68; Antoine (liq) A = 10.5793, B = 3883.9, C = 0, dev = 1.0, range = 293 to 313.

$C_3H_6N_6$ MW 126.12

Melamine, CA 108-78-1: Irritant; causes dermatitis; moderately soluble water; soluble hot ethanol. Tm = 620.15, decomposes; Antoine (sol) A = 11.645, B = 6440, C = 0, dev = 5.0, range = 417 to 614.

$C_3H_6N_6O_3$ MW 174.12

Triazine, hexahydro-1,3,5-trinitroso, CA 13980-04-6: Antoine (sol) A = 13.3119, B = 5879.8, C = 0, dev not specified, range = 325 to 360.

$C_3H_6N_6O_6$ MW 222.12

Cyclonite, hexahydro-1,3,5-trinitro-1,3,5-triazine, CA 121-82-4: Carcinogen; powerful explosive; rat poison; almost insoluble water; soluble methanol, ether, acetone, acetic acid. Tm = 476.65; Antoine I (sol) A = 13.305, B = 6799, C = 0, dev = 5.0, range = 328 to 371; Antoine II (liq) A = 9.67329, B = 4410.398, C = 0, dev = 1.0 to 5.0, range = 503 to 523.

C_3H_6O MW 58.08

Acetone, 2-propanone, CA 67-64-1: Moderately toxic vapor; highly flammable, flash point 253; soluble water, most organic solvents. Tm = 178.45; Tb = 329.25; Vm = 0.0735 at 293.15; Tc = 508.7; Pc = 4720; Vc = 0.209; Antoine I (liq) A = 6.24204, B = 1210.6, C = -43.49, dev = 0.1, range = 261 to 329; Antoine II (liq) A = 6.75622, B = 1566.69, C = 0.269, dev = 1,0, range = 329 to 488; Antoine III (liq) A = 3.6452, B = 469.5, C = -108.21, dev = 1.0, range = 178 to 243; Antoine IV (liq) A = 6.19735, B = 1190.382, C = -45.373, dev = 1.0, range = 203 to 269; Antoine V (liq) A = 6.26483, B = 1221.852, C = -42.338, dev = 0.1, range = 257 to 334; Antoine VI (liq) A = 6.24554, B = 1211.515, C = -43.471, dev = 0.02 to 0.1, range = 323 to 379; Antoine VII (liq) A = 6.69966, B = 1542.465, C = 0.447, dev = 0.1, range = 374 to 464; Antoine VIII (liq) A = 7.56948, B = 2457.295, C = 122.324, dev = 0.1 to 1.0, range = 457 to 508.

Allyl alcohol, CA 107-18-6: Highly toxic and irritant; lachrymator; flammable, flash point 294; soluble water, most organic solvents. Tm = 144.15; Tb = 370.05; Vm = 0.0677 at 288.15; Tc = 545.15; Antoine (liq) A = 7.40725, B = 1790.13, C = -38.295, dev = 1.0, range = 253 to 370.

Methyl vinyl ether, CA 107-25-5: Irritant; polymerizes easily; forms explosive peroxide; extremely flammable, flash point 217; *(continues)*

C_3H_6O *(continued)*

moderately soluble water; soluble ethanol, ether, acetone, benzene. Tm = 150.35; Tb = 278.65; Vm = 0.0773 at 293.15; Antoine (liq) A = 6.3882, B = 1221.5, C = 0, dev = 5.0, range = 278 to 412.

Propionaldehyde, propanal, CA 123-38-6: Irritant; moderately toxic; extremely flammable, flash point 243; stench; may form explosive peroxides; very soluble water; soluble ethanol, ether. Tm = 192.15; Tb = 321.15; Vm = 0.0729 at 298.15; Tc = 496, calculated; Pc = 4660, calculated; Vc = 0.210, calculated; Antoine I (liq) A = 6.2047, B = 1166.99, C = -43.15, dev = 1.0, range = 290 to 322; Antoine II (liq) A = 6.2336, B = 1180, C = -42.0, dev = 1.0, range = 250 to 330.

Propylene oxide, 1 2-epoxypropane, CA 16033-71-9: Moderately toxic by inhalation; suspected carcinogen; extremely flammable, flash point 236; moderately soluble water; soluble most organic solvents; Tm = 161.15; Tb = 307.35; Vm = 0.0676 at 273.15; Tc = 482.3; Pc = 4920; Vc = 0.186; Antoine (liq) A = 6.09487, B = 1065.27, C = -46.867, dev = 1.0, range = 225 to 308.

$C_3H_6O_2$ MW 74.08

1,3-Dioxolane, glycol methylene ether, CA 646-06-0: Irritant; highly flammable, flash point 275; may form explosive peroxides; soluble water, ethanol, ether, acetone. Tf = 178.15; Tb = 351.15; Vm = 0.0695 at 288.15; Antoine (liq) A = 6.10722, B = 1185.643, C = -59.824, dev = 0.1, range = 280 to 323.

2,3-Epoxy-1-propanol, glycidol, CA 556-52-5: Irritant; moderately toxic; absorbed through skin; polymerizes; flammable, flash point 354; soluble water, most organic solvents. Tb = 439.15 to 440.15, decomposes; Vm = 0.0665 at 298.15; Antoine (liq) A = 34.4207, B = 53298.34, C = 1227.306, dev = 1,0 to 5.0, range not specified.

Ethyl formate, CA 109-94-4: Irritant; extremely flammable, flash point 253; moderately soluble water with gradual hydrolysis; soluble ethanol, ether, benzene. Tf = 193.15; Tb = 327.45; Vm = 0.0808 at 293.15; Tc = 508.4; Pc = 4742; Vc = 0.229; Antoine I (liq) A = 6.1384, B = 1151.08, C = -48.94, dev = 1.0, range = 213 to 336; Antoine II (liq) A = 6.4206, B = 1326.4, C = -26.867, dev = 1.0, range = 327 to 498.

Methyl acetate, CA 79-20-9: Irritant; narcotic in high concentrations; flammable, flash point 263; solvent for gums and lacquers; soluble water, most organic solvents. Tm = 174.45; Tb = 329.45; Vm = 0.0793 at 293.15; Tc = 506.8; Pc = 4691; Vc = 0.228; Antoine I (liq) A = 6.190152, B = 1157.622, C = -53.426, dev = 0.1, range = 274 to 331; Antoine II (liq) A = 6.18771, B = 1156.219, C = -53.589, dev = 0.1, range = 274 to 331.

Propionic acid, propanoic acid, CA 79-09-4: Corrosive; irritant; causes burns; flammable, flash point 325; soluble water, ethanol, ether, chloroform. Tm = 251.15; Tb = 414.25; Vm = 0.0747 at 293.15; Tc = 612; Pc = 5370; Vc = 0.230; Antoine I (liq) A = 6.60267, B = 1577.96, C = -70.844, dev = 0.1, range = 343 to 419; Antoine II (liq) A = 9.24101, B = 2835.99, C = -23.07, dev = 1.0, range = 414 to 511.

$C_3H_6O_3$ MW 90.08

Methoxyacetic acid, CA 625-45-6: Corrosive; hygroscopic; flammable, flash point over 385; soluble water, ethanol, ether.   *(continues)*

$C_3H_6O_3$ *(continued)*

Tb = 476.15 to 477.15; Vm = 0.0765 at 293.15; Antoine (liq) A = 8.84007,
B = 3563.45, C = 43.98, dev = 1.0, range = 325 to 477.

Methyl glycolate, CA 96-35-5: Flammable; soluble water, ethanol, ether.
Tb = 424.15; Vm = 0.0771 at 291.15; Antoine (liq) A = 7.59063, B = 2369.19,
C = -0.58, dev = 1.0, range = 282 to 425.

Peroxypropionic acid, CA 4212-43-5: Flammable; may explode on heating.
Tf = 259.65; Antoine (liq) A = 7.748, B = 2256, C = 0, dev = 5.0, range =
273 to 393.

Propylene ozonide: Explosive at room temperature. Antoine (liq) A =
7.551, B = 1929, C = 0, dev = 5.0, range = 261 to 296.

1,3,5-Trioxane, trioxymethylene, CA 110-88-3: Irritant; flammable, flash
point 318; cyclic polymer of formaldehyde; soluble water, most organic
solvents. Tm = 334.15 to 335.15; Tb = 387.65; Vm = 0.0770 at 338.15;
Antoine (liq) A = 6.79843, B = 1687.5, C = -35.12, dev = 0.1, range = 329
to 387.

C_3H_6S ME 74.14

2-Methylthiirane, propylene sulfide, CA 1072-43-1: Highly toxic; stench;
flammable, flash point 283; slightly soluble water; soluble chloroform.
Tm = 182.15; Tb = 348.15 to 350.15; Vm = 0.0784 at 291.15; Antoine (liq)
A = 6.17838, B = 1249.3, C = -48.15, dev = 1.0 to 5.0, range = 272 to 423.

Thiacyclobutane, trimethylene sulfide, CA 287-27-4: Irritant; stench;
flammable, flash point 271; soluble ethanol, acetone, benzene. Tm = 199.95;
Tb = 366.95 to 367.35; Vm = 0.0721 at 293.15; Antoine (liq) A = 6.13786,
B = 1319.372, C = -48.822, dev = 0.02, range = 320 to 405.

C_3H_7Br MW 122.99

1-Bromopropane, propyl bromide, CA 106-94-5: Highly irritant; flammable,
flash point 298; slightly soluble water; soluble most organic solvents.
Tf = 163.15; Tb = 344.05; Vm = 0.0910 at 293.15; Antoine I (liq) A =
6.03555, B = 1194.889, C = -47.64, dev = 0.1. range = 250 to 368; Antoine
II (liq) A = 6.03823, B = 1193.612 C = -48.005, dev = 0.02 to 0.1, range =
301 to 344.

2-Bromopropane, isopropyl bromide, CA 75-26-3: Flammable, flash point 292;
slightly soluble water; soluble most organic solvents. Tf = 184.15; Tb =
332.45; Vm = 0.0936 at 293.15; Antoine I (liq) A = 5.28473, B = 858.03, C =
-71.18, dev = 1.0, range = 236 to 328: Antoine II (liq) A = 5.91155, B =
1098.573, C = -51.268, dev = 0.1 to 1.0, range = 299 to 332.

C_3H_7Cl MW 78.54

1-Chloropropane, propyl chloride, CA 540-54-5: Irritant; toxic vapor;
highly flammable, flash point below 255; slightly soluble water; soluble
ethanol, ether, benzene. Tm = 150.35; Tb = 319.75; Vm = 0.0876 at 288.15;
Antoine (liq) A = 6.09145, B = 1126.383, C = -43.78, dev = 0.1, range = 248
to 320.

2-Chloropropane, isopropyl chloride, CA 75-29-6: Eye irritant; highly
flammable, flash point 241; slightly soluble water;   *(continues)*

C_3H_7Cl *(continued)*

soluble ethanol, ether, benzene, chloroform. Tm = 156.15; Tb = 307.95;
Vm = 0.0905 at 288.15; Antoine (liq) A = 5.54823, B = 860.49, C = -66.25,
dev = 1.0, range = 239 to 310.

C_3H_7ClO MW 94.54

2-Chloro-1-propanol, propylene chlorohydrin, CA 78-89-7: Highly toxic;
flammable, flash point 325; soluble water, ethanol, ether, most organic
solvents; Tb = 406.15 to 407.15; Vm = 0.0852 at 293.15; Antoine (liq) A =
5.86512, B = 1163.6, C = -97.75, dev = 1.0, range = 316 to 399.

$C_3H_7ClO_2S$ MW 142.60

1-Propanesulfonyl chloride, CA 10147-36-1: Flammable. Tm = 227.15; Tb =
453.15, decomposes; Vm = 0.1112 at 288.15; Antoine (liq) A = 9.628, B =
3142.2, C = 0, dev = 1.0 to 5.0, range = 243 to 273.

C_3H_7ClS MW 110.60

Methyl(2-chloroethyl) sulfide, CA 542-81-4: Irritant; vesicant; stench;
flammable, flash point 315; soluble ethanol, ether, acetone. Tb = 413.15;
Vm = 0.0983 at 293.15; Antoine (liq) A = 7.41427, B = 2215.7, C = 0, dev =
1.0, range = 273 to 373.

C_3H_7F MW 62.08

1-Fluoropropane, propyl fluoride, CA 460-13-9: Flammable; soluble ethanol,
ether. Tm = 114.15; Tb = 269.95; Vm = 0.0780 at 293.15; Antoine (liq) A =
6.0782, B = 965.18, C = -33.65, dev = 1.0, range = 196 to 289.

2-Fluoropropane, isopropyl fluoride, CA 420-26-8: Flammable. Tm = 139.75;
Tb = 263.75; Antoine (liq) A = 6.043, B = 933.98, C = -32.41, dev = 1.0,
range = 190 to 264.

C_3H_7I MW 169.99

1-Iodopropane, propyl iodide, CA 107-08-4: Suspected carcinogen; light-
sensitive; flammable, flash point 317; slightly soluble water; soluble
ethanol, ether, benzene, chloroform. Tf = 171.85; Tb = 375.65; Vm = 0.0973
at 293.15; Antoine I (liq) A = 5.6036, B = 1160.5, C = -59.55, dev = 1.0,
range = 171 to 271; Antoine II (liq) A = 5.94093, B = 1267.062, C = -53.62,
dev = 0.1, range = 271 to 402.

2-Iodopropane, isopropyl iodide, CA 75-30-9: Irritant; suspected carcinogen; flammable, flash point 315; slightly soluble water; soluble ethanol,
ether, benzene, chloroform. Tf = 183.05; Tb = 362.15 to 363.15; Vm =
0.0998 at 293.15; Antoine I (liq) A = 6.2763, B = 1414.85, C = -31.45,
dev = 1.0, range = 261 to 363; Antoine II (liq) A = 5.2724, B = 989.55, C =
-69.18, dev = 1.0, range = 173 to 262.

C_3H_7N MW 57.10

Allylamine, 3-aminopropylene, CA 107-11-9: Irritant; lachrymator; highly
toxic; highly flammable, flash point 245; soluble water, *(continues)*

C_3H_7N *(continued)*

ethanol, ether, chloroform. Tm = 184.95; Tb = 326.05; Vm = 0.0749 at
293.15; Antoine I (liq) A = 7.305, B = 1725, C = 0, dev = 1.0, range = 273
to 303; Antoine II (liq) A = 7.05068, B = 1612.877, C = -7.635, dev = 1.0,
range = 273 to 324.

Azetidine, CA 503-29-7: Expected carcinogen; flammable; soluble water,
ethanol, ether, acetone. Tb = 336.15; Vm = 0.0677 at 293.15; Antoine (liq)
A = 7.15472, B = 1650.863, C = -12.88, dev = 1.0, range = 273 to 303.

C_3H_7NO MW 73.09

Acetone, oxime, CA 127-06-0: Flammable; soluble water, ethanol, ether,
light hydrocarbons. Tm = 334.15; Tb = 407.15 to 408.15; Vm = 0.0802 at
335.15; Antoine I (sol) A = 9.90438, B = 3113.245, C = 0, dev = 1.0 to 5.0,
range = 313 to 333; Antoine II (liq) A = 8.63144, B = 2683.893, C = 0,
dev = 1.0, range = 338 to 352.

Dimethyl formamide, CA 68-12-2: Highly toxic; reacts violently with many
materials; flammable, flash point 331; soluble water, most organic solvents.
Tm = 212.15; Tb = 426.15; Vm = 0.0774 at 298.15; Tc = 647.14, calculated;
Pc = 4480, calculated; Vc = 0.265, calculated; Antoine (liq) A = 5.37646,
B = 1049.26, C = -113.84, dev = 1.0, range = 301 to 426.

N-Methylacetamide, CA 79-16-3: Flammable, flash point 381; soluble water,
ethanol, ether, acetone, benzene. Tm = 303.7; Tb = 479.15; Antoine I (sol)
A = 7.864, B = 2823, C = 0, dev = 1.0 to 5.0, range = 288 to 301; Antoine
II (liq) A = 7.8259, B = 2793.3, C = 0, dev = 1.0, range = 353 to 479;
Antoine III (liq) A = 6.60575, B = 1868.206, C = -75.963, dev = 1.0, range
= 333 to 443.

Propionaldehyde, oxime, CA 627-39-4: Flammable. Tm = 313.15; Tb = 403.15
to 405.15; Antoine (liq) A = 7.23313, B = 1798.182, C = -58.676, dev = 1.0,
range = 313 to 339.

Propionamide, CA 79-05-0: Flammable; soluble water, ethanol, ether, acetone, chloroform. Tm = 352.15; Tb = 495.35; Vm = 0.0789 at 383.15; Antoine
I (sol) A = 11.1659, B = 4138.9, C = 0, dev = 1.0, range = 318 to 346;
Antoine II (liq) A = 6.86665, B = 1945.1, C = -92.63, dev = 1.0, range =
381 to 493.

$C_3H_7NO_2$ MW 89.09

Alanine, *levo*, CA 56-41-7: Very soluble water; slightly soluble ethanol;
insoluble ether. Tm = 570.15, decomposes; Antoine I (sol) A = 13.935, B =
7220, C = 0, dev = 5.0, range = 423 to 473; Antoine II (sol) A = 13.35743,
B = 6936.384, C = 0, dev = 1.0 to 5.0, range = 407 to 426.

Ethyl carbamate, urethane, CA 51-79-6: Expected carcinogen; flammable,
flash point 365; very soluble water; soluble ethanol, ether, benzene,
chloroform. Tm = 322.15 to 323.15; Tb = 457.15; Antoine I (sol) A =
13.215, B = 4646, C = 0, dev = 1.0, range = 292 to 307; Antoine II (liq) A
= 7.35655, B = 2179.29, C = -49.918, dev = 1.0, range = 338 to 457.

Isopropyl nitrite, CA 541-42-4: Toxic; can cause headaches and cyanosis;
flammable; decomposes at room temperature; soluble ethanol, ether. Tb =
312.15 to 313.15; Vm = 0.1041 at 273.15; Antoine (liq) A = 6.345, B = 1360,
C = 0, dev = 5.0, range = 253 to 268.   *(continues)*

$C_3H_7NO_2$ *(continued)*

1-Nitropropane, CA 108-03-2: Moderately toxic; may explode on heating; flammable, flash point 309; moderately soluble water; soluble most organic solvents. $Tm = 169.16$; $Tb = 404.33$; $Vm = 0.0890$ at 293.15; $Tc = 675.15$; Antoine (liq) $A = 6.252442$, $B = 1474.299$, $C = -57.164$, dev = 0.1, range = 293 to 405.

2-Nitropropane, CA 79-46-9: Suspected carcinogen; may explode on heating; flammable, flash point 297; moderately soluble water; soluble most organic solvents; $Tm = 181.83$; $Tb = 393.4$; $Vm = 0.0902$ at 293.15; $Tc = 617.15$; Antoine (liq) $A = 6.208143$, $B = 1422.898$, $C = -54.809$, dev = 0.1, range = 284 to 394.

Propyl nitrite, CA 543-67-9: Moderately toxic; flammable; soluble ethanol, ether. $Tb = 322.05$ to 322.55; $Vm = 0.1005$ at 293.15; Antoine (liq) $A = 6.595$, $B = 1480$, $C = 0$, dev = 5.0, range = 253 to 268.

$C_3H_7NO_3$ MW 105.09

Isopropyl nitrate, CA 1712-64-7: Flammable; can explode on heating or shock; soluble ethanol, ether. $Tb = 374.15$ to 375.15; $Vm = 0.1014$ at 292.15; Antoine (liq) $A = 6.4279$, $B = 1457$, $C = -45.65$, dev = 1.0, range = 273 to 343.

Propyl nitrate, CA 627-13-4: Toxic; highly flammable, flash point 293; can explode on heating or shock; used in rocket fuels; slightly soluble water; soluble ethanol, ether. $Tm = 173.15$; $Tb = 383.15$; $Vm = 0.0997$ at 293.15; Antoine (liq) $A = 6.3709$, $B = 1444$, $C = -52.35$, dev = 1.0, range = 273 to 343.

$C_3H_7N_3$ MW 85.11

Propane, 1-azido, CA 22293-25-0: Antoine (liq) $A = 6.09541$, $B = 1298.796$, $C = -28.314$, dev = 1.0, range = 253 to 298.

Propane, 2-azido, CA 691-57-6: Antoine (liq) $A = 6.5470$, $B = 1403.76$, $C = -26.765$, dev = 1.0, range = 253 to 298.

C_3H_8 MW 44.10

Propane, CA 74-98-6: Extremely flammable, flash point 169; moderately soluble organic solvents; soluble hydrocarbons at low temperatures. $Tm = 85.5$; $Tb = 231.1$; $Vm = 0.0753$ at 228.15; $Tc = 369.8$; $Pc = 4240$; $Vc = 0.203$; Antoine I (liq) $A = 5.92888$, $B = 803.81$, $C = -26.16$, dev = 0.1, range = 165 to 248; Antoine II (liq) $A = 6.757$, $B = 1045.3$, $C = -7.15$, dev = 1.0, range = 104 to 165; Antoine III (liq) $A = 5.97611$, $B = 822.497$, $C = -23.892$, dev = 0.1, range = 231 to 281; Antoine IV (liq) $A = 6.30221$, $B = 998.045$, $C = 2.129$, dev = 0.1, range = 278 to 332; Antoine V (liq) $A = 7.42085$, $B = 1889.081$, $C = 128.11$, dev = 0.1, range = 329 to 369.

$C_3H_8N_2$ MW 72.11

Dimethyl ammonium cyanide: Flammable. $Tm = 219.15$; Antoine (liq) $A = 9.86369$, $B = 2552.893$, $C = 0$, dev = 1.0, range = 250 to 296.

84

C_3H_8O MW 60.10

Ethyl methyl ether, methoxyethane, CA 540-67-0: Anesthetic in high concentrations; forms explosive peroxides; extremely flammable, flash point 236; soluble water, ethanol, ether, acetone. Tb = 283.95; Vm = 0.0829 at 273.15; Antoine I (liq) A = 5.00683, B = 504.49, C = -112.4, dev = 1.0, range = 216 to 299; Antoine II (liq) A = 5.30082, B = 523.34, C = -124.745, dev = 1.0, range = 281 to 433; Antoine III (liq) A = 6.13202, B = 988.801, C = -41.053, dev = 1.0 to 5.0, range = 281 to 438.

1-Propanol, propyl alcohol, CA 71-23-8: Moderately toxic; flammable, flash point 296; soluble water, ethanol, ether, acetone. Tm = 146.95; Tb = 370.35; Vm = 0.0748 at 293.15; Tc = 536.71; Pc = 5170; Vc = 0.219; Antoine I (liq) A = 6.86874, B = 1437.906, C = -74.621, dev = 0.02, range = 333 to 378; Antoine II (liq) A = 6.74195, B = 1364.911, C = -82.114, dev = 0.02, range = 356 to 378; Antoine III (liq) A = 8.7592, B = 2506, C = 0, dev = 1.0 to 5.0, range = 200 to 228; Antoine IV (liq) A = 6.74403, B = 1366.08, C = -81.994, dev = 0.02, range = 356 to 376; Antoine V (liq) A = 6.87377, B = 1440.743, C = -74.344, dev = 0.02 to 0.1, range = 333 to 376; Antoine VI (liq) A = 6.58415, B = 1273.365, C = -92.178, dev = 0.1, range = 369 to 407; Antoine VII (liq) A = 6.43938, B = 1185.921, C = -102.916, dev = 0.1 to 1.0, range = 401 to 482; Antoine VIII (liq) A = 6.89456, B = 1557.973, C = -47.251, dev = 1.0, range = 478 to 507.

2-Propanol, isopropyl alcohol, CA 67-63-0: Eye irritant; may cause headaches and nausea; flammable, flash point 285; may react explosively with some materials; soluble water, alcohols, ether, acetone, chloroform. Tm = 184.65; Tb = 355.45; Vm = 0.0765 at 293.15; Tc = 508.35; Pc = 4760; Vc = 0.220; Antoine I (liq) A = 6.86087, B = 1357.514, C = -75.786, dev = 0.02, range = 325 to 363; Antoine II (liq) A = 6.72348, B = 1282.26, C = -83.591, dev = 0.02, range = 347 to 363; Antoine III (liq) A = 9.6871, B = 2626, C = 0, dev = 1.0 to 5.0, range = 195 to 228; Antoine IV (liq) A = 6.73782, B = 1290.039, C = -82.771, dev = 0.02, range = 347 to 363; Antoine V (liq) A = 6.86451, B = 1359.473, C = -75.592, dev = 0.02 to 0.1, range = 325 to 363; Antoine VI (liq) A = 6.61939, B = 1225.439, C = -89.774, dev = 0.1 to 1.0, range = 350 to 383; Antoine VII (liq) A = 6.40823, B = 1107.303, C = -103.944, dev = 1.0, range = 379 to 461; Antoine VIII (liq) A = 7.02506, B = 1588.226, C = -33.839, dev = 1.0, range = 453 to 508.

$C_3H_8OS_2$ MW 124.22

3-Hydroxy-1,2-propanedithiol, CA 59-52-9: Highly irritant; stench; flammable, flash point above 385; moderately soluble water, with decomposition; soluble vegetable oils. Vm = 0.0997 at 293.15; Antoine (liq) A = 8.45143, B = 3196.8, C = 0, dev = 1.0, range = 353 to 413.

$C_3H_8O_2$ MW 76.10

2-Methoxyethanol, ethylene glycol, monomethyl ether, CA 109-86-4: Irritant; flammable, flash point 312; soluble water, ethanol, ether, benzene. Tm = 188.05; Tb = 397.65; Vm = 0.0788 at 293.15; Antoine (liq) A = 6.84907, B = 1715.47, C = -43.15, dev = 0.1 to 1.0, range = 333 to 423.

Methylal, dimethoxymethane, CA 109-87-5: Irritant; may form explosive peroxides; extremely flammable, flash point 241; very soluble water; soluble ethanol, ether, oils. Tm = 168.15; Tb = 314.15 to 315.15; Antoine I (liq) A = 5.50613, B = 804.78, C = -85.3, dev = 1.0, range = 273 to 316; Antoine II (liq) A = 7.06105, B = 1623.024, C = 5.834, dev = 1.0, range = 273 to 318.   *(continues)*

$C_3H_8O_2$ *(continued)*

1,2-Propanediol, propylene glycol, CA 57-55-6: May oxidize at high temperatures; flammable, flash point 372; soluble water, acetone, chloroform. Tm = 213.15; Tb = 460.75; Vm = 0.0733 at 293.15; Tc = 626, calculated; Pc = 6100, calculated; Vc = 0.239, calculated; Antoine (liq) A = 7.91179, B = 2554.9, C = -28.611, dev = 1.0, range = 318 to 461.

1,3-Propanediol, trimethylene glycol, CA 504-63-2: Flammable; soluble water, ethanol, ether. Tm = 246.15; Tb = 483.15 to 484.15; Vm = 0.0718 at 293.15; Tc = 658, calculated; Pc = 5920, calculated; Vc = 0.217, calculated; Antoine (liq) A = 8.34759, B = 3149.87, C = 9.144, dev = 1.0, range = 332 to 488.

$C_3H_8O_3$ MW 92.09

Glycerol, CA 56-81-5: Moderately irritant; hygroscopic; flammable, flash point 433; reacts violently with some oxidizing agents; soluble water, ethanol. Tm = 291.32; Tb = 563.15; Vm = 0.0730 at 298.15; Tc = 723, calculated; Pc = 4000, calculated; Vc = 0.264, calculated; Antoine I (liq) A = 5.13022, B = 990.45, C = -245.819, dev = 1.0, range = 469 to 563; Antoine II (liq) A = 10.39913, B = 4480.5, C = 0, dev = 1.0, range = 293 to 343.

C_3H_8S MW 76.16

Ethyl methyl sulfide, CA 624-89-5: Toxic; stench; flammable, flash point 258; soluble ethanol, ether. Tf = 168.35; Tb = 340.05; Vm = 0.0904 at 293.15; Antoine (liq) A = 6.05849, B = 1179.88, C = -48.666, dev = 0.02, range = 295 to 374.

1-Mercaptopropane, propyl mercaptan, CA 107-03-9: Moderately toxic; stench; highly flammable, flash point 253; slightly soluble water; soluble ethanol, ether, acetone, benzene. Tm = 160.02; Tb = 340.95; Vm = 0.0911 at 298.15; Antoine (liq) A = 6.05019, B = 1181.703, C = -48.687, dev = 0.02, range = 296 to 376.

2-Mercaptopropane, isopropyl mercaptan, CA 75-33-2: Toxic; stench; highly flammable, flash point 238; slightly soluble water; soluble ethanol, ether, acetone. Tm = 142.61; Tb = 325.75; Vm = 0.0935 at 293.15; Antoine (liq) A = 5.99782, B = 1111.471, C = -47.281, dev = 0.02, range = 282 to 359.

$C_3H_8S_2$ MW 108.22

1,3-Propanedithiol, CA 109-80-8: Irritant; toxic; stench; flammable, flash point 331; slightly soluble water; soluble ethanol, ether, benzene, chloroform. Tm = 194.15; Tb = 442.15 to 443.15; Vm = 0.1004 at 293.15; Antoine (liq) A = 6.8773, B = 2173, C = 0, dev = 1.0, range = 377 to 446.

C_3H_9N MW 59.11

Isopropyl amine, CA 75-31-0: Corrosive; highly irritant; extremely flammable, flash point 236; soluble water, ethanol, ether, acetone. Tm = 177.95; Tb = 305.55; Vm = 0.0858 at 293.15; Tc = 476.15; Pc = 4425; Vc = 0.235; Antoine (liq) A = 6.0104, B = 983.201, C = -59.395, dev = 0.02, range = 276 to 334.

(continues)

C_3H_9N *(continued)*

Propyl amine, CA 107-10-8: Corrosive; irritant; highly flammable, flash
point 236; soluble water, ethanol, ether, acetone. Tf = 190.15; Tb =
320.95; Vm = 0.0823 at 293.15; Tc = 496.95; Pc = 4742; Vc = 0.298; Antoine
(liq) A = 6.04693, B = 1041.715, C = -62.596, dev = 0.02, range = 295
to 351.

Trimethyl amine, CA 75-50-3: Corrosive; irritant; flammable, flash point
267; very soluble water; soluble ethanol, ether, benzene, chloroform. Tm =
156.05; Tb = 276.05; Vm = 0.0935 at 293.15; Tc = 433.25; Pc = 4073; Vc =
0.254; Antoine (liq) A = 6.01402, B = 968.978, C = -34.253, dev = 0.1,
range = 192 to 277.

C_3H_9NO MW 75.11

1-Amino-2-propanol, CA 78-96-6: Irritant; moderately toxic; flammable,
flash point 346; soluble water, ethanol, ether, acetone. Tm = 271.15; Tb =
433.15; Vm = 0.0772 at 293.15; Antoine (liq) A = 8.15363, B = 2626.73, C =
-4.193, dev = 1.0, range = 306 to 431.

Dimethylamine, *N*-methoxy, CA 5669-39-6: Irritant; flammable. Tm = 175.95;
Antoine (liq) A = 5.89071, B = 979.551, C = -50.971, dev = 1.0, range =
194 to 297.

Ethyl amine, 2-methoxy, CA 109-85-3: Corrosive; flammable, flash point
282; soluble water, ethanol. Tb = 368.15; Vm = 0.0869 at 293.15; Antoine
(liq) A = 6.3184, B = 1334.4, C = -55.01, dev = 1.0, range = 278 to 318.

$C_3H_9NO_3S$ MW 139.17

Trimethyl amine, sulfur dioxide complex, CA 3162-58-1: Irritant; moisture-
sensitive. Tm = 505.15, decomposes; Antoine (sol) A = 10.4009, B = 3165,
C = 0, dev = 1.0, range = 292 to 349.

$C_3H_9O_3P$ MW 124.08

Methylphosphonic acid, dimethyl ester, CA 756-79-6: Irritant; flammable,
flash point 316; soluble water. Tb = 454.15; Vm = 0.1070 at 293.15;
Antoine (liq) A = 4.23131, B = 573.682, C = -204.997, dev = 1.0, range =
336 to 408.

Trimethyl phosphite, CA 121-45-9: Irritant; moisture- and air-sensitive;
flammable, flash point 327; soluble ethanol, ether. Tm = 195.15; Tb =
384.15 to 385.15; Vm = 0.1179 at 293.15; Antoine (liq) A = 6.52092, B =
1714.26, C = 0, dev = 1.0, range = 422 to 494.

$C_3H_9O_4P$ MW 140.08

Trimethyl phosphate, CA 512-56-1: Suspected carcinogen; nonflammable; soluble water, ether. Tm = 227.05; Tb = 470.35; Vm = 0.1153 at 293.15;
Antoine (liq) A = 6.64252, B = 1992.12, C = -36.063, dev = 1.0, range = 296
to 466.

C_3H_9P MW 76.08

Trimethyl phosphine, CA 594-09-2: Irritant; toxic; highly flammable, may
ignite in air; insoluble water; soluble ether. Tm = 187.25; Tb = 311.15 to
314.15; Antoine (liq) A = 6.8578, B = 1512, C = 0, dev = 1.0, range = 248
to 310.

$C_3H_{10}N_2$ MW 74.13

1,2-Propanediamine, CA 10424-38-1: Corrosive; very hygroscopic; flammable,
flash point 306; very soluble water. Tb = 392.15 to 393.15; Vm = 0.0844 at
288.15; Antoine I (liq) A = 7.615, B = 2204, C = 0, dev = 5.0, range = 293
to 393; Antoine II (liq) A = 6.7097, B = 1595.12, C = -54.22, dev = 0.1,
range = 242 to 293.

Trimethylhydrazine, CA 1741-01-1: Toxic; flammable; used as rocket fuel.
Tm = 200.15; Tb = 333.15 to 334.15; Antoine (liq) A = 6.22105, B =
1184.891, C = -51.585, dev = 0.02, range = 256 to 292.

C_3N_2O MW 80.05

Carbonyl cyanide, mesoxalonitrile, CA 1115-12-4: Highly flammable; explodes with water. Tm = 237.15 to 238.15; Tb = 338.65; Vm = 0.0712 at
293.15; Antoine (liq) A = 7.7928, B = 1960, C = 0, dev = 1.0, range = 250
to 291.

C_3N_3P MW 109.03

Tricyanophosphine, phosphorus cyanide, CA 1116-01-4: Antoine (sol) A =
9.97978, B = 4088.62, C = 0, dev = 5.0, range = 293 to 323.

C_3O_2 MW 68.03

Carbon suboxide, CA 504-64-3: Highly toxic; lachrymator; attacks eyes and
lungs; highly flammable; polymerizes above 288 to a red solid; decomposes
with water; slightly soluble organic solvents. Tm = 161.85; Tb = 279.95;
Vm = 0.0611 at 273.15; Antoine (liq) A = 6.31391, B = 1100.94, C = -24.0,
dev = 0.1, range = 160 to 249.

C_3S_2 MW 100.15

Carbon subsulfide, CA 627-34-9: Extremely flammable; explosion hazard, explodes at 373; polymerizes rapidly in air and light. Tm = 272.65; Vm =
0.0759 at 288.15; Antoine (liq) A = 6.48321, B = 1881.8, C = -31.87, dev =
1.0, range = 287 to 383.

C_4BrClF_9N MW 348.39

1,1,2-Trifluoro-2-chloro-2-bromo-N,N-bis(trifluoromethyl) ethylamine:
Antoine (liq) A = 6.7239, B = 1730, C = 0, dev = 1.0, range = 329 to 364.

$C_4BrCl_2F_8N$ MW 364.85

Ethylamine, 2-bromo-1,2-dichloro-1,2-difluoro-*N,N*-bis(trifluoromethyl),
CA 4905-98-0: Antoine (liq) A = 6.7819, B = 1904, C = 0, dev = 1.0, range
= 358 to 394.

C_4BrF_6N MW 255.95

Ethynylamine, 2-bromo-*N,N*-bis(trifluoromethyl), CA 22130-38-7: Antoine
(liq) A = 6.7789, B = 1586, C = 0, dev = 1.0 to 5.0, range = 311 to 329.

C_4BrF_8N MW 293.94

N,N-Bis(trifluoromethyl)-2,2-difluoro-1-bromo vinylamine, CA 17725-57-4:
Antoine (liq) A = 7.0599, B = 1637, C = 0, dev = 1.0, range = 293 to 320.

$C_4BrF_{10}N$ MW 331.94

Ethylamine, 1,1,2,2-tetrafluoro-2-bromo-*N,N*-bis(trifluoromethyl): Antoine
(liq) A = 6.7179, B = 1574, C = 0, dev = 1.0, range = 289 to 329.

$C_4Br_2F_9N$ MW 392.84

Ethylamine, *N,N*-bis(trifluoromethyl)-1,2-dibromo-1,2,2-trifluoro,
CA 17725-58-5: Antoine (liq) A = 6.7489, B = 1790, C = 0, dev = 1.0, range
= 326 to 366.

C_4ClF_6NO MW 227.49

2-Chloro-2-isocyanato-hexafluoropropopane, CA 39095-53-9: Antoine (liq)
A = 6.935, B = 1594, C = 0, dev = 1.0 to 5.0, range not specified.

C_4ClF_8N MW 249.49

Vinylamine, *N,N*-bis(trifluoromethyl)-2-chloro-1,2-difluoro, CA 13747-22-3:
Antoine (liq) A = 6.8489, B = 1522, C = 0, dev = 1.0, range = 273 to 312.

C_4ClF_9S MW 286.54

1,1,2,2,3,3,4,4,4-Nonafluorobutanesulfenyl chloride, CA 42769-81-3:
Antoine (liq) A = 7.015, B = 1783, C = 0, dev = 1.0 to 5.0, range not
specified.

$C_4ClF_{10}N$ MW 287.49

N-Chloro-bis(pentafluoroethyl) amine, CA 54566-79-9: Antoine (liq) A =
6.945, B = 1606, C = 0, dev = 1.0 to 5.0, range not specified.

2-Propanamine, *N*-chloro-1,1,1,2,3,3,3-heptafluoro-*N*-(trifluoromethyl),
CA 53684-04-1: Antoine (liq) A = 7.255, B = 1704, C = 0, dev = 1.0 to 5.0,
range not specified.

$C_4ClF_{12}NS$ MW 357.54

Sulfur, chlorodifluoro(1,1,1,2,3,3,3-heptafluoro-2-propanaminato)(trifluoromethyl), CA 62609-69-2: Antoine (liq) A = 6.905, B = 1910, C = 0,
dev = 1.0 to 5.0, range not specified.

$C_4ClF_{12}S$ MW 343.54

Chlorotetrafluoro(nonafluorobutyl) sulfur: Antoine (liq) A = 7.285, B =
1940, C = 0, dev = 1.0 to 5.0, range not specified.

$C_4Cl_2F_6$ MW 232.94

1,4-Dichlorohexafluoro-2-butene, CA 20972-44-5: Antoine (liq) A = 7.3579,
B = 1774, C = 0, dev = 1.0 to 5.0, range = 279 to 330.

2,3-Dichlorohexafluoro-2-butene, cis, CA 2418-22-6: Tm = 205.85; Tb =
341.15 to 342.15; Vm = 0.1435 at 293.15; Antoine (liq) A = 6.987, B = 1699,
C = 0, dev = 1.0, range = 298 to 341.

2,3-Dichlorohexafluoro-2-butene, trans, CA 2418-21-5: Tm = 220.15; Antoine
(liq) A = 6.9681, B = 1684, C = 0, dev = 1.0, range = 298 to 340.

$C_4Cl_2F_7N$ MW 265.95

Azetidine, 2,3,4,4-tetrafluoro-2,3-dichloro-1-(trifluoromethyl): Antoine
(liq) A = 7.0019, B = 1702, C = 0, dev = 1.0, range = 273 to 333.

Methylamine, 1,1,1-trifluoro-N-(1,2,3,3-tetrafluoro-2,3-dichloropropyl-
idene): Antoine (liq) A = 6.0749, B = 1412, C = 0, dev = 1.0, range not
specified

$C_4Cl_3F_7$ MW 287.39

2,2,3-Trichloroheptafluorobutane, CA 335-44-4: Antoine (liq) A = 5.89497,
B = 1177.1, C = -67.33, dev = 1.0, range = 302 to 446.

$C_4Cl_4F_4$ MW 265.85

1,2,3,4-Tetrachlorotetrafluoro-1-butene: Antoine (liq) A = 6.9926, B =
2060.4, C = 0, dev = 1.0, range = 362 to 414.

$C_4Cl_4F_6O$ MW 319.85

Trichloromethyl 2-chloro-1,1,2,3,3,3-hexafluoropropyl ether, CA 61136-57-0:
Antoine (liq) A = 5.99553, B = 1334.607, C = -68.957, dev = 0.02 to 0.1,
range = 325 to 403.

C_4Cl_6 MW 260.76

Perchloro-1,3-butadiene, CA 87-68-3: Polymerizes; nonflammable; soluble
ethanol, ether. Tm = 252.15; Tb = 484.15 to 488.15; Vm = 0.1552 at 293.15;
Antoine (liq) A = 6.4781, B = 1783.8, C = -84.48, dev = 1.0, range = 343
to 484.

$C_4Cl_6O_3$ MW 308.76

Trichloroacetic acid, anhydride, CA 4124-31-6: Corrosive; nonflammable;
moisture-sensitive; soluble ether, acetic acid. Tb = 495.15 to 497.15,
decomposes; Vm = 0.1826 at 293.15; Antoine (liq) A = 7.1397, B = 2371.33,
C = -34.45, dev = 1.0, range = 329 to 496.

$C_4F_6O_3$ MW 210.03

Trifluoroacetic acid, anhydride, CA 407-25-0: Corrosive; nonflammable;
moisture-sensitive; moderately toxic; causes severe burns. Tm = 208.15;
Tb = 312.65; Vm = 0.1410 at 298.15; Antoine (liq) A = 6.64084, B = 1202.32,
C = -52.903, dev = 1.0, range = 271 to 312.

C_4F_7NO MW 211.04

1,2-Oxazetidine, 4,4-difluoro-3-(difluoromethylene)-2-(trifluoromethyl):
Antoine (liq) A = 7.415, B = 1625, C = 0, dev = 1.0 to 5.0, range = 238
to 283.

2H-1,4-Oxazine, 3,6-dihydro-2,2,3,3,5,6,6-heptafluoro, CA 4777-13-3:
Antoine (liq) A = 6.8109, B = 1428, C = 0, dev = 1.0 to 5.0, range = 249
to 293.

$C_4F_7NO_3S$ MW 275.10

Fluorosulfuric acid, ester with 3,3,3-trifluoro-2-(trifluoromethyl)lacto-
nitrile, CA 26404-53-5: Tm = 210.15; Antoine (liq) A = 6.3609, B = 1629,
C = 0, dev = 1.0, range = 262 to 320.

C_4F_8 MW 200.03

Perfluoro-1-butene, CA 357-26-6: Nonflammable; polymerizes. Tm = 155.15;
Tb = 277.95; Vm = 0.1295 at 273.15; Antoine (liq) A = 7.441, B = 1508,
C = 0, dev = 5.0, range = 203 to 279.

Perfluorocyclobutane, CA 115-25-3: Nontoxic; nonflammable; used as
refrigerant; soluble ether. Tm = 231.75; Tb = 267.05; Vm = 0.1322 at
294.15; Antoine I (liq) A = 5.98896, B = 883.59, C = -45.34, dev = 0.1,
range = 233 to 274; Antoine II (liq) A = 6.02418, B = 897.928, C = -43.723,
dev = 0.1, range = 233 to 291; Antoine III (liq) A = 6.22743, B = 1004.573,
C = -29.023, dev = 0.1, range = 289 to 348; Antoine IV (liq) A = 7.35602,
B = 1849.487, C = 84.325, dev = 0.1, range = 343 to 388.

$C_4F_8N_2O_3$ MW 276.04

Perfluoro-2-(tetrafluoro-2-nitroethyl)-1,2-oxazetidine: Antoine (liq) A =
6.6189, B = 1620, C = 0, dev = 1.0, range = 273 to 343.

C_4F_8OS MW 248.09

Octafluorotetrahydro-1-thiophene oxide, CA 42060-62-8: Tm = 230.15;
Antoine (liq) A = 7.555, B = 1939, C = 0, dev = 1.0 to 5.0, range not
specified.

$C_4F_8O_4S$ MW 296.09

Anhydride of heptafluorobutyric acid and fluorosulfuric acid, CA 6069-35-8: Antoine (liq) A = 6.49536, B = 1216.126, C = -78.769, dev = 5.0, range = 268 to 352.

C_4F_8S MW 232.09

Octafluorotetrahydrothiophene, CA 706-76-3: Antoine (liq) A = 6.505, B = 1406, C = 0, dev = 1.0 to 5.0, range not specified.

$C_4F_8S_2$ MW 264.15

2,2,3,3,5,5,6,6-Octafluoro-1,4-dithiane, CA 710-65-6: Antoine (liq) A = 7.025, B = 1724, C = 0, dev = 1.0 to 5.0, range not specified.

C_4F_9N MW 233.04

Methanamine, 1,1,1-trifluoro-N-[2,2,2-trifluoro-1-(trifluoromethyl)ethylidene], CA 453-22-5: Antoine (liq) A = 6.515, B = 1300, C = 0, dev = 1.0 to 5.0, range not specified.

Perfluoro[N,N-dimethyl(vinylamine)], CA 13821-49-3: Antoine (liq) A = 7.0609, B = 1437, C = 0, dev = 1.0, range = 257 to 280.

Perfluoro[N-methyl(propylideneamine)]: Antoine (liq) A = 6.8879, B = 1392, C = 0, dev = 1.0, range = 245 to 280.

Perfluoro[N-propyl(methyleneamine)]: Antoine (liq) A = 6.9519, B = 1478, C = 0, dev = 1.0, range = 250 to 291.

C_4F_9NO MW 249.04

Butyramide, nonafluoro, CA 32822-51-8: Antoine (liq) A = 7.035, B = 1540, C = 0, dev = 1.0 to 5.0, range not specified.

2,2,4,4,5,5-Hexafluoro-3-(trifluoromethyl) oxazolidine: Antoine (liq) A = 6.8369, B = 1429, C = 0, dev = 1.0, range = 253 to 293.

Perfluoro[2,4-bis(trifluoromethyl)-1,2-oxazetidine]: Antoine (liq) A = 6.5869, B = 1353, C = 0, dev = 1.0, range = 266 to 289.

C_4F_9NOS MW 281.10

Methanesulfinamide, 1,1,1-trifluoro-N-[2,2,2-trifluoro-1-(trifluoromethyl)ethylidene], CA 31340-35-9: Antoine (liq) A = 7.335, B = 1922, C = 0, dev = 1.0 to 5.0, range not specified.

$C_4F_9NO_2$ MW 265.04

O-(Trifluoroacetyl)-N,N-bis(trifluoromethyl) hydroxylamine, CA 15496-02-3: Antoine (liq) A = 7.2588, B = 1592.1, C = 0, dev = 1.0, range = 234 to 296.

$C_4F_9NO_2S$ MW 297.10

Sulfoxime, *N*-(trifluoroacetyl)-*S*,*S*-bis(trifluoromethyl), CA 34556-29-1:
Antoine (liq) A = 7.065, B = 1835.9, C = 0, dev = 1.0 to 5.0, range not
specified.

$C_4F_9NO_3$ MW 271.03

2-Propanol, 1,1,1,3,3,3-hexafluoro-2-(trifluoromethyl)nitrate,
CA 55064-78-3: Antoine (liq) A = 7.2728, B = 1751.5, C = 0, dev = 1.0,
range not specified.

C_4F_9NS MW 265.10

Methanesulfenamide, 1,1,1-trifluoro-*N*-[2,2,2-trifluoro-1-(trifluoromethyl)
ethylidene], CA 31340-34-8: Antoine (liq) A = 6.885, B = 1584, C = 0,
dev = 1.0 to 5.0, range not specified.

C_4F_{10} MW 238.03

Decafluorobutane, perfluorobutane, CA 355-25-9: Nonflammable; soluble benzene, chloroform. Tm = 145.15; Tb = 277.11; Vm = 0.1444 at 293.15; Antoine
I (liq) A = 6.1373, B = 1003.2, C = -28.07, dev = 1.0 to 5.0, range = 233
to 273; Antoine II (liq) A = 5.80191, B = 814.464, C = -56.463, dev = 1.0,
range = 272 to 327; Antoine III (liq) A = 7.08295, B = 1647.636, C =
56.969, dev = 1.0 to 5.0, range = 323 to 386.

$C_4F_{10}OS$ MW 286.09

Bis(pentafluoroethyl) sulfoxide, CA 33622-19-4: Antoine (liq) A = 7.475,
B = 1836, C = 0, dev = 1.0 to 5.0, range not specified.

(Trifluoromethyl)(heptafluoropropyl) sulfoxide, CA 33622-18-3: Antoine
(liq) A = 7.365, B = 1790, C = 0, dev = 1.0 to 5.0, range not specified.

$C_4F_{10}O_3S$ MW 318.09

Fluorosulfuric acid, perfluoro(1-methylpropyl) ester: Antoine (liq) A =
5.72387, B = 991.09, C = -77.17, dev = 1.0, range = 294 to 342.

$C_4F_{10}O_4S$ MW 334.09

Fluoroperoxysulfuric acid, 2,2,2-trifluoro-1,1-bis(trifluoromethyl)ethyl
ester, CA 55064-77-2: Antoine (liq) A = 7.5619, B = 1965, C = 0, dev =
1.0, range not specified.

$C_4F_{10}O_6S_2$ MW 398.14

1,1,1,2,3,4,4,4-Octafluoro-2,3-bis(fluorosulfato) butane: Antoine I (liq)
A = 7.19453, B = 2092.514, C = 0, dev = 5.0, range = 316 to 393; Antoine II
(liq) A = 5.47922, B = 1420.5, C = 0, dev = 1.0, range = 392 to 411.

$C_4F_{10}S$ MW 270.09

Perfluoro(1,1,2,3,4,5-hexahydrothiophene), CA 42060-60-6: Antoine (liq) A
= 8.265, B = 2171, C = 0, dev = 1.0 to 5.0, range not specified.

(Trifluoromethyl)(heptafluoropropyl) sulfide, CA 33547-11-4: Antoine (liq)
A = 6.625, B = 1439, C = 0, dev = 1.0 to 5.0, range not specified.

$C_4F_{11}NOS$ MW 319.09

Sulfur, difluoro(1,1,1,3,3,3-hexafluoro-2-propaniminato)oxo(trifluoromethyl), CA 62609-62-5: Antoine (liq) A = 6.635, B = 1833, C = 0, dev =
1.0 to 5.0, range not specified.

$C_4F_{11}NS$ MW 303.09

Sulfur, fluoro(trifluoromethyl)[2,2,2,1-tetrafluoro-1-(trifluoromethyl)-
ethyl]imino, CA 37826-43-0: Antoine (liq) A = 9.00573, B = 3144.777, C =
113.895, dev = 1.0, range = 300 to 333.

$C_4F_{12}N_2O$ MW 320.04

Perfluoro(2,3-dimethyl)-4-oxo-2,3-diazapentane: Antoine (liq) A = 7.3519,
B = 1671, C = 0, dev = 1.0, range = 276 to 308.

Perfluoro(2,4-dimethyl)-3-oxo-2,4-diazapentane: Antoine (liq) A = 6.8979,
B = 1574, C = 0, dev = 1.0, range = 288 to 318.

$C_4F_{12}N_2O_2S$ MW 368.10

Hydroxylamine, O,O'-thiobis[N,N-bis(trifluoromethyl)], CA 21951-03-1:
Antoine (liq) A = 8.085, B = 2099, C = 0, dev = 1.0 to 5.0, range not
specified.

$C_4F_{12}N_2O_2S_2$ MW 400.16

Hydroxylamine, O,O'-dithiobis[N,N-bis(trifluoromethyl)], CA 21951-02-0:
Antoine (liq) A = 9.005, B = 2633, C = 0, dev = 1.0 to 5.0, range not
specified.

$C_4F_{12}N_2O_4S$ MW 400.10

Hydroxylamine, O,O'-sulfonylbis[N,N-bis(trifluoromethyl)], CA 21950-98-1:
Antoine (liq) A = 8.005, B = 2172, C = 0, dev = 1.0 to 5.0, range not
specified.

$C_4F_{12}N_2S$ MW 336.10

Sulfur, difluorobis(1,1,2,2,2-pentafluoroethanaminato), CA 4101-59-1:
Antoine (liq) A = 7.635, B = 1934, C = 0, dev = 1.0 to 5.0, range not
specified.

$C_4F_{12}OS$ MW 324.08

Sulfur, difluorobis(pentafluoroethyl), CA 33564-25-9: Antoine (liq) A = 7.13648, B = 1726.719, C = -4.196, dev = 1.0 to 5.0, range = 284 to 341.

$C_4F_{12}O_2S$ MW 340.08

Sulfur, bis(trifluoromethyl)bis(trifluoromethoxy), CA 63465-11-2: Antoine (liq) A = 6.445, B = 1532, C = 0, dev = 5.0, range = 273 to 325.

$C_4F_{12}O_3S$ MW 356.08

Sulfur, oxobis(trifluoromethyl)bis(trifluoromethoxy), CA 66632-46-0: Antoine (liq) A = 6.725, B = 1746, C = 0, dev = 5.0, range = 273 to 335.

$C_4F_{12}P_4$ MW 399.92

1,2,3,4-Tetrakis(trifluoromethyl)tetraphosphetane, CA 393-02-2: Tm = 339.55; Antoine I (sol) A = 10.94466, B = 3413.129, C = 0, dev = 1.0, range = 292 to 339; Antoine II (liq) A = 7.29031, B = 2099.129, C = -11.623, dev = 1.0, range = 313 to 375.

$C_4F_{12}S$ MW 308.08

Dodecafluorooctahydrothiophene, CA 373-83-1: Antoine (liq) A = 6.915, B = 1644, C = 0, dev = 1.0 to 5.0, range not specified.

Sulfur, difluorobis(pentafluoroethyl), CA 33622-15-0: Antoine (liq) A = 6.925, B = 1685, C = 0, dev = 1.0 to 5.0, range not specified.

Sulfur, difluoro(trifluoromethyl)(heptafluoropropyl), CA 31206-31-2: Antoine (liq) A = 7.005, B = 1712, C = 0, dev = 1.0 to 5.0, range not specified.

$C_4F_{13}NOS$ MW 357.09

Sulfur, trifluoro(1,1,1,2,3,3,3-heptafluoro-2-propanaminato)(trifluoro-methanolato), CA 65844-09-9: Antoine (liq) A = 6.555, B = 1770, C = 0, dev = 1.0 to 5.0, range not specified.

$C_4F_{16}S_2$ MW 416.14

1,4-Dithiane, hexadecafluorooctahydro, CA 4556-31-4: Antoine (liq) A = 7.205, B = 2117, C = 0, dev = 1.0 to 5.0, range not specified.

C_4HBrF_7N MW 275.95

Vinylamine, 1-bromo-2-fluoro-N,N-bis(trifluoromethyl), *cis*, CA 25273-49-8: Antoine (liq) A = 6.420, B = 1559, C = 0, dev = 1.0 to 5.0, range = 321 to 342.

C_4HBrF_9N MW 313.95

Ethylamine, 2-bromo-1,1,2-trifluoro-N,N-bis(trifluoromethyl): Antoine
(liq) A = 6.7989, B = 1664, C = 0, dev = 1.0, range = 308 to 342.

Ethylamine, 2-bromo-1,2,2-trifluoro-N,N-bis(trifluoromethyl, CA 4905-96-8:
Antoine (liq) A = 7.2679, B = 1766, C = 0, dev = 1.0, range = 301 to 332.

$C_4HBr_2F_6N$ MW 336.86

Vinylamine, 1,2-dibromo-N,N-bis(trifluoromethyl), *trans*: Antoine (liq) A =
6.5429, B = 1746, C = 0, dev = 1.0 to 5.0, range = 355 to 382.

$C_4HCl_2F_5O_2$ MW 246.95

Butyric acid, 3,4-dichloro-2,2,3,4,4-pentafluoro, CA 375-07-5: Tm =
243.15; Antoine (liq) A = 8.281, B = 2863, C = 0, dev = 5.0, range = 373
to 456.

C_4HF_5 MW 144.04

3,3,4,4,4-Pentafluoro-1-butyne, CA 7096-51-7: Antoine (liq) A = 6.7201,
B = 1231.7, C = 0, dev = 5.0, range = 203 to 261.

C_4HF_6N MW 177.05

Ethynylamine, N,N-bis(trifluoromethyl), CA 13747-21-2: Antoine (liq) A =
6.9179, B = 1361, C = 0, dev = 1.0 to 5.0, range = 229 to 271.

C_4HF_7 MW 182.04

1,1,3,3,4,4,4-Heptafluoro-1-butene, CA 681-22-1: Antoine (liq) A = 7.2249,
B = 1450, C = 0, dev = 1.0 to 5.0, range not specified.

C_4HF_7N MW 196.05

Butane, 4-diazo-1,1,1,2,2,3,3-heptafluoro, CA 3937-92-6: Antoine (liq) A =
7.1309, B = 1680, C = 0, dev = 1.0 to 5.0, range not specified.

$C_4HF_7O_2$ MW 214.04

Perfluorobutyric acid, CA 375-22-4: Corrosive; stench; highly toxic; nonflammable; soluble water, ether, toluene. Tm = 255.65; Tb = 393.15; Vm =
0.1297 at 293.15; Antoine I (liq) A = 8.2722, B = 2469.4, C = 0, dev = 1.0,
range = 353 to 393; Antoine II (liq) A = 7.5694, B = 1988.193, C = -36.893,
dev = 1.0, range = 329 to 493.

C_4HF_8N MW 215.05

N,N-Bis(trifluoromethyl)-1,2-difluorovinylamine, CA 13747-24-5: Antoine
(liq) A = 7.0469, B = 1503, C = 0, dev = 1.0, range = 276 to 296.

N,N-Bis(trifluoromethyl)-2,2-difluorovinylamine, CA 13747-23-4: Antoine
(liq) A = 6.9419, B = 1447, C = 0, dev = 1.0, range = 274 to 291.

C_4HF_8NO MW 231.05

Morpholine, 2,2,3,3,5,5,6,6-octafluoro, CA 13580-54-6: Antoine (liq) A =
7.1519, B = 1706, C = 0, dev = 1.0 to 5.0, range = 273 to 323.

$C_4HF_9N_2OS$ MW 296.11

Methanesulfonimidamide, 1,1,1-trifluoro-N-[2,2,2-trifluoro-1-(trifluoromethyl)ethylidene], CA 62609-65-8: Antoine (liq) A = 7.025, B = 1948,
C = 0, dev = 5.0, range not specified.

$C_4HF_9O_2$ MW 252.04

2,2,2-Trifluoro-1,1-bis(trifluoromethyl)ethyl hydroperoxide, CA 64957-49-9:
Tm = 237.15; Antoine (liq) A = 7.5479, B = 1914, C = 0, dev = 1.0 to 5.0,
range not specified.

$C_4HF_9O_2S$ MW 284.10

Trifluoromethanesulfinic acid, 2,2,2-trifluoro-1-(trifluoromethyl)ethyl
ester, CA 52225-50-0: Antoine (liq) A = 7.705, B = 2061, C = 0, dev = 1.0
to 5.0, range not specified.

$C_4HF_{10}N$ MW 253.04

Bis(pentafluoroethyl) amine, CA 54566-81-3: Antoine (liq) A = 7.095, B =
1556, C = 0, dev = 1.0 to 5.0, range not specified.

Ethylamine, 1,2,2,2-tetrafluoro-N,1-bis(trifluoromethyl): Antoine (liq)
A = 7.345, B = 1650, C = 0, dev = 1.0 to 5.0, range not specified.

$C_4HF_{10}NOS$ MW 301.10

Sulfoxime, S,S-bis(pentafluoroethyl), CA 34556-24-6: Antoine (liq) A =
7.025, B = 1838.6, C = 0 dev = 1.0 to 5.0, range not specified.

C_4H_2 MW 50.06

1,3-Butadiyne, biacetylene, CA 460-12-8: Highly flammable; potentially
explosive; polymerizes; soluble ethanol, ether, acetone, chloroform. Tm =
237.15 to 238.15; Tb = 283.45; Vm = 0.0680 at 273.15; Antoine I (sol) A =
9.0789, B = 1897.56, C = 0, dev = 1.0, range = 188 to 237; Antoine II (liq)
A = 6.8271, B = 1361.42, C = 0, dev = 1.0, range = 237 to 283.

$C_4H_2BrF_6N$ MW 257.96

Vinylamine, 1-bromo-N,N-bis(trifluoromethyl), CA 19451-87-7: Antoine (liq)
A = 7.1429, B = 1714, C = 0, dev = 1.0 to 5.0, range = 288 to 327.

Vinylamine, 2-bromo-N,N-bis(trifluoromethyl), cis, CA 19483-21-7: Antoine
(liq) A = 6.4259, B = 1552, C = 0, dev = 1.0 to 5.0, range = 314 to 346.
                                                                (continues)

$C_4H_2BrF_6N$ *(continued)*

Vinylamine, 2-bromo-*N*,*N*-bis(trifluoromethyl), *trans*, CA 19483-20-6:
Antoine (liq) A = 6.5559, B = 1569, C = 0, dev = 1.0 to 5.0, range = 314
to 341.

Vinylamine, 2-bromo-*N*,*N*-bis(trifluoromethyl), mixed isomers: Antoine (liq)
A = 6.535, B = 1550, C = 0, dev = 5.0, range not specified.

$C_4H_2BrF_8N$ MW 295.96

Ethylamine, 2-bromo-1,2-difluoro-*N*,*N*-bis(trifluoromethyl), CA 6857-63-2:
Antoine (liq) A = 6.8389, B = 1690, C = 0, dev = 1.0 to 5.0, range = 314
to 348.

Ethylamine, 2-bromo-2,2-difluoro-*N*,*N*-bis(trifluoromethyl): Antoine (liq)
A = 6.9729, B = 1753, C = 0, dev = 1.0 to 5.0, range = 313 to 348.

$C_4H_2Br_2S$ MW 241.93

3,4-Dibromothiophene, CA 3141-26-2: Irritant; flammable, flash point above
385. Tm = 277.65; Tb = 494.15 to 495.15; Vm = 0.1106 at 293.15; Antoine
(liq) A = 5.0087, B = 1678.5, C = 0, dev = 1.0, range = 333 to 374.

$C_4H_2Cl_2O_2$ MW 152.96

Fumaroyl chloride, CA 627-63-4: Highly irritant; corrosive; lachrymator;
flammable, flash point 346; decomposes with water to give toxic fumes.
Tb = 431.15 to 433.15; Vm = 0.1086 at 293.15; Antoine (liq) A = 7.81719,
B = 2594.71, C = 13.131, dev = 1.0, range = 288 to 433.

$C_4H_2Cl_2S$ MW 153.03

2,5-Dichlorothiophene, CA 3172-52-9: Irritant; flammable, flash point 332;
soluble ethanol, ether. Tf = 229.75; Tb = 435.15; Vm = 0.1061 at 293.15;
Antoine (liq) A = 5.9702, B = 1762, C = 0, dev = 1.0, range = 333 to 374.

$C_4H_2F_4$ MW 126.05

1,1,4,4-Tetrafluoro-1,3-butadiene, CA 407-70-5: Antoine (liq) A = 6.1983,
B = 1168.7, C = 0, dev = 1.0, range = 239 to 271.

$C_4H_2F_6$ MW 164.05

1,1,2-Trifluoro-3-(trifluoromethyl)cyclopropane: Antoine (liq) A = 7.2539,
B = 1553, C = 0, dev = 1.0 to 5.0, range not specified.

$C_4H_2F_6OS$ MW 212.11

Trifluorothioacetic acid, *S*-(1,2,2-trifluoroethyl) ester, CA 35709-12-7:
Antoine (liq) A = 7.1732, B = 1791.97, C = 0, dev = 1.0 to 5.0, range =
282 to 322.

$C_4H_2F_6O_2$ MW 196.05

Trifluoroacetic acid, 2,2,2-trifluoroethyl ester, CA 407-38-5: Antoine
(liq) A = 7.045, B = 1663, C = 0, dev = 5.0, range not specified.

$C_4H_2F_7N$ MW 197.06

Vinylamine, 2-fluoro-N,N-bis(trifluoromethyl), *cis*, CA 25273-51-2: Antoine
(liq) A = 6.8459, B = 1521, C = 0, dev = 1.0 to 5.0, range = 289 to 311.

Vinylamine, 2-fluoro-N,N-bis(trifluoromethyl), *trans*, CA 25211-47-6:
Antoine (liq) A = 7.0189, B = 1491, C = 0, dev = 1.0 to 5.0, range = 273
to 295.

$C_4H_2N_2$ MW 78.07

Fumaronitrile, CA 764-42-1: Irritant; flammable; soluble ethanol, ether,
acetone, benzene. Tm = 369.15; Tb = 459.15; Vm = 0.0829 at 384.15;
Antoine (sol) A = 5.42433, B = 1368.6, C = -103.82, dev = 1.0, range = 245
to 282.

$C_4H_2N_2O_4S$ MW 174.13

2,4-Dinitrothiophene, CA 5347-12-6: Flammable; soluble ethanol. Tm =
329.15; Antoine (liq) A = 7.5099, B = 3116.1, C = 0, dev = 1.0, range = 388
to 523.

$C_4H_2O_3$ MW 98.06

Maleic anhydride, CA 108-31-6: Irritant; corrosive; causes burns; flammable, flash point 385; moisture-sensitive; polymer raw material; reacts
with water to form maleic acid; forms ester with ethanol; very soluble
acetone, benzene, chloroform. Tm = 326; Tb = 475.15; Vm = 0.0746 at
333.15; Tc = 710, calculated; Pc = 6300, calculated; Vc = 0.219, calculated;
Antoine I (sol) A = 13.190, B = 4461, C = 0, dev = 5.0, range = 308 to
326; Antoine II (liq) A = 7.49625, B = 2636.73, C = 4.881, dev = 1.0,
range = 336 to 475.

$C_4H_3BrF_7N$ MW 277.97

Ethylamine, 2-bromo-2-fluoro-N,N-bis(trifluoromethyl), CA 25237-12-1:
Antoine (liq) A = 6.4979, B = 1613, C = 0, dev = 1.0 to 5.0, range = 329
to 355.

C_4H_3BrS MW 163.03

2-Bromothiophene, CA 1003-09-4: Flammable, flash point 333; soluble ether,
acetone. Tb = 428.15 to 429.15; Vm = 0.0968 at 293.15; Antoine (liq) A =
5.3886, B = 1457.8, C = 0, dev = 1.0, range = 333 to 420.

3-Bromothiophene, CA 872-31-1: Flammable, flash point 329; soluble
acetone, benzene. Tb = 431.65 to 432.15; Vm = 0.0940 at 293.15; Antoine
(liq) A = 5.4016, B = 1508.3, C = 0, dev = 1.0, range = 333 to 374.

$C_4H_3ClF_6O_2S$ MW 264.57

Chlorosulfonic acid, 1,1-bis(trifluoromethyl)ethyl ester, CA 57169-82-1:
Antoine (liq) A = 7.165, B = 2071, C = 0, dev = 1.0 to 5.0, range not
specified.

C_4H_3ClS MW 118.58

2-Chlorothiophene, CA 96-43-5: Flammable, flash point 395; soluble
ethanol, ether. Tm = 201.25; Tb = 403.15; Vm = 0.0922 at 293.15; Antoine
(liq) A = 6.5366, B = 1798.9, C = 0, dev = 1.0, range = 333 to 374.

$C_4H_3F_5OS$ MW 194.12

Trifluorothioacetic acid, S-(2,2-difluoroethyl) ester, CA 35709-11-6:
Antoine (liq) A = 4.45122, B = 635.097, C = -131.354, dev = 1.0, range =
282 to 322.

$C_4H_3F_6NO_2$ MW 211.06

N,N-Bis(trifluoromethyl)acetamide-N-oxide, CA 22743-78-8: Antoine (liq)
A = 8.32614, B = 2160.6, C = 2.45, dev = 1.0, range = 268 to 336.

$C_4H_3F_7O$ MW 200.06

2,2,3,3,4,4,4-Heptafluoro-1-butanol, CA 375-01-9: Flammable, flash point
298; soluble ethanol, acetone. Tb = 368.15; Vm = 0.1250 at 293.15; Antoine
(liq) A = 8.298, B = 2277, C = 0, dev = 5.0, range = 273 to 298.

$C_4H_3F_7O_2S$ MW 248.12

Fluorosulfuric acid, 1,1-bis(trifluoromethyl)ethyl ester, CA 57169-83-2:
Antoine (liq) A = 7.175, B = 1898, C = 0, dev = 1.0 to 5.0, range not
specified.

C_4H_3IS MW 210.04

2-Iodothiophene, CA 3437-95-4: Flammable, flash point 344; soluble ethanol,
ether. Tm = 233.15; Tb = 453.15 to 455.15; Antoine (liq) A = 5.1221, B =
1517.2, C = 0, dev = 1.0, range = 333 to 374.

$C_4H_3NO_2S$ MW 129.13

2-Nitrothiophene, CA 609-40-5: Toxic; vesicant; light-sensitive; flammable,
flash point 366; soluble ethanol. Tm = 314.15; Tb = 497.15 to 498.15; Vm =
0.0946 at 316.15; Antoine (liq) A = 6.9316, B = 2367.65, C = -17.271, dev =
1.0, range = 321 to 498.

C_4H_4 MW 52.08

1-Buten-3-yne, vinyl acetylene, CA 689-97-4: Decomposes violently on
heating; forms explosive compounds if exposed to air; extremely *(continues)*

C_4H_4 *(continued)*

flammable, flash point below 268. Tb = 275.15 to 276.15; Vm = 0.0734 at 273.15; Antoine (liq) A = 6.15005, B = 999.11, C = -37.33, dev = 1.0, range = 180 to 278.

$C_4H_4BrF_6N$ MW 259.98

Ethylamine, 2-bromo-*N*,*N*-bis(trifluoromethyl), CA 1683-83-6: Antoine (liq) A = 6.5189, B = 1620, C = 0, dev = 1.0 to 5.0, range = 323 to 356.

$C_4H_4Cl_2$ MW 122.98

1,2-Dichloro-1,3-butadiene, CA 3574-40-1: Soluble carbon tetrachloride. Vm = 0.1026 at 293.15; Antoine (liq) A = 2.5541, B = 254.2, C = -168.95, dev = 1.0 to 5.0, range = 260 to 308.

2,3-Dichloro-1,3-butadiene, CA 1653-19-6: Soluble chloroform. Tb = 371.15; Vm = 0.1040 at 293.15; Antoine (liq) A = 7.1263, B = 2040.5, C = 23.45, dev = 1.0, range = 299 to 368.

$C_4H_4Cl_2O_2$ MW 154.98

Succinyl chloride, CA 543-20-4: Corrosive; highly irritant; lachrymator; flammable, flash point 349; soluble ethanol, acetone, benzene. Tm = 293.15; Tb = 466.45; Vm = 0.1127 at 293.15; Antoine (liq) A = 7.06279, B = 2127.04, C = -44.895, dev = 1.0, range = 312 to 466.

$C_4H_4Cl_2O_3$ MW 170.98

Chloroacetic acid, anhydride, CA 541-88-8: Corrosive; irritant; moisture-sensitive; soluble ethanol, ether, chloroform. Tm = 319.15; Tb = 476.15; Antoine (liq) A = 8.88774, B = 3458.26, C = 12.505, dev = 1.0, range = 340 to 490.

$C_4H_4Cl_4O_2S$ MW 257.95

3,3,4,4-Tetrachlorotetrahydrothiophene, 1,1-dioxide, CA 3737-41-5: Antoine (liq) A = 10.868, B = 4630.184, C = 0, dev = 1.0, range = 303 to 348.

$C_4H_4F_3NO_3$ MW 171.08

N-(Trifluoroacetyl)aminoacetic acid, CA 383-70-0: Tm = 392.65; Antoine (sol) A = 12.605, B = 5163, C = 0, dev = 5.0, range = 273 to 393.

$C_4H_4F_4OS$ MW 176.13

Trifluorothioacetic acid, *S*-(2-fluoroethyl) ester, CA 35709-10-5: Antoine (liq) A = 4.34949, B = 577.263, C = -142.945, dev = 1.0, range = 282 to 322.

$C_4H_4F_6N_2S$ MW 226.14

Ethanimidamide, 2,2,2-trifluoro-*N*-methyl-*N'*-[(trifluoromethyl)thio], CA 62067-10-1: Antoine (liq) A = 6.405, B = 1821.3, C = 0, dev = 5.0, range = 339 to 387.

$C_4H_4F_6O_2S$ MW 230.12

Trifluoromethanesulfinic acid, 2,2,2-trifluoro-1-methylethyl ester,
CA 52225-48-6: Antoine (liq) A = 7.135, B = 1921, C = 0, dev = 1.0 to 5.0,
range not specified.

$C_4H_4N_2$ MW 80.09

Succinonitrile, CA 110-61-2: Irritant; may emit toxic fumes on heating;
flammable, flash point over 385; slightly soluble water; soluble acetone,
chloroform. Tm = 327.65; Tb = 538.15 to 540.15; Vm = 0.0812 at 333.15;
Antoine (sol) A = 9.2722, B = 3656, C = 0, dev = 1.0, range = 279 to 299.

$C_4H_4N_4O_7$ MW 220.10

Furazandimethanol dinitrate, CA 57449-43-1: Antoine (liq) A = 7.215, B =
3067, C = 0, dev = 5.0, range = 399 to 433.

$C_4H_4N_4O_8$ MW 236.10

Furazandimethanol dinitrate, 2-oxide, CA 57449-44-2: Antoine (liq) A =
7.235, B = 3358.8, C = 0, dev = 5.0, range = 413 to 453.

$C_4H_4N_8O_{13}$ MW 372.12

Diethylamine, 2,2,2,2',2',2'-hexanitro-N-nitroso, CA 34882-73-0: Tm =
358.15; Antoine (sol) A = 11.465, B = 5104, C = 0, dev = 1.0 to 5.0,
range = 333 to 354.

$C_4H_4N_8O_{14}$ MW 388.12

Diethylamine, N,2,2,2,2',2',2'-heptanitro, CA 19836-28-3: Antoine (sol)
A = 13.925, B = 6140, C = 0, dev = 1.0 to 5.0, range = 340 to 356.

C_4H_4O MW 68.08

Furan, CA 110-00-9: Narcotic; absorbed through skin; extremely flammable,
flash point below 273; soluble ethanol, ether. Tm = 187.55; Tb = 305.15;
Vm = 0.0706 at 273.15; Antoine (liq) A = 6.10013, B = 1060.851, C = -45.41,
dev = 0.1, range = 238 to 363.

$C_4H_4O_2$ MW 84.07

Diketene, CA 674-82-8: Moderately toxic; lachrymator; slowly polymerizes
on standing; flammable, flash point 306; insoluble water; soluble organic
solvents. Tm = 266.65; Tb = 400.55; Vm = 0.0771 at 293.15; Antoine (liq)
A = 6.69663, B = 1677.09, C = -42.02, dev = 1.0, range = 297 to 388.

$C_4H_4O_3$ MW 100.07

Succinic anhydride, CA 108-30-5: Irritant; moisture-sensitive; slightly
soluble water; moderately soluble ethanol, chloroform, carbon *(continues)*

$C_4H_4O_3$ *(continued)*

tetrachloride. Tm = 392.75; Tb = 534.15; Antoine (liq) A = 7.8826, B = 3239.62, C = 16.732, dev = 1.0 to 5.0, range = 401 to 534.

$C_4H_4O_4$ MW 116.07

Fumaric acid, CA 110-17-8: Irritant; slightly soluble water, ether, acetone; moderately soluble ethanol; isomerizes to maleic acid at temperatures over 473; decomposes to maleic anhydride at temperatures over 560. Tm = 560.15; Tb = 563.15, decomposes; Antoine (sol) A = 13.21606, B = 6438.5, C = 0, dev = 1.0 to 5.0, range = 371 to 391.

Glycolide, 2,5-*para*-dioxanedione, CA 502-97-6: Moderately soluble ethanol, acetone. Tm = 356.15; Antoine (liq) A = 7.59624, B = 2755.49, C = -20.445, dev = 1.0, range = 376 to 413.

Maleic acid, CA 110-16-7: Strongly irritant; corrosive; very soluble water, methanol, ethanol, acetone; on heating above melting point, isomerizes partly to fumaric acid and dehydrates. Tm = 411.15 to 412.15; Tb = approximately 411.15, decomposes; Tc = 563, calculated; Pc = 4990, calculated; Vc = 0.297, calculated.

C_4H_4S MW 84.14

Thiophene, CA 110-02-1: Moderately toxic; stench; highly flammable, flash point 267; soluble ethanol, ether, acetone, benzene. Tm = 234.85; Tb = 357.31; Vm = 0.0790 at 293.15; Tc = 579.35; Pc = 5700; Vc = 0.219; Antoine I (sol) A = 9.84733, B = 2447.236, C = 0, dev = 1.0, range = 195 to 228; Antoine II (liq) A = 6.06132, B = 1232.35, C = -53.438, dev = 0.02, range = 311 to 393.

C_4H_5Cl MW 88.54

Chloroprene, 2-chloro-1,3-butadiene, CA 126-99-8: Irritant; moderately toxic; oxidizes to unstable peroxides; polymerizes; extremely flammable, flash point 253; slightly soluble water; soluble most organic solvents. Tm = 143.15; Tb = 332.55; Vm = 0.0924 at 293.15; Tc = 534.85; Antoine I (liq) A = 6.652, B = 1545.3, C = 0, dev = 1.0, range = 279 to 333; Antoine II (liq) A = 6.6519, B = 1545, C = 0, dev = 1.0, range = 243 to 263.

C_4H_5ClO MW 104.54

Methacryloyl chloride, CA 920-46-7: Corrosive; irritant; flammable, flash point 275; polymerizes; soluble ether, acetone, chloroform. Tm = 313.15; Tb = 369.15 to 371.15; Vm = 0.0962 at 298.15; Antoine (liq) A = 6.06582, B = 1273.251, C = -58.175, dev = 1.0, range = 313 to 372.

$C_4H_5Cl_3O_2$ MW 191.44

Ethyl trichloroacetate, CA 515-84-4: Irritant; corrosive; flammable, flash point 338; soluble ethanol, ether, benzene. Tb = 440.15 to 441.15; Vm = 0.1384 at 293.15; Antoine (liq) A = 7.29815, B = 2217.36, C = -21.202, dev = 1.0, range = 293 to 440.

$C_4H_5Cl_5$ MW 230.35

1,2,2,3,4-Pentachlorobutane, CA 2431-52-9: Soluble acetone, chloroform.
Vm = 0.1482 at 293.15; Antoine (liq) A = 4.8567, B = 901.07, C = -181.46,
dev = 1.0, range = 368 to 498.

$C_4H_5F_2I$ MW 217.98

1,1-Difluoro-4-iodo-1-butene: Antoine (liq) A = 7.3989, B = 2119.4, C = 0,
dev = 1.0 to 5.0, range = 318 to 342.

$C_4H_5F_3$ MW 110.08

(Trifluoromethyl)cyclopropane, CA 381-74-8: Antoine (liq) A = 6.8719, B =
1422, C = 0, dev = 1.0 to 5.0, range not specified.

$C_4H_5F_3O$ MW 126.08

Vinyl 2,2,2-trifluoroethyl ether, CA 406-90-6: Anesthetic; forms peroxides; potentially explosive; flammable. Tb = 315.65; Vm = 0.1111 at
293.15; Antoine (liq) A = 3.37764, B = 136.8463, C = -216.959, dev = 1.0,
range = 293 to 317.

$C_4H_5F_3OS$ MW 158.14

Trifluorothioacetic acid, S-ethyl ester, CA 383-64-2: Irritant; flammable,
flash point 276. Tb = 363.65; Vm = 0.1286 at 293.15; Antoine (liq) A =
8.28712, B = 2194.303, C = 0, dev = 5.0, range = 273 to 313.

$C_4H_5F_3O_2$ MW 142.08

Trifluoroacetic acid, ethyl ester, CA 383-63-1: Irritant; corrosive; flammable, flash point 272. Tb = 333.15 to 335.15; Vm = 0.1190 at 293.15;
Antoine (liq) A = 7.395, B = 1809, C = 0, dev = 5.0, range not specified.

C_4H_5N MW 67.09

Allyl cyanide, 3-butenenitrile, CA 109-75-1: Moderately toxic; flammable,
flash point 296. Tm = 186.15; Tb = 392.15; Vm = 0.0804 at 293.15; Antoine
(liq) A = 6.22803, B = 1423.7, C = -54.43, dev = 1.0, range = 293 to 417.

Crotononitrile, cis, CA 1190-76-7: Highly irritant; flammable, flash point
below 373. Tm = 200.55; Tb = 380.55; Antoine (liq) A = 6.28662, B =
1450.5, C = -41.84, dev = 1.0, range = 297 to 405.

Crotononitrile, trans, CA 627-26-9: Highly irritant; flammable, flash
point below 373; soluble ether, acetone. Tm = 221.65; Tb = 391.15 to
392.15; Vm = 0.0814 at 293.15; Antoine (liq) A = 6.18962, B = 1440.2, C =
-51.19, dev = 1.0, range = 292 to 420.

Methacrylonitrile, CA 126-98-7: Highly toxic; flammable, flash point 274;
polymerizes; moderately soluble water; soluble ethanol, ether, acetone.
Tm = 237.15; Tb = 363.15; Vm = 0.0839 at 293.15; Antoine (liq) A = 6.10511,
B = 1274.959, C = -52.416, dev = 0.1 to 1.0, range = 273 to 373.

(continues)

104

C_4H_5N *(continued)*

Pyrrole, CA 109-97-7: Irritant; emits toxic fumes on heating; flammable, flash point 312; moisture-sensitive; slightly soluble water; soluble ethanol, ether, acetone, benzene. Tm = 254.65; Tb = 403.15; Vm = 0.0692 at 293.15; Tc = 639.15; Antoine (liq) A = 6.42263, B = 1504.171, C = -62.39, dev = 0.1, range = 338 to 440.

$C_4H_5NO_2$ MW 99.09

Methyl cyanoacetate, CA 105-34-0: Highly toxic; flammable, flash point 383; slightly soluble water; soluble most organic solvents. Tm = 250.65; Tb = 479.15; Vm = 0.0890 at 293.15; Antoine (liq) A = 6.73114, B = 1914.22, C = -73.15, dev = 0.1 to 1.0, range = 385 to 573.

Succinimide, 2,5-pyrrolidinedione, CA 123-56-8: Flammable; very soluble water, ethanol. Tm = 399.15 to 400.15; Tb = 560.15 to 561.15, decomposes; Antoine (liq) A = 8.13603, B = 3401.67, C = -5.849, dev = 1.0, range = 416 to 561.

C_4H_5NS MW 99.15

Allyl isothiocyanate, mustard oil, CA 57-06-7: Highly toxic; suspected carcinogen; flammable, flash point 319; slightly soluble water; soluble ethanol, ether, benzene. Tm = 193.15; Tb = 424.15; Vm = 0.0977 at 288.15; Antoine (liq) A = 3.5137, B = 217.53, C = -279.74, dev = 1.0 to 5.0, range = 370 to 430.

2-Methylthiazole, CA 3581-87-1: Stench; flammable; soluble water, ethanol, acetone. Tm = 248.55; Tb = 402.15; Vm = 0.0893 at 298.15; Antoine I (liq) A = 6.16303, B = 1405.036, C = -63.996, dev = 0.02, range = 353 to 402; Antoine II (liq) A = 6.16599, B = 1406.419, C = -63.893, dev = 0.02, range = 342 to 404.

4-Methylthiazole, CA 693-95-8: Stench; flammable, flash point 305; soluble water, ethanol, ether. Tb = 403.15 to 406.15; Vm = 0.0893 at 298.15; Antoine (liq) A = 6.18588, B = 1423.787, C = -65.759, dev = 0.02 to 0.1, range = 346 to 408.

$C_4H_5N_7O_{12}$ MW 343.12

Ethanamine, 2,2,2-trinitro-N-(2,2,2-trinitroethyl), CA 34880-53-0: Tm = 386.15, decomposes; Antoine (sol) A = 9.015, B = 4223, C = 0, dev = 1.0 to 5.0, range = 337 to 349.

C_4H_6 MW 54.09

1,2-Butadiene, methylallene, CA 590-19-2: Highly flammable; polymerizes on heating; soluble ethanol, ether, benzene. Tm = 136.96; Tb = 284; Vm = 0.0800 at 273.15; Tc = 444; Pc = 4500; Vc = 0.219; Antoine I (liq) A = 6.52312, B = 1219.877, C = -13.374, dev = 0.1, range = 204 to 243; Antoine II (liq) A = 6.11873, B = 1041.117, C = -30.876, dev = 0.1, range = 243 to 291.

1,3-Butadiene, CA 106-99-0: Irritant; extremely flammable, flash point below 266; polymerizes; raw material for synthetic rubber and polymers; soluble ethanol, ether, acetone, benzene. Tm = 164.25;   *(continues)*

C_4H_6 *(continued)*

Tb = 268.74; Vm = 0.0871 at 293.15, Tc = 425.15; Pc = 4320; Vc = 0.221; Antoine I (liq) A = 6.16045, B = 998.106, C = -27.916, dev = 0.1, range = 193 to 213; Antoine II (liq) A = 5.97489, B = 930.546, C = -34.306, dev = 0.02, range = 213 to 276; Antoine III (liq) A = 5.99667, B = 940.687, C = -33.017, dev = 0.1, range = 270 to 318; Antoine IV (liq) A = 6.31615, B = 1130.927, C = -5.606, dev = 0.1, range = 315 to 382; Antoine V (liq) A = 8.86984, B = 3877.451, C = 315.612, dev = 0.1, range = 380 to 425.

1-Butyne, ethyl acetylene, CA 107-00-6: Extremely flammable, flash point below 266; polymerizes; soluble ethanol, ether. Tm = 147.28; Tb = 281.45; Vm = 0.0797 at 273.15; Antoine (liq) A = 6.10688, B = 988.75, C = -40.14, dev = 0.1, range = 205 to 289.

2-Butyne, dimethyl acetylene, CA 503-17-3: Irritant; extremely flammable, flash point below 253; polymerizes. Tm = 240.35; Tb = 300.35 to 300.75; Vm = 0.0757 at 273.15; Antoine I (sol) A = 6.16281, B = 896.91, C = -74.09, dev = 0.1, range = 222 to 240; Antoine II (liq) A = 6.19828, B = 1101.71, C = -37.34, dev = 1.0, range = 240 to 308.

Cyclobutene, cyclobutylene, CA 822-35-5: Extremely flammable, flash point below 263; soluble acetone, benzene. Tb = 275.15; Vm = 0.0738 at 273.15; Antoine (liq) A = 6.62604, B = 1261.44, C = -2.861, dev = 1.0 to 5.0, range = 206 to 275.

$C_4H_6ClFO_2$ MW 140.54

Fluoroacetic acid, 2-chloroethyl ester: Irritant; flammable. Tm = 264.45; Antoine (liq) A = 8.675, B = 2945 C = 0, dev = 5.0, range = 273 to 333.

$C_4H_6Cl_2$ MW 125.00

3,4-Dichloro-1-butene, CA 760-23-6: Irritant; corrosive; flammable, flash point 318; soluble ethanol, ether, benzene. Tm = 212.15; Tb = 396.15; Vm = 0.1079 at 293.15; Antoine (liq) A = 7.92992, B = 2647.465, C = 51.57, dev = 1.0 to 5.0, range = 320 to 396.

1,3-Dichloro-2-butene, *trans*, CA 7415-31-8: Highly toxic by inhalation; lachrymator; flammable, flash point 306; soluble ethanol, ether, acetone, benzene. Tm = below 198.15; Tb = 403.05; Vm = 0.1079 at 293.15; Antoine (liq) A = 7.0609, B = 2010.18, C = -3.54, dev = 0.1 to 1.0, range = 306 to 401.

1,4-Dichloro-2-butene, *trans*, CA 110-57-6: Irritant; corrosive; lachrymator; flammable, flash point 326; soluble ethanol, ether, acetone, benzene. Tm = 274.15; Tb = 428.65; Vm = 0.1057 at 298.15; Antoine (liq) A = 8.44412, B = 3052.95, C = 46.929, dev = 1.0 to 5.0, range = 340 to 379.

$C_4H_6Cl_2O_2$ MW 157.00

Chloroacetic acid, 2-chloroethyl ester, CA 3848-12-2: Irritant; soluble ethanol. Tb = 475.15; Vm = 0.1154 at 298.15; Antoine (liq) A = 7.22088, B = 2353.37, C = -26.927, dev = 1.0, range = 319 to 478.

Ethyl dichloroacetate, CA 535-15-9: Irritant; soluble ethanol, ether, acetone. Tb = 428.65; Vm = 0.1224 at 293.15; Antoine (liq) A = 7.43845, B = 2350.41, C = 2.772, dev = 1.0, range = 283 to 430.

$C_4H_6Cl_4$ MW 195.90

1,2,3,4-Tetrachlorobutane, CA 13138-51-7: Soluble ethanol, ether, acetone.
Tb = 476.15 to 479.15; Vm = 0.1380 at 293.15; Antoine (liq) A = 4.5475, B =
728.81, C = -178.94, dev = 1.0, range = 349 to 464.

C_4H_6FN MW 87.10

4-Fluorobutyronitrile, CA 407-83-0: Antoine (liq) A = 7.4753, B = 2362,
C = 0, dev = 1.0, range = 273 to 371.

$C_4H_6F_2O$ MW 124.09

Fluoroacetic acid, 2-fluoroethyl ester: Tm = 247.75; Antoine (liq) A =
8.838, B = 2876, C = 0, dev = 5.0, range = 273 to 333.

$C_4H_6F_3I$ MW 237.99

1,1,1-Trifluoro-3-iodobutane, CA 540-87-4: Antoine (liq) A = 6.5139, B =
1690, C = 0, dev = 1.0 to 5.0, range = 304 to 321.

1,1,1-Trifluoro-3-iodo-2-methylpropane, CA 26653-47-4: Antoine (liq) A =
6.3009, B = 1586, C = 0, dev = 1.0 to 5.0, range = 298 to 368.

$C_4H_6F_6N_2O$ MW 212.09

Hydrazine, 1,1-dimethyl-2,2-bis-(trifluoromethyl)-2-oxide, CA 30295-33-1:
Antoine (liq) A = 6.795, B = 1900, C = 0, dev = 1.0 to 5.0, range = 287
to 356.

$C_4H_6N_2O$ MW 98.10

Dimethylfurazan, CA 4975-21-7: Flammable; soluble ethanol, ether. Tm =
266.15; Tb = 429.15; Vm = 0.0932 at 287.15; Antoine (liq) A = 8.275, B =
2670, C = 0, dev = 1.0 to 5.0, range = 353 to 427.

$C_4H_6N_2O_2$ MW 114.10

Dimethylfurazan, 2-oxide, CA 2518-42-5: Flammable. Antoine (liq) A =
8.015, B = 2980, C = 0, dev = 1.0 to 5.0, range = 353 to 493.

2,5-Piperazinedione, CA 106-57-0: Sublimes at 553; moderately soluble
water; soluble hydrochloric acid. Tm = 584.15 to 585.15, decomposes;
Antoine (sol) A = 8.7755, B = 5423, C = 0, dev = 1.0, range = 413 to 450.

$C_4H_6N_4O_8$ MW 238.11

1,1,1,3-Tetranitro-2-methylpropane, CA 42216-58-0: Flammable. Tm = 305.15;
Antoine (liq) A = 10.185, B = 3956, C = 0, dev = 5.0, range = 304 to 327.

$C_4H_6N_4O_{11}$ MW 286.11

2-Nitro-2-hydroxymethyl-1,3-propanedioltrinitrate, CA 20820-44-4: Antoine
(liq) A = 7.72404, B = 3810, C = 0, dev = 1.0, range = 313 to 353.

C_4H_6O MW 70.09

Crotonaldehyde, CA 123-73-9: Strong irritant; toxic vapor; lachrymator; very dangerous to eyes; flammable, flash point 286; reacts with oxygen; polymerizes readily; very soluble water; soluble ethanol, ether, acetone, benzene. Tm = 204.15; Tb = 375.35; Vm = 0.0822 at 293.15; Antoine (liq) A = 6.10753, B = 1315.9653, C = -55.1465, dev = 1.0, range = 306 to 376.

Cyclobutanone, CA 1191-95-3: Flammable, flash point 283; soluble water, ethanol, ether, benzene, chloroform. Tb = 371.15 to 372.15; Vm = 0.0734 at 273.15; Antoine I (liq) A = 6.4845, B = 1464.75, C = -43.15, dev = 1.0 to 5.0, range = 283 to 313; Antoine II (liq) A = 6.23217, B = 1356.14, C = -51.113, dev = 0.02 to 0.1, range = 317 to 380.

Divinyl ether, CA 109-93-3: Anesthetic; extremely flammable, flash point below 243; forms explosive peroxides; decomposes and polymerizes in light; slightly soluble water; soluble ethanol, ether, acetone. Tm = 172.05; Tb = 301.15 to 304.15; Vm = 0.0914 at 298.15; Tc = 456.15, calculated; Pc = 3040, calculated; Antoine (liq) A = 6.12828, B = 1061.08, C = -44.15, dev = 1.0, range = 253 to 323.

Methacrolein, 2-methylpropenal, CA 78-85-3: Highly irritant; very toxic; lachrymator; forms peroxides and acids with air; highly flammable, flash point 258; must be stored below 283 to prevent dimer formation; polymerizes at room temperature and above; soluble water, ethanol, ether. Tm = 192.15; Tb = 341.15; Vm = 0.0837 at 293.15.

Methyl vinyl ketone, CA 78-94-4: Highly irritant; lachrymator; absorbed through skin; flammable, flash point 266; polymerizes on standing; soluble water, ethanol, methanol, ether, acetone. Tm = 267.15; Tb = 352.15 to 353.15; Vm = 0.0812 at 293.15; Antoine I (liq) A = 6.1392, B = 1273, C = -46.65, dev = 1.0 to 5.0, range = 300 to 355; Antoine II (liq) A = 7.14499, B = 1896.711, C = 14.641, dev = 1.0 to 5.0, range = 279 to 355.

C_4H_6OS MW 102.15

2(3H)-Dihydrothiophenone, CA 1003-10-7: Irritant; flammable. Vm = 0.0866 at 293.15; Antoine (liq) A = 5.885, B = 1946.3, C = 0, dev not specified, range not specified.

$C_4H_6O_2$ MW 86.09

2,3-Butanedione, CA 431-03-8: Moderately toxic; flammable, flash point 300; soluble water, ethanol, ether, acetone, benzene. Tm = 270.75; Tb = 361.15; Vm = 0.0878 at 291.65; Antoine (liq) A = 7.621, B = 2012.1, C = 0, dev = 1.0 to 5.0, range = 273 to 348.

2-Butenoic acid, cis, CA 503-64-0: Highly irritant; flammable; isomerizes in sunlight to trans isomer; soluble water, ethanol, acetone, carbon disulfide. Tm = 288.65; Tb = 442.15; Vm = 0.0839 at 293.15; Antoine (liq) A = 7.62029, B = 2297.25, C = -35.829, dev = 1.0, range = 306 to 445.

2-Butenoic acid, trans, CA 107-93-7: Irritant; corrosive; toxic; flammable, flash point 360; very soluble water; soluble ethanol, ether, acetone, toluene. Tm = 344.55 to 344.85; Tb = 457.85; Vm = 0.0893 at 353.15; Antoine (liq) A = 7.60186, B = 2333.67, C = -41.238, dev = 1.0, range = 353 to 458.

(continues)

$C_4H_6O_2$ *(continued)*

2-Butyne-1,4-diol, CA 110-65-6: Toxic; flammable, flash point 425; can explode on distillation; very soluble water; soluble methanol, ethanol, acetone. Tm = 331.15; Tb = 521.15; Antoine (liq) A = 8.8781, B = 3603, C = 0, dev = 1.0 to 5.0, range = 418 to 520.

gamma-Butyrolactone, CA 96-48-0: Expected carcinogen; flammable, flash point 371; soluble water, organic solvents. Tm = 229.15; Tb = 477.15; Vm = 0.0763 at 293.15; Tc = 709.15; Pc = 3430; Antoine (liq) A = 10.18937, B = 5483.794, C = 193.404, dev = 1.0, range = 392 to 474.

Methacrylic acid, 2-methylpropenoic acid, CA 79-41-4: Highly irritant; corrosive; flammable, flash point 350; causes burns; polymerizes on repeated heating; soluble water, ethanol, ether. Tm = 287.15; Tb = 433.65; Vm = 0.0848 at 298.15; Tc = 643, calculated; Pc = 4700, calculated; Vc = 0.270, calculated; Antoine I (liq) A = 8.10127, B = 2621.47, C = -4.216, dev = 1.0, range = 298 to 434; Antoine II (liq) A = 7.15818, B = 1942.119, C = -56.944, dev = 1.0, range = 321 to 435.

Methyl acrylate, CA 96-33-3: Irritant; moderately toxic; lachrymator; flammable, flash point 270; forms explosive peroxides and can polymerize violently; moderately soluble water; soluble most organic solvents. Tm = 196.65; Tb = 353.15; Vm = 0.0903 at 293.15; Tc = 536, calculated; Pc = 4250, calculated; Vc = 0.270, calculated; Antoine (liq) A = 6.5561, B = 1467.93, C = -30.849, dev = 5.0, range = 316 to 354.

Vinyl acetate, CA 108-05-4: Irritant; highly flammable, flash point 265; can polymerize violently; slightly soluble water; soluble ethanol, ether, acetone, benzene. Tm = 179.95; Tb = 345.85; Vm = 0.0922 at 293.15; Tc = 524, calculated; Pc = 4250, calculated; Vc = 0.270, calculated; Antoine I (liq) A = 6.3799, B = 1320.2716, C = -43.96, dev = 0.1, range = 293 to 346; Antoine II (liq) A = 6.34264, B = 1300.335, C = -46.041, dev = 0.1, range = 294 to 346.

$C_4H_6O_2S$ MW 118.15

Diacetyl sulfide, CA 3232-39-1: Irritant; flammable. Antoine (liq) A = 8.32266, B = 2659.162, C = 0, dev = 5.0, range = 325 to 355.

3-Sulfolene, 2,5-dihydrothiophene-1,1-dioxide, CA 77-79-2: Flammable, flash point 386; soluble water, ethanol, ether, benzene. Tm = 338.15; Vm = 0.0899 at 343.15.

$C_4H_6O_3$ MW 102.09

Acetic anhydride, CA 108-24-7: Corrosive; moderately toxic; lachrymator; causes burns; flammable, flash point 322; moderately soluble water, reacts; soluble ether, benzene, chloroform; forms ethyl acetate with ethanol. Tm = 199.02; Tb = 411.78; Vm = 0.0941 at 288.15; Tc = 569.15; Pc = 4681; Vc = 0.290, calculated; Antoine I (liq) A = 6.24655, B = 1427.77, C = -75.113, dev = 1.0, range = 320 to 413; Antoine II (liq) A = 5.38392, B = 2696.31, C = 17.794, dev = 1.0, range = 413 to 526.

Propylene carbonate, CA 108-32-7: Flammable, flash point 408. Tm = 218.15; Tb = 513.15; Vm = 0.0859 at 293.15; Antoine I (liq) A = 4.5875, B = 1767.4, C = 0, dev = 5.0, range = 323 to 370; Antoine II (liq) A = 6.19309, B = 1786.408, C = -88.452, dev = 0.1 to 1.0, range = 412 to 466.

$C_4H_6O_4$ MW 118.09

Dimethyl oxalate, CA 553-90-2: Flammable, flash point 348; moderately soluble water; soluble ethanol, ether, acetone. Tm = 327.15; Tb = 436.65;
Vm = 0.1008 at 333.15; Antoine I (sol) A = 7.5743, B = 2475.376, C = 0,
dev = 1.0, range = 289 to 306; Antoine II (liq) A = 7.50429, B = 2324.49,
C = -13.869, dev = 1.0, range = 293 to 437.

Succinic acid, butanedioic acid, CA 110-15-6: Irritant; flammable; used in
food industry; moderately soluble water; soluble ethanol, ether, acetone.
Tm = 461.25, decomposes; Antoine (sol) A = 13.193, B = 6132, C = 0, dev =
5.0, range = 372 to 401.

C_4H_6S MW 86.15

2-Vinylthiirane, CA 5954-75-6. Irritant; flammable. Antoine (liq) A =
10.46348, B = 4267.1, C = 130.05, dev = 1.0, range = 273 to 335.

$C_4H_6S_3$ MW 150.27

1,3-Dithiane-2-thion, CA 1748-15-8: Irritant; flammable. Tm = 351.15;
Antoine (sol) A = 10.55106, B = 4662.417, C = 0, dev = 0.1 to 1.0, range =
321 to 349.

C_4H_7Br MW 135.00

1-Bromo-1-butene, *cis*, CA 31849-78-2: Flammable; soluble ethanol, acetone,
benzene. Tb = 359.3; Vm = 0.1018 at 288.15; Antoine (liq) A = 6.08567, B =
1261.3, C = -50.15, dev = 0.1, range = 280 to 397.

1-Bromo-1-butene, *trans*, CA 32620-08-9: Flammable; soluble ethanol, acetone, benzene. Tm = 172.85; Tb = 367.85; Vm = 0.1022 at 288.15; Antoine
(liq) A = 6.52585, B = 1563.07, C = -22.205, dev = 1.0, range = 234 to 368.

2-Bromo-1-butene, CA 23074-36-4: Flammable; soluble ethanol, acetone, benzene. Tm = 139.75; Tb = 361.15; Vm = 0.1022 at 288.15; Antoine (liq) A =
6.0831, B = 1243.6, C = -49.15, dev = 1.0, range = 276 to 391.

2-Bromo-2-butene, *cis*, CA 3017-68-3: Flammable; soluble ethanol, ether,
benzene. Tm = 161.65; Tb = 364.15 to 365.65; Vm = 0.1006 at 288.15;
Antoine (liq) A = 6.40505, B = 1488.62, C = -28.759, dev = 1.0, range = 234
to 367.

2-Bromo-2-butene, *trans*, CA 3017-71-8: Flammable; soluble ethanol, ether,
benzene, chloroform. Tm = 158.55; Tb = 356.65 to 357.15; Vm = 0.1013 at
288.15; Antoine (liq) A = 6.27031, B = 1388.84, C = -33.132, dev = 1.0,
range = 228 to 359.

C_4H_7BrO MW 151.00

1-Bromo-2-butanone, CA 816-40-0: Irritant; flammable, flash point 341.
Vm = 0.1021 at 293.15; Antoine (liq) A = 6.1167, B = 1382.9, C = -91.25,
dev = 1.0 to 5.0, range = 322 to 428.

3-Bromo-2-butanone, CA 814-75-5: Irritant; flammable. Antoine (liq) A =
6.0126, B = 1294.8, C = -86.05, dev = 1.0 to 5.0, range = 306 to 409.

(continues)

110

C_4H_7BrO *(continued)*

Isobutyryl bromide, 2-methylpropionyl bromide, CA 2736-37-0: Irritant; flammable. Vm = 0.1073 at 288.15; Antoine (liq) A = 7.22323, B = 2219.68, C = -10.814, dev = 1.0, range = 286 to 436.

$C_4H_7Br_3$ MW 294.81

Propane, 1,3-dibromo-2-bromoethyl: Tm = 241.15; Antoine (liq) A = 6.5092, B = 2242.7, C = -95.15, dev = 5.0, range = 475 to 660.

1,1,2-Tribromobutane, CA 3675-68-1: Soluble ethanol, ether, chloroform. Tb = 489.35; Vm = 0.1350 at 293.15; Antoine (liq) A = 7.1426, B = 2473.6, C = -8.11, dev = 1.0, range = 361 to 490.

1,2,2-Tribromobutane, CA 3675-69-2: Soluble ethanol, ether, chloroform. Tb = 486.95; Vm = 0.1359 at 293.15; Antoine (liq) A = 6.756, B = 2192.8, C = -25.3, dev = 1.0, range = 356 to 487.

1,2,3-Tribromobutane, CA 632-05-3: Soluble ethanol, ether, chloroform. Tm = 254.15; Tb = 493.15; Vm = 0.1346 at 293.15; Antoine (liq) A = 6.4937, B = 1871.5, C = -76.15, dev = 5.0, range = 394 to 546.

1,2,4-Tribromobutane, CA 38300-67-3: Soluble ethanol, ether, chloroform. Tm = 255.15; Tb = 488.15; Vm = 0.1359 at 293.15; Antoine (liq) A = 6.49334, B = 1853.4, C = -75.13, dev = 5.0, range = 390 to 541.

2,2,3-Tribromobutane, CA 62127-47-3: Soluble ethanol, ether, chloroform. Tm = 275; Tb = 479.65, Vm = 0.1357 at 293.15; Antoine (liq) A = 6.50758, B = 1937.17, C = -49.69, dev = 1.0, range = 311 to 480.

C_4H_7Cl MW 90.55

1-Chloro-2-methyl-1-propene, CA 513-37-1: Irritant; lachrymator; anesthetic; flammable, flash point 272; soluble ethanol, ether, acetone. Tb = 341.15 to 342.15; Vm = 0.0986 at 293.15; Antoine (liq) A = 6.2687, B = 1271.243, C = -43.15, dev = 1.0, range = 285 to 343.

3-Chloro-2-methyl-1-propene, CA 563-47-3: Irritant; lachrymator; flammable, flash point 263; soluble ethanol, ether, acetone, chloroform. Tb = 345.25 to 345.35; Vm = 0.0988 at 293.15; Antoine (liq) A = 6.2303, B = 1275.366, C = -43.15, dev = 1.0, range = 285 to 348.

C_4H_7ClO MW 106.55

1-Chloro-2-butanone, CA 616-27-3: Irritant; flammable; soluble methanol. Tb = 410.65; Vm = 0.0982 at 293.15; Antoine (liq) A = 5.3177, B = 952.2, C = -125.57, dev = 5.0, range = 307 to 411.

3-Chloro-2-butanone, CA 4091-39-8: Irritant; flammable, flash point 294; soluble methanol. Tb = 388.15; Vm = 0.1010 at 273.15; Antoine (liq) A = 7.215, B = 2025, C = 0, dev = 5.0, range = 313 to 389.

3-Chloro-2-butene-1-ol, CA 40605-42-3: Irritant; flammable. Tm = 233.15; Tb = 434.15 to 435.15; Vm = 0.0973 at 293.15; Antoine (liq) A = 6.0577, B = 1330.9, C = -109.72, dev = 1.0, range = 345 to 437.

$C_4H_7ClO_2$ MW 122.55

Ethyl chloroacetate CA 105-39-5: Highly toxic; lachrymator; absorbed
through skin; flammable, flash point 337; decomposed by hot water or prolonged boiling; soluble ethanol, ether, acetone, benzene. Tf = 247.15;
Tb = 418.15 to 419.15; Vm = 0.1066 at 293.15; Antoine (liq) A = 7.15526,
B = 2051.27, C = -19.016, dev = 1.0, range = 274 to 418.

Isopropyl chloroformate, CA 108-23-6: Irritant, corrosive; flammable,
flash point 296; soluble ether. Tb = 378.15; Vm = 0.1137 at 293.15; Antoine (liq) A = 7.253, B = 1976, C = 0, dev = 5.0, range not specified.

C_4H_7ClS MW 122.61

2-Butene-3-chloro-1-thiol: Irritant; flammable. Antoine (liq) A = 8.184,
B = 2518, C = 0, dev = 1.0, range = 341 to 397.

$C_4H_7Cl_2O_4P$ MW 220.98

Dimethyl-(2,2-dichlorovinyl) phosphate, CA 62-73-7: Highly toxic by skin
absorption and inhalation; expected teratogen; insecticide; moderately soluble water; soluble ethanol. Vm = 0.1562 at 298.15; Antoine (liq) A =
9.3756, B = 3581.3, C = 1.2, dev = 1.0 to 5.0, range = 283 to 387.

$C_4H_7Cl_3$ MW 161.46

1,2,3-Trichlorobutane, CA 18338-40-4: Soluble ethanol, ether, chloroform.
Tb = 438.15 to 440.15; Vm = 0.1223 at 293.15; Antoine (liq) A = 6.01465,
B = 1618.85, C = -38.416, dev = 1.0, range = 273 to 442.

C_4H_7FOS MW 122.16

Thioacetic acid, 2-fluoroethyl ester: Irritant; flammable. Antoine (liq)
A = 7.6273, B = 2336.2, C = 0, dev = 1.0, range = 273 to 333.

$C_4H_7FO_2$ MW 106.10

Ethyl fluoroacetate, CA 459-72-3: Irritant; highly toxic; lachrymator;
flammable, flash point 303; soluble water, most organic solvents. Tb =
392.45; Vm = 0.0966 at 293.15; Antoine (liq) A = 7.2529, B = 2188.9, C = 0,
dev = 1.0, range = 273 to 333.

$C_4H_7F_3$ MW 112.09

1,1,1-Trifluorobutane, CA 460-34-4: Tm = 158.35; Antoine (liq) A =
6.15792, B = 1049.3, C = -37.15, dev = 1.0, range = 226 to 320.

$C_4H_7IO_2$ MW 214.00

Ethyl iodoacetate, CA 623-48-3: Irritant; corrosive; highly toxic; flammable, flash point 349; soluble ethanol, ether. Tb = 451.15 to 453.15;
Vm = 0.1184 at 293.15; Antoine (liq) A = 8.0924, B = 2723, C = 0, dev = 1.0
to 5.0, range = 301 to 362.

C_4H_7N MW 69.11

Butyronitrile, CA 109-74-0: Highly toxic by inhalation and skin absorption;
flammable, flash point 297; moderately soluble water; soluble ethanol,
ether, benzene. Tm = 161.15; Tb = 390.15; Vm = 0.0884 at 303.15; Antoine
(liq) A = 6.25397, B = 1452.076, C = -48.9645, dev = 0.02, range = 332 to
401.

C_4H_7NO MW 85.11

Acetone cyanohydrin, CA 75-86-5: Irritant; highly toxic; flammable, flash
point 307; decomposes to form hydrogen cyanide; soluble water, ethanol,
ether; raw material for methacrylic acid. Tm = 254.15; Tb = 393.15, decomposes; Vm = 0.0918 at 298.15; Antoine (liq) A = 16.159, B = 5564.35, C = 0,
dev = 5.0, range = 355 to 393.

2-Butenoic acid, amide, *cis*, CA 31110-30-2: Moderately soluble water; soluble ethanol, benzene. Tm = 388.15; Antoine (sol) A = 9.3066, B = 3550.7,
C = 0, dev = 5.0, range = 353 to 387.

2-Butenoic acid, amide, *trans*, CA 625-37-6: Moderately soluble water; soluble ethanol, benzene. Tm = 432.15 to 433.15; Antoine (sol) A = 5.4895,
B = 1297.9, C = -166.9, dev = 1.0, range = 363 to 413.

2-Hydroxybutyronitrile, CA 4476-02-2: Irritant; flammable. Antoine (liq)
A = 8.28203, B = 2734.75, C = -16.093, dev = 5.0, range = 314 to 452.

Methacrylamide, CA 79-39-0: Irritant; toxic; soluble ethanol. Tm = 383.15;
Antoine (liq) A = 11.63906, B = 4510.93, C = 0, dev = 1.0 to 5.0, range =
390 to 418.

3-Methoxypropionitrile, CA 110-67-8: Irritant; flammable; moisture-
sensitive; soluble ethanol, ether. Tm = 210.6; Tb = 436.15; Vm = 0.0900 at
293.15; Antoine I (liq) A = 7.7186, B = 2486.5, C = 0, dev = 5.0, range =
328 to 438; Antoine II (liq) A = 6.18085, B = 1498.691, C = -79.616, dev =
1.0 to 5.0, range = 293 to 436.

2-Pyrrolidone, CA 616-45-5: Flammable, flash point 402; hygroscopic; soluble water, most organic solvents; important high-boiling solvent. Tm =
298.75; Tb = 523.65; Vm = 0.0769 at 298.15; Antoine (liq) A = 8.0514, B =
3132.57, C = 0, dev = 5.0, range = 395 to 518.

$C_4H_7NO_2$ MW 101.10

Diacetamide, CA 625-77-4: Decomposes at 523; reacts with boiling water;
soluble ether. Tm = 352.15; Tb = 489.15 to 491.15; Antoine (liq) A =
7.89581, B = 2810.23, C = -19.283, dev = 1.0, range = 368 to 496.

2-Nitro-1-butene, CA 2783-12-2: Flammable. Antoine (liq) A = 7.7322, B =
2298.7, C = 0, dev = 1.0 to 5.0, range = 273 to 333.

$C_4H_7NO_3$ MW 117.10

N-Acetylglycine, CA 543-24-8: Moderately soluble water; soluble ethanol.
Tm = 479.15, decomposes; Antoine (sol) A = 13.65849, B = 6640.105, C = 0,
dev = 1.0 to 5.0, range = 383 to 400.

$C_4H_7N_3O_9$ MW 241.11

1,2,4-Butanetriol, trinitrate, CA 6659-60-5: Flammable; used as explosive.
Tm = 261.81; Antoine (liq) A = 9.8611, B = 3134.8, C = 0, dev = 1.0, range
= 293 to 313.

C_4H_8 MW 56.11

1-Butene, 1-butylene, CA 106-98-9: Narcotic in high concentrations; extremely flammable, flash point 193; slightly soluble water; moderately
soluble organic solvents. Tm = 87.8; Tb = 266.89; Vm = 0.1090 at 293.15;
Tc = 419.55; Pc = 4023; Vc = 0.240; Antoine I (liq) A = 5.9678, B = 926.1,
C = -33.15, dev = 0.1, range = 200 to 274; Antoine II (liq) A = 8.1706, B =
1601.52, C = 7.059, dev = 1.0, range = 126 to 192; Antoine III (liq) A =
6.05416, B = 970.771, C = -27.089, dev = 0.1, range = 267 to 345; Antoine
IV (liq) A = 6.77294, B = 1482.801, C = 48.073, dev = 0.1, range = 342 to
411; Antoine V (liq) A = 6.27411, B = 1097.171, C = -9.657, dev = 1.0,
range = 267 to 411.

2-Butene, *cis*, CA 590-18-1: Extremely flammable. Tm = 134.23; Tb = 276.87;
Vm = 0.0903 at 293.15; Tc = 435.58; Pc = 4205; Vc = 0.234; Antoine I (liq)
A = 5.99416, B = 960.10, C = -36.15, dev = 0.02, range = 221 to 290;
Antoine II (liq) A = 6.11733, B = 1022.247, C = -28.245, dev = 0.02, range
= 276 to 325; Antoine III (liq) A = 6.43156, B = 1209.771, C = -3.164,
dev = 0.1, range = 324 to 386; Antoine IV (liq) A = 7.28545, B = 1967.426,
C = 104.145, dev = 0.1 to 1.0, range = 383 to 431.

2-Butene, *trans*, CA 624-64-6: Extremely flammable. Tm = 167.62; Tb =
274.05; Vm = 0.0929 at 293.15; Tc = 428.63; Pc = 4104; Vc = 0.238; Antoine
I (liq) A = 5.99442, B = 960.8, C = -33.15, dev = 0.1, range = 205 to 287;
Antoine II (liq) A = 6.1576, B = 1042.105, C = -23.027, dev = 0.1, range =
273 to 315; Antoine III (liq) A = 6.54029, B = 1274.473, C = 7.499, dev =
1.0, range = 313 to 385; Antoine IV (liq) A = 6.94808, B = 1643.833, C =
64.733, dev = 1.0, range = 382 to 428.

Cyclobutane, CA 287-23-0: Extremely flammable; soluble ethanol, ether,
acetone, benzene. Tf = 182.45; Tb = 284.35 to 284.55; Vm = 0.0814 at
298.15; Antoine (liq) A = 6.03126, B = 1020.06, C = -32.27, dev = 0.1,
range = 198 to 287.

Methylcyclopropane, CA 594-11-6: Extremely flammable; soluble ethanol,
ether. Tm = 95.95; Tb = 273.85; Vm = 0.0812 at 253.15; Antoine (liq) A =
5.96539, B = 952.41, C = -37.157, dev = 1.0, range = 177 to 278.

2-Methylpropene, isobutylene, CA 115-11-7: Extremely flammable; soluble
ethanol, ether, benzene, sulfuric acid. Tm = 132.81; Tb = 266.25; Vm =
0.0944 at 293.15; Tc = 417.9, Pc = 4000, Vc = 0.239; Antoine I (liq) A =
5.96624, B = 923.2, C = -33.15, dev = 0.02, range = 212 to 279; Antoine II
(liq) A = 5.93211, B = 907.664, C = -35.082, dev = 0.1, range = 266 to 313;
Antoine III (liq) A = 6.27428, B = 1095.288, C = -9.441, dev = 1.0, range =
310 to 376; Antoine IV (liq) A = 7.64267, B = 2336.466, C = 160.311, dev =
1.0, range = 371 to 418.

C_4H_8BrClO MW 187.46

2-Bromoethyl-2-chloroethyl ether, CA 51070-66-7: Antoine (liq) A =
6.68372, B = 1945.77, C = -52.983, dev = 1.0, range = 309 to 469.

$C_4H_8Br_2$ MW 215.92

1,1-Dibromobutane, CA 62168-25-6: Irritant; flammable. Vm = 0.1200 at
298.15; Antoine (liq) A = 6.390, B = 1609, C = -64.15, dev = 5.0, range =
342 to 477.

1,2-Dibromobutane, CA 533-98-2: Irritant; narcotic; flammable, flash point
over 385; soluble ethanol, ether, chloroform. Tm = 207.75; Tb = 439.15;
Vm = 0.1200 at 293.15; Antoine I (liq) A = 6.88931, B = 2130.04, C = -3.55,
dev = 1.0, range = 281 to 439; Antoine II (liq) A = 6.03731, B = 1507.416,
C = -65.446, dev = 0.1, range = 338 to 425.

1,3-Dibromobutane, CA 107-80-2: Irritant; soluble ether, chloroform. Tb =
447.15; Vm = 0.1200 at 293.15; Antoine (liq) A = 6.15995, B = 1610.817, C =
-61.816, dev = 0.1, range = 351 to 450.

1,4-Dibromobutane, CA 110-52-1: Irritant; lachrymator; flammable, flash
point over 385; soluble chloroform. Tf = 253.15; Tb = 470.15 to 471.15;
Vm = 0.1193 at 293.15; Antoine (liq) A = 6.49664, B = 1791.9, C = -71.15,
dev = 5.0, range = 375 to 520.

2,3-Dibromobutane, *meso*, CA 5780-13-2: Irritant; soluble ether, chloroform.
Tm = 238.65; Tb = 430.45; Vm = 0.1205 at 288.15; Antoine (liq) A = 6.65637,
B = 1919.98, C = -17.725, dev = 1.0, range = 274 to 431.

2,3-Dibromobutane, ± *threo*, CA 598-71-0: Irritant; soluble ether, chloroform. Tm = 238.65; Tb = 430.45; Vm = 0.1207 at 295.15; Antoine (liq) A =
7.10146, B = 2250.44, C = 7.818, dev = 1.0, range = 278 to 434.

1,2-Dibromo-2-methylpropane, CA 594-34-3: Irritant; flammable, flash point
383; soluble ethanol, ether, benzene, chloroform. Tm = 282.15 to 285.15;
Tb = 422.15 to 424.15; Vm = 0.1213 at 293.15; Antoine (liq) A = 5.38956,
B = 1314.98, C = -33.653, dev = 1.0, range = 244 to 422.

1,3-Dibromo-2-methylpropane, CA 28148-04-1: Irritant. Tb = 450.15 to
451.15; Vm = 0.1211 at 293.15; Antoine (liq) A = 6.6170, B = 1933.2, C =
-28.484, dev = 1.0, range = 287 to 448.

$C_4H_8Br_2O$ MW 231.91

Bis(2-bromoethyl) ether, CA 5414-19-7: Irritant; forms explosive peroxides.
Vm = 0.1273 at 300.15; Antoine (liq) A = 6.65758, B = 1999.47, C = -55.785,
dev = 1.0, range = 320 to 486.

$C_4H_8Cl_2$ MW 127.01

1,1-Dichlorobutane, CA 541-33-3: Irritant; soluble ethanol, ether, chloroform. Tb = 387.15 to 388.15; Vm = 0.1176 at 298.15; Antoine I (liq) A =
6.160, B = 1376, C = -55.75, dev = 5.0, range = 303 to 428; Antoine II
(liq) A = 6.11538, B = 1367.111, C = -55.83, dev = 1.0, range = 304 to 386.

1,2-Dichlorobutane, CA 616-21-7: Irritant; soluble ether, chloroform.
Tb = 398.15 to 403.15; Vm = 0.1143 at 298.15; Antoine I (liq) A = 6.18945,
B = 1520.81, C = -33.204, dev = 1.0, range = 249 to 397; Antoine II (liq)
A = 6.02227, B = 1352.224, C = -60.482, dev = 0.1, range = 312 to 394.

(continues)

$C_4H_8Cl_2$ *(continued)*

1,3-Dichlorobutane, CA 1190-22-3:   Irritant, lachrymator; flammable, flash
point 303; soluble ether, chloroform.   Tb = 407.15; Vm = 0.1138 at 293.15;
Antoine (liq) A = 6.11577, B = 1429.065, C = -59.172, dev = 0.1, range =
318 to 407.

1,4-Dichlorobutane, CA 110-56-5:   Irritant; flammable, flash point 313;
soluble chloroform.   Tm = 235.85; Tb = 434.15 to 436.15; Vm = 0.1095 at
293.15; Antoine I (liq) A = 6.11603, B = 1495.7, C = -63.15, dev = 1.0,
range = 334 to 473; Antoine II (liq) A = 6.19263, B = 1530.168, C = -62.825,
dev = 0.1, range = 336 to 425.

2,2-Dichlorobutane, CA 4279-22-5:   Irritant; soluble chloroform.   Tm =
199.15; Tb = 377.15; Vm = 0.1150 at 298.15; Antoine (liq) A = 5.91028, B =
1233.893, C = -59.678, dev = 0.1 to 1.0, range = 293 to 376.

2,3-Dichlorobutane, mixed isomers, CA 2211-67-8:   Irritant; flammable,
flash point 291.   Tm = 192.75; Tb = 389.15 to 391.15; Vm = 0.1195 at
298.15; Antoine (liq) A = 6.2404, B = 1476.37, C = -40.433, dev = 1.0,
range = 247 to 389.

1,1-Dichloro-2-methylpropane, CA 598-76-5:   Irritant; soluble ethanol,
ether, benzene, chloroform.   Tb = 376.15 to 378.15, decomposes; Vm = 0.1256
at 285.15; Antoine (liq) A = 6.38106, B = 1504.81, C = -35.222, dev = 1.0,
range = 242 to 379.

1,2-Dichloro-2-methylpropane, CA 594-37-6:   Irritant; flammable, flash
point 288; soluble ethanol, ether, acetone, benzene.   Tb = 379.65; Vm =
0.1162 at 293.15; Antoine (liq) A = 6.49685, B = 1540.97, C = -38.008,
dev = 1.0, range = 247 to 381.

1,3-Dichloro-2-methylpropane, CA 616-19-3:   Irritant; soluble ethanol,
ether, benzene.   Tb = 409.15 to 413.15; Vm = 0.1122 at 298.15; Antoine
(liq) A = 7.16999, B = 1988.44, C = -23.142, dev = 1.0, range = 270 to 408.

$C_4H_8Cl_2O$ MW 143.01

Bis(2-chloroethyl) ether, CA 111-44-4:   Irritant; highly toxic; suspected
carcinogen; absorbed through skin; flammable, flash point 328; forms explosive peroxides; soluble most organic solvents.   Tf = 223.15; Tb = 450.15
to 451.15; Vm = 0.1172 at 293.15; Antoine (liq) A = 6.7637, B = 1948.62,
C = -41.974, dev = 1.0, range = 297 to 452.

$C_4H_8Cl_2S$ MW 159.07

Bis(2-chloroethyl) sulfide, mustard gas, CA 505-60-2:   Powerful vesicant
and poison, causing conjunctivitis and blindness; carcinogen; used as war
gas; slowly hydrolyzed by water; slightly soluble water; soluble most
organic solvents.   Tm = 286.15 to 287.15; Tb = 488.15 to 490.15; Vm =
0.1249 at 293.15; Antoine I (sol) A = 11.7813, B = 4030.9, C = 0, dev =
1.0, range = 263 to 287; Antoine II (liq) A = 8.6068, B = 3117.2, C = 0,
dev = 1.0, range = 288 to 358.

$C_4H_8Cl_2S_3$ MW 223.19

Bis(2-chloroethyl) trisulfide: Irritant.   Tm = 302.15; Antoine (liq) A =
8.35519, B = 3565.7, C = 0, dev = 5.0, range = 273 to 333.

$C_4H_8Cl_3O_4P$ MW 257.44

(1-Hydroxy-2,2,2-trichloroethyl) phosphonic acid, dimethyl ester,
CA 52-68-6: Toxic insecticide; roach killer, very soluble water; soluble
ethanol, chloroform. Tm = 357.15; Antoine (sol) A = 13.0497, B = 5588.7,
C = 0, dev = 1.0, range = 293 to 357.

$C_4H_8F_2$ MW 94.10

1,1-Difluorobutane, CA 2358-38-5: Antoine (liq) A = 6.182, B = 1136, C =
-42.15, dev = 5.0, range = 246 to 347.

2,2-Difluorobutane, CA 353-81-1: Tm = 155.65; Antoine (liq) A = 6.19827,
B = 1106.4, C = -40.15, dev = 1.0, range = 238 to 336.

$C_4H_8F_2O_4S$ MW 190.16

Bis(2-fluoroethyl) sulfate: Antoine (liq) A = 8.9366, B = 3335.6, C = 0,
dev = 1.0, range = 273 to 333.

$C_4H_8F_3N$ MW 127.11

N,N-Dimethyl-(2,2,2-trifluoroethyl) amine, CA 819-06-7: Tb = 320.15; Vm =
0.1238 at 297.15; Antoine (liq) A = 7.2309, B = 1671, C = 0, dev = 1.0 to
5.0, range not specified.

$C_4H_8N_2O_2$ MW 116.12

1,2-Diacetylhydrazine, CA 3148-73-0: Irritant. Tm = 413.15; Antoine (sol)
A = 11.3196, B = 5384.4, C = 0, dev = 1.0, range = 347 to 358.

Dimethyl glyoxime, CA 95-45-4: Almost insoluble water; soluble ethanol,
ether, acetone.   Tm = 511.15 to 513.15; Antoine (sol) A = 11.2449, B =
5057.7, C = 0, dev = 1.0, range = 331 to 352.

$C_4H_8N_2O_6$ MW 180.12

1,3-Butanediol, dinitrate, CA 6423-44-5: Flammable. Tm = 253.38; Antoine
(liq) A = 13.730, B = 3730, C = 0, dev = 1.0 to 5.0, range = 293 to 313.

1,4-Butanediol, dinitrate, CA 3457-91-8: Flammable. Tm = 285.45; Antoine
(liq) A = 10.3228, B = 3001.1, C = 0, dev = 1.0, range = 293 to 313.

$C_4H_8N_2O_7$ MW 196.12

Diethyleneglycol dinitrate, CA 693-21-0: Flammable. Tm = 261.75; Antoine
(liq) A = 13.548, B = 4926.2, C = 0, dev = 1.0 to 5.0, range = 293 to 333.

$C_4H_8N_4O_2$ MW 144.13

1,4-Dinitrosopiperazine, CA 140-79-4: Antoine (sol) A = 11.715, B = 5301,
C = 0, dev = 1.0 to 5.0, range = 325 to 360.

$C_4H_8N_4O_4$ MW 176.13

1,4-Dinitropiperazine, CA 4164-37-8: Antoine (sol) A = 11.925, B = 5821, C = 0, dev = 1.0 to 5.0, range = 325 to 360.

$C_4H_8N_8O_8$ MW 296.16

Octahydro-1,3,5,7-tetranitro-1,3,5,7-tetrazocine, CA 2691-41-0: Tm = 551.65, *delta* form; Antoine I (sol, *beta* form) A = 15.305, B = 9154, C = 0, dev = 5.0, range = 370 to 403; Antoine II (sol, *delta* form) A = 14.45166, B = 8636.401, C = 0, dev = 5.0, range = 461 to 487.

C_4H_8O MW 72.11

2-Butanone, methyl ethyl ketone, CA 78-93-3: Irritant; moderately toxic; flammable, flash point 264; expected teratogen; forms a highly toxic per-oxide; important low-boiling solvent; very soluble water; soluble most organic solvents. Tf = 187.25; Tb = 352.72; Vm = 0.0896 at 293.15; Tc = 535.5; Pc = 4154; Vc = 0.267; Antoine I (liq) A = 6.247219, B = 1294.53, C = -47.442, dev = 0.1, range = 294 to 352; Antoine II (liq) A = 6.18497, B = 1259.519, C = -51.359, dev = 0.02, range = 315 to 363; Antoine III (liq) A = 6.22518, B = 1286.794, C = -47.766, dev = 0.1, range = 353 to 403; Antoine IV (liq) A = 6.45545, B = 1456.517, C = -24.944, dev = 0.1, range = 397 to 479; Antoine V (liq) A = 8.56912, B = 4050.052, C = 282.032, dev = 1.0, range = 473 to 537.

3-Buten-2-ol ±, CA 6118-14-5: Irritant; flammable, flash point 289; sol-uble water. Tm = below 173.15; Tb = 369.15 to 371.15; Vm = 0.0867 at 293.15; Antoine (liq) A = 7.55576, B = 2049.048, C = 0, dev = 5.0, range = 304 to 370.

Butyraldehyde, CA 123-72-8: Irritant; moderately toxic; flammable, flash point 251; important industrial solvent; very soluble water; soluble ethanol, ether, acetone. Tm = 176.22; Tb = 348.85; Vm = 0.0896 at 293.15; Tc = 525, calculated; Pc = 4000, calculated; Vc = 0.263, calculated; Antoine I (liq) A = 5.68618, B = 994.1, C = -78.05, dev = 1.0, range = 293 to 349; Antoine II (liq) A = 5.40874, B = 1182.472, C = 0, dev = 5.0, range = 348 to 423.

1,2-Epoxybutane ±, CA 3760-95-0: Irritant; flammable, flash point 258; moisture-sensitive; soluble water, ethanol, ether, acetone. Tb = 336.15; Vm = 0.0868 at 293.15; Antoine (liq) A = 6.569, B = 1720.5, C = 41.3, dev = 1.0, range = 254 to 347.

1,2-Epoxy-2-methylpropane, CA 558-30-5: Flammable; soluble ethanol, ether. Tb = 325.15; Vm = 0.0834 at 273.15; Antoine (liq) A = 5.94797, B = 1173.2, C = -31.123, dev = 1.0, range = 204 to 329.

Ethyl vinyl ether, CA 109-92-2: Anesthetic; lachrymator; flammable, flash point below 227; forms explosive peroxides; may polymerize violently; slightly soluble water; soluble ethanol, ether. Tm = 157.75; Tb = 308.85; Vm = 0.0956 at 293.15; Antoine I (liq) A = 7.0133, B = 1543.2, C = 0, dev = 1.0, range = 223 to 309; Antoine II (liq) A = 6.06857, B = 1075.837, C = -43.943, dev = 1.0, range = 221 to 364.

Isobutyraldehyde, CA 78-84-2: Moderately toxic; stench; flammable, flash point 255; important chemical intermediate; moderately soluble water; sol-uble ether, acetone, chloroform. Tm = 208.15; Tb = 337.65; Vm = 0.0908 at 293.15; Tc = 507, calculated; Pc = 4100, calculated; *(continues)*

C_4H_8O *(continued)*

Vc = 0.263, calculated; Antoine I (liq) A = 5.83446, B = 1040.42, C = -65.53, dev = 0.1, range = 283 to 337; Antoine II (liq) A = 9.07587, B = 2379.61, C = 0, dev = 5.0, range = 336 to 373.

Methyl propenyl ether, *cis*, CA 4188-68-5: Flammable; polymerizes; forms explosive peroxides; soluble ethanol, ether, acetone. Tm = 158.75; Antoine (liq) A = 7.02565, B = 1598.9, C = 0, dev = 1.0, range = 293 to 318.

Methyl propenyl ether, *trans*, CA 4188-69-6: Flammable; polymerizes; forms explosive peroxides; soluble ethanol, ether, acetone. Tm = 173.46; Antoine (liq) A = 6.772, B = 1539.1, C = 0, dev = 1.0, range = 293 to 322.

Tetrahydrofuran, CA 109-99-9: Irritant; flammable, flash point 259; forms explosive peroxides; moderately soluble water; soluble most organic solvents. Tf = 208.15; Tb = 399.15; Vm = 0.0812 at 294.15; Antoine I (liq) A = 5.92617, B = 1101.47, C = -57.95, dev = 1.0, range = 273 to 339; Antoine II (liq) A = 6.12052, B = 1202.561, C = -46.863, dev = 0.02, range = 296 to 373; Antoine III (liq) A = 6.63507, B = 1626.656, C = 15.041, dev = 1.0 to 5.0, range = 399 to 479; Antoine IV (liq) A = 6.73137, B = 1702.922, C = 23.613, dev = 1.0 to 5.0, range = 467 to 541.

C_4H_8OS MW 104.17

1,4-Oxathiane, CA 15980-15-1: Highly toxic orally; stench; flammable, flash point 315; slightly soluble water. Tm = 256.15; Tb = 420.15; Vm = 0.0932 at 293.15; Antoine (liq) A = 5.30265, B = 979.58, C = -125.806, dev = 1.0, range = 342 to 411.

$C_4H_8O_2$ MW 88.11

2-Butene-1,4-diol, *cis*, CA 6117-80-2: Skin irritant; flammable, flash point 401; soluble water, ethanol. Tm = 277.15; Tb = 508.15; Vm = 0.0824 at 293.15; Antoine (liq) A = 8.6493, B = 3378, C = 0, dev = 5.0, range = 373 to 508.

Butyric acid, butanoic acid, CA 107-92-6: Irritant; corrosive; toxic; causes burns; flammable, flash point 345; smells of rancid butter; soluble water, ethanol, ether. Tm = 265.25; Tb = 436.65; Vm = 0.0919 at 293.15; Tc = 628; Pc = 4420, calculated; Vc = 0.283, calculated; Antoine I (liq) A = 6.50913, B = 1542.6, C = -94.15, dev = 1.0, range = 355 to 453; Antoine II (liq) A = 7.3554, B = 2180.05, C = -29.337, dev = 1.0, range = 437 to 592; Antoine III (liq) A = 11.53324, B = 5291.631, C = 128.778, dev = 1.0, range = 301 to 358.

1,3-Dioxane, CA 505-22-6: Moderately toxic; flammable; soluble water, ethanol, ether, acetone. Tm = 231.15; Tb = 378.15; Vm = 0.0852 at 293.15; Antoine (liq) A = 6.905, B = 1869, C = 0, dev = 5.0, range not specified.

1,4-Dioxane, CA 123-91-1: Moderately toxic by inhalation and skin absorption; flammable, flash point 285; carcinogen; may form explosive peroxides; soluble water, ethanol, ether, acetone. Tm = 284.95; Tb = 374.15; Vm = 0.0852 at 293.15; Tc = 587; Pc = 5208; Vc = 0.238; Antoine (liq) A = 6.40318, B = 1457.97, C = -42.888, dev = 0.02 to 0.1, range = 285 to 375.

Ethyl acetate, CA 141-78-6: Moderately irritant; flammable, flash point 269; slowly hydrolyzed by water; important low-boiling solvent; very soluble water; soluble most organic solvents; Tm = 189.15;   *(continues)*

$C_4H_8O_2$ *(continued)*

Tb = 350.15; Vm = 0.0979 at 293.15; Tc = 523.25; Pc = 3830; Vc = 0.286;
Antoine I (liq) A = 6.227229, B = 1245.239, C = -55.239, dev = 0.1, range =
288 to 351; Antoine II (liq) A = 6.38462, B = 1369.41, C = -37.675, dev =
1.0, range = 350 to 508; Antoine III (liq) A = 6.22825, B = 1245.68, C =
-55.193, dev = 0.1, range = 288 to 351.

3-Hydroxy-2-butanone, acetoin, CA 52217-02-4: Flammable; soluble water,
ethanol, acetone. Tm = 201.15; Tb = 421.15; Vm = 0.0876 at 293.15; Antoine
(liq) A = 7.3688, B = 2368.3, C = 24.85, dev = 5.0, range = 273 to 418.

Isobutyric acid, 2-methylpropanoic acid, CA 79-31-2: Irritant; corrosive;
toxic; flammable, flash point 329; very soluble water; soluble ethanol,
ether, chloroform. Tf = 227.05; Tb = 428.15; Vm = 0.0927 at 293.15;
Antoine I (liq) A = 7.20794, B = 2023.52, C = -38.649, dev = 1.0, range =
288 to 428; Antoine II (liq) A = 7.11635, B = 2006.61, C = -35.297, dev =
1.0, range = 428 to 562.

Isopropyl formate, CA 625-55-8: Flammable; slightly soluble water with
some hydrolysis; soluble ethanol, ether, acetone. Tb = 341.35; Vm = 0.1010
at 293.15; Antoine (liq) A = 6.79997, B = 1557.57, C = -16.632, dev = 1.0,
range = 221 to 342.

Methyl propionate, CA 554-12-1: Flammable, flash point 271; moderately
soluble water; soluble ethanol, ether. Tm = 185.65; Tb = 352.85; Vm =
0.0963 at 293.15; Tc = 530.6; Pc = 4004; Vc = 0.282; Antoine I (liq) A =
6.49537, B = 1393.26, C = -42.656, dev = 1.0, range = 231 to 353; Antoine
II (liq) A = 6.43771, B = 1414.65, C = -33.767, dev = 1.0, range = 353 to
486; Antoine III (liq) A = 6.05388, B = 1163.255, C = -65.163, dev = 0.1 to
1.0, range = 293 to 353.

Propyl formate, CA 110-74-7: Flammable, flash point 270; slightly soluble
water with some hydrolysis; soluble ethanol, ether. Tm = 180.25; Tb =
354.45; Vm = 0.0973 at 293.15; Tc = 538; Pc = 4063; Vc = 0.285; Antoine I
(liq) A = 6.73268, B = 1560.29, C = -24.287, dev = 1.0, range = 230 to 355;
Antoine II (liq) A = 6.2378, B = 1301.3, C = -46.767, dev = 1.0, range =
354 to 518.

$C_4H_8O_2S$ MW 120.17

Allylmethyl sulfone, CA 16215-14-8: Irritant; flammable. Antoine (liq)
A = 9.075, B = 3563, C = 0, dev = 5.0, range = 405 to 450.

Sulfolane, tetrahydrothiophene-1,1-dioxide, CA 126-33-0: Moderately toxic;
flammable, flash point 450; used to extract aromatics; soluble water, acetone, toluene. Tm = 301.65; Tb = 560.45; Vm = 0.0949 at 303.15; Tc = 855,
calculated; Pc = 5030, calculated; Vc = 0.300, calculated; Antoine I (liq)
A = 7.512, B = 3068, C = 0, dev = 5.0, range = 413 to 558; Antoine II (liq)
A = 3.22174, B = 1617.184, C = 0, dev = 5.0, range = 303 to 328.

$C_4H_8O_3$ MW 104.11

Ethoxyacetic acid, CA 627-03-2: Corrosive; flammable, flash point 370;
soluble ethanol, ether. Tb = 479.15 to 480.15; Vm = 0.0945 at 293.15;
Antoine (liq) A = 10.0982, B = 3612, C = 0, dev not specified, range = 280
to 310.

(continues)

120

$C_4H_8O_3$ *(continued)*

Ethyl glycolate, CA 623-50-7: Flammable; soluble ethanol, ether. Tb = 433.15; Vm = 0.0962 at 296.15; Antoine (liq) A = 7.47541, B = 2306.73, C = -9.705, dev = 1.0, range = 287 to 432.

2-Hydroxyisobutyric acid, acetonic acid, CA 594-61-6: Flammable; soluble water, ethanol, ether. Tm = 352.15; Tb = 485.15; Antoine (liq) A = 8.40826, B = 2863.62, C = -38.073, dev = 1.0, range = 371 to 485.

3-Hydroxypropionic acid, methyl ester, CA 6149-41-3: Flammable; soluble water, ethanol, ether. Tb = 450.15 to 457.15; Vm = 0.0931 at 298.15; Antoine (liq) A = 9.252, B = 3132, C = 0, dev = 5.0, range = 330 to 343.

Methoxyacetic acid, methyl ester, CA 6290-49-9: Flammable, flash point 308; soluble ethanol, ether, acetone. Tb = 404.15; Vm = 0.1020 at 293.15; Antoine (liq) A = 6.9973, B = 2053, C = 0, dev not specified, range = 285 to 310.

Methyl lactate ±, CA 547-64-8: Flammable, flash point 322; moisture-sensitive; soluble water, decomposes; soluble ethanol, ether. Tm = 207.15; Tb = 418.15; Vm = 0.0952 at 293.15; Antoine (liq) A = 7.6188, B = 2333.2, C = 0, dev = 5.0, range = 313 to 418.

Pyroxybutyric acid, CA 13122-71-9: Flammable. Tm = 263.15; Antoine (liq) A = 7.955, B = 2376, C = 0, dev = 5.0, range = 273 to 393.

C_4H_8S MW 88.17

2,2-Dimethylthiirane, CA 3772-13-2: Irritant; flammable. Antoine (liq) A = 6.17325, B = 1271.1, C = -54.15, dev = 1.0, range = 273 to 473.

2-Ethylthiirane, CA 3195-86-6: Irritant; flammable. Antoine (liq) A = 6.17445, B = 1309, C = -64.15, dev = 1.0, range = 298 to 450.

Tetrahydrothiophene, CA 110-01-0: Irritant; stench; flammable, flash point 285; slightly soluble water; soluble ethanol, ether, benzene. Tm = 177; Tb = 392.15 to 395.15; Vm = 0.0918 at 291.15; Antoine (liq) A = 6.10833, B = 1394.668, C = -54.326, dev = 0.02 to 0.1, range = 343 to 434.

$C_4H_8S_2$ MW 120.23

1,4-Dithiane, CA 505-29-3: Irritant; stench; flammable; slightly soluble water; soluble ethanol, ether, acetic acid. Tm = 385.15 to 386.15; Tb = 472.15 to 473.15; Antoine (liq) A = 5.30559, B = 1126.34, C = -132.475, dev = 1.0, range = 388 to 437.

C_4H_9Br MW 137.02

1-Bromobutane, butyl bromide, CA 109-65-9: Irritant; flammable, flash point 291; soluble ethanol, ether, acetone, chloroform. Tm = 160.45; Tb = 373.65; Vm = 0.1080 at 293.15; Antoine I (liq) A = 6.04744, B = 1298.608, C = -53.45, dev = 0.1, range = 273 to 400; Antoine II (liq) A = 6.1388, B = 1349.142, C = -48.003, dev = 0.1, range = 338 to 373.

2-Bromobutane ±, *sec*-butyl bromide, CA 5787-31-5: Irritant; suspected car-cinogen; flammable, flash point 294; soluble ethanol, ether,   *(continues)*

C_4H_9Br *(continued)*

acetone, chloroform. Tf = 161.25; Tb = 364.35; Vm = 0.1094 at 298.15; Antoine (liq) A = 5.87179, B = 1210.9, C = -51.15, dev = 0.1, range = 281 to 403.

1-Bromo-2-methylpropane, isobutyl bromide, CA 78-77-3: Irritant; suspected carcinogen; flammable, flash point 291; soluble ethanol, ether, acetone, benzene. Tf = 155.75; Tb = 363.65 to 364.15; Vm = 0.1094 at 293.15; Antoine I (liq) A = 5.87377, B = 1212.8, C = -51.15, dev = 1.0, range = 281 to 404; Antoine II (liq) A = 6.5623, B = 1572.161, C = -19.269, dev = 0.1, range = 305 to 363.

2-Bromo-2-methylpropane, *tert*-butyl bromide, CA 507-19-7: Irritant; suspected carcinogen; flammable, flash point 291; soluble most organic solvents. Tf = 256.85; Tb = 345.95; Vm = 0.1130 at 298.15; Antoine (liq) A = 6.52642, B = 1518.8, C = -10.0, dev = 1.0, range = 248 to 346.

C_4H_9BrO MW 153.02

1-Bromo-2-butanol, CA 2482-57-7: Flammable. Antoine (liq) A = 8.10004, B = 2298.97, C = -40.852, dev = 1.0, range = 296 to 418.

C_4H_9Cl MW 92.57

1-Chlorobutane, butyl chloride, CA 109-69-3: Moderately toxic; flammable, flash point 264; soluble ethanol, ether. Tf = 150.05; Tb = 351.65; Vm = 0.1051 at 298.15; Antoine (liq) A = 6.0628, B = 1227.433, C = -49.05, dev = 0.1, range = 257 to 389.

2-Chlorobutane ±, *sec*-butyl chloride, CA 53178-20-1: Moderately toxic; suspected carcinogen; flammable, flash point 258; soluble ethanol, ether, benzene, chloroform. Tm = 141.85; Tb = 341.15; Vm = 0.1063 at 293.15; Antoine (liq) A = 6.06958, B = 1195.8, C = -47.15, dev = 1.0, range = 266 to 377.

1-Chloro-2-methylpropane, isobutyl chloride, CA 513-36-0: Moderately toxic; flammable, flash point below 294; soluble ethanol, ether, acetone, chloroform. Tm = 141.95; Tb = 341.95; Vm = 0.1051 at 293.15; Antoine (liq) A = 6.01854, B = 1176.06, C = -49.025, dev = 1.0, range = 219 to 342.

2-Chloro-2-methylpropane, *tert*-butyl chloride, CA 507-20-0: Moderately toxic; flammable, flash point below 273; decomposes with hot water; soluble ethanol, ether, benzene, chloroform. Antoine I (liq) A = 5.75076, B = 972.6, C = -64.15, dev = 1.0, range = 253 to 358; Antoine II (liq) A = 6.48677, B = 1453.331, C = 0, dev = 1.0, range = 295 to 323.

$C_4H_9ClO_2$ MW 124.57

2-(2-Chloroethoxy) ethanol, CA 628-89-7: Irritant; lachrymator; flammable, flash point 363. Tb = 453.15 to 458.15; Vm = 0.1056 at 293.15; Antoine (liq) A = 8.07173, B = 2702.28, C = -23.702, dev = 1.0, range = 326 to 469.

$C_4H_9ClO_2S$ MW 156.63

Butyl sulfonyl chloride, CA 2386-60-9: Irritant; corrosive; lachrymator; flammable, flash point 352. Tm = 244.15; Vm = 0.1297 at 293.15; Antoine (liq) A = 9.161, B = 3145.6, C = 0, dev = 1.0, range = 253 to 283.

C_4H_9ClS MW 124.63

Ethyl(2-chloroethyl) sulfide, CA 693-07-2: Toxic; stench; mutagenic; soluble chloroform. Tb = 429.65; Vm = 0.1165 at 298.15; Antoine (liq) A = 7.41945, B = 2318.8, C = 0, dev = 5.0, range = 273 to 430.

C_4H_9F MW 76.11

1-Fluorobutane, butyl fluoride, CA 2366-52-1: Flammable; soluble ethanol, ether. Tm = 139.15; Tb = 305.65; Vm = 0.0977 at 293.15; Antoine (liq) A = 6.0830, B = 1081.71, C = -40.35, dev = 1.0, range = 222 to 326.

2-Fluorobutane, sec-butyl fluoride, CA 359-01-3: Flammable. Tm = 151.75; Tb = 298.15 to 299.15; Vm = 0.1007 at 298.15; Antoine (liq) A = 6.17833, B = 1081.1, C = -39.15, dev = 1.0, range = 233 to 329.

2-Fluoro-2-methylpropane, tert-butyl fluoride, CA 353-61-7: Flammable, flash point 261. Tm = 196.15; Tb = 285.25; Antoine (liq) A = 6.12738, B = 1022.6, C = -37.15, dev = 1.0, range = 222 to 315.

C_4H_9FO MW 92.11

4-Fluoro-1-butanol, CA 372-93-0: Flammable; soluble ethanol, ether, acetone. Antoine (liq) A = 10.3111, B = 3343, C = 0, dev = 1.0 to 5.0, range = 323 to 343.

C_4H_9I MW 184.02

1-Iodobutane, butyl iodide, CA 542-69-8: Suspected carcinogen; flammable, flash point 306; decomposes in light; soluble ethanol, ether. Tf = 170.15; Tb = 403.55 to 404.15; Vm = 0.1138 at 293.15; Antoine (liq) A = 5.94752, B = 1358.86, C = -58.95, dev = 0.1, range = 292 to 431.

1-Iodo-2-methylpropane, isobutyl iodide, CA 513-38-2: Irritant; flammable, flash point 285; decomposes in light; soluble ethanol, ether. Tm = 180.15; Tb = 393.15; Vm = 0.1151 at 293.15; Antoine (liq) A = 6.86675, B = 1803.89, C = -22.488, dev = 1.0, range = 256 to 393.

2-Iodo-2-methylpropane, tert-butyl iodide, CA 558-17-8: Irritant; flammable, flash point 266; decomposes in light; soluble ethanol. Tm = 234.95; Tb = 368.15 to 373.15; Vm = 0.1191 at 293.15; Antoine I (sol) A = 10.13066, B = 2600.591, C = 0, dev = 1.0 to 5.0, range = 202 to 223; Antoine II (liq) A = 6.79134, B = 1820.033, C = 0, dev = 1.0, range = 236 to 294.

C_4H_9N MW 71.12

Pyrrolidine, tetrahydropyrrole, CA 123-75-1: Irritant; corrosive; flammable, flash point 276; strong base; soluble water, ethanol, ether, chloroform. Tm = 215.34; Tb = 361.65 to 362.15; Vm = 0.0835 at 295.65; Antoine I (liq) A = 6.04739, B = 1179.072, C = -67.978, dev = 0.02, range = 316 to 394; Antoine II (liq) A = 6.16467, B = 1241.932, C = -61.074, dev = 0.1 to 1.0, range = 273 to 313.

C_4H_9NO MW 87.12

2-Butanone oxime, CA 96-29-7: Flammable; soluble water, ethanol, ether.
Tm = 243.65; Tb = 425.15; Vm = 0.0944 at 293.15; Antoine I (liq) A =
6.7944, B = 1692.8, C = -71.95, dev = 1.0, range = 308 to 425; Antoine II
(liq) A = 7.15637, B = 1819.534, C = -68.499, dev = 1.0, range = 318 to
343.

Butyraldehyde, oxime, CA 110-69-0: Flammable; can explode during vacuum
distillation; soluble ethanol, ether, acetone, benzene. Tm = 243.65; Tb =
425.15; Vm = 0.0944 at 293.15; Antoine (liq) A = 6.31358, B = 1391.561, C =
-101.195, dev = 1.0, range = 313 to 343.

Butyramide, CA 541-35-5: Soluble water, ethanol. Tm = 388.15 to 389.15;
Tb = 489.15; Antoine I (sol) A = 11.8639, B = 4546, C = 0, dev = 1.0 to
5.0, range = 298 to 341; Antoine II (sol) A = 11.7189, B = 4513.3, C = 0,
dev = 1.0, range = 336 to 382; Antoine III (liq) A = 6.64669, B = 1847.6,
C = -105.65, dev = 1.0, range = 397 to 504.

N,N-Dimethylacetamide, CA 127-19-5: Moderately toxic by skin absorption
and inhalation; expected teratogen; flammable, flash point 343; dipolar
solvent; soluble water, ethanol, ether, acetone, benzene. Tm = 253.15;
Tb = 439.25; Vm = 0.0930 at 298.15; Antoine I (liq) A = 7.6349, B = 2360,
C = 0, dev = 1.0 to 5.0, range = 303 to 363; Antoine II (liq) A = 6.14898,
B = 1494.747, C = -78.434, dev = 0.1, range = 371 to 423.

N-Methylpropionamide, CA 1187-58-2: Flammable. Tm = 242.25; Tb = 421.15;
Vm = 0.0936 at 298.15; Antoine (liq) A = 8.0949, B = 2840, C = 0, dev = 1.0
to 5.0, range = 303 to 363.

Morpholine, CA 110-91-8: Irritant; moderately toxic by inhalation; flammable, flash point 310; strong base; solvent for resins, waxes and dyes;
soluble water, most organic solvents. Tm = 268.25; Tb = 402.05; Vm =
0.0872 at 293.15; Antoine I (liq) A = 6.2852, B = 1447.7, C = -63.15, dev =
1.0, range = 317 to 443; Antoine II (liq) A = 6.4863, B = 1547.56, C =
-54.943, dev = 1.0 to 5.0, range = 273 to 318.

$C_4H_9NO_2$ MW 103.12

2-Aminobutyric acid ±, CA 2835-81-6: Soluble water; slightly soluble
ethanol. Tm = 577.15; Antoine (sol) A = 13.4675, B = 6895.258, C = 0,
dev = 1.0 to 5.0, range = 400 to 418.

2-Aminobutyric acid S-form, CA 1492-24-6: Soluble water. Tm = 565.15,
decomposes; Antoine (sol) A = 16.985, B = 8490, C = 0, dev = 5.0, range =
449 to 462.

2-Aminoisobutyric acid, 2-amino-2-methylpropionic acid, CA 62-57-7: Soluble water; moderately soluble ethanol. Tm = 610.15, sealed tube; Antoine
I (sol) A = 13.025, B = 6570, C = 0, dev = 5.0, range = 439 to 462; Antoine
II (sol) A = 13.58471, B = 7009.954, C = 0, dev = 1.0 to 5.0, range = 403
to 424.

sec-Butyl nitrite, CA 924-43-6: Flammable; soluble ethanol, ether, chloroform. Tb = 341.15 to 342.15; Vm = 0.1182 at 293.15; Antoine (liq) A =
6.455, B = 1545, C = 0, dev = 5.0, range = 267 to 287.

tert-Butyl nitrite, CA 540-80-7: Oxidizer; flammable, flash point 260;
slightly soluble water; soluble ethanol, ether, chloroform.   *(continues)*

124

$C_4H_9NO_2$ *(continued)*

Tb = 336.65; Vm = 0.1189 at 293.15; Antoine (liq) A = 6.785, B = 1610, C = 0, dev = 5.0, range = 267 to 337.

Lactic acid, *N*-methylamide: Flammable. Tm = 345.15; Antoine (liq) A = 9.28406, B = 3800.26, C = 0, dev = 1.0, range = 359 to 415.

N-Methylcarbamic acid, ethyl ester, CA 105-40-8: Narcotic; flammable; suspected teratogen; soluble water, ethanol. Tb = 443.15; Vm = 0.1019 at 293.15; Antoine (liq) A = 7.49671, B = 2300.91, C = -24.132, dev = 1.0, range = 299 to 443.

2-Methyl-1-nitropropane, CA 625-74-1: Flammable; soluble ethanol, ether. Tm = 196.33; Tb = 410.15 to 413.15; Vm = 0.1071 at 298.15; Antoine (liq) A = 6.199044, B = 1483.643, C = -61.055, dev = 0.1, range = 347 to 415.

2-Methyl-2-nitropropane, CA 594-70-7: Irritant; flammable, flash point 292; may explode on distillation; soluble ethanol, ether, benzene. Tm = 299.15; Tb = 399.15; Vm = 0.1085 at 301.15; Antoine (liq) A = 6.112625, B = 1396.948, C = -60.161, dev = 0.1, range = 334 to 401.

1-Nitrobutane, CA 627-05-4: Oxidizer; flammable, flash point 320; soluble ethanol, ether, dilute alkali. Tm = 191.82; Tb = 425.92; Vm = 0.1065 at 298.15; Antoine (liq) A = 6.220403, B = 1523.797, C = -64.372, dev = 0.1, range = 357 to 426.

2-Nitrobutane ±, CA 600-24-8: Flammable. Tf = 141.15; Tb = 412.65; Vm = 0.1074 at 298.15; Antoine (liq) A = 6.202795, B = 1494.318, C = -56.608, dev = 0.1, range = 345 to 413.

Propyl carbamate, CA 627-12-3: Flammable; soluble ethanol, ether, acetone. Tm = 333.15; Tb = 469.15; Antoine (liq) A = 7.65329, B = 2375.8, C = -47.656, dev = 1.0, range = 325 to 468.

$C_4H_9NO_3$ MW 119.12

Butyl nitrate, CA 928-45-0: Flammable. Tb = 409.15; Vm = 0.1165 at 303.15; Antoine (liq) A = 6.7339, B = 1742, C = -37.55, dev = 5.0, range = 273 to 406.

Isobutyl nitrate, CA 543-29-3: Flammable; soluble ethanol, ether. Tb = 396.65 to 397.65; Vm = 0.1174 at 293.15; Antoine (liq) A = 6.5489, B = 1597, C = -44.55, dev = 1.0, range = 273 to 397.

$C_4H_9N_3O_2$ MW 131.13

Bis(nitrosoethyl) amine: Irritant; flammable. Antoine (liq) A = 6.81889, B = 2046.41, C = -24.869, dev = 1.0, range = 291 to 450.

C_4H_9P MW 88.09

Allylmethylphosphine, CA 62778-93-2: Irritant. Antoine (liq) A = 7.09902, B = 1799.007, C = 0, dev = 1.0, range = 242 to 291.

Phospholane, CA 3466-00-0: Irritant. Tm = 185.15; Antoine (liq) A = 6.66982, B = 1665.748, C = -20.893, dev = 1.0, range = 257 to 347.

C_4H_{10} MW 58.12

Butane, CA 106-97-8: Extremely flammable, flash point 199; important fuel
and chemical raw material; slightly soluble water; moderately soluble
ethanol, ether, chloroform. Tm = 134.8; Tb = 272.65; Vm = 0.1004 at
293.15; Tc = 425.16; Pc = 3797; Vc = 0.255; Antoine I (liq) A = 5.93386,
B = 935.86, C = -34.42, dev = 0.1, range = 195 to 292; Antoine II (liq) A =
7.3327, B = 1409.73, C = 0, dev = 1.0, range = 135 to 213; Antoine III
(liq) A = 6.07512, B = 1007.247, C = -25.172, dev = 0.1 to 1.0, range = 273
to 321; Antoine IV (liq) A = 6.32267, B = 1161.10, C = -3.107, dev = 0.1,
range = 316 to 383; Antoine V (liq) A = 7.04942, B = 1770.348, C = 84.979,
dev = 0.1 to 1.0, range = 375 to 425.

Isobutane, CA 72-28-5: Extremely flammable, flash point 190; slightly soluble water; moderately soluble ethanol, ether, chloroform. Tm = 113.54;
Tb = 261.43; Vm = 0.1059 at 303.15; Tc = 408.13; Pc = 3648; Vc = 0.263;
Antoine I (liq) A = 6.03538, B = 946.35, C = -26.47, dev = 0.1, range = 186
to 280; Antoine II (liq) A = 7.83572, B = 1470.08, C = 3.99, dev = 1.0,
range = 121 to 187; Antoine III (liq) A = 5.93028, B = 907.164, C = -30.14,
dev = 0.1 to 1.0, range = 263 to 306; Antoine IV (liq) A = 6.26924, B =
1102.296, C = -2.12, dev = 0.1 to 1.0, range = 301 to 366; Antoine V (liq)
A = 6.95371, B = 1648.648, C = 77.939, dev = 0.1 to 1.0, range = 361 to 408.

$C_4H_{10}F_3NOS$ MW 177.18

Sulfur, (diethylaminato)trifluorooxo, CA 26458-94-6: Antoine (liq) A =
7.63974, B = 2587.66, C = 0, dev = 1.0, range = 329 to 354.

$C_4H_{10}F_3NS$ MW 161.18

Sulfur, (N-ethylethanaminato)trifluoro, CA 38078-09-0: Irritant; corrosive;
moisture-sensitive. Vm = 0.1221 at 293.15; Antoine (liq) A = 7.53514, B =
2363.597, C = 0, dev = 1.0, range = 318 to 340.

$C_4H_{10}N_2$ MW 86.14

Piperazine, CA 110-85-0: Irritant; corrosive; causes burns; hygroscopic;
flammable, flash point 354; strong base; very soluble water, ethanol, acetone, glycols. Tm = 382.75; Tb = 421.65; Antoine (sol) A = 10.4976, B =
3267.7, C = -21.928, dev = 1.0, range = 279 to 321.

Trimethylammonium cyanide: Toxic; flammable. Tm = 241.15; Antoine (sol)
A = 10.39227, B = 2350.249, C = 0, dev = 1.0, range = 219 to 236.

$C_4H_{10}O$ MW 74.12

1-Butanol, butyl alcohol, CA 71-36-3: Irritant; moderately toxic; flammable, flash point 310; important solvent; moderately soluble water; soluble ethanol, ether, acetone, benzene. Tm = 183.85; Tb = 390.85; Vm =
0.0915 at 293.15; Tc = 562.93; Pc = 4413; Vc = 0.275; Antoine I (liq) A =
6.41661, B = 1264.515, C = -104.202, dev = 0.02, range = 376 to 399;
Antoine II (liq) A = 6.54172, B = 1336.026, C = -96.348, dev = 0.02, range
= 323 to 413; Antoine III (liq) A = 7.05559, B = 1738.4, C = -46.544, dev =
1.0, range = 413 to 550; Antoine IV (liq) A = 8.9241, B = 2697, C = 0,
dev = 1.0, range = 209 to 251; Antoine V (liq) A = 6.41594, B = 1264.106,
C = -104.251, dev = 0.02, range = 376 to 397; Antoine VI (liq) A = 6.54723,
B = 1339.093, C = -96.03, dev = 0.1, range = 351 to 397;   *(continues)*

$C_4H_{10}O$ *(continued)*

Antoine VII (liq) A = 6.30192, B = 1197.531, C = -112.14, dev = 0.1 to 1.0, range = 391 to 429; Antoine VIII (liq) A = 6.20010, B = 1136.475, C = -119.98, dev = 1.0, range = 415 to 501; Antoine IX (liq) A = 6.82849, B = 1671.332, C = -38.175, dev = 1.0, range = 497 to 563.

2-Butanol, *sec*-butyl alcohol ±, CA 15892-23-6: Moderately toxic; flammable, flash point 297; old samples may explode on distillation; very soluble water; soluble ethanol, ether, acetone, benzene. Tm = 158.45; Tb = 372.65; Vm = 0.0919 at 293.15; Tc = 536.01; Pc = 4194; Vc = 0.268; Antoine I (liq) A = 6.26823, B = 1126.887, C = -108.291, dev = 0.02, range = 359 to 381; Antoine II (liq) A = 6.34976, B = 1169.754, C = -103.388, dev = 0.02, range = 303 to 403; Antoine III (liq) A = 5.74369, B = 735.87, C = -176.795, dev = 5.0, range = 372 to 524; Antoine IV (liq) A = 7.50959, B = 1751.931, C = -52.906, dev = 1.0, range = 210 to 303; Antoine V (liq) A = 6.2663, B = 1125.853, C = -108.414, dev = 0.02, range = 359 to 380; Antoine VI (liq) A = 6.35314, B = 1171.484, C = -103.199, dev = 0.02 to 0.1, range = 340 to 379; Antoine VII (liq) A = 6.18643, B = 1083.25, C = -113.557, dev = 0.1 to 1.0, range = 368 to 404; Antoine VIII (liq) A = 6.12622, B = 1050.17, C = -117.808, dev = 1.0, range = 395 to 485; Antoine IX (liq) A = 6.61842, B = 1439.696, C = -55.524, dev = 1.0, range = 476 to 536.

Diethyl ether, CA 60-29-7: Irritant; narcotic; flammable, flash point 228; forms explosive peroxides; important laboratory solvent; moderately soluble water; soluble most organic solvents. Tm = 156.85; Tb = 307.63; Vm = 0.1056 at 303.15; Tc = 466.7; Pc = 3638; Vc = 0.280; Antoine I (liq) A = 6.02962, B = 1051.432, C = -46.284, dev = 0.02, range = 286 to 329; Antoine II (liq) A = 6.05115, B = 1062.409, C = -44.967, dev = 0.1, range = 250 to 329; Antoine III (liq) A = 6.30714, B = 1236.75, C = -20.11, dev = 1.0, range = 307 to 457; Antoine IV (liq) A = 6.05933, B = 1067.576, C = -44.217, dev = 0.1, range = 305 to 360; Antoine V (liq) A = 6.37811, B = 1276.822, C = -14.869, dev = 0.1, range = 351 to 420; Antoine VI (liq) A = 6.98097, B = 1794.569, C = 57.993, dev = 0.1, range = 417 to 467.

Isopropyl methyl ether, 2-methoxypropane, CA 598-53-8: Flammable; forms explosive peroxides. Tb = 305.65; Vm = 0.1024 at 288.15; Antoine I (liq) A = 6.046, B = 1054.063, C = -43.038, dev = 0.02 to 0.1, range = 250 to 325; Antoine II (liq) A = 6.02934, B = 1045.588, C = -44.059, dev = 0.1, range = 260 to 325.

2-Methyl-1-propanol, isobutyl alcohol, CA 78-83-1: Moderately toxic; eye irritant; flammable, flash point 301; moderately soluble water; soluble ethanol, ether, acetone. Tm = 165.15; Tb = 381.25; Vm = 0.0924 at 293.15; Tc = 547.73; Pc = 4295; Vc = 0.272; Antoine I (liq) A = 6.34528, B = 1190.463, C = -106.712, dev = 0.02, range = 369 to 389; Antoine II (liq) A = 6.49241, B = 1271.027, C = -97.758, dev = 0.02, range = 313 to 411; Antoine III (liq) A = 7.05055, B = 1511.48, C = -81.634, dev = 1.0 to 5.0, range = 381 to 524; Antoine IV (liq) A = 9.8507, B = 2875, C = 0, dev = 1.0, range = 202 to 243; Antoine V (liq) A = 6.34606, B = 1190.8481, C = -106.673, dev = 0.02, range = 369 to 389; Antoine VI (liq) A = 6.50104, B = 1275.669, C = -97.269, dev = 0.1, range = 342 to 389; Antoine VII (liq) A = 6.27047, B = 1147.676, C = -111.933, dev = 0.1 to 1.0, range = 383 to 416; Antoine VIII (liq) A = 6.14833, B = 1077.094, C = -121.099, dev = 1.0, range = 401 to 493; Antoine IX (liq) A = 6.70286, B = 1525.5, C = -50.929, dev = 1.0, range = 483 to 548.

2-Methyl-2-propanol, *tert*-butyl alcohol, CA 75-65-0: Irritant; moderately toxic; flammable, flash point 284; soluble water, ethanol, ether. Tm = 298.65; Tb = 355.65; Vm = 0.0955 at 303.15; Tc = 506.2; Pc = 3972; Vc = 0.275; Antoine I (liq) A = 6.22619, B = 1042.416, C = -108.5, *(continues)*

$C_4H_{10}O$ *(continued)*

dev = 0.02, range = 347 to 363; Antoine II (liq) A = 6.35045, B = 1104.341, C = -101.315, dev = 0.02, range = 299 to 375; Antoine III (liq) A = 6.27388, B = 989.74, C = -124.966, dev = 1.0, range = 356 to 480; Antoine IV (liq) A = 6.23125, B = 1044.891, C = -108.211, dev = 0.02, range = 347 to 363; Antoine V (liq) A = 6.35498, B = 1106.556, C = -101.071, dev = 0.02 to 0.1, range = 329 to 363; Antoine VI (liq) A = 6.09542, B = 975.944, C = -116.864, dev = 0.1 to 1.0, range = 357 to 461; Antoine VII (liq) A = 6.87411, B = 1577.41, C = -24.596, dev = 1.0, range = 453 to 506.

Methyl propyl ether, CA 557-17-5: Anesthetic; flammable, flash point below 253; forms explosive peroxides; moderately soluble water; soluble ethanol, ether, acetone. Tb = 312.05; Vm = 0.1004 at 293.15; Antoine I (liq) A = 4.63212, B = 478.29, C = -130.18, dev = 0.1 to 1.0, range = 273 to 321; Antoine II (liq) A = 6.02543, B = 1071.218, C = -45.218, dev = 0.1, range = 253 to 328; Antoine III (liq) A = 6.22322, B = 1193.578, C = -28.564, dev = 0.1 to 1.0, range = 325 to 407; Antoine IV (liq) A = 7.4099, B = 2256.879, C = 112.962, dev = 1.0, range = 401 to 476.

$C_4H_{10}O_2$ MW 90.12

1,3-Butanediol ±, CA 18826-95-4: Irritant; flammable, flash point 394; very hygroscopic; soluble water, ethanol, acetone. Tm = 196.15; Tb = 480.65; Vm = 0.0909 at 313.15; Antoine (liq) A = 7.15953, B = 2005, C = -92.74, dev = 1.0, range = 362 to 483.

1,4-Butanediol, CA 110-63-4: Moderately toxic; flammable, flash point 394; soluble water, ethanol, acetone. Tm = 293.35; Tb = 501.15; Vm = 0.0888 at 298.15; Tc = 667, calculated; Pc = 4880, calculated; Vc = 0.297, calculated; Antoine (liq) A = 7.53422, B = 2292.1, C = -86.69, dev = 1.0, range = 380 to 510.

2,3-Butanediol ±, CA 6982-25-8: Hygroscopic; flammable, flash point 358; soluble water, ethanol, ether, acetone. Tm = 280.75; Tb = 455.65; Vm = 0.0898 at 293.15; Antoine (liq) A = 6.74297, B = 1675.5, C = -102.83, dev = 1.0, range = 348 to 457.

2,3-Butanediol *meso*, CA 5341-95-7: Hygroscopic; flammable, flash point 358; soluble water, ethanol, ether. Tm = 307.55; Tb = 454.85; Antoine (liq) A = 8.26902, B = 2852.58, C = 0, dev = 1.0, range not specified.

tert-Butylhydroperoxide, CA 75-91-2: Irritant; flammable, flash point below 300; stable to about 348; may explode on distillation; catalyst for polymerization reactions; soluble water, ethanol, ether, chloroform. Tm = 277.15 to 277.65; Tb = 362.15, decomposes; Vm = 0.1006 at 293.15; Antoine (liq) A = 8.469, B = 2507.3, C = 0, dev = 5.0, range not specified.

Diethylperoxide, CA 628-37-5: Flammable; may explode from heating or shock; slightly soluble water; soluble ethanol, ether. Tm = 205.15; Tb = 335.15 to 336.15; Vm = 0.1094 at 293.15; Antoine (liq) A = 6.481, B = 1517, C = 0, dev = 5.0, range = 253 to 333.

1,1-Dimethoxyethane, dimethyl acetal, CA 534-15-6: Moderately toxic; flammable, flash point 274; soluble water, ethanol, ether, acetone, chloroform. Tm = 159.95; Tb = 337.65; Vm = 0.1058 at 293.15; Antoine (liq) A = 5.7449, B = 1002.6, C = -69.54, dev = 1.0, range = 273 to 333.

1,2-Dimethoxyethane, ethylene glycol, dimethyl ether, CA 110-71-4: moderately irritant; flammable, flash point 271; may form explosive *(continues)*

$C_4H_{10}O_2$ *(continued)*

peroxides; soluble water, ethanol, ether, light hydrocarbons. Tm = 213.15; Tb = 356.15; Vm = 0.1048 at 306.15; Antoine I (liq) A = 6.4183, B = 1380.5, C = -44.93, dev = 1.0, range = 238 to 363; Antoine II (liq) A = 6.06187, B = 1229.09, C = -57.16, dev = 0.1 to 1.0, range = 238 to 298.

2-Ethoxyethanol, ethylene glycol, monoethyl ether, CA 110-80-5: Irritant; moderately toxic; flammable, flash point 312; may form explosive peroxides; solvent for lacquers, oils; soluble water, ethanol, ether, acetone. Tf = 203.15; Tb = 409.15; Vm = 0.0981 at 298.15; Antoine (liq) A = 6.944, B = 1801.9, C = -70.15, dev = 1.0, range = 336 to 408.

$C_4H_{10}O_2S$ MW 122.18

Bis(2-hydroxyethyl) sulfide, 2,2'-thiodiethanol, CA 111-48-8: Irritant; flammable, flash point 433; soluble water, ethanol. Tm = 257.15; Tb = 556.15; Vm = 0.1033 at 293.15; Antoine (liq) A = 3.72022, B = 1466.01, C = 6.713, dev = 1.0, range = 368 to 483.

$C_4H_{10}O_3$ MW 106.12

1,2,3-Butanetriol, CA 4435-50-1: Flammable; soluble water, ethanol. Antoine (liq) A = 10.1144, B = 5126.85, C = 94.904, dev = 1.0, range = 375 to 537.

Diethylene glycol, CA 111-46-6: Suspected carcinogen; flammable, flash point 397; soluble water, ethanol, ether, acetone. Tm = 262.7; Tb = 518.15; Vm = 0.0949 at 293.15; Tc = 680, calculated; Pc = 4600, calculated; Vc = 0.312, calculated; Antoine (liq) A = 11.9511, B = 7046.39, C = 190.015, dev = 1.0 to 5.0, range = 364 to 518.

Orthoformic acid, trimethyl ester, CA 149-73-5: Irritant; moderately toxic; flammable, flash point 288; soluble ethanol, ether. Tb = 376.15 to 378.15; Vm = 0.1103 at 298.15; Antoine (liq) A = 7.5134, B = 2039, C = 0, dev = 1.0 to 5.0, range = 273 to 358.

$C_4H_{10}O_3S$ MW 138.18

Diethyl sulfite, CA 623-81-4: Irritant; flammable, flash point 326; soluble ethanol, ether. Tb = 432.15 to 433.15; Vm = 0.1388 at 293.15; Antoine (liq) A = 7.1035, B = 2132.8, C = -12.62, dev = 1.0, range = 283 to 431.

$C_4H_{10}O_4$ MW 122.12

meso-Erythritol, CA 149-32-6: Irritant; very soluble water; soluble pyridine. Tm = 393.15; Tb = 602.15 to 604.15; Antoine I (sol) A = 15.241, B = 7057.1, C = 0, dev = 1.0 to 5.0, range = 379 to 392; Antoine II (liq) A = 9.705, B = 4873.9, C = 0, dev = 1.0 to 5.0, range = 394 to 401.

$C_4H_{10}O_4S$ MW 154.18

Diethyl sulfate, CA 64-67-5: Highly toxic by inhalation and skin absorption; suspected carcinogen; flammable, flash point 377; decomposed by hot water; slightly soluble water, decomposes; soluble ethanol,   *(continues)*

$C_4H_{10}O_4S$ *(continued)*

ether. Tm = 248.75; Tb = 481.15, decomposes; Vm = 0.1309 at 293.15; Antoine (liq) A = 7.62478, B = 2783, C = 13.36, dev = 1.0, range = 413 to 484.

$C_4H_{10}S$ MW 90.18

1-Butanethiol, butyl mercaptan, CA 109-79-5: Moderately toxic; stench; flammable, flash point 275; produced by skunks; slightly soluble water; soluble ethanol, ether. Tm = 157.48; Tb = 371.55; Vm = 0.1077 at 298.15; Antoine (liq) A = 6.05011, B = 1279.95, C = -55.132, dev = 0.02, range = 323 to 409.

2-Butanethiol ±, *sec*-butyl mercaptan, CA 513-53-1: Irritant; stench; flammable, flash point 250; slightly soluble water; soluble ethanol, ether. Tm = 133; Tb = 357.15 to 358.15; Vm = 0.1094 at 298.15; Antoine (liq) A = 6.00882, B = 1228.335, C = -51.282, dev = 0.02, range = 310 to 395.

Diethyl sulfide, CA 352-93-2: Moderately toxic; stench; flammable, flash point 264; slightly soluble water; soluble ethanol, ether. Tf = 171.1; Tb = 365.15; Vm = 0.1077 at 293.15; Antoine (liq) A = 6.04973, B = 1256.013, C = -54.664, dev = 0.02, range = 318 to 396.

Methyl isopropyl sulfide, CA 1551-21-9: Moderately toxic; stench; flammable; soluble ethanol, ether, acetone. Tm = 171.67; Tb = 366.15 to 368.15; Vm = 0.1088 at 293.15; Antoine (liq) A = 6.03231, B = 1235.05, C = -51.163, dev = 0.1, range = 298 to 368.

2-Methyl-1-propanethiol, isobutyl mercaptan, CA 513-44-0: Moderately toxic; stench; flammable, flash point 264; slightly soluble water; soluble ethanol, ether. Tf = 194.15; Tb = 361.15; Vm = 0.1081 at 293.15; Antoine (liq) A = 6.00927, B = 1235.697, C = -52.99, dev = 0.02, range = 314 to 399.

2-Methyl-2-propanethiol, *tert*-butyl mercaptan, CA 75-66-1: Moderately toxic; stench; flammable, flash point below 244; slightly soluble water; soluble ethanol, ether. Tm = 272.65; Tb = 336.85 to 337.35; Vm = 0.1135 at 298.15; Antoine (liq) A = 5.90891, B = 1113.534, C = -52.07, dev = 0.02, range = 292 to 373.

Methyl propyl sulfide, CA 3877-15-4: Moderately toxic; stench; flammable; soluble ethanol, ether, acetone. Tm = 160.15; Tb = 368.65; Vm = 0.1071 at 293.15; Antoine (liq) A = 6.07406, B = 1280.92, C = -53.84, dev = 0.1, range = 308 to 374.

$C_4H_{10}S_2$ MW 122.24

1,4-Butanedithiol, CA 1191-08-8: Moderately toxic; stench; flammable, flash point 343; soluble ethanol. Tm = 219.25; Tb = 468.15 to 469.15; Vm = 0.1175 at 293.15; Antoine (liq) A = 6.1944, B = 1636.95, C = -77.95, dev = 1.0, range = 347 to 469.

Diethyl disulfide, CA 110-81-6: Irritant; moderately toxic; stench; flammable, flash point 313; slightly soluble water. Tm = 171.65; Tb = 427.15; Vm = 0.1231 at 293.15; Antoine I (liq) A = 6.1434, B = 1513.63, C = -61.303, dev = 0.1, range = 287 to 434; Antoine II (liq) A = 6.09141, B = 1481.007, C = -64.671, dev = 0.02, range = 373 to 431.

$C_4H_{11}N$ MW 73.14

Butyl amine, CA 109-73-9: Irritant; corrosive; toxic by inhalation and
skin absorption; flammable, flash point 261; soluble water, ethanol, ether.
Tm = 224.05; Tb = 351.15; Vm = 0.0998 at 298.15; Tc = 531.9, Pc = 4200;
Vc = 0.313, calculated; Antoine (liq) A = 6.2635, B = 1258.745, C = -54.49,
dev = 0.1, range = 313 to 350.

sec-Butyl amine ±, 2-aminobutane, CA 33966-50-6: Highly irritant; flammable, flash point 264; soluble water, ethanol, ether, acetone. Tm = below
261.15; Tb = 336.15; Vm = 0.1010 at 293.15; Antoine (liq) A = 6.27559, B =
1238.3, C = -46.15, dev = 5.0, range = 264 to 371.

tert-Butyl amine, CA 75-64-9: Irritant; toxic; flammable, flash point 265;
soluble water, ethanol, ether. Tf = 200.5; Tb = 319.55; Vm = 0.1051 at
293.15; Antoine (liq) A = 5.90597, B = 992.175, C = -62.794, dev = 0.1,
range = 292 to 349.

Diethyl amine, CA 109-89-7: Irritant; corrosive; moderately toxic; flammable, flash point 250; soluble water, ethanol, ether. Tf = 223.15; Tb =
328.65; Vm = 0.1029 at 293.15; Tc = 496.65; Pc = 3695; Vc = 0.297; Antoine
I (liq) A = 5.96802, B = 1058.538, C = -61.331, dev = 0.1, range = 302 to
328; Antoine II (liq) A = 5.92687, B = 1028.405, C = -66.206, dev = 1.0,
range = 325 to 437; Antoine III (liq) A = 7.52503, B = 2484.979, C =
131.508, dev = 1.0, range = 431 to 496.

Isobutyl amine, CA 78-81-9: Highly irritant; flammable, flash point 264;
soluble water, ethanol, ether. Tf = 188.55; Tb = 341.15 to 342.15; Vm =
0.1010 at 298.15; Antoine (liq) A = 5.88848, B = 1050.547, C = -70.261,
dev = 0.1, range = 248 to 347.

$C_4H_{11}NO$ MW 89.14

2-(Dimethylamino)ethanol, CA 108-01-0: Irritant; corrosive; lachrymator;
flammable, flash point 314; soluble water, ethanol, ether. Tm = 214.15;
Tb = 408.15; Vm = 0.1005 at 293.15; Antoine I (liq) A = 6.82764, B =
1753.53, C = -43.15, dev = 0.1, range = 350 to 387; Antoine II (liq) A =
6.6682, B = 1697.94, C = -43.15, dev = 1.0, range = 323 to 408.

Propylamine, 3-methoxy, CA 5332-73-0: Irritant; flammable; soluble
methanol, acetone, benzene, chloroform. Tb = 389.15 to 392.15; Vm = 0.1021
at 293.15; Antoine (liq) A = 6.3822, B = 1438.84, C = -62.328, dev = 1.0,
range = 278 to 390.

$C_4H_{11}NO_2$ MW 105.14

Diethanolamine, CA 111-42-2: Irritant; corrosive; flammable, flash point
445; hygroscopic; strong base; acid gas absorbent; soluble water, methanol,
ethanol, acetone. Tm = 301.15; Tb = 541.15, decomposes; Vm = 0.0965 at
303.15; Tc = 715; Pc = 3270; Vc = 0.349; Antoine (liq) A = 7.26044, B =
2326.23, C = -98.907, dev = 0.1, range = 423 to 542.

$C_4H_{11}NO_2S$ MW 137.20

Ethanesulfonamide, N,N-dimethyl, CA 6338-68-7: Irritant; flammable. Tm =
277.65; Antoine (liq) A = 7.49481, B = 2837.483, C = 0, dev = 1.0 to 5.0,
range = 384 to 517.

$C_4H_{11}O_3P$ MW 138.10

Diethyl phosphite, CA 762-04-9: Irritant; moisture-sensitive; flammable, flash point 363. Tb = 460.15 to 461.15; Vm = 0.1286 at 293.15; Antoine (liq) A = 6.2989, B = 1988.5, C = 0, dev = 1.0, range = 338 to 471.

Ethylphosphonic acid, dimethyl ester, CA 6163-75-3: Irritant; flammable. Vm = 0.1252 at 303.15; soluble water, ethanol, benzene. Antoine (liq) A = 4.16476, B = 559.815, C = -211.243, dev = 1.0, range = 333 to 410.

$C_4H_{12}ClN$ MW 109.60

Butylammonium chloride, CA 3858-78-4: Soluble water. Tm = 468.15; Antoine (liq) A = 6.8439, B = 3244.9, C = 0, dev = 1.0, range = 489 to 508.

Diethyl amine, hydrochloride, CA 660-68-4: Soluble water, ethanol, chloroform. Tm = 496.65; Tb = 593.15 to 603.15; Antoine (liq) A = 18.5699, B = 9277.1, C = 0, dev = 1.0, range = 513 to 558.

$C_4H_{12}FN_2OP$ MW 154.12

Dimefox, bis(dimethylamido)fluorophosphate, CA 115-26-4: Toxic pesticide; aqueous solutions are stable; soluble water, most organic solvents. Vm = 0.1382 at 293.15; Antoine (liq) A = 7.66128, B = 2632.2, C = 0, dev = 1.0, range = 312 to 350.

$C_4H_{12}NP$ MW 105.12

Dimethyl(dimethylamino) phosphine: Irritant; flammable. Tm = 178.15; Antoine (liq) A = 6.0346, B = 1303.452, C = -48.971, dev = 1.0, range = 264 to 372.

$C_4H_{12}N_2$ MW 88.15

1,2-Butanediamine ±, CA 4426-48-6: Irritant; flammable. Antoine (liq) A = 6.34573, B = 1528.52, C = -65.468, dev = 0.1, range = 251 to 293.

1,2-Propanediamine, 2-methyl, CA 811-93-8: Irritant; flammable, flash point 296. Tm = 256.1; Tb = 332.15; Vm = 0.1048 at 293.15; Antoine (liq) A = 6.35027, B = 1460.03, C = -63.857, dev = 0.1, range = 256 to 293.

Tetramethylhydrazine, CA 6415-12-9: Irritant; flammable. Tm = 155.15; Tb = 345.65 to 346.15; Antoine (liq) A = 5.60644, B = 975.59, C = -75.03, dev = 1.0, range = 290 to 346.

$C_4H_{12}N_2O$ MW 104.15

N-(2-Hydroxyethyl) ethylene diamine, CA 111-41-1: Corrosive; lachrymator; flammable, flash point 402; soluble water, ethanol, acetone. Tb = 517.15; Vm = 0.1011 at 293.15; Antoine (liq) A = 8.3461, B = 3278.6, C = 0, dev = 1.0, range = 383 to 517.

$C_4H_{12}N_2OS$ MW 136.21

Sulfurous diamide, tetramethyl, CA 3768-60-3: Flammable; soluble ether. Tm = 304.15; Tm = 482.15; Antoine (liq) A = 6.55381, B = 2186.977, C = 0, dev = 1.0, range = 320 to 351.

$C_4H_{12}N_2O_2S$ MW 152.21

Sulfamide, tetramethyl, CA 3768-63-6: Flammable; soluble ethanol. Tm = 346.15; Tb = 498.15; Antoine (liq) A = 8.64261, B = 3636.96, C = 53.55, dev = 1.0 to 5.0, range = 358 to 495.

$C_4H_{12}N_2S$ MW 120.21

Sulfoxylic diamide, tetramethyl, CA 2129-20-6: Flammable. Tm = 293.15; Antoine (liq) A = 7.29017, B = 2110.389, C = 0, dev = 1.0, range = 301 to 326.

$C_4H_{13}N_3$ MW 103.17

Dimethylene triamine, CA 111-40-0: Irritant; corrosive; toxic; causes burns; flammable, flash point 371; soluble water, ethanol, light hydrocarbons. Tm = 234.15; Tb = 480.15; Vm = 0.1076 at 293.15; Antoine (liq) A = 7.27732, B = 2257.3, C = -43.15; dev = 1.0, range = 371 to 441.

C_4N_2 MW 76.06

2-Butyne dinitrile, dicyano acetylene, CA 1071-98-3: Toxic; flammable; polymerizes; potentially explosive. Tm = 294.15; Tb = 349.15; Vm = 0.0783 at 298.15; Antoine I (sol) A = 9.1364, B = 2312, C = 0, dev = 1.0, range = 263 to 273; Antoine II (liq) A = 5.63295, B = 1158.1, C = -30.38, dev = 1.0, range = 295 to 350.

$C_5BrF_{12}N$ MW 381.95

N,N-Bis(trifluoromethyl)propylamine, 1,1,2,3,3,3-hexafluoro-2-bromo: Antoine (liq) A = 6.4509, B = 1580, C = 0, dev = 1.0, range = 324 to 351.

C_5ClF_5 MW 190.50

1-Chloro-2,3,4,5,5-pentafluorocyclopentadiene, CA 30221-57-9: Antoine (liq) A = 6.775, B = 1620, C = 0, dev = 1.0 to 5.0, range = 273 to 303.

5-Chloro-1,2,3,4,5-pentafluorocyclopentadiene, CA 30221-56-8: Antoine (liq) A = 6.495, B = 1500, C = 0, dev = 1.0 to 5.0, range = 283 to 323.

$C_5ClF_{10}N$ MW 299.50

Ethanimidoyl chloride, 2,2,2-trifluoro-N-[1,2,2,2-tetrafluoro-1-(trifluoromethyl) ethyl], CA 54120-14-8: Antoine (liq) A = 7.095, B = 1686, C = 0, dev = 1.0 to 5.0, range not specified.

$C_5ClF_{12}N$ MW 337.50

2-Propanamine, N-chloro-1,1,1,2,3,3,3-heptafluoro-N-(pentafluoroethyl), CA 54566-78-8: Antoine (liq) A = 7.085, B = 1759, C = 0, dev = 1.0 to 5.0, range not specified.

$C_5Cl_2F_6$ MW 244.95

  1,2-Dichlorohexafluorocyclopentene, CA 706-79-6: Tm = 167.35; Tb = 363.85;
  Vm = 0.1480 at 293.15; Antoine (liq) A = 5.91394, B = 1163.9, C = -66.15,
  dev = 0.1 to 1.0, range not specified.

$C_5Cl_2F_9N$ MW 315.95

  Ethanamine, 1,1-dichloro-2,2,2-trifluoro-N-[2,2,2-trifluoro-1-(trifluoromethyl) ethylidene], CA 54566-77-7: Antoine (liq) A = 7.095, B = 1841,
  C = 0, dev = 1.0 to 5.0, range not specified.

$C_5Cl_3F_7$ MW 299.40

  1-Pentene, 1,1,5-trichloro-2,3,3,4,4,5,5-heptafluoro, CA 16327-67-6:
  Antoine (liq) A = 6.8824, B = 1938.1, C = 0, dev not specified, range not
  specified.

$C_5Cl_4F_8$ MW 353.85

  Pentane, 1,1,1,5-tetrachlorooctafluoro, CA 678-25-1: Antoine (liq) A =
  6.4651, B = 1914.0, C = 0, dev not specified, range not specified.

$C_5Cl_5F_7O$ MW 386.31

  (1,1,2-Trifluoro-2,2-dichloroethyl)(2,2,3,3-tetrafluoro-1,1,3-trichloro-
  propyl) ether, CA 61196-11-0: Antoine (liq) A = 5.99459, B = 1478.228, C =
  -78.894, dev = 0.02, range = 362 to 449.

C_5Cl_6 MW 272.77

  Hexachlorocyclopentadiene, CA 77-47-4: Highly toxic; corrosive; Tm =
  284.49; Tb = 512.15; Vm = 0.1603 at 298.15; Antoine (liq) A = 7.4813, B =
  2804.32, C = 0, dev = 1.0 to 5.0, range = 335 to 512.

C_5F_5N MW 169.05

  Perfluoropyridine, CA 700-16-3: Irritant; flammable, flash point 296; weak
  base. Tm = 231.65; Tb = 356.45; Vm = 0.1098 at 293.15; Antoine (liq) A =
  7.325, B = 1898, C = 0, dev = 5.0, range = 273 to 363.

C_5F_8 MW 212.04

  Perfluoro-1,2-pentadiene, CA 21972-01-0: Readily dimerizes. Antoine (liq)
  A = 6.625, B = 1363, C = 0, dev = 5.0, range = 262 to 276.

C_5F_9N MW 245.05

  1-Propylamine, 3,3,3-trifluoro-N,N-bis(trifluoromethyl), CA 19451-91-3:
  Antoine (liq) A = 6.4059, B = 1302, C = 0, dev = 1.0 to 5.0, range = 277
  to 293.

                                                                   (continues)

134

C_5F_9N *(continued)*

Pyridine, 2,3,4,5-tetrahydrononafluoro, CA 714-37-4: Antoine (liq) A = 6.8889, B = 1532, C = 0, dev = 1.0 to 5.0, range = 249 to 310.

C_5F_9NO
MW 261.05

Acetamide, 2,2,2-trifluoro-N-[2,2,2-trifluoro-1-(trifluoromethyl)ethyl-idene], CA 52225-57-7: Antoine (liq) A = 7.275, B = 1678, C = 0, dev = 1.0 to 5.0, range not specified.

$2H$-1,2-Oxazine, 3,3,4,5,6,6-hexafluoro-3,6-dihydro-2-(trifluoromethyl), CA 4827-67-2: Antoine (liq) A = 7.0357, B = 1638, C = 0, dev = 1.0, range = 263 to 323.

$C_5F_9NO_3S$
MW 325.11

1-Butanesulfonyl isocyanate, nonafluoro, CA 34805-64-6: Tm = 258.15; Antoine (liq) A = 5.68699, B = 1058.78, C = -113.641, dev = 1.0 to 5.0, range = 309 to 401.

C_5F_{10}
MW 250.04

Perfluorocyclopentane, CA 376-77-2: Tm = 283.55; Antoine I (sol) A = 7.70395, B = 1675.544, C = 0, dev = 0.1, range = 229 to 281; Antoine II (liq) A = 5.08451, B = 597.4, C = -101.61, dev = 0.1, range = 285 to 297.

$C_5F_{10}NP$
MW 295.02

Bis(pentafluoroethyl)phosphinocyanide, CA 35449-90-2: Antoine (liq) A = 6.9169, B = 1725, C = 0, dev = 1.0 to 5.0, range not specified.

$C_5F_{10}N_2O_2$
MW 310.05

Glutaramide, decafluoro, CA 32822-52-9: Antoine (liq) A = 7.075, B = 1865, C = 0, dev = 1.0 to 5.0, range not specified.

Piperidine, 1-nitrodecafluoro, CA 1840-07-9: Antoine (liq) A = 6.3019, B = 1545, C = 0, dev = 1.0 to 5.0, range = 283 to 343.

$C_5F_{10}O_2$
MW 282.04

Carbonofluoridic acid, 2,2,2-trifluoro-1,1-bis(trifluoromethyl)ethyl ester, CA 55064-79-4: Antoine (liq) A = 7.278, B = 1683.2, C = 0, dev = 1.0, range = 275 to 305.

$C_5F_{10}O_3$
MW 298.04

Carbonofluoridoperoxoic acid, 2,2,2-trifluoro-1,1-bis(trifluoromethyl)ethyl ester, CA 64957-47-7: Antoine (liq) A = 7.3904, B = 1798.9, C = 0, dev = 1.0, range not specified.

$C_5F_{10}O_6S_2$ MW 410.16

Octafluorocyclopentanediol, bis(fluorosulfate): Antoine (liq) A = 6.1399, B = 1395.1, C = -83.22, dev = 1.0, range = 334 to 423.

$C_5F_{11}N$ MW 283.04

Perfluoropiperidine, CA 836-77-1: Tm = 274.12; Antoine (liq) A = 5.97088, B = 1055.979, C = -56.442, dev = 0.02, range = 302 to 355.

Pyrrolidine, octafluoro-1-trifluoromethyl, CA 2344-10-7: Antoine (liq) A = 6.9679, B = 1535, C = 0, dev = 1.0 to 5.0, range = 249 to 306.

$C_5F_{11}NO$ MW 299.04

Acetamide, N,2,2,2-tetrafluoro-N-[1,2,2,2-tetrafluoro-1-(trifluoromethyl) ethyl], CA 52225-65-7: Antoine (liq) A = 7.125, B = 1701, C = 0, dev = 1.0 to 5.0, range not specified.

C_5F_{12} MW 288.04

Perfluoro-2-methylbutane, CA 594-91-2: Tm = 175.35; Antoine (liq) A = 6.11062, B = 1051.72, C = -46.95, dev = 1.0, range = 228 to 308.

Perfluoropentane, CA 678-26-2: Tm = 147.15; Tb = 302.35; Vm = 0.1778 at 293.15; Tc = 422.15; Pc = 2040; Vc = 0.512, calculated; Antoine (liq) A = 6.30553, B = 1131.8, C = -38.92, dev = 0.1 to 1.0, range = 221 to 303.

$C_5F_{12}N_2$ MW 316.05

Diazene, [2,2,2-trifluoro-1,1-bis(trifluoromethyl)ethyl](trifluoromethyl), CA 53684-06-3: Antoine (liq) A = 6.545, B = 1399, C = 0, dev = 1.0 to 5,0, range not specified.

$C_5F_{12}O_2$ MW 320.03

Bis(pentafluoroethoxy)difluoromethane, CA 20822-11-1: Tm = about 89.15; Antoine (liq) A = 5.59518, B = 854.939, C = -76.006, dev = 1.0, range = 246 to 299.

$C_5F_{12}O_2S$ MW 352.09

Trifluoromethanesulfinic acid, 2,2,2-trifluoro-1,1-bis(trifluoromethyl) ethyl ester, CA 52225-54-4: Antoine (liq) A - 7.715, B = 2025, C = 0, dev = 1.0 to 5.0, range not specified.

$C_5F_{12}O_3$ MW 336.03

Hypofluorous acid, difluoro[(2,2,2-trifluoro-1,1-bis[trifluoromethyl]ethyl) dioxy]methyl ester, CA 64957-50-2: Antoine (liq) A = 10.2449, B = 2859, C = 0, dev = 1.0 to 5.0, range not specified.

$C_5F_{12}O_4S$ MW 384.09

Sulfur, pentafluoro(2,2,3,3,4,4,5-heptafluoro-5-oxopentaneperoxoato),
CA 60672-63-1: Antoine (liq) A = 7.2659, B = 2040.4, C = 0, dev = 1.0 to
5.0, range not specified.

$C_5F_{13}N$ MW 321.04

N-(Trifluoromethyl)bis(pentafluoroethyl) amine, CA 758-48-5: Antoine (liq)
A = 5.93975, B = 1019.992, C = -60.317, dev = 0.02, range = 298 to 319.

$C_5F_{13}NS$ MW 353.10

Sulfilimine, N-[1,2,2-tetrafluoro-1-(trifluoromethyl)ethyl]-S,S-bis(trifluoromethyl), CA 37826-44-1: Antoine (liq) A = 6.90522, B = 1907.686,
C = 26.384, dev = 1.0, range = 314 to 360.

$C_5F_{14}N_2O$ MW 370.05

Hydrazine, 1-[difluoro(trifluoromethoxy)methyl]-1,2,2-tris(trifluoromethyl),
CA 17636-89-4: Antoine (liq) A = 7.3799, B = 1811, C = 0, dev = 1.0 to 5.0,
range = 302 to 331.

Methanediamine, 1,1-difluoro-N-(trifluoromethoxy)-N,N',N'-tris(trifluoromethyl), CA 17636-88-3: Antoine (liq) A = 7.3579, B = 1759, C = 0, dev =
1.0 to 5.0, range = 282 to 323.

$C_5F_{14}OS$ MW 374.09

Pentafluoro[(nonafluorocyclopentyl)oxy] sulfur: Antoine (liq) A = 6.50383,
B = 1447.486, C = -39.009, dev = 1.0, range = 300 to 361.

$C_5F_{15}NS$ MW 391.10

Sulfur, difluoro[1,1,1,2,3,3,3-heptafluoro-2-propanaminato]bis(trifluoromethyl), CA 65844-10-2: Antoine (liq) A = 6.575, B = 1718, C = 0, dev =
1.0 to 5.0, range not specified.

$C_5F_{15}P_5$ MW 499.90

1,2,3,4,5-Pentakis(trifluoromethyl)pentaphospholane: Tm = 240.15; Antoine
(liq) A = 6.40021, B = 1742.161, C = -65.869, dev = 1.0, range = 319 to 435.

$C_5HClF_8O_2$ MW 280.50

Trifluoroacetic acid, 2,2,2-trifluoro-(chlorodifluoromethyl)ethyl ester,
CA 52225-55-5: Antoine (liq) A = 7.695, B = 1921, C = 0, dev = 1.0 to 5.0,
range not specified.

C_5HF_9 MW 232.05

Nonafluorocyclopentane, CA 376-65-8: Antoine (liq) A = 6.06239, B =
1047.552, C = -53.556, dev = 0.02 to 0.1, range = 289 to 348.

C_5HF_9IN MW 372.96

Propenylamine, 3,3,3-trifluoro-1-iodo-*N*,*N*-bis(trifluoromethyl), *cis*,
CA 20257-34-5: Antoine (liq) A = 6.4519, B = 1636, C = 0, dev = 1.0 to
5.0, range = 343 to 366.

Propenylamine, 3,3,3-trifluoro-1-iodo-*N*,*N*-bis(trifluoromethyl), *trans*,
CA 20257-35-6: Antoine (liq) A = 6.9239, B = 1826, C = 0, dev = 1.0 to
5.0, range = 345 to 368.

$C_5HF_9O_2$ MW 264.05

Trifluoroacetic acid, 2,2,2-trifluoro-1-(trifluoromethyl)ethyl ester,
CA 42031-15-2: Antoine (liq) A = 6.635, B = 1487, C = 0, dev = 1.0 to 5.0,
range not specified.

$C_5HF_{10}N$ MW 265.05

Piperidine, 2,2,3,3,4,4,5,5,6,6-decafluoro, CA 559-31-9: Antoine (liq) A =
6.9289, B = 1706, C = 0, dev = 1.0, range = 273 to 313.

$C_5HF_{10}NO$ MW 281.05

Acetamide, 2,2,2-trifluoro-*N*-[1,2,2,2-tetrafluoro-1-(trifluoromethyl)ethyl],
CA 52225-63-5: Antoine (liq) A = 8.035, B = 2211, C = 0, dev = 1.0 to 5.0,
range not specified.

$C_5HF_{12}N$ MW 303.05

2-Propanamine, 1,1,1,2,3,3,3-heptafluoro-*N*-(pentafluoroethyl),
CA 54566-8-02: Antoine (liq) A = 7.135, B = 1667, C = 0, dev = 1.0 to 5.0,
range not specified.

C_5HN_3 MW 103.08

Ethylenetricarbonitrile, CA 997-76-2: Antoine (liq) A = 4.17653, B =
910.337, C = -159, dev = 1.0, range = 313 to 343.

$C_5H_2BrF_8N$ MW 307.97

Allylamine, 2-bromo-3,3-difluoro-*N*,*N*-bis(trifluoromethyl), CA 19451-93-5:
Antoine (liq) A = 6.7599, B = 1765, C = 0, dev = 1.0 to 5.0, range = 336
to 367.

$C_5H_2F_9N$ MW 247.06

Propenylamine, 3,3,3-trifluoro-*N*,*N*-bis(trifluoromethyl), *trans,*
CA 19451-88-8: Antoine (liq) A = 6.5849, B = 1471, C = 0, dev = 1.0 to
5.0, range = 287 to 319.

$C_5H_2F_9NOS$ MW 295.12

Ethanimidic acid, 2,2,2-trifluoro-*N*-[(trifluoromethyl)thio]-2,2,2-tri-
fluoroethyl ester, CA 62067-07-6: Antoine (liq) A = 7.045, B = 1878.3,
C = 0, dev = 1.0 to 5.0, range not specified.

$C_5H_3BrF_9N$ MW 327.98

Propylamine, 2-bromo-3,3,3-trifluoro-N,N-bis(trifluoromethyl),
CA 19451-92-4: Antoine (liq) A = 6.8569, B = 1785, C = 0, dev = 1.0 to
5.0, range = 342 to 365.

$C_5H_3F_6N$ MW 191.08

1-Propynylamine, N,N-bis(trifluoromethyl), CA 25237-11-0: Antoine (liq)
A = 7.1679, B = 1625, C = 0, dev = 1.0 to 5.0, range = 295 to 312.

$C_5H_3F_9N_2OS$ MW 310.14

Methanesulfonimidamide, 1,1,1-trifluoro-N'-methyl-N-[2,2,2-trifluoro-1-
(trifluoromethyl)ethylidene], CA 62609-63-6: Antoine (liq) A = 6.505, B =
1922, C = 0, dev = 5.0, range not specified.

$C_5H_3F_9O_2S$ MW 298.12

Trifluoromethanesulfinic acid, 2,2,2-trifluoro-1-methyl-1-(trifluoromethyl)
ethyl ester, CA 52225-51-1: Antoine (liq) A = 6.635, B = 1784, C = 0,
dev = 5.0, range not specified.

$C_5H_4BrF_6N$ MW 271.99

Propenylamine, 2-bromo-N,N-bis(trifluoromethyl), *cis*, CA 25273-47-6:
Antoine (liq) A = 6.9549, B = 1846, C = 0, dev = 1.0 to 5.0, range = 346
to 367.

Propenylamine, 2-bromo-N,N-bis(trifluoromethyl), *trans*, CA 25273-48-7:
Antoine (liq) A = 6.8479, B = 1742, C = 0, dev = 1.0 to 5.0, range = 336
to 360.

C_5H_4BrN MW 158.00

3-Bromopyridine, CA 626-55-1: Irritant; highly toxic; flammable, flash
point 324; moderately soluble water. Tm = 245.85; Tb = 446.15 to 447.15;
Vm = 0.0968 at 283.15; Antoine (liq) A = 6.85249, B = 2090.41, C = -15.439,
dev = 1.0, range = 289 to 447.

C_5H_4ClN MW 113.55

2-Chloropyridine, CA 109-09-1: Irritant; highly toxic; flammable, flash
point 338; soluble ethanol, ether. Tb = 443.15; Vm = 0.0942 at 288.15;
Antoine (liq) A = 6.88198, B = 2085.18, C = -15.951, dev = 1.0, range = 286
to 444.

$C_5H_4F_7I$ MW 323.98

1,1,1,2,2,3,3-Heptafluoro-5-iodopentane, CA 1513-88-8: Tm = 268.15;
Antoine (liq) A = 7.1558, B = 2019.8, C = 0, dev = 1.0, range = 317 to 386.

$C_5H_4F_9N$ MW 249.08

Propylamine, 3,3,3-trifluoro-*N*,*N*-bis(trifluoromethyl), CA 19451-89-9:
Antoine (liq) A = 6.851, B = 1622, C = 0, dev = 1.0 to 5.0, range = 290
to 333.

$C_5H_4N_2$ MW 92.10

2-Methyl-2-butenedinitrile, *cis*, CA 37580-43-1: Tm = 379.45 to 379.65;
Antoine (liq) A = 7.9341, B = 3055.5, C = 0, dev = 1.0, range = 395 to 467.

2-Methyl-2-butenedinitrile, *trans*, CA 37580-44-2: Antoine (liq) A = 7.557,
B = 2504, C = 0, dev = 1.0 to 5.0, range = 339 to 411.

$C_5H_4O_2$ MW 96.09

Furfural, CA 98-01-1: Irritant; moderately toxic; flammable, flash point
333; polymerizes on standing; very soluble water; soluble most organic solvents. Tm = 236.65; Tb = 434.85; Vm = 0.0829 at 293.15; Tc = 657, calculated; Pc = 5512; Vc = 0.252, calculated; Antoine (liq) A = 6.41784, B =
1663.16, C = -57.88, dev = 1.0, range = 357 to 435.

$C_5H_4O_2S$ MW 128.15

2-Thiophene carboxylic acid, CA 527-72-0: Very soluble hot water; very
soluble ethanol, ether, chloroform. Tm = 402.15 to 403.15; Tb = 533.15,
decomposes; Antoine (sol) A = 12.655, B = 5065, C = 0, dev = 1.0, range =
314 to 323.

$C_5H_4O_3$ MW 112.08

Citraconic anhydride, CA 616-02-4: Moisture-sensitive; flammable, flash
point 374; soluble ethanol, ether, acetone. Tm = 281.15; Tb = 486.15; Vm =
0.0899 at 289.15; Antoine (liq) A = 7.01294, B = 2289.59, C = -29.502,
dev = 1.0, range = 320 to 487.

2-Furancarboxylic acid, CA 88-14-2: Very soluble hot water; soluble ether,
ethanol. Tm = 406.15 to 407.15; Tb = 503.15 to 505.15; Antoine (sol) A =
14.745, B = 5667, C = 0, dev = 5.0, range = 317 to 328.

$C_5H_5F_6NO$ MW 209.09

N,*N*-Bis(trifluoromethyl)allylamine-*N*-oxide, CA 22743-77-7: Antoine (liq)
A = 7.235, B = 1730, C = 0, dev = 5.0, range = 254 to 328.

Vinylamine, 1-methoxy-*N*,*N*-bis(trifluoromethyl), CA 22130-39-8: Antoine
(liq) A = 6.894, B = 1692, C = 0, dev = 1.0 to 5.0, range = 321 to 343.

Vinylamine, 2-methoxy-*N*,*N*-bis(trifluoromethyl), *cis*, CA 22298-35-7:
Antoine (liq) A = 6.634, B = 1700, C = 0, dev = 1.0 to 5.0, range = 341
to 362.

$C_5H_5F_6NO_2$ MW 225.09

N,*N*-Bis(trifluoromethyl)propionamide-*N*-oxide, CA 22743-66-4: Antoine (liq)
A = 7.985, B = 2197, C = 0, dev = 5.0, range = 278 to 361.

C_5H_5N MW 79.10

2,4-Pentadienenitrile, *cis*, CA 2180-69-0: Flammable; soluble ether, acetone. Tm = 209.15; Vm = 0.0926 at 299.15; Antoine (liq) A = 6.5009, B =
1639, C = -43.15, dev = 1.0, range = 318 to 383.

Pyridine, CA 110-86-1: Irritant; toxic by skin absorption and inhalation;
flammable, flash point 293; powerful solvent; soluble water, most organic
solvents. Tm = 231.55; Tb = 388.45; Vm = 0.0809 at 298.15; Tc = 619.95;
Pc = 5630; Vc = 0.254; Antoine I (liq) A = 6.17372, B = 1379.353, C =
-57.456, dev = 0.02, range = 323 to 426; Antoine II (liq) A = 6.30308, B =
1448.781, C = -50.948, dev = 0.1, range = 296 to 353; Antoine III (liq) A =
6.16446, B = 1373.263, C = -58.18, dev = 0.02, range = 348 to 434; Antoine
IV (liq) A = 6.284, B = 1455.584, C = -48.272, dev = 0.1 to 1.0, range =
431 to 558; Antoine V (liq) A = 7.25663, B = 2578.625, C = 115.604, dev =
1.0, range = 552 to 620.

$C_5H_5NO_2$ MW 111.10

2-Cyanoacrylic acid, methyl ester, CA 137-05-3: Toxic; flammable. Antoine
(liq) A = 9.0785, B = 3018.8, C = 0, dev = 1.0, range = 258 to 283.

2-Pyrrole carboxylic acid, CA 634-97-9: Soluble water, ethanol, ether.
Tm = 481.65, decomposes; Antoine (sol) A = 15.725, B = 6633, C = 0, dev =
5.0, range = 349 to 359.

$C_5H_5N_3O$ MW 123.11

Pyrazine carboxamide, CA 98-96-4: Suspected carcinogen; soluble water,
methanol, ethanol, chloroform. Tm = 462.15 to 464.15; Antoine (sol) A =
10.285, B = 4592, C = 0, dev = 5.0, range = 353 to 383.

$C_5H_5N_5$ MW 135.13

Adenine, 6-aminopurine, CA 73-24-5: Expected teratogen; moderately soluble
hot water. Tm = 633.15 to 638.15, decomposes; Antoine (sol) A = 9.562, B =
5705, C = 0, dev = 5.0, range = 448 to 473.

C_5H_6 MW 66.10

1,3-Cyclopentadiene, CA 542-92-7: Toxic vapor; flammable, flash point
below 298; can dimerize explosively at room temperature; forms unstable
peroxides; soluble ethanol, ether, benzene, acetone. Tm = 188.15; Tb =
314.65; Vm = 0.0835 at 303.15; Tc = 507, calculated; Pc = 5150, calculated;
Vc = 0.225, calculated; Antoine (liq) A = 6.16935, B = 1168.2, C = -32.77,
dev = 1.0, range = 291 to 314.

Ethynyl cyclopropane, CA 6746-94-7: Flammable; polymerizes. Antoine (liq)
A = 7.010, B = 1627, C = 0, dev not specified, range = 290 to 320.

C_5H_6ClN MW 115.56

3-Pentenenitrile, 4-chloro, CA 32366-08-8: Flammable; polymerizes.
Antoine (liq) A = 9.70558, B = 3338.666, C = 0, dev = 5.0, range = 349 to
433.

$C_5H_6Cl_2O_2$ MW 169.01

Glutaryl chloride, CA 2873-74-7: Corrosive; lachrymator; flammable, flash
point 379; soluble ether. Tb = 489.15 to 491.15; Vm = 0.1278 at 295.15;
Antoine (liq) A = 7.48697, B = 2574.56, C = -20.82, dev = 1.0, range = 329
to 490.

$C_5H_6F_3NO_3$ MW 185.10

Glycine, *N*-(trifluoroacetyl), methyl ester, CA 383-72-2: Antoine (sol) A =
8.145, B = 2994, C = 0, dev = 5.0, range = 293 to 463.

$C_5H_6F_6N_2S$ MW 240.17

Ethanimidamide, 2,2,2-trifluoro-*N*,*N*-dimethyl-*N*'-[(trifluoromethyl)thio],
CA 62067-11-2: Antoine (liq) A = 7.985, B = 2392.3, C = 0, dev = 1.0 to
5.0, range not specified.

Sulfoxylic diamide, dimethyl[2,2,2-trifluoro-1-(trifluoromethyl)ethylidene],
CA 38005-19-5: Antoine (liq) A = 7.445, B = 2081, C = 0, dev = 1.0 to 5.0,
range not specified.

$C_5H_6F_6O_2S$ MW 244.15

Trifluoromethanesulfinic acid, 2,2,2-trifluoro-1,1-dimethylethyl ester,
CA 52225-49-7: Antoine (liq) A = 6.785, B = 1854, C = 0, dev = 1.0 to 5.0,
range not specified.

$C_5H_6F_6O_5S_2$ MW 324.21

Methanesulfinic acid, trifluoro-2-hydroxy-1,3-propanediyl ester,
CA 61915-97-7: Antoine (liq) A = 5.325, B = 1712, C = 0, dev = 5.0, range
= 333 to 418.

$C_5H_6N_2$ MW 94.12

Dimethylmalanonitrile, CA 7321-55-3: Irritant; flammable; soluble water.
Tm = 304.15 to 305.15; Tb = 442.65; Antoine (liq) A = 6.20280, B = 1573.99,
C = -68.4, dev = 1.0, range = 322 to 413.

Glutaronitrile, CA 544-13-8: Irritant; flammable, flash point above 385;
soluble water, ethanol, chloroform. Tm = 243.65; Tb = 559.15; Vm = 0.0952
at 296.15; Antoine I (liq) A = 6.7318, B = 2431.32, C = -45.286, dev = 1.0,
range = 364 to 560; Antoine II (liq) A = 8.7426, B = 3490, C = 0, dev =
1.0, range = 277 to 303.

C_5H_6O MW 82.10

2-Methylfuran, CA 534-22-5: Highly toxic; flammable, flash point 243;
slightly soluble water; soluble ethanol, ether. Tm = 185.15; Tb = 336.15
to 336.65; Vm = 0.0993 at 293.15; Antoine (liq) A = 5.95585, B = 1107.3,
C = -56.88, dev = 1.0, range = 251 to 338.

142

$C_5H_6O_2$ MW 98.10

2(3H)-Furanone, 5-methyl, CA 591-12-8: Carcinogenic in animals; flammable;
readily isomerizes to 2(5H)-isomer; soluble water. Tm = 291.15; Tb =
440.15; Vm = 0.0899 at 293.15; Antoine (liq) A = 8.98983, B = 3993.752, C =
127.959, dev = 1.0 to 5.0, range = 324 to 442.

2(5H)-Furanone, 5-methyl ±, CA 591-11-7: Flammable; forms a dimer; soluble
water, ethanol, ether. Tm = below 193.15; Tb = 481.15 to 482.15; Vm =
0.0907 at 293.15; Antoine (liq) A = 8.0922, B = 3224.894, C = 48.624, dev =
1.0 to 5.0, range = 356 to 481.

Furfuryl alcohol, CA 98-00-0: Irritant; moderately toxic; flammable, flash
point 348; aqueous solutions decompose on standing; soluble water, ethanol,
ether. Tm = 259.15; Tb = 443.15 to 444.15; Vm = 0.0870 at 296.15; Antoine
(liq) A = 8.81987, B = 3223.12, C = 29.705, dev = 1.0, range = 304 to 443.

3-Pentyn-2-one-5-hydroxy, CA 15441-65-3: Flammable. Antoine (liq) A =
9.2974, B = 3362.6, C = 0, dev = 1.0, range = 273 to 333.

$C_5H_6O_3$ MW 114.10

Glutaric anhydride, CA 108-55-4: Irritant; moderately toxic; moisture-
sensitive; flammable, flash point above 385; soluble water, reacts; moderately soluble ethanol. Tm = 329.15; Antoine (liq) A = 7.89186, B =
3362.14, C = 10.996, dev = 1.0, range = 373 to 560.

Monomethylsuccinate anhydride ±, CA 4100-80-5: Irritant; moisture-
sensitive; flammable, flash point above 385; moderately soluble chloroform.
Tm = 306.15; Tb = 517.15 to 521.15; Antoine (liq) A = 6.60291, B = 2105.73,
C = -62.441, dev = 1.0, range = 342 to 521.

C_5H_6S MW 98.16

2-Methylthiophene, CA 554-14-3: Toxic; flammable, flash point 280; soluble
ethanol, ether, acetone, benzene. Tm = 209.75; Tb = 383.15 to 385.15; Vm =
0.0963 at 293.15; Antoine (liq) A = 6.0598, B = 1324.362, C = -59.038,
dev = 0.1, range = 324 to 391.

3-Methylthiophene, CA 616-44-4: Toxic; flammable, flash point 284; soluble
ethanol, ether, acetone, benzene. Tm = 204.15; Tb = 388.15; Vm = 0.0958 at
293.15; Antoine (liq) A = 6.10694, B = 1361.798, C = -56.548, dev = 0.1,
range = 327 to 399.

$C_5H_7ClO_3$ MW 150.56

Acetic acid, chlorooxo, propyl ester, CA 54166-91-5: Flammable. Antoine
(liq) A = 7.08543, B = 1848.09, C = -51.262, dev = 1.0, range = 282 to 396.

$C_5H_7FO_2$ MW 118.11

Allyl fluoroacetate: Flammable. Tb = 409.15 to 410.15; Antoine (liq) A =
8.3486, B = 2554.3, C = 0, dev = 1.0, range = 273 to 333.

C_5H_7N MW 81.12

Angelic acid, nitrile, CA 20068-02-4: Flammable. Antoine (liq) A =
6.37576, B = 1626.55, C = -40.936, dev = 1.0, range = 265 to 413. *(continues)*

C_5H_7N *(continued)*

2-Ethylacrylonitrile: Flammable. Antoine (liq) A = 6.33791, B = 1563.46, C = -26.287, dev = 1.0, range = 244 to 387.

1-Methylpyrrole, CA 96-54-8: Flammable, flash point 289; soluble ethanol, ether. Tm = 216.15; Tb = 385.15 to 386.15; Vm = 0.0887 at 288.15; Antoine (liq) A = 6.2088, B = 1368.437, C = -60.325, dev = 0.02, range = 321 to 423.

Tiglic acid, nitrile, CA 30574-97-1: Flammable. Antoine (liq) A = 6.18163, B = 1524.04, C = -30.387, dev = 1.0, range = 247 to 395.

C_5H_7NO MW 97.12

Pentanenitrile, 4-oxo, CA 927-56-0: Flammable. Antoine (liq) A = 7.365, B = 2732, C = 0, dev = 5.0, range = 393 to 473.

$C_5H_7NO_2$ MW 113.12

Ethyl cyanoacetate, CA 105-56-6: Irritant; flammable, flash point 383; lachrymator; slightly soluble water; soluble most organic solvents. Tm = 250.65; Tb = 481.15 to 483.15; Vm = 0.1071 at 298.15; Antoine (liq) A = 8.2774, B = 2690.39, C = -50.198, dev = 1.0, range = 340 to 479.

$C_5H_7NO_3$ MW 129.12

2-Pyrrolidenecarboxylic acid, 5-oxo ±, CA 149-87-1: Moderately soluble water. Tm = 451.15 to 452.15; Antoine (sol) A = 13.75981, B = 6956.113, C = 0, dev = 1.0 to 5.0, range = 394 to 416.

C_5H_7NS MW 113.18

Isothiocyanic acid, 3-butenyl ester, CA 34424-44-7: Flammable. Antoine (liq) A = 7.3630, B = 2360, C = 0, dev = 1.0, range = 342 to 443.

Thiazole, 2,4-dimethyl, CA 541-58-2: Flammable; hygroscopic; soluble cold water, less soluble hot; soluble ethanol, ether. Antoine (liq) A = 6.09804, B = 1415.23, C = -72.969, dev = 0.02 to 0.1, range = 357 to 421.

C_5H_8 MW 68.12

Bicyclo [2,1,0] pentane, CA 185-94-4: Flammable. Tb = 318.65; Vm = 0.0862 at 293.15; Antoine (liq) A = 5.95539, B = 1078.706, C = -46.01, dev = 0.02 to 0.1, range = 296 to 315.

Cyclopentene, CA 142-29-0: Moderately toxic; flammable, flash point 244; can polymerize on standing; soluble most organic solvents. Tf = 138.07; Tb = 317.39; Vm = 0.0882 at 293.15; Tc = 506.15; Pc = 4790, calculated; Vc = 0.247, calculated; Antoine (liq) A = 6.01607, B = 1105.926, C = -41.615, dev = 0.02, range = 249 to 318.

3-Methyl-1,2-butadiene, CA 598-25-4: Irritant; flammable, flash point 261; readily polymerizes; soluble most organic solvents. Tm = 159.55; Tb = 313.15 to 313.65; Vm = 0.1001 at 298.15; Antoine I (liq) A = 6.27685, B = 1194.537, C = -33.685, dev = 0.02, range = 227 to 253; Antoine II (liq) A = 6.06840, B = 1103.901, C = -42.26, dev = 0.02, range = 252 to 323.

(continues)

C_5H_8 *(continued)*

2-Methyl-1,3-butadiene, isoprene, CA 78-79-5: Moderately irritant; flammable, flash point 219; forms explosive peroxides in air; readily polymerizes; monomer for both natural and synthetic rubber; soluble most organic solvents. Tf = 127.2; Tb = 307.22; Vm = 0.1016 at 303.15; Antoine I (liq) A = 6.13677, B = 1126.159, C = -34.266, dev = 0.02, range = 221 to 254; Antoine II (liq) A = 6.01054, B = 1071.578, C = -39.637, dev = 0.02, range = 254 to 316.

3-Methyl-1-butyne, isopropyl acetylene, CA 598-23-2: Flammable; readily polymerizes; soluble ethanol, ether. Tm = 183.15; Tb = 301.15; Vm = 0.1023 at 293.15; Antoine (liq) A = 6.0097, B = 1014.81, C = -46.04, dev = 5.0, range = 218 to 320.

Methylenecyclobutane, CA 1120-56-5: Flammable, flash point 239; polymerizes. Tf = 138.47; Tb = 315.15; Vm = 0.0926 at 293.15; Antoine I (liq) A = 6.3291, B = 1362.2, C = 0, dev = 1.0, range = 290 to 316; Antoine II (liq) A = 8.28852, B = 2485.737, C = 83.248, dev = 0.1 to 1.0, range = 292 to 306.

1,2-Pentadiene, ethyl allene, CA 591-95-7: Flammable; polymerizes; soluble ethanol, ether, acetone, benzene. Tm = 135.85; Tb = 317.15 to 318.15; Vm = 0.0984 at 293.15; Antoine I (liq) A = 6.3848, B = 1250.293, C = -31.187, dev = 0.1, range = 231 to 249; Antoine II (liq) A = 6.04296, B = 1104.991, C = -44.299, dev = 0.02, range = 249 to 331.

1,3-Pentadiene, *cis*, CA 1574-41-0: Irritant; flammable, flash point 245; readily polymerizes; soluble ethanol, ether, benzene. Tm = 132.25; Tb = 316.65; Vm = 0.0986 at 293.15; Antoine I (liq) A = 6.31877, B = 1223.602, C = -32.526, dev = 0.02, range = 230 to 255; Antoine II (liq) A = 6.03579, B = 1101.923, C = -43.783, dev = 0.02, range = 255 to 326.

1,3-Pentadiene, *trans*, CA 2004-70-8: Irritant; flammable, flash point 245; readily polymerizes; soluble ethanol, ether, benzene. Tf = 185.75; Tb = 315.15; Vm = 0.1015 at 298.15; Antoine I (liq) A = 6.22702, B = 1185.389, C = -33.744, dev = 0.02, range = 228 to 256; Antoine II (liq) A = 6.03807, B = 1103.84, C = -41.426, dev = 0.02, range = 256 to 324.

1,4-Pentadiene, CA 591-93-5: Flammable, flash point 277; readily polymerizes; soluble most organic solvents. Tf = 124.85; Tb = 298.95 to 299.35; Vm = 0.1031 at 293.15; Antoine I (liq) A = 6.29891, B = 1155.378, C = -28.852, dev = 0.02, range = 216 to 236; Antoine II (liq) A = 5.96033, B = 1017.995, C = -41.689, dev = 0.02, range = 236 to 307.

2,3-Pentadiene ±, CA 28383-16-6: Flammable; polymerizes; soluble most organic solvents. Tm = 147.55; Tb = 323.15; Vm = 0.0980 at 293.15; Antoine I (liq) A = 6.32743, B = 1231.768, C = -35.588, dev = 0.02, range = 234 to 258; Antoine II (liq) A = 6.08706, B = 1126.837, C = -45.309, dev = 0.02, range = 258 to 330.

1-Pentyne, CA 627-19-0: Irritant; flammable, flash point below 239; polymerizes; soluble ethanol, ether, benzene. Tf = 167.15; Tb = 321.15 to 322.15; Vm = 0.0987 at 293.15; Antoine (liq) A = 6.09224, B = 1092.52, C = -45.97, dev = 1.0, range = 229 to 315.

2-Pentyne, CA 627-21-4: Flammable, flash point 243; polymerizes; soluble most organic solvents. Tf = 163.83; Tb = 329.27; Vm = 0.0965 at 298.15; Antoine (liq) A = 6.1933, B = 1200.1, C = -42.38, dev = 1.0 to 5.0, range = 240 to 329.

(continues)

C_5H_8 *(continued)*

Spiropentane, CA 157-40-4: Flammable. Tf = 166.15; Tb = 311.45 to 311.65; Vm = 0.0902 at 293.15; Antoine (liq) A = 6.03502, B = 1086.269, C = -42.522, dev = 0.02 to 0.1, range = 276 to 344.

Vinyl cyclopropane: Flammable; readily polymerizes. Tf = 150.55; Tb = 313.15 to 315.15; Antoine (liq) A = 6.8141, B = 1509, C = 0, dev not specified, range = 289 to 310.

$C_5H_8Br_2$ MW 227.93

1,2-Dibromocyclopentane, *trans*, CA 10230-26-9: Tm = 260.65; Vm = 0.1227 at 293.15; Antoine (liq) A = 7.5369, B = 2500.5, C = 0, dev = 1.0, range = 273 to 332.

$C_5H_8Br_4$ MW 387.73

Pentaerythritol tetrabromide, CA 3229-00-3: Soluble ethanol, benzene. Tm = 436.15; Tb = 578.15 to 579.15; Antoine I (sol) A = 10.330, B = 4387, C = 0, dev = 1.0, range = 384 to 434; Antoine II (liq) A = 7.5514, B = 3185.8, C = 0, dev = 1.0, range = 439 to 466.

$C_5H_8ClFO_2$ MW 154.57

3-Chloro-4-fluorobutyric acid, methyl ester: Flammable. Antoine (liq) A = 8.5882, B = 2845.4, C = 0, dev = 1.0 to 5.0, range = 273 to 333.

$C_5H_8Cl_4$ MW 209.93

1,1,1,5-Tetrachloropentane, CA 2467-10-9: Vm = 0.1565 at 298.15; Antoine (liq) A = 5.31153, B = 1168.208, C = -140.949, dev = 1.0, range = 340 to 432.

$C_5H_8F_2O_3$ MW 154.11

Bis(2-fluoroethyl) carbonate: Flammable. Antoine (liq) A = 9.037, B = 3214, C = 0, dev = 1.0 to 5.0, range = 273 to 333.

$C_5H_8N_4O_{12}$ MW 316.14

Pentaerythritol tetranitrate, CA 78-11-5: Toxic; can explode from heating or shock; used as an explosive; slightly soluble water; moderately soluble ethanol, ether, acetone. Tm = 411.15 to 413.15; Antoine (sol) A = 16.855, B = 7750, C = 0, dev = 5.0, range = 370 to 411.

C_5H_8O MW 84.12

Cyclopentanone, CA 120-92-3: Moderately toxic; flammable, flash point 299; may explode if mixed with oxidizing agents; polymerizes easily, particularly if acids are present; slightly soluble water; soluble ethanol, ether, acetone. Tm = 221.85; Tb = 403.15; Vm = 0.0885 at 293.15; Antoine I (liq) A = 5.72768, B = 1196.9, C = -81.97, dev = 1.0, range = 293 to 404; Antoine II (liq) A = 6.19474, B = 1450.534, C = -57.43, dev = 0.02, range = 338 to 416.

(continues)

146

C_5H_8O *(continued)*

Cyclopropyl methyl ketone, CA 765-43-5: Flammable, flash point 294; soluble water, ethanol, ether. Tm = 154.15; Tb = 387.15; Vm = 0.0991 at 293.15; Antoine (liq) A = 7.0929, B = 1963, C = 0, dev = 1.0 to 5.0, range = 361 to 387.

$2H$-Pyran, dihydro, CA 25512-65-6: Irritant; flammable, flash point 258; soluble water, ethanol. Tm = 203.15; Tb = 359.15 to 360.15; Vm = 0.0912 at 293.15; Antoine (liq) A = 6.695, B = 1680.98, C = 0, dev = 5.0, range = 273 to 358.

2-Methyl-2-butenal, *trans*, tiglaldehyde, CA 497-03-0: Flammable; polymerizes; moderately soluble water; soluble ethanol, ether. Tb = 389.65 to 390.65; Vm = 0.0966 at 293.15; Antoine (liq) A = 6.57807, B = 1661.94, C = -26.057, dev = 1.0, range = 248 to 390.

3-Methyl-3-buten-2-one, isopropenyl methyl ketone, CA 814-78-8: Flammable; soluble ethanol. Tm = 219.15; Tb = 371.15 to 373.15; Vm = 0.0984 at 293.15; Antoine (liq) A = 6.08048, B = 1270.26, C = -59.13, dev = 0.1, range = 313 to 371.

2-Methyl-3-butyn-2-ol, CA 115-19-5: Irritant; flammable, flash point 298; polymerizes; soluble water, most organic solvents. Tf = 275.75; Tb = 376.75; Vm = 0.0976 at 293.15; Antoine (liq) A = 8.089, B = 2291, C = 0, dev = 1.0 to 5.0, range = 294 to 380.

1-Penten-3-one, ethyl vinyl ketone, CA 1629-58-9: Irritant; lachrymator; flammable, flash point 267; polymerizes. Tb = 375.15; Vm = 0.0993 at 293.15; Antoine (liq) A = 7.078, B = 1918, C = 0, dev = 5.0, range = 303 to 376.

C_5H_8OS MW 116.18

Tetrahydro-$4H$-thiopyran-4-one, CA 1072-72-6: Flammable; soluble ethanol. Tm = 332.15; Antoine (sol) A = 11.055, B = 3746.1, C = 0, dev not specified, range not specified.

$C_5H_8O_2$ MW 100.12

Acetic acid, isopropenyl ester, CA 108-22-5: Irritant; flammable, flash point 291; polymerizes. Tm = 180.25; Tb = 369.15; Vm = 0.1101 at 293.15; Antoine (liq) A = 7.1002, B = 1884.4, C = 0, dev = 1.0, range not specified.

Acetylacetone, 2,4-pentanedione, CA 123-54-6: Irritant; moderately toxic; flammable, flash point 307; very soluble water; soluble most organic solvents. Tm = 249.65; Tb = 413.55; Vm = 0.1030 at 298.15; Tc = 601.15; Pc = 3900; Antoine I (liq) A = 5.98985, B = 1377.34, C = -64.8, dev = 1.0, range = 288 to 378; Antoine II (liq) A = 6.4771, B = 1840, C = 0, dev = 1.0, range = 378 to 411.

Allyl acetate, CA 591-87-7: Highly toxic; flammable, flash point 279; readily polymerizes; intermediate in synthesis of 1,4-butanediol; soluble ethanol, ether, acetone. Tb = 376.15 to 377.15; Vm = 0.1079 at 293.15.

Angelic acid, 2-methyl-2-butenoic acid, *cis*, CA 565-63-9: Irritant; vesicant; flammable; isomerizes to *trans* form at boiling point; moderately soluble hot water; soluble ethanol, ether. Tm = 318.15; Tb = 458.15; Vm = 0.1019 at 320.15; Antoine (liq) A = 9.0544, B = 3229.35, C = 0, dev = 1.0 to 5.0, range = 361 to 458.   *(continues)*

$C_5H_8O_2$ *(continued)*

Ethyl acrylate, CA 140-88-5: Highly irritant; toxic; lachrymator; flammable, flash point 283; polymerizes on standing or on distillation; moderately soluble water; soluble ethanol, ether. Tm = 201.95; Tb = 372.15 to 373.15; Vm = 0.1092 at 298.15; Tc = 553, calculated; Pc = 3680, calculated; Vc = 0.323, calculated; Antoine (liq) A = 6.25041, B = 1354.65, C = -53.603, dev = 1.0, range = 244 to 373.

2-Ethylacrylic acid, CA 3586-58-1: Flammable; polymerizes. Tm = 257.15; Tb = 453.15; Antoine (liq) A = 8.25967, B = 2610.46, C = -34.774, dev = 5.0, range = 320 to 453.

Levulinaldehyde, 4-oxovaleraldehyde, CA 626-96-0: Flammable; polymerizes; soluble water, ethanol, ether. Tm = below 252.15; Tb = 459.15 to 461.15, decomposes; Vm = 0.0983 at 294.15; Antoine (liq) A = 7.06445, B = 2221.13, C = -21.078, dev = 1.0, range = 301 to 460.

3-Methylcrotonic acid, CA 541-47-9: Flammable; moderately soluble water. Tm = 343.15; Tb = 472.15; Antoine (liq) A = 6.94867, B = 1947.7, C = -73.88, dev = 1.0, range = 363 to 473.

Methyl methacrylate, CA 80-62-6: Irritant; lachrymator; flammable, flash point 283; polymerizes readily; moderately soluble water; soluble most organic solvents. Tm = 225.15; Tb = 373.15 to 374.15; Vm = 0.1066 at 298.15; Tc = 564, calculated; Pc = 3680, calculated; Vc = 0.323, calculated; Antoine I (liq) A = 6.63751, B = 1597.9, C = -28.76, dev = 1.0, range = 293 to 374; Antoine II (liq) A = 6.43088, B = 1461.197, C = -43.15, dev = 1.0, range = 293 to 373.

Propionic acid, vinyl ester, CA 105-38-4: Irritant; flammable, flash point 274; polymerizes. Tb = 364.35; Vm = 0.1215 at 364.35; Antoine (liq) A = 6.85612, B = 1782.604, C = 0, dev = 1.0, range not specified.

Tiglic acid, 2-methyl-2-butenoic acid, *trans*, CA 80-59-1: Irritant; corrosive; vesicant; flammable; moderately soluble hot water; soluble ethanol, ether. Tm = 338.65; Tb = 471.65; Vm = 0.1038 at 349.15; Antoine (liq) A = 6.6985, B = 1793.4, C = -91.483, dev = 1.0 to 5.0, range = 350 to 453.

gamma-Valerolactone ±, CA 108-29-2: Irritant; flammable, flash point 354. Tm = 242.15; Tb = 478.15 to 480.15; Vm = 0.0947 at 293.15; Antoine (liq) A = 6.35886, B = 1837.91, C = -58.49, dev = 5.0, range = 310 to 481.

$C_5H_8O_3$ MW 116.12

Levulinic acid, 4-oxopentanoic acid, CA 123-76-2: Corrosive; flammable, flash point 410; very soluble water; soluble ethanol, ether. Tm = 306.65; Tb = 518.15 to 519.15, decomposes; Antoine (liq) A = 9.07997, B = 3552.05, C = -17.123, dev = 5.0, range = 375 to 519.

Methyl acetoacetate, CA 105-45-3: Moderately toxic; flammable, flash point 343; soluble water, ethanol, ether. Tm = 193.15; Tb = 445.15; Vm = 0.1080 at 293.15; Antoine (liq) A = 7.3348, B = 2371, C = 0, dev = 1.0, range = 289 to 446.

$C_5H_8O_4$ MW 132.12

Dimethyl malonate, CA 108-59-8: Flammable, flash point 363; slightly soluble water; very soluble organic solvents. Tm = 211.25;   *(continues)*

$C_5H_8O_4$ *(continued)*

Tb = 454.55; Vm = 0.1146 at 298.15; Antoine (liq) A = 7.50022, B = 2339.5, C = -27.959, dev = 1.0, range = 308 to 454.

Glutaric acid, pentanedioic acid, CA 110-94-1: Flammable; very soluble most organic solvents. Tm = 370.15 to 371.15; Tb = 575.15 to 577.15, decomposes; Antoine (liq) A = 9.0442, B = 3521.97, C = -75.657, dev = 1.0, range = 428 to 576.

Methylene diacetate, diacetoxy methane, CA 628-51-3: Flammable; moderately soluble water; soluble ethanol, ether. Tm = 250.15; Tb = 442.15 to 444.15; Vm = 0.1163 at 293.15; Antoine (liq) A = 6.9085, B = 1915.5, C = -51.91, dev = 1.0, range = 334 to 443.

C_5H_9BrO MW 165.03

3-Bromo-2-pentanone, CA 815-48-5: Irritant; flammable. Antoine (liq) A = 7.5505, B = 2359.4, C = 0, dev = 1.0 to 5.0, range = 273 to 333.

C_5H_9Cl MW 104.58

Cyclopentyl chloride, CA 930-28-9: Flammable, flash point 255. Tb = 387.65 to 388.15; Vm = 0.1041 at 293.15; Antoine (liq) A = 6.13744, B = 1370.14, C = -54.65, dev = 0.1, range = 322 to 387.

$C_5H_9ClO_2$ MW 136.58

2-Chloropropionic acid, ethyl ester, CA 535-13-7: Corrosive; lachrymator; flammable, flash point 311; slightly soluble water; soluble ethanol, ether. Tb = 420.15 to 421.15; Vm = 0.1257 at 293.15; Antoine (liq) A = 7.28171, B = 2109.02, C = -19.881, dev = 1.0, range = 279 to 420.

3-Chloropropionic acid, ethyl ester, CA 623-71-2: Irritant; lachrymator; flammable, flash point 327; soluble ethanol, ether. Tb = 435.15; Vm = 0.1232 at 293.15; Antoine (liq) A = 9.073, B = 2923, C = 0, dev = 5.0, range = 316 to 358.

Isopropyl chloroacetate, CA 105-48-6: Irritant; flammable. Tb = 422.15; Vm = 0.1254 at 293.15; Antoine (liq) A = 7.86227, B = 2605.84, C = 23.271, dev = 1.0, range = 276 to 422.

C_5H_9ClS MW 136.64

(2-Chloroethyl) allyl sulfide, CA 19155-35-2: Irritant; flammable. Antoine (liq) A = 8.00836, B = 2619.9, C = 0, dev = 1.0, range = 273 to 341.

$C_5H_9Cl_3O$ MW 191.48

3-Chloro-2,2-bis(chloromethyl)-1-propanol, CA 813-99-0: Irritant; flammable. Antoine (liq) A = 5.06777, B = 885.017, C = -225.23, dev = 1.0, range = 404 to 450.

C_5H_9FOS MW 136.18

4-Fluorothiobutyric acid, methyl ester: Irritant; flammable. Antoine
(liq) A = 8.358, B = 2735, C = 0, dev = 5.0, range = 273 to 333.

$C_5H_9FO_2$ MW 120.12

4-Fluorobutyric acid, methyl ester: Flammable; slightly soluble water.
Tb = 408.15 to 410.15; Antoine (liq) A = 8.2014, B = 2471.7, C = 0, dev =
1.0, range = 273 to 351.

Isopropyl fluoroacetate, CA 406-06-4: Flammable. Antoine (liq) A = 7.875,
B = 2316, C = 0, dev = 5.0, range = 273 to 395.

$C_5H_9FO_3$ MW 136.12

3-Fluoro-2-hydroxybutyric acid, methyl ester: Flammable. Antoine (liq)
A = 9.2285, B = 3252.1, C = 0, dev = 1.0 to 5.0, range = 273 to 353.

C_5H_9N MW 83.13

Trimethylacetonitrile, pivalonitrile, CA 630-18-2: Irritant; lachrymator;
flammable, flash point 277. Tm = 288.15 to 289.15; Tb = 378.15 to 379.15;
Vm = 0.1096 at 298.15; Antoine (liq) A = 5.9551, B = 1248.639, C =
-62.237, dev = 0.1, range = 313 to 371.

Valeronitrile, CA 110-59-8: Irritant; flammable, flash point 313. Tm =
177.15; Tb = 414.25 to 414.35; Vm = 0.1038 at 293.15; Antoine (liq) A =
6.16961, B = 1476.88, C = -59.73, dev − 1.0, range = 313 to 418.

C_5H_9NO MW 99.13

Butyl isocyanate, CA 111-36-4: Highly irritant and toxic; flammable, flash
point below 300. Tm = below 203.15; Tb = 388.15; Vm = 0.1109 at 293.15;
Antoine (liq) A = 6.1758, B = 1296.761, C = -77.902, dev = 1.0 to 5.0,
range = 273 to 389.

Isobutyl isocyanate, CA 15585-98-5: Highly irritant and toxic; flammable.
Tb = 374.65; Antoine (liq) A = 5.89869, B = 1121.599, C = -86.876, dev =
5.0, range = 273 to 376.

Methacrylamide, *N*-methyl, CA 3887-02-3: Irritant; flammable. Tm = 273.95;
Antoine (liq) A = 6.74546, B = 1922.399, C = -82.192, dev = 1.0 to 5.0,
range = 355 to 489.

1-Methyl-2-pyrrolidinone, CA 872-50-4: Flammable, flash point 368; soluble
water; dipolar solvent used to extract aromatics from lube oils. Tm =
248.75; Tb = 475.15; Vm = 0.0964 at 298.15; Antoine I (liq) A = 7.4038, B =
2570.3, C = 0, dev = 1.0, range = 361 to 477; Antoine II (liq) A = 6.43532,
B = 1846.874, C = -59.019, dev = 1.0, range = 291 to 299.

2-Pentenoic acid, *cis*, amide, CA 15856-96-9: Flammable. Tm = 341.15;
Antoine I (sol) A = 15.1815, B = 5562, C = 0, dev = 1.0, range = 323 to
333; Antoine II (liq) A = 10.0136, B = 3856, C = 0, dev = 1.0, range = 343
to 384.

(continues)

150

C$_5$H$_9$NO *(continued)*

2-Pentenoic acid, *trans*, amide, CA 15856-96-9: Flammable. Tm = 421.15;
Antoine (sol) A = 7.1208, B = 2784.5, C = -14.95, dev = 5.0, range = 353
to 383.

Propanenitrile, 2-ethoxy, CA 14631-45-9: Flammable. Antoine (liq) A =
7.15516, B = 2201.391, C = -18.087, dev = 1.0 to 5.0, range = 348 to 445.

2-Piperidone, *delta*-valerolactone, CA 675-20-7: Hygroscopic; flammable,
flash point above 385; soluble water, ethanol, ether. Tb = 529.15; Antoine
(liq) A = 9.817, B = 3891, C = 0, dev = 5.0, range = 293 to 312.

C$_5$H$_9$NO$_2$ MW 115.13

L-Proline, CA 147-85-3: Very soluble water, ethanol. Tm = 493.15 to
495.15, decomposes; Antoine I (sol) A = 9.445, B = 5040, C = 0, dev = 5.0,
range = 442 to 467; Antoine II (sol) A = 12.98089, B = 6654.076, C = 0,
dev = 1.0 to 5.0, range = 396 to 416.

C$_5$H$_9$NO$_3$ MW 131.13

L-Proline, 4-hydroxy, *trans*, CA 51-35-4: Very soluble water; moderately
soluble ethanol. Tm = 547.15; Antoine (sol) A = 14.64075, B = 8494.402,
C = 0, dev = 1.0 to 5.0, range = 461 to 481.

C$_5$H$_9$N$_3$O$_7$ MW 223.14

Propane, 2-ethoxy-1,1,1-trinitro, CA 26459-85-8: Flammable; explosive.
Antoine (liq) A = 9.4570, B = 3016, C = 0, dev = 1.0, range = 293 to 310.

C$_5$H$_9$N$_3$O$_9$ MW 255.14

2-Hydroxymethyl-2-methyl-1,3-propanediol, trinitrate, CA 3032-55-1: Flammable; explosive. Antoine (liq) A = 10.7486, B = 4603.4, C = 0, dev = 1.0,
range = 299 to 345.

1,2,5-Pentanetriol, trinitrate: Flammable; explosive. Antoine (liq) A =
5.758, B = 2177, C = 0, dev = 5.0, range = 293 to 313.

C$_5$H$_{10}$ MW 70.13

Cyclopentane, CA 287-92-3: Narcotic at high concentrations; flammable,
flash point below 266; soluble most organic solvents. Tm = 179.28; Tb =
322.41; Vm = 0.0934 at 293.15; Tc = 511.76; Pc = 4502; Vc = 0.258; Antoine
I (liq) A = 6.0080, B = 1122.21, C = -42.011, dev = 0.02, range = 280 to
331; Antoine II (liq) A = 6.08918, B = 1174.132, C = -34.864, dev = 0.1,
range = 322 to 384; Antoine III (liq) A = 6.41769, B = 1415.096, C = -0.66,
dev = 0.1, range = 381 to 455; Antoine IV (liq) A´= 6.77782, B = 1749.65,
C = 48.533, dev = 1.0, range = 452 to 511.

2-Methyl-1-butene, CA 563-46-2: Flammable, flash point 253; polymerizes;
soluble most organic solvents. Tf = 135.59; Tb = 304.2; Vm = 0.1079 at
293.15; Antoine (liq) A = 5.99292, B = 1050.937, C = -40.727, dev = 0.02,
range = 240 to 336.

(continues)

C_5H_{10} *(continued)*

3-Methyl-1-butene, CA 563-45-1: Flammable, flash point 217; polymerizes; soluble most organic solvents. Tf = 104.65; Tb = 293.25; Vm = 0.1110 at 288.15; Antoine (liq) A = 5.94656, B = 1010.866, C = -36.694, dev = 0.02 to 0.1, range = 237 to 324.

2-Methyl-2-butene, CA 513-35-9: Flammable, flash point 228; polymerizes on standing; soluble most organic solvents. Tf = 139.15; Tb = 311.55; Vm = 0.1059 at 293.15; Antoine (liq) A = 6.04475, B = 1097.501, C = -39.985, dev = 0.02, range = 271 to 343.

1-Pentene, CA 109-67-1: Moderately toxic; narcotic; flammable, flash point 255; polymerizes; soluble most organic solvents. Tm = 108.15; Tb = 303.1; Vm = 0.1094 at 293.15; Tc = 464.78; Pc = 3529; Vc = 0.296; Antoine (liq) A = 5.96999, B = 1043.962, C = -39.767, dev = 0.02, range = 218 to 311.

2-Pentene, *cis*, CA 627-20-3: Irritant; flammable, flash point 228; polymerizes; soluble most organic solvents. Tf = 122.15; Tb = 310.03; Vm = 0.1071 at 293.15; Tc = 475.93; Pc = 3654; Vc = 0.302; Antoine (liq) A = 5.99069, B = 1064.178, C = -43.035, dev = 0.02 to 0.1, range = 234 to 318.

2-Pentene, *trans*, CA 646-04-8: Irritant; flammable, flash point 225; polymerizes; soluble most organic solvents. Tm = 132.91; Tb = 309.49; Vm = 0.1082 at 293.15; Tc = 475.37; Pc = 3654; Vc = 0.302; Antoine (liq) A = 6.02797, B = 1082.41, C = -40.388, dev = 0.02 to 0.1, range = 251 to 341.

$C_5H_{10}Br_2$ MW 299.94

1,1-Dibromopentane, CA 13320-56-4: Irritant; flammable. Antoine (liq) A = 6.432, B = 1704, C = -68.15, dev = 5.0, range = 360 to 501.

1,2-Dibromopentane ±, CA 3234-49-9: Irritant; flammable. Tb = 463.15 to 464.15; Vm = 0.1377 at 291.15; Antoine I (liq) A = 6.62929, B = 1867.23, C = -44.249, dev = 5.0, range = 292 to 448; Antoine II (liq) A = 6.11629, B = 1618.433, C = -66.658, dev = 0.1, range = 348 to 465.

1,4-Dibromopentane, CA 626-87-9: Irritant; flammable, flash point above 385. Tm = 238.73; Vm = 0.1418 at 293.15; Antoine (liq) A = 6.49219, B = 1799.1, C = -72.15, dev = 5.0, range = 377 to 524.

1,5-Dibromopentane, CA 111-24-0: Irritant; flammable, flash point 352. Tm = 233.65; Tb = 495.45; Vm = 0.1351 at 293.15; Antoine (liq) A = 6.499, B = 1884, C = -76.15, dev = 1.0, range = 396 to 549.

$C_5H_{10}Cl_2$ MW 141.04

1,1-Dichloropentane, CA 820-55-3: Irritant; flammable. Antoine (liq) A = 6.202, B = 1478, C = -60.75, dev = 1.0 to 5.0, range = 325 to 457.

1,2-Dichloropentane ±, CA 1674-33-5: Irritant. flammable. Tb = 421.15 to 422.15; Vm = 0.1310 at 298.15; Antoine (liq) A = 6.07545, B = 1454.915, C = -63.909, dev = 0.1, range = 332 to 418.

1,4-Dichloropentane ±, CA 626-92-6: Irritant; flammable. Tb = 434.15 to 436.15; Vm = 0.1301 at 293.15; Antoine (liq) A = 5.98145, B = 1448.414, C = -77.935, dev = 0.1, range = 348 to 443.

(continues)

$C_5H_{10}Cl_2$ *(continued)*

1,5-Dichloropentane, CA 628-76-2: Irritant; flammable, flash point above 300. Tm = 200.35; Tb = 452.15 to 453.15; Vm = 0.1372 at 293.15; Antoine I (liq) A = 6.1319, B = 1586.2, C = -68.15, dev = 5.0, range = 355 to 503; Antoine II (liq) A = 6.0623, B = 1523.731, C = -80.493, dev = 0.1, range = 362 to 453.

$C_5H_{10}Cl_2O$ MW 157.04

(2-Chloroethyl)-(2-chloroisopropyl) ether, CA 52250-75-6: Irritant; flammable. Antoine (liq) A = 6.78921, B = 1976.88, C = -39.721, dev = 5.0, range = 297 to 453.

(2-Chloroethyl)-(2-chloropropyl) ether, CA 42434-29-7: Irritant; flammable. Antoine (liq) A = 6.6422, B = 1989.12, C = -38.23, dev = 1.0, range = 302 to 467.

$C_5H_{10}Cl_2O_2$ MW 173.04

Bis-(2-chloroethoxy) methane, CA 111-91-1: Irritant; flammable. Antoine (liq) A = 7.54778, B = 2641.33, C = -11.518, dev = 1.0, range = 326 to 488.

$C_5H_{10}F_2$ MW 108.13

1,1-Difluoropentane, CA 62127-40-6: Flammable. Antoine (liq) A = 6.226, B = 1245, C = -47.15, dev = 5.0, range = 268 to 378.

2,2-Difluoropentane, CA 371-65-3: Flammable. Tm = 175.05; Antoine (liq) A = 6.28916, B = 1232.4, C = -45.15, dev = 1.0 to 5.0, range = 262 to 367.

3,3-Difluoropentane, CA 358-03-2: Flammable. Tm = 179.15; Antoine (liq) A = 6.29255, B = 1235.9, C = -45.15, dev = 1.0 to 5.0, range = 262 to 368.

$C_5H_{10}F_2O_2$ MW 140.13

Bis-(2-fluoroethoxy) methane, CA 373-40-0: Flammable. Antoine (liq) A = 8.4133, B = 2730.6, C = 0, dev = 1.0, range = 273 to 333.

$C_5H_{10}F_3NOS$ MW 189.20

Methanesulfenamide, *N,N*-diethyl-1,1,1-trifluoro, CA 14674-10-3: Flammable. Antoine (liq) A = 6.465, B = 1988, C = 0, dev = 1.0 to 5.0, range not specified.

$C_5H_{10}N_2$ MW 98.15

3-(Dimethylamino) propionitrile, CA 1738-25-6: Highly toxic; flammable, flash point 335. Tm = 228.95; Tb = 444.15; Vm = 0.1128 at 293.15; Antoine I (liq) A = 7.39234, B = 2396.9, C = 0, dev = 1.0, range = 330 to 445; Antoine II (liq) A = 5.25814, B = 1032.97, C = -133.025, dev = 1.0, range = 331 to 407.

$C_5H_{10}N_2O$ MW 114.15

Piperidine, 1-nitroso, CA 100-75-4: Suspected carcinogen; flammable.
Antoine (liq) A = 7.0879, B = 2492, C = 0, dev = 1.0, range = 333 to 383.

$C_5H_{10}N_2O_2$ MW 130.15

Acetylglycine, N-methylamide: Tm = 430.35; Antoine (sol) A = 10.4929, B =
5109.2, C = 0, dev = 1.0, range = 348 to 363.

$C_5H_{10}N_2O_4$ MW 162.14

1,1-Dinitropentane, CA 3759-56-6: Flammable; explosive. Antoine (liq) A =
8.2709, B = 3000, C = 0, dev not specified, range = 293 to 323.

$C_5H_{10}N_2O_6$ MW 194.14

1,5-Pentanediol, dinitrate, CA 3457-92-9: Flammable; explosive. Tm =
256.57; Antoine (liq) A = 13.853, B = 4122, C = 0, dev = 5.0, range = 293
to 313.

2,4-Pentanediol, dinitrate, CA 25385-63-1: Flammable; explosive. Tm =
254.68; Antoine (liq) A = 11.396, B = 3166, C = 0, dev = 5.0, range = 293
to 313.

Propane, 1-(methoxymethoxy)-2,2-dinitro, CA 67727-92-8: Flammable; explosive. Antoine (liq) A = 10.1296, B = 3725, C = 0, dev = 1.0, range = 293
to 333.

$C_5H_{10}O$ MW 86.13

Allyl ethyl ether, CA 557-31-3: Highly irritant; highly toxic; flammable,
flash point 253; may form explosive peroxides; slightly soluble water; soluble ethanol, ether, acetone. Tb = 340.75; Vm = 0.1126 at 293.15; Antoine
(liq) A = 6.09495, B = 1197.801, C = -47.875, dev = 1.0, range = 244 to 401.

3-Buten-3-methyl-1-ol, CA 763-32-6: Flammable, flash point 309. Vm =
0.1010 at 293.15; Antoine (liq) A = 5.012, B = 656.6, C = -184.66, dev =
5.0, range = 338 to 409.

3-Buten-3-methyl-2-ol ±, CA 10473-14-0: Flammable. Tm = 245.15; Tb =
389.15 to 390.15; Vm = 0.1031 at 290.15; Antoine (liq) A = 10.436, B =
4531.7, C = 167.11, dev = 5.0, range = 358 to 379.

Cyclopentanol, CA 96-41-3: Flammable, flash point 324; slightly soluble
water; soluble ethanol, ether, acetone. Tf = 254.15; Tb = 412.15 to
413.15; Vm = 0.0917 at 303.15; Antoine I (liq) A = 7.77346, B = 2093.02,
C = -46.12, dev = 0.1, range = 283 to 321; Antoine II (liq) A = 7.87826,
B = 2136.338, C = -44.233, dev = 0.1 to 1.0, range = 283 to 323.

3-Methyl-2-butanone, methyl isopropyl ketone, CA 563-80-4: Toxic; flammable, flash point 279; very soluble water; soluble ethanol, ether, acetone.
Tf = 181.15; Tb = 367.35; Vm = 0.1071 at 293.15; Antoine I (liq) A =
5.2823, B = 865.6, C = -103.15, dev = 1.0, range = 311 to 369; Antoine II
(liq) A = 6.08847, B = 1264.546, C = -57.754, dev = 0.02 to 0.1, range =
328 to 390; Antoine III (liq) A = 6.13272, B = 1292.675,   *(continues)*

154

$C_5H_{10}O$ *(continued)*

C = -54.259, dev = 0.1 to 1.0, range = 363 to 415; Antoine IV (liq) A = 6.42540, B = 1515.454, C = -24.034, dev = 1.0, range = 405 to 500; Antoine V (liq) A = 6.98323, B = 2079.602, C = 55.808, dev = 1.0, range = 490 to 555.

2-Methyltetrahydrofuran, CA 96-47-9: Irritant; flammable, flash point 262; moderately soluble cold water, less soluble hot; soluble ethanol, benzene, ether. Tm = 135.95; Tb = 353.15; Vm = 0.1007 at 293.15; Antoine (liq) A = 6.19376, B = 1298.63, C = -43.15, dev = 1.0, range = 283 to 353.

2-Pentanone, methyl propyl ketone, CA 107-87-9: Irritant; flammable, flash point 280; moderately soluble water; soluble ethanol, ether. Tm = 195.15 to 195.35; Tb = 375.55; Vm = 0.1066 at 293.15; Antoine I (liq) A = 6.14243, B = 1311.145, C = -58.457, dev = 0.1, range = 329 to 385; Antoine II (liq) A = 6.1404, B = 1310.269, C = -58.514, dev = 0.02 to 0.1, range = 336 to 422; Antoine III (liq) A = 6.47975, B = 1569.596, C = -24.035, dev = 0.1 to 1.0, range = 416 to 501; Antoine IV (liq) A = 7.34104, B = 2487.843, C = 98.19, dev = 0.1 to 1.0, range = 487 to 561.

3-Pentanone, diethyl ketone, CA 96-22-0: Moderately toxic; flammable, flash point 286; forms explosive peroxides with oxidizing agents; moderately soluble water; soluble ethanol, acetone. Tm = 233.75; Tb = 374.95; Vm = 0.1056 at 293.15; Tc = 560.95; Pc = 3739; Vc = 0.336; Antoine I (liq) A = 6.14917, B = 1309.653, C = -59.032, dev = 0.1, range = 329 to 384; Antoine II (liq) A = 6.14635, B = 1308.327, C = -59.137, dev = 0.02 to 0.1, range = 329 to 426; Antoine III (liq) A = 6.45505, B = 1544.596, C = -27.379, dev = 0.1 to 1.0, range = 421 to 502; Antoine IV (liq) A = 7.14424, B = 2259.87, C = 71.059, dev = 0.1 to 1.0, range = 494 to 561.

Tetrahydropyran, pentamethylene oxide, CA 142-68-7: Moderately irritant; flammable, flash point 253; can form explosive peroxides; very soluble water; soluble ethanol, ether, benzene. Tm = 228.95; Tb = 361.15; Vm = 0.0978 at 293.15; Antoine (liq) A = 6.01171, B = 1249.062, C = -49.943, dev = 1.0, range = 273 to 362.

Valeraldehyde, pentanal, CA 110-62-3: Irritant; moderately toxic; flammable, flash point 285; slightly soluble water; soluble ethanol, ether. Tm = 182.15; Tb = 376.15; Vm = 0.1064 at 293.15; Antoine (liq) A = 6.01144, B = 1072.24, C = -109.14, dev = 1.0, range = 290 to 385.

$C_5H_{10}OS$ MW 118.19

1-(Methylthio)-2-(vinyloxy)-ethane, CA 6607-53-0: Irritant; flammable. Antoine (liq) A = 7.982, B = 2483.2, C = 0, dev = 1.0, range = 316 to 347.

$C_5H_{10}O_2$ MW 102.13

Butyl formate, CA 592-84-7: Irritant; flammable, flash point 291; slightly soluble water; soluble ethanol, ether, acetone. Tm = 181.25; Tb = 379.65 to 380.45; Vm = 0.1149 at 293.15; Tc = 559, calculated; Pc = 3510, calculated; Vc = 0.336, calculated; Antoine (liq) A = 6.84067, B = 1739.2, C = -19.49, dev = 1.0, range = 295 to 380.

sec-Butyl formate ±, CA 589-40-2: Irritant; flammable; soluble ethanol, ether, acetone. Tb = 370.15; Vm = 0.1155 at 293.15; Antoine (liq) A = 6.92732, B = 1727.71, C = -15.856, dev = 1.0, range = 238 to 367.

(continues)

$C_5H_{10}O_2$ *(continued)*

tert-Butyl formate, CA 762-75-4: Irritant; corrosive; flammable, flash point 264. Tb = 355.15 to 356.15; Vm = 0.1171 at 293.15; Antoine (liq) A = 6.889, B = 1738, C = 0, dev = 1.0, range not specified.

Ethyl propionate, CA 105-37-3: Flammable, flash point 285; moderately soluble water; soluble ethanol, ether, acetone. Tf = 200.15; Tb = 372.15; Vm = 0.1145 at 293.15; Tc = 546; Pc = 3362; Vc = 0.345; Antoine I (liq) A = 6.134869, B = 1268.942, C = -64.849, dev = 0.1, range = 306 to 372; Antoine II (liq) A = 6.4443, B = 1507.82, C = -32.549, dev = 1.0, range = 372 to 538.

3-Hydroxy-3-methyl-2-butanone, CA 115-22-0: Flammable, flash point 315. Tb = 413.15 to 414.15; Vm = 0.1052 at 293.15; Antoine (liq) A = 7.166, B = 2148, C = 0, dev = 5.0, range = 317 to 419.

4-Hydroxy-3-methyl-2-butanone, CA 3393-64-4: Flammable, flash point 351. Vm = 0.1028 at 293.15; Antoine (liq) A = 7.34764, B = 2072.75, C = -68.15, dev = 1.0, range = 375 to 528.

Isobutyl formate, CA 542-55-2: Irritant; corrosive; flammable, flash point 283; slightly soluble water; soluble ethanol, ether, acetone. Tf = 178.15; Tb = 371.35; Vm = 0.1153 at 293.15; Antoine I (liq) A = 6.63347, B = 1586.26, C = -28.659, dev = 1.0, range = 240 to 372; Antoine II (liq) A = 6.04040, B = 1149.53, C = -86.645, dev = 1.0 to 5.0, range = 371 to 507.

Isopropyl acetate, CA 108-21-4: Irritant; flammable, flash point 275; very soluble water; soluble ethanol, ether, acetone; solvent for cellulose derivatives. Tm = 199.75; Tb = 361.15 to 364.15; Vm = 0.1171 at 293.15; Antoine (liq) A = 6.45885, B = 1436.53, C = -39.485, dev = 1.0, range = 235 to 362.

Isovaleric acid, CA 503-74-2: Irritant; corrosive; toxic; flammable, flash point 343; smells of rancid cheese; moderately soluble water; soluble ether, ethanol, chloroform. Tm = 235.55; Tb = 449.15; Vm = 0.1100 at 293.15; Antoine (liq) A = 7.46656, B = 2209.72, C = -43.525, dev = 1.0 to 5.0, range = 307 to 448.

1-Methoxy-2-butanone, CA 50741-70-3: Flammable. Antoine (liq) A = 6.23677, B = 1429.5, C = -68.15, dev = 1.0, range = 297 to 408.

Methyl butyrate, CA 623-42-7: Irritant; flammable, flash point 287; moderately soluble water; soluble ethanol, ether. Tm = 188.35; Tb = 375.45; Vm = 0.1137 at 293.15; Antoine I (liq) A = 6.27187, B = 1351.36, C = -58.739, dev = 1.0, range = 246 to 375; Antoine II (liq) A = 6.62592, B = 1678.76, C = -12.021, dev = 1.0 to 5.0, range = 375 to 545.

Methyl isobutyrate, CA 547-63-7: Irritant; flammable, flash point 276; slightly soluble water; soluble ethanol, ether, acetone. Tm = 188.45; Tb = 366.15; Vm = 0.1146 at 293.15; Antoine I (liq) A = 6.51181, B = 1459.48, C = -41.822, dev = 1.0, range = 239 to 366; Antoine II (liq) A = 6.36875, B = 1432.58, C = -37.32, dev = 1.0 to 5.0, range = 366 to 533.

Propyl acetate, CA 109-60-4: Irritant; flammable, flash point 286; moderately soluble water; soluble ethanol, ether; solvent used for lacquers, waxes, nitrocellulose, insecticides. Tm = 180.15; Tb = 374.15; Vm = 0.1150 at 293.15; Tc = 549.4; Pc = 3364; Vc = 0.345; Antoine I (liq) A = 6.142106, B = 1282.873, C = -64.486, dev = 0.1, range = 312 to 374; Antoine II (liq) A = 6.48937, B = 1544.31, C = -30.623, dev = 1.0, range = 374 to 542; Antoine III (liq) A = 6.13951, B = 1281.682, C = -64.58, dev = 0.1, range = 312 to 374. *(continues)*

$C_5H_{10}O_2$ *(continued)*

Tetrahydrofurfuryl alcohol, CA 97-99-4: Irritant; flammable, flash point 348; hygroscopic; soluble water, most organic solvents; solvent for fats, waxes, resins. Tm = below 193.15; Tb = 451.15; Vm = 0.0969 at 293.15; Antoine (liq) A = 7.3609, B = 2412, C = 0, dev = 1.0, range = 393 to 453.

Valeric acid, pentanoic acid, CA 109-52-4: Irritant; corrosive; flammable, flash point 361; moderately soluble water; soluble ethanol, ether. Tm = 238.65; Tb = 460.15; Vm = 0.1084 at 293.15; Antoine (liq) A = 6.69856, B = 1694.37, C = -98.15, dev = 1.0, range = 375 to 523.

$C_5H_{10}O_3$ MW 118.13

Diethyl carbonate, CA 105-58-8: Irritant; flammable, flash point 298; moisture-sensitive; slightly soluble water; soluble ethanol, ether. Tm = 230.15; Tb = 399.95; Vm = 0.1212 at 293.15; Antoine (liq) A = 6.64355, B = 1685.3, C = -36.13; dev = 1.0, range = 308 to 400.

Ethylene glycol, monomethylether acetate, CA 110-49-6: Moderately toxic by inhalation and skin contact; flammable, flash point 322; soluble water, most organic solvents; solvent for oils, gums, resins. Tm = 208.15; Tb = 418.15; Vm = 0.1175 at 293.15; Antoine (liq) A = 6.7832, B = 1789.3, C = -43.15, dev = 1.0, range = 343 to 417.

Ethyl lactate ±, CA 2676-33-7: Flammable, flash point 319; soluble water, ethanol, ether. Tm = 248.15; Tb = 427.15; Vm = 0.1142 at 293.15; Antoine I (liq) A = 6.60606, B = 1673.8, C = -62.21, dev = 1.0 to 5.0, range = 308 to 426; Antoine II (liq) A = 6.2975, B = 1441.066, C = -90.17, dev = 1.0 to 5.0, range = 324 to 427.

3-Hydroxyproprionic acid, ethyl ester, CA 623-72-3: Flammable. Tm = 458.15 to 463.15; Vm = 0.1115 at 293.15; Antoine (liq) A = 9.429, B = 3249, C = 0, dev = 5.0, range = 338 to 356.

3-Methoxypropionic acid, methyl ester, CA 3852-09-3: Flammable. Tb = 415.75 to 415.95; Vm = 0.1164 at 288.15; Antoine (liq) A = 7.4215, B = 2265, C = 0, dev = 5.0, range = 355 to 438.

Glycerol, 1-monoacetate, CA 106-61-6: Flammable; very hygroscopic; soluble water, ethanol; used in manufacture of smokeless powder. Vm = 0.1112 at 293.15; Antoine (liq) A = 9.23008, B = 3864.8, C = 0, dev = 1.0, range = 385 to 458.

$C_5H_{10}S$ MW 102.19

Allyl ethyl sulfide, CA 5296-62-8: Irritant; stench; flammable. Antoine (liq) A = 7.2278, B = 2011.115, C = 0, dev = 1.0, range = 300 to 327.

Cyclopentanethiol, CA 1679-07-8: Irritant; stench; flammable. Tm = 155.39, Tb = 404.15 to 405.15; Antoine (liq) A = 6.03809, B = 1387.975, C = -61.121, dev = 0.02, range = 348 to 446.

2-Methylthiophene, tetrahydro, CA 1795-09-1: Toxic; flammable; soluble ethanol, ether, acetone, benzene. Tm = 172.46; Tb = 407.15; Vm = 0.1071 at 293.15; Antoine (liq) A = 6.06613, B = 1408.115, C = -58.848, dev = 0.02, range = 335 to 447.

(continues)

157

$C_5H_{10}S$ *(continued)*

3-Methylthiophene, tetrahydro, CA 4740-00-5: Toxic; flammable; soluble ethanol, ether, acetone, benzene. Tm = 192.02; Tb = 410.65 to 411.65; Vm = 0.1062 at 293.15; Antoine (liq) A = 6.07359, B = 1432.084, C = -59.452, dev = 0.02, range = 340 to 453.

2H-Thiopyran, tetrahydro, pentamethylene sulfide, CA 1613-51-0: Irritant; stench; flammable, flash point 294; soluble most organic solvents. Tm = 292.22; Tb = 415.15; Vm = 0.1037 at 293.15; Antoine (liq) A = 6.02831, B = 1421.852, C = -61.445, dev = 0.1, range = 347 to 423.

$C_5H_{11}Br$ MW 151.05

1-Bromo-2,2-dimethylpropane, neopentyl bromide, CA 630-17-1: Irritant; flammable, flash point 279; soluble ethanol, ether, acetone, benzene. Tb = 379.15; Vm = 0.1259 at 293.15; Antoine (liq) A = 5.92621, B = 1274.2, C = -54.15, dev = 5.0, range = 293 to 420.

1-Bromo-2-methylbutane ±, CA 5973-11-5: Irritant; flammable, flash point 295; slightly soluble water; soluble ethanol, ether, chloroform. Tb = 389.65 to 391.15; Vm = 0.1235 at 293.15; Antoine (liq) A = 5.97649, B = 1336.2, C = -57.15, dev = 1.0, range = 306 to 436.

1-Bromo-3-methylbutane, isopentyl bromide, CA 107-82-4: Irritant; flammable, flash point 305; slightly soluble water; soluble ethanol, ether, chloroform. Tf = 161.15; Tb = 393.15 to 394.15; Vm = 0.1251 at 293.15; Antoine (liq) A = 5.9759, B = 1335.6, C = -57.15, dev = 1.0, range = 306 to 436.

2-Bromo-2-methylbutane, *tert*-pentyl bromide, CA 507-36-8: Irritant; flammable; soluble ethanol, ether. Tb = 381.15 to 382.15, decomposes; Vm = 0.1261 at 291.65; Antoine (liq) A = 5.90074, B = 1262, C = -57.15, dev = 5.0, range = 295 to 422.

2-Bromo-3-methylbutane ±, CA 18295-25-5: Irritant; flammable; soluble ethanol, ether. Tb = 388.45; Antoine (liq) A = 5.95778, B = 1313.3, C = -56.15, dev = 1.0, range = 301 to 430.

1-Bromopentane, pentyl bromide, CA 110-53-2: Irritant; highly toxic; flammable, flash point 304; slightly soluble water; soluble ethanol, benzene, ether, chloroform. Tf = 117.9; Tb = 402.85; Vm = 0.1240 at 293.15; Antoine (liq) A = 6.0807, B = 1401.634, C = -58.77, dev = 0.1, range = 293 to 443.

2-Bromopentane ±, *sec*-pentyl bromide, CA 107-81-3: Irritant; flammable, flash point 293; soluble ethanol, ether, benzene, chloroform. Tf = 177.65; Tb = 390.55; Vm = 0.1251 at 293.15; Antoine (liq) A = 5.97038, B = 1325.8, C = -56.15, dev = 1.0 to 5.0, range = 303 to 432.

3-Bromopentane, CA 1809-10-5: Irritant; flammable, flash point 293; soluble ethanol, ether, benzene, chloroform. Tm = 146.95; Tb = 391.75; Vm = 0.1238 at 293.15; Antoine (liq) A = 5.96517, B = 1324.8, C = -57.15, dev = 1.0 to 5.0, range = 304 to 434.

$C_5H_{11}Cl$ MW 106.59

1-Chloro-2,2-dimethylpropane, neopentyl chloride, CA 753-89-9: Irritant; flammable, flash point 264; soluble ethanol, ether, benzene, chloroform. Tm = 253.15; Tb = 357.45; Vm = 0.1231 at 293.15; Antoine (liq) A = 6.07978, B = 1252, C = -50.15, dev = 1.0, range = 279 to 395.   *(continues)*

158

$C_5H_{11}Cl$ (continued)

1-Chloro-2-methylbutane ±, CA 616-13-7: Irritant; flammable; soluble ethanol, ether. Tb = 369.15 to 372.15; Vm = 0.1209 at 288.15; Antoine (liq) A = 6.39554, B = 1505.8, C = -30.56, dev = 1.0, range = 300 to 374.

2-Chloro-2-methylbutane, *tert*-pentyl chloride, CA 594-36-5: Irritant; flammable; soluble ethanol, ether. Tm = 200.15; Tb = 359.15; Vm = 0.1232 at 293.15; Antoine (liq) A = 6.08392, B = 1258.5, C = -50.15, dev = 1.0, range = 280 to 396.

2-Chloro-3-methylbutane ±, CA 631-65-2: Irritant; flammable; soluble ethanol, ether. Tb = 363.15 to 366.15; Vm = 0.1237 at 293.15; Antoine (liq) A = 6.07928, B = 1278.3, C = -52.15, dev = 1.0, range = 285 to 405.

1-Chloropentane, pentyl chloride, CA 543-59-9: Irritant; flammable, flash point 284; soluble ethanol, ether, benzene, chloroform. Tm = 174.15; Tb = 381.55; Vm = 0.1222 at 303.15; Antoine (liq) A = 6.09107, B = 1332.89, C = -54.65, dev = 0.1, range = 277 to 421.

2-Chloropentane ±, CA 625-29-6: Irritant; flammable; soluble ethanol, benzene, ether, chloroform. Tm = 136.15; Tb = 368.15 to 370.15; Vm = 0.1225 at 293.15; Antoine (liq) A = 6.09187, B = 1298.8, C = -52.15, dev = 0.1 to 1.0, range = 289 to 409.

3-Chloropentane, CA 616-20-6: Irritant; flammable; soluble ethanol, ether, benzene, chloroform. Tm = 168.15; Tb = 368.55 to 369.45; Vm = 0.1222 at 293.15; Antoine (liq) A = 6.08132, B = 1295.2, C = -53.15, dev = 0.1 to 1.0, range = 289 to 410.

$C_5H_{11}ClO_2S$ MW 170.65

1-Pentanesulfonyl chloride, CA 6303-18-0: Irritant; flammable. Tm = 251.15; Antoine (liq) A = 8.772, B = 3159.2, C = 0, dev = 1.0, range = 263 to 293.

$C_5H_{11}Cl_2N$ MW 156.05

N-Methyl-bis(2-chloroethyl) amine, CA 51-75-2: Very toxic; vesicant; flammable; suspected carcinogen; mutagen; slightly soluble water. Antoine (liq) A = 8.12188, B = 2850.4, C = 0, dev = 1.0, range = 273 to 333.

$C_5H_{11}F$ MW 90.14

1-Fluoro-2-methylbutane, CA 33965-74-1: Flammable. Tb = 329.05; Antoine (liq) A = 6.07668, B = 1152.56, C = -45.96, dev = 1.0, range = 287 to 329.

2-Fluoro-2-methylbutane, *tert*-pentyl fluoride, CA 661-53-0: Flammable. Tm = 152.15; Tb = 317.95; Antoine (liq) A = 6.24224, B = 1166.3, C = -42.65, dev = 1.0, range = 249 to 341.

1-Fluoropentane, pentyl fluoride, CA 592-50-7: Irritant; flammable; soluble ethanol, ether. Tm = 153.15; Tb = 335.15 to 335.65; Vm = 0.1140 at 293.15; Antoine (liq) A = 6.1106, B = 1190.03, C = -46.05, dev = 1.0, range = 245 to 373.

$C_5H_{11}I$ MW 198.05

1-Iodo-2-methylbutane ±, CA 616-14-8: Irritant; flammable; light-sensitive; soluble ethanol, ether. Tb = 416.15 to 417.15; Vm = 0.1326 at 293.15; Antoine (liq) A = 7.1849, B = 2254.04, C = 14.69, dev = 1.0, range = 339 to 406.

1-Iodo-3-methylbutane, isopentyl iodide, CA 541-28-6: Irritant; flammable; light-sensitive; slightly soluble water; soluble ethanol, ether. Tb = 422.15 to 423.15; Vm = 0.1310 at 293.15; Antoine (liq) A = 6.39137, B = 1670.43, C = -40.582, dev = 1.0, range = 270 to 422.

2-Iodo-2-methylbutane, *tert*-pentyl iodide, CA 594-38-7: Irritant; flammable; light-sensitive; rapidly hydrolyzed by water; soluble ethanol, ether. Tb = 401.15; Vm = 0.1326 at 293.15; Antoine (liq) A = 7.3170, B = 2113.1, C = 0, dev = 1.0, range = 308 to 398.

1-Iodopentane, pentyl iodide, CA 628-17-1: Irritant; flammable, flash point 324; light-sensitive; soluble ethanol, ether. Tm = 187.55; Tb = 430.15; Vm = 0.1306 at 293.15; Antoine (liq) A = 5.97662, B = 1454.028, C = -63.98, dev = 1.0, range = 312 to 473.

$C_5H_{11}N$ MW 85.15

Cyclopentyl amine, CA 1003-03-8: Irritant; flammable, flash point 290; soluble water, acetone, benzene. Tf = 187.45; Tb = 380.15 to 381.15; Vm = 0.0980 at 293.15; Antoine (liq) A = 6.03827, B = 1266.748, C = -67.544, dev = 0.02, range = 317 to 419.

1-Methylpyrrolidine, CA 120-94-5: Irritant; flammable, flash point 259; soluble water, ethanol. Tb = 353.15 to 354.15; Vm = 0.1040 at 293.15; Antoine (liq) A = 6.51764, B = 1473.185, C = -24.635, dev = 0.1, range = 273 to 315.

Piperidine, CA 110-89-4: Highly toxic by inhalation and skin absorption; causes burns; flammable, flash point 289; strong base; soluble water, most organic solvents. Tm = 266.15; Tb = 379.15; Vm = 0.0988 at 293.15; Antoine (liq) A = 5.97847, B = 1237.885, C = -67.781, dev = 0.02, range = 315 to 417.

$C_5H_{11}NO$ MW 101.15

N,N-Diethylformamide, CA 617-84-5: Irritant; flammable, flash point 333; soluble water, most organic solvents. Tb = 450.15 to 451.15; Vm = 0.1114 at 292.15; Antoine (liq) A = 7.8569, B = 2557, C = 0, dev = 1.0 to 5.0, range = 303 to 363.

N-Methylmorpholine, 4-methylmorpholine, CA 109-02-4: Irritant; flammable, flash point 297; soluble water, most organic solvents; solvent for resins, waxes, and dyes. Tm = 207.25; Tb = 388.45; Vm = 0.1098 at 293.15; Antoine I (liq) A = 7.15751, B = 2004.5, C = 0, dev = 1.0 to 5.0, range = 297 to 389; Antoine II (liq) A = 6.0459, B = 1336.45, C = -58.125, dev = 1.0, range = 276 to 390.

3-Pentanone, oxime, CA 1188-11-0: Flammable. Antoine (liq) A = 6.87621, B = 1677.712, C = -80.087, dev = 1.0, range = 318 to 425.

(continues)

$C_5H_{11}NO$ *(continued)*

Propionamide, *N,N*-dimethyl, CA 758-96-3: Irritant; flammable, flash point 335. Tm = 228.15; Tb = 438.15 to 451.15; Vm = 0.1099 at 293.15; Antoine (liq) A = 5.54447, B = 1166.969, C = -120.379, dev = 1.0, range = 326 to 424.

Valeramide, pentanamide, CA 626-97-1: Flammable; very soluble water; soluble ethanol, ether. Tm = 379.15; Vm = 0.1158 at 383.15; Antoine (sol) A = 11.9709, B = 4665.8, C = 0, dev = 1.0, range = 333 to 374.

$C_5H_{11}NO_2$ MW 117.15

2-Aminopentanoic acid ±, norvaline, CA 760-78-1: Very soluble water; slightly soluble organic solvents. Tm = 576.15; Antoine (sol) A = 12.015, B = 6270, C = 0, dev = 5.0, range = 439 to 461.

Butyl carbamate, CA 592-35-8: Flammable, flash point 381; soluble ethanol. Tm = 324.15 to 329.15; Tb = 476.15 to 477.15, decomposes; Antoine (sol) A = 13.707, B = 4919, C = 0, dev = 5.0, range = 292 to 316.

Isobutyl carbamate, CA 543-28-2: Flammable; soluble ethanol, ether. Tm = 340.15; Tb = 479.15; Antoine (liq) A = 7.88293, B = 2673.89, C = -24.729, dev = 5.0, range = 356 to 479.

Lactamide, *N,N*-dimethyl, CA 31502-31-5: Flammable. Tm = 318.15; Antoine (liq) A = 9.36635, B = 3850.293, C = 0, dev = 1.0, range = 351 to 417.

L-Valine, CA 72-18-4: Very soluble water; slightly soluble ethanol, ether. Tm = 588.15, sealed tube; Antoine (sol) A = 17.115, B = 8490, C = 0, dev = 5.0, range = 438 to 456.

$C_5H_{11}NO_2S$ MW 149.21

L-Methionine, CA 63-68-3: Soluble water; slightly soluble organic solvents. Tm = 556.15, decomposes; Antoine (sol) A = 11.795, B = 6530, C = 0, dev = 5.0, range = 463 to 485.

$C_5H_{11}NO_3$ MW 133.15

Isopentyl nitrate, CA 543-87-3: Flammable; slightly soluble water; soluble ethanol, ether. Tb = 420.15 to 421.15; Vm = 0.1337 at 295.15; Antoine (liq) A = 6.82681, B = 1838.17, C = -39.343, dev = 1.0, range = 278 to 421.

$C_5H_{11}P$ MW 102.12

Phosphorinane, CA 4743-40-2: Flammable. Tm = 290.45; Antoine I (sol) A = 7.7909, B = 2260, C = 0, dev = 1.0, range = 250 to 291; Antoine II (liq) A = 7.1939, B = 2086, C = 0, dev = 1.0, range = 294 to 345.

C_5H_{12} MW 72.15

2,2-Dimethylpropane, neopentane, CA 463-82-1: Flammable, flash point 198; slightly soluble water; soluble most organic solvents. Tm = 256.58; Tb = 282.65; Vm = 0.1176 at 293.15; Tc = 433.78; Pc = 3199; Vc = 0.304; Antoine I (sol) A = 6.3283, B = 1020.7, C = -43.15, dev = 1.0 to 5.0, *(continues)*

C_5H_{12} *(continued)*

range = 223 to 256; Antoine II (sol) A = 7.07825, B = 1372.459, C = -8.39, dev = 0.1, range = 223 to 256; Antoine III (liq) A = 5.76532, B = 900.545, C = -43.111, dev = 0.02, range = 268 to 313; Antoine IV (liq) A = 5.83935, B = 937.641, C = -38.071, dev = 0.1, range = 257 to 315; Antoine V (liq) A = 6.08953, B = 1080.237, C = -17.896, dev = 0.1, range = 312 to 385; Antoine VI (liq) A = 7.26795, B = 2114.713, C = 128.175, dev = 0.1, range = 382 to 433.

2-Methylbutane, isopentane, CA 78-78-4: Flammable, flash point 213; slightly soluble water; soluble most organic solvents. Tm = 113.25; Tb = 301.0; Vm = 0.1164 at 293.15; Tc = 460.39; Pc = 3381; Vc = 0.306; Antoine I (liq) A = 5.95805, B = 1040.73, C = -37.705, dev = 0.02, range = 216 to 323; Antoine II (liq) A = 6.32287, B = 1279.08, C = -4.481, dev = 1.0 to 5.0, range = 300 to 460; Antoine III (liq) A = 6.39629, B = 1325.048, C = 1.244, dev = 0.1 to 1.0, range = 320 to 391; Antoine IV (liq) A = 6.22589, B = 1212.803, C = -12.958, dev = 1.0, range = 385 to 416; Antoine V (liq) A = 8.09160, B = 3167.01, C = 233.708, dev = 1.0, range = 412 to 460.

Pentane, CA 109-66-0: Narcotic in high concentrations; hygroscopic; flammable, flash point 224; slightly soluble water; soluble most organic solvents. Tm = 143.42; Tb = 309.22; Vm = 0.1153 at 293.15; Tc = 469.7; Pc = 3369; Vc = 0.312; Antoine I (liq) A = 7.6922, B = 1686.65, C = 0, dev = 1.0, range = 143 to 223; Antoine II (liq) A = 5.99466, B = 1073.139, C = -40.188, dev = 0.02, range = 223 to 352; Antoine III (liq) A = 5.98799, B = 1070.14, C = -40.485, dev = 0.02, range = 269 to 335; Antoine IV (liq) A = 6.28417, B = 1260.973, C = -14.031, dev = 0.1, range = 350 to 422; Antoine V (liq) A = 7.47436, B = 2414.137, C = 141.919, dev = 0.1, range = 418 to 470.

$C_5H_{12}ClF_3N_2OS$ MW 240.67

Sulfur, chlorobis(*N*-methylmethanaminato)oxo(trifluoromethyl), CA 63265-73-6: Flammable. Antoine (liq) A = 6.475, B = 2133, C = 0, dev = 1.0 to 5.0, range not specified.

$C_5H_{12}ClF_3N_2S$ MW 224.67

Sulfur, chlorobis(*N*-methylmethanaminato)(trifluoromethyl), CA 63265-71-4: Flammable. Antoine (liq) A = 6.245, B = 1984, C = 0, dev = 1.0 to 5.0, range not specified.

$C_5H_{12}NO_3PS_2$ MW 229.25

Phosphorodithioic acid, *O,O*-dimethyl-*S*-[2-(methylamino)-2-oxoethyl] ester, CA 60-51-5: Flammable; slightly soluble water; soluble ethanol, acetone, ether; contact insecticide. Tm = 325.15; Vm = 0.1795 at 338.15; Antoine (liq) A = 8.9588, B = 3832.6, C = -36.03, dev = 1.0, range = 283 to 390.

$C_5H_{12}N_2$ MW 100.16

Piperazine, *N*-methyl, CA 109-01-3: Irritant; flammable, flash point 315; soluble water, ethanol, ether. Tb = 411.15; Vm = 0.1109 at 293.15; Antoine (liq) A = 6.343, B = 1508.38, C = -61.56, dev = 0.1 to 1.0, range = 274 to 319.

162

$C_5H_{12}O$ MW 88.15

Butyl methyl ether, 1-methoxybutane, CA 628-28-4: Narcotic in high concentrations; flammable, flash point 263; forms explosive peroxides; slightly soluble water; soluble ethanol, ether, acetone. Tm = 157.65; Tb = 344.15; Vm = 0.1184 at 293.15; Antoine I (liq) A = 6.008786, B = 1160.272, C = -53.444, dev not specified, range = 295 to 342; Antoine II (liq) A = 6.02963, B = 1171.617, C = -52.131, dev = 0.1, range = 293 to 367.

2,2-Dimethyl-1-propanol, neopentyl alcohol, CA 75-84-3: Flammable, flash point 310; moderately soluble water; soluble ethanol, ether. Tm = 326.15; Tb = 386.15 to 387.15; Antoine (liq) A = 6.05629, B = 1092.234, C = -115.87, dev = 0.1, range = 330 to 387.

Ethyl propyl ether, 1-ethoxypropane, CA 628-32-0: Narcotic in high concentrations; flammable, flash point below 253; forms explosive peroxides; slightly soluble water; soluble ethanol, ether, acetic acid. Tm = 146.45; Tb = 336.75; Vm = 0.1193 at 293.15; Antoine I (liq) A = 6.121381, B = 1194.642, C = -46.015, dev = 1.0, range = 2 3 to 336; Antoine II (liq) A = 6.04017, B = 1152.298, C = -50.755, dev = 0.02 to 0.1, range = 264 to 359.

2-Methyl-1-butanol, CA 137-32-6: Flammable, flash point 323; moderately soluble water; soluble ethanol, ether, acetone. Tm = below 203.15; Tb = 401.15; Vm = 0.1080 at 293.15; Antoine I (liq) A = 7.05829, B = 1682, C = -69.15, dev = 1.0, range = 338 to 402; Antoine II (liq) A = 7.1983, B = 1766.724, C = -63.028, dev = 1.0, range = 249 to 319; Antoine III (liq) A = 6.18669, B = 1202.782, C = -114.73, dev = 0.1 to 1.0, range = 317 to 403.

3-Methyl-1-butanol, isopentyl alcohol, CA 123-51-3: Moderately toxic; expected carcinogen; flammable, flash point 316; slightly soluble water; soluble most organic solvents; solvent for fats, resins; major component of fusel oil. Tm = 155.95; Tb = 403.65; Vm = 0.1086 at 293.15; Antoine (liq) A = 6.17508, B = 1201.77, C = -115.83, dev = 1.0, range = 303 to 412.

2-Methyl-2-butanol, *tert*-pentanol, CA 75-85-4: Moderately toxic; flammable, flash point 292; moderately soluble water; soluble ethanol, ether, acetone, benzene. Tm = 261.25; Tb = 374.95; Vm = 0.1094 at 298.15; Antoine I (liq) A = 7.02635, B = 1557.51, C = -64.738, dev = 1.0, range = 280 to 375; Antoine II (liq) A = 6.07488, B = 1075.367, C = -111.143, dev = 0.1, range = 323 to 376.

3-Methyl-2-butanol ±, CA 598-75-4: Flammable, flash point 299; moderately soluble water; soluble ethanol, ether, acetone, benzene. Tb = 385.15; Vm = 0.1076 at 293.15; Antoine (liq) A = 8.30313, B = 2425.45, C = 0, dev = 1.0 to 5.0, range = 293 to 385.

Methyl *tert*-butyl ether, CA 1634-04-4: Flammable, flash point 245; forms explosive peroxides; slightly soluble water; soluble ethanol, ether; antiknock additive for gasoline. Tm = 164.15; Tb = 328.25; Vm = 0.1190 at 293.15; Tc = 497.1; Pc = 3430; Vc = 0.329, calculated; Antoine (liq) A = 6.09111, B = 1171.54, C = -41.542, dev = 0.1, range = 287 to 351.

1-Pentanol, pentyl alcohol, CA 71-41-0: Moderately toxic vapor; irritant; flammable, flash point 306; moderately soluble water; soluble ethanol, ether, acetone. Tm = 194.95; Tb = 410.95; Vm = 0.1087 at 298.15; Tc = 586.15; Pc = 3880; Vc = 0.326; Antoine I (liq) A = 6.14668, B = 1195.924, C = -122.348, dev = 0.02, range = 388 to 420; Antoine II (liq) A = 6.30990, B = 1290.23, C = -111.419, dev = 0.1, range = 347 to 429; Antoine III (liq) A = 6.1490, B = 1197.233, C = -122.194, dev = 0.02, range = 388 to 420; Antoine IV (liq) A = 6.3975, B = 1337.613, C = -106.567, dev = 0.1, range = 326 to 411; Antoine V (liq) A = 6.28069, B = 1277.413,   *(continues)*

$C_5H_{12}O$ *(continued)*

C = -112.34, dev = 0.1, range = 408 to 441; Antoine VI (liq) A = 6.00943,
B = 1110.258, C = -133.942, dev = 1.0, range = 433 to 514.

2-Pentanol ±, CA 6032-29-7: Moderately toxic vapor; irritant; flammable,
flash point 314; moderately soluble water; soluble ethanol, ether. Tb =
392.45; Vm = 0.1062 at 293.15; Antoine (liq) A = 6.34413, B = 1252.4, C =
-104.179, dev = 1.0, range = 274 to 393.

3-Pentanol, CA 584-02-1: Irritant; flammable, flash point 314; moderately
soluble water; soluble ethanol, ether, acetone; used as flotation agent.
Tm = below 198.15; Tb = 388.75; Vm = 0.1082 at 298.15; Antoine I (liq) A =
6.14986, B = 1137.7, C = -114.58, dev = 1.0, range = 297 to 390; Antoine II
(liq) A = 6.61558, B = 1392.387, C = -86.248, dev = 1.0, range = 245 to 390.

$C_5H_{12}O_2$ MW 104.15

Ethylene glycol, monoisopropyl ether, CA 109-59-1: Moderately toxic; flammable; soluble water, ethanol, ether, acetone. Tb = 417.15; Vm = 0.1153 at
293.15; Antoine (liq) A = 6.8967, B = 1817.1, C = -43.15, dev = 1.0, range
= 341 to 413.

Ethylene glycol, monopropyl ether, CA 2807-30-9: Moderately toxic; flam-
mable; soluble water, ethanol, ether; anti-icing additive for gasoline.
Tb = 423.15; Vm = 0.1143 at 293.15; Antoine (liq) A = 6.9542, B = 1878.9,
C = -43.15, dev = 1.0, range = 350 to 422.

Formaldehyde, diethyl acetal, CA 462-95-3: Irritant; flammable; soluble
water, ethanol, ether, acetone, benzene. Tm = 206.65; Tb = 362.15; Vm =
0.1252 at 293.15; Antoine (liq) A = 6.1765, B = 1307.1, C = -47.93, dev =
1.0, range = 273 to 361.

2-Methyl-1,3-butanediol, CA 684-84-4: Flammable; soluble water, ethanol,
ether. Vm = 0.1061 at 303.15; Antoine (liq) A = 7.35262, B = 2195.8, C =
-74.15, dev = 1.0 to 5.0, range = 399 to 561.

3-Methyl-1,3-butanediol, CA 2568-33-4: Flammable; soluble water, ethanol,
ether. Tb = 475.15 to 476.15; Vm = 0.1102 at 293.15; Antoine (liq) A =
8.6399, B = 3147.8, C = 0, dev = 1.0, range = 346 to 475.

Neopentyl glycol, 2,2-dimethyl-1,3-propanediol, CA 126-30-7: Irritant;
flammable, flash point 402; soluble water, most organic solvents. Tm =
402.15; Tb = 483.15; Antoine (liq) A = 10.6606, B = 4150.57, C = 0, dev =
1.0 to 5.0, range = 400 to 480.

1,5-Pentanediol, CA 111-29-5: Irritant; flammable, flash point 402; soluble water, methanol, ethanol, acetone. Tm = 255.15; Tb = 511.15 to
512.15; Vm = 0.1050 at 293.15; Antoine (liq) A = 6.25865, B = 1524.3, C =
-158.38, dev = 0.1, range = 391 to 479.

$C_5H_{12}O_3$ MW 120.15

Diethylene glycol, monomethyl ether, CA 111-77-3: Flammable, flash point
356; soluble water, ethanol, ether, acetone. Tm = 197.15; Tb = 467.15;
Vm = 0.1177 at 293.15; Antoine (liq) A = 7.0934, B = 2155.4, C = -43.15,
dev = 1.0, range = 385 to 466.

$C_5H_{12}O_4$ MW 136.15

Pentaerythritol, tetramethylol methane, CA 115-77-5: Hygroscopic; moderately soluble water; soluble ethanol, glycerol; used in resins, paints, and
varnishes. Tm = 533.15; Antoine (sol) A = 15.295, B = 75.28, C = 0, dev =
1.0 to 5.0, range = 379 to 409.

$C_5H_{12}S$ MW 104.21

Butyl methyl sulfide, CA 628-29-5: Stench; toxic; flammable; soluble
ethanol. Tm = 175.35; Tb = 396.35; Vm = 0.1237 at 293.15; Antoine I (liq)
A = 6.0691, B = 1363.237, C = -61.091, dev = 0.02, range = 343 to 436;
Antoine II (liq) A = 6.61672, B = 1845.917, C = 0, dev = 1.0, range = 301
to 330.

tert-Butyl methyl sulfide, CA 6163-64-0: Stench; toxic; flammable; soluble
ethanol. Tm = 190.87; Antoine (liq) A = 5.9665, B = 1256.624, C = -54.778,
dev = 0.02, range = 305 to 411.

2,2-Dimethyl-1-propanethiol, neopentyl mercaptan, CA 1679-08-9: Stench;
toxic; flammable. Tm = 205.85; Antoine (liq) A = 5.9640, B = 1274.117,
C = -54.947, dev = 0.02, range = 292 to 416.

Ethyl isopropyl sulfide, CA 5145-99-3: Stench; toxic; flammable. Tm =
150.96; Antoine (liq) A = 6.01338, B = 1291.716, C = -58.221, dev = 0.1,
range = 319 to 391.

Ethyl propyl sulfide, CA 4110-50-3: Stench; toxic; flammable; soluble
ethanol. Tm = 156.15; Tb = 391.65; Vm = 0.1245 at 293.15; Antoine (liq) A
= 6.05033, B = 1336.821, C = -61.135, dev = 0.1, range = 331 to 398.

2-Methyl-1-butanethiol, CA 1878-18-8: Stench; toxic; flammable, flash
point 292; soluble ethanol, ether. Tb = 392.15 to 394.15; Vm = 0.1238 at
296.15; Antoine (liq) A = 6.03776, B = 1347.127, C = -58.051, dev = 0.02,
range = 324 to 432.

3-Methyl-1-butanethiol, isopentyl mercaptan, CA 541-31-1: Stench; toxic;
flammable, flash point 291; soluble ethanol, ether. Tb = 389.15; Vm =
0.1248 at 293.15; Antoine (liq) A = 6.03692, B = 1341.11, C = -58.819,
dev = 0.02, range = 323 to 431.

2-Methyl-2-butanethiol, CA 1679-09-0: Stench; toxic; flammable, flash
point 272; soluble ethanol, ether. Tm = 169.38; Tb = 372.15 to 376.15;
Vm = 0.1238 at 293.15; Antoine (liq) A = 5.9499, B = 1253.118, C = -54.567,
dev = 0.02, range = 320 to 411.

3-Methyl-2-butanethiol, CA 2084-18-6: Stench; toxic; flammable, flash
point 276; soluble ethanol, ether. Tb = 383.15; Antoine (liq) A = 5.99785,
B = 1306.129, C = -55.739, dev = 0.02, range = 315 to 422.

1-Pentanethiol, pentyl mercaptan, CA 110-66-7: Stench; allergen; toxic;
flammable, flash point 291; slightly soluble water; soluble ethanol, ether.
Tm = 197.45; Tb = 399.65; Vm = 0.1238 at 293.15; Antoine (liq) A = 6.05703,
B = 1369.359, C = -61.794, dev = 0.02, range = 347 to 430.

$C_5H_{12}S_2$ MW 136.27

Ethyl isopropyl disulfide, CA 53966-36-2: Stench; toxic; flammable. Antoine (liq) A = 6.08945, B = 1520.006, C = -66.442, dev = 0.1, range = 363
to 427.   *(continues)*

$C_5H_{12}S_2$ *(continued)*

Ethyl propyl disulfide, CA 30453-31-7: Stench; toxic; flammable. Antoine (liq) A = 6.14352, B = 1572.62, C = -66.814, dev = 0.1, range = 373 to 414.

1,5-Pentanedithiol, CA 928-98-3: Stench; toxic; flammable, flash point 368. Tm = 200.65; Vm = 0.1341 at 293.15; Antoine (liq) A = 6.7819, B = 2150.3, C = -40.24, dev = 1.0, range = 363 to 491.

$C_5H_{13}N$ MW 87.16

Diethylmethylamine, CA 616-39-7: Irritant; flammable, flash point 246; soluble water, ethanol, ether. Tm = 158.55; Tb = 339.15 to 340.15; Vm = 0.1240 at 298.15; Antoine (liq) A = 5.9422, B = 1130.178, C = -52.007, dev = 0.1, range = 283 to 339.

N-Methyl butyl amine, CA 110-68-9: Irritant; corrosive; flammable, flash point 274. Tb = 363.65 to 364.65; Vm = 0.1184 at 291.25; Antoine (liq) A = 7.54635, B = 1992.9, C = 0, dev = 1.0, range = 283 to 313.

Pentyl amine, CA 110-58-7: Strong irritant; moderately toxic; flammable, flash point 272; very soluble water; soluble ethanol, ether, acetone, benzene. Tf = 218.15; Tb = 377.15; Vm = 0.1155 at 293.15; Antoine (liq) A = 6.3229, B = 1396.9, C = -53.65, dev = 1.0, range = 298 to 417.

$C_5H_{13}NO_2$ MW 119.16

Methyl diethanolamine, CA 105-59-9: Irritant; flammable, flash point 400; soluble water. Tf = 252.15; Tb = 520.15; Vm = 0.1148 at 293.15; Antoine (liq) A = 9.33371, B = 3815.32, C = 0, dev = 1.0 to 5.0, range = 390 to 520.

$C_5H_{13}NO_2S$ MW 151.22

Methanesulfonamide, *N,N*-diethyl, CA 2374-61-0: Irritant; flammable. Tm = 284.65; Antoine (liq) A = 7.16246, B = 2720.698, C = 0, dev = 1.0 to 5.0, range = 384 to 528.

$C_5H_{13}NS$ MW 119.22

tert-Butylsulfinic acid, monomethylamide: Irritant; flammable. Antoine (liq) A = 7.50923, B = 2187.3, C = 0, dev = 1.0, range = 329 to 397.

$C_5H_{13}O_3P$ MW 152.13

Methylphosphonic acid, diethyl ester, CA 683-08-9: Irritant; flammable; soluble water, ethanol, ether. Tb = 465.15 to 467.15; Vm = 0.1447 at 296.15; Antoine (liq) A = 7.9222, B = 2706.8, C = 0, dev = 1.0, range = 343 to 402.

$C_5H_{14}N_2$ MW 102.18

N,N-Dimethyl-1,3-propanediamine, CA 109-55-7: Irritant; flammable, flash point 318. Tf = 173.15; Tb = 408.05; Vm = 0.1258 at 293.15; Antoine I (liq) A = 7.39835, B = 2196.8, C = 0, dev = 1.0, range = 303 to 408; Antoine II (liq) A = 5.92247, B = 1254.734, C = -87.13, dev = 1.0, range = 303 to 366.   *(continues)*

$C_5H_{14}N_2$ *(continued)*

Methane, bis(dimethylamino), CA 51-80-9: Irritant; flammable, flash point 261; soluble water. Tm = below 133.15; Tb = 355.65; Vm = 0.1364 at 293.15; Antoine (liq) A = 5.39197, B = 926.13, C = -84.046, dev = 1.0, range = 301 to 348.

C_5O_2 MW 92.05

Pentacarbondioxide, CA 51799-36-1: Tm = about 173.15; Antoine (liq) A = 2.635, B = 240.2, C = 0, dev = 5.0, range = 186 to 273.

C_6BrF_5 MW 246.96

Bromopentafluorobenzene, CA 344-04-7: Irritant; flammable, flash point 360. Tm = 243.25; Tb = 408.75; Vm = 0.1268 at 298.15; Antoine (liq) A = 6.20068, B = 1475.78, C = -57.82, dev = 1.0, range = 400 to 522.

$C_6BrF_{15}N_2S$ MW 497.02

Diimidosulfuryl bromide fluoride, bis[1,2,2,2-tetrafluoro-1-(trifluoro-methyl)ethyl], CA 62977-74-6: Antoine (liq) A = 6.505, B = 2142, C = 0, dev = 5.0, range not specified.

C_6ClF_5 MW 202.51

Chloropentafluorobenzene, CA 344-07-0: Nonflammable. Tb = 395.15 to 396.15; Vm = 0.1292 at 298.15; Antoine I (liq) A = 6.19210, B = 1388.701, C = -59.393, dev = 0.02, range = 348 to 402; Antoine II (liq) A = 6.18873, B = 1386.456, C = -59.659, dev = 0.02 to 0.1, range = 307 to 417.

$C_6ClF_{13}N_2$ MW 382.51

Azoethane, 1-chloro-1',2,2,2,2',2',2'-heptafluoro-1,1'-bis(trifluoromethyl), CA 33757-14-1: Antoine (liq) A = 6.88502, B = 1742.156, C = 0, dev = 1.0 to 5.0, range = 297 to 355.

$C_6ClF_{15}N_2S$ MW 452.57

Diimidosulfuryl chloride fluoride, bis[1,2,2,2-tetrafluoro-1-(trifluoro-methyl)ethyl], CA 62977-72-4: Antoine (liq) A = 6.255, B = 1945, C = 0, dev = 5.0, range not specified.

$C_6Cl_2F_{12}N_2S$ MW 431.03

Sulfur diimide, bis[1-chloro-2,2,2-trifluoro-1-(trifluoromethyl)ethyl], CA 38005-17-3: Antoine (liq) A = 7.625, B = 2274, C = 0, dev = 1.0 to 5.0, range not specified.

$C_6Cl_3F_3$ MW 235.42

1,3,5-Trichloro-2,4,6-trifluorobenzene, CA 319-88-0: Antoine (liq) A = 6.26803, B = 1712.481, C = -69.744, dev = 0.1, range = 364 to 496.

$C_6Cl_3N_3O_6$ MW 316.44

1,3,5-Trichloro-2,4,6-trinitrobenzene, CA 2631-68-7: Explosive. Tm =
466.15; Antoine (liq) A = 8.08354, B = 3601.7, C = 0, dev = 1.0, range =
503 to 543.

$C_6Cl_4O_2$ MW 245.88

Tetrachloro-1,4-benzoquinone, chloranil, CA 118-75-2: Irritant; flammable;
soluble ether. Tm = 563.15, sealed tube; Tb = sublimes; Antoine (sol) A =
11.185, B = 5170, C = 0, dev = 5.0, range = 333 to 356.

C_6Cl_6 MW 284.78

Hexachlorobenzene, perchlorobenzene, CA 118-74-1: Moderately toxic; flammable; moisture-sensitive; suspected carcinogen; soluble ether, benzene,
chloroform. Tm = 501.85; Tb = 592.45; Tc = 824.15; Pc = 2847; Vc = 0.518;
Antoine I (sol) A = 7.00706, B = 2831.85, C = -28.25, dev = 1.0, range =
387 to 502; Antoine II (liq) A = 7.35248, B = 2786.78, C = -61.33, dev =
1.0, range = 502 to 589.

C_6F_6 MW 186.06

Hexafluorobenzene, perfluorobenzene, CA 392-56-3: Flammable, flash point
283; soluble most organic solvents. Tm = 278.25; Tb = 353.41; Vm = 0.1158
at 298.15; Tc = 516.72; Pc = 3304; Antoine I (sol) A = 11.49514, B =
3518.13, C = 44.44, dev = 0.1, range = 215 to 278; Antoine II (liq) A =
6.15233, B = 1224.974, C = -57.984, dev = 0.02, range = 278 to 354; Antoine
III (liq) A = 6.19544, B = 1251.177, C = -54.775, dev = 0.1, range = 348 to
389; Antoine IV (liq) A = 6.43336, B = 1418.987, C = -32.607, dev = 0.1,
range = 384 to 462; Antoine V (liq) A = 7.64375, B = 2643.629, C = 123.522,
dev = 0.1, range = 458 to 517.

C_6F_8 MW 224.05

Perfluoro(2-methyl-3-methylene-cyclobutene): Antoine (liq) A = 7.125, B =
1619, C = 0, dev = 1.0 to 5.0, range = 243 to 306.

$C_6F_{10}O_4$ MW 326.05

Hexafluoro-4-(fluoroformyl)peroxybutyric acid, trifluoromethyl ester,
CA 32750-98-4: Antoine (liq) A = 8.1335, B = 2287.7, C = 0, dev not specified, range not specified.

$C_6F_{11}NO$ MW 311.05

Propanamide, 2,2,3,3,3-pentafluoro-N-[2,2,2-trifluoro-1-(trifluoromethyl)
ethylidene], CA 52225-58-8: Antoine (liq) A = 7.065, B = 1707, C = 0,
dev = 1.0 to 5.0, range not specified.

C_6F_{12} MW 300.05

Perfluorocyclohexane, CA 355-68-0: Nonflammable. Tm = 321.15 to 322.15;
Tb = 325.15; Antoine I (sol) A = 7.6741, B = 1809.3,   *(continues)*

C_6F_{12} *(continued)*

C = -6.55, dev = 1.0, range = 252 to 322; Antoine II (liq) A = 6.3199, B = 1297.8, C = -21.58, dev = 1.0, range = 350 to 451.

Perfluoro(1,2-dimethylcyclobutane), CA 2994-71-0: Tm = 231.85; Antoine (liq) A = 6.00841, B = 1060.41, C = -52.57, dev = 0.1, range = 242 to 318.

$C_6F_{12}N_2$ MW 328.06

1,2-Ethynediamine, *N,N,N',N'*-tetrakis(trifluoromethyl), CA 19451-96-8: Antoine (liq) A = 7.0899, B = 1676, C = 0, dev = 1.0 to 5.0, range = 305 to 328.

$C_6F_{12}N_2OS$ MW 376.12

2-Propanamine, 1,1,1,3,3,3-hexafluoro-2-isothiocyanato-*N*-[2,2,2-trifluoro-1-(trifluoromethyl)ethylidene], CA 34619-84-6: Antoine (liq) A = 7.505, B = 2068, C = 0, dev = 1.0 to 5.0, range not specified.

$C_6F_{12}N_2O_2S$ MW 392.12

Methanesulfonimidamide, 1,1,1-trifluoro-*N'*-(trifluoroacetyl)-*N*-[2,2,2-trifluoro-1-(trifluoromethyl)ethylidene], CA 62609-66-9: Antoine (liq) A = 6.745, B = 1915, C = 0, dev = 1.0 to 5.0, range not specified.

$C_6F_{12}N_2S$ MW 360.12

Sulfoxylic diamide, bis[2,2,2-trifluoro-1-(trifluoromethyl)ethylidene], CA 31340-33-7: Antoine (liq) A = 7.285, B = 2068, C = 0, dev = 1.0 to 5.0, range not specified.

$C_6F_{12}N_2S_2$ MW 392.18

2-Propanimine, *N,N'*-dithiobis(1,1,1,3,3,3-hexafluoro), CA 38005-16-2: Antoine (liq) A = 7.785, B = 2415, C = 0, dev = 1.0 to 5.0, range not specified.

$C_6F_{12}O$ MW 316.05

Perfluoro(methoxycyclopentane), CA 788-40-9: Tm = 159.15; Antoine (liq) A = 5.0742, B = 677.31, C = -109.2, dev = 1.0, range = 246 to 330.

$C_6F_{12}O_2$ MW 332.05

Trifluoroacetic acid, 2,2,2-trifluoro-1,1-bis(trifluoromethyl)ethyl ester, CA 24165-10-4: Antoine (liq) A = 7.4729, B = 1793.8, C = 0, dev = 1.0, range = 264 to 298.

$C_6F_{12}O_4$ MW 364.04

Carbonoperoxoic acid, *O*-[2,2,2-trifluoro-1,1-bis(trifluoromethyl)ethyl]-*OO*-(trifluoromethyl) ester, CA 55100-93-1: Antoine (liq) A = 6.9465, B = 1748.3, C = 0, dev = 1.0, range = 273 to 315.

$C_6F_{13}NS$ MW 365.11

Ethanimidothioic acid, 2,2,2-trifluoro-*N*-[1,2,2,2-tetrafluoro-1-(trifluoromethyl)ethyl]-trifluoromethyl ester, CA 54120-07-9: Antoine (liq) A =
7.125, B = 1841, C = 0, dev = 1.0 to 5.0, range not specified.

C_6F_{14} MW 338.04

Perfluoro-2,3-dimethylbutane, CA 354-96-1: Tm = 258.05; Antoine (liq) A =
6.35129, B = 1281.85, C = -37.88, dev = 0.1, range = 262 to 333.

Perfluorohexane, CA 355-42-0: Soluble ether, chloroform, benzene. Tm =
187.15; Tb = 331.15; Vm = 0.2012 at 298.15; Tc = 447.65; Pc = 1900; Vc =
0.602, calculated; Antoine I (liq) A = 5.72424, B = 949.53, C = -74.97,
dev = 0.1, range = 261 to 334; Antoine II (liq) A = 7.15614, B = 1743.756,
C = 0, dev = 1.0, range = 433 to 449.

Perfluoro-2-methylpentane, CA 355-04-4: Soluble benzene. Tm = 102.45;
Tb = 330.85; Vm = 0.1951 at 293.15; Antoine (liq) A = 5.90039, B = 1037.59,
C = -64.27, dev = 0.1, range = 253 to 329.

Perfluoro-3-methylpentane, CA 865-71-4: Soluble benzene. Tm = 157.75;
Antoine (liq) A = 5.98448, B = 1094.68, C = -56.29, dev = 0.1, range = 282
to 333.

$C_6F_{14}N_2S$ MW 398.12

Sulfur diimide, bis[1,2,2,2-tetrafluoro-1-(trifluoromethyl)ethyl],
CA 34451-12-2: Antoine (liq) A = 6.93637, B = 1776.112, C = -20.278, dev =
1.0, range = 325 to 378.

$C_6F_{15}N$ MW 371.05

Perfluorotriethylamine, CA 359-70-6: Non-polar solvent; insoluble water,
other polar solvents. Tb = 344.15; Vm = 0.2145 at 298.15; Antoine I (liq)
A = 5.65060, B = 937.41, C = -86.51, dev = 1.0, range = 317 to 349; Antoine
II (liq) A = 6.01894, B = 1133.101, C = -61.157, dev = 0.1, range = 320
to 334.

2-Propanamine, 1,1,1,2,3,3,3-heptafluoro-*N*-(pentafluoroethyl)-*N*-(trifluoromethyl), CA 54566-82-4: Antoine (liq) A = 6.745, B = 1601, C = 0, dev =
1.0 to 5.0, range not specified.

$C_6F_{16}N_2S$ MW 436.11

Diimidosulfuryl fluoride, bis[1,2,2,2-tetrafluoro-1-(trifluoromethyl)ethyl],
CA 59617-31-1: Antoine (liq) A = 7.115, B = 1936, C = 0, dev = 1.0 to 5.0,
range not specified.

$C_6F_{16}O_4S_2$ MW 504.16

1,3-Dithietane, 2,2,4,4-tetrafluoro-1,1,3,3-tetrahydro-1,1,3,3-tetrakis
(trifluoromethoxy), CA 63441-15-6: Antoine (liq) A = 6.795, B = 1935,
C = 0, dev = 5.0, range not specified.

$C_6F_{16}S$ MW 408.10

Difluorobis[1,2,2,2-tetrafluoro-1-(trifluoromethyl)ethyl] sulfur, CA 1423-
18-3: Tm = 195.15; Antoine (liq) A = 6.975, B = 1911, C = 0, dev = 1.0 to
5.0, range = 273 to 383.

$C_6F_{17}N_3$ MW 437.06

Hydrazine, 1-[(bis[trifluoromethyl]amino)difluoromethyl]-1,2,2-tris(trifluoromethyl), CA 18015-82-2: Antoine (liq) A = 7.3779, B = 1920, C = 0,
dev = 1.0 to 5.0, range not specified.

$C_6HBrF_{12}N_2$ MW 408.97

Vinylenediamine, 1-bromo-N,N,N',N'-tetrakis(trifluoromethyl), CA 19451-
95-7: Antoine (liq) A = 6.5159, B = 1683, C = 0, dev = 1.0 to 5.0, range =
348 to 371.

$C_6HClF_{11}NO$ MW 347.52

Propanamide, N-[1-chloro-2,2,2-trifluoro-1-(trifluoromethyl)ethyl]-
2,2,3,3,3-pentafluoro, CA 52225-62-4: Antoine (liq) A = 7.605, B = 2132,
C = 0, dev = 1.0 to 5.0, range not specified.

$C_6HCl_2N_3O_6$ MW 282.00

1,3-Dichloro-2,4,6-trinitrobenzene, CA 1630-09-7: Explosive; soluble
ethanol. Tm = 402.15; Antoine (liq) A = 5.4544, B = 2452, C = 0, dev =
5.0, range = 504 to 563.

$C_6HCl_3F_8O_2$ MW 363.42

Octafluoro-3,5,6-trichlorohexanoic acid, CA 2106-54-9: Antoine (liq) A =
8.6507, B = 3356, C = 0, dev = 1.0 to 5.0, range = 373 to 505.

$C_6HCl_3O_2$ MW 211.43

Trichloro-1,4-benzoquinone, CA 634-85-5: Flammable; soluble ethanol, ether.
Tm = 442.15 to 443.15; Antoine (sol) A = 11.155, B = 4630, C = 0, dev =
5.0, range = 300 to 328.

C_6HCl_5 MW 250.34

Pentachlorobenzene, CA 608-93-5: Flammable; suspected teratogen; soluble
ethanol. Tm = 358.15; Tb = 549.15; Antoine (liq) A = 8.00795, B = 3325.33,
C = 4.814, dev = 1.0, range = 371 to 549.

C_6HCl_5O MW 266.34

Pentachlorophenol, CA 87-86-5: Irritant; highly toxic by inhalation and
skin absorption; almost insoluble water; soluble ethanol, ether, benzene;
used as antifungal agent. Tm = 463.15; Tb = 582.15 to 583.15, decomposes;
Antoine (liq) A = 8.198, B = 3606, C = 0, dev = 1.0, range = 463 to 507.

C$_6$HF$_5$ MW 168.07

Pentafluorobenzene, CA 363-72-4: Flammable, flash point 283. Tm = 225.15;
Tb = 358.15; Vm = 0.1110 at 293.15; Antoine I (liq) A = 6.15571, B =
1250.946, C = -57.457, dev = 0.02, range = 322 to 368; Antoine II (liq) A =
6.20712, B = 1282.574, C = -53.619, dev = 0.1, range = 358 to 397; Antoine
III (liq) A = 6.47368, B = 1477.401, C = -27.81, dev = 0.1, range = 393 to
479; Antoine IV (liq) A = 7.65942, B = 2731.601, C = 133.448, dev = 0.1,
range = 473 to 531.

C$_6$HF$_5$O   MW 184.07

Pentafluorophenol, CA 771-61-9: Irritant; flammable, flash point 345.
Tm = 307.15 to 309.15; Tb = 416.15; Antoine I (sol) A = 11.291, B = 3523,
C = 0, dev = 1.0, range = 273 to 296; Antoine II (liq) A = 6.18665, B =
1377.011, C = -89.435, dev = 0.02, range = 378 to 428.

C$_6$HF$_{12}$NO   MW 331.06

Propanamide, 2,2,3,3,3-pentafluoro-N-[1,2,2,2-tetrafluoro-1-(trifluoromethyl)ethyl], CA 52225-64-6: Antoine (liq) A = 7.865, B = 2158, C = 0,
dev = 1.0 to 5.0, range not specified.

C$_6$HF$_{12}$NOS   MW 363.12

2,2,2-Trifluoro-N-[(trifluoromethyl)thio]ethanimidic acid, 2,2,2-trifluoro-
1-(trifluoromethyl)ethyl ester, CA 62067-08-7: Antoine (liq) A = 6.825,
B = 1752.4, C = 0, dev = 1.0 to 5.0, range not specified.

C$_6$H$_2$BrCl$_3$O   MW 276.34

3-Bromo-2,4,6-trichlorophenol: Irritant; flammable. Tm = 336.15; Antoine
(liq) A = 7.37465, B = 2918.99, C = -35.025, dev = 1.0, range = 385 to 579.

C$_6$H$_2$ClN$_3$O$_6$ MW 247.55

1-Chloro-2,4,6-trinitrobenzene, picryl chloride, CA 88-88-0: Irritant; explosive; soluble ethanol, acetone, benzene. Tm = 356.15; Antoine (liq) A =
7.0949, B = 3298, C = 0, dev = 5.0, range = 473 to 543.

C$_6$H$_2$Cl$_2$O$_2$ MW 176.99

2,6-Dichloro-1,4-benzoquinone, CA 697-91-6: Flammable; soluble ethanol,
chloroform. Tm = 394.15; Antoine (sol) A = 8.975, B = 3670, C = 0, dev =
5.0, range = 274 to 315.

C$_6$H$_2$Cl$_3$F   MW 199.44

1-Fluoro-2,4,6-trichlorobenzene, CA 400-04-4: Irritant; toxic. Tm =
284.35; Antoine (liq) A = 8.2712, B = 3725.8, C = 113.74, dev = 1.0, range
= 344 to 489.

$C_6H_2Cl_3NO_2$ MW 226.45

2,4,5-Trichloro-1-nitrobenzene, CA 89-69-0: Irritant; flammable, flash
point above 385; Tm = 330.15; Tb = 561.15; Antoine (liq) A = 7.315, B =
2963, C = 0, dev = 5.0, range = 427 to 560.

$C_6H_2Cl_4$ MW 215.89

1,2,3,4-Tetrachlorobenzene, CA 634-66-2: Irritant; flammable, flash point
above 385; soluble most organic solvents. Tm = 319.15; Tb = 528.05; Vm =
0.127 at 319.15; Tc = 723.15; Pc = 3380; Vc = 0.400; Antoine (liq) A =
5.7082, B = 1517.2, C = -117.384, dev = 1.0, range = 331 to 527.

1,2,3,5-Tetrachlorobenzene, CA 634-90-2: Irritant; flammable, flash point
above 385; soluble most organic solvents. Tm = 324.15; Tb = 519.15;
Antoine (liq) A = 6.7756, B = 2394.0, C = -17.85, dev = 1.0, range = 331
to 519.

1,2,4,5-Tetrachlorobenzene, CA 95-94-3: Irritant; nonflammable under most
conditions; soluble most organic solvents. Tm = 412.65; Tb = 521.15; Tc =
762.95; Pc = 3380; Vc = 0.475; Antoine (liq) A = 9.1357, B = 4642.36, C =
132.952, dev = 1.0, range = 419 to 518.

$C_6H_2Cl_4O$ MW 231.89

2,3,4,6-Tetrachlorophenol, CA 58-90-2: Highly toxic by inhalation or skin
absorption; soluble ethanol, benzene, chloroform, light hydrocarbons; used
as a wood preservative. Tm = 342.15 to 343.15; Antoine (liq) A = 7.96172,
B = 3227.92, C = -6.121, dev = 1.0, range = 373 to 548.

$C_6H_2Cl_4O_2$ MW 247.89

Tetrachlorohydroquinone, CA 87-87-6: Soluble ethanol, ether. Tm = 516.15
to 518.15; Antoine (sol) A = 9.205, B = 4650, C = 0, dev = 5.0, range =
298 to 359.

$C_6H_2F_4$ MW 150.08

1,2,3,4-Tetrafluorobenzene, CA 551-62-2: Irritant; flammable, flash point
277. Tm = 231.15; Tb = 368.15; Vm = 0.1055 at 298.15; Antoine I (liq) A =
6.31876, B = 1396.067, C = -44.277, dev = 0.1, range = 279 to 323; Antoine
II (liq) A = 6.16042, B = 1292.174, C = -56.495, dev = 0.02, range = 300
to 392.

1,2,3,5-Tetrafluorobenzene, CA 2367-82-0: Irritant; flammable, flash point
289. Tm = 225.15; Tb = 356.15; Vm = 0.1077 at 293.15; Antoine I (liq) A =
6.20248, B = 1290.984, C = -50.295, dev = 0.1, range = 279 to 323; Antoine
II (liq) A = 6.15119, B = 1253.771, C = -55.168, dev = 0.02, range = 287 to
382; Antoine III (liq) A = 6.24644, B = 1317.349, C = -46.898, dev = 0.02,
range = 385 to 416.

1,2,4,5-Tetrafluorobenzene, CA 327-54-8: Irritant; flammable, flash point
277. Tm = 277.15; Tb = 363.15; Vm = 0.1054 at 298.15; Antoine I (liq) A =
6.17614, B = 1278.93, C = -56.748, dev = 0.02 to 0.1, range = 293 to 390;
Antoine II (liq) A = 6.42009, B = 1454.406, C = -33.675, dev = 0.1 to 1.0,
range = 390 to 488; Antoine III (liq) A = 7.88521, B = 3090.851, C =
174.387, dev = 1.0, range = 488 to 543.

$C_6H_2F_{12}O_3S$ MW 382.12

2-Propanol, 1,1,1,3,3,3-hexafluorosulfite, CA 53517-89-8: Antoine (liq)
A = 7.485, B = 2216, C = 0, dev = 1.0 to 5.0, range not specified.

$C_6H_3BrCl_2O$ MW 241.90

2-Bromo-4,6-dichlorophenol, CA 4524-77-0: Irritant; toxic; flammable; soluble ethanol, benzene, chloroform. Tm = 341.15; Antoine (liq) A = 7.41375,
B = 2886.48, C = -7.484, dev = 1.0, range = 357 to 541.

$C_6H_3ClN_2O_4$ MW 202.55

1-Chloro-2,4-dinitrobenzene, CA 97-00-7: Irritant; very toxic; causes
severe dermatitis; flammable, flash point 459; has been used as explosive;
soluble ethanol, benzene. Tm = 324.15; Tb = 588.15, decomposes; Vm =
0.1352 at 348.15; Antoine (liq) A = 9.1525, B = 4203.39, C = 0, dev = 1.0
to 5.0, range = 430 to 590.

$C_6H_3ClO_2$ MW 142.54

2-Chloro-1,4-benzoquinone, CA 695-99-8: Irritant; flammable; soluble
water, ethanol, ether, chloroform. Tm = 330.15; Antoine (sol) A = 9.865,
B = 3620, C = 0, dev = 5.0, range = 264 to 298.

$C_6H_3Cl_2NO_2$ MW 192.00

3,4-Dichloro-1-nitrobenzene, CA 99-54-7: Irritant; flammable, flash point
396; soluble ether, benzene, hot ethanol. Tm = 315.15 to 316.15; Tb =
528.15 to 529.15; Vm = 0.1319 at 348.15; Antoine (liq) A = 7.385, B = 2894,
C = 0, dev = 5.0, range = 417 to 515.

$C_6H_3Cl_3$ MW 181.45

1,2,3-Trichlorobenzene, CA 87-61-6: Irritant; flammable, flash point 399;
soluble ether, benzene, carbon disulfide. Tm = 326.65; Tb = 491.65;
Antoine I (sol) A = 9.787, B = 3440, C = 0, dev = 5.0, range = 289 to 303;
Antoine II (liq) A = 7.23008, B = 2624.09, C = 10.506, dev = 1.0, range =
343 to 492.

1,2,4-Trichlorobenzene, CA 120-82-1: Irritant; flammable, flash point 378;
soluble ether, benzene, carbon disulfide. Tm = 290.3; Tb = 486.95; Vm =
0.1251 at 299.15; Tc = 726.45; Pc = 3718; Vc = 0.447; Antoine I (sol) A =
9.570, B = 3254, C = 0, dev = 1.0 to 5.0, range = 279 to 290; Antoine II
(liq) A = 6.31998, B = 1827, C = -63.15, dev = 0.1, range = 383 to 553;
Antoine III (liq) A = 6.6802, B = 2064.4, C = -43.05, dev = 1.0, range =
293 to 383.

1,3,5-Trichlorobenzene, CA 108-70-3: Moderately toxic and irritant; flammable, flash point 380; soluble ether, benzene, carbon disulfide, light
hydrocarbons. Tm = 336.65; Tb = 481.65; Antoine I (sol) A = 8.301, B =
2956, C = 0, dev = 5.0, range = 282 to 301; Antoine II (liq) A = 6.43345,
B = 1932.26, C = -45.268, dev = 1.0, range = 336 to 482.

$C_6H_3Cl_3O$ MW 197.45

2,4,5-Trichlorophenol, CA 95-95-4: Irritant; toxic; flammable; soluble
most organic solvents; used to manufacture herbicides. Tm = 341.15; Tb =
517.15 to 521.15; Antoine (liq) A = 7.38197, B = 2812.25, C = -2.091, dev =
1.0, range = 345 to 525.

2,4,6-Trichlorophenol, CA 88-06-2: Irritant; suspected carcinogen; nonflam-
mable under most conditions; soluble most organic solvents. Tm = 341.15;
Tb = 519.15; Vm = 0.1325 at 348.15; Antoine (liq) A = 7.67323, B = 2876.7,
C = -11.682, dev = 1.0, range = 349 to 519.

$C_6H_3Cl_3O_2$ MW 213.45

Trichlorohydroquinone, CA 608-94-6: Moderately soluble water. Tm = 411.15;
Antoine (sol) A = 12.695, B = 5300, C = 0, dev = 5.0, range = 298 to 336.

$C_6H_3F_3$ MW 132.08

1,3,5-Trifluorobenzene, CA 372-38-3: Irritant; flammable, flash point 266.
Tm = 267.65; Tb = 348.65; Vm = 0.1034 at 298.15; Antoine (liq) A = 6.04363,
B = 1196.385, C = -54.131, dev = 1.0, range = 279 to 350.

$C_6H_3F_9O_2$ MW 278.07

Acetic acid, 2,2,2-trifluoro-1,1-bis(trifluoromethyl)ethyl ester, CA 24165-
09-1: Antoine (liq) A = 7.8467, B = 2092.3, C = 0, dev = 1.0, range = 273
to 328.

Trifluoroacetic acid, 1,1-bis(trifluoromethyl)ethyl ester, CA 42031-16-3:
Antoine (liq) A = 7.165, B = 1743, C = 0, dev = 1.0 to 5.0, range not
specified.

$C_6H_3F_{10}NS$ MW 311.14

Ethanimidothioic acid, 2,2,2-trifluoro-N-[1,2,2,2-tetrafluoro-1-(trifluoromethyl)ethyl], methyl ester, CA 54120-08-0: Antoine (liq) A = 6.325, B =
1651, C = 0, dev = 1.0 to 5.0, range not specified.

$C_6H_3N_3O_6$ MW 213.11

1,2,3-Trinitrobenzene, CA 603-13-4: Explosive when heated rapidly; soluble
ethanol, ether, acetone, benzene. Tm = 400.65; Antoine (liq) A = 6.635,
B = 3148, C = 0, dev = 5.0, range = 523 to 573.

1,2,4-Trinitrobenzene, CA 610-31-1: Explosive when heated rapidly; soluble
ethanol, ether, acetone, benzene. Tm = 334.15 to 335.15; Antoine (liq) A =
9.013, B = 4314, C = 0, dev = 5.0, range = 523 to 573.

1,3,5-Trinitrobenzene, CA 99-35-4: Explosive when heated rapidly; soluble
methanol, ethanol, ether, acetone, benzene. Tm = 395.65, higher melting
form; Tb = 588.15; Vm = 0.1442 at 425.15; Antoine (liq) A = 7.876, B =
3671, C = 0, dev = 5.0, range = 475 to 585.

$C_6H_3N_3O_7$ MW 229.11

2,4,6-Trinitrophenol, picric acid, CA 88-89-1: Toxic; can explode from
heating or shock; moderately soluble water; soluble ethanol, ether, acetone, benzene; has been used as military explosive. Tm = 395.65, melting
point of stable form; Tb = above 573.15, explodes; Antoine (liq) A = 11.319,
B = 5560, C = 0, dev = 5.0, range = 468 to 598.

$C_6H_3N_3O_8$ MW 245.11

Resorcinol, 2,4,6-trinitro, CA 82-71-3: Explodes on rapid heating or shock;
soluble ethanol, ether; dibasic acid moderately soluble water. Tm = 452.15
to 453.15; Antoine (sol) A = 13.03, B = 6301, C = 0, dev not specified,
range = 325 to 436.

C_6H_4BrCl MW 191.45

1-Bromo-3-chlorobenzene, CA 108-37-2: Irritant; flammable, flash point 353;
soluble ethanol, ether, benzene, chloroform. Tm = 252.15; Tb = 469.15;
Vm = 0.1174 at 293.15; Antoine (liq) A = 6.4972, B = 1912.7, C = -43.15,
dev = 0.1 to 1.0, range = 252 to 469.

1-Bromo-4-chlorobenzene, CA 106-39-8: Irritant; flammable; soluble ethanol,
ether, benzene, chloroform. Tm = 340.55; Tb = 469.15; Vm = 0.1215 at
344.15; Antoine I (sol) A = 10.478, B = 3548.4, C = 0, dev = 1.0, range =
294 to 337; Antoine II (liq) A = 6.71377, B = 2074.22, C = -35.248, dev =
1.0, range = 333 to 470.

$C_6H_4BrNO_2$ MW 202.01

4-Bromo-1-nitrobenzene, CA 586-78-7: Irritant; flammable; soluble ethanol,
ether, benzene. Tm = 400.15; Tb = 529.15; Antoine (sol) A = 11.8319, B =
4615, C = 0, dev = 1.0 to 5.0, range = 293 to 313.

$C_6H_4Br_2$ MW 235.91

1,2-Dibromobenzene, CA 583-53-9: Irritant; flammable, flash point 364;
soluble ether, benzene, chloroform. Tm = 280.25; Tb = 498.15; Vm = 0.1201
at 293.15; Antoine (liq) A = 6.22755, B = 1825.77, C = -66.15, dev = 0.1,
range = 388 to 568.

1,3-Dibromobenzene, CA 108-36-1: Irritant; flammable, flash point 366;
soluble ether, benzene, chloroform. Tm = 266.15; Tb = 492.65; Vm = 0.1208
at 293.15; Antoine (liq) A = 5.971, B = 1603.4, C = -87.55, dev = 1.0,
range = 417 to 500.

1,4-Dibromobenzene, CA 106-37-6: Irritant; flammable; soluble ether, benzene, chloroform. Tm = 360.45; Tb = 491.15 to 492.15; Antoine I (sol) A =
10.717, B = 3826.2, C = 0, dev = 1.0 to 5.0, range = 298 to 354; Antoine II
(liq) A = 6.5732, B = 2057.2, C = -43.15, dev = 0.1 to 1.0, range = 373
to 493.

C_6H_4ClF MW 130.55

1-Chloro-3-fluorobenzene, CA 625-98-9: Irritant; flammable, flash point
291; soluble benzene. Tm = below 195.15; Tb = 400.75;   *(continues)*

C_6H_4ClF *(continued)*

Vm = 0.1069 at 298.15; Antoine (liq) A = 6.9669, B = 2003.85, C = 3.47, dev = 1.0, range = 273 to 403.

C_6H_4ClI MW 238.46

1-Chloro-4-iodobenzene, CA 637-87-6: Irritant; flammable, flash point 381; soluble ethanol, benzene. Tm = 330.15; Tb = 500.15; Vm = 0.1264 at 300.15; Antoine I (sol) A = 8.944, B = 3200, C = 0, dev = 5.0, range = 303 to 323; Antoine II (liq) A = 5.63678, B = 1465.651, C = -102.487, dev = 5.0, range = 333 to 500.

$C_6H_4ClNO_2$ MW 157.56

1-Chloro-2-nitrobenzene, CA 88-73-3: Irritant; highly toxic by inhalation or skin absorption; flammable, flash point above 385; soluble ether, acetone, ethanol, benzene. Tm = 306.15; Tb = 518.15; Antoine (liq) A = 6.7806, B = 2280.1, C = -39.85, dev = 1.0, range not specified.

1-Chloro-4-nitrobenzene, CA 100-00-5: Irritant; highly toxic by inhalation or skin absorption; flammable, flash point 383; soluble ethanol, acetone, ether, benzene. Tm = 356.15; Tb = 515.15; Vm = 0.1214 at 363.65; Antoine I (sol) A = 12.0579, B = 4345, C = 0, dev = 1.0 to 5.0, range = 283 to 303; Antoine II (liq) A = 7.245, B = 2680, C = 0, dev = 5.0, range = 385 to 515.

$C_6H_4Cl_2$ MW 147.00

1,2-Dichlorobenzene, *ortho*-dichlorobenzene, CA 95-50-1: Irritant; moderately toxic by inhalation; flammable, flash point 339; soluble ethanol, ether, acetone, benzene. Tm = 256.18; Tb = 453.55; Vm = 0.1126 at 293.15; Tc = 690.35; Pc = 4031; Vc = 0.411; Antoine (liq) A = 6.26918, B = 1705.55, C = -53.56, dev = 0.1, range = 373 to 453.

1,3-Dichlorobenzene, *meta*-dichlorobenzene, CA 541-73-1: Irritant; moderately toxic by inhalation; flammable, flash point 336; soluble ethanol, ether, acetone, benzene. Tf = 248.39; Tb = 446.15; Vm = 0.1141 at 293.15; Tc = 688.45; Pc = 4864; Vc = 0.458; Antoine (liq) A = 6.00535, B = 1496.2, C = -72.15, dev = 0.1 to 1.0, range = 348 to 513.

1,4-Dichlorobenzene, *para*-dichlorobenzene, CA 106-46-7: Irritant; moderately toxic by inhalation; flammable, flash point 339; soluble ethanol, ether, acetone, benzene. Tm = 326.19; Tb = 447.25; Tc = 680.65; Pc = 4109; Vc = 0.411; Antoine I (sol) A = 10.472, B = 3382.9, C = 0, dev = 1.0 to 5.0, range = 293 to 313; Antoine II (sol) A = 10.181, B = 3290.4, C = 0, dev = 1.0 to 5.0, range = 310 to 326; Antoine III (liq) A = 6.12695, B = 1578.51, C = -64.22, dev = 0.1, range = 341 to 448.

$C_6H_4Cl_2O$ MW 163.00

2,4-Dichlorophenol, CA 120-83-2: Irritant; moderately toxic by skin absorption; causes burns; flammable, flash point 387; soluble ethanol, ether, benzene, chloroform. Tm = 316.15 to 317.15; Tb = 483.15 to 484.15; Antoine I (liq) A = 6.75941, B = 1945.1, C = -73.987, dev = 1.0, range = 326 to 483; Antoine II (liq) A = 6.32554, B = 1807.32, C = -69.17, dev = 1.0, range = 391 to 474.

(continues)

$C_6H_4Cl_2O$ *(continued)*

2,6-Dichlorophenol, CA 87-65-0: Irritant; moderately toxic by skin absorption; causes burns; flammable; soluble ethanol, ether, benzene, chloroform. Tm = 340.15; Tb = 492.15 to 493.15; Antoine I (liq) A = 7.32845, B = 2436.59, C = -35.584, dev = 1.0, range = 333 to 493; Antoine II (liq) A = 5.2254, B = 1106.4, C = -151.42, dev = 1.0, range = 353 to 493.

3,5-Dichlorophenol, CA 591-35-5: Irritant; moderately toxic by skin absorption; causes burns; flammable; soluble ethanol, ether, benzene, chloroform. Tm = 341.15; Tb = 506.15; Antoine (sol) A = 9.63028, B = 3750.828, C = 0, dev = 1.0, range = 273 to 295.

$C_6H_4Cl_2O_2$ MW 179.00

2,6-Dichlorohydroquinone, CA 20103-10-0: Flammable; moderately soluble water; soluble ethanol, ether, acetone. Tm = 437.15; Antoine (sol) A = 10.665, B = 4800, C = 0, dev = 5.0, range = 323 to 345.

$C_6H_4Cl_2O_3$ MW 195.00

Vinyl mucochlorate: Antoine (liq) A = 8.9542, B = 3340.3, C = 0, dev = 1.0 to 5.0, range = 273 to 333.

$C_6H_4Cl_3N$ MW 196.46

2,4,6-Trichloroaniline, CA 634-93-5: Irritant; soluble ethanol, ether. Tm = 351.65; Tb = 535.15; Antoine (liq) A = 11.085, B = 4864.09, C = 0.469, dev = 1.0, range = 407 to 535.

$C_6H_4F_2$ MW 114.09

1,2-Difluorobenzene, CA 367-11-3: Flammable, flash point 280; soluble acetone, benzene, chloroform. Tm = 239.15; Tb = 364.15 to 365.15; Vm = 0.0992 at 298.15.

1,3-Difluorobenzene, CA 372-18-9: Flammable, flash point 262; soluble acetone, benzene, chloroform. Tm = 213.85; Tb = 355.15 to 356.15; Vm = 0.0986 at 293.15.

1,4-Difluorobenzene, CA 540-36-3: Flammable, flash point 261; soluble acetone, benzene, chloroform. Tm = 260.15; Tb = 361.15 to 362.15; Vm = 0.0974 at 293.15.

$C_6H_4INO_2$ MW 249.01

2-Iodo-1-nitrobenzene, CA 609-73-4: Irritant; flammable, flash point above 385; soluble ethanol, ether. Tm = 327.15; Tb = 561.15 to 562.15; Vm = 0.1298 at 348.15; Antoine (liq) A = 7.555, B = 3129, C = 0, dev = 5.0, range = 433 to 563.

$C_6H_4I_2$ MW 329.91

1,4-Diiodobenzene, CA 624-38-4: Irritant; flammable; soluble ethanol, ether. Tm = 402.15; Tb = 558.15; Antoine I (sol) A = 8.29504, *(continues)*

178

$C_6H_4I_2$ *(continued)*

B = 3311.326, C = 0, dev = 5.0, range = 372 to 401; Antoine II (liq) A = 6.8859, B = 2747.48, C = 0, dev = 1.0 to 5.0, range = 402 to 560.

$C_6H_4N_2$ MW 104.11

Nicotinic acid, nitrile, CA 100-54-9: Irritant; flammable, flash point 357; soluble water, ethanol, ether, benzene. Tm = 323.15; Tb = 513.15 to 518.15; Antoine (liq) A = 5.9485, B = 1544.73, C = -88.25, dev = 1.0, range = 453 to 479.

$C_6H_4N_2O$ MW 120.11

Benzofurazan, CA 273-09-6: Irritant; flammable; insoluble water; soluble ethanol. Tm = 328.15; Antoine (sol) A = 11.025, B = 3728, C = 0, dev = 5.0, range = 278 to 298.

$C_6H_4N_2O_2$ MW 136.11

Benzofurazan-1-oxide, CA 480-96-6: Irritant; flammable; soluble ethanol. Tm = 343.15 to 344.15; Antoine (sol) A = 11.095, B = 4291.1, C = 0, dev = 5.0, range = 288 to 318.

$C_6H_4N_2O_3$ MW 152.11

1-Nitro-2-nitrosobenzene, CA 612-29-3: Flammable; probably dimeric in solid form; soluble ethanol, acetone, benzene, chloroform. Tm = 399.15 to 399.65; Antoine (sol) A = 9.695, B = 4989.81, C = 0, dev not specified, range = 323 to 343.

$C_6H_4N_2O_4$ MW 168.11

1,2-Dinitrobenzene, CA 528-29-0: Highly toxic by inhalation and skin absorption; explosive; soluble ethanol, ether, benzene, chloroform. Tm = 391.15; Tb = 592.15; Vm = 0.1281 at 393.15; Antoine I (sol) A = 9.19067, B = 4330.64, C = 0, dev = 5.0, range = 343 to 387; Antoine II (liq) A = 7.3029, B = 3135, C = 0, dev = 5.0, range = 454 to 593.

1,3-Dinitrobenzene, CA 99-65-0: Highly toxic by inhalation and skin absorption; explosive; soluble ethanol, ether, benzene, chloroform. Tm = 362.72; Tb = 564.15; Antoine I (sol) A = 8.43083, B = 3974.218, C = 0, dev = 5.0, range = 335 to 356; Antoine II (liq) A = 11.45398, B = 5051.427, C = 0, dev = 1.0 to 5.0, range = 366 to 379; Antoine III (liq) A = 7.6779, B = 3244, C = 0, dev = 5.0, range = 456 to 576.

1,4-Dinitrobenzene, CA 100-25-4: Highly toxic by inhalation and skin absorption; explosive; soluble ethanol, ether, benzene, chloroform. Tm = 446.15 to 447.15; Tb = 571.15; Antoine I (sol) A = 10.4219, B = 4888.291, C = 0, dev = 5.0, range = 339 to 398; Antoine II (liq) A = 7.5439, B = 3152, C = 0, dev = 5.0, range = 445 to 572.

$C_6H_4N_2O_5$ MW 184.11

2,3-Dinitrophenol, CA 66-56-8: Highly toxic; potentially explosive; slightly soluble water; soluble ethanol, ether, benzene,   *(continues)*

$C_6H_4N_2O_5$ *(continued)*

acetone, chloroform. Tm = 417.15 to 418.15; Antoine (sol) A = 11.705, B = 5171, C = 0, dev = 5.0, range = 303 to 343.

2,4-Dinitrophenol, CA 51-28-5: Highly toxic; potentially explosive; slightly soluble water; soluble ethanol, ether, benzene, chloroform, acetone. Tm = 385.15 to 387.15; Antoine (sol) A = 13.075, B = 5466, C = 0, dev = 5.0, range = 293 to 333.

2,5-Dinitrophenol, CA 329-71-5: Highly toxic; potentially explosive; slightly soluble water; soluble ethanol, ether, benzene, acetone, chloroform. Tm = 381.15; Antoine (sol) A = 11.575, B = 4876, C = 0, dev = 5.0, range = 278 to 333.

2,6-Dinitrophenol, CA 573-56-8: Highly toxic; potentially explosive; slightly soluble water; soluble ethanol, ether, benzene, acetone, chloroform. Tm = 336.15 to 337.15; Antoine (sol) A = 14.505, B = 5860, C = 0, dev = 5.0, range = 293 to 333.

3,4-Dinitrophenol, CA 577-71-9: Highly toxic; potentially explosive; slightly soluble water; soluble ethanol, ether, benzene, acetone, chloroform. Tm = 407.15; Antoine (sol) A = 13.375, B = 6451, C = 0, dev = 5.0, range = 328 to 383.

3,5-Dinitrophenol, CA 586-11-8: Highly toxic; potentially explosive; slightly soluble water; soluble ethanol, ether, benzene, acetone, chloroform. Tm = 399.15; Antoine (sol) A = 4.89345, B = 3257.944, C = 0, dev = 5.0, range = 327 to 368.

$C_6H_4N_4O_6$ MW 228.12

2,4,6-Trinitroaniline, picramide, CA 489-98-5: Explosive from heating or shock; soluble benzene, acetone. Tm = 465.15 to 468.15; Antoine (sol) A = 11.225, B = 6056, C = 0, dev = 5.0, range = 328 to 371.

$C_6H_4O_2$ MW 108.10

1,4-Benzoquinone, CA 106-51-4: Irritant; highly toxic; causes dermatitis and conjunctivitis; flammable; slightly soluble water; soluble ethanol, ether. Tm = 390.15; Antoine I (sol) A = 10.296, B = 3577, C = 0, dev = 1.0 to 5.0, range = 284 to 386; Antoine II (liq) A = 7.498, B = 2498.7, C = 0, dev = 1.0 to 5.0, range = 388 to 402.

C_6H_5Br MW 157.01

Bromobenzene, phenyl bromide, CA 108-86-1: Irritant; moderately toxic; flammable, flash point 324; soluble ether, benzene, chloroform, light hydrocarbons. Tm = 242.55; Tb = 429.35; Vm = 0.1060 at 303.15; Tc = 670.15; Pc = 4518; Vc = 0.486; Antoine I (liq) A = 6.03934, B = 1474.06, C = -63.75, dev = 0.1, range = 333 to 463; Antoine II (liq) A = 6.40524, B = 1776.58, C = -25.639, dev = 1.0, range = 429 to 633.

C_6H_5BrO MW 173.01

2-Bromophenol, CA 95-56-7: Irritant; moderately toxic; flammable, flash point 315; moderately soluble water; soluble ethanol, ether,   *(continues)*

180

C_6H_5BrO *(continued)*

chloroform. Tm = 278.75; Tb = 467.15; Vm = 0.1159 at 293.15; Antoine (liq) A = 6.4341, B = 1752, C = -70.95, dev = 1.0 to 5.0, range not specified.

3-Bromophenol, CA 591-20-8: Irritant; moderately toxic; flammable, flash point above 385; moderately soluble water; soluble ethanol, ether, chloroform. Tm = 306.15; Tb = 508.15 to 509.15; Antoine (liq) A = 9.556, B = 3840.2, C = 0. dev = 5.0, range = 410 to 510.

4-Bromophenol, CA 106-41-2: Irritant; moderately toxic; flammable; moderately soluble water; soluble ethanol, ether, chloroform. Tm = 339.15; Tb = 511.15; Vm = 0.0940 at 288.15; Antoine I (sol) A = 12.492, B = 4560.8, C = 0, dev = 1.0, range = 260 to 302; Antoine II (liq) A = 8.0145, B = 3071.41, C = 0, dev = 5.0, range = 390 to 511.

C_6H_5Cl MW 112.56

Chlorobenzene, phenyl chloride, CA 108-90-7: Irritant; toxic by inhalation; flammable, flash point 301; soluble ethanol, ether, benzene, chloroform. Tm = 227.81; Tb = 404.85; Vm = 0.1018 at 293.15; Tc = 632.35; Pc = 4519; Vc = 0.3655; Antoine I (liq) A = 6.11512, B = 1438.86, C = -54.72, dev = 0.1, range = 333 to 405; Antoine II (liq) A = 6.62988, B = 1897.41, C = 5.21, dev = 1.0, range = 405 to 597.

C_6H_5ClO MW 128.56

2-Chlorophenol, CA 95-57-8: Irritant; corrosive; toxic; causes burns; flammable, flash point 337; slightly soluble water; soluble ethanol, ether, benzene. Tm = 281.85; Tb = 448.15 to 449.15; Vm = 0.1018 at 293.15; Antoine I (liq) A = 5.78693, B = 1314.9, C = -101.05, dev = 1.0, range = 333 to 449; Antoine II (liq) A = 5.3685, B = 1096.98, C = -122.58, dev = 1.0, range = 354 to 448.

3-Chlorophenol, CA 108-43-0: Irritant; corrosive; toxic; causes burns; flammable, flash point above 385; slightly soluble water; soluble ethanol, ether, benzene. Tm = 305.95; Tb = 488.15 to 490.15; Vm = 0.1033 at 318.15; Antoine I (sol) A = 7.67412, B = 3178.132, C = 0, dev = 1.0, range = 252 to 293; Antoine II (liq) A = 6.54908, B = 1978.86, C = -51.572, dev = 1.0, range = 317 to 487.

4-Chlorophenol, CA 106-48-9: Irritant; corrosive; toxic; causes burns; flammable, flash point 394; slightly soluble water; soluble ethanol, ether, benzene. Tm = 313.15 to 314.15; Tb = 492.15; Vm = 0.1016 at 313.15; Antoine (liq) A = 5.83238, B = 1385.1, C = -131.1, dev = 1.0, range = 373 to 493.

$C_6H_5ClO_2$ MW 144.56

Chlorohydroquinone, 2-chloro-1,4-benzenediol, CA 615-67-8: Irritant; flammable; soluble ethanol, ether, benzene, chloroform. Tm = 379.15; Tb = 536.15; Antoine (sol) A = 13.455, B = 5370, C = 0, dev = 5.0, range = 298 to 334.

$C_6H_5ClO_2S$ MW 176.62

Benzene sulfonylchloride, CA 98-09-9: Irritant; corrosive; toxic; causes burns; flammable, flash point above 385; soluble ethanol,   *(continues)*

$C_6H_5ClO_2S$ *(continued)*

ether. Tm = 287.65; Tb = 524.15 to 525.15, decomposes; Vm = 0.1277 at 288.15; Antoine (liq) A = 6.62109, B = 2219.13, C = -43.826, dev = 1.0, range = 338 to 525.

$C_6H_5Cl_2N$ MW 162.02

3,4-Dichloroaniline, CA 95-76-1: Irritant; highly toxic; soluble ethanol, ether, benzene; used to manufacture herbicides. Tm = 344.65; Tb = 545.15; Antoine (liq) A = 7.6189, B = 3060.03, C = 0, dev = 1.0 to 5.0, range = 420 to 545.

$C_6H_5Cl_2O_2P$ MW 210.98

Phenyl dichlorophosphate, CA 770-12-7: Corrosive; moisture-sensitive; flammable, flash point above 385. Tb = 514.15 to 516.15; Vm = 0.1494 at 293.15; Antoine (liq) A = 6.30021, B = 1800.71, C = -93.092, dev = 1.0, range = 339 to 513.

C_6H_5F MW 96.10

Fluorobenzene, CA 462-06-6: Highly toxic vapor; flammable, flash point 258; soluble ethanol, ether, benzene. Tm = 230.93; Tb = 357.88; Vm = 0.0944 at 298.15; Tc = 560.09; Pc = 4551; Vc = 0.269; Antoine I (liq) A = 6.07234, B = 1245.564, C = -51.587, dev = 0.02, range = 312 to 394; Antoine II (liq) A = 5.4113, B = 1398.61, C = -31.295, dev = 1.0, range = 358 to 530; Antoine III (liq) A = 6.14135, B = 1291.116, C = -45.664, dev = 0.1, range = 373 to 419; Antoine IV (liq) A = 6.37857, B = 1478.807, C = -18.847, dev = 0.1, range = 414 to 501; Antoine V (liq) A = 7.59182, B = 2876.741, C = 171.093, dev = 0.1 to 1.0, range = 497 to 561.

C_6H_5FO MW 112.10

2-Fluorophenol, CA 367-12-4: Irritant; toxic; flammable, flash point 319; moderately soluble water; soluble acetone, light hydrocarbons. Tm = 289.15; Tb = 427.15; Vm = 0.1001 at 298.15; Antoine (liq) A = 6.6295, B = 1667.3, C = -62.65, dev = 1.0 to 5.0, range not specified.

3-Fluorophenol, CA 372-20-3: Irritant; toxic; flammable, flash point 344; moderately soluble water; soluble acetone, light hydrocarbons. Tm = 286.85; Tb = 451.15; Vm = 0.0905 at 293.15; Antoine (liq) A = 7.83338, B = 2629.15, C = 0, dev = 1.0 to 5.0, range = 373 to 451.

4-Fluorophenol, CA 371-41-5: Irritant; toxic; flammable, flash point 341; moderately soluble water; soluble acetone, light hydrocarbons. Tm = 321.15; Tb = 458.65; Antoine (liq) A = 7.56041, B = 2547.66, C = 0, dev = 1.0 to 5.0, range = 360 to 460.

C_6H_5I MW 204.01

Iodobenzene, phenyl iodide, CA 591-50-4: Flammable, flash point 347; decomposes in light; soluble ethanol, ether, benzene, chloroform. Tm = 241.75; Tb = 461.75; Vm = 0.1114 at 293.15; Antoine I (liq) A = 6.01996, B = 1562.87, C = -72.15, dev = 0.1, range = 358 to 543;   *(continues)*

182

C_6H_5I *(continued)*

Antoine II (liq) A = 6.46493, B = 1967.69, C = -20.202, dev = 1.0, range = 462 to 679; Antoine III (liq) A = 6.36129, B = 1765.99, C = -54.15, dev = 1.0, range = 273 to 358.

C_6H_5NO MW 107.11

Nitrosobenzene, CA 586-96-9: Flammable; dimeric in solid form; soluble ethanol, ether, benzene, light hydrocarbons. Tm = 340.65 to 341.15; Antoine (sol) A = 4.42787, B = 508.51, C = -205.51, dev = 1.0, range = 297 to 339.

$C_6H_5NO_2$ MW 123.11

Nitrobenzene, CA 98-95-3: Moderately toxic; causes cyanosis; moderate fire and explosion hazard; flammable, flash point 361; slightly soluble water; soluble ethanol, ether, benzene. Tm = 279; Tb = 484.05; Vm = 0.1027 at 298.15; Tc = 719, calculated; Pc = 4340, calculated; Vc = 0.349, calculated; Antoine I (liq) A = 6.22069, B = 1732.222, C = -72.886, dev = 0.1, range = 407 to 484; Antoine II (liq) A = 6.6699, B = 2064, C = -43.15, dev = 1.0, range = 279 to 296.

$C_6H_5NO_3$ MW 139.11

2-Nitrophenol, CA 88-75-5: Irritant; moderately toxic by inhalation and skin absorption; flammable; very soluble hot water; soluble ethanol, ether, benzene, acetone. Tm = 318.05; Tb = 489.15; Antoine I (sol) A = 7.8446, B = 2864.6, C = 0, dev = 1.0, range = 273 to 292; Antoine II (liq) A = 6.04963, B = 1571.7, C = -101.17, dev = 1.0, range = 366 to 490.

3-Nitrophenol, CA 554-84-7: Irritant; moderately toxic by inhalation and skin absorption; flammable; soluble ethanol, ether, benzene, acetone. Tm = 370.15; Vm = 0.1087 at 373.15; Antoine (sol) A = 8.93697, B = 3981.386, C = 0, dev = 1.0, range = 305 to 334.

4-Nitrophenol, CA 100-02-7: Irritant; moderately toxic by inhalation and skin absorption; flammable; soluble ethanol, ether, benzene, acetone. Tm = 387.15; Tb = 552.15, decomposes; Antoine (sol) A = 11.9529, B = 5159.7, C = 0, dev = 1.0, range = 304 to 352.

$C_6H_5NO_4$ MW 155.11

1,3-Dihydroxy-2-nitrobenzene, 2-nitro-1,3-benzenediol, CA 601-89-8: Flammable; soluble ethanol. Tm = 357.15 to 358.15; Antoine (sol) A = 10.175, B = 3892, C = 0, dev = 5.0, range = 253 to 293.

$C_6H_5N_3$ MW 119.13

Triazobenzene, phenyl azide, CA 622-37-7: Explosive by heating or shock; explodes at normal boiling point. Tm = 245.65; Vm = 0.1093 at 298.15; Antoine (liq) A = 7.4323, B = 2361, C = 0, dev = 1.0, range = 348 to 368.

$C_6H_5N_5O_6$ MW 243.14

1,3-Diamino-2,4,6-trinitrobenzene, CA 28930-29-2: Explosive by heating or
shock. Tm = 556.15; Antoine (sol) A = 12.855, B = 7314, C = 0, dev = 5.0,
range = 335 to 382.

C_6H_6 MW 78.11

Benzene, CA 71-43-2: Highly toxic by inhalation and skin contact; suspected carcinogen; flammable, flash point 262; soluble most organic solvents; important chemical raw material. Tm = 278.68; Tb = 353.25; Vm =
0.0889 at 293.15; Tc = 562.6; Pc = 4898; Vc = 0.259; Antoine I (sol) A =
10.0091, B = 2836.0, C = 25.31, dev = 0.1, range = 223 to 279; Antoine II
(sol) A = 8.45261, B = 1986.69, C = -23.089, dev = 0.1, range = 218 to 279;
Antoine III (liq) A = 6.01907, B = 1204.682, C = -53.072, dev = 0.02,
range = 279 to 377; Antoine IV (liq) A = 6.06832, B = 1236.034, C = -48.99,
dev = 0.02, range = 353 to 422; Antoine V (liq) A = 6.3607, B = 1466.083,
C = -15.44, dev = 0.1, range = 420 to 502; Antoine VI (liq) A = 7.51922,
B = 2809.514, C = 171.489, dev = 0.1, range = 501 to 562.

1,5-Hexadien-3-yne, divinyl acetylene, CA 821-08-9: Flammable, flash point
below 253; forms explosive polymers; soluble benzene. Tm = 185.15; Tb =
356.65; Vm = 0.0995 at 293.15; Antoine (liq) A = 5.950, B = 1169.97, C =
-60.51, dev = 1.0, range = 223 to 357.

1,3-Hexadien-5-yne, CA 10420-90-3: Flammable; forms explosive polymers;
soluble benzene. Tm = 192.15; Tb = 356.15 to 357.15; Vm = 0.1001 at
293.15; Antoine (liq) A = 4.6695, B = 698.74, C = -106.15, dev = 1.0,
range = 223 to 303.

C_6H_6ClN MW 127.57

2-Chloroaniline, CA 95-51-2: Irritant; highly toxic by inhalation and skin
absorption; flammable, flash point 370; slightly soluble water; soluble
most organic solvents. Tm = 269.65; Tb = 481.15 to 483.15; Vm = 0.1052 at
293.15; Antoine (liq) A = 5.84228, B = 1432.2, C = -108.81, dev = 1.0,
range = 397 to 482.

3-Chloroaniline, CA 108-42-9: Irritant; highly toxic by inhalation and
skin absorption; flammable, flash point 396; slightly soluble water; soluble most organic solvents. Tm = 263.15; Tb = 503.15 to 504.15; Vm =
0.1050 at 295.15; Antoine (liq) A = 6.36093, B = 1857.75, C = -76.51, dev =
1.0, range = 398 to 573.

4-Chloroaniline, CA 106-47-8: Irritant; highly toxic by inhalation and
skin absorption; flammable; slightly soluble water; soluble most organic
solvents. Tm = 345.65; Tb = 505.15; Antoine I (sol) A = 13.448, B = 4736,
C = 0, dev = 5.0, range = 283 to 303; Antoine II (liq) A = 7.3489, B =
2729, C = 0, dev = 5.0, range = 363 to 505.

$C_6H_6Cl_4$ MW 219.93

alpha-3,4,5,6-Tetrachlorocyclohexene, CA 41992-55-6: Irritant; flammable.
Tm = 306.35 to 306.65; Antoine (liq) A = 7.89772, B = 3031.907, C = 0,
dev = 1.0, range = 353 to 399.

184

$C_6H_6Cl_6$ MW 290.83

alpha-1,2,3,4,5,6-Hexachlorocyclohexane, benzene hexachloride, CA 319-84-6:
Toxic; carcinogen; soluble acetone, ethyl acetate, dioxane. Tm = 432.35;
Tb = 561.15; Antoine (sol) A = 11.5547, B = 4996.9, C = 0, dev = 5.0,
range = 313 to 363.

beta-Hexachlorocyclohexane, CA 319-85-7: Toxic; carcinogen; soluble acetone, dioxane. Tm = 584.85; Antoine (sol) A = 11.5908, B = 5590, C = 0,
dev = 1.0, range = 313 to 363.

gamma-Hexachlorocyclohexane, CA 58-99-9: Irritant; toxic; carcinogen;
powerful insecticide, not presently used; soluble ether, acetone, benzene,
chloroform. Tm = 386.05; Tb = 596.55; Antoine (sol) A = 12.2883, B =
5180.3, C = 0, dev = 5.0, range = 313 to 363.

delta-Hexachlorocyclohexane, CA 319-86-8: Irritant; carcinogen; violent
reaction with iron; soluble acetone, ether, benzene, ethyl acetate. Tm =
413.95; Antoine (sol) A = 11.7109, B = 5081.4, C = 0, dev = 1.0, range =
313 to 363.

$C_6H_6F_9N_3S$ MW 323.18

2-Propanimine, *N*-[*N,N'*-dimethyl-*S*-(trifluoromethyl)sulfonodiimidoyl]-
1,1,1,3,3,3-hexafluoro, CA 63265-76-9: Antoine (liq) A = 5.015, B = 1710,
C = 0, dev = 1.0 to 5.0, range not specified.

$C_6H_6N_2$ MW 106.13

3-Hexene, dinitrile, CA 1119-85-3: Irritant; flammable. Tm = 349.15;
Antoine (liq) A = 6.360, B = 2582, C = 0, dev = 5.0, range = 353 to 448.

$C_6H_6N_2O$ MW 122.13

2-Pyridinecarboxamide, picolinamide, CA 1452-77-3: Irritant; flammable;
soluble water, ethanol, benzene. Tm = 379.65; Antoine (sol) A = 11.945,
B = 4862, C = 0, dev = 1.0 to 5.0, range = 323 to 373.

3-Pyridinecarboxamide, nicotinamide, CA 98-92-0: Irritant; flammable,
flash point above 423; soluble water, ethanol, benzene. Tm = 402.15 to
403.15; Antoine (sol) A = 13.005, B = 5841, C = 0, dev = 1.0 to 5.0, range
= 363 to 393.

4-Pyridinecarboxamide, isonicotinamide, CA 1453-82-3: Irritant; flammable;
soluble water, ethanol, benzene. Tm = 428.65 to 429.15; Antoine (sol) A =
11.305, B = 5219, C = 0, dev = 5.0, range = 383 to 412.

$C_6H_6N_2O_2$ MW 138.13

2-Nitroaniline, CA 88-74-4: Highly toxic by inhalation and skin absorption; flammable; moderately soluble water; soluble ethanol, ether, benzene;
used in manufacture of azo dyes. Tm = 344.15 to 345.15; Tb = 557.15;
Antoine I (sol) A = 11.625, B = 4701, C = 0, dev = 5.0, range = 273 to 323;
Antoine II (liq) A = 11.3629, B = 7444.3, C = 240.83, dev = 1.0, range =
423 to 553.

(continues)

$C_6H_6N_2O_2$ *(continued)*

3-Nitroaniline, CA 99-09-2: Highly toxic by inhalation and skin absorption; flammable; moderately soluble water; soluble methanol, ethanol, ether; used in manufacture of azo dyes. Tm = 387.15; Tb = 578.15 to 580.15, decomposes; Vm = 0.1176 at 433.15; Antoine I (sol) A = 12.125, B = 5095, C = 0, dev = 5.0, range = 288 to 343; Antoine II (liq) A = 9.1005, B = 4653.8, C = 78.62, dev = 1.0, range = 443 to 578.

4-Nitroaniline, CA 100-01-6: Highly toxic by inhalation and skin absorption; flammable, flash point 472; moderately soluble hot water; soluble ethanol, ether, acetone, chloroform; used in manufacture of azo dyes. Tm = 421.15 to 422.15; Tb = 605.15; Antoine I (sol) A = 11.1109, B = 5093, C = 0, dev = 5.0, range = 346 to 366; Antoine II (liq) A = 8.7688, B = 4071.3, C = 0, dev = 1.0, range = 473 to 538.

$C_6H_6N_6O_6$ 　　　　　　　　　　　　　　　　　　　　　　　　MW 258.15

2,4,6-Trinitro-1,3,5-benzenetriamine, CA 3058-38-6: Antoine (sol) A = 13.855, B = 8787, C = 0, dev = 5.0, range = 402 to 451.

C_6H_6O 　　　　　　　　　　　　　　　　　　　　　　　　　　MW 94.11

Phenol, carbolic acid, CA 108-95-2: Highly toxic by skin contact; corrosive; causes burns; flammable, flash point 352; very soluble water; soluble ethanol, ether, acetone, benzene, carbon tetrachloride; used to manufacture resins and polymers. Tm = 314.06; Tb = 454.99; Vm = 0.0904 at 333.15; Tc = 694.25; Pc = 6130; Vc = 0.229; Antoine I (sol) A = 10.6887, B = 3586.36, C = 0, dev = 1.0, range = 282 to 313; Antoine II (sol) A = 10.71099, B = 3594.703, C = 0, dev = 1.0, range = 273 to 313; Antoine III (liq) A = 6.25947, B = 1516.072, C = -98.581, dev = 0.1, range = 383 to 473; Antoine IV (liq) A = 6.34757, B = 1482.82, C = -113.862, dev = 1.0, range = 455 to 655; Antoine V (liq) A = 6.57957, B = 1710.287, C = -80.273, dev = 0.1 to 1.0, range = 314 to 395; Antoine VI (liq) A = 6.26694, B = 1522.07, C = -97.834, dev = 0.1, range = 387 to 456; Antoine VII (liq) A = 6.30177, B = 1548.368, C = -94.612, dev = 0.1 to 1.0, range = 449 to 526; Antoine VIII (liq) A = 6.92874, B = 2146.053, C = -17.025, dev = 1.0 to 5.0, range = 520 to 625.

$C_6H_6O_2$ 　　　　　　　　　　　　　　　　　　　　　　　　　MW 110.11

Catechol, 1,2-benzenediol, CA 120-80-9: Irritant; toxic by skin absorption; causes burns; flammable, flash point 400; soluble water, ethanol, ether, acetone, benzene. Tm = 378.15; Tb = 518.15; Antoine (liq) A = 6.61028, B = 1954.6, C = -94.25, dev = 1.0, range = 395 to 519.

Hydroquinone, 1,4-benzenediol, CA 123-31-9: Irritant; toxic; may cause dermatitis; vapor can damage eyes; flammable, flash point 438; soluble water, ethanol, ether, acetone, benzene. Tm = 445.15; Tb = 560.15; Antoine I (sol) A = 12.585, B = 5420, C = 0, dev = 5.0, range = 298 to 346; Antoine II (liq) A = 7.00575, B = 2321.92, C = -95.235, dev = 1.0, range = 448 to 559.

Resorcinol, 1,3-benzenediol, CA 108-46-3: Irritant; moderately toxic; flammable, flash point 400; soluble water, ethanol, ether, acetone, benzene. Tm = 383.15; Tb = 550.15; Antoine I (sol) A = 11.425, B = 4876, C = 0, dev = 5.0, range = 283 to 323; Antoine II (liq) A = 　　*(continues)*

$C_6H_6O_2$ *(continued)*

6.52635, B = 1918.1, C = -128.65, dev = 1.0 to 5.0, range = 419 to 550;
Antoine III (liq) A = 6.1041, B = 1745.2, C = -133.81, dev = 1.0 to 5.0,
range = 392 to 463.

$C_6H_6O_3$ MW 126.11

1,2,3-Trihydroxybenzene, pyrogallol, CA 87-66-1: Irritant; highly toxic;
suspected carcinogen; easily absorbed through skin; causes kidney and liver
damage; readily absorbs oxygen; soluble water, ethanol, ether. Tm = 406.15
to 407.15; Tb = 582.15, decomposes; Antoine (liq) A = 6.90499, B = 2477.24,
C = -76.412, dev = 1.0, range = 425 to 582.

$C_6H_6O_3S$ MW 158.17

Benzenesulfonic acid, CA 98-11-3: Highly irritant; corrosive; causes
burns; flammable; soluble water, ethanol, acetic acid. Tm = 338.15 to
339.15; Tb = 383.75.

$C_6H_6O_4$ MW 142.11

Butynedioic acid, dimethyl ester, CA 762-42-5: Corrosive; lachrymator;
flammable, flash point 359; potentially explosive in pure state or concentrated solutions; soluble ethanol, ether. Tb = 468.15 to 471.15, decomposes; Vm = 0.1229 at 293.15; Antoine (liq) A = 8.33305, B = 2941.4, C = 0,
dev = 1.0 to 5.0, range = 273 to 460.

C_6H_6S MW 100.71

Thiophenol, benzenethiol, CA 108-98-5: Irritant; toxic; stench; causes
dermatitis; flammable, flash point 323; soluble ethanol, ether, carbon disulfide, benzene. Tm = 258.25; Tb = 442.25; Vm = 0.1027 at 298.15; Antoine
(liq) A = 6.11531, B = 1530.286, C = -69.948, dev = 0.02, range = 385 to
486.

$C_6H_7Cl_2N$ MW 164.03

2-Chloroaniline, hydrochloride, CA 137-04-2: Soluble water, ethanol. Tm =
508.15; Antoine (sol) A = 10.02662, B = 3964.413, C = -4.206, dev = 1.0,
range = 373 to 473.

3-Chloroaniline, hydrochloride, CA 141-85-5: Soluble water, ethanol. Tm =
495.15; Antoine (sol) A = 9.59062, B = 3534.651, C = -10.27, dev = 1.0,
range = 383 to 473.

4-Chloroaniline, hydrochloride, CA 20265-96-7: Soluble water, ethanol.
Antoine (sol) A = 9.82487, B = 4000.971, C = -3.044, dev = 1.0, range = 373
to 483.

$C_6H_7F_3N_2O_4$ MW 228.13

Glycine, N-[N-(trifluoroacetyl)glycyl], CA 400-58-8: Tm = 460.55; Antoine
(sol) A = 6.135, B = 3501, C = 0, dev = 5.0, range = 273 to 423.

C_6H_7N MW 93.13

Aniline, CA 62-53-3: Toxic by inhalation and skin absorption; suspected
carcinogen; flammable, flash point 343; very soluble water; soluble most
organic solvents; used in manufacture of rubber chemicals. Tm = 267; Tb =
457.55; Vm = 0.0912 at 293.15; Tc = 699.0, Pc = 5309; Vc = 0.270; Antoine I
(liq) A = 6.40627, B = 1702.817, C = -70.155, dev = 0.1, range = 304 to
458; Antoine II (liq) A = 8.1019, B = 2728, C = 0, dev = 1.0, range = 273
to 338; Antoine III (liq) A = 6.41147, B = 1708.239, C = -69.454, dev =
0.1, range = 373 to 458; Antoine IV (liq) A = 6.44338, B = 1682.148, C =
-78.065, dev = 1.0, range = 455 to 523.

2-Methylpyridine, 2-picoline, CA 109-06-8: Irritant; toxic; flammable,
flash point 299; soluble water, ethanol, ether, acetone. Tm = 209.15; Tb =
402.65; Vm = 0.0984 at 293.15; Antoine I (liq) A = 5.2309, B = 1164.1, C =
-71.0, dev = 1.0 to 5.0, range = 209 to 245; Antoine II (liq) A = 6.1558,
B = 1415.29, C = -61.521, dev = 0.02, range = 352 to 442; Antoine III (liq)
A = 6.15522, B = 1414.906, C = -61.566, dev = 0.02, range = 352 to 442;
Antoine IV (liq) A = 6.32369, B = 1546.248, C = -44.271, dev = 0.1 to 1.0,
range = 429 to 537; Antoine V (liq) A = 7.32144, B = 2667.496, C = 107.978,
dev = 1.0, range = 521 to 621.

3-Methylpyridine, 3-picoline, CA 108-99-6: Irritant; toxic; flammable,
flash point 309; soluble water, ethanol, ether, acetone. Tm = 254.85; Tb =
417.05; Vm = 0.0978 at 298.15; Antoine I (sol) A = 11.245, B = 3246.9, C =
0, dev = 1.0 to 5.0, range = 225 to 255; Antoine II (liq) A = 6.17593, B =
1482.943, C = -61.705, dev = 0.02, range = 347 to 458; Antoine III (liq)
A = 6.17791, B = 1484.285, C = -61.554, dev = 0.02 to 0.1, range = 347 to
458; Antoine IV (liq) A = 6.18988, B = 1491.897, C = -60.745, dev = 0.02,
range = 347 to 381; Antoine V (liq) A = 6.16648, B = 1476.25, C = -62.502,
dev = 0.02, range = 374 to 458; Antoine VI (liq) A = 6.38586, B = 1659.184,
C = -38.176, dev = 1.0, range = 450 to 570; Antoine VII (liq) A = 7.57549,
B = 3151.52, C = 161.352, dev = 1.0 to 5.0, range = 561 to 645.

4-Methylpyridine, 4-picoline, CA 108-99-4: Irritant; toxic; flammable,
flash point 329; soluble water, ethanol, ether. Tm = 276.85; Tb = 418.05;
Vm = 0.0974 at 293.15; Antoine I (sol) A = 12.246, B = 3277.6, C = 0, dev =
1.0 to 5.0, range = 213 to 239; Antoine II (liq) A = 6.16752, B = 1481.88,
C = -62.459, dev = 0.02, range = 348 to 459; Antoine III (liq) A = 6.16906,
B = 1482.916, C = -62.343, dev = 0.02 to 0.1, range = 348 to 460; Antoine
IV (liq) A = 6.17965, B = 1489.655, C = -61.626, dev = 0.02, range = 348
to 387; Antoine V (liq) A = 6.15555, B = 1473.357, C = -63.481, dev = 0.02,
range = 381 to 460; Antoine VI (liq) A = 6.32903, B = 1613.866, C =
-45.037, dev = 0.1 to 1.0, range = 452 to 573; Antoine VII (liq) A =
7.34093, B = 2808.114, C = 117.355, dev = 1.0, range = 564 to 646.

C_6H_7NO MW 109.13

4-Aminophenol, CA 123-30-8: Irritant; allergen; highly toxic; decomposes
in air and light; moderately soluble hot water; soluble ethanol, methyl
ethyl ketone. Tm = 459.15; Tb = 557.15, decomposes; Antoine (sol) A =
13.239, B = 5801, C = 0, dev = 5.0, range = 423 to 459.

2-Methoxypyridine, CA 1628-89-3: Irritant; flammable, flash point 305;
soluble benzene. Tb = 414.15 to 415.15; Antoine (liq) A = 7.08839, B =
2117.863, C = 0, dev not specified, range = 304 to 338.

2(1H)-Pyridone, 1-methyl, CA 694-85-9: Flammable; soluble ethanol, ether,
benzene, chloroform. Tm = 280.15; Tb = 523.15; Vm = 0.0981 at 293.15;
Antoine (liq) A = 8.13699, B = 3146.384, C = 0, dev not specified, range =
353 to 399.

C_6H_7NS MW 125.19

Pyridine, 4-(methylthio), CA 22581-72-2: Toxic; flammable; soluble light
hydrocarbons. Tm = 320.15; Antoine (liq) A = 7.7704, B = 2913.134, C = 0,
dev not specified, range = 346 to 383.

4(1H)-Pyridinethion, 1-methyl, CA 6887-59-8: Toxic; flammable; soluble
ethanol. Tm = 434.15 to 436.15; Antoine (liq) A = 19.71379, B = 8942.552,
C = 0, dev not specified, range = 440 to 465.

$C_6H_7N_5$ MW 149.15

9-Methyladenine, CA 700-00-5: Tm = 582.15; Antoine (sol) A = 12.471, B =
6355, C = 0, dev = 5.0, range = 413 to 458.

C_6H_8 MW 80.13

1,3-Cyclohexadiene, CA 592-57-4: Forms explosive peroxides in contact with
air; polymerizes; flammable, flash point 299; soluble most organic solvents.
Tm = 184.15; Tb = 356.15 to 357.15; Vm = 0.0953 at 293.15; Antoine (liq)
A = 5.990348, B = 1205.995, C = -50.8724, dev = 0.02, range = 307 to 364.

1,4-Cyclohexadiene, CA 628-41-1: Forms explosive peroxides; polymerizes;
flammable, flash point 267; soluble most organic solvents. Tm = 223.95;
Tb = 359.15 to 360.15; Vm = 0.0946 at 293.15; Antoine (liq) A = 6.41736,
B = 1475.149, C = -28.108, dev = 1.0, range = 304 to 360.

1,3,5-Hexatriene, cis, CA 2612-46-6: Toxic; polymerizes; flammable, flash
point 310; soluble most organic solvents. Tm = 261.15; Tb = 351.65; Vm =
0.1117 at 293.15; Antoine (liq) A = 6.28031, B = 1363.971, C = -35.919,
dev = 0.1, range = 306 to 323.

C_6H_8ClN MW 129.59

Aniline hydrochloride, CA 142-04-1: Highly toxic; suspected carcinogen;
decomposes in light; flammable, flash point 466; soluble water, ethanol.
Tm = 471.15; Tb = 518.15; Antoine (sol) A = 11.21944, B = 4570.798, C = 0,
dev = 1.0, range = 383 to 471.

Pyridine hydrochloride, 3-methyl, CA 14401-92-4: Flammable; soluble water.
Antoine (liq) A = 7.31696, B = 2177.2, C = -95.99, dev = 1.0, range = 420
to 471.

Pyridine hydrochloride, 4-methyl, CA 14401-93-5: Flammable; soluble water.
Antoine (liq) A = 13.0498, B = 8659.23, C = 271.089, dev = 0.1, range = 437
to 473.

$C_6H_8Cl_2O_4$ MW 215.03

Ethylene glycol, bis(chloroacetate), CA 6941-69-1: Irritant; flammable;
soluble ethanol. Tm = 319.15; Antoine (liq) A = 7.50867, B = 2704.76, C =
-65.131, dev = 1.0 to 5.0, range = 385 to 557.

$C_6H_8N_2$ MW 108.14

Adiponitrile, CA 111-69-3: High oral toxicity; flammable, flash point 366;
very soluble water; soluble methanol, ethanol, benzene,   (continues)

$C_6H_8N_2$ *(continued)*

chloroform. Tm = 275.64; Tb = 568.15; Vm = 0.1121 at 293.15; Tc = 780.15; Pc = 2800; Antoine I (liq) A = 7.6729, B = 3236, C = 0, dev = 5.0, range = 392 to 523; Antoine II (liq) A = 7.3562, B = 3067.5, C = 0, dev = 1.0 to 5.0, range = 348 to 523.

2-(Methylamino) pyridine, CA 4597-87-9: Irritant; flammable, flash point 360; soluble water, ethanol, ether, benzene. Tm = 288.15; Tb = 473.15 to 474.15; Vm = 0.1028 at 302.15; Antoine (liq) A = 7.235, B = 2559, C = 0, dev = 5.0, range = 308 to 323.

3-(Methylamino) pyridine, CA 18364-47-1: Irritant; flammable; soluble water, ethanol, ether, benzene. Antoine (liq) A = 8.165, B = 2987, C = 0, dev = 5.0, range = 313 to 343.

4-(Methylamino) pyridine, CA 1121-58-0: Irritant; flammable; soluble water, acetone, ethanol, ether, benzene. Tm = 388.15 to 391.15; Antoine (liq) A = 7.515, B = 2827, C = 0, dev = 5.0, range = 313 to 343.

1,3-Phenylenediamine, 1,3-diaminobenzene, CA 108-45-2: Irritant; highly toxic by inhalation and skin absorption; flammable; very soluble hot water; soluble ethanol, ether, acetone, benzene, chloroform. Tm = 336.15; Tb = 560.15; Vm = 0.1011 at 331.15; Antoine (liq) A = 7.16114, B = 2674.75, C = -39.859, dev = 1.0, range = 372 to 559.

Phenyl hydrazine, CA 100-63-0: Irritant; toxic by inhalation and skin absorption; suspected carcinogen; flammable, flash point 361; slightly soluble water; soluble ethanol, ether, benzene, chloroform. Tm = 292.75; Tb = 516.15, decomposes; Vm = 0.0988 at 298.15; Antoine (liq) A = 6.53614, B = 1975.68, C = -80.15, dev = 1.0, range = 413 to 518.

$C_6H_8O_3$ MW 128.13

2,2-Dimethylsuccinic acid, anhydride, CA 17347-61-4: Flammable; moisture-sensitive. Tm = 305.15; Antoine (liq) A = 7.59319, B = 2620.89, C = -23.719, dev = 1.0, range = 334 to 493.

2-Methylglutaric acid, anhydride, CA 31468-33-4: Flammable; moisture-sensitive. Tb = 545.15 to 548.15; Antoine (liq) A = 7.12822, B = 2694.89, C = -29.617, dev = 1.0, range = 366 to 556.

$C_6H_8O_4$ MW 144.13

Dimethyl fumarate, CA 624-49-7: Irritant; flammable; soluble acetone, chloroform. Tm = 375.15; Tb = 466.15; Antoine (liq) A = 7.955, B = 2807, C = 0, dev = 5.0, range = 361 to 466.

Dimethyl maleate, CA 624-48-6: Toxic; irritant; flammable, flash point 364; soluble ethanol. Tm = 254.15; Tb = 478.15; Vm = 0.1242 at 293.15; Antoine (liq) A = 7.780, B = 2715, C = 0, dev = 5.0, range = 385 to 421.

C_6H_8S MW 112.19

2,3-Dimethylthiophene, CA 632-16-6: Toxic; stench; flammable; soluble ethanol, ether, benzene. Tf = 224.15; Tb = 413.15 to 414.15; Vm = 0.1120 at 293.15; Antoine (liq) A = 6.0498, B = 1430, C = -61.15, dev = 1.0, range = 353 to 473.   *(continues)*

C_6H_8S *(continued)*

2,4-Dimethylthiophene, CA 638-00-6: Toxic; stench; flammable; soluble ethanol, ether, benzene. Tb = 413.85; Vm = 0.1127 at 293.15; Antoine (liq) A = 6.1188, B = 1450.7, C = -61.15, dev = 1.0, range = 323 to 493.

2,5-Dimethylthiophene, CA 638-02-8: Toxic; stench; flammable, flash point 296; soluble ethanol, ether, benzene. Tf = 210.55; Tb = 409.95; Vm = 0.1139 at 293.15; Antoine (liq) A = 6.8735, B = 1925.8, C = -12.75, dev = 1.0, range = 333 to 374.

3,4-Dimethylthiophene, CA 632-15-5: Toxic; stench; flammable; soluble ethanol, ether, benzene. Tb = 417.15; Vm = 0.1130 at 298.15; Antoine (liq) A = 6.0638, B = 1446.7, C = -61.25, dev = 5.0, range = 328 to 478.

2-Ethylthiophene, CA 872-55-9: Toxic; stench; flammable, flash point 294; soluble ethanol, ether, benzene. Tb = 407.15; Vm = 0.1130 at 293.15; Antoine (liq) A = 9.379, B = 3933.1, C = 130.95, dev = 1.0, range = 333 to 374.

3-Ethylthiophene, CA 1795-01-3: Toxic; stench; flammable; soluble ethanol, ether, benzene. Tf = 184.05; Tb = 409.15; Vm = 0.1124 at 293.15; Antoine (liq) A = 6.0779, B = 1422.0, C = -59.95, dev = 5.0, range = 318 to 473.

$C_6H_9F_3O_2$ MW 170.13

Butyl trifluoroacetate, CA 367-64-6: Flammable; soluble chloroform. Tb = 373.35; Vm = 0.1657 at 295.15; Antoine (liq) A = 6.8041, B = 1676.2, C = -28.15, dev = 1.0, range = 343 to 377.

C_6H_9N MW 95.14

2,5-Dimethylpyrrole, CA 625-84-3: Flammable, flash point 327; slightly soluble water; soluble ethanol, ether. Tb = 438.15; Vm = 0.1018 at 293.15, Antoine (liq) A = 6.33355, B = 1514.044, C = -90.841, dev = 0.1, range = 373 to 473.

C_6H_{10} MW 82.15

Bicyclo(3,1,0)hexane, *cis*, CA 285-58-5: Flammable. Antoine (liq) A = 7.0057, B = 1759.9, C = 0, dev = 1.0 to 5.0, range = 273 to 300.

Cyclohexene, CA 110-83-8: Irritant; flammable, flash point 261; polymerizes; soluble ethanol, ether, acetone, benzene. Tm = 169.65; Tb = 356.15; Vm = 0.1014 at 293.15; Antoine (liq) A = 5.997323, B = 1221.899, C = -49.978, dev = 0.02, range = 309 to 365.

2,3-Dimethyl-1,3-butadiene, CA 513-81-5: Flammable, flash point 251; polymerizes on heating; may form explosive peroxides; soluble most organic solvents. Tf = 197.15; Tb = 342.15 to 343.15; Vm = 0.1137 at 298.15; Antoine (liq) A = 6.3266, B = 1346, C = -30.15, dev = 1.0, range = 273 to 342.

1,3-Dimethylcyclobutene, CA 1489-61-8: Flammable; polymerizes. Antoine (liq) A = 6.9809, B = 1633, C = 0, dev = 1.0, range = 269 to 296.

1,3-Hexadiene, *trans*, CA 20237-34-7: Flammable, flash point 270; polymerizes; soluble most organic solvents. Tf = 170.75; Tb = 346.35; *(continues)*

C_6H_{10} *(continued)*

Vm = 0.1174 at 298.15; Antoine (liq) A = 5.69364, B = 1010.68, C = -71.364, dev = 0.1, range = 299 to 319.

1,4-Hexadiene, *trans*, CA 592-45-0: Flammable, flash point 252; polymerizes; soluble most organic solvents. Tf = 134.45; Tb = 338.15; Vm = 0.1182 at 298.15; Antoine (liq) A = 6.04268, B = 1188.124, C = -44.033, dev = 0.1, range = 304 to 323.

1,5-Hexadiene, CA 592-42-7: Flammable, flash point 246; polymerizes; soluble most organic solvents. Tf = 132.45; Tb = 332.75; Vm = 0.1194 at 298.15; Antoine I (liq) A = 5.7368, B = 1032, C = -56.15, dev = 1.0, range = 273 to 333; Antoine II (liq) A = 5.98314, B = 1159.908, C = -40.998, dev = 0.1, range = 299 to 333.

2,4-Hexadiene, CA 592-46-1: Flammable, flash point 266; readily polymerizes; soluble ethanol, ether, chloroform. Tm = 194.15; Tb = 350.95 to 354.35; Vm = 0.1142 at 293.15; Antoine (liq) A = 6.16354, B = 1263.15, C = -49.35, dev = 5.0, range = 276 to 401.

2,4-Hexadiene, *trans, trans*, CA 5194-51-4: Flammable; readily polymerizes; soluble ethanol, ether, chloroform. Tf = 228.25; Tb = 354.15; Vm = 0.1157 at 298.15; Antoine (liq) A = 5.99492, B = 1208.584, C = -52.769, dev = 0.1 to 1.0, range = 304 to 354.

1-Hexyne, CA 693-02-7: Irritant; flammable, flash point 252; polymerizes; may form explosive polymers; soluble most organic solvents. Tf = 141.06; Tb = 344.5; Vm = 0.1148 at 293.15; Antoine (liq) A = 6.03702, B = 1194.6, C = -48.15, dev = 1.0, range = 265 to 391.

3-Hexyne, CA 928-49-4: Irritant; flammable; polymerizes; may form explosive polymers; soluble most organic solvents. Tm = 167.62; Tb = 354.15; Vm = 0.1136 at 293.15; Antoine (liq) A = 8.31264, B = 2774.8, C = 85.19, dev = 1.0, range = 253 to 354.

1-Methylcyclopentene, CA 693-89-0: Flammable, flash point 256; soluble most organic solvents. Tm = 146.15; Tb = 349.15; Vm = 0.1060 at 298.15; Antoine (liq) A = 5.99374, B = 1199.6, C = -48.15, dev not specified, range = 268 to 403.

3-Methylcyclopentene ±, CA 39750-75-9: Flammable; soluble most organic solvents. Tb = 339.65 to 340.15; Vm = 0.1085 at 298.15; Antoine (liq) A = 5.99749, B = 1165.6, C = -46.15, dev not specified, range = 263 to 392.

4-Methylcyclopentene, CA 1759-81-5: Flammable; soluble most organic solvents. Tf = 112.35; Tb = 348.15; Vm = 0.1056 at 298.15; Antoine (liq) A = 5.99505, B = 1197.6, C = -48.15, dev not specified, range = 271 to 403.

$C_6H_{10}Br_2$ MW 241.95

1,2-Dibromocyclohexane, *trans*, CA 7429-37-0: Irritant; flammable, flash point above 385; soluble ethanol, ether, acetone, benzene. Tm = 268.15; Vm = 0.1362 at 293.15; Antoine (liq) A = 6.56169, B = 1964.941, C = -58.15, dev = 0.1 to 1.0, range = 350 to 416.

$C_6H_{10}ClFO_2$ MW 168.60

3-Fluorobutyric acid, 2-chloroethyl ester: Irritant; flammable. Antoine (liq) A = 8.8131, B = 3152.8, C = 0, dev = 1.0, range = 273 to 333.

$C_6H_{10}Cl_2$ MW 153.05

1,1-Dichlorocyclohexane, CA 2108-92-1: Irritant; flammable; soluble benzene. Tf = 226.15; Tb = 444.15; Vm = 0.0982 at 293.15; Antoine (liq) A = 5.76762, B = 1382.86, C = -76.642, dev = 1.0, range = 335 to 444.

1,2-Dichlorocyclohexane, *cis*, CA 10498-35-8: Irritant; flammable; soluble benzene. Tf = 271.65; Tb = 479.15 to 482.15; Vm = 0.1273 at 293.15; Antoine (liq) A = 6.82084, B = 2258.14, C = -10.915, dev = 1.0, range = 364 to 480.

1,2-Dichlorocyclohexane, *trans*, CA 822-86-6: Irritant; flammable, flash point 339; soluble benzene. Tf = 266.85; Tb = 462.15; Vm = 0.1293 at 293.15; Antoine (liq) A = 6.11854, B = 1651.207, C = -60.685, dev = 1.0, range = 344 to 462.

1,4-Dichlorocyclohexane, *cis*, CA 16749-11-4: Irritant; flammable; soluble benzene. Tm = 291.15; Tb = 466.15, decomposes; Vm = 0.1286 at 293.15; Antoine (liq) A = 7.28994, B = 2498.144, C = 0, dev = 1.0, range = 353 to 406.

$C_6H_{10}Cl_2O_2$ MW 185.05

Isobutyl dichloroacetate, CA 37079-08-6: Irritant; flammable. Antoine (liq) A = 6.91236, B = 2038.88, C = -40.442, dev = 1.0 to 5.0, range = 301 to 456.

$C_6H_{10}F_2O_2$ MW 152.14

3-Fluorobutyric acid, 2-fluoroethyl ester: Irritant; flammable. Antoine (liq) A = 8.411, B = 2861, C = 0, dev = 5.0, range = 273 to 333.

$C_6H_{10}N_6O_9$ MW 310.18

Dipropylamine, 2,2,2',2'-tetranitro-*N*-nitroso, CA 28464-26-8: Flammable. Tm = 403.15; Antoine (sol) A = 13.825, B = 5798, C = 0, dev = 1.0 to 5.0, range = 323 to 336.

$C_6H_{10}N_6O_{10}$ MW 326.18

Dipropylamine, *N*,2,2,2',2'-pentanitro, CA 28464-24-6: Flammable; explosive. Tm = 460.15; Antoine (sol) A = 9.075, B = 5175.4, C = 0, dev = 1.0 to 5.0, range = 397 to 423.

$C_6H_{10}O$ MW 98.14

Cyclohexanone, CA 108-94-1: Irritant; toxic vapor; flammable, flash point 317; may form explosive peroxides; very soluble cold water; soluble most organic solvents. Tm = 242.0; Tb = 429.85; Vm = 0.1062 at 298.15; Tc = 629.15; Pc = 3850; Vc = 0.311, calculated; Antoine I (sol) A = 8.434, B = 2576.6, C = 0, dev = 5.0, range not specified; Antoine II (liq) A = 6.103304, B = 1495.511, C = -63.5983, dev = 0.02 to 0.1, range = 362 to 439.

2,3-Dihydro-4-methyl-2*H*-pyran, CA 12655-16-2: Flammable. Antoine (liq) A = 7.0979, B = 1992.4, C = 0, dev = 1.0, range = 304 to 392.

(continues)

$C_6H_{10}O$ *(continued)*

5-Hexen-2-one, allylacetone, CA 109-49-9: Flammable, flash point 396; polymerizes. Tb = 402.75; Vm = 0.1178 at 300.15; Antoine (liq) A = 6.78088, B = 1996.2, C = 23.29, dev = 1.0, range = 449 to 561.

Methylenetetrahydro-2*H*-pyran, CA 35656-02-1: Flammable. Antoine (liq) A = 4.15464, B = 416.062, C = -188.682, dev = 1.0 to 5.0, range = 339 to 382.

2-Methyl-1-penten-4-one, isomesityl oxide, CA 3744-02-3: Irritant; lachrymator; flammable; soluble most organic solvents. Tm = 200.58; Tb = 394.15; Antoine I (liq) A = 6.13791, B = 1361.5, C = -65.15, dev = 1.0, range = 306 to 398; Antoine II (liq) A = 6.24543, B = 1440.437, C = -54.91, dev not specified, range = 389 to 461.

4-Methyl-3-penten-2-one, mesityl oxide, CA 141-79-7: Moderately toxic by inhalation and skin absorption; lachrymator; flammable, flash point 304; moderately soluble water; soluble most organic solvents; solvent for nitro-cellulose, gums, resins. Tm = 226.75; Tb = 402.95; Vm = 0.1153 at 298.15; Tc = 600.15; Pc = 3400; Antoine I (liq) A = 6.13756, B = 1399.09, C = -64.3, dev = 1.0, range = 313 to 405; Antoine II (liq) A = 6.25018, B = 1470.963, C = -56.375, dev = 1.0, range = 399 to 471.

$C_6H_{10}O_2$ MW 114.14

Allyl glycidyl ether, CA 106-92-3: Corrosive; toxic; flammable, flash point 330; forms explosive peroxides; polymer raw material. Tb = 427.15; Vm = 0.1177 at 293.15; Antoine (liq) A = 6.26935, B = 1508.678, C = -72.565, dev = 1.0, range = 323 to 420.

Butyric acid, vinyl ester, CA 123-20-6: Irritant; lachrymator; flammable. Tb = 389.85; Vm = 0.1267 at 293.15; Antoine (liq) A = 7.29103, B = 2051.021, C = 0, dev = 1.0, range = 365 to 387.

5,5-Dimethyldihydro-2(3*H*)-furanone, CA 3123-97-5: Flammable; soluble water. Tm = 273.15; Vm = 0.1132 at 298.15; Antoine (liq) A = 6.47306, B = 1898.86, C = -55.081, dev = 1.0, range = 311 to 480.

Ethyl crotonate, CA 10544-63-5: Irritant; lachrymator; flammable, flash point 275; soluble ethanol, ether. Tb = 411.15; Vm = 0.1244 at 293.15; Antoine (liq) A = 5.62336, B = 1092.6, C = -114.47, dev = 1.0, range = 329 to 420.

Ethyl methacrylate, CA 97-63-2: Irritant; lachrymator; flammable, flash point 293; soluble ethanol, ether. Tb = 391.15 to 392.15; Vm = 0.1256 at 298.15; Antoine (liq) A = 7.137, B = 2003, C = 0, dev = 5.0, range = 285 to 390.

2,5-Hexanedione, acetonyl acetone, CA 110-13-4: Highly toxic by inhalation; flammable, flash point 352; soluble water, ethanol, ether, acetone. Tm = 267.75; Tb = 465.45; Vm = 0.1175 at 293.15; Tc = 663.35; Pc = 3400; Antoine (liq) A = 7.56515, B = 2586.5, C = -2.53, dev = 1.0, range = 386 to 474.

Propyl acrylate, CA 925-60-0: Irritant; lachrymator; flammable. Tb = 396.15; Vm = 0.1257 at 293.15; Antoine (liq) A = 7.021, B = 1982, C = 0, dev = 5.0, range = 287 to 395.

4-Vinyl-1,3-dioxane, CA 1072-96-4: Irritant; lachrymator; flammable. Antoine (liq) A = 4.8903, B = 748.2, C = -155.95, dev = 1.0 to 5.0, range = 306 to 416.

$C_6H_{10}O_3$ MW 130.14

Cyclohexene ozonide: Flammable; potentially explosive. Antoine (liq) A = 2.49825, B = 387.846, C = 0, dev = 5.0, range = 276 to 311.

Ethyl acetoacetate, CA 141-97-9: Irritant; moderately toxic; flammable, flash point 330; very soluble water; soluble most organic solvents. Tm = 234.15; Tb = 454.15; Vm = 0.1274 at 298.15; Antoine (liq) A = 6.91605, B = 2003.6, C = -45.751, dev = 1.0 to 5.0, range = 301 to 454.

Methyl levulinate, CA 624-45-3: Flammable; soluble ethanol, ether, acetone, benzene. Tb = 469.15; Vm = 0.1238 at 293.15; Antoine (liq) A = 7.43787, B = 2521.12, C = -7.039, dev = 5.0, range = 312 to 471.

Propionic acid, anhydride, CA 123-62-6: Corrosive; lachrymator; flammable, flash point 336; moisture-sensitive; soluble methanol, ethanol, chloroform, ether. Tm = 228.15; Tb = 441.55 to 441.95; Vm = 0.1315 at 313.15; Antoine (liq) A = 7.4913, B = 2364.27, C = -9.448, dev = 5.0, range = 293 to 440.

$C_6H_{10}O_4$ MW 146.14

2-Acetoxypropionic acid, methyl ester, CA 6284-75-9: Flammable. Antoine (liq) A = 6.31995, B = 1527.888, C = -90.15, dev = 1.0 to 5.0, range = 337 to 445.

3-Acetoxypropionic acid, methyl ester, CA 38003-42-8: Flammable. Antoine (liq) A = 10.147, B = 3550, C = 0, dev = 5.0, range = 343 to 358.

Adipic acid, hexanedioic acid, CA 124-04-9: Irritant; flammable, flash point 469; very soluble hot water; soluble methanol, ethanol, acetone; used in manufacture of nylon and other polymers. Tm = 426.15 to 426.25; Tb = 610.65, decomposes; Vm = 0.133 at 426.15; Tc = 809, calculated; Pc = 3530, calculated; Vc = 0.400; Antoine I (sol) A = 14.588, B = 6757, C = 0, dev = 5.0, range = 359 to 406; Antoine II (liq) A = 6.97589, B = 2377.36, C = -132.475, dev = 1.0, range = 432 to 611.

Diethyl oxalate, CA 95-92-1: Irritant; high oral toxicity; flammable, flash point 348; slightly soluble water; soluble most organic solvents. Tm = 232.55; Tb = 458.55; Vm = 0.1355 at 293.15; Antoine I (liq) A = 7.61183, B = 2259.46, C = -55.688, dev = 1.0, range = 320 to 459; Antoine II (liq) A = 6.51656, B = 1699.399, C = -79.741, dev = 1.0 to 5.0, range = 343 to 457.

Dimethyl succinate, CA 106-65-0: Flammable, flash point 358; slightly soluble water; moderately soluble ethanol, ether, acetone. Tm = 292.15; Tb = 473.15; Vm = 0.1305 at 293.15; Antoine (liq) A = 7.4959, B = 2575, C = 0, dev = 1.0, range = 349 to 470.

Ethylene glycol, diacetate, CA 111-55-7: Moderately toxic; flammable, flash point 361; soluble water, ethanol, ether, acetone, benzene. Tf = 242.15; Tb = 459.15 to 460.15; Vm = 0.1323 at 293.15; Antoine (liq) A = 6.98671, B = 2057.72, C = -50.488, dev = 1.0, range = 311 to 464.

Ethylidene diacetate, 1,1-diacetoxyethane, CA 542-10-9: Flammable; slightly soluble water; soluble ethanol, ether. Tm = 291.15 to 292.15; Tb = 441.15; Vm = 0.1366 at 298.15; Antoine (liq) A = 7.9408, B = 2597.4, C = 0, dev = 1.0, range = 343 to 438.

$C_6H_{10}O_5$ MW 162.14

levo-Glucosan, CA 498-07-7: Soluble water. Tm = 457.15; Antoine (liq) A = 10.215, B = 4815, C = 0, dev = 5.0, range = 468 to 528.

Malic acid, dimethyl ester, *levo*, CA 617-55-0: Decomposed by water; on heating, isomerizes to dimethyl fumarate. Tb = 515.15; Antoine (liq) A = 7.67876, B = 2854.84, C = -12.721, dev = 1.0, range = 348 to 516.

Methyl[1-(methoxycarbonyl)ethyl] carbonate: Toxic; flammable. Antoine (liq) A = 6.5753, B = 1797, C = -80.15, dev = 5.0, range = 358 to 483.

$C_6H_{10}O_6$ MW 178.14

Tartaric acid, dimethyl ester, *dextro*, CA 608-68-4: Flammable, flash point above 385; soluble water, ethanol, ether, acetone. Tm = 321.15; Tb = 553.15; Antoine I (sol) A = 15.7353, B = 5903.2, C = 0, dev = 5.0, range = 308 to 317; Antoine II (liq) A = 9.8545, B = 3993.1, C = 0, dev = 1.0, range = 322 to 365; Antoine III (liq) A = 7.47348, B = 2812.2, C = -38.793, dev = 1.0, range = 375 to 553.

Tartaric acid, dimethyl ester ±, CA 608-69-5: Flammable, flash point above 385; soluble water, ethanol, ether, acetone. Tm = 363.15; Tb = 555.15; Vm = 0.1413 at 363.15; Antoine I (sol) A = 15.2517, B = 5941.3, C = 0, dev = 5.0, range = 315 to 358; Antoine II (liq) A = 8.14012, B = 3546.8, C = 22.833, dev = 1.0, range = 373 to 555.

$C_6H_{10}S$ MW 114.20

Allyl sulfide, CA 592-88-1: Toxic; stench; lachrymator; flammable, flash point 319; slightly soluble water; soluble ethanol, ether, chloroform, carbon tetrachloride; found in garlic. Tm = 190.15; Tb = 412.15; Vm = 0.1286 at 300.15; Antoine (liq) A = 6.23844, B = 1539.03, C = -48.133, dev = 1.0, range = 263 to 412.

$C_6H_{11}BrO_2$ MW 195.06

2-Bromo-2-methylpropionate, ethyl ester, CA 600-00-0: Corrosive; lachrymator; flammable, flash point 333; soluble ethanol, ether. Tb = 435.15 to 437.15; Vm = 0.1471 at 293.15; Antoine (liq) A = 6.94308, B = 2077.51, C = -16.044, dev = 1.0, range = 283 to 437.

$C_6H_{11}Cl$ MW 118.61

Chlorocyclohexane, cyclohexyl chloride, CA 542-18-7: Irritant; flammable, flash point 305; soluble ethanol, ether, acetone, benzene. Tm = 230.15; Tb = 415.15; Vm = 0.1186 at 293.15; Antoine I (liq) A = 6.01819, B = 1422.52, C = -61.24, dev = 0.1 to 1.0, range = 350 to 416; Antoine II (liq) A = 5.69855, B = 1200.629, C = -90.518, dev = 1.0, range = 352 to 416.

$C_6H_{11}ClO$ MW 134.61

Diethylacetyl chloride, CA 2736-40-5: Irritant; flammable; moisture-sensitive. Tb = 407.15 to 410.15; Antoine (liq) A = 7.0185, B = 2058.5, C = 0, dev = 1.0 to 5.0, range = 313 to 412.

$C_6H_{11}ClO_2$ MW 150.60

Chloroacetic acid, *sec*-butyl ester, CA 17696-64-9: Irritant; flammable; soluble ethanol, ether. Tb = 436.15 to 437.15; Vm = 0.1421 at 293.15; Antoine (liq) A = 6.66579, B = 1829.03, C = -48.541, dev = 1.0, range = 290 to 441.

Chloroacetic acid, isobutyl ester, CA 13361-35-8: Irritant; flammable; soluble ethanol, ether, acetone. Tb = 443.15; Vm = 0.1419 at 293.15; Antoine (liq) A = 5.8295, B = 1410.923, C = -66.142, dev = 1.0, range = 293 to 323.

$C_6H_{11}F$ MW 102.15

Fluorocyclohexane, cyclohexyl fluoride, CA 372-46-3: Flammable; soluble pyridine. Tm = 286.15; Tb = 375.15 to 377.15; Vm = 0.1101 at 293.15; Antoine (liq) A = 6.899, B = 1825.8, C = 0, dev = 1.0, range = 316 to 373.

$C_6H_{11}FO_2$ MW 134.15

2-Fluorohexanoic acid, CA 1578-57-0: Irritant; flammable. Antoine (liq) A = 10.79603, B = 4225.211, C = 0, dev = 5.0, range = 387 to 411.

$C_6H_{11}F_3N_2$ MW 168.16

Hexanamidine, *N,N,N'*-trifluoro, CA 31330-22-0: Irritant; flammable. Antoine (liq) A = 8.0377, B = 2429.3, C = 0, dev not specified, range not specified.

$C_6H_{11}I$ MW 210.06

Iodocyclohexane, cyclohexyl iodide, CA 626-62-0: Irritant; flammable, flash point 344; light-sensitive; soluble ethanol, ether, acetone, benzene. Tb = 453.15, decomposes; Vm = 0.1293 at 293.15; Antoine (liq) A = 6.84464, B = 2245.195, C = 0, dev = 1.0, range = 358 to 408.

$C_6H_{11}N$ MW 97.16

Capronitrile, hexanenitrile, CA 628-73-9: Irritant; flammable, flash point 316; soluble ethanol, ether. Tm = 192.85; Tb = 433.15; Vm = 0.1207 at 293.15; Antoine (liq) A = 6.195123, B = 1553.811, C = -65.8114, dev = 0.02 to 0.1, range = 371 to 442.

Diallyl amine, CA 124-02-7: Highly toxic; lachrymator; flammable, flash point 288; moderately soluble water; soluble ethanol, ether. Tm = 184.75; Tb = 383.55; Vm = 0.1234 at 293.15.

4-Methylvaleronitrile, isocapronitrile, CA 542-54-1: Irritant; flammable; insoluble water; soluble ethanol, ether. Tm = 222.15; Tb = 429.15 to 430.15; Vm = 0.1210 at 293.15; Antoine (liq) A = 6.3552, B = 1865.5, C = 0, dev = 1.0 to 5.0, range = 332 to 430.

$C_6H_{11}NO$ MW 113.16

Caprolactam, CA 105-60-2: Irritant; toxic by inhalation; moisture-sensitive; flammable, flash point 398; very soluble water,   *(continues)*

$C_6H_{11}NO$ *(continued)*

reacts; soluble most organic solvents; used to produce nylon-6. Tm = 342.45; Tb = 543.15; Vm = 0.1109 at 350.15; Tc = 805, calculated; Pc = 4770, calculated; Vc = 0.402, calculated; Antoine I (sol) A = 9.855, B = 4049, C = 0, dev = 5.0, range = 258 to 308; Antoine II (liq) A = 6.5562, B = 2154, C = -72.15, dev = 5.0, range = 373 to 543.

2-Hexenoic acid, *cis*, amide, CA 820-99-5: Irritant; flammable. Tm = 341.75; Antoine I (sol) A = 11.0114, B = 4179, C = 0, dev = 1.0, range = 323 to 333; Antoine II (liq) A = 8.1634, B = 3225.1, C = 0, dev = 1.0, range = 343 to 383.

2-Hexenoic acid, *trans*, amide, CA 820-99-5: Irritant; flammable. Tm = 397.95; Antoine (sol) A = 7.155, B = 2914.3, C = 0, dev = 5.0, range = 353 to 393.

2-Piperidone, 1-methyl, CA 931-20-4: Irritant; flammable; soluble water, ethanol, ether. Vm = 0.1103 at 298.15; Antoine (liq) A = 7.99278, B = 2894.295, C = 0, dev not specified, range = 341 to 385.

Pyridine, 2,3,4,5-tetrahydro-6-methoxy, CA 5693-62-9: Toxic; flammable; soluble water. Antoine (liq) A = 7.38134, B = 2237.376, C = 0, dev not specified, range = 292 to 338.

$C_6H_{11}NO_2$ MW 129.16

1-Aminocyclopentanecarboxylic acid, CA 52-52-8: Tm = 601.15 to 602.15, decomposes; Antoine (sol) A = 12.485, B = 6440, C = 0, dev = 5.0, range = 443 to 468.

Lactic acid, *N*-allylamide: Irritant; flammable. Antoine (liq) A = 9.78073, B = 4084.174, C = 0, dev = 1.0, range = 359 to 419.

$C_6H_{11}NO_3$ MW 145.16

Ethyl acetamidoacetate, CA 1906-82-7: Flammable, flash point above 385; soluble water, ethanol, ether, benzene, chloroform. Tm = 321.15; Tb = 533.15; Antoine (liq) A = 8.8737, B = 3627, C = 0, dev = 1.0, range = 383 to 466.

$C_6H_{11}NS$ MW 129.22

2-Piperidinethione, 1-methyl, CA 13070-07-0: Irritant; flammable. Antoine (liq) A = 7.84065, B = 3305.124, C = 0, dev not specified, range = 363 to 370.

Pyridine, 2,3,4,5-tetrahydro-4-(methylthio), CA 19766-29-1: Toxic; flammable; soluble water. Antoine (liq) A = 7.94199, B = 2747.67, C = 0, dev not specified, range = 313 to 351.

C_6H_{12} MW 84.16

Cyclohexane, CA 110-82-7: Flammable, flash point 253; soluble most organic solvents; important chemical raw material. Tm = 279.7; Tb = 353.89; Vm = 0.1080 at 293.15; Tc = 554.15; Pc = 4075; Vc = 0.308; Antoine I (sol) A = 7.2778, B = 1747.2, C = 26.84, dev = 1.0, range = 223 to 280; *(continues)*

198

C_6H_{12} *(continued)*

Antoine II (liq) A = 5.9662, B = 1201.531, C = -50.503, dev = 0.02, range = 293 to 355; Antoine III (liq) A = 6.03245, B = 1244.124, C = -44.911, dev = 0.1, range = 353 to 414; Antoine IV (liq) A = 6.36849, B = 1519.732, C = -4.032, dev = 0.1, range = 412 to 491; Antoine V (liq) A = 7.37347, B = 2683.075, C = 159.31, dev = 0.1, range = 489 to 553.

2,3-Dimethyl-1-butene, CA 563-78-0: Flammable, flash point below 253; polymerizes; soluble most organic solvents. Tf = 115.85; Tb = 329.15; Vm = 0.1237 at 293.15; Antoine (liq) A = 5.97875, B = 1129.944, C = -44.352, dev = 0.02, range = 267 to 335.

3,3-Dimethyl-1-butene, CA 558-37-2: Flammable, flash point 245; polymerizes; soluble most organic solvents. Tf = 157.95; Tb = 314.35; Vm = 0.1289 at 293.15; Antoine (liq) A = 5.8624, B = 1041.236, C = -44.405, dev = 0.02, range = 254 to 316.

2,3-Dimethyl-2-butene, CA 563-79-1: Flammable, flash point below 253; polymerizes; soluble most organic solvents. Tm = 198.85; Tb = 345.15 to 346.15; Vm = 0.1189 at 293.15; Antoine (liq) A = 6.07133, B = 1213.182, C = -47.947, dev = 0.02, range = 282 to 348.

1-Hexene, CA 592-41-6: Moderate irritant; flammable, flash point 247; polymerizes; soluble most organic solvents. Tm = 133.33; Tb = 336.64; Vm = 0.1250 at 293.15; Tc = 504.03; Pc = 3140; Vc = 0.354; Antoine (liq) A = 5.98336, B = 1149.029, C = -47.755, dev = 0.02, range = 273 to 343.

2-Hexene, *cis*, CA 7688-21-3: Flammable, flash point 244; polymerizes; soluble most organic solvents. Tm = 131.85; Tb = 341.95; Vm = 0.1225 at 293.15; Antoine (liq) A = 6.14811, B = 1250.582, C = -40.138, dev = 0.02, range = 278 to 343.

2-Hexene, *trans*, CA 4050-45-7: Flammable, flash point 244; polymerizes; soluble most organic solvents. Tf = 140.15; Tb = 341.15; Vm = 0.1250 at 298.15; Antoine (liq) A = 5.9507, B = 1137.488, C = -52.714, dev = 0.02, range = 283 to 342.

3-Hexene, *cis*, CA 7642-09-3: Flammable, flash point 261; polymerizes; soluble most organic solvents. Tm = 138.15; Tb = 339.95 to 340.05; Vm = 0.1238 at 303.15; Antoine (liq) A = 5.99142, B = 1157.506, C = -49.177, dev = 0.02, range = 276 to 348.

3-Hexene, *trans*, CA 13269-52-8: Flammable, flash point 261; polymerizes; soluble most organic solvents. Tf = 160.15; Tb = 340.55 to 340.75; Vm = 0.1243 at 293.15; Antoine (liq) A = 6.06139, B = 1190.215, C = -46.741, dev = 0.02, range = 278 to 341.

Methylcyclopentane, CA 96-37-7: Flammable, flash point 244; soluble most organic solvents. Tf = 130.7; Tb = 344.96; Vm = 0.1124 at 293.15; Antoine (liq) A = 5.98551, B = 1184.874, C = -47.232, dev = 0.02, range = 288 to 346.

2-Methyl-1-pentene, CA 763-29-1: Flammable, flash point 245; polymerizes; soluble most organic solvents. Tf = 137.45; Tb = 334.65 to 335.15; Vm = 0.1238 at 293.15; Antoine (liq) A = 5.97842, B = 1140.282, C = -48.224, dev = 0.02, range = 272 to 341.

3-Methyl-1-pentene, CA 760-20-3: Flammable, flash point 245; polymerizes; soluble most organic solvents. Tf = 120.15; Tb = 326.75 to 327.15; Vm = 0.1261 at 293.15; Antoine I (liq) A = 5.89056, B = 1091.679, *(continues)*

C_6H_{12} *(continued)*

C = -46.306, dev = 0.02, range = 265 to 333; Antoine II (liq) A = 5.99434, B = 1148.616, C = -49.853, dev = 0.02, range = 275 to 344.

4-Methyl-1-pentene, CA 691-37-2: Flammable, flash point 242; polymerizes; soluble most organic solvents. Tm = 116.52; Tb = 326.95; Vm = 0.1267 at 293.15; Antoine (liq) A = 5.94694, B = 1114.082, C = -44.332, dev = 0.02, range = 265 to 333.

2-Methyl-2-pentene, CA 625-27-4: Flammable, flash point 250; polymerizes; soluble most organic solvents. Tf = 138.15; Tb = 340.35 to 340.65; Vm = 0.1226 at 293.15; Antoine (liq) A = 6.02657, B = 1171.612, C = -49.064, dev = 0.02, range = 277 to 346.

3-Methyl-2-pentene, *cis*, CA 922-62-3: Flammable, flash point 267; polymerizes; soluble most organic solvents. Tm = 134.75; Tb = 341.15 to 342.15; Vm = 0.1205 at 293.15; Antoine (liq) A = 6.02143, B = 1178.519, C = -47.367, dev = 0.02, range = 277 to 347.

3-Methyl-2-pentene, *trans*, CA 616-12-6: Flammable, flash point 267; polymerizes; soluble most organic solvents. Tf = 134.65; Tb = 343.15 to 344.15; Vm = 0.1214 at 298.15; Antoine (liq) A = 6.04046, B = 1188.556, C = -49.0, dev = 0.02, range = 280 to 349.

4-Methyl-2-pentene, *cis*, CA 691-38-3: Flammable, flash point below 244; polymerizes; soluble most organic solvents. Tm = 138.75; Tb = 329.45; Vm = 0.1258 at 293.15; Antoine (liq) A = 5.9581, B = 1116.268, C = -47.098, dev = 0.02, range = 267 to 330.

4-Methyl-2-pentene, *trans*, CA 674-76-0: Flammable, flash point below 244; polymerizes; soluble most organic solvents. Tm = 132.35; Tb = 331.65; Vm = 0.1259 at 293.15; Antoine (liq) A = 5.9953, B = 1137.435, C = -46.651, dev = 0.02, range = 269 to 337.

$C_6H_{12}Br_2$ MW 243.97

1,1-Dibromohexane, CA 58133-26-9: Irritant; flammable; soluble ethanol, acetone. Antoine (liq) A = 6.458, B = 1794, C = -72.15, dev = 5.0, range = 378 to 526.

$C_6H_{12}ClNO$ MW 149.62

4-(2-Chloroethyl) morpholine, CA 3240-94-6: Irritant; flammable. Antoine (liq) A = 8.04461, B = 2808.7, C = 0, dev = 5.0, range = 273 to 333.

$C_6H_{12}Cl_2$ MW 155.07

1,1-Dichlorohexane, CA 62017-16-7: Irritant; flammable; soluble chloroform, ether. Antoine (liq) A = 6.248, B = 1578, C = -65.15, dev = 5.0, range = 345 to 484.

1,2-Dichlorohexane ±, CA 2162-92-7: Irritant; flammable; soluble ether, chloroform. Tb = 445.15 to 447.15; Vm = 0.1429 at 288.15; Antoine (liq) A = 6.08007, B = 1526.369, C = -70.762, dev = 0.1, range = 352 to 442.

$C_6H_{12}Cl_2O$ MW 171.07

Ether, bis(2-chloro-1-methylethyl), CA 108-60-1: Irritant; flammable;
forms explosive peroxides; soluble ethanol, ether, acetone, benzene. Tb =
461.15; Vm = 0.1551 at 293.15; Antoine (liq) A = 6.68233, B = 1856.14, C =
-58.793, dev = 1.0 to 5.0, range = 302 to 456.

$C_6H_{12}Cl_2O_2$ MW 187.07

Acetaldehyde, bis(2-chloroethyl) acetal, CA 14689-97-5: Irritant; flammable; soluble ethanol, ether. Tb = 467.15 to 469.15; Vm = 0.1594 at
293.15; Antoine (liq) A = 7.52856, B = 2482.34, C = -36.122, dev = 5.0,
range = 329 to 486.

$C_6H_{12}Cl_3N$ MW 204.53

Tris(2-chloroethyl) amine, nitrogen mustard gas, CA 555-77-1: Powerful
vesicant; carcinogen; flammable; decomposes on standing; slightly soluble
water; soluble ethanol, ether, carbon tetrachloride, carbon disulfide;
chemical warfare agent. Tm = 269.15; Antoine (liq) A = 8.54111, B =
3393.4, C = 0, dev = 1.0, range = 273 to 367.

$C_6H_{12}Cl_3O_4P$ MW 285.49

Tris(2-chloroethyl) phosphate, CA 115-96-8: Irritant; flammable, flash
point 505. Tb = 603.15; Vm = 0.2054 at 293.15; Antoine (liq) A = 4.345,
B = 1917, C = 0, dev = 5.0, range = 293 to 445.

$C_6H_{12}F_2$ MW 122.16

1,1-Difluorohexane, CA 62127-41-7: Irritant; flammable. Antoine (liq) A =
6.287, B = 1353, C = -52.15, dev = 5.0, range = 290 to 407.

$C_6H_{12}F_4N_2$ MW 188.17

N,N,N',N'-Tetrafluoro-2-methyl-1,2-pentanediamine, CA 16096-76-7: Irritant; flammable. Antoine (liq) A = 6.18432, B = 1499.818, C = -58.041,
dev = 0.1, range = 253 to 293.

$C_6H_{12}N_2$ MW 112.17

1,4-Diazabicyclo(2,2,2)octane, triethylene diamine, CA 280-57-9: Corrosive; irritant; extremely hygroscopic; flammable, flash point above 323;
soluble water, ethanol, acetone, benzene. Tf = 431.15; Tb = 447.15;
Antoine I (sol) A = 9.5981, B = 3233.63, C = 0, dev = 1.0, range = 324 to
352; Antoine II (sol) A = 8.1518, B = 2722.18, C = 0, dev = 1.0, range =
353 to 369.

$C_6H_{12}N_2O_6$ MW 208.17

2,5-Hexanediol, dinitrate: Flammable; potentially explosive. Tm = 323.51;
Antoine (sol) A = 20.445, B = 6217, C = 0, dev = 5.0, range = 293 to 313.

$C_6H_{12}N_2O_8$ MW 240.17

Triethylene glycol, dinitrate, CA 111-22-8: Flammable; used as explosive.
Antoine (liq) A = 10.6041, B = 4614.7, C = 0, dev = 1.0, range = 303 to 348.

$C_6H_{12}N_4$ MW 140.19

Hexamethylene tetramine, methenamine, CA 100-97-0: Irritant; flammable,
flash point 523; soluble water, ethanol, acetone, chloroform. Tm = 553.15,
sublimes; Antoine (sol) A = 9.3109, B = 4012, C = 0, dev = 1.0 to 5.0,
range = 298 to 453.

$C_6H_{12}O$ MW 100.16

Allyl isopropyl ether, CA 6140-80-3: Irritant; lachrymator; polymerizes;
flammable; forms explosive peroxides; soluble ethanol, ether, acetone.
Tb = 355.15 to 356.15; Vm = 0.1290 at 293.15; Antoine I (liq) A = 7.01253,
B = 1690.5, C = -14.994, dev = 1.0, range = 229 to 353; Antoine II (liq)
A = 6.06433, B = 1218.691, C = -52.409, dev = 1.0, range = 253 to 415.

Allyl propyl ether, CA 1471-03-0: Irritant; lachrymator; polymerizes;
flammable; forms explosive peroxides; soluble ethanol, ether, acetone.
Tb = 363.15 to 365.15; Vm = 0.1290 at 293.15; Antoine I (liq) A = 6.63777,
B = 1584.1, C = -21.779, dev = 1.0, range = 234 to 364; Antoine II (liq)
A = 6.05785, B = 1249.871, C = -55.238, dev = 1.0, range = 261 to 428.

Butyl vinyl ether, CA 111-34-2: Irritant; lachrymator; polymerizes; flammable, flash point 264; forms explosive peroxides; soluble ethanol, ether,
acetone, benzene. Tm = 181.25; Tb = 366.65; Vm = 0.1270 at 293.15; Antoine
(liq) A = 7.145, B = 1884.9, C = 0, dev = 1.0, range = 269 to 368.

Cyclohexanol, CA 108-93-0: Irritant; toxic vapor; narcotic in high concen-
trations; flammable, flash point 341; moderately soluble water; soluble
most organic solvents. Tm = 298.3; Tb = 434.25; Vm = 0.1041 at 293.15;
Tc = 625.15; Pc = 3749; Vc = 0.322, calculated; Antoine I (sol) A = 9.631,
B = 3173.1, C = 0, dev = 1.0 to 5.0, range = 272 to 298; Antoine II (liq)
A = 6.1634, B = 1318.5, C = -116.55, dev = 1.0, range = 318 to 434; Antoine
III (liq) A = 6.27792, B = 1381.8, C = -110.132, dev = 0.1 to 1.0, range =
300 to 434.

3,3-Dimethyl-2-butanone, pinacolone, CA 75-97-8: Flammable, flash point
296; moderately soluble water; soluble ethanol, ether, acetone. Tm =
223.35; Tb = 379.35; Vm = 0.1250 at 298.15; Antoine I (liq) A = 6.3188, B =
1451.4, C = -43.15, dev = 1.0, range = 311 to 381; Antoine II (liq) A =
6.05794, B = 1296.563, C = -59.301, dev = 0.02, range = 363 to 400; Antoine
III (liq) A = 6.07608, B = 1307.382, C = -58.071, dev = 0.1 to 1.0, range =
289 to 402; Antoine IV (liq) A = 6.36848, B = 1527.032, C = -28.989, dev =
1.0, range = 396 to 509; Antoine V (liq) A = 7.08726, B = 2274.685, C =
74.296, dev = 1.0 to 5.0, range = 491 to 567.

2,5-Dimethyltetrahydrofuran ±, CA 38484-59-2: Flammable, flash point 299;
soluble ethanol, ether, benzene. Tb = 365.15 to 367.15; Vm = 0.1202 at
293.15; Antoine (liq) A = 6.44476, B = 1470.95, C = -31.613, dev = 1.0 to
5.0, range = 278 to 370.

1,2-Epoxyhexane, CA 1436-34-6: Flammable, flash point 288; moisture-
sensitive. Tb = 391.15 to 393.15; Vm = 0.1205 at 293.15; Antoine (liq) A =
5.2257, B = 883.99, C = -117.25, dev = 1.0, range = 300 to 390.

(continues)

202

$C_6H_{12}O$ *(continued)*

2-Hexanone, butyl methyl ketone, CA 591-78-6: Irritant; moderately toxic; flammable, flash point 298; slightly soluble water; soluble ethanol, ether, acetone. Tm = 216.25; Tb = 400.15; Vm = 0.1235 at 293.15; Antoine I (liq) A = 6.4127, B = 1575.5, C = -43.15, dev = 5.0, range = 293 to 411; Antoine II (liq) A = 6.14816, B = 1392.968, C = -64.465, dev = 0.02, range = 279 to 423; Antoine III (liq) A = 6.15949, B = 1399.959, C = -63.704, dev = 0.1, range = 310 to 427; Antoine IV (liq) A = 6.48416, B = 1653.979, C = -31.154, dev = 1.0, range = 421 to 523; Antoine V (liq) A = 7.40894, B = 2670.75, C = 99.934, dev = 1.0, range = 513 to 587.

3-Hexanone, ethyl propyl ketone, CA 589-38-8: Moderately toxic; flammable, flash point 308; soluble ethanol, ether, acetone. Tm = 217.53; Tb = 396.15 to 396.65; Vm = 0.1234 at 293.15; Antoine I (liq) A = 6.12573, B = 1365.798, C = -65.143, dev = 1.0, range = 292 to 406; Antoine II (liq) A = 6.12591, B = 1366.412, C = -65.018, dev = 0.02, range = 348 to 413; Antoine III (liq) A = 6.36209, B = 1542.134, C = -42.563, dev = 0.1, range = 408 to 517; Antoine IV (liq) A = 7.39423, B = 2668.487, C = 106.115, dev = 0.1 to 1.0, range = 511 to 583.

Isobutyl vinyl ether, CA 109-53-5: Irritant; lachrymator; polymerizes; flammable, flash point 264; forms explosive peroxides; soluble ethanol, ether, acetone, benzene. Tf = 161.05; Tb = 356.15; Vm = 0.1304 at 293.15; Antoine (liq) A = 7.522, B = 1954, C = 0, dev = 5.0, range = 266 to 357.

1-Methylcyclopentanol, CA 1462-03-9: Flammable, flash point 313; moderately soluble water; soluble most organic solvents. Tm = 308.15 to 309.15; Tb = 408.15 to 409.15; Antoine (liq) A = 7.8527, B = 2386.1, C = 0, dev = 1.0, range = 354 to 407.

2-Methyl-2,3-epoxypentane, CA 1192-22-9: Flammable. Antoine (liq) A = 4.3439, B = 466.3, C = -169.85, dev = 1.0 to 5.0, range = 306 to 369.

3-Methyl-2-pentanone ±, *sec*-butyl methyl ketone, CA 565-61-7: Moderately toxic; flammable, flash point 285; soluble ethanol, ether, chloroform. Tb = 391.15; Vm = 0.1232 at 293.15; Antoine I (liq) A = 6.3986, B = 1526.1, C = -43.15, dev = 5.0, range = 286 to 400; Antoine II (liq) A = 6.19488, B = 1386.759, C = -59.533, dev = 1.0, range = 283 to 395; Antoine III (liq) A = 6.23833, B = 1421.785, C = -54.658, dev = 1.0, range = 385 to 455.

4-Methyl-2-pentanone, methyl isobutyl ketone, CA 108-10-1: Irritant; moderately toxic; flammable, flash point 291; moderately soluble water; soluble ethanol, ether, acetone, benzene; solvent for cellulose esters. Tf = 189.15; Tb = 389.35; Vm = 0.1251 at 293.15; Tc = 571.15; Pc = 3300; Vc = 0.369, calculated; Antoine I (liq) A = 5.79768, B = 1168.443, C = -81.202, dev = 1.0, range = 294 to 390; Antoine II (liq) A = 5.83311, B = 1188.115, C = -79.265, dev = 1.0, range = 281 to 400.

2-Methyl-3-pentanone, ethyl isopropyl ketone, CA 565-69-5: Irritant; moderately toxic; flammable, flash point 286; soluble ethanol, ether, acetone, benzene. Tb = 388.15 to 389.15; Vm = 0.1230 at 291.15; Antoine I (liq) A = 5.42548, B = 944.0, C = -111.46, dev = 1.0, range = 300 to 387; Antoine II (liq) A = 6.20686, B = 1380.785, C = -57.899, dev = 1.0, range = 280 to 387; Antoine III (liq) A = 6.22276, B = 1398.029, C = -55.063, dev = 1.0, range = 377 to 450.

$C_6H_{12}O_2$ MW 116.16

Butyl acetate, CA 123-86-4: Irritant; flammable, flash point 295; moderately soluble water; soluble most organic solvents.   *(continues)*

$C_6H_{12}O_2$ *(continued)*

Tf = 196.15; Tb = 399.15; Vm = 0.1329 at 298.15; Tc = 579.15; Pc = 3160, calculated; Vc = 0.389, calculated; Antoine I (liq) A = 6.176, B = 1385.8, C = -67.05, dev = 1.0, range = 332 to 399; Antoine II (liq) A = 6.13505, B = 1355.816, C = -70.705, dev = 0.1, range = 341 to 399.

Caproic acid, hexanoic acid, CA 142-62-1: Corrosive; toxic; flammable, flash point 375; slightly soluble water; soluble ethanol, ether; has a typical goat-like odor. Tm = 269.75; Tb = 478.95; Vm = 0.1250 at 293.15; Antoine (liq) A = 7.08241, B = 2009.93, C = -82.69, dev = 1.0, range = 335 to 487.

1,2-Cyclohexanediol, *cis*, CA 1792-81-0: Flammable; soluble water, ethanol, ether, benzene. Tm = 371.15; Vm = 0.1128 at 374.15; Antoine (sol) A = 5.8079, B = 2282.1, C = 0, dev = 1.0, range = 289 to 320.

1,2-Cyclohexanediol, *trans*, CA 54383-22-1: Flammable; soluble water, ethanol, ether, benzene. Tm = 377.15; Antoine (sol) A = 5.6642, B = 2218.9, C = 0, dev = 1.0, range = 289 to 320.

Diacetone alcohol, CA 123-42-2: Irritant; moderately toxic; flammable, flash point 337; decomposes when distilled; all physical properties affected by impurity, particularly presence of free acetone; soluble water, polar solvents. Tf = 228.95; Tb = 442.35; Vm = 0.1248 at 298.15; Tc = 605.15; Pc = 3500; Antoine (liq) A = 7.6801, B = 2482.93, C = 0, dev = 1.0 to 5.0, range = 301 to 440.

4,4-Dimethyl-1,3-dioxane, CA 766-15-4: Flammable; soluble ethanol, ether. Antoine (liq) A = 8.1205, B = 3006.6, C = 85.19, dev = 1.0, range = 333 to 407.

4,5-Dimethyl-1,3-dioxane, *cis*, CA 2391-24-4: Flammable; soluble ethanol, ether. Antoine (liq) A = 4.0788, B = 384.83, C = -223.945, dev = 1.0 to 5.0, range = 353 to 410.

4,5-Dimethyl-1,3-dioxane, *trans*, CA 1121-20-6: Flammable; soluble ethanol, ether. Antoine (liq) A = 5.0183, B = 808.16, C = -136.2, dev = 1.0 to 5.0, range = 353 to 408.

Ethyl butyrate, CA 105-54-4: Irritant in high concentrations; flammable, flash point 297; slightly soluble water; soluble ethanol, ether; has a typical pineapple odor; used in foods and perfumes. Tf = 179.85; Tb = 393.15 to 394.15; Vm = 0.1322 at 293.15; Antoine (liq) A = 5.79321, B = 1154.21, C = -89.57, dev = 1.0, range = 263 to 404.

2-Ethylbutyric acid, CA 88-09-5: Corrosive; toxic; flammable, flash point 372; slightly soluble water; soluble ethanol, ether. Tf = 258.15; Tb = 467.15 to 468.15; Vm = 0.1257 at 293.15; Antoine (liq) A = 8.528, B = 3040, C = 0, dev = 5.0, range = 373 to 466.

4-Ethyl-1,3-dioxane, CA 1121-61-5: Flammable; soluble ethanol, ether. Antoine (liq) A = 6.2195, B = 1511.5, C = -53.45, dev = 1.0, range = 362 to 412.

Ethyl isobutyrate, CA 97-62-1: Irritant; flammable, flash point 286; slightly soluble water; soluble ethanol, ether. Tm = 184.95; Tb = 383.15; Vm = 0.1336 at 293.15; Antoine I (liq) A = 6.06445, B = 1285.96, C = -66.535, dev = 1.0, range = 249 to 393; Antoine II (liq) A = 6.49953, B = 1565.55, C = -34.996, dev = 1.0, range = 383 to 483.

(continues)

204

$C_6H_{12}O_2$ *(continued)*

2-Hydroxymethyltetrahydropyran, CA 100-72-1: Irritant; moderately toxic; flammable, flash point 366; soluble ethanol, ether. Tb = 460.15; Vm = 0.1122 at 298.15; Antoine (liq) A = 7.5681, B = 2559, C = 0, dev = 5.0, range = 344 to 460.

Isobutyl acetate, CA 110-19-0: Moderately toxic; flammable, flash point 291; slightly soluble water; soluble ethanol, ether, acetone. Tf = 174.15; Tb = 391.15; Vm = 0.1328 at 293.15; Antoine (liq) A = 6.66966, B = 1709.03, C = -24.779, dev = 1.0, range = 252 to 391.

Isocaproic acid, 4-methylvaleric acid, CA 646-07-1: Irritant; flammable; slightly soluble water; soluble ethanol, ether. Tm = 240.15; Tb = 472.25; Vm = 0.1259 at 293.15; Antoine (liq) A = 4.49814, B = 639.69, C = -223.885, dev = 5.0, range = 339 to 481.

Isopentyl formate, CA 110-45-2: Irritant; moderately toxic; flammable; slightly soluble water; soluble ethanol, ether. Tm = 179.65; Tb = 397.35; Vm = 0.1312 at 293.15; Antoine (liq) A = 7.17776, B = 2060.14, C = 1.929, dev = 1.0, range = 255 to 397.

Methyl isovalerate, methyl 3-methylbutanoate, CA 556-24-1: Irritant; flammable; slightly soluble water; soluble ethanol, ether. Tb = 389.15 to 390.15; Vm = 0.1319 at 293.15; Antoine (liq) A = 6.67921, B = 1675.41, C = -31.431, dev = 1.0, range = 254 to 390.

Methyl valerate, CA 624-24-8: Irritant; flammable, flash point 295; soluble ethanol, ether, acetone. Tm = 182.15; Tb = 399.65; Vm = 0.1298 at 293.15; Antoine (liq) A = 6.62646, B = 1658.4, C = -42.09, dev = 1.0, range = 297 to 411.

tert-Pentyl formate, CA 757-88-0: Flammable. Antoine (liq) A = 6.825, B = 1859, C = 0, dev = 1.0, range not specified.

Propyl propionate, CA 106-36-5: Flammable, flash point 352; slightly soluble water; soluble ethanol, ether, acetone. Tm = 197.25; Tb = 395.15 to 397.15; Vm = 0.1319 at 293.15; Antoine (liq) A = 6.58895, B = 1623.45, C = -41.317, dev = 1.0, range = 259 to 396.

$C_6H_{12}O_3$ MW 132.16

sec-Butyl glycolate: Flammable. Antoine (liq) A = 7.02974, B = 2052.59, C = -42.005, dev = 1.0, range = 301 to 451.

Ethoxyacetic acid, ethyl ester, CA 817-95-8: Flammable; soluble ethanol, ether, acetone. Tb = 431.15; Vm = 0.1362 at 293.15; Antoine (liq) A = 7.6109, B = 2408.269, C = 0, dev = 1.0 to 5.0, range = 330 to 430.

3-Ethoxypropionic acid, methyl ester, CA 14144-33-3: Flammable. Antoine (liq) A = 7.352, B = 2314, C = 0, dev = 5.0, range = 320 to 432.

2-Ethoxyethanol, acetate, CA 111-15-9: Flammable, flash point 320; very soluble water; soluble ethanol, ether. Tm = 211.15; Tb = 429.15; Vm = 0.1358 at 293.15; Antoine (liq) A = 5.5777, B = 1107.8, C = -119.15, dev = 1.0, range = 322 to 430.

Glycerol, 1-monoallyl ether: Irritant; flammable; polymerizes; forms explosive peroxides. Tm = 183.15; Antoine (liq) A = 6.6503, B = 1945.5, C = -99.15, dev = 1.0, range = 323 to 383.   *(continues)*

$C_6H_{12}O_3$ *(continued)*

1-Hexene ozonide: Flammable; potentially explosive. Antoine (liq) A = 1.53615, B = 170.387, C = 0, dev = 5.0, range = 335 to 414.

3-Hydroxypropionic acid, propyl ester: Flammable. Antoine (liq) A = 8.911, B = 3182, C = 0, dev = 5.0, range = 350 to 375.

3-Methoxypropionic acid, ethyl ester, CA 10606-42-5: Flammable. Antoine (liq) A = 7.4011, B = 2329, C = 0, dev = 5.0, range = 313 to 432.

Propyl lactate, CA 616-09-1: Flammable; soluble water, ethanol, ether, benzene. Tm = 233.15; Antoine (liq) A = 6.25936, B = 1496.05, C = -90.15, dev = 1.0 to 5.0, range = 334 to 442.

2,4,6-Trimethyl-1,3,5-trioxane, CA 123-63-7: Flammable, flash point 309; narcotic; can lead to addiction; moderately soluble water; soluble ethanol, ether, chloroform. Tm = 285.75; Tb = 398.5; Vm = 0.1333 at 298.15; Antoine (liq) A = 7.465, B = 2168, C = 0, dev = 1.0 to 5.0, range = 323 to 396.

$C_6H_{12}O_4$ MW 148.16

Glycerol, 1-propionate ±, CA 624-47-5: Flammable. Antoine (liq) A = 9.42172, B = 3961.3, C = 0, dev = 1.0, range = 388 to 456.

$C_6H_{12}S$ MW 116.22

Cyclohexanethiol, CA 1569-69-3: Irritant; stench; flammable, flash point 316; soluble ethanol, ether, acetone, benzene, chloroform. Tb = 431.15; Vm = 0.1188 at 293.15; Antoine (liq) A = 6.01115, B = 1476.976, C = -63.234, dev = 0.02, range = 355 to 476.

Cyclopentyl methyl sulfide, CA 7133-36-0: Irritant; stench; flammable. Antoine (liq) A = 6.06738, B = 1482.501, C = -64.412, dev = 0.02, range = 354 to 473.

2,5-Dimethyltetrahydrothiophene, *cis*, CA 5161-13-7: Flammable; soluble ethanol, ether, benzene. Tm = 183.75; Antoine (liq) A = 6.01964, B = 1414.739, C = -63.243, dev = 0.1, range = 349 to 427.

2,5-Dimethyltetrahydrothiophene, *trans*, CA 5161-14-8: Flammable; soluble ethanol, ether, benzene. Tm = 196.83; Antoine (liq) A = 6.05769, B = 1445.091, C = -58.505, dev = 0.1, range = 348 to 396.

2-Ethyltetrahydrothiophene, CA 1551-32-2: Flammable; soluble ethanol, ether, benzene. Antoine (liq) A = 6.04642, B = 1478.9, C = -64.15, dev = 5.0, range = 333 to 488.

3-Ethyltetrahydrothiophene, CA 62184-67-2: Flammable; soluble ethanol, ether, benzene. Antoine (liq) A = 6.04485, B = 1506.6, C = -65.15, dev = 1.0, range = 343 to 503.

2-Methyltetrahydro-2*H*-thiopyran, CA 5161-16-0: Flammable; soluble ethanol, ether, benzene. Tm = 215.04; Antoine (liq) A = 5.97765, B = 1440.674, C = -63.509, dev = 0.1, range = 356 to 438.

3-Methyltetrahydro-2*H*-thiopyran, CA 5258-50-4: Flammable; soluble ethanol, ether, benzene. Tm = 213.01; Antoine (liq) A = 6.07754, B = 1520.204, C = -57.862, dev = 0.1, range = 361 to 435.   *(continues)*

$C_6H_{12}O_4$ *(continued)*

4-Methyltetrahydro-2*H*-thiopyran, CA 5161-17-1: Flammable; soluble ethanol, ether, benzene. Tm = 245.05; Antoine (liq) A = 6.00297, B = 1474.243, C = -63.0, dev = 0.1, range = 361 to 441.

$C_6H_{13}Br$ MW 165.07

2-Bromo-3,3-dimethylbutane, CA 26356-06-9: Irritant; flammable; soluble ethanol, ether, acetone, chloroform. Tm = 298.15; Antoine (liq) A = 6.0194, B = 1388.7, C = -59.15, dev = 5.0, range = 315 to 449.

1-Bromohexane, hexyl bromide, CA 111-25-1: Irritant; flammable, flash point 330; soluble ethanol, ether, acetone, chloroform. Tf = 188.45; Tb = 428.45; Vm = 0.1406 at 293.15; Antoine (liq) A = 6.1272, B = 1503.52, C = -63.65, dev = 1.0, range = 333 to 456.

2-Bromohexane ±, CA 3577-86-4: Irritant; flammable; soluble ethanol, acetone, ether, chloroform. Tb = 417.15; Vm = 0.1416 at 293.15; Antoine (liq) A = 6.0731, B = 1443, C = -61.15, dev = 5.0, range = 303 to 416.

2-Bromo-4-methylpentane, CA 30310-22-6: Irritant; flammable; soluble ethanol, ether, acetone, benzene. Tm = 178.95; Antoine (liq) A = 6.01362, B = 1382.7, C = -59.15, dev = 5.0, range = 315 to 448.

$C_6H_{13}Cl$ MW 120.62

2-Chloro-2,3-dimethylbutane, CA 594-57-0: Irritant; flammable; soluble ethanol, ether, acetone, benzene. Tm = 263.15 to 264.15; Tb = 383.15 to 385.15; Vm = 0.1374 at 293.15; Antoine (liq) A = 6.09787, B = 1350.4, C = -55.15, dev = 1.0, range = 301 to 426.

2-Chloro-3,3-dimethylbutane ±, CA 5750-00-5: Irritant; flammable; soluble ethanol, ether, acetone, benzene. Tm = 272.25; Tb = 386.15 to 387.15; Vm = 0.1391 at 293.15; Antoine (liq) A = 6.0948, B = 1345.4, C = -55.15, dev = 5.0, range = 300 to 425.

1-Chlorohexane, hexyl chloride, CA 544-10-5: Irritant; flammable, flash point 308; soluble ethanol, ether, acetone, benzene. Tm = 179.15; Tb = 408.15 to 409.15; Vm = 0.1377 at 293.15; Antoine (liq) A = 5.89376, B = 1304.968, C = -73.092, dev = 0.1 to 1.0, range = 288 to 409.

2-Chlorohexane ±, CA 638-28-8: Irritant; flammable; soluble ethanol, acetone, ether, benzene. Tb = 396.15; Vm = 0.1387 at 294.15; Antoine (liq) A = 6.1721, B = 1419, C = -58.15, dev = 5.0, range = 300 to 399.

$C_6H_{13}ClO_2S$ MW 184.68

1-Hexanesulfonyl chloride, CA 14532-24-2: Irritant; flammable. Tm = 350.15; Antoine (sol) A = 8.403, B = 3169.1, C = 0, dev = 1.0, range = 273 to 303.

$C_6H_{13}Cl_2N$ MW 170.08

Triethylamine, 2,2-dichloro, CA 13426-57-8: Highly toxic; nitrogen mustard; deadly vesicant; flammable; insoluble water; soluble most organic solvents. Tm = 239.15; Vm = 0.1566 at 296.15; Antoine (liq) A = 8.14382, B = 2868.9, C = 0, dev = 1.0, range = 372 to 333.

$C_6H_{13}F$ MW 104.17

1-Fluorohexane, hexyl fluoride, CA 373-14-8: Flammable; soluble ethanol, benzene. Tm = 170.15; Tb = 364.15 to 365.15; Vm = 0.1303 at 293.15; Antoine (liq) A = 6.1554, B = 1299.19, C = -51.55, dev = 1.0, range = 273 to 388.

3-Fluorohexane, CA 52688-75-2: Flammable; soluble ethanol, benzene. Tm = 169.15; Antoine (liq) A = 6.34002, B = 1325.9, C = -50.15, dev = 1.0, range = 281 to 393.

$C_6H_{13}I$ MW 212.07

1-Iodohexane, hexyl iodide, CA 638-45-9: Irritant; light-sensitive; flammable, flash point 334. Tf = 198.95; Tb = 453.15; Vm = 0.1471 at 293.15; Antoine (liq) A = 6.0203, B = 1549.17, C = -68.6, dev = 1.0, range = 331 to 485.

$C_6H_{13}N$ MW 99.18

Cyclohexylamine, CA 108-91-8: Irritant; strong base; flammable, flash point 304; soluble water, most organic solvents. Tm = 255.15; Tb = 408.15; Vm = 0.1143 at 298.15; Antoine I (liq) A = 5.7521, B = 1203.2, C = -86.45, dev = 1.0, range = 333 to 408; Antoine II (liq) A = 4.86161, B = 706.036, C = -159.693, dev = 0.1 to 1.0, range = 363 to 407.

Hexamethyleneimine, hexahydro-1H-azepine, CA 111-49-9: Irritant; corrosive; flammable, flash point 291; very soluble water. Tb = 409.15 to 410.15; Vm = 0.1148 at 295.15; Antoine I (liq) A = 7.161, B = 2111, C = 0, dev = 5.0, range = 312 to 411; Antoine II (liq) A = 6.7851, B = 1968.5, C = 0, dev = 1.0, range = 348 to 423; Antoine III (liq) A = 6.1879, B = 1457.86, C = -61.132, dev = 1.0, range not specified.

N-Methylpiperidine, CA 626-67-5: Irritant; flammable, flash point 276; soluble water, ethanol, ether. Tb = 380.15; Vm = 0.1216 at 293.15; Antoine (liq) A = 4.96366, B = 864.007, C = -95.821, dev = 5.0, range = 273 to 380.

2-Methylpiperidine ±, CA 3000-79-1: Irritant; carcinogen; flammable, flash point 281; soluble water, ethanol, ether. Tm = 268.25; Tb = 390.15 to 391.15; Vm = 0.1175 at 297.15; Antoine (liq) A = 5.93994, B = 1272.811, C = -67.917, dev = 0.02, range = 323 to 431.

$C_6H_{13}NO$ MW 115.17

Butyramide, N,N-dimethyl, CA 760-79-2: Irritant; flammable; soluble ethanol, ether, benzene, acetone. Tm = 233.15; Tb = 458.15 to 461.15; Vm = 0.1271 at 298.15; Antoine (liq) A = 6.38232, B = 1709.66, C = -72.195, dev = 1.0, range = 351 to 432.

Caproamide, hexanamide, CA 628-02-4: Irritant; flammable; moderately soluble water; soluble ethanol, ether, benzene, chloroform. Tm = 373.15; Tb = 528.15; Vm = 0.1153 at 293.15; Antoine (sol) A = 12.4525, B = 4967.6, C = 0, dev = 1.0, range = 338 to 368.

$C_6H_{13}NO_2$ MW 131.17

Isoleucine ±, CA 443-79-8: Amino acid; very soluble water. Tm = 565.15, decomposes; Antoine (sol) A = 12.175, B = 6270, C = 0, dev = 5.0, range = 442 to 461. *(continues)*

208

$C_6H_{13}NO_2$ *(continued)*

Lactamide, *N*-isopropyl: Irritant; flammable. Tm = 324.65; Antoine (liq)
A = 8.99257, B = 3650.668, C = 0, dev = 1.0, range = 369 to 407.

Lactamide, *N*-propyl: Irritant; flammable. Tm = 293.15; Antoine (liq) A =
9.26539, B = 3867.83, C = 0, dev = 1.0, range = 373 to 423.

Leucine, *levo*, CA 7005-03-0: Amino acid; very soluble water. Tm = 566.15
to 568.15, decomposes; Antoine (sol) A = 15.405, B = 7860, C = 0, dev =
5.0, range = 446 to 464.

Norleucine ±, 2-aminohexanoic acid, CA 616-06-8: Nonessential amino acid;
very soluble water. Tm = 570.15 to 573.15, decomposes; Antoine (sol) A =
11.175, B = 5980, C = 0, dev = 5.0, range = 435 to 469.

C_6H_{14} MW 86.18

2,2-Dimethylbutane, neohexane, CA 75-83-2: Flammable, flash point 225;
soluble most organic solvents. Tm = 173.28; Tb = 322.89; Vm = 0.1337 at
298.15; Antoine (liq) A = 5.87731, B = 1079.798, C = -43.978, dev = 0.02,
range = 293 to 324.

2,3-Dimethylbutane, CA 79-29-8: Flammable, flash point 244; soluble most
organic solvents. Tm = 144.61; Tb = 331.14; Vm = 0.1303 at 293.15; Antoine
(liq) A = 5.95181, B = 1136.355, C = -43.159, dev = 0.02, range = 278 to
332.

Hexane, CA 110-54-3: Moderately toxic vapor; narcotic in high concentrations; flammable, flash point 251; soluble most organic solvents. Tm =
177.83; Tb = 341.89; Vm = 0.1316 at 298.15; Tc = 507.43; Pc = 3012; Vc =
0.370; Antoine I (liq) A = 6.00431, B = 1172.04, C = -48.747, dev = 0.02,
range = 293 to 343; Antoine II (liq) A = 6.15142, B = 1224.492, C =
-45.358, dev = 0.02, range = 238 to 298; Antoine III (liq) A = 8.47892, B =
1800.89, C = -4.115, dev = 0.1, range = 189 to 259; Antoine IV (liq) A =
5.99521, B = 1167.388, C = -49.272, dev = 0.02, range = 298 to 343; Antoine
V (liq) A = 5.91942, B = 1123.687, C = -54.776, dev = 0.1, range = 341 to
377; Antoine VI (liq) A = 6.4106, B = 1469.286, C = -7.702, dev = 0.1,
range = 374 to 451; Antoine VII (liq) A = 7.30814, B = 2367.155, C =
111.016, dev = 0.1, range = 445 to 508.

2-Methylpentane, isohexane, CA 107-83-5: Flammable, flash point 250; soluble most organic solvents. Tf = 119.49; Tb = 333.42; Vm = 0.1319 at
293.15; Antoine (liq) A = 5.97783, B = 1142.992, C = -45.657, dev = 0.02,
range = 293 to 335.

3-Methylpentane, CA 96-14-0: Flammable, flash point below 266; soluble
most organic solvents. Tm = 155.15; Tb = 336.43; Vm = 0.1297 at 293.15;
Antoine (liq) A = 5.97897, B = 1155.28, C = -45.659, dev = 0.02, range =
293 to 338.

$C_6H_{14}FO_3P$ MW 184.15

Fluorophosphoric acid, diisopropyl ester, CA 55-91-4: Highly toxic; suspected carcinogen; causes myosis; slightly soluble water, reacts to form
HF; soluble ether, vegetable oils. Tm = 191.15; Tb = 456.15; Vm = 0.1745
at 293.15; Antoine (liq) A = 9.8475, B = 3344.6, C = 0, dev = 1.0, range =
273 to 348.

$C_6H_{14}N_2$ MW 114.19

Cyclohexane, 1,4-diamino, CA 3114-70-3: Irritant; flammable. Tm = 345.15
to 346.15; Antoine (liq) A = 6.6877, B = 2000, C = -43.15, dev = 1.0,
range = 383 to 473.

1,4-Dimethylpiperazine, CA 106-58-1: Corrosive; irritant; flammable, flash
point 291; soluble water, ethanol, chloroform. Tb = 404.15 to 405.15; Vm =
0.133 at 293.15; Antoine (liq) A = 5.9462, B = 1368.65, C = -59.803, dev =
1.0, range = 276 to 405.

2,5-Dimethylpiperazine, *cis*, CA 6284-84-0: Irritant; flammable, flash
point 331; soluble water, ethanol, chloroform. Tm = 387.15; Tb = 435.15;
Antoine (liq) A = 4.89797, B = 588.864, C = -233.455, dev = 1.0, range =
437 to 609.

$C_6H_{14}O$ MW 102.18

Butyl ethyl ether, 1-ethoxybutane, CA 628-81-9: Irritant; flammable, flash
point 277; forms explosive peroxides; soluble ethanol, ether, acetone.
Tm = 149.15; Tb = 364.65; Vm = 0.1364 at 293.15; Antoine (liq) A =
6.062575, B = 1252.485, C = -56.685, dev = 0.1, range = 311 to 365.

tert-Butyl ethyl ether, 2-ethoxy-2-methylpropane, CA 637-92-3: Irritant;
flammable, flash point 254; forms explosive peroxides; soluble ethanol,
ether, acetone. Tm = 179.15; Tb = 346.25; Vm = 0.1359 at 298.15; Antoine I
(liq) A = 6.07456, B = 1195.382, C = -51.873, dev = 1.0, range = 284 to
346; Antoine II (liq) A = 6.09524, B = 1213.284, C = -49.453, dev = 1.0 to
5.0, range = 248 to 350; Antoine III (liq) A = 6.13244, B = 1241.854, C =
-45.22, dev = 1.0 to 5.0, range = 340 to 407.

Diisopropyl ether, isopropyl ether, CA 108-20-3: Irritant; flammable,
flash point 245; readily forms explosive peroxides; slightly soluble water;
soluble ethanol, ether, acetone. Tm = 187.65; Tb = 341.65; Vm = 0.1411 at
293.15; Tc = 500.05; Pc = 2878; Vc = 0.386; Antoine I (liq) A = 5.966496,
B = 1135.034, C = -54.92, dev = 0.1, range = 296 to 342; Antoine II (liq)
A = 5.97661, B = 1142.985, C = -53.82, dev = 0.1, range = 284 to 365;
Antoine III (liq) A = 6.26597, B = 1334.298, C = -28.271, dev = 0.1 to 1.0,
range = 360 to 440; Antoine IV (liq) A = 7.13537, B = 2140.415, C = 80.78,
dev = 0.1 to 1.0, range = 436 to 500.

2,2-Dimethyl-1-butanol, CA 1185-33-7: Moderately toxic; flammable;
slightly soluble water; soluble ethanol, ether, acetone. Tm = below
258.15; Tb = 410.15; Vm = 0.1234 at 293.15; Antoine (liq) A = 6.14386,
B = 1260.41, C = -105.38, dev = 0.1, range = 356 to 415.

2,3-Dimethyl-1-butanol ±, CA 20281-85-0: Moderately toxic; flammable;
slightly soluble water; soluble ethanol, ether, acetone. Tb = 417.15 to
418.15; Vm = 0.1232 at 293.65; Antoine (liq) A = 7.398, B = 2044, C =
-43.15, dev = 5.0, range = 324 to 431.

3,3-Dimethyl-1-butanol, CA 624-95-3: Moderately toxic; flammable, flash
point 320; slightly soluble water; soluble ethanol, ether, acetone. Tm =
213.15; Tb = 423.15; Vm = 0.1211 at 288.15; Antoine (liq) A = 7.396, B =
2010, C = -43.15, dev = 5.0, range = 319 to 424.

2,3-Dimethyl-2-butanol, CA 594-60-5: Moderately toxic; flammable, flash
point 302; slightly soluble water; soluble ethanol, ether, acetone. Tm =
259.15; Tb = 390.15 to 391.15; Vm = 0.1241 at 293.15; Antoine (liq) A =
6.33417, B = 1312.61, C = -88.49, dev = 1.0, range = 299 to 400. *(continues)*

$C_6H_{14}O$ *(continued)*

3,3-Dimethyl-2-butanol ±, CA 20281-91-8: Moderately toxic; flammable, flash point 301; slightly soluble water; soluble ethanol, ether, acetone. Tm = 278.75; Tb = 394.15; Vm = 0.1258 at 298.15; Antoine (liq) A = 7.386, B = 1883, C = -43.15, dev = 5.0, range = 302 to 401.

Dipropyl ether, propyl ether, CA 111-43-3: Irritant; flammable, flash point 294; forms explosive peroxides; slightly soluble water; soluble ethanol, ether, acetone. Tf = 151.15; Tb = 363.65; Vm = 0.1388 at 293.15; Antoine I (liq) A = 6.019715, B = 1227.468, C = -57.449, dev = 0.1, range = 312 to 371; Antoine II (liq) A = 6.0361, B = 1236.828, C = -56.358, dev = 0.1, range = 292 to 389; Antoine III (liq) A = 6.50897, B = 1579.466, C = -12.142, dev = 0.1 to 1.0, range = 385 to 467; Antoine IV (liq) A = 8.20381, B = 3494.323, C = 209.259, dev = 0.1 to 1.0, range = 465 to 530.

2-Ethyl-1-butanol, CA 97-95-0: Irritant; moderately toxic; flammable, flash point 330; slightly soluble water; soluble ethanol, ether. Tm = 159.15; Tb = 420.15; Vm = 0.1226 at 293.15; Antoine I (liq) A = 6.1956, B = 1327.8, C = -103.25, dev = 1.0, range = 321 to 426; Antoine II (liq) A = 11.51831, B = 3865.368, C = 17.884, dev = 1.0, range = 262 to 295.

1-Hexanol, hexyl alcohol, CA 111-27-3: Irritant; moderately toxic by skin absorption; flammable, flash point 336; moderately soluble water; soluble ethanol, ether, acetone, benzene. Tm = 228.15; Tb = 430.15; Vm = 0.1256 at 293.15; Tc = 611.35; Pc = 3510; Vc = 0.381; Antoine I (liq) A = 6.41271, B = 1422.031, C = -107.706, dev = 0.1 to 1.0, range = 325 to 431; Antoine II (liq) A = 10.60355, B = 3986.406, C = 46.713, dev = 1.0, range = 298 to 343.

2-Hexanol ±, CA 20281-86-1: Irritant; moderately toxic; flammable, flash point 331; slightly soluble water; soluble ethanol, ether, acetone. Tb = 409.15; Vm = 0.1252 at 293.15; Antoine I (liq) A = 6.76591, B = 1595.34, C = -77.9, dev = 0.1, range = 360 to 415; Antoine II (liq) A = 7.19139, B = 1898.882, C = -46.663, dev = 0.1 to 1.0, range = 351 to 412.

3-Hexanol ±, CA 17015-11-1: Irritant; moderately toxic; flammable, flash point 314; slightly soluble water; soluble ethanol, ether, acetone. Tb = 407.65 to 408.65; Vm = 0.1249 at 293.15; Antoine I (liq) A = 7.15606, B = 1886.9, C = -42.21, dev = 0.1, range = 354 to 410; Antoine II (liq) A = 7.33693, B = 1801.99, C = -66.255, dev = 1.0, range = 280 to 320.

2-Methyl-1-pentanol ±, CA 17092-54-5: Irritant; moderately toxic; flammable, flash point 327; slightly soluble water; soluble ethanol, ether, acetone. Tm = 166.15; Tb = 422.15 to 423.15; Vm = 0.1237 at 293.15; Antoine I (liq) A = 6.80909, B = 1662.71, C = -75.01, dev = 0.1, range = 367 to 423; Antoine II (liq) A = 7.06975, B = 1673.745, C = -82.699, dev = 1.0, range = 261 to 294.

3-Methyl-1-pentanol ±, CA 20281-83-8: Irritant; moderately toxic; flammable, flash point 332; slightly soluble water; soluble ethanol, acetone, ether. Tb = 424.15 to 425.15; Vm = 0.1242 at 297.15; Antoine (liq) A = 6.54045, B = 1501.77, C = -94.41, dev = 0.1, range = 328 to 427.

4-Methyl-1-pentanol, isohexyl alcohol, CA 626-89-1: Irritant; moderately toxic; flammable, flash point 324; slightly soluble water; soluble ethanol, ether, acetone. Tb = 425.15 to 426.15; Vm = 0.1260 at 298.15; Antoine I (liq) A = 6.00689, B = 1187.31, C = -128.22, dev = 0.1, range = 371 to 427; Antoine II (liq) A = 6.20701, B = 1293.791, C = -117.419, dev = 0.1, range = 357 to 427.

(continues)

$C_6H_{14}O$ *(continued)*

2-Methyl-2-pentanol, CA 590-36-3: Irritant; moderately toxic; flammable, flash point 294; slightly soluble water; soluble ethanol, ether, acetone. Tm = 170.15; Tb = 395.15; Vm = 0.1224 at 289.15; Antoine I (liq) A = 6.01861, B = 1175.42, C = -101.73, dev = 0.1, range = 341 to 396; Antoine II (liq) A = 5.90931, B = 1070.24, C = -121.462, dev = 0.1, range = 330 to 397.

3-Methyl-2-pentanol ±, CA 74497-29-3: Irritant; moderately toxic; flammable, flash point 313; slightly soluble water; soluble ethanol, ether, acetone. Tb = 407.45; Vm = 0.1230 at 293.15; Antoine I (liq) A = 5.87451, B = 1094.51, C = -124.5, dev = 0.1, range = 314 to 409; Antoine II (liq) A = 6.34728, B = 1367.055, C = -95.283, dev = 0.1, range = 255 to 295.

4-Methyl-2-pentanol ±, CA 108-11-2: Irritant; moderately toxic; flammable, flash point 314; slightly soluble water; soluble ethanol, ether, acetone. Tb = 405.15; Vm = 0.1265 at 293.15; Antoine I (liq) A = 7.07349, B = 1751.56, C = -57.93, dev = 1.0, range = 353 to 404; Antoine II (liq) A = 7.10533, B = 1651.468, C = -75.737, dev = 1.0, range = 249 to 295.

2-Methyl-3-pentanol ±, CA 20281-90-7: Irritant; moderately toxic; flammable, flash point 319; slightly soluble water; soluble ethanol, acetone, ether. Tm = 249.55; Tb = 399.15 to 400.15; Vm = 0.1240 at 293.15; Antoine I (liq) A = 5.82962, B = 1069.34, C = -120.17, dev = 0.1, range = 307 to 401; Antoine II (liq) A = 7.07746, B = 1800.133, C = -46.033, dev = 0.1 to 1.0, range = 342 to 400.

3-Methyl-3-pentanol, CA 77-74-7: Irritant; moderately toxic; flammable, flash point 319; slightly soluble water; soluble ethanol, ether, acetone. Tm = 249.55; Tb = 395.25 to 396.05; Vm = 0.1233 at 293.15; Antoine (liq) A = 7.0213, B = 1903.2, C = -15.95, dev = 1.0, range = 322 to 397.

$C_6H_{14}OS$ MW 134.24

2-Methyl-2-propanesulfonic acid, ethyl ester: Irritant; flammable. Antoine (liq) A = 3.2371, B = 730.04, C = 0, dev = 1.0, range = 337 to 343.

$C_6H_{14}O_2$ MW 118.18

2-Butoxyethanol, CA 111-76-2: Irritant; moderately toxic; may form explosive peroxides; flammable, flash point 335; very soluble water; soluble most organic solvents; important industrial solvent and emulsifier. Tm = 203.15; Tb = 443.25; Vm = 0.1311 at 293.15; Antoine (liq) A = 6.9697, B = 1988.9, C = -43.15; dev = 1.0, range = 366 to 443.

Diethyl acetal, 1,1-diethoxyethane, CA 105-57-7: Irritant; narcotic vapor; flammable, flash point 252; readily polymerizes; moderately soluble water; soluble ethanol, ether, acetone. Tb = 376.35; Vm = 0.1432 at 293.15; Antoine (liq) A = 6.71087, B = 1552.05, C = -45.6, dev = 1.0, range = 281 to 384.

2,3-Dimethyl-2,3-butanediol, pinacol, CA 76-09-5: Irritant; flammable, flash point 350; soluble hot water, ethanol, ether. Tm = 311.15; Tb = 447.55; Antoine (liq) A = 6.16442, B = 1355.65, C = -121.56, dev = 1.0, range = 346 to 448.

Ethylene glycol, diethyl ether, 1,2-diethoxyethane, CA 629-14-1: Irritant; flammable, flash point 308; may form explosive peroxides;   *(continues)*

$C_6H_{14}O_2$ *(continued)*

very soluble water; soluble ethanol, ether, acetone, benzene. Tm = 199.15;
Tb = 394.15; Vm = 0.1393 at 293.15; Antoine (liq) A = 5.45472, B = 1145.24,
C = -60.442, dev = 1.0 to 5.0, range = 239 to 393.

2-Isobutoxyethanol, CA 4439-24-1: Highly toxic by skin absorption; flammable, flash point 331; soluble water, ethanol, ether. Tb = 430.15 to
431.15; Vm = 0.1328 at 293.15; Antoine (liq) A = 6.9997, B = 1945.5, C =
-43.15, dev = 1.0, range = 344 to 432.

3-Methyl-1,5-pentanediol, CA 4457-71-0: Irritant; flammable; hygroscopic;
soluble water, ethanol, ether. Antoine (liq) A = 6.91687, B = 1917.3, C =
-128.74, dev = 1.0, range = 402 to 485.

2-Methyl-2,4-pentanediol ±, CA 107-41-5: Irritant; flammable, flash point
369; hygroscopic; soluble water, most organic solvents. Tb = 471.15; Vm =
0.1277 at 290.15; Antoine (liq) A = 8.4533, B = 3037.3, C = 0, dev = 5.0,
range = 373 to 473.

3-Methyl-2,4-pentanediol ±, CA 25618-01-3: Irritant; flammable; hygroscopic; soluble water, ethanol, ether. Tb = 484.15 to 485.15; Vm = 0.1236
at 293.15; Antoine (liq) A = 6.8387, B = 1852.3, C = -100.38, dev = 1.0,
range = 368 to 424.

$C_6H_{14}O_3$ MW 134.17

Diethylene glycol, dimethyl ether, diglyme, CA 111-96-6: Irritant; flammable, flash point 340; forms explosive peroxides on exposure to air and
light; soluble water, most organic solvents. Tm = 209.15; Tb = 435.15;
Vm = 0.1420 at 293.15; Antoine (liq) A = 7.02673, B = 2032.01, C = -28.261,
dev = 1.0, range = 286 to 433.

Diethylene glycol, monoethyl ether, CA 111-90-0: Irritant; moderately
toxic orally; hygroscopic; may form explosive peroxides; soluble water,
most organic solvents. Tm = 195.15; Tb = 475.85; Vm = 0.1361 at 298.15;
Antoine (liq) A = 8.13351, B = 3019.21, C = 17.729, dev = 1.0, range = 318
to 475.

Dipropylene glycol, CA 25265-71-8: Irritant; hygroscopic; flammable, flash
point 394; soluble water, ethanol. Tm = 233.15; Tb = 505.15; Vm = 0.1312
at 293.15; Antoine (liq) A = 8.2924, B = 3156.61, C = -2.912, dev = 1.0,
range = 423 to 505.

2-Ethyl-2-hydroxymethyl-1,3-propanediol, trimethylol propane, CA 77-99-6:
Flammable; essentially nontoxic; soluble water, ethanol, acetone. Tm =
331.95; Tb = 568.15; Antoine (liq) A = 6.982, B = 2204, C = -125.25, dev =
5.0, range = 433 to 570.

1,2,6-Hexanetriol, 1,2,6-trihydroxyhexane, CA 106-69-4: Irritant; hygroscopic; flammable, flash point 464; soluble water, ethanol, acetone; used
as diuretic. Tm = 240.35; Vm = 0.1213 at 293.15; Antoine (liq) A = 4.5998,
B = 1154.2, C = -213.05, dev = 1.0, range = 393 to 433.

$C_6H_{14}O_4$ MW 150.17

Ethane, 1,1,2,2-tetramethoxy, CA 2517-44-4: Irritant; flammable; soluble
ethanol, ether. Antoine (liq) A = 7.1894, B = 2239.7, C = 0, dev = 1.0,
range = 351 to 432.   *(continues)*

$C_6H_{14}O_4$ *(continued)*

Triethylene glycol, CA 112-27-6: Irritant; very hygroscopic; flammable, flash point 450; soluble water, ethanol, benzene; high temperature solvent. Tm = 268.85; Tb = 551.46; Vm = 0.1332 at 288.15; Tc = 700, calculated; Pc = 3320, calculated; Vc = 0.443, calculated; Antoine I (liq) A = 8.82922, B = 3778.12, C = 2.204, dev = 1.0, range = 387 to 552; Antoine II (liq) A = 8.1182, B = 3534, C = 0, dev = 1.0 to 5.0, range = 288 to 303.

$C_6H_{14}S$ MW 118.24

Butyl ethyl sulfide, CA 638-46-0: Irritant; stench; flammable; soluble ethanol, ether, chloroform. Tm = 178.05; Tb = 417.35; Vm = 0.1412 at 293.15; Antoine (liq) A = 6.06592, B = 1421.32, C = -67.34, dev = 0.1, range = 354 to 424.

sec-Butyl ethyl sulfide, CA 5008-72-0: Irritant; stench; flammable; soluble ethanol, ether, chloroform. Antoine (liq) A = 6.06029, B = 1399.88, C = -61.55, dev = 0.1, range = 345 to 409.

tert-Butyl ethyl sulfide, CA 14290-92-7: Irritant; stench; flammable; soluble ethanol, ether, chloroform. Tm = 184.23; Antoine (liq) A = 5.97576, B = 1323.24, C = -60.26, dev = 0.1, range = 332 to 400.

Diisopropyl sulfide, CA 625-80-9: Irritant; stench; flammable, flash point 280; soluble ethanol, ether, chloroform. Tm = 195.1; Tb = 393.15 to 394.15; Vm = 0.1452 at 293.15; Antoine (liq) A = 5.99561, B = 1326.755, C = -60.635, dev = 0.02, range = 324 to 433.

2,3-Dimethyl-2-butanethiol, CA 1639-01-6: Irritant; stench; flammable; soluble ethanol, ether, chloroform. Antoine (liq) A = 5.96241, B = 1353.374, C = -57.244, dev = 0.02, range = 328 to 441.

Dipropyl sulfide, CA 111-47-7: Irritant; stench; flammable; soluble ethanol, ether, chloroform. Tm = 171.25; Tb = 414.15 to 415.15; Vm = 0.1411 at 293.15; Antoine (liq) A = 6.06067, B = 1413.44, C = -67.42, dev = 0.1, range = 353 to 427.

Ethyl isobutyl sulfide, CA 1613-45-2: Irritant; stench; flammable; soluble ethanol, ether, chloroform. Tb = 407.35; Vm = 0.1424 at 293.15; Antoine (liq) A = 6.04818, B = 1391.97, C = -63.05, dev = 0.1, range = 345 to 414.

1-Hexanethiol, hexyl mercaptan, CA 111-31-9: Irritant; stench; flammable, flash point 293; soluble ethanol, ether, chloroform. Tm = 192.66; Tb = 425.75; Vm = 0.1404 at 293.15; Antoine (liq) A = 6.07016, B = 1453.659, C = -68.181, dev = 0.02, range = 352 to 468.

2-Hexanethiol, CA 1679-06-7: Irritant; stench; flammable; soluble ethanol, ether, chloroform, benzene. Tm = 126.15; Tb = 415.15; Vm = 0.1417 at 293.15; Antoine (liq) A = 5.8888, B = 1303.31, C = -76.45, dev = 1.0, range = 328 to 423.

Isopropyl propyl sulfide, CA 5008-73-1: Irritant; stench; flammable; soluble ethanol, ether, chloroform. Antoine (liq) A = 6.04754, B = 1383.97, C = -62.8, dev = 0.1, range = 343 to 416.

2-Methyl-2-pentanethiol, CA 1633-97-2: Irritant; stench; flammable; soluble ethanol, ether, chloroform. Antoine (liq) A = 5.98146, B = 1343.026, C = -60.388, dev = 0.02, range = 327 to 439.

$C_6H_{14}S_2$ MW 150.30

Diisopropyl disulfide, CA 4253-89-8: Irritant; stench; flammable, flash
point 291; soluble ethanol, ether. Tm = 204.1; Tb = 450.15; Vm = 0.1593 at
293.15; Antoine (liq) A = 6.03442, B = 1524.917, C = -71.875, dev = 0.1,
range = 377 to 447.

Dipropyl disulfide, CA 629-19-6: Irritant; stench; flammable, flash point
339; soluble ethanol, ether; occurs in garlic and onion flavors. Tm =
187.69; Tb = 467.15; Vm = 0.1566 at 293.15; Antoine (liq) A = 6.09923, B =
1603.846, C = -77.239, dev = 0.02 to 0.1, range = 389 to 447.

Ethyl (1,1-dimethylethyl) disulfide, CA 4151-69-3: Irritant; stench; flammable; soluble ethanol, ether. Tm = 206.47; Antoine (liq) A = 6.06471, B =
1550.607, C = -66.824, dev = 0.1, range = 373 to 461.

1,6-Hexanedithiol, CA 1191-43-1: Irritant; stench; flammable, flash point
363; soluble ethanol, ether. Tm = 252.15; Vm = 0.1529 at 293.15; Antoine
(liq) A = 6.28256, B = 1835.7, C = -81.04, dev = 1.0, range = 379 to 511.

Isopropyl propyl disulfide, CA 33672-51-4: Irritant; stench; flammable;
soluble ethanol, ether. Antoine (liq) A = 6.10315, B = 1584.422, C =
-72.431, dev = 0.1, range = 383 to 433.

$C_6H_{14}S_3$ MW 182.36

Trithiodiethylene glycol, dimethyl ether: Irritant; flammable. Tm =
297.45; Antoine (liq) A = 13.042, B = 5414, C = 0, dev = 5.0, range = 391
to 418.

$C_6H_{15}N$ MW 101.19

Butyl ethyl amine, CA 13360-63-9: Irritant; flammable; very soluble water;
soluble ethanol, ether, acetone, benzene. Tb = 381.15 to 382.15; Vm =
0.1368 at 293.15; Antoine (liq) A = 5.63094, B = 1085.2, C = -86.52, dev =
1.0, range = 283 to 382.

sec-Butyl ethyl amine ±, CA 21035-44-9: Irritant; flammable; soluble ethanol, ether, acetone, benzene. Tm = 168.85; Tb = 371.95; Vm = 0.1375 at
293.15; Antoine (liq) A = 6.20766, B = 1331.9, C = -53.5, dev = 1.0, range
= 283 to 372.

Diisopropyl amine, CA 108-18-9: Irritant; flammable, flash point 272;
strongly basic; soluble water, ethanol, ether, acetone, benzene. Tm =
212.15; Tb = 357.05; Vm = 0.1411 at 293.15; Antoine I (liq) A = 6.21541,
B = 1329.67, C = -41.71, dev = 1.0, range = 260 to 412; Antoine II (liq)
A = 6.1498, B = 1275.7, C = -49.15, dev = 1.0, range = 273 to 367.

Dimethyl tert-butyl amine, CA 918-02-5: Irritant; flammable; soluble ethanol, ether, acetone, benzene. Antoine (liq) A = 5.91391, B = 1205.37, C =
-55.232, dev = 0.1, range = 283 to 318.

Dipropyl amine, CA 142-84-7: Irritant; toxic; flammable, flash point 290;
soluble water, ethanol, ether, acetone. Tm = 209.55; Tb = 382.45; Vm =
0.1367 at 293.15; Antoine (liq) A = 6.33153, B = 1415.4, C = -55.15, dev =
1.0 to 5.0, range = 302 to 422.

Hexylamine, CA 111-26-2: Irritant; highly toxic; flammable, flash point
302; soluble ethanol, ether. Tf = 254.15; Tb = 402.15 to   (continues)

$C_6H_{15}S_3$ *(continued)*

403.15; Vm = 0.1321 at 293.15; Antoine (liq) A = 6.31049, B = 1513.5, C = -54.35, dev = 1.0, range = 303 to 406.

Triethylamine, CA 121-44-8: Irritant; flammable, flash point 266; very soluble water; soluble ethanol, ether, acetone. Tm = 158.45; Tb = 361.95; Vm = 0.1395 at 298.15; Antoine I (liq) A = 5.90069, B = 1176.7, C = -60.37, dev = 1.0, range = 283 to 363; Antoine II (liq) A = 5.9365, B = 1200.887, C = -57.286, dev = 0.1 to 1.0, range = 298 to 324.

$C_6H_{15}NO$ MW 117.19

Diethyl amine, *N*-(methoxymethyl), CA 5888-29-9: Irritant; flammable; soluble water, ethanol, ether. Tb = 390.15; Antoine (liq) A = 7.2535, B = 1984, C = 0, dev = 1.0, range = 293 to 318.

2-Diethylaminoethanol, CA 100-37-8: Irritant; flammable, flash point 333; soluble water, ethanol, ether, acetone, benzene. Tb = 436.15; Vm = 0.1332 at 298.15; Antoine (liq) A = 7.8425, B = 2534.3, C = 0, dev = 1.0, range = 328 to 433.

$C_6H_{15}NO_2$ MW 133.19

Diisopropanolamine, CA 110-97-4: Irritant; flammable, flash point 400; soluble water, ethanol, ether. Tm = 320.15; Tb = 521.15; Vm = 0.1845 at 293.15; Antoine (liq) A = 7.96099, B = 2847.259, C = -42.402, dev = 1.0, range = 390 to 521.

2-[2-(Dimethylamino)ethoxy]ethanol, CA 1704-62-7: Irritant; flammable; soluble water, ethanol, ether. Antoine (liq) A = 7.34992, B = 2297.71, C = -43.15, dev = 0.1 to 1.0, range = 412 to 452.

$C_6H_{15}NO_2S$ MW 165.25

Ethanesulfonamide, *N,N*-diethyl, CA 33718-39-7: Irritant; flammable. Tm = 295.15; Antoine (liq) A = 7.51469, B = 2895.084, C = 0, dev = 1.0 to 5.0, range = 392 to 526.

$C_6H_{15}NO_3$ MW 149.19

Triethanolamine, CA 102-71-6: Irritant; very hygroscopic; flammable, flash point 469; soluble water, ethanol, acetone, chloroform. Tm = 291.05; Tb = 608.55; Vm = 0.1333 at 298.15; Tc = 787; Pc = 2450; Vc = 0.472, calculated; Antoine (liq) A = 9.53861, B = 4951.87, C = 49.99, dev = 0.1 to 1.0, range = 523 to 579.

$C_6H_{15}NS$ MW 133.25

N,N-Dimethyl-*S-tert*-butyl thiohydroxylamine: Irritant; flammable. Antoine (liq) A = 5.54214, B = 1480, C = 0, dev = 1.0, range = 328 to 334.

$C_6H_{15}N_3$ MW 129.20

1-(2-Aminoethyl) piperazine, CA 140-31-8: Irritant; flammable, flash point 366; soluble water, ethanol; curing agent for epoxy resins.   *(continues)*

216

$C_6H_{15}N_3$ *(continued)*

Tm = 255.55; Tb = 495.15; Vm = 0.1309 at 293.15; Antoine (liq) A = 7.810, B = 2827.4, C = 0, dev not specified, range not specified.

$C_6H_{15}O_2PS_3$ MW 246.34

O,O-Dimethyl-*S*-[2-(ethylthio)ethyl] dithiophosphate, CA 640-15-3: Toxic liquid; insecticide; flammable; slightly soluble water; soluble most organic solvents. Vm = 0.2038 at 293.15; Antoine (liq) A = 9.2361, B = 3980.8, C = -1.21, dev = 1.0, range = 283 to 394.

$C_6H_{15}O_3P$ MW 166.16

Phosphonic acid, dipropyl ester, CA 1809-21-8: Irritant; flammable. Antoine (liq) A = 6.1869, B = 1991.1, C = 0, dev = 1.0, range = 318 to 467.

$C_6H_{15}O_3PS$ MW 198.22

O,O,O-Triethylthiophosphate, CA 126-68-1: Highly toxic insecticide; flammable. Vm = 0.1842 at 293.15; Antoine (liq) A = 13.1259, B = 4570, C = 0, dev = 1.0 to 5.0, range = 305 to 335.

O,O,S-Triethylthiophosphate, CA 1186-09-0: Highly toxic insecticide; flammable. Tb = 510.15, decomposes; Vm = 0.1789 at 293.15; Antoine (liq) A = 10.5879, B = 3985, C = 0, dev = 1.0 to 5.0, range = 312 to 352.

$C_6H_{15}O_3PS_2$ MW 230.28

Phosphorothioic acid, *O*-[2-(ethylthio)ethyl]-*O,O*-dimethyl ester, CA 867-27-6: Toxic insecticide; flammable; slightly soluble water; soluble most organic solvents. Vm = 0.1935 at 293.15; Antoine (liq) A = 6.66585, B = 2428.6, C = -56.73, dev = 1.0, range = 283 to 379.

Phosphorothioic acid, *S*-[2-(ethylthio)ethyl]-*O,O*-dimethyl ester, CA 919-86-8: Toxic insecticide; flammable; soluble most organic solvents. Antoine (liq) A = 8.48854, B = 3410, C = -26.72, dev = 1.0, range = 283 to 407.

$C_6H_{15}O_4P$ MW 182.16

Triethyl phosphate, CA 78-40-0: Irritant; flammable; moderately soluble water, decomposes; soluble ethanol, ether. Tm = 216.75; Tb = 488.15 to 489.15; Vm = 0.1703 at 293.15; Antoine (liq) A = 8.13559, B = 3387.7, C = 68.426, dev = 1.0, range = 312 to 484.

$C_6H_{15}P$ MW 118.16

Triethyl phosphine, CA 554-70-1: Toxic; vile odor; flammable; forms explosive products by reaction with oxygen; insoluble water; soluble ethanol, ether. Tm = 185.15; Tb = 402.15 to 403.15; Vm = 0.1477 at 293.15; Antoine (liq) A = 6.985, B = 2000, C = 0, dev = 1.0, range = 291 to 402.

$C_6H_{16}FN_2OP$ MW 182.18

Phosphorodiamidic fluoride, *N,N'*-diisopropyl, mipafox, CA 371-86-8: Toxic
insecticide; flammable; slightly soluble water. Tm = 338.15; Antoine (liq)
A = 6.874, B = 3033, C = 0, dev = 5.0, range = 278 to 398.

$C_6H_{16}N_2$ MW 116.21

1,6-Hexanediamine, CA 124-09-4: Irritant; flammable, flash point 358; soluble water, ethanol, benzene. Tm = 314.05; Tb = 477.65; Tc = 663, calculated; Pc = 3290, calculated; Vc = 0.475, calculated; Antoine I (liq) A =
7.6767, B = 2680, C = 0, dev = 5.0, range = 338 to 473; Antoine II (liq)
A = 7.4439, B = 2577.3, C = 0, dev = 1.0, range = 348 to 474.

$C_6H_{16}N_2O_2$ MW 148.20

Diisopropyl ammonium nitrite, CA 3129-93-9: Tm = 413.15; Antoine (sol) A =
3.735, B = 2035, C = 0, dev = 5.0, range = 288 to 299.

$C_6H_{18}N_4$ MW 146.24

Triethylenetetramine, CA 112-24-3: Irritant; corrosive; causes burns;
flammable, flash point 408; soluble water, ethanol. Tm = 234.15; Tb =
550.55; Vm = 0.1450 at 293.15; Antoine (liq) A = 7.07504, B = 2549, C =
-43.15, dev = 5.0, range = 431 to 550.

C_6N_2 MW 100.08

Butadiyne, dicyano, CA 16419-78-6: Flammable; polymerizes. Tm = 337.65;
Antoine I (sol) A = 6.38095, B = 1794.5, C = 0, dev = 1.0, range = 294 to
335; Antoine II (liq) A = 5.7423, B = 1579.2, C = 0, dev not specified,
range = 341 to 369.

C_6N_4 MW 128.09

Tetracyanoethylene, CA 670-54-2: Soluble acetone; forms complexes with
olefins and aromatics. Tm = 473.15 to 475.15; Tb = 496.16; Antoine (sol)
A = 10.795, B = 4250, C = 0, dev = 1.0 to 5.0, range = 333 to 371.

C_6N_6 MW 156.11

1,3,5-Triazine, 2,4,6-tricarbonitrile, CA 7615-57-8: Reacts with water;
soluble benzene. Tm = 393.15; Tb = 535.15; Antoine (liq) A = 9.925, B =
4040, C = 0, dev = 1.0 to 5.0, range not specified.

$C_7ClF_{17}N_2S$ MW 502.58

Chloro(trifluoromethyl)bis-[*N*-(1,2,2,2-tetrafluoro-1-[trifluoromethyl]
ethyl)-imino] sulfur: Antoine (liq) A = 6.335, B = 2022, C = 0, dev = 1.0
to 5.0, range not specified.

$C_7Cl_4F_{12}$ MW 453.87

1,1,1,7-Tetrachloroperfluoroheptane, CA 1550-21-6: Antoine (liq) A =
5.8708, B = 1907, C = 0, dev not specified, range not specified.

218

$C_7F_6O_2$ MW 230.07

Fluoroformic acid, pentafluorophenyl ester, CA 59483-82-8: Irritant; flammable. Tm = 266.55; Antoine (liq) A = 9.2289, B = 2881.6, C = 0, dev = 5.0, range not specified.

C_7F_8 MW 236.06

Perfluorotoluene, CA 434-64-0: Flammable, flash point 293. Tb = 376.15; Vm = 0.1417 at 293.15; Antoine (liq) A = 7.7325, B = 2138.35, C = 0, dev = 1.0 to 5.0, range = 285 to 376.

C_7F_{10} MW 274.06

3,3-Difluoro-1,2-bis(trifluoromethyl)-4-(difluoromethylene)-cyclobutene, CA 14451-74-2: Antoine (liq) A = 6.925, B = 1647, C = 0, dev = 1.0 to 5.0, range = 272 to 316.

$C_7F_{12}O_6$ MW 408.05

Hexafluoroperoxyglutaric acid, bis(trifluoromethyl) ester, CA 32751-20-5: Tm = 197.15; Antoine (liq) A = 8.3571, B = 2472.7, C = 0, dev not specified, range = 200 to 390.

C_7F_{14} MW 350.05

Perfluoromethylcyclohexane, CA 355-02-2: Nonflammable; soluble acetone, benzene. Tm = 231.15; Tb = 349.25; Vm = 0.1958 at 298.15; Antoine (liq) A = 5.94674, B = 1132.493, C = -62.084, dev = 0.02, range = 305 to 385.

C_7F_{16} MW 388.05

Perfluoroheptane, CA 335-57-9: Soluble ethanol, ether, acetone, chloroform. Tm = 222.15; Tb = 355.65; Vm = 0.2239 at 293.15; Tc = 474.75; Pc = 1620; Vc = 0.664; Antoine (liq) A = 6.08406, B = 1193.014, C = -63.12, dev = 0.1, range = 271 to 379.

$C_7F_{16}N_2OS$ MW 464.12

Methanesulfonimidamide, 1,1,1-trifluoro-N'-[1,2,2,2-tetrafluoro-1-(trifluoromethyl)ethyl]-N-[2,2,2-trifluoro-1-(trifluoromethyl)ethylidene], CA 62609-64-7: Antoine (liq) A = 6.085, B = 1851, C = 0, dev = 1.0 to 5.0, range not specified.

$C_7HF_5O_2$ MW 212.08

Pentafluorobenzoic acid, CA 602-94-8: Tm = 376.15; Tb = 493.15; Antoine (sol) A = 12.483, B = 4781, C = 0, dev not specified, range = 334 to 359.

C_7HF_{15} MW 370.06

1,1,1,2,2,3,3,4,4,5,5,6,6,7,7-Pentadecafluoroheptane, CA 375-83-7: Tb = 369.15 to 370.15; Vm = 0.2145 at 298.15; Antoine (liq) A = 7.263, B = 1935.7, C = 0, dev = 5.0, range = 292 to 370.

$C_7H_2F_{13}NO$ MW 363.08

2-Propanamine, 1,1,1,2,3,3,3-heptafluoro-N-[2,2,2-trifluoro-1-(2,2,2-tri-fluoroethoxy)ethylidene], CA 54181-88-3: Antoine (liq) A = 7.065, B = 1857, C = 0, dev = 1.0 to 5.0, range not specified.

$C_7H_3ClF_3NO_2$ MW 225.55

Benzene, 1-(trifluoromethyl)-2-chloro-5-nitro, CA 777-37-7: Irritant; flammable, flash point 371. Tf = 295.15; Tb = 505.15; Vm = 0.1477 at 298.15; Antoine (liq) A = 6.27899, B = 1779.91, C = -88.51, dev = 0.1, range = 364 to 508.

Benzene, 1-(trifluoromethyl)-4-chloro-3-nitro, CA 121-17-5: Irritant; flammable, flash point 374. Tm = 270.61; Tb = 495.15; Vm = 0.1493 at 293.15; Antoine (liq) A = 6.28268, B = 1738.71, C = -89.2, dev = 0.1, range = 358 to 495.

$C_7H_3Cl_2F_3$ MW 215.00

Benzene, 1-(trifluoromethyl)-3,4-dichloro, CA 328-84-7: Irritant; flam-mable, flash point 338. Tm = 261.05; Tb = 446.15 to 447.15; Vm = 0.1455 at 293.15; Antoine (liq) A = 7.3898, B = 2474.7, C = 13.35, dev = 1.0, range = 353 to 453.

$C_7H_3Cl_2NO$ MW 188.01

Phenyl isocyanate, 3,4-dichloro, CA 102-36-3: Irritant; toxic; lachry-mator; flammable, flash point 415; moisture-sensitive. Tm = 315.15; Vm = 0.1353 at 323.15; Antoine (liq) A = 7.80421, B = 3312.281, C = 60.837, dev = 1.0, range = 373 to 473.

$C_7H_3Cl_5$ MW 264.37

Benzene, 1-(trichloromethyl)-3,4-dichloro, CA 13014-24-9: Irritant; flam-mable. Tf = 298.95; Tb = 556.25; Antoine (liq) A = 6.11016, B = 1868.905, C = -101.15, dev = 1.0, range = 438 to 663.

$C_7H_3F_5$ MW 182.09

2,3,4,5,6-Pentafluorotoluene, CA 771-56-2: Irritant; flammable. Tm = 243.15; Tb = 390.15 to 391.15; Vm = 0.1265 at 298.15; Antoine I (liq) A = 6.19445, B = 1382.934, C = -60.494, dev = 0.02 to 0.1, range = 312 to 416; Antoine II (liq) A = 6.21241, B = 1394.345, C = -59.194, dev = 0.02, range = 348 to 401.

$C_7H_4ClF_3$ MW 180.56

Benzene, 1-(trifluoromethyl)-2-chloro, CA 88-16-4: Irritant; flammable, flash point 331. Tm = 267.15; Tb = 421.35 to 421.85; Vm = 0.1440 at 303.15; Antoine (liq) A = 6.1393, B = 1490.4, C = -64.91, dev = 0.1 to 1.0, range = 310 to 426.

Benzene, 1-(trifluoromethyl)-3-chloro, CA 98-15-7: Irritant; flammable, flash point 309. Tm = 217.15; Tb = 411.15 to 412.15;   *(continues)*

$C_7H_4ClF_3$ *(continued)*

Vm = 0.1356 at 298.15; Antoine (liq) A = 6.1804, B = 1458.9, C = -61.33, dev = 0.1 to 1.0, range = 302 to 411.

Benzene, 1-(trifluoromethyl)-4-chloro, CA 98-56-6: Irritant; flammable, flash point 320. Tm = 240.15; Tb = 408.15 to 409.15; Vm = 0.1335 at 293.15; Antoine (liq) A = 6.2889, B = 1536.3, C = -52.35, dev = 0.1 to 1.0, range = 302 to 412.

C_7H_4ClNO MW 153.57

Phenyl isocyanate, 3-chloro, CA 2909-38-8: Highly toxic; lachrymator; flammable, flash point 359; reacts violently with water. Tm = 268.75; Vm = 0.1219 at 293.15; Antoine (liq) A = 5.70529, B = 1384.1, C = -105.27, dev = 1.0, range = 344 to 432.

Phenyl isocyanate, 4-chloro, CA 104-12-1: Highly toxic; lachrymator; flammable, flash point 383; can explode during distillation; reacts violently with water. Tm = 301.15; Tb = 476.15 to 477.15; Vm = 0.1229 at 313.15; Antoine (liq) A = 11.3908, B = 6532.551, C = 226.438, dev = 1.0, range = 363 to 443.

$C_7H_4ClNO_3$ MW 185.57

Benzoyl chloride, 3-nitro, CA 121-90-4: Corrosive; moderately toxic; flammable, flash point above 385; moisture-sensitive; can explode from heating or shock; soluble ether. Tm = 306.15; Tb = 548.15 to 551.15; Antoine (liq) A = 7.935, B = 3260, C = 0, dev = 5.0, range = 428 to 551.

$C_7H_4Cl_2O$ MW 175.01

Benzoyl chloride, 2-chloro, CA 609-65-4: Corrosive; lachrymator; flammable, flash point 383; reacts with water. Tm = 269.15; Tb = 508.15 to 511.15; Vm = 0.1266 at 293.15; Antoine (liq) A = 7.615, B = 2790, C = 0, dev = 5.0, range = 374 to 395.

Benzoyl chloride, 3-chloro, CA 618-46-2: Corrosive; lachrymator; flammable, flash point 383; reacts with water. Tb = 498.15; Antoine (liq) A = 7.145, B = 2579, C = 0, dev = 5.0, range = 367 to 500.

Benzoyl chloride, 4-chloro, CA 122-01-0: Corrosive; lachrymator; flammable, flash point 378; reacts with water. Tm = 289.15; Tb = 493.15 to 495.15; Vm = 0.1271 at 293.15; Antoine (liq) A = 8.015, B = 2909, C = 0, dev = 5.0, range = 370 to 490.

$C_7H_4Cl_4$ MW 229.92

Benzene, 1-(trichloromethyl)-2-chloro, CA 2136-89-2: Corrosive; lachrymator; flammable, flash point 371; soluble ether, acetone. Tm = 302.55; Tb = 537.45; Vm = 0.1514 at 293.15; Antoine (liq) A = 6.24284, B = 1951.37, C = -76.88, dev = 1.0, range = 423 to 588.

2,3,5,6-Tetrachlorotoluene, CA 1006-31-1: Toxic; flammable; soluble methanol, ethanol, ether; used as herbicide. Tm = 366.15 to 367.15; Antoine (liq) A = 7.3449, B = 3034, C = 20.85, dev = 1.0, range = 399 to 548.

$C_7H_4F_3NO_2$ MW 191.11

Benzene, 1-(trifluoromethyl)-3-nitro, CA 98-46-4: Irritant; highly toxic;
flammable, flash point 360. Tm = 271.83; Tb = 475.95; Vm = 0.1331 at
288.15; Antoine (liq) A = 6.30515, B = 1710.6, C = -78.03, dev = 0.1,
range = 341 to 475.

$C_7H_4F_4$ MW 164.10

Benzene, 1-(trifluoromethyl)-3-fluoro, CA 401-80-9: Irritant; flammable,
flash point 280. Tm = 191.65; Tb = 374.15 to 375.15; Vm = 0.1260 at
290.15; Antoine (liq) A = 6.13285, B = 1305.692, C = -57.384, dev = 0.02
to 0.1, range = 313 to 410.

Benzene, 1-(trifluoromethyl)-4-fluoro, CA 402-44-8: Irritant; flammable,
flash point 283. Tm = 231.45; Tb = 375.15 to 378.15; Vm = 0.1269 at
293.15; Antoine (liq) A = 6.5794, B = 1627.35, C = -20.1, dev = 1.0, range
= 286 to 381.

$C_7H_4F_{12}O$ MW 332.09

2,2,3,3,4,4,5,5,6,6,7,7-Dodecafluoro-1-heptanol, CA 335-99-9: Tm = 303.15;
Antoine (liq) A = 8.297, B = 2791, C = 0, dev = 5.0, range = 355 to 446.

C_7H_5BrO MW 185.02

Benzoyl bromide, CA 618-32-6: Corrosive; lachrymator; flammable, flash
point 363; reacts with water; soluble ether, benzene. Tf = 249.15; Tb =
491.15 to 492.15; Vm = 0.1178 at 288.15; Antoine (liq) A = 6.82147, B =
2204.91, C = -33.861, dev = 1.0, range = 320 to 492.

C_7H_5ClO MW 140.57

Benzoyl chloride, CA 98-88-4: Corrosive; lachrymator; causes burns; flammable, flash point 345; fumes in air; reacts with water and ethanol; soluble ether, benzene, carbon disulfide. Tf = 272.15; Tb = 470.35; Vm =
0.1165 at 298.15; Antoine (liq) A = 6.65026, B = 2006.37, C = -38.409,
dev = 1.0, range = 305 to 470.

2-Chlorobenzaldehyde, CA 89-98-5: Irritant; causes burns; flammable, flash
point 360; soluble ethanol, ether, acetone, benzene. Tf = 284.15; Tb =
486.15 to 487.15; Vm = 0.1126 at 293.15; Antoine (liq) A = 6.18706, B =
1718.1, C = -74.15, dev = 0.1 to 1.0, range = 382 to 563.

$C_7H_5ClO_2$ MW 156.57

2-Chlorobenzoic acid, CA 118-91-2: Irritant; flammable; soluble ethanol,
ether, acetone, benzene; preservative and chemical intermediate.

$C_7H_5Cl_2N$ MW 174.03

Phenylcarbonimidic dichloride, CA 622-44-6: Corrosive; toxic; lachrymator;
flammable, flash point 352. Tb = 482.15 to 484.15; Vm = 0.1360 at 288.15;
Antoine (liq) A = 8.032, B = 2820, C = 0, dev = 5.0, range = 273 to 378.

$C_7H_5Cl_3$ MW 195.48

Benzene, 1-(chloromethyl)-2,4-dichloro, CA 94-99-5: Irritant; toxic; flammable; soluble ethanol, ether. Tm = 270.55; Tb = 521.15; Vm = 0.1389 at 293.15; Antoine (liq) A = 6.27225, B = 1881.38, C = -80.22, dev = 0.1, range = 413 to 578.

(Trichloromethyl)benzene, CA 98-07-7: Corrosive; toxic; suspected carcinogen; flammable, flash point 400; reacts with water and ethanol; soluble ether, benzene, chloroform. Tm = 268.4; Tb = 493.75; Vm = 0.1424 at 293.15; Antoine (liq) A = 6.95923, B = 2268.82, C = -28.669, dev = 1.0 to 5.0, range = 318 to 487.

2,3,6-Trichlorotoluene, CA 2077-46-5: Irritant; toxic; flammable; soluble ethanol, acetone. Tm = 318.15 to 319.15; Antoine (liq) A = 5.6249, B = 1323, C = -144.15, dev = 1.0, range = 384 to 509.

$C_7H_5Cl_3F_8$ MW 347.46

1,1,2,2,3,3,4,4-Octafluoro-1,5,5-trichloroheptane, CA 16395-71-4: Irritant. Antoine (liq) A = 6.2443, B = 1972.2, C = 0, dev not specified, range not specified.

$C_7H_5FN_2O_4$ MW 200.13

Benzene, (fluorodinitromethyl), CA 17003-70-2: Irritant; flammable; potentially explosive. Antoine (liq) A = 7.845, B = 2760.3, C = 0, dev = 1.0 to 5.0, range = 328 to 363.

$C_7H_5FO_2$ MW 140.11

3-Fluorobenzoic acid, CA 455-38-9: Lachrymator; flammable; moderately soluble water; soluble ether. Tm = 397.15; Antoine (sol) A = 11.562, B = 4771, C = 0, dev = 1.0, range = 358 to 382.

$C_7H_5F_3$ MW 146.11

Benzene, (trifluoromethyl), CA 98-08-8: Toxic by inhalation and skin absorption; corrosive; flammable, flash point 285; reacts with water; soluble most organic solvents. Tm = 244.13; Tb = 375.2; Vm = 0.1237 at 298.15; Antoine (liq) A = 6.09081, B = 1303.905, C = -56.013, dev = 0.02, range = 328 to 413.

$C_7H_5F_{10}NS$ MW 325.17

Ethanimidothioic acid, 2,2,2-trifluoro-N-[1,2,2,2-tetrafluoro-1-(trifluoromethyl)ethyl, ethyl ester, CA 54120-09-1: Antoine (liq) A = 6.515, B = 1778, C = 0, dev = 1.0 to 5.0, range not specified.

C_7H_5N MW 103.12

Benzonitrile, phenyl cyanide, CA 100-47-0: Toxic; flammable, flash point 352; moderately soluble hot water; soluble most organic solvents. Tm = 260.35; Tb = 464.25; Vm = 0.1020 at 288.15; Antoine (liq) A = 6.79506, B = 2066.71, C = -32.19, dev = 1.0, range = 301 to 464.   *(continues)*

C$_7$H$_5$N *(continued)*

Phenyl isocyanide, CA 931-54-4: Irritant; unstable. Vm = 0.1052 at 288.15;
Antoine (liq) A = 6.70002, B = 1898.49, C = -33.748, dev = 1.0, range = 285
to 438.

C$_7$H$_5$NO   MW 119.12

Phenyl isocyanate, CA 103-71-9: Irritant; lachrymator; highly toxic; flammable, flash point 328; moisture-sensitive; reacts violently with water;
soluble ether. Tm = 243.15; Tb = 438.15; Vm = 0.1094 at 299.05; Antoine
(liq) A = 6.72286, B = 1942.36, C = -27.037, dev = 1.0, range = 283 to 439.

C$_7$H$_5$NO$_3$ MW 151.12

2-Nitrobenzaldehyde, CA 552-89-6: Irritant; flammable, flash point above
385; potentially explosive; moderately soluble water; soluble ethanol, benzene, ether. Tm = 316.15 to 317.15; Vm = 0.1177 at 293.15; Antoine (liq)
A = 6.99614, B = 2550.68, C = -35.44, dev = 1.0, range = 390 to 547.

3-Nitrobenzaldehyde, CA 99-61-6: Irritant; flammable; potentially explosive; moderately soluble water; soluble ethanol, ether, benzene, chloroform. Tm = 331.15; Vm = 0.1181 at 293.15; Antoine (liq) A = 7.62933, B =
3015.68, C = -14.57, dev = 1.0, range = 401 to 552.

C$_7$H$_5$NS   MW 135.18

Phenyl isothiocyanate, phenyl mustard oil, CA 103-72-0: Irritant; corrosive; lachrymator; highly toxic orally; flammable, flash point 360; soluble
ethanol, ether. Tf = 252.15; Tb = 494.15; Vm = 0.1198 at 298.15; Antoine
(liq) A = 7.06159, B = 2409.59, C = -15.196, dev = 1.0, range = 320 to 492.

C$_7$H$_5$N$_3$O$_6$ MW 227.13

2,4,6-Trinitrotoluene, TNT, CA 118-96-7: Toxic by inhalation and skin absorption; explodes above 513; slightly soluble water; soluble ether, acetone, benzene, pyridine. Tm = 353.9; Antoine I (sol) A = 13.596, B =
5874.238, C = 0, dev = 1.0 to 5.0, range = 293 to 353; Antoine II (sol) A =
12.2025, B = 5400.536, C = 0, dev = 1.0, range = 337 to 350; Antoine III
(liq) A = 6.40336, B = 2191.85, C = -121.43, dev = 1.0, range = 353 to 523.

C$_7$H$_5$N$_3$O$_7$ MW 243.13

Anisole, 2,4,6-trinitro, CA 606-35-9: Toxic; potentially explosive; soluble ethanol, ether, benzene. Tm = 341.15; Vm = 0.1627 at 353.15; Antoine
I (sol) A = 16.3749, B = 6918, C = 0, dev = 1.0 to 5.0, range = 334 to 342;
Antoine II (liq) A = 10.1909, B = 4803, C = 0, dev = 1.0, range = 342 to
363; Antoine III (liq) A = 21.605, B = 9780, C = 0, dev = 5.0, range = 457
to 473.

Phenol, 3-methyl-2,4,6-trinitro, CA 602-99-3: Toxic; explodes above 423;
soluble ethanol, ether, acetone, benzene. Tm = 382.15 to 383.15; Antoine
(sol) A = 12.59, B = 5808, C = 0, dev not specified, range = 310 to 366.

$C_7H_5N_5O_8$ MW 287.15

Aniline, *N*-methyl-*N*,2,4,6-tetranitro, CA 479-45-8: Irritant; explodes at
453 to 463; soluble ethanol, ether, benzene; used as explosive sensitizer.
Tm = 404.15 to 405.15; Antoine (sol) A = 14.14, B = 6987, C = 0, dev not
specified, range = 335 to 405.

$C_7H_6Cl_2$ MW 161.03

Benzene, (dichloromethyl), CA 98-87-3: Highly irritant; corrosive; lachrymator; flammable, flash point 365; soluble most organic solvents. Tm =
256.75; Tb = 478.35; Vm = 0.1281 at 287.15; Antoine (liq) A = 6.17925, B =
1815.31, C = -52.238, dev = 1.0 to 5.0, range = 308 to 487.

2,4-Dichlorotoluene, CA 95-73-8: Irritant; toxic; flammable, flash point
352; soluble ethanol, ether, benzene. Tm = 259.65; Tb = 474.25; Vm =
0.1286 at 293.15; Antoine (liq) A = 5.6199, B = 1330.4, C = -104.65, dev =
1.0, range = 346 to 475.

3,4-Dichlorotoluene, CA 95-75-0: Irritant; toxic; flammable, flash point
358; soluble ethanol, ether, benzene. Tm = 257.85; Tb = 482.05; Vm =
0.1279 at 293.15; Antoine (liq) A = 6.10415, B = 1655.44, C = -78.15, dev =
0.1, range = 378 to 543.

$C_7H_6F_3N$ MW 161.13

Benzene, 1-(trifluoromethyl)-3-amino, CA 98-16-8: Irritant; highly toxic;
flammable, flash point 358; soluble ethanol, ether. Tm = 278.15 to 279.15;
Tb = 462.15 to 464.15; Vm = 0.1249 at 293.15; Antoine (liq) A = 6.2952, B =
1650.21, C = -79.57, dev = 0.1, range = 334 to 464.

$C_7H_6F_3NS$ MW 193.19

Aniline, *N*-(trifluoromethyl)thio: Toxic; flammable. Antoine (liq) A =
7.295, B = 2454, C = 0, dev = 1.0 to 5.0, range = 333 to 413.

$C_7H_6N_2O_4$ MW 182.14

Benzene, (dinitromethyl), CA 611-38-1: Toxic; flammable; potentially explosive. Tm = 351.15 to 353.15; Antoine (sol) A = 10.225, B = 3975, C = 0,
dev = 5.0, range = 312 to 323.

2,4-Dinitrotoluene, CA 121-14-2: Highly toxic by inhalation and skin absorption; suspected carcinogen; flammable, flash point 480; potentially
explosive on prolonged heating; decomposes above 573; slightly soluble
water; soluble ethanol, ether, acetone, benzene. Tm = 343.15 to 344.15;
Vm = 0.1379 at 344.15; Antoine I (sol) A = 12.27361, B = 5009.432, C = 0,
dev = 5.0, range = 277 to 343; Antoine II (sol) A = 11.7426, B = 5139.058,
C = 0, dev = 1.0, range = 331 to 342; Antoine III (liq) A = 7.1423, B =
3039, C = 0, dev = 1.0 to 5.0, range = 473 to 572; Antoine IV (liq) A =
6.04898, B = 1956.095, C = -108.183, dev = 5.0, range = 344 to 572.

2,6-Dinitrotoluene, CA 606-20-2: Highly toxic; flammable; potentially explosive; slightly soluble water; soluble ethanol, ether, acetone, benzene.
Tm = 339.15; Vm = 0.1419 at 384.15; Antoine I (sol) A = 11.9436, B =
4446.22, C = -21.279, dev = 5.0, range = 277 to 323; Antoine II (liq) A =
7.329, B = 2971, C = 0, dev = 5.0, range = 423 to 533;   *(continues)*

$C_7H_6N_2O_4$ *(continued)*

Antoine III (liq) A = 6.70024, B = 2160.968, C = -93.282, dev = 5.0, range = 330 to 533.

3,5-Dinitrotoluene, CA 618-85-9: Highly toxic; flammable; potentially explosive; soluble ethanol, ether, benzene, chloroform. Tm = 366.15; Vm = 0.1426 at 384.15; Antoine (liq) A = 7.5674, B = 3271, C = 0, dev = 5.0, range = 493 to 543.

$C_7H_6N_2O_5$ MW 198.13

2-Methyl-4,6-dinitrophenol, CA 534-52-1: Highly toxic; irritant; potentially explosive; slightly soluble water; soluble ethanol, ether, acetone; contact insecticide. Tm = 359.65; Antoine (sol) A = 13.265, B = 5400, C = 0, dev = 1.0, range = 290 to 324.

$C_7H_6N_4$ MW 146.15

1H-Tetrazole, 1-phenyl, CA 5378-52-9: Flammable; soluble aqueous ethanol, carbon tetrachloride. Tm = 338.15; Antoine (sol) A = 13.917, B = 5421, C = 0, dev not specified, range not specified.

1H-Tetrazole, 5-phenyl, CA 3999-10-8: Flammable; soluble aqueous ethanol, carbon tetrachloride. Tm = 488.15, decomposes; Antoine (sol) A = 12.843, B = 5999, C = 0, dev not specified, range not specified.

C_7H_6O MW 106.12

Benzaldehyde, CA 100-52-7: Moderately toxic orally; flammable, flash point 336; slowly oxidizes in air; slightly soluble water; soluble most organic solvents. Tm = 247.15; Tb = 452.15; Vm = 0.1017 at 298.15; Antoine I (liq) A = 5.56823, B = 1197.54, C = -115.829, dev = 1.0, range = 348 to 452; Antoine II (liq) A = 7.4764, B = 2455.4, C = 0, dev = 1.0, range = 273 to 373; Antoine III (liq) A = 6.20256, B = 1611.255, C = -67.979, dev = 0.02, range = 409 to 481; Antoine IV (liq) A = 6.48764, B = 1782.204, C = -52.863, dev = 0.1, range = 311 to 376; Antoine V (liq) A = 6.22745, B = 1629.229, C = -65.993, dev = 0.1, range = 370 to 475; Antoine VI (liq) A = 6.28780, B = 1682.466, C = -58.948, dev = 0.1 to 1.0, range = 465 to 541; Antoine VII (liq) A = 6.52485, B = 1916.921, C = -26.699, dev = 1.0, range = 529 to 599.

2,4,6-Cycloheptatriene-1-one, tropone, CA 539-80-0: Hygroscopic oil; flammable; soluble benzene, chloroform. Tm = 265.15 to 268.15; Vm = 0.0969 at 295.15; Antoine (liq) A = 7.569, B = 2829, C = 0, dev not specified, range = 273 to 323.

$C_7H_6O_2$ MW 122.12

Benzene, 1,2-methylenedioxy, CA 274-09-9: Toxic; flammable. Tb = 445.15 to 446.15; Vm = 0.1148 at 293.15; Antoine (liq) A = 6.815, B = 2164.1, C = 0, dev = 5.0, range = liquid to 447.

Benzoic acid, CA 65-85-0: Irritant; toxic by skin absorption; flammable, flash point 394; moderately soluble water; soluble ethanol, ether, acetone, chloroform; used for plastics and as a preservative in food industry. Tm = 395.55; Tb = 522.35; Vm = 0.1136 at 403.15;   *(continues)*

226

$C_7H_6O_2$ *(continued)*

Tc = 751, calculated; Pc = 4470, calculated; Vc = 0.344, calculated; Antoine I (sol) A = 15.94025, B = 7420.596, C = 74.333, dev = 1.0, range = 294 to 321; Antoine II (sol) A = 11.8285, B = 4719.5, C = 0, dev = 5.0, range = 343 to 373; Antoine III (liq) A = 7.80991, B = 2776.12, C = -43.978, dev = 1.0, range = 405 to 523.

Formic acid, phenyl ester, CA 1864-94-4: Irritant; flammable. Tb = 446.15, decomposes; Vm = 0.1122 at 273.15; Antoine (liq) A = 20.6099, B = 6358.1, C = 0, dev = 5.0, range = 287 to 305.

4-Hydroxybenzaldehyde, CA 123-08-0: Irritant; flammable; moderately soluble hot water; soluble ethanol, ether, benzene. Tm = 389.15; Vm = 0.1082 at 403.15; Antoine I (sol) A = 11.955, B = 5129.7, C = 0, dev = 1.0, range = 303 to 336; Antoine II (liq) A = 6.98689, B = 2538.01, C = -73.504, dev = 1.0, range = 394 to 583.

2-Hydroxy-2,4,6-cycloheptatriene-1-one, tropolone, CA 533-75-5: Hygroscopic; flammable, flash point above 385; soluble water, ether, acetone, light hydrocarbons. Tm = 322.15 to 323.15; Antoine (sol) A = 12.213, B = 4380, C = 0, dev = 5.0, range = 273 to 323.

Salicylaldehyde, 2-hydroxybenzaldehyde, CA 90-02-8: Irritant; flammable, flash point 351; moderately soluble water; soluble ethanol, ether, acetone, benzene; used in perfumes. Tm = 275.15; Tb = 470.15; Vm = 0.1046 at 293.15; Antoine (liq) A = 6.6349, B = 1971.2, C = 44.03, dev = 1.0, range = 383 to 470.

$C_7H_6O_3$ MW 138.12

4-Hydroxybenzoic acid, CA 99-96-7: Flammable; very soluble hot water; soluble methanol, ethanol, ether, acetone. Tm = 487.65 to 488.65; Antoine (sol) A = 12.7477, B = 6063.5, C = 0, dev = 1.0, range = 398 to 433.

Salicylic acid, 2-hydroxybenzoic acid, CA 69-72-7: Irritant; moisture-sensitive; flammable, flash point 430; moderately soluble hot water; soluble methanol, ethanol, ether, acetone; used in drug and food industry. Tm = 431.75; Vm = 0.1474 at 431.75; Tc = 729, calculated; Pc = 5180, calculated; Vc = 0.364, calculated; Antoine I (sol) A = 11.9834, B = 4968.7, C = 0, dev = 1.0, range = 368 to 408; Antoine II (liq) A = 5.53812, B = 1049.95, C = -228.144, dev = 5.0, range = 445 to 504.

$C_7H_6O_5$ MW 170.12

3,4,5-Trihydroxybenzoic acid, gallic acid, CA 149-91-7: Irritant; hygroscopic; light-sensitive; very soluble hot water; soluble ethanol, acetone. Tm = 526.15, decomposes; Antoine (sol) A = 8.851, B = 3926, C = 0, dev = 5.0, range = 391 to 421.

C_7H_7Br MW 171.04

Benzene, (bromomethyl), CA 100-39-0: Corrosive; lachrymator; causes burns; flammable, flash point 359; reacts slowly with water; soluble ethanol, ether. Tf = 269.25; Tb = 471.15 to 472.15; Vm = 0.1189 at 298.15; Antoine I (liq) A = 7.09196, B = 2374.88, C = -5.048, dev = 1.0, range = 305 to 472; Antoine II (liq) A = 7.36905, B = 2627.88, C = 16.624, dev = 1.0, range = 340 to 407.   *(continues)*

C_7H_7Br *(continued)*

2-Bromotoluene, CA 95-46-5: Irritant; moderately toxic; flammable, flash point 352; soluble ethanol, ether, benzene, acetone. Tm = 246.15; Tb = 454.55; Vm = 0.1202 at 293.15; Antoine (liq) A = 6.03337, B = 1549.39, C = -70.15, dev = 0.1 to 1.0, range = 353 to 518.

3-Bromotoluene, CA 591-17-3: Irritant; moderately toxic; flammable, flash point 333; soluble ethanol, ether, benzene, acetone. Tm = 233.15; Tb = 457.15; Vm = 0.1213 at 293.15; Antoine (liq) A = 6.86228, B = 2085.49, C = -27.333, dev = 1.0, range = 287 to 457.

4-Bromotoluene, CA 106-38-7: Irritant; moderately toxic; flammable, flash point 358; soluble ethanol, ether, benzene, acetone. Tm = 301.15; Tb = 457.15; Vm = 0.1218 at 308.15; Antoine (liq) A = 6.13252, B = 1612.35, C = -66.79, dev = 0.1 to 1.0, range = 358 to 523.

C_7H_7BrO MW 187.04

4-Bromoanisole, CA 104-92-7: Flammable, flash point 367; soluble ethanol, ether, chloroform. Tm = 284.15; Tb = 496.15; Vm = 0.1284 at 293.15; Antoine (liq) A = 6.6234, B = 2091.43, C = -43.339, dev = 1.0, range = 321 to 496.

C_7H_7Cl MW 126.59

Benzene, (chloromethyl), CA 100-44-7: Highly irritant; corrosive; lachrymator; suspected carcinogen; causes burns; flammable, flash point 340; decomposes with hot water; soluble most organic solvents. Tm = 233.95; Tb = 452.55; Vm = 0.1148 at 293.15; Antoine I (liq) A = 6.7176, B = 1954.13, C = -38.02, dev = 1.0, range = 295 to 453; Antoine II (liq) A = 7.73903, B = 2642.08, C = 12.819, dev = 1.0, range = 320 to 390.

2-Chlorotoluene, CA 95-49-8: Toxic vapor; narcotic; flammable, flash point 320; soluble most organic solvents. Tm = 237.55; Tb = 432.35; Vm = 0.1180 at 303.15; Antoine (liq) A = 6.07253, B = 1497.2, C = -64.15, dev = 0.1, range = 338 to 493.

3-Chlorotoluene, CA 108-41-8: Toxic vapor; narcotic; flammable, flash point 323; soluble most organic solvents. Tm = 225.35; Tb = 434.85; Vm = 0.1181 at 293.15; Antoine (liq) A = 6.78821, B = 2028.13, C = -11.629, dev = 1.0, range = 277 to 436.

4-Chlorotoluene, CA 106-43-4: Toxic vapor; narcotic; flammable, flash point 322; soluble most organic solvents. Tm = 280.65; Tb = 435.55; Vm = 0.1183 at 293.15; Antoine (liq) A = 6.90317, B = 2107.52, C = -5.373, dev = 1.0, range = 304 to 436.

C_7H_7ClO MW 142.58

2-Chloroanisole, CA 766-51-8: Irritant; flammable, flash point 349; soluble ethanol, ether. Tm = 246.35; Tb = 468.15 to 469.15; Vm = 0.1197 at 293.15; Antoine (liq) A = 6.66563, B = 2012.4, C = -43.15, dev = 0.1, range = 388 to 460.

C_7H_7F MW 110.13

Benzene, (fluoromethyl), CA 350-50-5: Flammable; soluble ethanol, ether. Tm = 238.15; Tb = 413.15; Vm = 0.1077 at 298.15; Antoine (liq) A = 7.24465, B = 2066.216, C = -15.2, dev = 1.0 to 5.0, range = 297 to 410. *(continues)*

C_7H_7F *(continued)*

2-Fluorotoluene, CA 95-52-3: Flammable, flash point 285; soluble ethanol, ether. Tm = 211.15; Tb = 388.15; Vm = 0.1097 at 286.15; Antoine (liq) A = 6.0981, B = 1356.8, C = -56.0, dev = 1.0, range = 295 to 388.

3-Fluorotoluene, CA 352-70-5: Flammable, flash point 282; soluble ethanol, ether. Tf = 162.35; Tb = 389.15; Vm = 0.1103 at 293.15; Antoine (liq) A = 6.1344, B = 1382.7, C = -54.81, dev = 1.0, range = 293 to 390.

4-Fluorotoluene, CA 352-32-9: Flammable, flash point 313; soluble ethanol, ether. Tm = 217.15; Tb = 390.15; Vm = 0.1101 at 289.15; Antoine (liq) A = 6.11911, B = 1374.021, C = -55.755, dev = 0.02 to 0.1, range = 340 to 429.

$C_7H_7F_2N$ MW 143.14

Benzylamine, *N,N*-difluoro, CA 23162-99-4: Irritant; flammable. Antoine (liq) A = 9.61889, B = 4065.956, C = 0, dev = 1.0, range = 313 to 333.

C_7H_7I MW 218.04

Benzene, (iodomethyl), CA 620-05-3: Moderately irritant; flammable; light-sensitive; soluble methanol, ethanol, ether, benzene. Tm = 297.65; Vm = 0.1258 at 298.15; Antoine (liq) A = 6.90329, B = 2476.303, C = 0, dev = 1.0, range = 358 to 403.

2-Iodotoluene, CA 615-37-2: Irritant; flammable, flash point 363; soluble ethanol, ether, benzene. Tb = 484.15 to 485.15; Vm = 0.1273 at 293.15; Antoine (liq) A = 6.48132, B = 1986.16, C = -40.589, dev = 1.0, range = 310 to 484.

C_7H_7IO MW 234.04

4-Iodoanisole, CA 696-62-8: Flammable, flash point above 385; light-sensitive; soluble ethanol, ether, chloroform. Tm = 324.15 to 325.15; Antoine (liq) A = 6.42412, B = 2003.4, C = -66.762, dev = 1.0 to 5.0, range = 401 to 520.

C_7H_7NO MW 121.14

Benzamide, CA 55-21-0: Irritant; flammable; moderately soluble hot water; soluble ethanol, benzene, pyridine. Tm = 403.15; Tb = 561.15, decomposes; Vm = 0.1122 at 403.15; Antoine (sol) A = 11.69847, B = 5062.899, C = 0, dev = 1.0, range = 325 to 342.

2,4,6-Cycloheptatriene-1-one, 2-amino, CA 6264-93-3: Irritant; flammable; soluble ethanol, benzene, chloroform. Tm = 379.15 to 380.15; Antoine (sol) A = 7.7089, B = 3718, C = 0, dev = 1.0 to 5.0, range = 273 to 323.

Formanilide, CA 103-70-8: Irritant; flammable, flash point above 385; moderately soluble water; soluble ethanol, benzene. Tm = 323.15; Tb = 544.15; Vm = 0.1070 at 323.15; Antoine (sol) A = 9.61889, B = 4065.956, C = 0, dev = 1.0, range = 298 to 318.

$C_7H_7NO_2$ MW 137.14

2-Aminobenzoic acid, anthranilic acid, CA 118-92-3: Irritant; moderately toxic; flammable; soluble hot water, ethanol, ether, chloroform; corrosion inhibitor and dye intermediate. Tm = 417.15 to 419.15; Antoine (sol) A = 12.02961, B = 5222.086, C = 0, dev = 1.0 to 5.0, range = 331 to 349.

3-Aminobenzoic acid, CA 99-05-8: Irritant; suspected carcinogen; flammable; soluble hot water, ethanol, ether, acetone; dye intermediate. Tm = 447.15; Antoine (sol) A = 13.60972, B = 6373.631, C = 0, dev = 1.0 to 5.0, range = 366 to 385.

4-Aminobenzoic acid, CA 150-13-0: Irritant; moderately toxic; flammable; soluble hot water, ethanol, ether, ethyl acetate. Tm = 460.15 to 460.65; Antoine (sol) A = 12.33406, B = 5866.164, C = 0, dev = 1.0 to 5.0, range = 364 to 384.

Benzene, (nitromethyl), CA 622-42-4: Irritant; toxic; flammable; soluble ethanol, acetone. Tb = 498.15 to 500.15; Antoine (liq) A = 7.545, B = 2810, C = 0, dev = 5.0, range = 363 to 413.

2-Nitrotoluene, CA 88-72-2: Irritant; highly toxic by inhalation and skin absorption; flammable, flash point 379; soluble most organic solvents. Tm = 269.3; Tb = 494.85; Vm = 0.1179 at 293.15; Antoine (liq) A = 6.32043, B = 1827.66, C = -71.63, dev = 0.1, range = 402 to 496.

3-Nitrotoluene, CA 99-08-1: Irritant; highly toxic by inhalation and skin absorption; flammable, flash point 379; soluble most organic solvents. Tm = 289.25; Tb = 505.05; Vm = 0.1185 at 293.15; Antoine (liq) A = 7.00458, B = 2481.45, C = -8.58, dev = 1.0, range = 353 to 505.

4-Nitrotoluene, CA 99-99-0: Irritant; highly toxic by inhalation and skin absorption; flammable, flash point 379; soluble most organic solvents. Tm = 324.85; Tb = 511.65; Vm = 0.1242 at 348.15; Antoine I (sol) A = 10.6673, B = 4130.07, C = 0, dev = 1.0, range = 296 to 310; Antoine II (liq) A = 7.40605, B = 2889.12, C = 23.37, dev = 1.0, range = 423 to 512.

$C_7H_7NO_3$ MW 153.14

2-Nitroanisole, CA 91-23-6: Irritant; flammable, flash point above 385; insoluble water; soluble ethanol, ether. Tm = 283.15; Tb = 545.15; Vm = 0.1221 at 293.15; Antoine (liq) A = 7.615, B = 3060, C = 0, dev = 5.0, range = 424 to 545.

$C_7H_7N_3$ MW 133.15

Benzene, azidomethyl, CA 622-79-7: Flammable; potentially explosive. Antoine (liq) A = 7.365, B = 2506.0, C = 0, dev = 1.0 to 5.0, range = 333 to 363.

C_7H_8 MW 92.14

Bicyclo(2,2,1)hepta-2,5-diene, 2,5-norbornadiene, CA 121-46-0: Flammable, flash point 252; polymerizes; soluble most organic solvents. Tf = 254.05; Tb = 363.45; Vm = 0.1017 at 293.15; Antoine (liq) A = 5.96613, B = 1247.078, C = -49.397, dev = 1.0, range = 300 to 364.

(continues)

C_7H_8 *(continued)*

1,3,5-Cycloheptatriene, tropilidene, CA 544-25-2: Toxic; flammable, flash point 299; polymerizes in air; soluble most organic solvents. Tm = 193.65; Tb = 390.15; Vm = 0.1038 at 292.15; Antoine I (liq) A = 6.09522, B = 1374.656, C = -52.612, dev = 1.0, range = 273 to 390; Antoine II (liq) A = 6.12574, B = 1390.771, C = -53.069, dev = 0.1, range = 273 to 390.

Tetracyclo[3,2,0,0^{2,7},0^{4,6}]heptane, quadricyclane, CA 278-06-8: Flammable; may isomerize to norbornadiene. Tb = 371.15; Antoine (liq) A = 6.56659, B = 1621.163, C = -27.958, dev = 1.0, range = 302 to 372.

Toluene, CA 108-88-3: Irritant; moderately toxic vapor; flammable, flash point 277; reacts violently with many oxidizing agents; soluble most organic solvents. Tm = 178.19; Tb = 383.78; Vm = 0.1069 at 298.15; Tc = 591.79; Pc = 4109; Vc = 0.316; Antoine I (liq) A = 6.08627, B = 1349.122, C = -53.154, dev = 0.02, range = 308 to 386; Antoine II (liq) A = 6.1528, B = 1376.81, C = -51.1, dev = 0.1, range = 210 to 279; Antoine III (liq) A = 6.12072, B = 1374.901, C = -49.657, dev = 0.02, range = 383 to 445; Antoine IV (liq) A = 6.40851, B = 1615.834, C = -15.897, dev = 0.1, range = 440 to 531; Antoine V (liq) A = 7.65383, B = 3153.235, C = 188.566, dev = 0.1, range = 530 to 592; Antoine VI (liq) A = 6.16273, B = 1391.005, C = -48.974, dev = 0.02, range = 273 to 295.

C_7H_8O MW 108.14

Anisole, methoxybenzene, CA 100-66-3: Moderately toxic; flammable, flash point 325; forms explosive peroxides. Tm = 235.65; Tb = 428.15; Vm = 0.1086 at 293.15; Antoine (liq) A = 6.17662, B = 1489.957, C = -69.525, dev = 0.1, range = 382 to 437.

Benzyl alcohol, CA 100-51-6: Irritant; moderately toxic; flammable, flash point 213; moderately soluble water; soluble most organic solvents; used in foods and pharmaceuticals. Tm = 258.15; Tb = 478.15; Vm = 0.1036 at 298.15; Antoine I (liq) A = 6.7069, B = 1904.3, C = -73.15, dev = 1.0, range = 385 to 573; Antoine II (liq) A = 8.963, B = 3214, C = 0, dev = 5.0, range = 293 to 313.

2-Hydroxytoluene, *ortho*-cresol, CA 95-48-7: Highly toxic; corrosive; causes burns; flammable, flash point 354; soluble ethanol, ether, benzene, chloroform. Tm = 303.15; Tb = 464.15 to 465.15; Vm = 0.1044 at 303.15; Tc = 697.55; Pc = 5006; Vc = 0.282; Antoine I (sol) A = 11.68858, B = 3909.409, C = 0, dev = 1.0, range = 273 to 303; Antoine II (liq) A = 6.19545, B = 1542.299, C = -96.04, dev = 0.1, range = 383 to 473; Antoine III (liq) A = 6.47616, B = 1714.489, C = -79.841, dev = 0.1 to 1.0, range = 304 to 409; Antoine IV (liq) A = 6.19561, B = 1543.097, C = -95.902, dev = 0.1, range = 399 to 470; Antoine V (liq) A = 6.24893, B = 1584.403, C = -90.794, dev = 1.0, range = 463 to 526; Antoine VI (liq) A = 6.82237, B = 2134.352, C = -19.536, dev = 1.0 to 5.0, range = 517 to 630.

3-Hydroxytoluene, *meta*-cresol, CA 108-39-4: Highly toxic; corrosive; causes burns; flammable, flash point 359; soluble ethanol, ether, benzene, chloroform. Tm = 284.15 to 285.15; Tb = 475.15; Vm = 0.1046 at 293.15; Antoine I (sol) A = 8.0462, B = 2930.845, C = 0, dev = 1.0 to 5.0, range = 273 to 285; Antoine II (liq) A = 6.28394, B = 1603.811, C = -100.504, dev = 0.1, range = 383 to 473; Antoine III (liq) A = 9.0902, B = 3223.45, C = 0, dev = 1.0, range = 284 to 313; Antoine IV (liq) A = 7.150, B = 2123.548, C = -59.018, dev = 1.0, range = 285 to 416; Antoine V (liq) A = 6.28579, B = 1605.855, C = -100.232, dev = 0.1, range = 410 to 477; Antoine VI (liq) A = 5.80987, B = 1293.277, C = -135.465, dev = 1.0, *(continues)*

C_7H_8 *(continued)*

range = 471 to 531; Antoine VII (liq) A = 6.64135, B = 2069.208, C = -26.534, dev = 1.0 to 5.0, range = 523 to 633.

4-Hydroxytoluene, *para*-cresol, CA 106-44-5: Highly toxic; corrosive; causes burns; flammable, flash point 359; soluble ethanol, ether, benzene, chloroform. Tm = 309.15; Tb = 475.65; Antoine I (sol) A = 12.0298, B = 3861.98, C = 0, dev = 1.0, range = 273 to 307; Antoine II (sol) A = 11.16859, B = 3868.314, C = 0, dev = 1.0, range = 277 to 307; Antoine III (liq) A = 6.24257, B = 1566.029, C = -105.47, dev = 0.1, range = 383 to 473; Antoine IV (liq) A = 6.83697, B = 1930.688, C = -73.442, dev = 0.1 to 1.0, range = 308 to 393; Antoine V (liq) A = 6.2376, B = 1563.08, C = -105.776, dev = 0.1, range = 385 to 477; Antoine VI (liq) A = 6.19164, B = 1533.535, C = -108.781, dev = 1.0, range = 463 to 533; Antoine VII (liq) A = 6.99685, B = 2310.405, C = -10.362, dev = 1.0 to 5.0, range = 523 to 635.

$C_7H_8O_2$ MW 124.14

2,4-Dihydroxytoluene, CA 496-73-1: Corrosive; toxic; flammable; soluble water, ethanol, ether, benzene. Tm = 377.15 to 378.15; Tb = 540.15 to 543.15; Antoine (liq) A = 5.8914, B = 1711.3, C = -132.23, dev = 1.0, range = 391 to 459.

2,6-Dihydroxytoluene, CA 608-25-3: Corrosive; toxic; flammable; soluble water, ethanol, ether, benzene. Tm = 392.15 to 393.15; Tb = 537.15; Antoine (liq) A = 5.5882, B = 1521.6, C = -140.28, dev = 1.0, range = 398 to 434.

3,4-Dihydroxytoluene, CA 452-86-8: Corrosive; toxic; flammable; soluble water, ethanol, ether, benzene. Tm = 338.15; Tb = 524.15; Vm = 0.1100 at 347.15; Antoine (liq) A = 11.765, B = 4700, C = 0, dev = 5.0, range = 387 to 415.

3,5-Dihydroxytoluene, CA 504-15-4: Corrosive; toxic; flammable; soluble water, ethanol, ether, benzene. Tm = 380.65; Tb = 560.15 to 563.15; Antoine (liq) A = 6.0777, B = 1798.5, C = -137.25, dev = 1.0, range = 402 to 468.

2-Methoxyphenol, guiaicol, CA 90-05-1: Irritant; toxic; flammable, flash point 355; light-sensitive; moderately soluble water; soluble ethanol, ether, chloroform. Tm = 305.15; Tb = 478.65; Antoine (liq) A = 6.44572, B = 1786.15, C = -76.43, dev = 1.0, range = 378 to 479.

3-Methoxyphenol, CA 150-19-6: Irritant; toxic; flammable, flash point above 385; moderately soluble water; soluble ethanol, ether, chloroform. Tm = below 256.15; Tb = 517.15; Antoine (liq) A = 6.12536, B = 1572.51, C = -136.16, dev = 1.0, range = 413 to 518.

4-Methoxyphenol, CA 150-76-5: Irritant; toxic; flammable, flash point above 385; moderately soluble water; soluble ethanol, ether, benzene. Tm = 326.15; Tb = 516.15; Antoine I (sol) A = 12.27865, B = 4631.266, C = 0, dev = 1.0, range = 278 to 300; Antoine II (liq) A = 6.8462, B = 2111.03, C = -81.56, dev = 1.0, range = 418 to 518.

2*H*-Pyran-2-one, 3,5-dimethyl, CA 63233-31-8: Flammable; soluble water, ethanol, ether. Tm = 324.65; Antoine (liq) A = 6.9785, B = 2232.48, C = -69.112, dev = 1.0, range = 352 to 518.

$C_7H_8O_2S$ MW 156.20

6-Methyl-4-methoxy-2H-pyran-2-thion, CA 52911-98-5: Irritant; flammable.
Tm = 399.15; Antoine (liq) A = 12.80165, B = 5685.948, C = 0, dev not
specified, range = 401 to 415.

2-Methyl-6-(methylthio)-4H-pyran-4-one, CA 52911-99-6: Irritant; flammable.
Tm = 345.15; Antoine (liq) A = 7.52416, B = 3274.416, C = 0, dev not specified, range = 387 to 432.

$C_7H_8O_3$ MW 140.14

2-Furancarboxylic acid, ethyl ester, CA 614-99-3: Flammable, flash point
343; soluble ethanol. Tm = 307.15; Tb = 468.15; Antoine (liq) A = 6.96778,
B = 2124.04, C = -40.161, dev = 1.0, range = 354 to 389.

2-Methoxy-6-methyl-4H-pyran-4-one, CA 4225-42-7: Flammable. Tm = 367.65;
Antoine (liq) A = 9.01232, B = 3801.846, C = 0, dev not specified, range =
370 to 384.

4-Methoxy-6-methyl-2H-pyran-2-one, CA 672-89-9: Flammable; soluble light
hydrocarbons. Tm = 362.15 to 363.15; Antoine (liq) A = 7.26392, B =
2999.702, C = 0, dev not specified, range = 385 to 434.

C_7H_8S MW 124.20

2-Methylbenzenethiol, 2-mercaptotoluene, CA 137-06-4: Irritant; stench;
flammable, flash point 336; soluble ethanol, ether. Tm = 288.15; Tb =
467.45; Vm = 0.1193 at 293.15; Antoine (liq) A = 6.1908, B = 1675.3, C =
-67.05, dev = 1.0, range = 379 to 470.

3-Methylbenzenethiol, 3-mercaptotoluene, CA 108-40-7: Irritant; stench;
flammable, flash point 345; soluble ethanol, ether. Tm = below 253.15;
Tb = 468.15 to 473.15; Vm = 0.1190 at 293.15; Antoine (liq) A = 6.1670, B =
1651.7, C = -71.35, dev = 1.0, range = 380 to 471.

4-Methylbenzenethiol, 4-mercaptotoluene, CA 106-45-6: Irritant; stench;
flammable, flash point 341; soluble ethanol, ether. Tm = 316.15 to 317.15;
Tb = 468.15; Antoine (liq) A = 6.0955, B = 1619.3, C = -72.15, dev = 1.0,
range = 379 to 471.

Methyl phenyl sulfide, CA 100-68-5: Irritant; stench; flammable; soluble
ethanol, benzene. Tb = 467.15 to 469.15; Vm = 0.1179 at 298.15; Antoine
(liq) A = 6.13467, B = 1603.382, C = -79.166, dev = 0.02, range = 389
to 475.

$C_7H_8S_3$ MW 188.32

4,5,6,7-Tetrahydro-1,3-benzodithiole-2-thion, CA 698-42-0: Irritant;
stench; flammable. Tm = 356.15; Antoine (sol) A = 11.54996, B = 5127.087,
C = 0, dev = 1.0, range = 340 to 352.

$C_7H_9F_3N_2O_4$ MW 242.15

Glycine, N-[N-(trifluoroacetyl)glycyl], methyl ester, CA 433-33-0: Tm =
419.85; Antoine I (sol) A = 14.975, B = 6682, C = 0, dev = 5.0, range = 323
to 419; Antoine II (liq) A = 10.735, B = 4902, C = 0, dev = 5.0, range =
420 to 443.

$C_7H_9F_5O_2$ MW 220.14

Pentafluoropropionic acid, butyl ester, CA 680-28-4: Flammable. Antoine
(liq) A = 6.6904, B = 1669.7, C = -33.35, dev = 1.0, range = 354 to 389.

C_7H_9N MW 107.15

Benzyl amine, CA 100-46-9: Highly irritant; causes burns; flammable, flash
point 333; strong base; soluble water, ethanol, ether. Tm = 283.15; Tb =
458.15; Vm = 0.1092 at 293.15; Antoine (liq) A = 6.71728, B = 1921.37, C =
-49.746, dev = 1.0, range = 302 to 458.

2,3-Dimethylpyridine, 2,3-lutidine, CA 583-61-9: Irritant; toxic; flammable; very soluble water; soluble ethanol, ether. Tm = 257.65; Tb =
434.65; Vm = 0.1150 at 298.15; Antoine (liq) A = 6.18881, B = 1538.772,
C = -66.477, dev = 0.02, range = 372 to 436.

2,4-Dimethylpyridine, 2,4-lutidine, CA 108-47-4: Irritant; toxic; flammable, flash point 310; soluble water, ethanol, ether. Tm = 209.15; Tb =
431.85; Vm = 0.1149 at 293.15; Antoine (liq) A = 6.25215, B = 1572.844, C =
-61.184, dev = 0.02, range = 349 to 433.

2,5-Dimethylpyridine, 2,5-lutidine, CA 589-93-5: Irritant; toxic; hygro-
scopic; flammable, flash point 320; soluble cold water, ether, ethanol.
Tm = 257.45; Tb = 430.15; Vm = 0.1148 at 293.15; Antoine (liq) A = 6.20695,
B = 1541.012, C = -63.376, dev = 0.02, range = 358 to 431.

2,6-Dimethylpyridine, 2,6-lutidine, CA 108-48-5: Irritant; toxic; flammable, flash point 306; very soluble water; soluble ethanol, ether. Tm =
267.05; Tb = 416.85; Vm = 0.1161 at 293.15; Antoine (liq) A = 6.18368, B =
1472.109, C = -64.861, dev = 0.02 to 0.1, range = 352 to 418.

3,4-Dimethylpyridine, 3,4-lutidine, CA 583-58-4: Irritant; highly toxic;
hygroscopic; flammable, flash point 326; very soluble water; soluble ethanol, ether. Tm = 262.55; Tb = 452.25; Vm = 0.1155 at 293.15; Antoine (liq)
A = 6.19734, B = 1611.09, C = -67.957, dev = 0.02, range = 385 to 454.

3,5-Dimethylpyridine, 3,5-lutidine, CA 591-22-0: Irritant; toxic; hygroscopic; flammable, flash point 326; moderately soluble water; soluble ethanol, ether. Tm = 266.55; Tb = 445.85; Vm = 0.1142 at 298.15; Antoine (liq)
A = 6.21786, B = 1598.682, C = -65.548, dev = 0.02 to 0.1, range = 373
to 446.

2-Ethylpyridine, CA 100-71-0: Irritant; stench; flammable, flash point
302; moderately soluble water; soluble ethanol, ether, acetone. Tm =
210.05; Tb = 421.75; Vm = 0.1144 at 290.15; Antoine (liq) A = 5.81102, B =
1295.6, C = -83.15, dev = 1.0, range = 323 to 373.

3-Ethylpyridine, CA 536-78-7: Irritant; stench; flammable, flash point
321; moderately soluble water; soluble ethanol, ether, acetone. Tm =
196.25; Tb = 435.15 to 438.15; Vm = 0.1140 at 296.15; Antoine (liq) A =
6.00312, B = 1472.5, C = -71.56, dev = 1.0, range = 334 to 373.

4-Ethylpyridine, CA 536-75-4: Irritant; stench; flammable, flash point
320; moderately soluble water; soluble ethanol, ether, acetone. Tm =
182.65; Tb = 442.75 to 443.15; Vm = 0.1138 at 293.15; Antoine (liq) A =
5.97423, B = 1456.3, C = -74.84, dev = 1.0, range = 333 to 442.

N-Methylaniline, CA 100-61-8: Irritant; toxic by inhalation and skin absorption; flammable, flash point 351; slightly soluble water;   (continues)

C_7H_9N *(continued)*

soluble ethanol, ether. Tf = 216.15; Tb = 469.25; Vm = 0.1083 at 293.15; Antoine I (liq) A = 6.59793, B = 1885.03, C = -58.155, dev = 1.0, range = 309 to 469; Antoine II (liq) A = 6.33055, B = 1685.664, C = -78.688, dev = 0.1 to 1.0, range = 353 to 470.

ortho-Toluidine, 2-aminotoluene, CA 95-53-4: Irritant; toxic by inhalation and skin absorption; suspected carcinogen; flammable, flash point 358; slightly soluble water; soluble ethanol, ether. Tm = 257.65; Tb = 473.15 to 475.15; Vm = 0.1073 at 293.15; Antoine (liq) A = 6.26948, B = 1672.87, C = -81.11, dev = 1.0, range = 391 to 474.

meta-Toluidine, 3-aminotoluene, CA 108-44-1: Irritant; toxic by inhalation and skin absorption; flammable, flash point 358; slightly soluble water; soluble ethanol, ether, acetone, benzene. Tm = 241.65; Tb = 476.15 to 477.15; Vm = 0.1084 at 293.15; Antoine (liq) A = 6.27299, B = 1669.26, C = -85.339, dev = 0.1, range = 394 to 477.

para-Toluidine, 4-aminotoluene, CA 106-49-0: Irritant; toxic by inhalation and skin absorption; flammable, flash point 360; soluble ethanol, acetone, ether, pyridine. Tm = 317.15 to 318.15; Tb = 473.15 to 474.15; Vm = 0.1114 at 293.15; Antoine (liq) A = 6.17451, B = 1585.0, C = -93.44, dev = 1.0, range = 393 to 474.

C_7H_9NO MW 123.15

2-Methoxyaniline, CA 90-04-0: Irritant; flammable; highly toxic by inhalation and skin absorption; almost insoluble water; soluble ethanol, ether, acetone, benzene. Tm = 278.15; Tb = 498.15; Vm = 0.1127 at 293.15; Antoine (liq) A = 7.93885, B = 2901.19, C = -2.898, dev = 1.0 to 5.0, range = 334 to 492.

C_7H_{10} MW 94.16

Bicyclo(2,2,1)hept-2-ene, 2-norbornene, CA 498-66-8: Flammable, flash point 258; polymerizes; soluble most organic solvents. Tm = 317.15 to 317.65; Tb = 367.15 to 370.15; Antoine (liq) A = 5.37005, B = 953.45, C = -85.282, dev = 1.0, range = 301 to 350.

Bicyclo(4,1,0)hept-3-ene, CA 16554-83-9: Flammable; soluble most organic solvents. Antoine (liq) A = 6.05576, B = 1350.394, C = -55.805, dev = 0.1, range = 333 to 384.

Tricyclo[2,2,1,0^{2,6}] heptane, nortricyclene, CA 279-19-6: Flammable; soluble most organic solvents. Tm = 329.15; Tb = 379.15 to 380.15; Antoine (liq) A = 6.05065, B = 1222.751, C = -69.091, dev = 1.0, range = 302 to 337.

Tricyclo[4,1,0,0^{2,4}] heptane, CA 187-26-8: Flammable; soluble most organic solvents. Antoine (liq) A = 6.04798, B = 1315.522, C = -52.24, dev = 0.02, range = 322 to 373.

$C_7H_{10}N_2$ MW 122.17

Diallyl cyanamide, CA 538-08-9: Irritant; lachrymator; flammable; insoluble water; soluble ethanol, ether, benzene, acetone. Tm = 203.15; Antoine (liq) A = 7.52552, B = 2730.2, C = 0, dev = 1.0, range = 369 to 495.

(continues)

$C_7H_{10}N_2$ *(continued)*

2,4-Diaminotoluene, CA 95-80-7: Irritant; toxic; suspected carcinogen;
flammable; soluble water, ethanol, ether, benzene. Tm = 372.15; Tb =
557.15; Vm = 0.1167 at 373.15; Tc = 804, calculated; Pc = 4380, calculated;
Vc = 0.430, calculated; Antoine (liq) A = 8.07251, B = 3288.08, C = -11.132,
dev = 1.0, range = 379 to 553.

Pimelonitrile, 1,5-dicyanopentane, CA 646-20-8: Irritant; toxic; flammable,
flash point above 385; soluble ethanol, ether, chloroform. Tm = 241.75;
Vm = 0.1287 at 291.15; Antoine (liq) A = 9.1538, B = 3891, C = 0, dev = 1.0,
range = 306 to 331.

4-Tolylhydrazine, CA 539-44-6: Toxic; flammable; moderately soluble water;
soluble ethanol, ether, benzene. Tm = 338.15 to 339.15; Tb = 513.15 to
517.15, decomposes; Antoine (liq) A = 7.60172, B = 2628.55, C = -45.383,
dev = 1.0 to 5.0, range = 355 to 515.

$C_7H_{10}O_2$ MW 126.15

5-Methyl-5-hexene-2,4-dione, CA 20583-46-4: Flammable; polymerizes. Tm =
248.15; Antoine (liq) A = 4.5556, B = 1381, C = 0, dev = 1.0, range = 323
to 363.

$C_7H_{10}O_3$ MW 142.15

3-Acetyl-2,4-pentanedione, triacetylmethane, CA 815-68-9: Flammable, flash
point 360; unstable, decomposes in water; soluble acetone. Vm = 0.1297 at
293.15; Antoine (liq) A = 8.0284, B = 2870.5, C = 0, dev = 1.0 to 5.0,
range = 369 to 477.

Trimethylsuccinic acid, anhydride: Flammable; moisture-sensitive; soluble
benzene. Tm = 311.65; Tb = 500.15; Antoine (liq) A = 6.65509, B = 2163.88,
C = -39.036, dev = 5.0, range = 326 to 504.

$C_7H_{10}O_4$ MW 158.15

Dimethyl citraconate, CA 617-54-9: Flammable; soluble ethanol, ether, acetone. Tb = 483.15 to 484.15; Vm = 0.1418 at 293.15; Antoine (liq) A =
7.2685, B = 2370.26, C = -33.311, dev = 1.0, range = 324 to 484.

Dimethyl itaconate, CA 617-52-7: Flammable, flash point 373; soluble ethanol, ether, acetone. Tm = 309.15; Tb = 481.15; Antoine (liq) A = 8.39414,
B = 2853.73, C = -34.615, dev = 1.0 to 5.0, range = 342 to 481.

Dimethyl mesaconate, CA 617-53-8: Flammable; slightly soluble water; soluble ethanol, ether, acetone. Tb = 478.65 to 479.65; Vm = 0.1449 at
293.15; Antoine (liq) A = 7.11704, B = 2250.93, C = -38.845, dev = 1.0,
range = 319 to 479.

$C_7H_{10}S_3$ MW 190.34

Trithiocarbonic acid, cyclic 1,2-cyclohexylene ester, CA 2164-87-6: Tm =
440.15; Antoine (sol) A = 11.16842, B = 5148.959, C = 0, dev = 1.0, range =
353 to 369.

$C_7H_{11}BrO_2$ MW 207.07

4-Bromo-3-methylcrotonic acid, ethyl ester, CA 26918-14-9: Irritant; flammable. Antoine (liq) A = 6.1599, B = 2252.752, C = 0, dev = 5.0, range =
346 to 381.

$C_7H_{11}ClO_5$ MW 210.61

(2-Chloroethyl)[(1-methoxycarbonyl)ethyl] carbonate: Toxic; flammable.
Antoine (liq) A = 8.6623, B = 3491, C = 0, dev = 5.0, range = 365 to 525.

$C_7H_{11}Cl_3O_2$ MW 233.52

Trichloroacetic acid, neopentyl ester, CA 57392-56-0: Irritant; flammable.
Antoine (liq) A = 8.455, B = 3013, C = 0, dev = 5.0, range = 378 to 473.

$C_7H_{11}NO_2$ MW 141.17

2-Methyl-2-acetoxybutyronitrile: Irritant; flammable. Antoine (liq) A =
6.9878, B = 2036.73, C = -59.454, dev = 1.0 to 5.0, range = 315 to 469.

5-Oxo-2-pyrrolidinecarboxylic acid, ethyl ester: Irritant; flammable.
Antoine (liq) A = 8.619, B = 3849, C = 0, dev = 1.0, range = 418 to 511.

C_7H_{12} MW 96.17

Bicyclo(2,2,1)heptane, norbornane, CA 279-23-2: Flammable; soluble most
organic solvents. Tm = 359.15 to 360.15; Antoine I (sol) A = 7.604, B =
2097.5, C = 0, dev = 1.0, range = 284 to 327; Antoine II (sol) A = 4.80833,
B = 981.971, C = -75.114, dev = 0.02, range = 218 to 247.

Bicyclo(4,1,0)heptane, *cis*, CA 286-08-8: Flammable; soluble most organic
solvents. Tb = 383.15; Antoine (liq) A = 7.146, B = 1987.0, C = 0, dev =
1.0 to 5.0, range = 298 to 385.

Bicyclo(4,1,0)heptane ±, norcarane, CA 286-08-8: Flammable; soluble most
organic solvents. Tb = 383.15; Antoine (liq) A = 6.01901, B = 1338.564,
C = -56.396, dev = 0.02, range = 333 to 385.

Cycloheptene, suberene, CA 628-92-2: Flammable, flash point 267; soluble
most organic solvents. Tm = 217.15; Tb = 389.15; Vm = 0.1169 at 293.15;
Antoine (liq) A = 7.2743, B = 2011.9, C = 0, dev = 0.1 to 1.0, range = 251
to 313.

1,2-Dimethylcyclopentene, CA 765-47-9: Flammable; soluble most organic
solvents. Tf = 182.75; Tb = 376.25 to 376.35; Vm = 0.1213 at 298.15;
Antoine (liq) A = 5.97984, B = 1290.8, C = -54.15, dev = 1.0, range = 294
to 431.

1,3-Dimethylcyclopentene ±, CA 62184-82-1: Flammable; soluble most organic
solvents. Tb = 365.55 to 369.15; Vm = 0.1255 at 293.15; Antoine (liq) A =
5.99297, B = 1252, C = -51.15, dev = 5.0, range = 283 to 410.

1,4-Dimethylcyclopentene ±, CA 57426-81-0: Flammable; soluble most organic
solvents. Tb = 365.85 to 365.95; Vm = 0.1235 at 293.15; Antoine (liq) A =
5.90895, B = 1226.4, C = -52.15, dev = 1.0, range = 273 to 413.

(continues)

C_7H_{12} *(continued)*

1,5-Dimethylcyclopentene ±, CA 16491-15-9: Flammable; soluble most organic solvents. Tf = 155.15; Tb = 366.75 to 366.85; Vm = 0.1233 at 293.15; Antoine (liq) A = 6.00571, B = 1288, C = -53.15, dev = 5.0, range = 273 to 423.

3,3-Dimethylcyclopentene, CA 58049-91-5: Flammable; soluble most organic solvents. Tb = 350.75 to 351.15; Vm = 0.1255 at 298.15; Antoine (liq) A = 5.94216, B = 1220.3, C = -51.15, dev = 5.0, range = 278 to 403.

4,4-Dimethylcyclopentene, CA 19037-72-0: Flammable; soluble most organic solvents. Tb = 353.15; Vm = 0.1255 at 298.15; Antoine (liq) A = 5.94216, B = 1220.3, C = -51.15, dev = 5.0, range = 278 to 403.

1-Ethylcyclopentene, CA 2146-38-5: Flammable; soluble most organic solvents. Tf = 154.65; Tb = 379.45; Vm = 0.1212 at 298.15; Antoine (liq) A = 5.98603, B = 1294.8, C = -54.15, dev = 1.0, range = 293 to 433.

3-Ethylcyclopentene, CA 694-35-9: Flammable; soluble most organic solvents. Tb = 370.95; Vm = 0.1235 at 298.15; Antoine (liq) A = 5.97458, B = 1262.5, C = -53.15, dev = 1.0, range = 283 to 423.

4-Ethylcyclopentene, CA 3742-38-9: Flammable; soluble most organic solvents. Tb = 371.45; Vm = 0.1236 at 298.15; Antoine (liq) A = 5.95709, B = 1284.2, C = -54.15, dev = 5.0, range = 288 to 435.

1-Heptyne, CA 628-71-7: Irritant; polymerizes; flammable, flash point 271; soluble most organic solvents. Tf = 192.22; Tb = 372.93; Vm = 0.1312 at 293.15; Antoine (liq) A = 6.4039, B = 1392.4, C = -56.55, dev = 0.1, range = 336 to 373.

2-Heptyne, CA 1119-65-9: Flammable; polymerizes; soluble most organic solvents. Tb = 383.15 to 386.15; Vm = 0.1293 at 293.15; Antoine (liq) A = 6.3335, B = 1413.4, C = -58.55, dev = 0.1, range = 346 to 385.

3-Heptyne, CA 2586-89-2: Flammable; polymerizes; soluble most organic solvents. Tm = 142.65; Tb = 379.15 to 380.15; Vm = 0.1311 at 298.15; Antoine (liq) A = 6.4560, B = 1433.5, C = -57.95, dev = 0.1, range = 343 to 380.

1-Methylbicyclo(3,1,0)hexane, CA 4625-24-5: Flammable; soluble most organic solvents. Antoine (liq) A = 6.00288, B - 1257.954, C = -51.527, dev = 0.02 to 0.1, range = 312 to 362.

1-Methylcyclohexene, CA 591-49-1: Flammable, flash point 270; soluble most organic solvents. Tf = 152.75; Tb = 383.45 to 383.55; Vm = 0.1187 at 293.15; Antoine (liq) A = 6.0101, B = 1311.087, C = -56.045, dev = 0.02, range = 333 to 384.

3-Methylcyclohexene ±, CA 56688-75-6: Flammable, flash point 270; soluble most organic solvents. Tm = 157.65; Tb = 377.15; Vm = 0.1204 at 298.15; Antoine (liq) A = 6.0253, B = 1289.7, C = -55.15, dev = 0.1, range = 335 to 376.

4-Methylcyclohexene ±, CA 26293-22-1: Flammable, flash point 272; soluble most organic solvents. Tm = 157.65; Tb = 375.05 to 375.75; Vm = 0.1203 at 293.15; Antoine (liq) A = 5.99371, B = 1283.1, C = -54.15, dev not specified, range = 292 to 429.

Methylene cyclohexane, CA 1192-37-6: Flammable, flash point 267; soluble most organic solvents. Tm = 166.45; Tb = 373.15 to 375.15;   *(continues)*

238

C_7H_{12} *(continued)*

Vm = 0.1191 at 293.15; Antoine (liq) A = 5.92827, B = 1253.192, C = -57.094, dev = 0.02, range = 331 to 387.

$C_7H_{12}Br_2$ MW 255.98

1,2-Dibromocycloheptane, CA 29974-68-3: Flammable. Antoine (liq) A = 7.1454, B = 2628, C = 0, dev = 1.0, range = 292 to 353.

$C_7H_{12}ClNO$ MW 161.63

6-Chlorohexyl isocyanate, CA 13654-91-6: Highly toxic; flammable; reacts with water. Antoine (liq) A = 7.1071, B = 2541.63, C = -14.12, dev = 1.0, range = 363 to 453.

$C_7H_{12}ClN_5$ MW 201.66

Simazine, CA 122-34-9: Almost insoluble water; slightly soluble dioxane; used as herbicide. Tm = 501.15 to 502.15; Antoine (sol) A = 14.292, B = 6833, C = 0, dev = 5.0, range = 323 to 403.

$C_7H_{12}Cl_2O_2$ MW 199.08

Dichloroacetic acid, neopentyl ester: Irritant; flammable. Antoine (liq) A = 8.515, B = 2996, C = 0, dev = 5.0, range = 368 to 463.

$C_7H_{12}Cl_2S$ MW 199.14

(2-Chloroethyl)(2-chlorocyclopentyl) sulfide: Toxic; stench; flammable. Antoine (liq) A = 8.85376, B = 3444.6, C = 0, dev = 1.0, range = 273 to 333.

$C_7H_{12}Cl_4$ MW 237.98

1,1,1,7-Tetrachloroheptane, CA 3922-36-9: Irritant; flammable. Antoine (liq) A = 4.3655, B = 815.981, C = -202.46, dev = 5.0, range = 370 to 454.

$C_7H_{12}O$ MW 112.17

Cycloheptanone, suberone, CA 502-42-1: Flammable, flash point 328; slightly soluble water; soluble ethanol, ether. Tm = 252.15; Tb = 452.15 to 454.15; Vm = 0.1180 at 293.15; Antoine I (liq) A = 6.1388, B = 1588.06, C = -68.15, dev = 1.0, range = 313 to 453; Antoine II (liq) A = 6.13093, B = 1591.359, C = -67.794, dev = 0.02 to 0.1, range = 373 to 465.

$C_7H_{12}O_2$ MW 128.17

Butyl acrylate, CA 141-32-2: Irritant; lachrymator; polymerizes; flammable, flash point 312; slightly soluble water; soluble ethanol, acetone, ether. Tm = 208.55; Tb = 417.15 to 422.15; Vm = 0.1434 at 298.15; Tc = 598, calculated; Pc = 2630, calculated; Vc = 0.428, calculated; Antoine (liq) A = 6.45509, B = 1676.02, C = -43.868, dev = 1.0, range = 272 to 421.

(continues)

$C_7H_{12}O_2$ *(continued)*

Isobutyl acrylate, CA 106-63-8: Irritant; lachrymator; polymerizes, flammable, flash point 303; soluble methanol, ethanol, ether. Tm = 212.15; Tb = 405.15; Vm = 0.1441 at 293.15; Antoine (liq) A = 7.653, B = 2288.06, C = 0, dev = 1.0 to 5.0, range = 330 to 410.

Propyl methacrylate, CA 2210-28-8: Irritant; lachrymator; polymerizes; flammable; soluble ethanol, ether. Tb = 414.15; Vm = 0.1421 at 293.15; Antoine (liq) A = 7.2687, B = 2175, C = 0, dev = 5.0, range = 304 to 413.

$C_7H_{12}O_3$ MW 144.17

2-Acetoxy-2-methyl-3-butanone, CA 10235-71-9: Flammable. Antoine (liq) A = 8.6472, B = 2864, C = 0, dev = 5.0, range = 337 to 368.

Ethyl levulinate, CA 539-88-8: Flammable, flash point 363; soluble water, ethanol. Tb = 478.15 to 479.15; Vm = 0.1426 at 293.15; Antoine (liq) A = 6.62446, B = 1885.16, C = -71.165, dev = 1.0, range = 320 to 480.

$C_7H_{12}O_4$ MW 160.17

2-Acetoxypropionic acid, ethyl ester, CA 2985-28-6: Flammable. Tb = 451.55; Vm = 0.1532 at 290.15; Antoine (liq) A = 6.3592, B = 1586.226, C = -90.15, dev = 1.0 to 5.0, range = 313 to 454.

3-Acetoxypropionic acid, ethyl ester, CA 40326-37-2: Flammable. Antoine (liq) A = 10.556, B = 3768, C = 0, dev = 5.0, range = 350 to 367.

Diethyl malonate, CA 105-53-3: Flammable, flash point 348; moderately soluble water; soluble ethanol, ether, acetone, benzene. Tm = 223.35; Tb = 472.05; Vm = 0.1518 at 293.15; Antoine I (liq) A = 7.8903, B = 2833.41, C = 9.566, dev = 1.0, range = 313 to 472; Antoine II (liq) A = 9.1982, B = 3307, C = 0, dev = 1.0 to 5.0, range = 293 to 318; Antoine III (liq) A = 6.07665, B = 1339.66, C = -137.721, dev = 1.0 to 5.0, range = 384 to 468.

Glutaric acid, dimethyl ester, CA 1119-40-0: Flammable, flash point 386; soluble ethanol, ether. Tm = 230.65; Tb = 487.15; Vm = 0.1473 at 293.15; Antoine (liq) A = 7.9259, B = 2860, C = 0, dev = 1.0, range = 366 to 483.

Monomethyl adipate, CA 627-91-8: Flammable, flash point above 385; soluble ethanol. Tm = 276.15; Vm = 0.1508 at 293.15; Antoine (liq) A = 10.045, B = 4330, C = 0, dev = 5.0, range = 453 to 503.

Pimelic acid, heptanedioic acid, CA 111-16-0: Flammable; very soluble water; soluble ethanol, ether. Tm = 377.15 to 378.15; Antoine (liq) A = 7.97118, B = 3228.95, C = -74.19, dev = 1.0, range = 436 to 615.

Spiro[5,5]-undecane, 2,4,8,10-tetraoxa, CA 180-43-8: Flammable; soluble water, ethanol, ether, acetone. Tm = 319.65; Tb = 486.95; Antoine (liq) A = 7.9149, B = 2942, C = 0, dev = 1.0 to 5.0, range not specified.

$C_7H_{12}O_5$ MW 176.17

Ethyl[(1-methoxycarbonyl)ethyl] carbonate: Toxic; flammable. Antoine (liq) A = 6.671, B = 1884, C = -80.17, dev = 5.0, range = 343 to 473.

(continues)

$C_7H_{12}O_5$ *(continued)*

2-(Lactyloxy)propionic acid, methyl ester: Flammable. Antoine (liq) A = 6.90295, B = 1992.175, C = -90.15, dev = 1.0 to 5.0, range = 317 to 384.

$C_7H_{13}ClO$ MW 148.63

Heptanoyl chloride, CA 2528-61-2: Corrosive; lachrymator; flammable, flash point 331; soluble ethanol, light hydrocarbons. Tf = 189.35; Tb = 398.35; Vm = 0.1550 at 293.15; Antoine (liq) A = 9.5793, B = 3072.13, C = -12.562, dev = above 5.0, range = 307 to 418.

$C_7H_{13}ClO_2$ MW 164.63

Chloroacetic acid, neopentyl ester: Irritant; flammable. Antoine (liq) A = 8.488, B = 2904, C = 0, dev = 5.0, range = 378 to 448.

$C_7H_{13}F_3O_3$ MW 202.17

Tris(2-fluoroethyl)orthoformate: Antoine (liq) A = 8.058, B = 3118, C = 0, dev = 5.0, range = 273 to 333.

$C_7H_{13}N$ MW 111.19

Heptanonitrile, CA 629-08-3: Irritant; flammable; soluble ethanol, acetone, benzene. Tm = 210.55; Tb = 456.15 to 457.15; Vm = 0.1371 at 293.15; Antoine (liq) A = 6.68234, B = 2007.6, C = -28.07, dev = 1.0, range = 313 to 473.

Quinuclidine, CA 100-76-5: Flammable; sublimes readily; very soluble water; soluble most organic solvents. Tm = 431.15; Antoine (sol) A = 8.2279, B = 2630, C = 0, dev = 5.0, range = 273 to 363.

$C_7H_{13}NO$ MW 127.19

2-Butoxypropionitrile: Irritant; flammable. Antoine (liq) A = 7.5429, B = 2439.1, C = 0, dev = 1.0, range = 373 to 423.

6-Heptenoic acid, *trans*, amide: Tm = 398.15; Antoine (sol) A = 12.2801, B = 5076.4, C = 0, dev = 1.0 to 5.0, range = 362 to 393.

$C_7H_{13}NO_2$ MW 143.19

Lactic acid, *N*-(methylallyl) amide: Antoine (liq) A = 10.10044, B = 4274.04, C = 0, dev = 1.0, range = 360 to 428.

N-Lactylmorpholine: Antoine (liq) A = 7.8151, B = 3274.243, C = 0, dev = 1.0, range = 371 to 423.

$C_7H_{13}NO_3$ MW 159.18

N-Acetylalanine ±, ethyl ester, CA 5143-72-6: Irritant; flammable. Tm = 312.15 to 313.15; Antoine (liq) A = 8.4896, B = 3408, C = 0, dev = 1.0, range = 372 to 460.

$C_7H_{13}O_6P$ MW 224.15

Mevinphos, CA 7786-34-7: Highly toxic by inhalation or skin absorption;
soluble water, most organic solvents. Vm = 0.181 at 293.15; Antoine (liq)
A = 8.6307, B = 3560, C = 0, dev = 1.0, range = 293 to 383.

C_7H_{14} MW 98.19

Cycloheptane, suberane, CA 291-64-5: Flammable, flash point 279; soluble
most organic solvents. Tm = 261.15; Tb = 391.15 to 393.15; Vm = 0.1213 at
293.15; Antoine I (liq) A = 5.98143, B = 1333.833, C = -56.458, dev = 0.02,
range = 341 to 433; Antoine II (liq) A = 6.12682, B = 1417.738, C =
-47.665, dev = 0.1, range = 282 to 333; Antoine III (liq) A = 5.97596, B =
1329.98, C = -56.968, dev = 0.02, range = 333 to 398; Antoine IV (liq) A =
7.05325, B = 2475.271, C = 108.922, dev = 1.0 to 5.0, range = 476 to 604.

1,1-Dimethylcyclopentane, CA 1638-26-2: Flammable; soluble most organic
solvents. Tf = 203.35; Tb = 361.0; Vm = 0.1309 at 298.15; Antoine (liq)
A = 5.93757, B = 1217.016, C = -51.464, dev = 0.02, range = 284 to 363.

1,2-Dimethylcyclopentane, cis, CA 1192-18-3: Flammable; soluble most organic solvents. Tm = 220.65; Tb = 372.38; Vm = 0.1278 at 298.15; Antoine
(liq) A = 5.96947, B = 1266.083, C = -53.267, dev = 0.02, range = 293
to 375.

1,2-Dimethylcyclopentane, ±trans, CA 822-50-4: Flammable; soluble most organic solvents. Tm = 154.15; Tb = 364.93; Vm = 0.1315 at 298.15; Antoine
(liq) A = 5.96743, B = 1241.79, C = -51.547, dev = 0.02, range = 295 to 367.

1,3-Dimethylcyclopentane, cis, CA 2532-58-3: Flammable; soluble most organic solvents. Tf = 139.35; Tb = 363.75 to 363.95; Vm = 0.1327 at 298.15;
Antoine (liq) A = 5.96114, B = 1239.17, C = -51.591, dev = 0.02, range =
295 to 366.

1,3-Dimethylcyclopentane, ±trans, CA 1759-58-6: Flammable; soluble most
organic solvents. Tm = 138.15; Tb = 364.65; Vm = 0.1319 at 298.15; Antoine
(liq) A = 5.98762, B = 1250.884, C = -49.72, dev = 0.02, range = 295 to 367.

2,3-Dimethyl-1-pentene ±, CA 29841-08-5: Flammable; soluble most organic
solvents. Tf = 139.15; Tb = 358.15; Vm = 0.1393 at 293.15; Antoine (liq)
A = 6.02306, B = 1222.4, C = -53.15, dev = 1.0, range = 311 to 382.

2,4-Dimethyl-1-pentene, CA 2213-32-3: Flammable; soluble most organic solvents. Tm = 149.35; Tb = 354.05 to 354.45; Vm = 0.1414 at 293.15; Antoine
(liq) A = 5.96946, B = 1206.617, C = -50.337, dev = 0.02, range = 311 to
361.

3,3-Dimethyl-1-pentene, CA 3404-73-7: Flammable; soluble most organic solvents. Tm = 138.85; Tb = 350.05; Vm = 0.1411 at 293.15; Antoine (liq) A =
6.03691, B = 1199.2, C = -53.15, dev = 1.0, range = 306 to 374.

3,4-Dimethyl-1-pentene ±, CA 29841-08-5: Flammable; soluble most organic
solvents. Tb = 353.95; Vm = 0.1416 at 298.15; Antoine (liq) A = 6.02998,
B = 1210.5, C = -53.15, dev = 1.0, range = 309 to 378.

4,4-Dimethyl-1-pentene, CA 762-62-9: Flammable; soluble most organic solvents. Tf = 135.35; Tb = 345.55; Vm = 0.1438 at 293.15; Antoine (liq) A =
5.88352, B = 1147.958, C = -49.628, dev = 0.02, range = 299 to 347.
(continues)

242

C_7H_{14} *(continued)*

2,3-Dimethyl-2-pentene, CA 10574-37-5: Flammable; soluble most organic solvents. Tf = 155.15; Tb = 370.15 to 371.15; Vm = 0.1349 at 293.15; Antoine (liq) A = 5.99846, B = 1267.3, C = -53.15, dev = 1.0, range = 322 to 396.

2,4-Dimethyl-2-pentene, CA 625-65-0: Flammable; soluble most organic solvents. Tf = 145.45; Tb = 356.15 to 357.15; Vm = 0.1412 at 293.15; Antoine (liq) A = 6.01564, B = 1213.898, C = -53.714, dev = 0.02, range = 286 to 363.

3,4-Dimethyl-2-pentene, *cis*, CA 4914-91-4: Flammable; soluble most organic solvents. Tm = 159.75; Tb = 362.45; Vm = 0.1385 at 298.15; Antoine (liq) A = 6.01347, B = 1239.4, C = -53.15, dev = 1.0, range = 316 to 387.

3,4-Dimethyl-2-pentene, *trans*, CA 4914-92-5: Flammable; soluble most organic solvents. Tf = 148.95; Tb = 362.35; Vm = 0.1385 at 298.15; Antoine (liq) A = 6.00924, B = 1247.1, C = -53.15, dev = 1.0, range = 317 to 390.

4,4-Dimethyl-2-pentene, *cis*, CA 762-63-0: Flammable; soluble most organic solvents. Tm = 137.75; Tb = 353.55; Vm = 0.1413 at 298.15; Antoine (liq) A = 5.91872, B = 1184.006, C = -50.994, dev = 0.02, range = 303 to 355.

4,4-Dimethyl-2-pentene, *trans*, CA 690-08-4: Flammable; soluble most organic solvents. Tf = 137.65; Tb = 349.15 to 349.25; Vm = 0.1425 at 293.15; Antoine (liq) A = 5.9934, B = 1190.659, C = -51.299, dev = 0.02, range = 295 to 352.

Ethylcyclopentane, CA 1640-89-7: Flammable, flash point 288; soluble most organic solvents. Tm = 135.25; Tb = 376.15 to 377.15; Vm = 0.1281 at 293.15; Antoine I (liq) A = 6.02171, B = 1304.525, C = -51.781, dev = 0.02, range = 308 to 387; Antoine II (liq) A = 6.15104, B = 1396.62, C = -39.666, dev = 1.0, range = 386 to 507; Antoine III (liq) A = 7.4518, B = 2858.104, C = 159.371, dev = 1.0, range = 499 to 569.

2-Ethyl-3-methyl-1-butene, CA 7357-93-9: Flammable; soluble most organic solvents. Tb = 362.15; Vm = 0.1373 at 293.15; Antoine (liq) A = 5.96725, B = 1206.183, C = -55.036, dev = 0.02, range = 303 to 361.

2-Ethyl-1-pentene, CA 3404-71-5: Flammable; soluble most organic solvents. Tb = 367.15; Vm = 0.1387 at 293.15; Antoine (liq) A = 6.00475, B = 1255.7, C = -53.15, dev = 5.0, range = 267 to 392.

3-Ethyl-1-pentene, CA 4038-04-4: Flammable; soluble most organic solvents. Tf = 145.65; Tb = 357.25; Vm = 0.1420 at 298.15; Antoine (liq) A = 6.02366, B = 1221.9, C = -53.15, dev = 1.0, range = 311 to 382.

3-Ethyl-2-pentene, CA 816-79-5: Flammable; soluble most organic solvents. Tb = 369.15; Vm = 0.1366 at 293.15; Antoine (liq) A = 6.00115, B = 1262.6, C = -53.15, dev = 1.0, range = 321 to 395.

1-Heptene, 1-heptylene, CA 592-76-7: Flammable, flash point 245; soluble most organic solvents. Tf = 154.12; Tb = 366.79; Vm = 0.1409 at 293.15; Tc = 537.29; Pc = 2830; Vc = 0.413; Antoine (liq) A = 5.99079, B = 1237.44, C = -56.26, dev = 0.02 to 0.1, range = 311 to 368.

2-Heptene, *cis*, CA 6443-92-1: Irritant; flammable; soluble most organic solvents. Tb = 371.65; Vm = 0.1387 at 293.15; Antoine (liq) A = 6.0565, B = 1284.7, C = -54.45, dev = 0.1 to 1.0, range = 315 to 372.

(continues)

C_7H_{14} *(continued)*

2-Heptene, *trans*, CA 14686-13-6: Flammable, flash point 272; soluble most organic solvents. Tm = 163.65; Tb = 371.65; Vm = 0.1400 at 293.15; Antoine (liq) A = 6.0559, B = 1282.5, C = -54.45, dev = 0.1 to 1.0, range = 314 to 373.

3-Heptene, *cis*, CA 7642-10-6: Flammable, flash point 266; soluble most organic solvents. Tb = 368.95; Vm = 0.1397 at 293.15; Antoine (liq) A = 6.0478, B = 1271.9, C = -54.1, dev = 0.1 to 1.0, range = 312 to 369.

3-Heptene, *trans*, CA 14686-14-7: Irritant; flammable, flash point 266; soluble most organic solvents. Tm = 136.45; Tb = 368.95; Vm = 0.1407 at 293.15; Antoine (liq) A = 6.0526, B = 1273.8, C = -54.1, dev = 0.1 to 1.0, range = 312 to 369.

Methylcyclohexane, CA 108-87-2: Irritant; flammable, flash point 269; soluble most organic solvents. Tm = 146.55; Tb = 373.55; Vm = 0.1276 at 293.15; Antoine I (liq) A = 5.9428, B = 1266.954, C = -52.282, dev = 0.02, range = 308 to 368; Antoine II (liq) A = 6.14677, B = 1413.495, C = -32.726, dev = 1.0, range = 373 to 511; Antoine III (liq) A = 7.29186, B = 2700.205, C = 147.549, dev = 1.0, range = 501 to 573.

2-Methyl-1-hexene, CA 6094-02-6: Flammable, flash point 267; soluble most organic solvents. Tf = 170.35; Tb = 365.15 to 366.15; Vm = 0.1403 at 293.15; Antoine (liq) A = 6.00827, B = 1248.8, C = -53.15, dev = 1.0, range = 318 to 390.

3-Methyl-1-hexene ±, CA 29841-10-9: Flammable, flash point 267; soluble most organic solvents. Tb = 357.15; Vm = 0.1429 at 298.15; Antoine (liq) A = 6.02414, B = 1221.2, C = -53.15, dev = 1.0, range = 311 to 381.

4-Methyl-1-hexene ±, CA 13643-03-3: Flammable; soluble most organic solvents. Tf = 131.65; Tb = 360.35 to 360.65; Vm = 0.1414 at 298.15; Antoine (liq) A = 6.01836, B = 1230.8, C = -53.15, dev = 1.0, range = 313 to 384.

5-Methyl-1-hexene, CA 3524-73-0: Flammable; soluble most organic solvents. Tb = 357.85; Vm = 0.1428 at 298.15; Antoine (liq) A = 6.0213, B = 1226, C = -53.15, dev = 1.0, range = 313 to 393.

2-Methyl-2-hexene, CA 2738-19-4: Flammable; soluble most organic solvents. Tf = 142.75; Tb = 367.55 to 367.75; Vm = 0.1395 at 298.15; Antoine (liq) A = 6.0021, B = 1260.5, C = -53.15, dev = 1.0, range = 322 to 394.

3-Methyl-2-hexene, *cis*, CA 10574-36-4: Flammable; soluble most organic solvents. Tm = 154.65; Tb = 370.45; Vm = 0.1307 at 298.15; Antoine (liq) A = 5.99865, B = 1266.8, C = -53.15, dev = 1.0, range = 322 to 396.

3-Methyl-2-hexene, *trans*, CA 20710-38-7: Flammable; soluble most organic solvents. Tf = 143.85; Tb = 368.35; Vm = 0.1383 at 298.15; Antoine (liq) A = 6.00247, B = 1259.7, C = -53.15, dev = 1.0, range = 321 to 394.

4-Methyl-2-hexene, *cis*, CA 3683-19-0: Flammable; soluble most organic solvents. Tb = 360.25 to 360.75; Vm = 0.1412 at 298.15; Antoine (liq) A = 6.01929, B = 1229.4, C = -53.15, dev = 1.0, range = 313 to 384.

4-Methyl-2-hexene, *trans*, CA 3683-22-5: Flammable; soluble most organic solvents. Tf = 147.45; Tb = 358.25 to 358.75; Vm = 0.1418 at 298.15; Antoine (liq) A = 6.01696, B = 1233.7, C = -53.15, dev = 1.0, range = 314 to 385.

(continues)

C_7H_{14} *(continued)*

5-Methyl-2-hexene, *cis*, CA 13151-17-2: Flammable; soluble most organic solvents. Tb = 364.25 to 364.75; Vm = 0.1409 at 298.15; Antoine (liq) A = 6.01314, B = 1240.3, C = -53.15, dev = 1.0, range = 354 to 372.

5-Methyl-2-hexene, *trans*, CA 7385-82-2: Flammable; soluble most organic solvents. Tf = 148.85; Tb = 358.75 to 359.25; Vm = 0.1427 at 298.15; Antoine (liq) A = 6.01564, B = 1235.5, C = -53.15, dev = 1.0, range = 315 to 386.

2-Methyl-3-hexene, *cis*, CA 15840-60-5: Flammable; soluble most organic solvents. Tb = 359.15 to 361.15; Vm = 0.1243 at 298.15; Antoine (liq) A = 6.01976, B = 1228.3, C = -53.15, dev = 5.0, range = 262 t0 383.

2-Methyl-3-hexene, *trans*, CA 692-24-0: Flammable; soluble most organic solvents. Tm = 131.55; Tb = 361.25; Vm = 0.1433 at 298.15; Antoine (liq) A = 6.02009, B = 1228, C = -53.15, dev = 1.0, range = 313 to 383.

3-Methyl-3-hexene, *cis*, CA 4914-89-0: Flammable; soluble most organic solvents. Tb = 368.55; Vm = 0.1387 at 298.15; Antoine (liq) A = 5.99033, B = 1233.289, C = -59.038, dev = 0.02, range = 307 to 375.

3-Methyl-3-hexene, *trans*, CA 3899-36-3: Flammable; soluble most organic solvents. Tb = 367.15 to 368.15; Vm = 0.1393 at 298.15; Antoine (liq) A = 6.02806, B = 1254.791, C = -54.732, dev = 0.02, range = 310 to 368.

2,3,3-Trimethyl-1-butene, CA 594-56-9: Flammable, flash point 256; soluble most organic solvents. Tm = 163.25; Tb = 349.15 to 351.15; Vm = 0.1393 at 293.15; Antoine (liq) A = 5.91745, B = 1177.947, C = -49.903, dev = 0.02, range = 288 to 353.

$C_7H_{14}Br_2$ MW 258.00

1,1-Dibromoheptane, CA 59104-79-9: Irritant; flammable; soluble ether, acetone, benzene. Antoine (liq) A = 6.498, B = 1882, C = -76.15, dev = 5.0, range = 395 to 548.

1,2-Dibromoheptane ±, CA 42474-21-5: Irritant; flammable; soluble ether, acetone, benzene. Tb = 500.15 to 502.15, decomposes; Vm = 0.1710 at 293.15; Antoine (liq) A = 7.596, B = 2761, C = 0, dev = 1.0 to 5.0, range = 295 to 500.

$C_7H_{14}Cl_2$ MW 169.09

1,1-Dichloroheptane, CA 821-25-0: Irritant; flammable; soluble ether, benzene, chloroform. Tb = 464.15; Vm = 0.1673 at 293.15; Antoine (liq) A = 6.306, B = 1677, C = -70.15, dev = 5.0, range = 364 to 510.

$C_7H_{14}F_2$ MW 136.18

1,1-Difluoroheptane, CA 407-96-5: Flammable. Tm = 191.15; Antoine (liq) A = 6.349, B = 1458, C = -57.15, dev = 1.0, range = 311 to 424.

$C_7H_{14}N_2$ MW 126.20

3-(Diethylamino)propionitrile, CA 5351-04-2: Irritant; flammable. Antoine (liq) A = 8.035, B = 2806, C = 0, dev = 5.0, range = 338 to 470.

$C_7H_{14}N_2O_2S$ MW 190.26

2-Methyl-2-(methylthio)propanal, O-[(methylamino)carbonyl] oxime, CA 116-06-3: Flammable; slightly soluble water; soluble benzene, methylene chloride; used as insecticide. Tm = 372.15 to 373.15; Antoine (sol) A = 8.96742, B = 4180.621, C = 0, dev = 5.0, range = 298 to 323.

$C_7H_{14}O$ MW 114.19

Cycloheptanol, suberol, CA 502-41-0: Flammable, flash point 344; slightly soluble water; soluble ethanol, ether. Tm = 275.15; Tb = 457.15 to 458.15; Vm = 0.1195 at 293.15; Antoine (liq) A = 10.75074, B = 3987.83, C = 25.772, dev = 1.0, range = 284 to 323.

2,4-Dimethyl-3-pentanone, diisopropyl ketone, CA 565-80-0: Flammable, flash point 288; soluble ethanol, ether, benzene. Tm = 204.15; Tb = 397.15 to 398.15; Vm = 0.1408 at 293.15; Antoine (liq) A = 6.40937, B = 1564.4, C = -43.15, dev = 1.0, range = 321 to 399.

1,2-Epoxyheptane, CA 5063-65-0: Flammable. Antoine (liq) A = 5.7877, B = 1250.42, C = -87.55, dev = 1.0, range = 305 to 414.

1-Ethyl-1-cyclopentanol, CA 1462-96-0: Flammable; soluble ethanol, ether. Tm = 265.15; Antoine (liq) A = 4.0916, B = 383.2, C = -242.45, dev = 5.0, range = 347 to 426.

Heptanal, heptaldehyde, CA 111-71-7: Irritant; flammable, flash point 308; slightly soluble water; soluble ethanol, ether; used in perfumery. Tm = 231.15; Tb = 428.15; Vm = 0.1411 at 303.15; Antoine (liq) A = 3.77351, B = 375.82, C = -215.348, dev = 1.0 to 5.0, range = 285 to 428.

2-Heptanone, methyl pentyl ketone, CA 110-43-0: Moderately toxic; flammable, flash point 320; slightly soluble water; soluble ethanol, ether. Tf = 237.65; Tb = 424.6; Vm = 0.1415 at 303.15; Antoine I (liq) A = 6.07656, B = 1408.13, C = -78.31, dev = 1.0, range = 303 to 424; Antoine II (liq) A = 6.15178, B = 1464.092, C = -71.076, dev = 0.02 to 0.1, range = 327 to 457; Antoine III (liq) A = 6.56718, B = 1810.283, C = -26.944, dev = 1.0, range = 449 to 580.

4-Heptanone, dipropyl ketone, CA 123-19-3: Moderately toxic; flammable, flash point 321; soluble ethanol, ether. Tm = 240.15; Tb = 415.15 to 417.15; Vm = 0.1397 at 293.15; Antoine I (liq) A = 8.16186, B = 2344.65, C = -36.058, dev = 1.0, range = 296 to 417; Antoine II (liq) A = 6.15729, B = 1454.275, C = -69.274, dev = 1.0, range = 304 to 490.

1-Methylcyclohexanol, CA 590-67-0: Moderately toxic by inhalation; flammable, flash point 340; slightly soluble water; soluble ethanol, ether. Tm = 297.15 to 298.15; Tb = 428.15; Vm = 0.1242 at 293.15; Antoine (liq) A = 7.992, B = 2563.1, C = 0, dev = 1.0 to 5.0, range = 340 to 430.

2-Methylcyclohexanol, mixture of *cis* and *trans* isomers, CA 583-59-5: Moderately toxic by inhalation; flammable, flash point 331; slightly soluble water; soluble ethanol, ether. Tm = 263.65 to 263.95; Tb = 436.15 to 439.15; Vm = 0.1228 at 293.15; Antoine (liq) A = 5.91177, B = 1257.6, C = -117.391, dev = 1.0, range = 311 to 418.

3-Methylcyclohexanol $\pm cis$, CA 24965-90-0: Moderately toxic by inhalation; flammable, flash point 335; slightly soluble water; soluble ethanol, ether. Tm = 268.45; Tb = 447.15 to 448.15; Vm = 0.1247 at 293.15; Antoine (liq) A = 8.339, B = 2835.1, C = 0, dev = 1.0 to 5.0, range = 340 to 450.

(continues)

$C_7H_{14}O$ *(continued)*

3-Methylcyclohexanol ±*trans*, CA 23068-71-5: Moderately toxic by inhalation; flammable, flash point 335; slightly soluble water; soluble ethanol, ether. Tm = 272.15; Tb = 446.15 to 447.15; Vm = 0.1239 at 293.15; Antoine (liq) A = 7.8533, B = 2611.87, C = 0, dev = 1.0 to 5.0, range = 350 to 450.

4-Methylcyclohexanol, *cis*, CA 7731-28-4: Moderately toxic by inhalation; flammable, flash point 343; slightly soluble water; soluble ethanol, ether. Tm = 263.95; Tb = 446.15 to 447.15; Vm = 0.1245 at 293.15; Antoine (liq) A = 7.8396, B = 2605.77, C = 0, dev = 1.0 to 5.0, range = 340 to 450.

4-Methylcyclohexanol, *trans*, CA 7731-29-5: Moderately toxic by inhalation; flammable, flash point 343; slightly soluble water; soluble ethanol, ether. Tb = 446.15 to 447.65; Vm = 0.1252 at 294.15; Antoine (liq) A = 8.09465, B = 2721.19, C = 0, dev = 1.0 to 5.0, range = 340 to 450.

2-Methyl-3-hexanone, propyl isopropyl ketone, CA 7379-12-6: Flammable; soluble ethanol, ether, acetone, chloroform. Tb = 408.15 to 409.15; Vm = 0.1411 at 293.15; Antoine (liq) A = 7.3179, B = 2159.5, C = 0, dev = 1.0 to 5.0, range = liquid to 406.

$C_7H_{14}O_2$ MW 130.19

Butyl propionate, CA 590-01-2: Flammable, flash point 305; slightly soluble water; soluble ethanol, ether. Tm = 183.65; Tb = 418.65; Vm = 0.1487 at 293.15; Antoine (liq) A = 5.86919, B = 1232.3, C = -97.91, dev = 1.0, range = 305 to 417.

Ethyl isovalerate, CA 108-64-5: Flammable, flash point 299; soluble ether, ethanol, benzene. Tm = 173.85; Tb = 407.85; Vm = 0.1504 at 293.15; Antoine (liq) A = 6.71902, B = 1751.38, C = -36.37, dev = 1.0, range = 301 to 418.

Formic acid, 1-ethyl-1-methylpropyl ester: Flammable. Antoine (liq) A = 6.943, B = 2041, C = 0, dev = 1.0, range not specified.

Heptanoic acid, CA 111-14-8: Highly toxic orally; corrosive; flammable, flash point above 385; soluble ethanol, ether, acetone. Tm = 262.65; Tb = 496.15; Vm = 0.1431 at 303.15; Antoine (liq) A = 7.35678, B = 2246.07, C = -75.375, dev = 1.0, range = 351 to 495.

Isobutyl propionate, CA 540-42-1: Flammable; soluble ethanol, ether. Tm = 201.75; Tb = 409.95; Vm = 0.1499 at 293.15; Antoine (liq) A = 6.96821, B = 1889.1, C = -29.238, dev = 1.0, range = 271 to 410.

Isopentyl acetate, CA 123-92-2: Flammable, flash point 298; slightly soluble water; soluble ethanol, ether, ethyl acetate. Tm = 195.15; Tb = 415.15; Vm = 0.1502 at 293.15; Antoine (liq) A = 6.76468, B = 1791.26, C = -38.64, dev = 1.0, range = 308 to 424.

Isopropyl isobutyrate, CA 617-50-5: Flammable; soluble ethanol, ether, acetone. Tb = 393.85; Vm = 0.1537 at 294.15; Antoine (liq) A = 6.31943, B = 1473.28, C = -52.202, dev = 1.0, range = 257 to 394.

4-Methoxy-4-methyl-2-pentanone, CA 107-70-0: Flammable. Antoine (liq) A = 6.09563, B = 1467.71, C = -74.94, dev = 0.1, range = 343 to 423.

Methyl caproate, methyl hexanoate, CA 106-70-7: Irritant; causes burns; flammable, flash point 318; soluble ethanol, ether, acetone,   *(continues)*

$C_7H_{14}O_2$ *(continued)*

benzene.   Tm = 202.15; Tb = 423.15; Vm = 0.1472 at 293.15; Antoine (liq)
A = 6.9549, B = 1935.93, C = -31.379, dev = 0.1, range = 315 to 383.

Neopentyl acetate, CA 926-41-0:   Flammable.   Tb = 398.65 to 399.15; Antoine
(liq) A = 8.4377, B = 2564, C = 0, dev = 5.0, range = 301 to 400.

Pentyl acetate, amyl acetate, CA 628-63-7:   Irritant; flammable, flash
point 289; soluble ethanol, ether, acetone.   Tm = 202.35; Tb = 422.35; Vm =
0.1487 at 293.15; Antoine (liq) A = 7.356, B = 2258.3, C = 0, dev = 1.0,
range = 329 to 423.

Propyl butyrate, CA 105-66-8:   Flammable; slightly soluble water; soluble
ethanol, ether.   Tm = 177.95; Tb = 416.15; Vm = 0.1491 at 293.15; Antoine
(liq) A = 6.7473, B = 1818.61, C = -32.407, dev = 1.0, range = 271 to 416.

Propyl isobutyrate, CA 644-49-5:   Flammable; soluble ethanol, ether, acetone.   Tb = 408.15 to 409.15; Vm = 0.1472 at 273.15; Antoine (liq) A =
5.75368, B = 1175.31, C = -93.388, dev = 1.0, range = 267 to 407.

$C_7H_{14}O_3$ MW 146.19

2-Butoxypropionic acid, CA 14620-87-2:   Irritant; flammable.   Antoine (liq)
A = 7.633, B = 2759.3, C = 0, dev = 5.0, range = 373 to 473.

Butyl lactate ±, CA 138-22-7:   Flammable; moderately soluble water; soluble
ethanol, ether.   Tm = 230.15; Tb = 460.15; Vm = 0.1491 at 295.15; Antoine
(liq) A = 6.67105, B = 1711.4, C = -89.45, dev = 1.0, range = 339 to 456.

3-Ethoxypropionic acid, ethyl ester, CA 763-69-9:   Flammable.   Vm = 0.1540
at 293.15; Antoine (liq) A = 7.337, B = 2379, C = 0, dev = 5.0, range = 312
to 446.

1-Heptene ozonide: Flammable; potentially explosive.   Antoine (liq) A =
1.60137, B = 186.434, C = 0, dev = 5.0, range = 326 to 392.

4-(2-Hydroxyethyl)-4-methyl-1,3-dioxane: Flammable.   Antoine (liq) A =
6.32549, B = 1622.7, C = -77.26, dev = 1.0, range = 329 to 455.

3-Hydroxypropionic acid, butyl ester: Flammable.   Antoine (liq) A = 8.541,
B = 3149, C = 0, dev = 5.0, range = 361 to 382.

3-Methoxypropionic acid, propyl ester: Flammable.   Antoine (liq) A =
7.4818, B = 2455, C = 0, dev = 5.0, range = 323 to 433.

3-Propoxypropionic acid, methyl ester, CA 14144-39-9:   Flammable.   Antoine
(liq) A = 7.4233, B = 2435, C = 0, dev = 5.0, range = 323 to 453.

$C_7H_{14}O_4$ MW 162.19

Butyric acid, 2,3-dihydroxypropyl ester ±, CA 557-25-5:   Flammable; soluble
water, ethanol.   Tb = 542.15 to 544.15; Vm = 0.1437 at 291.15; Antoine
(liq) A = 9.90582, B = 4200.8, C = 0, dev = 5.0, range = 392 to 449.

$C_7H_{14}S$ MW 130.25

Allyl *tert*-butyl sulfide: Toxic; lachrymator; stench; flammable.   Antoine
(liq) A = 7.44686, B = 2222.554, C = 0, dev = 1.0, range = 319 to 339.

$C_7H_{15}Br$ MW 179.10

1-Bromoheptane, heptyl bromide, CA 629-04-9: Irritant; flammable, flash
point 333; soluble ethanol, ether, chloroform. Tf = 215.05; Tb = 451.15 to
452.15; Vm = 0.1571 at 293.15; Antoine (liq) A = 6.1831, B = 1603.71, C =
-68.15, dev = 1.0, range = 333 to 483.

2-Bromoheptane ±, CA 1974-04-5: Irritant; flammable, flash point 320; soluble ethanol, ether, benzene. Tm = 320.15; Tb = 438.15 to 440.15; Vm =
0.1588 at 293.15; Antoine (liq) A = 6.1255, B = 1541, C = -66.15, dev =
5.0, range = 333 to 440.

$C_7H_{15}Cl$ MW 134.65

1-Chloroheptane, CA 629-06-1: Irritant; flammable, flash point 314; soluble ethanol, ether, chloroform. Tm = 203.65; Tb = 432.65; Vm = 0.1537 at
293.15; Antoine (liq) A = 5.9631, B = 1410.064, C = -77.511, dev = 1.0,
range = 307 to 434.

2-Chloroheptane ±, CA 1001-89-4: Irritant; flammable; soluble ethanol,
ether, chloroform. Vm = 0.1553 at 293.15; Antoine (liq) A = 6.2294, B =
1524, C = -63.15, dev = 5.0, range = 313 to 424.

$C_7H_{15}Cl_2N$ MW 184.11

N-Methyl-bis(2-chloropropyl) amine: Irritant; flammable. Antoine (liq)
A = 8.12188, B = 2850.4, C = 0, dev = 5.0, range = 273 to 333.

N-Propyl-bis(2-chloroethyl) amine: Irritant; flammable. Antoine (liq) A =
8.14374, B = 2966.7, C = 0, dev = 1.0, range = 273 to 369.

$C_7H_{15}F$ MW 118.19

1-Fluoroheptane, heptyl fluoride, CA 661-11-0: Flammable; soluble acetone,
ether, benzene. Tf = 200.15; Tb = 392.15; Vm = 0.1466 at 293.15; Antoine
(liq) A = 6.2084, B = 1405.79, C = -56.55, dev = 1.0, range = 287 to 417.

$C_7H_{15}I$ MW 226.10

1-Iodoheptane, heptyl iodide, CA 4282-40-0: Irritant; light-sensitive;
flammable, flash point 351; soluble ethanol, ether, acetone. Tm = 224.95;
Tb = 477.15; Vm = 0.1641 at 293.15; Antoine (liq) A = 6.0737, B = 1644.29,
C = -72.9, dev = 0.1 to 1.0, range = 373 to 513.

$C_7H_{15}N$ MW 113.20

Azocine, octahydro, CA 1121-92-2: Flammable. Tm = 240.15; Vm = 0.1263 at
294.15; Antoine (liq) A = 6.86775, B = 1940.165, C = -30.653, dev = 0.1 to
1.0, range = 273 to 313.

$C_7H_{15}NO$ MW 129.20

Heptanamide, CA 628-62-6: Flammable. Tm = 369.15; Tb = 523.15 to 531.15;
Vm = 0.1516 at 383.15; Antoine (sol) A = 12.7419, B = 5182.3, C = 0, dev =
1.0, range = 345 to 366.

$C_7H_{15}NO_2$ MW 145.20

Hexyl carbamate, CA 2114-20-7: Flammable. Tm = 329.95; Antoine (sol) A =
13.8729, B = 5018, C = 0, dev = 1.0 to 5.0, range = 291 to 314.

Lactic acid, N-butylamide, CA 30220-58-7: Flammable. Tm = 298.65; Antoine
(liq) A = 9.4919, B = 4043.436, C = 0, dev = 1.0, range = 365 to 433.

Lactic acid, N-sec-butylamide: Flammable. Tm = 297.15; Antoine (liq) A =
9.38012, B = 3895.414, C = 0, dev = 1.0, range = 368 to 418.

Lactic acid, N-isobutylamide: Flammable. Tm = 321.15; Antoine (liq) A =
9.1069, B = 3836.978, C = 0, dev = 1.0, range = 388 to 418.

Leucine, levo, methyl ester: Flammable. Antoine (liq) A = 6.03254, B =
2057.7, C = 0, dev = 1.0, range = 320 to 353.

C_7H_{16} MW 100.20

2,2-Dimethylpentane, CA 590-35-2: Flammable, flash point 288; soluble most
organic solvents. Tf = 149.34; Tb = 352.35; Vm = 0.1487 at 293.15; Antoine
I (liq) A = 5.93117, B = 1185.376, C = -50.37, dev = 0.02, range = 277 to
354; Antoine II (liq) A = 6.2280, B = 1399.333, C = -20.934, dev = 1.0,
range = 353 to 483.

2,3-Dimethylpentane ±, CA 565-59-3: Flammable, flash point below 266; soluble most organic solvents. Tb = 363.75; Vm = 0.1442 at 293.15; Antoine I
(liq) A = 5.97125, B = 1233.903, C = -51.775, dev = 0.02, range = 286 to
365; Antoine II (liq) A = 6.42613, B = 1450.517, C = -32.515, dev = 0.02 to
0.1, range = 208 to 286.

2,4-Dimethylpentane, CA 108-08-7: Flammable, flash point 261; soluble most
organic solvents. Tf = 153.91; Tb = 353.65; Vm = 0.1490 at 293.15; Antoine
(liq) A = 5.95921, B = 1196.516, C = -50.993, dev = 0.02, range = 284 to
355.

3,3-Dimethylpentane, CA 562-49-2: Flammable, flash point 267; soluble most
organic solvents. Tf = 138.15; Tb = 359.15 to 360.15; Vm = 0.1445 at
293.15; Antoine I (liq) A = 5.95139, B = 1228.138, C = -47.819, dev = 0.02,
range = 285 to 360; Antoine II (liq) A = 6.35011, B = 1415.316, C =
-31.302, dev = 0.1, range = 213 to 281; Antoine III (liq) A = 5.94685, B =
1225.973, C = -48.144, dev = 0.02, range = 280 to 360.

3-Ethylpentane, CA 617-78-7: Flammable; soluble most organic solvents.
Tm = 154.55; Tb = 366.45; Vm = 0.1435 at 293.15; Antoine (liq) A = 5.99985,
B = 1251.609, C = -53.262, dev = 0.02, range = 291 to 368.

Heptane, CA 142-82-5: Irritant; narcotic in high concentrations; flammable,
flash point 269; soluble most organic solvents. Tm = 182.15; Tb = 371.45;
Vm = 0.1466 at 293.15; Tc = 540.26; Pc = 2736, Vc = 0.432; Antoine (liq)
A = 6.02633, B = 1268.583, C = -56.054, dev = 0.02, range = 297 to 375.

2-Methylhexane, isoheptane, CA 591-76-4: Flammable, flash point below 255;
soluble most organic solvents. Tf = 154.05; Tb = 363.15; Vm = 0.1476 at
293.15; Antoine (liq) A = 6.00513, B = 1240.11, C = -53.123, dev = 0.02,
range = 296 to 365.

3-Methylhexane ±, CA 589-34-4: Flammable, flash point 269; soluble most
organic solvents. Tm = 100.15; Tb = 364.95; Vm = 0.1459 at 293.15; Antoine
(liq) A = 5.99126, B = 1239.57, C = -53.979, dev = 0.02, range = 289 to 366.

(continues)

C_7H_{16} (continued)

2,2,3-Trimethylbutane, triptane, CA 464-06-2: Flammable, flash point 267; soluble most organic solvents. Tm = 248.95; Tb = 354.05; Vm = 0.1452 at 293.15; Antoine I (liq) A = 5.9181, B = 1201.098, C = -47.026, dev = 0.02, range = 284 to 355; Antoine II (liq) A = 6.18145, B = 1390.726, C = -20.97, dev = 1.0, range = 353 to 483.

$C_7H_{16}O$ MW 116.20

2,4-Dimethyl-3-pentanol, CA 600-36-2: Flammable, flash point 322; slightly soluble water; soluble ethanol, ether. Tm = below 203.15; Tb = 412.15; Vm = 0.1402 at 293.15; Antoine (liq) A = 5.64812, B = 1045.596, C = -124.885, dev = 1.0 to 5.0, range = 307 to 412.

3-Ethyl-3-pentanol, CA 597-49-9: Flammable, flash point 313; slightly soluble water; soluble ethanol, ether. Tm = 260.82, Tb = 413.15 to 415.15; Vm = 0.1382 at 295.15; Antoine (liq) A = 8.5656, B = 2678.1, C = 0, dev = 5.0, range = 317 to 408.

1-Heptanol, heptyl alcohol, CA 111-70-6: Flammable, flash point 346; slightly soluble water; soluble ethanol, ether. Tf = 239.05; Tb = 449.45; Vm = 0.1419 at 298.15; Antoine I (liq) A = 5.9794, B = 1256.783, C = -133.487, dev = 1.0, range = 336 to 450; Antoine II (liq) A = 6.10408, B = 1322.62, C = -126.87, dev = 1.0, range = 335 to 450.

2-Heptanol ±, CA 52390-72-4: Flammable, flash point 344; slightly soluble water; soluble ethanol, ether, benzene. Tb = 431.15 to 433.15; Vm = 0.1423 at 293.15; Antoine (liq) A = 6.96023, B = 1808.607, C = -67.246, dev = 0.1 to 1.0, range = 357 to 431.

3-Heptanol ±, CA 40617-58-1: Flammable, flash point 333; slightly soluble water; soluble ethanol, ether. Tm = 203.15; Tb = 429.65 to 430.15; Vm = 0.1412 at 293.15; Antoine I (liq) A = 11.98108, B = 4275.711, C = 34.738, dev = 1.0, range = 263 to 295; Antoine II (liq) A = 5.46754, B = 964.547, C = -151.386, dev = 1.0, range = 325 to 430.

4-Heptanol, CA 589-55-9: Flammable; slightly soluble water; soluble ethanol, ether. Tf = 231.65; Tb = 427.15; Vm = 0.1420 at 293.15; Antoine I (liq) A = 6.47046, B = 1473.51, C = -98.047, dev = 1.0, range = 282 to 320; Antoine II (liq) A = 5.67603, B = 1071.152, C = -135.848, dev = 1.0, range = 320 to 428.

2-Methyl-2-hexanol, CA 625-23-0: Flammable, flash point 313; slightly soluble water; soluble ethanol, ether. Tb = 414.15 to 415.15; Vm = 0.1431 at 293.15; Antoine (liq) A = 6.24166, B = 1325.086, C = -102.632, dev = 1.0, range = 311 to 415.

$C_7H_{16}O_2$ MW 132.20

2,2-Diethoxypropane, acetone, diethyl acetal, CA 126-84-1: Flammable, flash point 280; soluble ethanol, ether, acetone, benzene. Tb = 387.15; Vm = 0.1612 at 294.15; Antoine (liq) A = 5.71529, B = 1473.016, C = 0, dev = 1.0, range = 286 to 304.

4-Methyl-4-methoxy-2-pentanol, CA 141-73-1: Flammable; moderately soluble water; soluble ethanol, ether. Antoine (liq) A = 6.3216, B = 1623.5, C = -65.55, dev = 5.0, range = 343 to 423.

$C_7H_{16}O_3$ MW 148.20

Diethylene glycol, monopropyl ether, CA 6881-94-3: Flammable; soluble
water, ethanol, ether. Antoine (liq) A = 9.3029, B = 3410.5, C = 0, dev
not specified, range = 369 to 404.

Orthoformic acid, triethyl ester, triethoxymethane, CA 122-51-0: Moderately toxic; irritant; flammable, flash point 303; slightly soluble water,
decomposes; soluble ethanol, ether. Tm = below 255.15; Tb = 418.15 to
420.15; Vm = 0.1673 at 298.15; Antoine I (liq) A = 6.95814, B = 1899.93,
C = -35.582, dev = 1.0, range = 278 to 419; Antoine II (liq) A = 8.5676,
B = 2558, C = 0, dev = 1.0 to 5.0, range = 293 to 323.

$C_7H_{16}S$ MW 132.26

1-Heptanethiol, CA 1639-09-4: Toxic; stench; flammable, flash point 319;
soluble ethanol, ether. Tm = 229.92; Tb = 450.05; Vm = 0.1569 at 293.15;
Antoine (liq) A = 6.07616, B = 1525.109, C = -75.423, dev = 0.02, range =
373 to 472.

2-Heptanethiol, CA 628-00-2: Toxic; stench; flammable; soluble ethanol,
ether. Tm = 259.05; Antoine (liq) A = 6.1559, B = 1542.5, C = -64.95,
dev = 1.0, range = 341 to 443.

$C_7H_{16}S_2$ MW 164.32

1,7-Heptanedithiol, CA 62224-02-6: Toxic; stench; flammable. Tm = 235.05;
Antoine (liq) A = 6.2802, B = 1856.4, C = -91.06, dev = 1.0, range = 392
to 526.

$C_7H_{17}N$ MW 115.22

Heptylamine, CA 111-68-2: Irritant; corrosive; flammable, flash point 308;
soluble ethanol, ether. Tf = 250.15; Tb = 428.4; Vm = 0.1486 at 293.15;
Antoine (liq) A = 5.66704, B = 1210.3, C = -100.03, dev = 1.0, range = 326
to 430.

$C_7H_{17}NO$ MW 131.22

Diethylamine, N-(ethoxymethyl), CA 7352-03-6: Irritant; flammable; soluble
ethanol, ether, acetone. Tb = 409.15; Antoine (liq) A = 7.1564, B = 2071,
C = 0, dev = 5.0, range = 285 to 400.

$C_7H_{17}O_2PS_3$ MW 260.36

Phorate, O,O-diethyl-S-[(ethylthio)methyl] dithiophosphate, CA 298-02-2:
Highly toxic liquid; flammable; slightly soluble water; soluble most or-
ganic solvents; used as insecticide. Vm = 0.2252 at 298.15; Antoine (liq)
A = 8.6498, B = 3697, C = 0, dev = 5.0, range = 283 to 387.

$C_7H_{18}N_2$ MW 130.23

N,N-Diethyl-1,3-propanediamine, CA 104-78-9: Toxic; corrosive; flammable,
flash point 331; soluble ethanol, ether. Tb = 432.15; Vm = 0.1577 at
293.15; Antoine (liq) A = 7.494, B = 2422.3, C = 0, dev = 1.0 to 5.0,
range = 329 to 443.   *(continues)*

$C_7H_{18}N_2$ *(continued)*

1,7-Heptanediamine, CA 646-19-5: Irritant; flammable, flash point 360; soluble ethanol, ether, acetone, benzene. Tm = 301.15 to 302.15; Tb = 496.15 to 498.15; Antoine (liq) A = 7.05686, B = 2040.712, C = -23.953, dev not specified, range = 273 to 313.

$C_7H_{18}N_2O$ MW 146.23

1,3-Bis(dimethylamino)-2-propanol, CA 5966-51-8: Irritant; flammable, flash point above 385; soluble water, ethanol, ether. Tb = 451.15 to 458.15; Vm = 0.1664 at 293.15; Antoine (liq) A = 7.88511, B = 2629.9, C = 0, dev = 1.0 to 5.0, range = 355 to 450.

$C_7H_{19}N_3$ MW 145.25

Ethanamine, *N,N*-diethyl-2-(1-methylhydrazino), CA 67727-91-7: Irritant; flammable. Antoine (liq) A = 9.7565, B = 3230, C = 0, dev = 1.0, range = 283 to 313.

$C_8Cl_4N_2$ MW 265.91

1,3-Benzenedicarbonitrile, 2,4,5,6-tetrachloro, chlorothalonil, CA 1897-45-6: Irritant; soluble xylenes, cyclohexane. Tm = 524.15; Tb = 623.15; Antoine (sol) A = 10.18458, B = 4477.93, C = -42.803, dev = 1.0, range = 363 to 418.

$C_8F_8O_2$ MW 280.07

Trifluoroacetic acid, pentafluorophenyl ester, CA 14533-84-7: Tm = 276.65; Antoine (liq) A = 6.1989, B = 1704.5, C = 0, dev not specified, range not specified.

$C_8F_8O_4$ MW 312.07

Monoperoxycarbonic acid, *O*-(pentafluorophenyl)-*OO*-(trifluoromethyl) ester, CA 59483-83-9: Antoine (liq) A = 13.046, B = 4616.4, C = 0, dev not specified, range not specified.

C_8F_{16} MW 400.06

Perfluoroethylcyclohexane, CA 335-21-7: Antoine (liq) A = 5.97684, B = 1217.729, C = -67.476, dev = 0.1, range = 311 to 411.

C_8F_{18} MW 438.06

Perfluorooctane, CA 307-34-6: Tm = 248.15; Tb = 376.15; Vm = 0.2475 at 298.15; Antoine (liq) A = 5.97984, B = 1199.76, C = -77.1, dev = 0.1, range = 310 to 379.

$C_8F_{18}N_2OS$ MW 514.13

Sulfur, bis(1,1,1,3,3,3-hexafluoro-2-propanaminato)oxobis(trifluoromethyl), CA 66632-47-1: Irritant. Antoine (liq) A = 6.645, B = 2069, C = 0, dev = 1.0 to 5.0, range = 273 to 333.

$C_8F_{18}N_2S$ MW 498.13

Sulfilimine, S,S-bis(trifluoromethyl)-N-[2,2,2-trifluoro-1-(trifluoromethyl)]-1-[(2,2,2-trifluoro-1-[trifluoromethyl]ethylidene)amino]ethyl,
CA 37826-45-2: Irritant. Antoine (liq) A = 8.92561, B = 3467.27, C =
93.153, dev = 1.0, range = 329 to 373.

$C_8F_{18}O$ MW 454.06

Bis(nonafluorobutyl) ether, CA 308-48-5: Irritant; forms explosive peroxides. Antoine I (liq) A = 5.99268, B = 1191.989, C = -75.34, dev = 0.02
to 0.1, range = 343 to 375; Antoine II (liq) A = 8.0796, B = 2207, C = 0,
dev = 1.0 to 5.0, range = 288 to 313; Antoine III (liq) A = 9.85214, B =
2942.979, C = 0, dev = 5.0, range = 374 to 413.

$C_8F_{18}O_2$ MW 470.06

Perfluoro[1,6-bis(methoxy)hexane]: Antoine (liq) A = 6.6119, B = 1755,
C = 0, dev = 1.0 to 5.0, range not specified.

$C_8F_{18}O_3S$ MW 518.12

Sulfuric acid, bis[2,2,2-trifluoro-1,1-bis(trifluoromethyl)ethyl] ester:
Antoine (liq) A = 6.875, B = 2018, C = 0, dev = 1.0 to 5.0, range not
specified.

$C_8F_{18}S_2$ MW 502.18

Bis(nonafluorobutyl) disulfide, CA 42060-69-5: Irritant. Antoine (liq)
A = 7.425, B = 2345, C = 0, dev = 1.0 to 5.0, range not specified.

$C_8F_{20}N_2S$ MW 536.13

Sulfur, difluoro[1,1,1,3,3,3-hexafluoro-N-(2,2,2-trifluoro-1-[trifluoro-
methyl]-ethylidene)-2,2-propanediaminato]-N-bis(trifluoromethyl), CA 65844-
11-3: Irritant. Antoine (liq) A = 7.285, B = 2059, C = 0, dev = 1.0 to
5.0, range not specified.

$C_8HCl_4F_{11}O_2$ MW 479.89

3,5,7,8-Tetrachloro-2,2,3,4,4,5,6,6,7,8,8-undecafluorooctanoic acid,
CA 2923-68-4: Irritant. Antoine (liq) A = 8.6783, B = 3690, C = 0, dev =
5.0, range = 373 to 553.

$C_8HF_{16}NO$ MW 431.08

2-Propanamine, 1,1,1,2,3,3,3-heptafluoro-N-[2,2,2-trifluoro]-1-[2,2,2-tri-
fluoro-1-(trifluoromethyl)ethoxy]ethylidene, CA 54181-87-2: Antoine (liq)
A = 7.165, B = 1879, C = 0, dev = 1.0 to 5.0, range not specified.

$C_8H_2F_{14}$ MW 364.08

1,1,1,2,2,3,3,6,6,7,7,8,8,8-Tetradecafluoro-4-octene, CA 35709-14-9:
Antoine (liq) A = 7.1149, B = 1930, C = 0, dev = 1.0 to 5.0, range not
specified.

254

$C_8H_2F_{16}$ MW 402.08

1,1,2,2,3,3,4,4,5,5,6,6,7,7,8,8-Hexadecafluorooctane, CA 307-99-3: Antoine (liq) A = 7.221, B = 2145.9, C = 0, dev = 5.0, range = 298 to 323.

$C_8H_3ClF_6$ MW 248.56

4-Chloro-1,3-bis(trifluoromethyl)benzene, CA 327-76-4: Irritant; toxic; flammable. Tm = 214.15; Antoine (liq) A = 7.106, B = 1959.4, C = -33.6, dev = 1.0, range = 275 to 353.

5-Chloro-1,3-bis(trifluoromethyl)benzene, CA 328-72-3: Irritant; toxic; flammable. Tm = 251.15; Antoine (liq) A = 7.244, B = 1992.7, C = -26.53, dev = 1.0, range = 275 to 353.

$C_8H_3Cl_4F_3$ MW 297.92

1,1,1-Trifluoro-2,2-dichloro-2-(3,4-dichlorophenyl)ethane: Irritant; flammable. Antoine (liq) A = 7.81798, B = 2972.982, C = 0, dev = 1.0 to 5.0, range = 417 to 461.

$C_8H_3F_5O_2$ MW 226.10

Acetic acid, pentafluorophenyl ester: Flammable. Tm = 303.15; Antoine (liq) A = 7.74818, B = 2513.566, C = 0, dev = 1.0, range = 283 to 322.

$C_8H_3F_{15}O$ MW 400.09

2,2,3,3,4,4,5,5,6,6,7,7,8,8,8-Pentadecafluoro-1-octanol, CA 307-30-2: Irritant. Tm = 310.15; Tb = 436.15 to 438.15; Antoine (liq) A = 8.4288, B = 2795, C = 0, dev = 5.0, range = 350 to 437.

$C_8H_4ClF_3O$ MW 208.57

Trifluoromethyl 3-chlorophenyl ketone, CA 321-31-3: Irritant; flammable. Antoine (liq) A = 7.70908, B = 2571.164, C = 0, dev = 1.0, range = 366 to 405.

$C_8H_4Cl_2O_2$ MW 203.02

Isophthaloyl chloride, CA 99-63-8: Corrosive; lachrymator; flammable, flash point 453; soluble ether. Tm = 316.15 to 317.15; Tb = 549.15; Antoine (liq) A = 7.905, B = 3211, C = 0, dev = 1.0 to 5.0, range = 443 to 550.

Phthaloyl chloride, CA 88-95-9: Corrosive; lachrymator; flammable, flash point above 385; soluble ether; decomposed by water or ethanol. Tm = 288.15 to 289.15; Tb = 554.15; Vm = 0.1441 at 293.15; Antoine (liq) A = 6.98493, B = 2576.14, C = -31.567, dev = 1.0 to 5.0, range = 391 to 549.

Terephthaloyl chloride, CA 100-20-9: Corrosive; lachrymator; flammable, flash point 453; soluble ether. Tm = 356.15 to 357.15; Tb = 532.15; Antoine (liq) A = 7.505, B = 2937, C = 0, dev = 1.0, range = 454 to 473.

$C_8H_4Cl_3F_3$ MW 263.47

1,1,1-Trifluoro-2,2-dichloro-2-(3-chlorophenyl) ethane: Irritant; flammable. Antoine (liq) A = 7.47098, B = 2633.442, C = 0, dev = 1.0 to 5.0,
range = 387 to 474.

$C_8H_4F_6$ MW 214.11

1,3-Bis(trifluoromethyl)benzene, CA 402-31-3: Corrosive; toxic; flammable,
flash point 299. Tb = 389.15; Vm = 0.1553 at 298.15; Antoine (liq) A =
6.440, B = 1488.8, C = -52.15, dev = 1.0, range = 275 to 390.

1,4-Bis(trifluoromethyl)benzene, CA 433-19-2: Corrosive; toxic; flammable.
Tm = 275.93; Antoine (liq) A = 6.1624, B = 1351.6, C = -64.26, dev = 1.0,
range = 287 to 390.

$C_8H_4O_3$ MW 148.12

Phthalic anhydride, CA 85-44-9: Flammable, flash point 424; moisture-
sensitive; hydrolyzed by hot water; soluble ethanol; important polymer raw
material. Tm = 404.15; Tb = 557.65, sublimes; Tc = 791, calculated; Pc =
4720, calculated; Vc = 0.421, calculated; Antoine I (sol) A = 11.2912, B =
4592.957, C = 0, dev = 1.0, range = 333 to 363; Antoine II (liq) A =
7.74204, B = 3542.32, C = 59.561, dev = 1.0 to 5.0, range = 407 to 558.

$C_8H_5Cl_2F_3$ MW 229.03

1,1,1-Trifluoro-2,2-dichloro-2-phenylethane, CA 309-10-4: Irritant; flammable. Antoine (liq) A = 7.45028, B = 2466.902, C = 0, dev = 1.0 to 5.0,
range = 365 to 446.

$C_8H_5Cl_2N$ MW 186.04

Phenylacetonitrile, *alpha*, *alpha*-dichloro, CA 40626-45-7: Irritant; flammable. Tb = 496.15 to 497.15; Antoine (liq) A = 7.06319, B = 2340.7, C =
-33.917, dev = 1.0, range = 329 to 497.

$C_8H_5Cl_5$ MW 278.39

Pentachloroethylbenzene, CA 606-07-5: Irritant; toxic; flammable. Antoine
(liq) A = 6.98091, B = 2739.5, C = -21.405, dev = 1.0, range = 369 to 572.

$C_8H_5F_3O$ MW 174.12

Acetophenone, 2,2,2-trifluoro, CA 434-45-7: Irritant; lachrymator; flammable, flash point 314; Tm = 233.15; Tb = 425.15; Vm = 0.1361 at 293.15;
Antoine (liq) A = 7.28578, B = 2250.274, C = 0, dev = 1.0, range = 342
to 425.

C_8H_5NO MW 131.13

Phenyl glyoxylonitrile, benzoyl cyanide, CA 613-90-1: Highly toxic;
moisture-sensitive; flammable, flash point 357; soluble ethanol, ether.
Tm = 305.15 to 306.15; Tb = 479.15 to 481.15; Antoine I (sol)   *(continues)*

C_8H_5NO *(continued)*

A = 10.9699, B = 4108, C = 0, dev = 1.0 to 5.0, range = 292 to 304; Antoine
II (liq) A = 7.1071, B = 2330, C = -24.615, dev = 1.0 to 5.0, range = 318
to 481.

$C_8H_5NO_2$ MW 147.13

Phthalimide, CA 85-41-6: Slightly soluble water; soluble ether, chloroform, hot ethanol. Tm = 511.15; Antoine (sol) A = 9.139, B = 4326, C = 0,
dev = 1.0 to 5.0, range = 378 to 418.

$C_8H_5N_3$ MW 143.15

Pyridinium dicyanomethylide, CA 27032-01-5: Antoine (sol) A = 12.080, B =
6549, C = 0, dev = 1.0, range = 403 to 440.

C_8H_6BrN MW 196.05

Phenylacetonitrile, *alpha*-bromo ±, CA 5798-79-8: Highly toxic; powerful
lachrymator; soluble most organic solvents; used as war gas. Tm = 302.15;
Tb = 515.15, decomposes; Vm = 0.1274 at 302.15; Antoine (liq) A = 8.10052,
B = 3025.17, C = -16.58, dev = 1.0 to 5.0, range = 293 to 515.

$C_8H_6ClNO_3$ MW 199.59

Benzeneacetyl chloride, 2-nitro, CA 22751-23-1: Irritant; toxic; flammable; soluble methanol, ether. Tm = 328.15; Antoine (sol) A = 13.3639,
B = 5413, C = 0, dev = 1.0, range = 296 to 327.

Benzeneacetyl chloride, 3-nitro, CA 99-47-8: Irritant; toxic; flammable;
soluble ether. Tm = 376.15; Antoine (sol) A = 13.2049, B = 5700, C = 0,
dev = 1.0, range = 299 to 343.

$C_8H_6Cl_2$ MW 173.04

2,3-Dichlorostyrene, CA 2123-28-6: Toxic; lachrymator; flammable; polymerizes. Vm = 0.1344 at 293.15; Antoine (liq) A = 7.13499, B = 2497.13, C =
-21.355, dev = 1.0, range = 334 to 508.

2,4-Dichlorostyrene, CA 2123-27-5: Toxic; lachrymator; flammable; polymerizes. Vm = 0.1392 at 298.15; Antoine (liq) A = 6.72763, B = 2127.2, C =
-47.716, dev = 1.0, range = 327 to 498.

2,5-Dichlorostyrene, CA 1123-84-8: Toxic; lachrymator; flammable; polymerizes. Vm = 0.1232 at 293.15; Antoine (liq) A = 6.88872, B = 2264.89, C =
-36.51, dev = 1.0, range = 328 to 500.

2,6-Dichlorostyrene, CA 28469-92-3: Toxic; lachrymator; flammable, flash
point 344; polymerizes. Tm = 361.15 to 363.15; Vm = 0.1370 at 293.15;
Antoine (liq) A = 7.29544, B = 2575.68, C = -3.519, dev = 1.0, range = 321
to 490.

3,4-Dichlorostyrene, CA 2039-83-0: Toxic; lachrymator; flammable; polymerizes. Vm = 0.1378 at 293.15; Antoine (liq) A = 7.10717, B = 2465.11, C =
-20.261, dev = 1.0, range = 330 to 503.   *(continues)*

$C_8H_6Cl_2$ *(continued)*

3,5-Dichlorostyrene, CA 2155-42-2: Toxic; lachrymator; flammable; polymerizes. Vm = 0.1413 at 298.15; Antoine (liq) A = 6.72763, B = 2127.2, C = -47.716, dev = 1.0, range = 326 to 498.

$C_8H_6Cl_2O$ MW 189.04

3-(Chloromethyl)benzoyl chloride, CA 63024-77-1: Corrosive; toxic; lachrymator; flammable. Antoine (liq) A = 7.06653, B = 2856.596, C = 0, dev = 1.0, range = 424 to 464.

4-(Chloromethyl)benzoyl chloride, CA 876-08-4: Corrosive; toxic; lachrymator; flammable, flash point 366. Tm = 306.15 to 308.15; Antoine (liq) A = 8.71529, B = 3570.318, C = 0, dev = 1.0 to 5.0, range = 440 to 466.

$C_8H_6Cl_4$ MW 243.95

2,3,4,6-Tetrachloro-1-ethylbenzene, CA 61911-56-6: Irritant; toxic; flammable. Antoine (liq) A = 7.25438, B = 2879.33, C = 5.225, dev = 1.0, range = 350 to 543.

3,4,5,6-Tetrachloro-1,2-dimethylbenzene, CA 877-08-7: Irritant; toxic; flammable. Antoine (liq) A = 7.22158, B = 2632.85, C = -41.998, dev = 1.0, range = 367 to 547.

$C_8H_6N_2O_2$ MW 162.15

3-Aminophthalimide, CA 2518-24-3: Soluble water, ethanol. Tm = 539.15 to 540.15; Antoine (sol) A = 10.990, B = 5655, C = 0, dev = 1.0 to 5.0, range = 386 to 459.

4-Aminophthalimide, CA 3676-85-5: Soluble water, ethanol. Tm = 567.15; Antoine (sol) A = 13.033, B = 7067, C = 0, dev = 1.0 to 5.0, range = 444 to 498.

C_8H_6O MW 118.13

Benzofuran, coumaron, CA 271-89-6: Flammable, flash point 329; polymerizes on standing; soluble ethanol, ether, benzene, carbon tetrachloride. Tm = 244.3; Tb = 447.15; Vm = 0.1082 at 298.15; Antoine (liq) A = 6.27117, B = 1645.2, C = -58.71, dev = 0.1 to 1.0, range = 323 to 403.

Phenol, 2-ethynyl, CA 5101-44-0: Flammable; polymerizes. Tm = 292.15; Antoine (liq) A = 3.816, B = 1175.6, C = 0, dev = 5.0, range = 300 to 373.

$C_8H_6O_2$ MW 134.13

Phenyl glyoxal, benzoyl formaldehyde, CA 1074-12-0: Irritant; flammable; moderately soluble water; soluble ethanol, ether, acetone, benzene. Antoine (liq) A = 7.44944, B = 2239.75, C = -55.213, dev = 1.0, range = 348 to 467.

Phthalide, CA 87-41-2: Flammable; soluble hot water, ethanol, ether. Tm = 348.15; Tb = 563.15; Vm = 0.1153 at 372.15; Antoine (liq) A = 7.07014, B = 2735.55, C = -22.994, dev = 1.0, range = 368 to 563.

$C_8H_6O_3$ MW 150.13

Piperonal, CA 120-57-0: Flammable, flash point above 385; soluble ethanol,
ether, acetone; used in perfumery and soft drinks. Tm = 310.15; Tb =
536.15; Antoine I (sol) A = 12.725, B = 4734, C = 0, dev = 1.0, range = 293
to 310; Antoine II (liq) A = 7.69393, B = 3016.04, C = -6.049, dev = 1.0,
range = 360 to 536.

$C_8H_6O_4$ MW 166.13

Isophthalic acid, CA 121-91-5: Irritant; slightly soluble water; soluble
ethanol, acetic acid. Tm = 621.15, sublimes; Antoine (sol) A = 10.072,
B = 4993, C = -43.15, dev = 5.0, range = 493 to 563.

Terephthalic acid, CA 100-21-0: Irritant; slightly soluble water; soluble
hot ethanol; important polymer material; sublimes without melting. Tm =
700.15; Tb = approximately 670, sublimes; Antoine (sol) A = 11.822, B =
6150, C = -43.15, dev = 5.0, range = 523 to 633.

C_8H_7Br MW 183.05

2-Bromostyrene, CA 2039-88-5: Irritant; flammable, flash point 359; polymerizes readily; soluble ethanol, ether, benzene. Tf = 220.35; Tb =
479.35; Vm = 0.1293 at 293.15; Antoine (liq) A = 6.03528, B = 1631.2, C =
-78.15, dev = 0.1, range = 378 to 543.

4-Bromostyrene, CA 2039-82-9: Irritant; flammable, flash point 348; polymerizes readily; soluble ethanol, ether, benzene. Tf = 265.45; Tb =
485.15; Vm = 0.1309 at 293.15; Antoine (liq) A = 6.1398, B = 1682.5, C =
-78.15, dev = 0.1, range = 383 to 543.

C_8H_7Cl MW 138.60

2-Chlorostyrene, CA 2039-87-4: Toxic; lachrymator; flammable, flash point
331; polymerizes readily; soluble most organic solvents. Tm = 210.0; Tb =
461.75; Vm = 0.1260 at 293.15; Antoine (liq) A = 5.99134, B = 1541.1, C =
-75.15, dev = 0.1, range = 363 to 523.

3-Chlorostyrene, CA 2039-85-2: Toxic; lachrymator; flammable, flash point
335; polymerizes readily; soluble most organic solvents. Vm = 0.1241 at
293.15; Antoine (liq) A = 6.83847, B = 2156.77, C = -16.882, dev = 1.0,
range = 298 to 463.

4-Chlorostyrene, CA 1073-67-2: Toxic; lachrymator; flammable, flash point
333; polymerizes readily; soluble most organic solvents. Tm = 257.25;
Tb = 465.15; Vm = 0.1275 at 293.15; Antoine (liq) A = 5.96738, B = 1545,
C = -75.15, dev = 0.1, range = 363 to 523.

C_8H_7ClO MW 154.60

2-Chloroacetophenone, phenacyl chloride, CA 532-27-4: Strongly irritant;
lachrymator; moisture-sensitive; flammable; soluble ethanol, ether, acetone, benzene; used as riot-control agent. Tm = 327.15; Tb = 520.15;
Antoine (sol) A = 12.904, B = 4740, C = 0, dev = 1.0 to 5.0, range = 278
to 323.

(continues)

C_8H_7ClO *(continued)*

4'-Chloroacetophenone, CA 99-91-2: Strongly irritant; flammable, flash point 363; soluble ethanol, ether. Tm = 293.15; Tb = 509.15; Vm = 0.1297 at 293.15; Antoine (liq) A = 6.30237, B = 1852.9, C = -79.15, dev = 1.0, range = 404 to 623.

Phenylacetyl chloride, CA 103-80-0: Corrosive; irritant; flammable, flash point 375; moisture-sensitive; soluble ether. Vm = 0.1323 at 293.15; Antoine (liq) A = 6.75779, B = 2020.19, C = -57.905, dev = 1.0, range = 321 to 483.

C_8H_7FO MW 138.14

2-Fluoroacetophenone, phenacyl fluoride, CA 450-95-3: Flammable; moisture-sensitive. Tm = 301.65 to 302.15; Antoine (liq) A = 8.8782, B = 3236.4, C = 0, dev = 1.0 to 5.0, range = 273 to 333.

C_8H_7N MW 117.15

Indole, CA 120-72-9: Moderately toxic; stench; flammable, flash point above 385; soluble hot water, most organic solvents; strong fecal odor; used in perfumery. Tm = 325.15 to 327.15; Tb = 527.15, decomposes; Antoine (sol) A = 10.3289, B = 3916, C = 0, dev = 1.0, range = 291 to 319.

Phenyl acetonitrile, benzyl cyanide, CA 140-29-4: Toxic by inhalation and skin absorption; eye irritant; flammable, flash point 384; soluble ethanol, ether, acetone. Tm = 249.35; Tb = 507.15; Vm = 0.1153 at 293.15; Antoine (liq) A = 6.98913, B = 2369.13, C = -31.329, dev = 1.0, range = 333 to 507.

2-Toluonitrile, CA 529-19-1: Irritant; flammable, flash point 357; soluble ethanol, ether. Tm = 259.15 to 260.15; Tb = 478.35; Vm = 0.1177 at 293.15; Antoine (liq) A = 6.56182, B = 1977.07, C = -44.337, dev = 1.0, range = 309 to 479.

4-Toluonitrile, CA 104-85-8: Irritant; flammable, flash point 358; soluble ethanol, ether. Tm = 302.15; Tb = 490.75; Vm = 1197 at 303.15; Antoine (liq) A = 7.02705, B = 2444.85, C = -4.194, dev = 1.0, range = 315 to 491.

2-Tolylisocyanide, CA 10468-64-1: Irritant; flammable. Antoine (liq) A = 6.7856, B = 2023, C = -33.348, dev = 1.0, range = 298 to 457.

C_8H_7NO MW 133.15

Benzyl isocyanate, CA 3173-56-6: Strongly irritant; toxic; lachrymator; moisture-sensitive; flammable, flash point 318. Vm = 0.1235 at 293.15; Antoine (liq) A = 6.2991, B = 2208.594, C = 0, dev = 1.0, range = 333 to 393.

$C_8H_7NO_3$ MW 165.15

2'-Nitroacetophenone, CA 577-59-3: Toxic; flammable, flash point above 385; soluble ethanol, ether. Tm = 301.15 to 302.15; Antoine (liq) A = 13.35022, B = 5409.87, C = 0, dev = 0.1, range = 293 to 333.

(continues)

$C_8H_7NO_3$ *(continued)*

3'-Nitroacetophenone, CA 121-89-1: Toxic; flammable; soluble ethanol, ether. Tm = 354.15; Tb = 475.15; Antoine (sol) A = 13.3718, B = 5746.6, C = 0, dev = 1.0, range = 293 to 343.

$C_8H_7NO_4$ MW 181.15

2-Nitrobenzoic acid, methyl ester, CA 606-27-9: Toxic; flammable, flash point above 385. Tm = 260.15; Tb = 548.15; Vm = 0.1409 at 293.15; Antoine (liq) A = 7.095, B = 2930, C = 0, dev = 5.0, range = 423 to 453.

(2-Nitrophenyl) acetate, CA 610-69-5: Toxic; flammable, flash point above 385; soluble ethanol, ether, acetone, benzene. Tm = 314.15; Tb = 526.15, decomposes; Antoine (liq) A = 8.92922, B = 3604.13, C = -5.824, dev = 1.0 to 5.0, range = 373 to 526.

C_8H_7NS MW 149.21

Benzyl isothiocyanate, benzyl mustard oil, CA 622-78-6: Lachrymator; moisture-sensitive; flammable, flash point above 385; soluble ether, ethanol. Tb = 516.15; Vm = 0.1327 at 289.15; Antoine (liq) A = 7.87754, B = 2944.06, C = -14.814, dev = 1.0, range = 352 to 516.

2-Methylbenzothiazole, CA 120-75-2: Flammable, flash point 375; soluble ethanol. Tm = 287.15; Tb = 511.15; Vm = 0.1268 at 292.15; Antoine (liq) A = 7.93713, B = 2831.45, C = -21.45, dev = 1.0 to 5.0, range = 343 to 499.

$C_8H_7N_3O_2$ MW 177.16

Phthalimide, 3,6-diamino, CA 1660-15-7: Antoine (sol) A = 8.8559, B = 5146.5, C = 0, dev = 1.0, range = 461 to 508.

$C_8H_7N_3O_6$ MW 241.16

Benzene, 2,4-dimethyl-1,3,5-trinitro, CA 632-92-8: Flammable; explosive; soluble benzene, chloroform, pyridine. Tm = 455.15; Antoine (sol) A = 14.73, B = 6779, C = 0, dev not specified, range = 318 to 412.

(2,2,2-Trinitroethyl)benzene, CA 38677-56-4: Flammable; explosive. Antoine (sol) A = 11.245, B = 4396.5, C = 0, dev = 1.0, range = 293 to 308.

$C_8H_7N_3O_7$ MW 257.16

2,4,6-Trinitrophenetole, CA 4732-14-3: Flammable; potentially explosive. Antoine I (sol) A = 13.519, B = 6292.2, C = 0, dev = 5.0, range = 342 to 351; Antoine II (liq) A = 8.353, B = 4124.5, C = 0, dev = 5.0, range = 352 to 364.

C_8H_8 MW 104.15

Cyclooctatetraene, CA 629-20-9: Flammable, flash point below 251; polymerizes; soluble most organic solvents. Tm = 268.45; Tb = 415.15 to 416.15; Vm = 0.1131 at 293.15; Antoine (liq) A = 6.19416, B = 1504.036, C = -54.616, dev = 0.1, range = 273 to 348.   *(continues)*

C_8H_8 *(continued)*

1,5,7-Octatriene-3-yne, CA 16607-77-5: Flammable; polymerizes; soluble most organic solvents. Antoine (liq) A = 6.2805, B = 1833.5, C = 0, dev = 1.0, range = 313 to 429.

Pentacyclooctane, cubane, CA 277-10-1: Flammable; decomposes above 473. Tm = 403.15 to 404.15; Antoine (sol) A = 13.12, B = 4190, C = 0, dev = 5.0, range = 239 to 262.

Styrene, CA 100-42-5: Irritant; lachrymator; flammable, flash point 304; light-sensitive; polymerizes readily; slightly soluble water; soluble most organic solvents; important polymer raw material. Tm = 242.55; Tb = 418.15; Vm = 0.1150 at 293.15; Tc = 648; Pc = 4000; Vc = 0.352; Antoine I (liq) A = 7.3945, B = 2221.3, C = 0, dev = 1.0, range = 245 to 334; Antoine II (liq) A = 6.08201, B = 1445.58, C = -63.72, dev = 0.1, range = 334 to 419.

$C_8H_8Br_2$ MW 263.96

(1,2-Dibromoethyl)benzene ±, CA 93-52-7: Corrosive; lachrymator; flammable; soluble most organic solvents. Tm = 347.15 to 347.65; Tb = 531.15 to 534.15; Antoine (liq) A = 7.38363, B = 2578.45, C = -47.645, dev = 1.0 to 5.0, range = 359 to 527.

$C_8H_8Cl_2$ MW 175.06

2,5-Dichloro-1,4-dimethylbenzene, CA 1124-05-6: Irritant; toxic; flammable; soluble benzene. Tm = 344.15; Tb = 497.15; Antoine (liq) A = 6.29217, B = 1797.4, C = -78.15, dev = 0.1, range = 393 to 573.

2,3-Dichloro-1-ethylbenzene, CA 54484-61-6: Irritant; toxic; flammable; soluble benzene, Tm = 232.35; Antoine (liq) A = 7.15654, B = 2549.62, C = -0.35, dev = 1.0, range = 319 to 495.

2,5-Dichloro-1-ethylbenzene, CA 54484-63-8: Irritant; toxic; flammable; soluble benzene. Tm = 211.95; Tb = 486.65; Vm = 0.1413 at 273.15; Antoine (liq) A = 7.34785, B = 2730.63, C = 21.535, dev = 1.0, range = 311 to 490.

3,4-Dichloro-1-ethylbenzene, CA 6623-59-2: Irritant; toxic; flammable; soluble benzene. Tm = 196.75; Antoine (liq) A = 7.13529, B = 2558.53, C = -1.227, dev = 1.0, range = 320 to 500.

$C_8H_8Cl_2O_2$ MW 207.06

2-(2,4-Dichlorophenoxy)ethanol, CA 120-67-2: Toxic; flammable; used as herbicide. Tm = 327.15 to 330.65; Antoine (liq) A = 6.40272, B = 2035.22, C = -122.779, dev = 1.0 to 5.0, range = 484 to 560.

$C_8H_8Cl_3O_3PS$ MW 321.54

O,O-Dimethyl-O-(2,4,5-trichlorophenyl) thiophosphate, CA 299-84-3: Toxic insecticide; flammable; soluble most organic solvents. Tm = 313.15 to 315.15; Antoine (liq) A = 5.722, B = 2967, C = 0, dev = 5.0, range = 298 to 373.

$C_8H_8N_2O_3$ MW 180.16

2'-Nitroacetanilide, CA 552-32-9: Irritant; flammable; soluble hot water,
ethanol, chloroform, benzene. Tm = 367.15; Antoine (liq) A = 5.855, B =
2300, C = 0, dev = 5.0, range = 473 to 593.

C_8H_8O MW 120.15

Acetophenone, methyl phenyl ketone, CA 98-86-2: Irritant; hypnotic; flammable, flash point 350; slightly soluble water; soluble ethanol, benzene,
ether, chloroform. Tm = 292.15 to 293.15; Tb = 474.85; Vm = 0.1168 at
293.15; Antoine (liq) A = 6.28228, B = 1723.46, C = -72.15, dev = 1.0,
range = 375 to 603.

Phenylacetaldehyde, CA 122-78-1: Flammable, flash point 344; polymerizes
on standing; slightly soluble water; soluble ethanol, ether. Tm = 306.15
to 307.15; Tb = 467.15; Antoine (liq) A = 8.263, B = 2846, C = 0, dev = 1.0
to 5.0, range = 283 to 333.

$C_8H_8O_2$ MW 136.15

1,4-Benzodioxan, CA 493-09-4: Flammable, flash point 360; soluble ethanol,
ether, benzene, chloroform. Tb = 485.15 to 487.15; Antoine (liq) A =
7.395, B = 2633.7, C = 0, dev = 5.0, range = liquid to 486.

Benzyl formate, CA 104-57-4: Irritant; flammable; soluble ethanol, ether,
acetone. Tb = 475.15 to 476.15; Vm = 0.1259 at 293.15; Antoine (liq) A =
7.675, B = 2697, C = 0, dev = 5.0, range = 298 to 357.

2,5-Dimethyl-1,4-benzoquinone, CA 137-18-8: Flammable; soluble ethanol,
ether, benzene. Tm = 398.15; Antoine (sol) A = 10.655, B = 4030, C = 0,
dev = 5.0, range = 273 to 293.

2'-Hydroxyacetophenone, CA 118-93-4: Irritant; flammable, flash point
above 385; slightly soluble water; soluble ethanol, ether, acetic acid.
Tm = 277.15 to 279.15; Tb = 491.15; Vm = 0.1204 at 293.15; Antoine (liq)
A = 5.78219, B = 1380.6, C = -125.11, dev = 1.0, range = 369 to 491.

4'-Hydroxyacetophenone, CA 99-93-4: Irritant; flammable; slightly soluble
water; soluble ethanol, ether. Tm = 382.15; Vm = 0.1228 at 382.15; Antoine
(sol) A = 11.32842, B = 4999.068, C = 0, dev = 1.0, range = 320 to 351.

4-Methoxybenzaldehyde, CA 123-11-5: Flammable; soluble ethanol, benzene,
ether. Tm = 273.15; Tb = 521.15; Vm = 0.1217 at 288.15; Antoine I (liq)
A = 8.182, B = 3153, C = 0, dev = 5.0, range = 283 to 323; Antoine II (liq)
A = 7.32783, B = 2671.13, C = -19.293, dev = 1.0, range = 346 to 521.

Methyl benzoate, CA 93-58-3: Moderately toxic; flammable, flash point 355;
soluble ethanol, ether. Tf = 260.85; Tb = 471.15 to 473.15; Vm = 0.1250 at
293.15; Antoine I (liq) A = 8.183, B = 2816.6, C = 0, dev = 1.0, range =
283 to 323; Antoine II (liq) A = 6.20322, B = 1656.25, C = -77.92, dev =
1.0, range = 373 to 533.

3-Methylbenzoic acid, meta-toluic acid, CA 99-04-7: Flammable; moderately
soluble hot water; soluble ethanol, ether. Tm = 384.15; Tb = 536.15; Vm =
0.1292 at 385.15; Antoine (liq) A = 8.1472, B = 3280.8, C = 0, dev = 1.0,
range = 473 to 533.

(continues)

$C_8H_8O_2$ *(continued)*

Phenyl acetate, CA 122-79-2: Flammable, flash point 349; slightly soluble water; soluble ethanol, ether, chloroform. Tb = 468.85; Vm = 0.1263 at 293.15; Antoine (liq) A = 7.19875, B = 2307.79, C = -24.719, dev = 1.0, range = 311 to 469.

Phenylacetic acid, CA 103-82-2: Highly toxic orally; flammable; soluble hot water, ethanol, ether, acetone. Tm = 350.15 to 351.65; Tb = 538.65; Vm = 0.1248 at 350.15; Antoine (liq) A = 8.00148, B = 3144.95, C = -14.408, dev = 1.0 to 5.0, range = 370 to 539.

$C_8H_8O_3$ MW 152.15

4-Hydroxybenzoic acid, methyl ester, CA 99-76-3: Allergen; flammable; moderately soluble water; soluble ethanol, ether, acetone. Tm = 404.15; Tb = 543.15 to 553.15, decomposes; Antoine (liq) A = 5.23662, B = 1159.34, C = -220.03, dev = 1.0, range = 446 to 517.

2-Hydroxy-3-methoxybenzaldehyde, *ortho*-vanillin, CA 148-53-8: Flammable, flash point above 385; soluble ethanol, ether, light hydrocarbons. Tm = 317.15 to 318.15; Tb = 538.15 to 539.15; Antoine (sol) A = 1.96587, B = 785.515, C = -138.009, dev = 1.0, range = 282 to 303.

2-Methylbenzoic acid, CA 579-75-9: Flammable; soluble ethanol, ether. Tm = 379.15; Antoine I (sol) A = 10.9955, B = 4746.1, C = 0, dev = 5.0, range = 353 to 369; Antoine II (sol) A = 12.4158, B = 5288.205, C = 0, dev = 1.0 to 5.0, range = 318 to 353.

3-Methoxybenzoic acid, CA 586-38-9: Flammable; soluble ethanol, ether. Tm = 383.65; Antoine (sol) A = 13.42611, B = 5611.721, C = 0, dev = 0.1 to 1.0, range = 317 to 336.

4-Methoxybenzoic acid, CA 100-09-4: Flammable; soluble ethanol, ether. Tm = 457.15; Tb = 548.15 to 553.15; Antoine (sol) A = 12.86054, B = 5738.274, C = 0, dev = 1.0, range = 334 to 356.

Methyl salicylate, CA 119-36-8: Irritant; moderately toxic; flammable, flash point 369; slightly soluble water; soluble methanol, ether, chloroform; used in perfumery and flavorings. Tm = 264.55; Tb = 496.45; Vm = 0.1291 at 298.15; Antoine I (liq) A = 5.91298, B = 1543.5, C = -101.547, dev = 1.0, range = 327 to 497; Antoine II (liq) A = 6.03559, B = 1620.399, C = -93.687, dev = 1.0, range = 329 to 496; Antoine III (liq) A = 8.19074, B = 2969.941, C = 0, dev = 1.0 to 5.0, range = 288 to 333.

Vanillin, 4-hydroxy-3-methoxybenzaldehyde, CA 121-33-5: Flammable; oxidizes in air; moisture-sensitive; light-sensitive; moderately soluble hot water; soluble ethanol, ether, chloroform. Tm = 354.15 to 356.15; Tb = 557.15, decomposes; Antoine (sol) A = 10.997, B = 4623, C = 0, dev = 5.0, range = 288 to 333; Antoine II (sol) A = 10.93562, B = 4535.023, C = 0, dev = 5.0, range = 297 to 328; Antoine III (liq) A = 7.91734, B = 3198.18, C = -17.047, dev = 1.0, range = 380 to 558.

$C_8H_8O_4$ MW 168.15

Dehydroacetic acid, CA 520-45-6: Irritant; highly toxic; flammable; slightly soluble water; soluble acetone, benzene; used as fungicide. Tm = 382.15; Tb = 543.15; Antoine (liq) A = 7.57554, B = 2933.36, C = -15.72, dev = 1.0 to 5.0, range = 364 to 542.

264

C_8H_9Br MW 185.06

1-Bromo-2,5-dimethylbenzene, CA 553-94-6: Irritant; flammable, flash point
352; soluble ethanol, benzene. Tm = 282.15; Tb = 471.15 to 472.15; Vm =
0.1363 at 291.15; Antoine (liq) A = 6.1701, B = 1699.71, C = -71.631, dev =
1.0, range = 310 to 480.

(2-Bromoethyl)benzene, CA 103-63-9: Irritant; flammable, flash point 362;
soluble ethanol, benzene. Tm = 217.25; Tb = 490.15 to 491.15, decomposes;
Vm = 0.1362 at 293.15; Antoine (liq) A = 7.06927, B = 2359.07, C = -22.96,
dev = 0.1, range = 348 to 401.

1-Bromo-2-ethylbenzene, CA 1973-22-4: Irritant; flammable, flash point
344; soluble ethanol, ether, acetone, benzene. Tm = 205.25; Tb = 472.45;
Vm = 0.1366 at 293.15; Antoine (liq) A = 6.0864, B = 1621.24, C = -75.15,
dev = 0.1, range = 368 to 523.

1-Bromo-4-ethylbenzene, CA 1585-07-5: Irritant; flammable, flash point
336; soluble ethanol, ether, acetone, benzene. Tm = 229.65; Tb = 473.15
to 476.15; Vm = 0.1386 at 298.15; Antoine (liq) A = 6.10699, B = 1632.6,
C = -80.15, dev = 0.1, range = 378 to 533.

C_8H_9Cl MW 140.61

(1-Chloroethyl)benzene ±, CA 38661-82-4: Irritant; lachrymator; flammable;
soluble ethanol, ether, benzene. Tb = 467.15 to 468.15, decomposes; Vm =
0.1324 at 293.15; Antoine (liq) A = 6.5437, B = 1933.8, C = -40.05, dev =
1.0, range = 342 to 378.

(2-Chloroethyl)benzene, CA 622-24-2: Irritant; lachrymator; flammable,
flash point 339; soluble ethanol, ether, acetone, benzene. Tb = 470.15 to
471.15; Vm = 0.1315 at 298.15; Antoine (liq) A = 2.496, B = 200.3, C =
-267.2, dev = 1.0 to 5.0, range = 356 to 380.

1-Chloro-2-ethylbenzene, CA 89-96-3: Irritant; lachrymator; flammable;
soluble ethanol, acetone, benzene, chloroform. Tm = 190.45; Tb = 452.15;
Vm = 0.1330 at 293.15; Antoine (liq) A = 6.10659, B = 1556.0, C = -72.15,
dev = 0.1, range = 353 to 503.

1-Chloro-3-ethylbenzene, CA 620-16-6: Irritant; lachrymator; flammable;
soluble ethanol, ether, acetone, benzene. Tf = 218.15; Tb = 456.95; Vm =
0.1335 at 293.15; Antoine (liq) A = 6.11572, B = 1577.3, C = -73.15, dev =
0.1, range = 358 to 508.

1-Chloro-4-ethylbenzene, CA 622-98-0: Irritant; lachrymator; flammable;
soluble ethanol, ether, benzene. Tm = 210.55; Tb = 457.65; Vm = 0.1345 at
293.15; Antoine (liq) A = 6.10799, B = 1577, C = -73.15; dev = 0.1, range
= 358 to 508.

1-(Chloromethyl)-4-methylbenzene, CA 104-82-5: Highly irritant; lachrymator; flammable, flash point 348; soluble ethanol, ether, benzene. Tm =
268.65; Tb = 473.15 to 475.15; Vm = 0.1338 at 293.15; Antoine (liq) A =
8.0843, B = 3357.3, C = 76.55, dev = 1.0, range = 376 to 457.

$C_8H_9ClNO_5PS$ MW 297.65

O,O-Dimethyl-O-(3-chloro-4-nitrophenyl) thiophosphate, chlorthion, CA 500-
28-7: Toxic insecticide; flammable; soluble ethanol, ether, benzene. Tm =
294.15; Antoine (liq) A = 10.2115, B = 4807.8, C = 0, dev = 1.0, range =
283 to 409.

C_8H_9ClO MW 156.61

1-Chloro-2-ethoxybenzene, CA 614-72-2: Irritant; toxic; flammable; soluble
ethanol, ether, benzene. Tb = 483.15; Vm = 0.1387 at 288.15; Antoine (liq)
A = 7.23273, B = 2408.53, C = -20.635, dev = 1.0 to 5.0, range = 318 to
481.

4-Chlorophenethyl alcohol, CA 1875-88-3: Irritant; flammable, flash point
above 385. Tb = 532.15; Vm = 0.1327 at 293.15; Antoine (liq) A = 6.54533,
B = 2039, C = -83.15; dev = 1.0, range = 426 to 673.

4-Chlorophenetole, 4-chloro-1-ethoxybenzene, CA 622-61-7: Irritant; flammable; soluble ethanol, ether, benzene. Tm = 294.15; Tb = 485.15 to 487.15;
Antoine (liq) A = 6.6896, B = 2070, C = -43.15; dev = 1.0, range = 395 to
485.

$C_8H_9ClO_2$ MW 172.61

Ethylene glycol, 4-chlorophenyl ether, CA 7477-64-7: Irritant; forms explosive peroxides; flammable. Tm = 301.15; Antoine (liq) A = 6.3106, B =
1889.56, C = -116.02, dev = 1.0, range = 410 to 554.

$C_8H_9Cl_3O_4$ MW 275.52

2-Acetyl-4,4,4-trichloro-3-oxobutyric acid, ethyl ester: Irritant; flammable. Antoine (liq) A = 7.0255, B = 2776.298, C = 0, dev = 1.0 to 5.0,
range = 374 to 409.

C_8H_9N MW 119.17

2-Methyl-5-vinylpyridine, CA 140-76-1: Toxic; flammable; polymerizes.
Antoine (liq) A = 5.28099, B = 1022.917, C = -143.974, dev = 1.0, range =
342 to 457.

C_8H_9NO MW 135.17

Acetamide, 2-phenyl, CA 103-81-1: Flammable; slightly soluble water; soluble ethanol, ether. Tm = 430.15 to 431.15; Antoine (sol) A = 11.10735,
B - 5036.675, C = 0, dev = 1.0, range - 329 to 352.

Acetanilide, CA 103-84-4: Irritant; flammable, flash point 446; moderately
soluble hot water; soluble ethanol, acetone, chloroform. Tm = 387.45; Tb =
577.15; Antoine I (sol) A = 9.3689, B = 4208.1, C = 0, dev = 1.0 to 5.0,
range = 303 to 324; Antoine II (sol) A = 10.36243, B = 4556.937, C = 0,
dev = 1.0, range = 317 to 336; Antoine III (liq) A = 7.22624, B = 2769.31,
C = -46.481, dev = 1.0, range = 473 to 577.

Acetophenone, 4'-amino, CA 99-92-3: Flammable; soluble hot water, ethanol,
ether. Tm = 379.15; Tb = 566.15 to 568.15; Antoine (sol) A = 11.04842, B =
4840.361, C = 0, dev = 1.0, range = 314 to 338.

N-Methylbenzamide, CA 613-93-4: Flammable; soluble ethanol, acetone. Tm =
355.15; Tb = 564.15; Antoine I (sol) A = 8.7679, B = 3917.2, C = 0, dev =
1.0, range = 297 to 321; Antoine II (sol) A = 10.50734, B = 4477.763,
C = 0, dev = 1.0, range = 307 to 329.

266

$C_8H_9NO_2$ MW 151.16

Anthranilic acid, methyl ester, CA 134-20-3: Flammable, flash point 377;
light-sensitive; slightly soluble water; soluble ethanol, ether; used in
essential oils. Tm = 297.15 to 298.15; Tb = 529.15; Antoine I (sol) A =
11.113, B = 4094, C = 0, dev = 1.0 to 5.0, range = 287 to 298; Antoine II
(liq) A = 8.285, B = 3252, C = 0, dev = 5.0, range = 299 to 520.

2-Nitro-1,3-dimethylbenzene, CA 81-20-9: Highly toxic; irritant; flammable, flash point 360; soluble ethanol, ether, benzene. Tm = 286.15; Tb =
498.15; Vm = 0.1359 at 288.15; Antoine (liq) A = 7.23058, B = 2595.182,
C = 0, dev = 1.0 to 5.0, range = 373 to 498.

4-Nitro-1,3-dimethylbenzene, CA 89-87-2: Highly toxic; irritant; flammable, flash point 380; soluble ethanol, ether, acetone, benzene. Tm =
275.15; Tb = 519.15; Vm = 0.1332 at 288.15; Antoine (liq) A = 6.25896, B =
1864.16, C = -79.026, dev = 1.0, range = 368 to 518.

2-Nitroethylbenzene, CA 612-22-6: Irritant; toxic; flammable; soluble acetone, ethanol, ether, benzene. Tf = 250.15; Tb = 500.15 to 501.15; Vm =
0.1342 at 297.65; Antoine (liq) A = 7.0207, B = 2289, C = -43.15, dev =
1.0, range = 353 to 433.

4-Nitroethylbenzene, CA 100-12-9: Irritant; toxic; flammable; soluble acetone, ethanol, ether, benzene. Tm = 260.85; Tb = 518.15 to 519.15; Vm =
0.1345 at 298.15; Antoine (liq) A = 7.0772, B = 2417, C = -43.15, dev =
1.0, range = 353 to 433.

C_8H_{10} MW 106.17

1,2-Dimethylbenzene, *ortho*-xylene, CA 95-47-6: Irritant; moderately toxic
by inhalation; suspected carcinogen; flammable, flash point 305; soluble
most organic solvents; used to manufacture phthalic anhydride. Tm =
247.97; Tb = 417.56; Vm = 0.1206 at 298.15; Tc = 630.3; Pc = 3730; Vc =
0.369; Antoine I (liq) A = 6.13132, B = 1480.155; C = -58.804, dev = 0.02,
range = 333 to 419; Antoine II (liq) A = 6.15921, B = 1502.949, C =
-55.725, dev = 0.02, range = 416 to 473; Antoine III (liq) A = 6.46119, B =
1772.963, C = -18.84, dev = 0.1, range = 471 to 571; Antoine IV (liq) A =
7.91427, B = 3735.582, C = 229.953, dev = 0.1, range = 567 to 630.

1,3-Dimethylbenzene, *meta*-xylene, CA 108-38-3: Irritant; moderately toxic
by inhalation; flammable, flash point 300; soluble most organic solvents.
Tm = 225.28; Tb = 412.27; Vm = 0.1229 at 298.15; Tc = 617.05; Pc = 3535;
Vc = 0.376; Antoine I (liq) A = 6.14083, B = 1467.244, C = -57.442, dev =
0.02, range = 331 to 415; Antoine II (liq) A = 5.76037, B = 1292.224, C =
-72.052, dev = 0.1, range = 267 to 301; Antoine III (liq) A = 6.17035, B =
1490.184, C = -54.448, dev = 0.02, range = 412 to 462; Antoine IV (liq) A =
6.42535, B = 1710.901, C = -24.591, dev = 0.1, range = 461 to 554; Antoine
V (liq) A = 7.59221, B = 3163.74, C = 165.278, dev = 0.1, range = 550 to
617.

1,4-Dimethylbenzene, *para*-xylene, CA 106-42-3: Irritant; moderately toxic
by inhalation; flammable, flash point 300; soluble most organic solvents;
used to manufacture terephthalic acid. Tm = 286.41; Tb = 411.52; Vm =
0.1233 at 298.15; Tc = 616.2; Pc = 3511; Vc = 0.379; Antoine I (sol) A =
15.50091, B = 6327.014, C = 115.724, dev = 0.1, range = 247 to 286; Antoine
II (liq) A = 6.14779, B = 1475.767, C = -55.241, dev = 0.1, range = 286 to
453; Antoine III (liq) A = 6.14049, B = 1472.733, C = -55.342, dev = 0.1,
range = 411 to 463; Antoine IV (liq) A = 6.44333, B = 1735.196, *(continues)*

C_8H_{10} *(continued)*

C = -19.846, dev = 0.1, range = 460 to 553; Antoine V (liq) A = 7.84182, B = 3543.356, C = 208.522, dev = 0.1, range = 551 to 616.

Ethylbenzene, CA 100-41-4: Irritant; moderately toxic; flammable, flash point 294; soluble most organic solvents; used to manufacture styrene. Tm = 178.17; Tb = 409.34; Vm = 0.1224 at 298.15; Tc = 616.2; Pc = 3701; Vc = 0.374; Antoine I (liq) A = 6.06991, B = 1416.922, C = -60.716, dev = 0.02, range = 298 to 420; Antoine II (liq) A = 6.10898, B = 1445.262, C = -57.128, dev = 0.02, range = 409 to 459; Antoine III (liq) A = 6.36656, B = 1665.991, C = -26.716, dev = 0.1, range = 457 to 554; Antoine IV (liq) A = 7.49119, B = 3056.747, C = 159.496, dev = 0.1, range = 549 to 617.

$C_8H_{10}F_3NO_3$ MW 225.17

N-Trifluoroacetyl-*levo*-proline, methyl ester, CA 61274-28-0: Flammable. Tm = 305.25; Antoine (liq) A = 7.675, B = 3026, C = 0, dev = 5.0, range = 303 to 523.

$C_8H_{10}F_3NO_5$ MW 257.17

N-Trifluoroacetyl-*levo*-aspartic acid, dimethyl ester, CA 688-09-5: Flammable. Tm = 306.45; Antoine (liq) A = 7.505, B = 3040, C = 0, dev = 5.0, range = 303 to 423.

$C_8H_{10}NO_5PS$ MW 263.20

O,O-Dimethyl-*O*-(4-nitrophenyl) thiophosphate, methylparathion, CA 298-00-0: Toxic insecticide; flammable; soluble most organic solvents. Tm = 309.15; Antoine (liq) A = 9.999, B = 4646.3, C = 0, dev = 1.0, range = 293 to 427.

$C_8H_{10}N_2O_2$ MW 166.18

3-Nitro-*N,N*-dimethylaniline, CA 619-31-8: Irritant; highly toxic; flammable; soluble ethanol, ether. Tm = 333.15 to 334.15; Tb = 553.15 to 558.15; Antoine (liq) A = 6.895, B = 2730, C = 0, dev = 5.0, range = 427 to 558.

4-Nitro-*N,N*-dimethylaniline, CA 100-23-2: Irritant; highly toxic; flammable; soluble ethanol, ether, acetic acid. Tm = 436.15 to 439.15; Antoine (sol) A = 11.210, B = 5163, C = 0, dev = 1.0, range = 344 to 366.

$C_8H_{10}O$ MW 122.17

2,3-Dimethylphenol, CA 526-75-0: Highly toxic; corrosive; flammable; soluble ethanol, ether. Tm = 348.15; Tb = 491.15; Antoine I (sol) A = 12.29616, B = 4394.694, C = 0, dev = 1.0, range = 282 to 323; Antoine II (liq) A = 6.17998, B = 1619.086, C = -102.197, dev = 0.02 to 0.1, range = 433 to 492.

2,4-Dimethylphenol, CA 105-67-9: Highly toxic; corrosive; suspected carcinogen; flammable, flash point above 385; soluble ethanol, ether. Tm = 300.15 to 301.15; Tb = 483.15; Antoine I (liq) A = 9.65613, B = 3442.574, C = 0, dev = 0.02 to 0.1, range = 282 to 318; Antoine II (liq) A = 6.1672, B = 1578.685, C = -104.772, dev = 1.0, range = 429 to 486.   *(continues)*

268

$C_8H_{10}O$ *(continued)*

2,5-Dimethylphenol, CA 95-87-4: Highly toxic; corrosive; flammable; soluble ethanol, ether. Tm = 344.15 to 346.15; Tb = 484.65; Antoine I (sol) A = 12.51064, B = 4445.233, C = 0, dev = 1.0, range = 282 to 323; Antoine II (liq) A = 6.17702, B = 1593.804, C = -102.241, dev = 0.02 to 0.1, range = 427 to 485.

2,6-Dimethylphenol, CA 576-26-1: Highly toxic; corrosive; suspected carcinogen; flammable, flash point 346; soluble ethanol, ether. Tm = 322.15; Tb = 476.15; Antoine I (sol) A = 11.6308, B = 3950.681, C = 0, dev = 1.0, range = 277 to 313; Antoine II (liq) A = 6.19644, B = 1629.621, C = -85.358, dev = 0.02 to 0.1, range = 417 to 476.

3,4-Dimethylphenol, CA 95-65-8: Highly toxic; corrosive; suspected carcinogen; flammable; soluble ethanol, ether. Tm = 335.15 to 337.15; Tb = 498.15; Antoine I (sol) A = 12.31521, B = 4485.592, C = 0, dev = 1.0, range = 282 to 323; Antoine II (liq) A = 6.20617, B = 1623.592, C = -113.623, dev = 0.02 to 0.1, range = 444 to 502.

3,5-Dimethylphenol, CA 108-68-9: Highly toxic; corrosive; suspected carcinogen; flammable; soluble ethanol, ether. Tm = 334.15; Tb = 492.65; Antoine I (sol) A = 11.97153, B = 4336.025, C = 0, dev = 1.0, range = 282 to 323; Antoine II (liq) A = 6.25292, B = 1638.564, C = -109.095, dev = 0.1, range = 427 to 497.

Ethoxybenzene, phenetole, CA 103-73-1: Flammable, flash point 330; soluble ethanol, ether. Tm = 240.15; Tb = 445.15; Vm = 0.1264 at 293.15; Antoine (liq) A = 6.14524, B = 1508.326, C = -78.613, dev = 0.1, range = 390 to 454.

2-Ethylphenol, CA 90-00-6: Irritant; toxic; flammable, flash point 351; soluble ethanol, ether, acetone, benzene. Tm = 269.75, stable solid form; Tb = 479.65 to 480.65; Vm = 0.1178 at 273.15; Antoine I (liq) A = 6.13344, B = 1550.409, C = -102.103, dev = 0.02 to 0.1, range = 423 to 491; Antoine II (liq) A = 9.44878, B = 3318.181, C = 0, dev = 1.0, range = 277 to 318.

3-Ethylphenol, CA 620-17-7: Irritant; toxic; flammable, flash point 367; soluble ethanol, ether, benzene. Tm = 269.15; Tb = 487.15; Vm = 0.1188 at 293.15; Antoine I (liq) A = 6.16692, B = 1573.226, C = -113.548, dev = 0.02 to 0.1, range = 445 to 503; Antoine II (liq) A = 9.76773, B = 3558.911, C = 0, dev = 1.0, range = 277 to 323.

4-Ethylphenol, CA 123-07-9: Irritant; toxic; flammable, flash point 373; soluble ethanol, ether, acetone, benzene. Tm = 320.15 to 321.15; Tb = 491.65 to 492.65; Antoine I (sol) A = 11.74364, B = 4188.624, C = 0, dev = 1.0, range = 278 to 317; Antoine II (liq) A = 6.13939, B = 1550.479, C = -116.1, dev = 0.02 to 0.1, range = 444 to 503.

4-Methyl benzyl alcohol, CA 589-18-4: Irritant; flammable; soluble ethanol, ether. Tm = 334.15 to 335.15; Tb = 490.15; Antoine (liq) A = 9.033, B = 3352, C = 0, dev = 1.0 to 5.0, range = 338 to 376.

1-Phenylethanol ±, CA 13323-81-4: Irritant; moderately toxic by skin absorption; flammable, flash point 358; soluble ethanol, ether. Tm = 293.15; Tb = 476.75; Vm = 0.1205 at 293.15; Antoine (liq) A = 7.02762, B = 2177, C = -43.15, dev = 1.0, range = 353 to 480.

2-Phenylethanol, CA 60-12-8: Irritant; moderately toxic; flammable, flash point 369; moderately soluble water; soluble ethanol, ether; used in perfumery. Tf = 246.15; Tb = 494.15 to 495.15; Vm = 0.1197 at *(continues)*

$C_8H_{10}O$ *(continued)*

298.15; Antoine I (liq) A = 6.59416, B = 1905.1, C = -76.15, dev = 1.0, range = 394 to 613; Antoine II (liq) A = 9.836, B = 3573, C = 0, dev = 5.0, range = 283 to 318.

$C_8H_{10}O_2$ MW 138.17

Benzyl alcohol, 4-methoxy, anisyl alcohol, CA 105-13-5: Moderately toxic; flammable, flash point above 385; soluble water, ethanol, ether. Tm = 298.15; Tb = 532.25; Antoine (liq) A = 12.341, B = 4995, C = 0, dev = 5.0, range = 394 to 424.

1,3-Dihydroxy-2,5-dimethylbenzene, CA 488-87-9: Irritant; flammable; soluble water, ethanol, ether. Tm = 436.15; Tb = 550.15 to 553.15; Antoine (liq) A = 6.1309, B = 1789.1, C = -131.45, dev = 1.0, range = 393 to 459.

1,3-Dihydroxy-4,5-dimethylbenzene, CA 527-55-9: Irritant; flammable; soluble water, ethanol, ether, benzene. Tm = 408.15 to 410.15; Antoine (liq) A = 4.3014, B = 1005.9, C = -203.93, dev = 1.0, range = 424 to 453.

1,4-Dihydroxy-2,5-dimethylbenzene, CA 615-90-7: Irritant; flammable; soluble water, ethanol, ether. Tm = 490.15; Antoine (sol) A = 11.485, B = 5280, C = 0, dev = 5.0, range = 331 to 361.

1,3-Dihydroxy-5-ethylbenzene, CA 4299-72-3: Irritant; flammable; soluble ethanol, ether. Tm = 371.15; Antoine (liq) A = 6.2492, B = 1852.4, C = -143.42, dev = 1.0, range = 408 to 479.

1,2-Dimethoxybenzene, veratrole, CA 91-16-7: Irritant; flammable, flash point 360; soluble ethanol, ether, benzene. Tm = 295.15 to 296.15; Tb = 479.85; Vm = 0.1271 at 298.15; Antoine (liq) A = 8.705, B = 3492, C = 0, dev = 5.0, range not specified.

1,3-Dimethoxybenzene, CA 151-10-0: Irritant; flammable, flash point 360; soluble ethanol, ether, benzene. Tm = 221.15; Tb = 489.65 to 490.85; Vm = 0.1306 at 298.15; Antoine (liq) A = 4.5012, B = 769.2, C = -188.95, dev = 1.0 to 5.0, range = 358 to 500.

1,4-Dimethoxybenzene, CA 150-78-7: Irritant; flammable; soluble ethanol, ether, benzene. Tm = 331.15 to 333.15; Tb = 485.75; Antoine (liq) A = 9.2954, B = 3244, C = 0, dev = 5.0, range = 298 to 357.

Ethylene glycol, monophenyl ether, 2-phenoxyethanol, CA 122-99-6: Irritant; flammable, flash point 394; very soluble water; soluble ethanol, ether. Tm = 287.15; Tb = 518.05; Vm = 0.1254 at 293.15; Antoine (liq) A = 6.69552, B = 2033.96, C = -84.761, dev = 1.0, range = 351 to 519.

3-Methoxy-4-hydroxytoluene, CA 93-51-6: Irritant; flammable; slightly soluble water; soluble ethanol, ether, benzene, chloroform. Tm = 278.65; Tb = 494.15; Vm = 0.1258 at 293.15; Antoine (liq) A = 7.6429, B = 2780.3, C = 0, dev = 5.0, range = 356 to 495.

$C_8H_{10}O_2S$ MW 170.23

Benzyl methyl sulfone, CA 3112-90-1: Flammable; soluble water. Tm = 400.15; Antoine (liq) A = 7.585, B = 3392, C = 0, dev = 1.0 to 5.0, range = 455 to 529.

$C_8H_{10}O_6$ MW 202.16

Butanedioic acid, dioxo, diethyl ester, CA 59743-08-7: Flammable. Tb = 506.15 to 507.15, decomposes; Antoine (liq) A = 7.56539, B = 2678.15, C = -25.033, dev = 1.0, range = 343 to 507.

$C_8H_{10}S$ MW 138.23

Benzyl methyl sulfide, CA 766-92-7: Toxic; stench; flammable, flash point 346; liquid with powerful horseradish odor. Tb = 479.15 to 483.15; Antoine (liq) A = 7.53546, B = 2652.709, C = 0, dev = 1.0, range = 336 to 368.

Ethyl phenyl sulfide, CA 622-38-8: Toxic; stench; flammable; soluble ethanol. Tb = 477.15; Vm = 0.1354 at 293.15; Antoine (liq) A = 7.77379, B = 2699.151, C = 0, dev = 1.0 to 5.0, range = 338 to 477.

$C_8H_{11}F_3O_2$ MW 196.17

Trifluoroacetic acid, cyclohexyl ester, CA 1549-45-7: Irritant; flammable. Antoine (liq) A = 6.8594, B = 1905.8, C = -28.35, dev = 1.0, range = 345 to 420.

$C_8H_{11}N$ MW 121.18

N,N-Dimethylaniline, CA 121-69-7: Highly toxic by inhalation and skin absorption; flammable, flash point 336; soluble ethanol, ether, acetone, benzene, chloroform. Tm = 275.15; Tb = 465.15 to 467.15; Vm = 0.1268 at 293.15; Antoine I (liq) A = 7.07329, B = 2301.63, C = -12.001, dev = 1.0, range = 302 to 467; Antoine II (liq) A = 6.55663, B = 1864.075, C = -55.854, dev = 1.0, range = 363 to 418.

2,4-Dimethylaniline, CA 95-68-1: Highly toxic by inhalation and skin absorption; flammable, flash point 363; soluble ethanol, ether, benzene. Tm = 258.85; Tb = 488.15 to 489.15; Vm = 0.1246 at 293.15; Antoine (liq) A = 6.85218, B = 2037.51, C = -64.116, dev = 1.0, range = 383 to 485.

2,6-Dimethylaniline, CA 87-62-7: Highly toxic by inhalation and skin absorption; flammable, flash point 364; soluble ethanol, ether. Tm = 284.15; Tb = 487.15; Vm = 0.1231 at 293.15; Antoine (liq) A = 7.04335, B = 2438.77, C = -7.329, dev = 1.0, range = 373 to 490.

N-Ethylaniline, CA 103-69-5: Irritant; highly toxic by inhalation and skin absorption; flammable, flash point 358; soluble ethanol, ether, acetone, benzene. Tm = 209.65; Tb = 477.85; Vm = 0.1261 at 298.15; Antoine (liq) A = 6.95175, B = 2194.84, C = -33.495, dev = 1.0, range = 311 to 477.

4-Ethylaniline, CA 589-16-2: Irritant; highly toxic by inhalation and skin absorption; flammable, flash point 358; soluble ethanol, ether. Tm = 268.15; Tb = 486.15 to 487.15; Vm = 0.1252 at 293.15; Antoine (liq) A = 6.04815, B = 1578.65, C = -100.0, dev = 1.0, range = 393 to 491.

5-Ethyl-2-methylpyridine, CA 104-90-5: Corrosive; moderately toxic; hygroscopic; flammable, flash point 339; moderately soluble water; soluble ethanol, ether, benzene. Tm = 202.25; Tb = 451.45; Vm = 0.1319 at 296.15; Antoine (liq) A = 6.86137, B = 2068.33, C = -23.81, dev = 1.0, range = 348 to 451.

(continues)

$C_8H_{11}N$ *(continued)*

alpha-Methyl benzylamine ±, CA 618-36-0: Corrosive; toxic; flammable, flash point 352; strong base; very soluble water; soluble ethanol, ether. Tb = 458.15; Vm = 0.1290 at 288.15; Antoine (liq) A = 6.59008, B = 1917.31, C = 0, dev not specified, range = 292 to 318.

4-Methyl benzylamine, CA 104-84-7: Irritant; toxic; flammable, flash point 348; very soluble water; soluble ethanol, ether. Tm = 285.75 to 286.35; Tb = 477.15; Vm = 0.1273 at 293.15; Antoine (liq) A = 8.1142, B = 2844, C = 0, dev = 5.0, range = 353 to 466.

2-Propylpyridine, CA 622-39-9: Irritant; flammable, flash point 329; soluble ethanol, ether, acetone. Tm = 275.15; Tb = 443.15; Vm = 0.1325 at 298.15; Antoine (liq) A = 5.47395, B = 1172.22, C = -107.9, dev = 1.0, range = 338 to 445.

3-Propylpyridine, CA 4673-31-8: Irritant; flammable; soluble ethanol, ether. Tb = 443.15; Antoine (liq) A = 9.8173, B = 4256, C = 101.41, dev = 1.0, range = 350 to 450.

4-Propylpyridine, CA 1122-81-2: Irritant; flammable; soluble ethanol, ether. Tb = 457.15 to 459.15; Vm = 0.1292 at 288.15; Antoine (liq) A = 5.4121, B = 1195.9, C = -113.5, dev = 1.0 to 5.0, range = 354 to 465.

$C_8H_{11}NO$ MW 137.18

2-Anilinoethanol, 2-phenylamino ethanol, CA 122-98-5: Irritant; highly toxic; flammable, flash point above 385; slightly soluble water; soluble ethanol, ether, chloroform. Tb = 559.15; Vm = 0.1253 at 293.15; Antoine (liq) A = 7.22165, B = 2544.81, C = -64.746, dev = 1.0 to 5.0, range = 377 to 553.

2-Ethoxyaniline, *ortho*-phenetidine, CA 94-70-2: Highly toxic and irritant; flammable; soluble ethanol, ether. Tm = below 252.15; Tb = 505.65; Antoine (liq) A = 7.92085, B = 2952.11, C = -2.626, dev = 1.0, range = 373 to 458.

4-Ethoxyaniline, *para*-phenetidine, CA 156-43-4: Highly toxic and irritant; flammable, flash point 388; light-sensitive; soluble ethanol, ether. Tm = 276.15 to 277.15; Tb = 527.15 to 528.15; Vm = 0.1288 at 289.15; Antoine (liq) A = 6.29024, B = 1750.62, C = -113.15; dev = 1.0 to 5.0, range = 421 to 523.

C_8H_{12} MW 108.18

Cyclooctadiene, mixed isomers: Flammable; polymerizes; soluble most organic solvents. Antoine (liq) A = 6.424, B = 1809, C = 0, dev = above 5.0, range = 290 to 474.

1,5-Cyclooctadiene ±, CA 10092-71-4: Flammable, flash point 318; polymerizes rapidly; soluble most organic solvents. Tm = 211.15; Tb = 423.95; Vm = 0.1227 at 298.15; Tc = 648.15, calculated; Pc = 3825, calculated; Vc = 0.365, calculated; Antoine (liq) A = 7.05145, B = 2138.134, C = 0, dev = 1.0, range = 348 to 386.

1,2-Divinylcyclobutane ±*trans*, CA 6553-48-6: Flammable; soluble most organic solvents. Tb = 385.15 to 386.15; Vm = 0.1379 at 293.15; Antoine (liq) A = 7.3369, B = 2040, C = 0, dev not specified, range = 350 to 385.

(continues)

C_8H_{12} *(continued)*

4-Vinyl-1-cyclohexene ±, CA 100-40-3: Irritant; flammable, flash point 289; polymerizes; soluble most organic solvents. Tf = 164.25; Tb = 402.65 to 403.65; Vm = 0.1304 at 293.15; Antoine I (liq) A = 6.01674, B = 1382.743, C = -57.472, dev = 1.0 to 5.0, range = 292 to 405; Antoine II (liq) A = 6.02159, B = 1385.61, C = -57.171, dev = 0.1, range not specified.

$C_8H_{12}Cl_2O_5$ MW 259.09

Diethylene glycol, bis(chloroacetate): Irritant; flammable. Antoine (liq) A = 9.3505, B = 4161.84, C = -19.363, dev = 1.0, range = 421 to 586.

$C_8H_{12}N_2$ MW 136.20

Suberic acid, dinitrile, CA 629-40-3: Irritant; flammable, flash point above 385. Tm = 269.65; Vm = 0.1428 at 293.15; Antoine (liq) A = 9.3935, B = 4036, C = 0, dev = 1.0, range = 303 to 339.

$C_8H_{12}O_4$ MW 172.18

Diethyl fumarate, CA 623-91-6: Flammable, flash point 364; soluble acetone, chloroform. Tm = 274.15 to 275.15; Tb = 486.15 to 488.15; Vm = 0.1644 at 298.15; Antoine (liq) A = 7.42655, B = 2609.73, C = -10.518, dev = 1.0 to 5.0, range = 326 to 492.

Diethyl maleate, CA 141-05-9: Flammable, flash point 366; soluble ethanol, ether. Tm = 264.35; Tb = 496.15; Vm = 0.1615 at 293.15; Antoine (liq) A = 7.11957, B = 2392.18, C = -30.693, dev = 1.0 to 5.0, range = 330 to 498.

C_8H_{14} MW 110.2

Bicyclo(2,2,2)-octane, CA 280-33-1: Flammable; soluble most organic solvents. Tm = 441.15 to 444.15; Antoine (sol) A = 7.753, B = 2416.4, C = 0, dev = 1.0, range = 323 to 363.

Bicyclo(3,3,0)-octane, *cis*: Flammable; soluble most organic solvents. Tm = 223.82; Tb = 409.15 to 409.65; Vm = 0.1276 at 298.15; Antoine (liq) A = 9.6545, B = 2867.6, C = 0, dev = 5.0, range = 298 to 318.

Bicyclo(3,3,0)-octane, *trans*: Flammable; soluble most organic solvents. Tm = 244.15; Tb = 405.15; Vm = 0.1278 at 291.15; Antoine (liq) A = 8.744, B = 2593, C = 0, dev = 5.0, range = 298 to 320.

Bicyclo(4,2,0)-octane, *cis*, CA 28282-35-1: Flammable; soluble most organic solvents. Tb = 409.15; Vm = 0.1285 at 293.15; Antoine (liq) A = 8.913, B = 2671.3, C = 0, dev = 5.0, range = 298 to 347.

Bicyclo(5,1,0)-octane, *cis*, CA 16526-90-2: Flammable; soluble most organic solvents. Antoine (liq) A = 7.664, B = 2277.9, C = 0, dev = 1.0, range = 297 to 322.

Cyclooctene, *cis*, CA 931-87-3: Flammable, flash point 298; soluble most organic solvents. Tm = 257.65 to 258.65; Tb = 411.15; Vm = 0.1304 at 298.15; Antoine (liq) A = 7.3641, B = 2194.3, C = 0, dev = 1.0, range = 273 to 411.

(continues)

C_8H_{14} *(continued)*

2,5-Dimethyl-1,5-hexadiene, CA 627-58-7: Flammable, flash point 286; polymerizes; soluble most organic solvents. Tf = 197.55; Tb = 389.15; Vm =
0.1398 at 298.15; Antoine (liq) A = 4.65684, B = 624.521, C = -153.168,
dev = 1.0, range = 330 to 388.

3,3-Dimethyl-1,5-hexadiene, CA 24253-25-6: Flammable; polymerizes; soluble
most organic solvents. Tb = 374.75; Antoine (liq) A = 6.960, B = 1840,
C = 0, dev = 5.0, range = 293 to 371.

1-Ethylcyclohexene, CA 1453-24-3: Flammable; soluble most organic solvents. Tf = 163.25; Tb = 410.15; Vm = 0.1348 at 298.15; Antoine (liq) A =
6.01221, B = 1381.846, C = -65.259, dev = 0.02, range = 353 to 412.

1-Methylbicyclo(4,1,0)-heptane, CA 2439-79-4: Flammable; soluble most organic solvents. Antoine (liq) A = 5.98593, B = 1355.31, C = -58.286, dev =
0.02 to 0.1, range = 340 to 394.

1-Octyne, CA 629-05-0: Irritant; flammable, flash point 290; polymerizes;
soluble most organic solvents. Tf = 193.67; Tb = 399.4; Vm = 0.1477 at
293.15; Antoine (liq) A = 6.1798, B = 1418.6, C = -59.55, dev = 0.1,
range = 340 to 400.

2-Octyne, CA 2809-67-8: Irritant; flammable; polymerizes; soluble most organic solvents. Tf = 211.55; Tb = 411.15; Vm = 0.1451 at 293.15; Antoine
(liq) A = 6.2112, B = 1473.7, C = -60.95, dev = 0.1, range = 350 to 412.

3-Octyne, CA 15232-76-5: Irritant; flammable; polymerizes; soluble most
organic solvents. Tf = 169.25; Tb = 406.29; Vm = 0.1464 at 293.15; Antoine
(liq) A = 6.2361, B = 1463.8, C = -60.35, dev = 0.1, range = 363 to 406.

4-Octyne, CA 1942-45-6: Irritant; flammable, flash point 293; polymerizes;
soluble most organic solvents. Tf = 170.65; Tb = 404.72; Vm = 0.1468 at
293.15; Antoine (liq) A = 6.2382, B = 1459.9, C = -60.05, dev = 0.1,
range = 362 to 405.

$C_8H_{14}Br_2$ MW 270.01

1,2-Dibromocyclooctane, CA 29974-69-4: Irritant; flammable. Antoine (liq)
A = 6.8549, B = 2629.6, C = 0, dev = 1.0 to 5.0, range = 292 to 354.

$C_8H_{14}ClN_5$ MW 215.69

2-Chloro-4-ethylamino-6-isopropylamino-1,3,5-triazine, CA 1912-24-9: Irritant; soluble methanol, chloroform; used as herbicide. Tm = 448.15 to
450.15; Antoine (sol) A = 12.8909, B = 5945, C = 0, dev = 1.0 to 5.0,
range = 323 to 403.

$C_8H_{14}Cl_2S$ MW 213.16

(2-Chlorocyclohexyl)(2-chloroethyl) sulfide, CA 16660-53-0: Toxic;
stench; flammable. Antoine (liq) A = 7.8709, B = 3265.5, C = 0, dev = 1.0
to 5.0, range = 273 to 333.

$C_8H_{14}N_2O_2$ MW 170.21

Acetylproline, *N*-methylamide: Tm = 371.15; Antoine I (sol, *alpha*) A =
7.7719, B = 3608.9, C = 0, dev = 1.0, range = 308 to 318; Antoine II (sol,
beta), A = 6.3929, B = 3169.6, C = 0, dev = 1.0, range = 319 to 335.

$C_8H_{14}O$ MW 126.20

Cyclohexyl methyl ketone, CA 823-76-7: Irritant; flammable, flash point
335; soluble ether. Tb = 453.15 to 454.15; Vm = 0.1375 at 293.15; Antoine
(liq) A = 7.3487, B = 2418.7, C = 0, dev = 1.0, range not specified.

Cyclooctanone, CA 502-49-8: Flammable, flash point 345; soluble ethanol,
acetone, benzene. Tm = 315.15; Tb = 468.15 to 470.15; Antoine I (liq) A =
5.7364, B = 1514.62, C = -73.15, dev = 1.0, range = 323 to 403; Antoine II
(liq) A = 6.10508, B = 1643.839, C = -73.617, dev = 0.1, range = 394 to
484.

2-Ethyl-2-hexenal, CA 645-62-5: Irritant; flammable; polymerizes. Tb =
448.15; Antoine (liq) A = 6.0276, B = 1485.64, C = -79.33, dev = 1.0,
range = 326 to 448.

2-Ethyl-4-methyl-2-pentenal, CA 28419-86-5: Irritant; flammable; polymerizes. Antoine (liq) A = 6.2657, B = 1590.55, C = -62.65, dev = 1.0,
range = 311 to 436.

2-Propylcyclopentanone ±, CA 1193-70-0: Flammable. Tm = 204.93; Tb =
456.25 to 456.35; Vm = 0.1400 at 293.15; Antoine (liq) A = 7.2707, B =
2401.2, C = 0, dev = 5.0, range = 332 to 457.

$C_8H_{14}O_2$ MW 142.20

Acrylic acid, neopentyl ester: Irritant; flammable; polymerizes. Antoine
(liq) A = 7.80872, B = 2388.368, C = 0, dev = 1.0, range = 301 to 325.

1,4-Butanediol, divinyl ether, CA 3891-33-6: Irritant; flammable; polymerizes; forms explosive peroxides. Tb = 463.15 to 464.15; Vm = 0.1464 at
293.15; Antoine (liq) A = 7.8229, B = 2558, C = 0, dev = 5.0, range = 335
to 440.

Butyl methacrylate, CA 97-88-1: Irritant; lachrymator; flammable, flash
point 325; soluble ethanol, ether. Tb = 436.65 to 443.65; Vm = 0.1600 at
298.15; Antoine I (liq) A = 6.9199, B = 1913.9, C = -43.15, dev = 1.0,
range = 343 to 373; Antoine II (liq) A = 6.63397, B = 1820.716, C = -43.15,
dev = 1.0, range = 344 to 437.

Cyclohexyl acetate, CA 622-45-7: Moderately toxic by inhalation; flammable, flash point 331; soluble ethanol, ether. Tb = 448.15; Vm = 0.1466
at 293.15; Antoine (liq) A = 6.8224, B = 1959.7, C = -39.65, dev = 1.0,
range = 368 to 446.

Methacrylic acid, *tert*-butyl ester, CA 585-07-9: Irritant; flammable;
lachrymator; polymerizes. Tm = 225.15; Antoine (liq) A = 6.10987, B =
1390.621, C = -69.471, dev = 1.0, range = 313 to 410.

Pentyl acrylate, CA 2998-23-4: Irritant; lachrymator; flammable; polymerizes. Vm = 0.1594 at 293.15; Antoine (liq) A = 7.3344, B = 2345, C = 0,
dev = 5.0, range = 325 to 440.

$C_8H_{14}O_3$ MW 158.20

Butyric anhydride, CA 106-31-0: Irritant; corrosive; causes burns; flammable, flash point 355; moisture-sensitive; soluble water, ethanol, with decomposition; soluble ether. Tm = 198.15; Tb = 471.15; Vm = 0.1636 at 293.15; Antoine (liq) A = 7.4603, B = 2564.2, C = 0, dev = 5.0, range = 349 to 470.

Diethylene glycol, divinyl ether, CA 764-99-8: Irritant; lachrymator; flammable; polymerizes; forms explosive peroxides. Antoine (liq) A = 7.575, B = 2610, C = 0, dev = 5.0, range = 336 to 470.

2-Ethylacetoacetic acid, ethyl ester, CA 607-97-6: Flammable; moderately soluble water; soluble ethanol, ether. Tb = 471.15; Vm = 0.1623 at 298.15; Antoine (liq) A = 7.10079, B = 2224.53, C = -34.605, dev = 1.0, range = 313 to 471.

Isopropyl levulinate, CA 21884-26-4: Flammable; soluble ethanol, acetone, ether, benzene. Tb = 482.45; Vm = 0.1607 at 293.15; Antoine (liq) A = 6.62251, B = 1924.16, C = -64.804, dev = 1.0, range = 321 to 481.

Propyl levulinate, CA 645-67-0: Flammable; soluble ethanol, ether, acetone, benzene. Tb = 494.35; Vm = 0.1599 at 293.15; Antoine (liq) A = 7.18953, B = 2385.63, C = -34.319, dev = 1.0, range = 332 to 495.

$C_8H_{14}O_4$ MW 174.20

2-Acetoxypropionic acid, propyl ester: Flammable. Antoine (liq) A = 6.36139, B = 1650.733, C = -90.15, dev = 1.0 to 5.0, range = 318 to 469.

3-Acetoxypropionic acid, propyl ester: Flammable. Antoine (liq) A = 10.615, B = 3904, C = 0, dev = 5.0, range = 361 to 373.

Diethyl succinate, CA 123-25-1: Flammable, flash point 363; soluble ethanol, ether; used in flavorings. Tm = 252.15; Tb = 490.85; Vm = 0.1675 at 293.15; Antoine (liq) A = 7.35157, B = 2458.47, C = -29.813, dev = 1.0, range = 327 to 490.

Diisopropyl oxalate, CA 615-81-6: Toxic; flammable; soluble ethanol, ether. Tb = 462.15; Vm = 0.1740 at 293.15; Antoine (liq) A = 7.06613, B = 2078.99, C = -55.733, dev = 1.0, range = 316 to 467.

Dimethyl adipate, CA 627-93-0: Flammable, flash point 380; soluble ether, ethanol. Tm = 281.15; Vm = 0.1643 at 293.15; Antoine I (liq) A = 8.035, B = 3020, C = 0, dev = 5.0, range = 418 to 501; Antoine II (liq) A = 8.1459, B = 3070, C = 0, dev = 1.0 to 5.0, range = 382 to 500.

Dipropyl oxalate, CA 615-98-5: Toxic; flammable; soluble ethanol, ether. Tm = 226.85; Tb = 487.15 to 488.15; Vm = 0.1710 at 293.15; Antoine (liq) A = 6.95982, B = 2145.57, C = -53.507, dev = 1.0, range = 326 to 487.

2-Methylmalonic acid, diethyl ester, CA 609-08-5: Flammable, flash point 349; soluble ethanol, ether, acetone, chloroform. Tb = 474.35 to 474.55; Vm = 0.1704 at 293.15; Antoine (liq) A = 6.90665, B = 2137.13, C = -38.335, dev = 1.0, range = 312 to 475.

Octanedioic acid, suberic acid, CA 505-48-6: Flammable; slightly soluble water; soluble ethanol. Tm = 417.15; Antoine I (sol) A = 16.062, B = 7472, C = 0, dev = 1.0 to 5.0, range = 379 to 407; Antoine II (liq) A = 8.86785, B = 3965.03, C = -40.8, dev = 1.0, range = 445 to 619.

$C_8H_{14}O_4S$ MW 206.26

Thiodiacetic acid, diethyl ester, CA 925-47-3: Irritant; flammable. Tb =
540.15 to 541.15; Antoine (liq) A = 5.17254, B = 1120.871, C = -188.403,
dev = 1.0, range = 384 to 448.

$C_8H_{14}O_5$ MW 190.20

Isopropyl[1-(methoxycarbonyl)ethyl] carbonate: Toxic; flammable. Antoine
(liq) A = 7.8888, B = 2897, C = 0, dev = 5.0, range = 330 to 493.

2-(Lactyloxy)propionic acid, ethyl ester: Flammable. Antoine (liq) A =
6.92521, B = 2031.393, C = -90.15, dev = 1.0 to 5.0, range = 321 to 389.

Malic acid, diethyl ester ±, CA 626-11-9: Flammable. Tb = 528.15; Antoine
(liq) A = 7.5684, B = 2837.99, C = -16.52, dev = 1.0, range = 353 to 527.

Propyl[1-(methoxycarbonyl)ethyl] carbonate: Toxic; flammable. Antoine
(liq) A = 8.1498, B = 3032.2, C = 0, dev = 5.0, range = 373 to 495.

$C_8H_{14}O_6$ MW 206.20

Tartaric acid, diethyl ester, *dextro*, CA 13811-71-7: Flammable, flash
point 366; moderately soluble water; soluble ethanol, ether, acetone. Tm =
290.15; Tb = 553.15; Vm = 0.1713 at 293.15; Antoine (liq) A = 7.34502, B =
2694.63, C = -48.406, dev = 1.0, range = 375 to 553.

Tartaric acid, diethyl ester ±, CA 87-91-2: Flammable; moderately soluble
water; soluble ethanol, ether, acetone. Tm = 291.85; Tb = 554.15; Vm =
0.1712 at 293.15; Antoine (liq) A = 7.18438, B = 2565.01, C = -57.713,
dev = 1.0, range = 375 to 553.

$C_8H_{14}O_6S$ MW 238.26

Sulfonyldiacetic acid, diethyl ester, CA 29771-87-7: Irritant; flammable.
Antoine (liq) A = 5.78528, B = 1603.171, C = -177.234, dev = 1.0 to 5.0,
range = 421 to 494.

$C_8H_{15}Br$ MW 191.11

(2-Bromoethyl) cyclohexane, CA 1647-26-3: Irritant; flammable. Tf =
216.15; Tb = 485.15; Vm = 0.1547 at 293.15; Antoine (liq) A = 5.81865, B =
1524.18, C = -86.514, dev = 1.0, range = 311 to 486.

$C_8H_{15}ClO$ MW 162.66

Caprylyl chloride, octanoyl chloride, CA 111-64-8: Corrosive; lachrymator;
flammable, flash point 355; soluble ether. Tm = 210.15; Tb = 468.15 to
469.15; Vm = 0.1706 at 288.15; Antoine (liq) A = 11.1832, B = 3889.9,
C = 0, dev = 1.0 to 5.0, range = 343 to 373.

5-Methylheptanoyl chloride: Irritant; lachrymator; flammable. Antoine
(liq) A = 10.1568, B = 3462.8, C = 0, dev = 1.0, range = 338 to 373.

$C_8H_{15}N$ MW 125.21

3-Azabicyclo(3,2,2)-nonane, CA 283-24-9: Corrosive; flammable, flash point
336. Tm = 460.15 to 461.15; Antoine (sol) A = 8.032, B = 2727, C = 0,
dev = 5.0, range = 303 to 443.

Caprylonitrile, octanenitrile, CA 124-12-9: Irritant; flammable, flash
point 346; soluble ether. Tm = 227.55; Tb = 471.15 to 473.15; Vm = 0.1539
at 293.15; Antoine (liq) A = 6.36566, B = 1796.28, C = -66.08, dev = 1.0,
range = 373 to 480.

$C_8H_{15}NO$ MW 141.21

Heptyl isocyanate, CA 4747-81-3: Highly toxic; flammable; reacts violently
with water. Antoine (liq) A = 7.39356, B = 2479.717, C = 0, dev = 1.0 to
5.0, range = 326 to 461.

Methacrylic acid, N-tert-butyl amide, CA 6554-73-0: Irritant; lachrymator;
polymerizes; flammable. Tm = 332.15; Antoine (liq) A = 7.55357, B =
2593.19, C = 0, dev = 5.0, range = 340 to 467.

2-Octenoic acid, amide, trans: Tm = 397.85 to 398.25; Antoine (sol) A =
9.111, B = 3840.5, C = 0, dev = 1.0, range = 373 to 393.

$C_8H_{15}NO_2$ MW 157.21

Methacrylic acid, 2-(dimethylamino)ethyl ester, CA 2867-47-2: Corrosive;
lachrymator; flammable, flash point 347. Tm = 243.15; Tb = 455.15 to
465.15; Vm = 0.1685 at 298.15; Antoine (liq) A = 6.82635, B = 2009.43, C =
-43.15, dev = 1.0, range = 372 to 460.

Piperidine, 1-lactoyl: Antoine (liq) A = 8.0622, B = 3244.362, C = 0,
dev = 1.0, range = 346 to 408.

$C_8H_{15}NO_3$ MW 173.21

N,N-Diethyloxamic acid, ethyl ester, CA 5411-58-5: Flammable. Antoine
(liq) A = 7.02404, B = 2400.13, C = -46.78, dev = 1.0, range = 349 to 525.

$C_8H_{15}N_5O$ MW 197.24

2-Methoxy-4,6-bis(ethylamino)-1,3,5-triazine, CA 673-04-1: Antoine (sol)
A = 11.0189, B = 5130, C = 0, dev = 1.0 to 5.0, range = 323 to 403.

$C_8H_{15}N_5S$ MW 213.30

2-Methylthio-4,6-bis(ethylamino)-1,3,5-triazine, CA 1014-70-6: Used as
herbicide. Tm = 355.15 to 356.15; Antoine (sol) A = 11.0389, B = 5293,
C = 0, dev = 1.0 to 5.0, range = 323 to 355.

2-Methylthio-4-methylamino-6-isopropylamino-1,3,5-triazine, CA 1014-69-3:
Used as herbicide. Tm = 357.15 to 359.15; Antoine (sol) A = 11.2259, B =
5302, C = 0, dev = 1.0 to 5.0, range = 323 to 357.

C_8H_{16} MW 112.21

Cyclooctane, CA 292-64-8: Flammable, flash point 303; soluble most organic
solvents. Tm = 287.45; Tb = 421.75 to 422.75; Vm = 0.1344 at 293.15;
Antoine I (liq) A = 5.98663, B = 1437.682, C = -63.147, dev = 0.1, range =
369 to 467; Antoine II (liq) A = 5.9899, B = 1440.707, C = -62.701, dev =
0.1, range = 369 to 468; Antoine III (liq) A = 6.20474, B = 1564.985, C =
-50.842, dev = 0.1, range = 289 to 369.

1,1-Dimethylcyclohexane, CA 590-66-9: Flammable, flash point 280; soluble
most organic solvents. Tf = 239.65; Tb = 393.15; Vm = 0.1437 at 293.15;
Antoine (liq) A = 5.92575, B = 1323.358, C = -55.112, dev = 0.02, range =
313 to 395.

1,2-Dimethylcyclohexane, *cis*, CA 2207-01-4: Flammable, flash point 288;
soluble most organic solvents. Tm = 222.95; Tb = 403.15; Vm = 0.1409 at
293.15; Antoine (liq) A = 5.96232, B = 1367.306, C = -57.314, dev = 0.02,
range = 322 to 405.

1,2-Dimethylcyclohexane, ±*trans*, CA 6876-23-9: Flammable, flash point 288;
soluble most organic solvents. Tm = 183.15; Tb = 397.15; Vm = 0.1446 at
293.15; Antoine (liq) A = 5.95748, B = 1353.605, C = -54.046, dev = 0.02,
range = 316 to 399.

1,3-Dimethylcyclohexane, *cis*, CA 638-04-0: Flammable, flash point 283;
soluble most organic solvents. Tm = 197.55; Tb = 392.95 to 393.45; Vm =
0.1465 at 293.15; Antoine (liq) A = 5.96898, B = 1341.733, C = -54.702,
dev = 0.02, range = 318 to 396.

1,3-Dimethylcyclohexane ±*trans*, CA 2207-03-6: Flammable, flash point 283;
soluble most organic solvents. Tf = 183.15; Tb = 397.15; Vm = 0.1430 at
293.15; Antoine (liq) A = 5.96258, B = 1345.645, C = -57.53, dev = 0.02,
range = 314 to 400.

1,4-Dimethylcyclohexane, *cis*, CA 624-29-3: Flammable, flash point 289;
soluble most organic solvents. Tf = 186.15; Tb = 397.15 to 398.15; Vm =
0.1433 at 293.15; Antoine (liq) A = 5.95775, B = 1345.647, C = -56.985,
dev = 0.02, range = 317 to 400.

1,4-Dimethylcyclohexane, ±*trans*, CA 2207-04-7: Flammable, flash point 284;
soluble most organic solvents. Tf = 236.15; Tb = 392.15 to 393.15; Vm =
0.1471 at 293.15; Antoine (liq) A = 5.94449, B = 1331.612, C = -54.43,
dev = 0.02, range = 313 to 395.

2,2-Dimethyl-3-hexene, *cis*, CA 690-92-6: Flammable; polymerizes; soluble
most organic solvents. Tf = 135.75; Tb = 378.65; Vm = 0.1584 at 298.15;
Antoine (liq) A = 5.8977, B = 1239.206, C = -60.194, dev = 0.02, range =
319 to 380.

2,2-Dimethyl-3-hexene, *trans*, CA 690-93-7: Flammable; polymerizes; soluble
most organic solvents. Tb = 373.95; Vm = 0.1604 at 298.15; Antoine (liq)
A = 5.94901, B = 1232.574, C = -61.471, dev = 0.02, range = 306 to 376.

Ethylcyclohexane, CA 1678-91-7: Flammable, flash point 308; soluble most
organic solvents. Tf = 161.85; Tb = 404.95; Vm = 0.1424 at 293.15; Antoine
(liq) A = 5.99043, B = 1381.396, C = -58.271, dev = 0.02, range = 323
to 407.

1-Ethyl-1-methylcyclopentane, CA 16747-50-5: Flammable; soluble most organic solvents. Tf = 129.35; Tb = 394.65; Vm = 0.1445 at 298.15; Antoine I
(liq) A = 5.98238, B = 1347.293, C = -55.883, dev = 0.02,   *(continues)*

C_8H_{16} *(continued)*

range = 331 to 397; Antoine II (liq) A = 6.34694, B = 1530.266, C = -39.889, dev = 0.1, range = 238 to 288.

1-Ethyl-2-methylcyclopentane ±cis, CA 930-89-2: Flammable; soluble most organic solvents. Tf = 167.15; Tb = 394.55; Vm = 0.1429 at 293.15; Antoine I (liq) A = 6.02721, B = 1386.726, C = -56.383, dev = 0.02, range = 303 to 403; Antoine II (liq) A = 6.41176, B = 1584.799, C = -39.199, dev = 0.1, range = 238 to 304.

3-Ethyl-2-methyl-1-pentene, CA 19780-66-6: Flammable; polymerizes; soluble most organic solvents. Tm = 160.17; Tb = 382.15 to 383.15; Vm = 0.1545 at 293.15; Antoine (liq) A = 5.99601, B = 1306.261, C = -55.047, dev = 0.02, range = 307 to 389.

Isopropylcyclopentane, CA 3875-51-2: Flammable; soluble most organic solvents. Tm = 161.75; Tb = 399.55; Vm = 0.1445 at 293.15; Antoine (liq) A = 6.00635, B = 1376.846, C = -55.423, dev = 0.02, range = 320 to 403.

2-Methyl-2-heptene, CA 627-97-4: Flammable; polymerizes; soluble most organic solvents. Tb = 395.15 to 396.15; Vm = 0.1558 at 298.15; Antoine (liq) A = 6.7207, B = 1734.82, C = -27.797, dev = 1.0, range = 257 to 396.

1-Octene, CA 111-66-0: Flammable, flash point 294; polymerizes; soluble most organic solvents. Tm = 171.41; Tb = 394.45; Vm = 0.1578 at 298.15; Tc = 566.6; Pc = 2550; Vc = 0.472; Antoine (liq) A = 6.05178, B = 1350.245, C = -60.716, dev = 0.02, range = 317 to 400.

2-Octene, *cis*, CA 7642-04-8: Flammable, flash point 294; polymerizes; soluble most organic solvents. Tf = 169.15; Tb = 397.75; Vm = 0.1549 at 293.15; Antoine (liq) A = 6.02004, B = 1337.146, C = -65.706, dev = 0.1, range = 356 to 400.

2-Octene, *trans*, CA 13389-42-9: Flammable, flash point 294; polymerizes; soluble most organic solvents. Tf = 185.15; Tb = 396.5; Vm = 0.1559 at 293.15; Antoine (liq) A = 6.04801, B = 1347.058, C = -64.866, dev = 0.1, range = 356 to 399.

3-Octene, *trans*, CA 14919-01-8: Flammable; polymerizes; soluble most organic solvents. Tf = 162.75; Tb = 395.55; Vm = 0.1569 at 293.15; Antoine (liq) A = 6.06043, B = 1355.147, C = -62.216, dev = 0.1, range = 354 to 396.

4-Octene, *cis*, CA 7642-15-1: Flammable, flash point 294; polymerizes; soluble most organic solvents. Tf = 155.15; Tb = 394.85; Vm = 0.1556 at 293.15; Antoine (liq) A = 6.04891, B = 1355.346, C = -60.477, dev = 0.1, range = 353 to 395.

4-Octene, *trans*, CA 14850-23-8: Flammable, flash point 294; polymerizes; soluble most organic solvents. Tf = 179.15; Tb = 394.55; Vm = 0.1571 at 293.15; Antoine (liq) A = 6.0543, B = 1351.136, C = -61.737, dev = 0.1, range = 353 to 396.

Propylcyclopentane, CA 2040-96-2: Flammable; soluble most organic solvents. Tf = 154.45; Tb = 403.95; Vm = 0.1445 at 293.15; Antoine (liq) A = 6.04236, B = 1393.284, C = -58.949, dev = 0.02, range = 323 to 406.

1,1,2-Trimethylcyclopentane ±, CA 4259-00-1: Flammable; soluble most organic solvents. Tb = 386.85; Vm = 0.1465 at 293.15; Antoine (liq) A = 5.94266, B = 1307.357, C = -54.81, dev = 0.02, range = 309 to 389.

(continues)

280

C_8H_{16} *(continued)*

1,1,3-Trimethylcyclopentane, CA 4516-69-2: Flammable; soluble most organic solvents. Tm = 130.75; Tb = 378.05; Vm = 0.1457 at 293.15; Antoine (liq) A = 5.93036, B = 1273.902, C = -53.454, dev = 0.02, range = 301 to 379.

cis-1,2-*trans*-4-Trimethylcyclopentane, CA 4850-28-6: Flammable; soluble most organic solvents. Tf = 140.85; Tb = 389.95; Vm = 0.1480 at 298.15; Antoine (liq) A = 5.97533, B = 1331.787, C = -54.392, dev = 0.02, range = 311 to 392.

trans-1,2-*cis*-4-Trimethylcyclopentane, CA 13398-35-1: Flammable; soluble most organic solvents. Tm = 142.36; Antoine (liq) A = 5.96994, B = 1303.665, C = -53.586, dev = 0.02, range = 305 to 385.

2,4,4-Trimethyl-1-pentene, diisobutylene, CA 107-39-1: Moderately irritant; flammable, flash point 268; polymerizes; soluble most organic solvents. Tf = 179.55; Tb = 374.35; Vm = 0.1569 at 293.15; Antoine (liq) A = 5.93872, B = 1261.611, C = -53.843, dev = 0.02, range = 343 to 381.

2,4,4-Trimethyl-2-pentene, CA 107-40-4: Flammable, flash point 275; polymerizes; soluble most organic solvents. Tf = 166.85; Tb = 377.65; Vm = 0.1555 at 293.15; Antoine (liq) A = 5.98282, B = 1272.235, C = -58.173, dev = 0.02, range = 319 to 380.

$C_8H_{16}Br_2$ MW 272.02

1,1-Dibromooctane: Flammable; soluble ether, chloroform. Antoine (liq) A = 6.528, B = 1967, C = -80.15, dev = 5.0, range = 412 to 571.

$C_8H_{16}Cl_2$ MW 183.12

1,1-Dichlorooctane, CA 20395-24-8: Irritant; flammable. Antoine (liq) A = 6.362, B = 1773, C = -74.15, dev = 5.0, range = 382 to 533.

$C_8H_{16}F_2$ MW 150.21

1,1-Difluorooctane, CA 61350-03-6: Flammable. Antoine (liq) A = 6.410, B = 1559, C = -61.15, dev = 5.0, range = 329 to 459.

$C_8H_{16}O$ MW 128.21

Caprylaldehyde, octanal, CA 124-13-0: Moderately toxic; flammable, flash point 325; slightly soluble water; soluble ethanol, ether, acetone, benzene. Tb = 436.55; Vm = 0.1561 at 293.15; Antoine (liq) A = 7.1191, B = 2269.6, C = 0, dev = 1.0 to 5.0, range = 293 to 438.

2,5-Dimethyl-3-hexanone, CA 1888-57-9: Flammable; slightly soluble water; soluble ethanol, ether, acetone. Tb = 420.15 to 421.15; Vm = 0.1550 at 273.15; Antoine (liq) A = 7.3122, B = 2221.9, C = 0, dev = 1.0, range = liquid to 418.

1-Ethylcyclohexanol, CA 1940-18-7: Flammable; slightly soluble water; soluble benzene, light hydrocarbons. Tm = 307.15 to 308.15; Tb = 439.15; Vm = 0.1390 at 298.15; Antoine (liq) A = 12.323, B = 6899, C = 229.45, dev = 5.0, range = 324 to 440.

(continues)

$C_8H_{16}O$ *(continued)*

2-Methyl-3-heptanone, CA 13019-20-0: Flammable; slightly soluble water; soluble ethanol, ether, acetone. Tb = 432.15 to 433.15; Vm = 0.1571 at 293.15; Antoine (liq) A = 7.3076, B = 2271.9, C = 0, dev = 1.0, range = liquid to 428.

6-Methyl-3-hepten-2-ol, CA 51500-48-2: Flammable; soluble ethanol. Antoine (liq) A = 8.03727, B = 2497.38, C = -34.468, dev = 5.0, range = 314 to 449.

6-Methyl-5-hepten-2-ol ±, CA 4630-06-2: Flammable, flash point 340; soluble ethanol. Tb = 447.15 to 449.15; Vm = 0.1500 at 293.15; Antoine (liq) A = 9.4987, B = 3594.5, C = 32.38, dev = 5.0, range = 314 to 448.

2-Octanone, methyl hexyl ketone, CA 111-13-7: Flammable, flash point 335; soluble ethanol, ether, acetone; used in flavorings and perfumery. Tf = 257.15; Tb = 446.45; Vm = 0.1563 at 293.15; Antoine I (liq) A = 6.99302, B = 2033.16, C = -38.283, dev = 1.0 to 5.0, range = 317 to 446; Antoine II (liq) A = 6.17264, B = 1542.746, C = -75.962, dev = 1.0, range = 324 to 520.

3-Octanone, ethyl pentyl ketone, CA 106-68-3: Flammable, flash point 319; soluble ethanol, ether, acetone; used as food flavoring. Tf = 227.15; Tb = 440.15 to 441.15; Vm = 0.1560 at 293.15; Antoine (liq) A = 9.0260, B = 3603.9, C = 78.46, dev = 1.0, range = 293 to 348.

4-Octanone, butyl propyl ketone, CA 589-63-9: Flammable; soluble ethanol, ether, acetone. Tb = 437.15 to 440.15; Vm = 0.1574 at 298.15; Antoine (liq) A = 6.65994, B = 2105.7, C = 15.71, dev = 1.0, range = 288 to 433.

1-Propylcyclopentanol, CA 1604-02-0: Flammable; soluble ethanol. Tm = 235.65; Tb = 446.65; Vm = 0.1418 at 298.15; Antoine (liq) A = 5.070, B = 813.6, C = -181.65, dev = 1.0, range = 344 to 447.

2,2,4-Trimethyl-3-pentanone, CA 5857-36-3: Flammable; soluble ether, acetone. Tm = 244.15; Tb = 407.15 to 408.15; Vm = 0.1590 at 293.15; Antoine (liq) A = 7.95531, B = 2191.89, C = -39.701, dev = 1.0, range = 287 to 408.

$C_8H_{16}O_2$ MW 144.21

3-Butoxy-2-butanone: Flammable; soluble acetone. Antoine (liq) A = 8.48478, B = 3258.7, C = 102.34, dev = 1.0, range = 323 to 398.

Caprylic acid, octanoic acid, CA 124-07-2: Irritant; corrosive; flammable, flash point 383; moderately soluble hot water; soluble ethanol, chloroform, ether, carbon disulfide; used in perfumery. Tm = 289.85; Tb = 512.85; Vm = 0.1587 at 293.15; Antoine I (liq) A = 6.70211, B = 1861, C = -115.393, dev = 1.0, range = 360 to 512; Antoine II (liq) A = 3.1122, B = 525.9, C = -203.26, dev = 1.0, range = 296 to 331.

Ethyl caproate, ethyl hexanoate, CA 123-66-0: Flammable, flash point 322; soluble ethanol, ether. Tm = 206.15; Tb = 439.15 to 440.15; Vm = 0.1656 at 293.15; Antoine I (liq) A = 7.03448, B = 1989.21, C = -44.25, dev = 1.0, range = 296 to 449; Antoine II (liq) A = 7.87441, B = 2538.838, C = 0, dev = 1.0 to 5.0, range = 300 to 376.

2-Ethylhexanoic acid ±, CA 149-57-5: Irritant; moderately toxic; flammable, flash point above 385; soluble ether. Tm = below 203.15; Tb = 498.15 to 501.15; Vm = 0.1597 at 298.15; Antoine (liq) A = 8.4588, B = 3227.8, C = 0, dev = 1.0, range = 403 to 500.   *(continues)*

$C_8H_{16}O_2$ *(continued)*

Formic acid, 1,1-dimethylpentyl ester: Flammable. Antoine (liq) A = 7.066, B = 2174, C = 0, dev = 1.0, range not specified.

Hexyl acetate, CA 142-92-7: Flammable, flash point 318; soluble ethanol, ether. Tm = 192.25; Tb = 442.35; Vm = 0.1643 at 288.15; Antoine (liq) A = 7.8442, B = 2556.973, C = 0, dev = 1.0, range = 304 to 381.

Isobutyl butyrate, CA 539-90-2: Flammable, flash point 323; slightly soluble water; soluble ethanol, ether. Tb = 430.15; Vm = 0.1724 at 291.15; Antoine (liq) A = 7.32254, B = 2347.42, C = 11.157, dev = 1.0 to 5.0, range = 277 to 430.

Isobutyl isobutyrate, CA 97-85-8: Flammable, flash point 311; soluble ethanol, ether, acetone. Tm = 192.55; Tb = 420.15; Vm = 0.1648 at 273.15; Antoine (liq) A = 6.74347, B = 1794.65, C = -41.828, dev = 1.0, range = 277 to 421.

Isopentyl propionate, CA 105-68-0: Flammable; soluble ethanol, ether. Tb = 433.85; Vm = 0.1658 at 293.15; Antoine (liq) A = 7.14879, B = 2191.13, C = -7.374, dev = 1.0, range = 281 to 434.

Methyl heptanoate, CA 106-73-0: Flammable, flash point 330; soluble ethanol, ether, acetone. Tm = 217.15; Tb = 445.25; Vm = 0.1636 at 293.15; Antoine (liq) A = 6.57638, B = 1768.74, C = -58.373, dev = 1.0, range = 332 to 402.

4-Methylvaleric acid, ethyl ester, CA 25415-67-2: Flammable. Tb = 433.15 to 437.15; Vm = 0.1657 at 293.15; Antoine (liq) A = 7.03303, B = 2088.15, C = -18.426, dev = 1.0, range = 284 to 434.

Propyl isovalerate, CA 557-00-6: Flammable; soluble ethanol, ether. Tb = 428.65; Vm = 0.1674 at 293.15; Antoine (liq) A = 7.22048, B = 2200.75, C = -7.221, dev = 1.0, range = 281 to 429.

$C_8H_{16}O_3$ MW 160.21

2-Butoxypropionic acid, methyl ester; Flammable. Antoine (liq) A = 8.462, B = 2712.3, C = 0, dev = 5.0, range = 348 to 417.

3-Butoxypropionic acid, methyl ester: Flammable. Antoine (liq) A = 7.6999, B = 2670.2, C = 0, dev = 1.0 to 5.0, range = 311 to 469.

3-Ethoxypropionic acid, propyl ester: Flammable. Antoine (liq) A = 7.4929, B = 2538, C = 0, dev = 5.0, range = 343 to 461.

Ethylene glycol, monobutylether acetate, CA 112-07-2: Flammable, flash point 344; moderately soluble water; soluble ethanol. Tf = 209.15; Tb = 465.15; Vm = 0.1700 at 293.15; Antoine (liq) A = 7.8740, B = 2711, C = 0, dev = 5.0, range = 293 to 465.

2-Hydroxyisobutyric acid, butyl ester, CA 816-50-2: Flammable. Antoine (liq) A = 6.7725, B = 1981, C = -43.15, dev = 1.0, range = 384 to 458.

3-Methoxypropionic acid, butyl ester, CA 4195-88-4: Flammable. Antoine (liq) A = 7.6708, B = 2657.7, C = 0, dev = 1.0, range = 311 to 469.

Pentyl lactate, CA 6382-06-5: Flammable. Tm = 251.15; Vm = 0.1657 at 293.15; Antoine (liq) A = 10.3487, B = 3859, C = 0, dev = 1.0 to 5.0, range = 288 to 370.

$C_8H_{16}O_4$ MW 176.21

Diethylene glycol, monoethylether acetate, CA 112-15-2: Flammable; soluble
water, ethanol. Tm = 248.15; Tb = 490.15; Vm = 0.1745 at 293.15; Antoine
(liq) A = 7.5096, B = 2698.2, C = 0, dev = 1.0, range = 293 to 491.

$C_8H_{17}Br$ MW 193.13

1-Bromooctane, octyl bromide, CA 111-83-1: Irritant; flammable, flash
point 351; soluble ethanol, ether. Tf = 218.15; Tb = 474.65; Vm = 0.1743
at 298.15; Antoine (liq) A = 6.2428, B = 1701.61, C = -72.35, dev = 1.0,
range = 373 to 475.

2-Bromooctane ±, CA 60251-57-2: Irritant; flammable; soluble ethanol,
ether. Tb = 461.15 to 462.15; Vm = 0.1775 at 298.15; Antoine (liq) A =
6.1728, B = 1634, C = -70.15, dev = 5.0, range = 343 to 463.

$C_8H_{17}Cl$ MW 148.68

1-Chlorooctane, octyl chloride, CA 111-85-3: Irritant; flammable, flash
point 343; soluble ethanol, ether. Tm = 215.35; Tb = 454.65; Vm = 0.1702
at 293.15; Antoine (liq) A = 5.968, B = 1469.829, C = -85.993, dev = 1.0,
range = 327 to 457.

2-Chlorooctane ±, CA 51261-14-4: Irritant; flammable; soluble ethanol,
ether. Tb = 444.15 to 446.15; Vm = 0.1708 at 288.15; Antoine (liq) A =
6.2696, B = 1619, C = -67.15, dev = 5.0, range = 330 to 446.

Heptane, 3-(chloromethyl), CA 123-04-6: Irritant; flammable; soluble ethanol, ether, acetone, benzene. Tb = 447.15; Vm = 0.1696 at 293.15; Antoine
(liq) A = 7.1764, B = 2306.6, C = 0, dev = 1.0, range = 371 to 443.

$C_8H_{17}ClO_4$ MW 212.67

Triethylene glycol, mono(2-chloroethyl) ether, CA 5197-66-0: Irritant;
flammable. Tm = 219.15; Antoine (liq) A = 8.48725, B = 3601.42, C = 1.025,
dev = 1.0, range = 383 to 555.

$C_8H_{17}Cl_2N$ MW 198.14

N-Butyl bis(2-chloroethyl) amine, CA 42520-97-8: Irritant; flammable.
Antoine (liq) A = 8.40851, B = 3169.8, C = 0, dev = 1.0, range = 273 to 380.

N-sec-Butyl bis(2-chloroethyl) amine: Irritant; flammable. Antoine (liq)
A = 8.29174, B = 3109.5, C = 0, dev = 1.0, range = 273 to 373.

N-tert-Butyl bis(2-chloroethyl) amine, CA 10125-86-7: Irritant; flammable.
Antoine (liq) A = 8.2592, B = 3050.9, C = 0, dev = 1.0, range = 273 to 345.

N-Isobutyl bis(2-chloroethyl) amine: Irritant; flammable. Antoine (liq)
A = 8.54732, B = 3152.5, C = 0, dev = 1.0, range = 273 to 345.

$C_8H_{17}F$ MW 132.22

1-Fluorooctane, octyl fluoride, CA 463-11-6: Flammable. Tm = 209.15; Tb =
415.15 to 416.15; Vm = 0.1632 at 293.15; Antoine (liq) A = 6.266, B =
1509.34, C = -61.15, dev = 1.0, range = 307 to 446.

$C_8H_{17}I$ MW 240.13

1-Iodooctane, octyl iodide, CA 629-27-6: Irritant; light-sensitive; flammable, flash point 368; soluble ethanol, ether. Tf = 227.45; Tb = 498.65;
Vm = 0.1806 at 293.15; Antoine (liq) A = 6.1319, B = 1738.53, C = -76.92,
dev = 0.1, range = 391 to 554.

$C_8H_{17}NO$ MW 143.23

Butyric acid, N,N-diethylamide, CA 1114-76-7: Irritant; flammable; soluble
water, ethanol. Tb = 479.15; Antoine (liq) A = 5.8587, B = 2022.9, C = 0,
dev = 1.0, range = 298 to 373.

Caprylaldehyde oxime, CA 929-55-5: Flammable; soluble ethanol, acetone.
Tm = 333.15; Antoine (liq) A = 9.785, B = 3726, C = 0, dev = 5.0, range =
313 to 400.

Capryl amide, CA 629-01-6: Flammable; soluble ethanol, ether, acetone.
Tm = 379.15 to 383.15; Tb = 512.15; Vm = 0.1695 at 383.15; Antoine (sol)
A = 14.045, B = 5783, C = 0, dev = 5.0, range = 325 to 374.

2-Octanone, oxime, CA 7207-49-0: Flammable. Tm = 266.75; Tb = 487.55 to
488.55; Antoine (liq) A = 9.245, B = 3526, C = 0, dev = 5.0, range = 293
to 487.

3-Octanone, oxime, CA 7207-50-3: Flammable. Antoine (liq) A = 9.275, B =
3508, C = 0, dev = 5.0, range = 293 to 400.

4-Octanone, oxime, CA 7207-51-4: Flammable. Antoine (liq) A = 9.635, B =
3595, C = 0, dev = 5.0, range = 293 to 400.

$C_8H_{17}NO_2$ MW 159.23

Lactic acid, N-isopentylamide: Irritant; flammable. Antoine (liq) A =
9.42102, B = 4067.555, C = 0, dev = 1.0, range = 386 to 433.

Lactic acid, N-pentylamide: Irritant; flammable. Tm = 302.65; Antoine
(liq) A = 9.77049, B = 4272.074, C = 0, dev = 1.0, range = 373 to 448.

Leucine, *levo*, ethyl ester, CA 2743-60-4: Flammable. Tb = 469.15; Antoine
(liq) A = 11.3495, B = 6637.3, C = 246.05, dev = 1.0 to 5.0, range = 333
to 449.

(1-Methylheptyl)nitrite: Flammable. Antoine (liq) A = 7.2522, B = 2345.6,
C = 0, dev = 5.0, range = 303 to 338.

C_8H_{18} MW 114.23

2,2-Dimethylhexane, CA 590-73-8: Flammable; soluble most organic solvents.
Tf = 152.15; Tb = 379.15 to 380.15; Vm = 0.1643 at 293.15; Antoine (liq)
A = 5.9595, B = 1272.349, C = -58.187, dev = 0.02, range = 302 to 381.

2,3-Dimethylhexane ±, CA 584-94-1: Flammable, flash point 280; soluble
most organic solvents. Tb = 388.95; Vm = 0.1604 at 293.15; Antoine (liq)
A = 5.99154, B = 1313.797, C = -59.146, dev = 0.02, range = 310 to 390.
(continues)

C_8H_{18} *(continued)*

2,4-Dimethylhexane ±, CA 589-43-5: Flammable, flash point 283; soluble most organic solvents. Tb = 383.15; Vm = 0.1631 at 293.15; Antoine (liq) A = 5.97958, B = 1288.984, C = -58.211, dev = 0.02, range = 305 to 385.

2,5-Dimethylhexane, CA 592-13-2: Flammable, flash point 299; soluble most organic solvents. Tf = 182.15; Tb = 381.15 to 383.15; Vm = 0.1647 at 293.15; Antoine (liq) A = 5.98696, B = 1288.764, C = -58.542, dev = 0.02, range = 307 to 383.

3,3-Dimethylhexane, CA 563-16-6: Flammable; soluble most organic solvents. Tf = 147.05; Tb = 384.15 to 385.15; Vm = 0.1609 at 293.15; Antoine (liq) A = 5.9720, B = 1305.752, C = -55.911, dev = 0.02, range = 308 to 386.

3,4-Dimethylhexane, CA 583-48-2: Flammable; soluble most organic solvents. Tb = 390.15 to 392.15; Vm = 0.1597 at 298.15; Antoine (liq) A = 6.00003, B = 1327.531, C = -58.527, dev = 0.02, range = 313 to 392.

3-Ethylhexane, CA 619-99-8: Flammable; soluble most organic solvents. Tb = 389.15 to 392.15; Vm = 0.1601 at 293.15; Antoine (liq) A = 6.01964, B = 1330.492, C = -60.22, dev = 0.02, range = 314 to 393.

3-Ethyl-2-methylpentane, CA 609-26-7: Flammable; soluble most organic solvents. Tm = 158.15; Tb = 388.75; Vm = 0.1588 at 293.15; Antoine (liq) A = 5.97587, B = 1310.709, C = -58.667, dev = 0.02, range = 311 to 390.

3-Ethyl-3-methylpentane, CA 1067-08-9: Flammable; soluble most organic solvents. Tm = 182.25; Tb = 391.35; Vm = 0.1570 at 293.15; Antoine (liq) A = 5.98793, B = 1344.948, C = -53.678, dev = 0.02, range = 312 to 393.

2-Methylheptane, CA 592-27-8: Flammable, flash point 277; soluble most organic solvents. Tf = 161.85; Tb = 390.75; Vm = 0.1637 at 293.15; Antoine I (liq) A = 6.05858, B = 1346.996, C = -58.436, dev = 0.02, range = 285 to 392; Antoine II (liq) A = 6.81199, B = 1703.6, C = -30.648, dev = 0.1, range = 233 to 286.

3-Methylheptane ±, CA 589-81-1: Flammable; soluble most organic solvents. Tm = 152.65; Tb = 393.15 to 395.15; Vm = 0.1618 at 293.15; Antoine I (liq) A = 6.02047, B = 1329.42, C = -60.945, dev = 0.02, range = 286 to 393; Antoine II (liq) A = 6.50909, B = 1567.46, C = -40.786, dev = 0.1, range = 238 to 286.

4-Methylheptane, CA 589-53-7: Flammable; soluble most organic solvents. Tf = 152.15; Tb = 391.15; Vm = 0.1621 at 293.15; Antoine (liq) A = 6.02799, B = 1329.347, C = -60.367, dev = 0.02, range = 312 to 392.

Octane, CA 111-65-9: Flammable, flash point 286; soluble most organic solvents. Tm = 216.35; Tb = 398.81; Vm = 0.1626 at 293.15; Tc = 568.83; Pc = 2486; Vc = 0.492; Antoine I (liq) A = 6.04231, B = 1351.497, C = -64.014, dev = 0.02, range = 297 to 400; Antoine II (liq) A = 7.90115, B = 2238.9, C = -4.53, dev = 1.0, range = 216 to 278; Antoine III (liq) A = 6.16936, B = 1440.32, C = -52.894, dev = 0.1, range = 396 to 432; Antoine IV (liq) A = 6.23406, B = 1492.068, C = -45.851, dev = 0.1, range = 428 to 510; Antoine V (liq) A = 7.66614, B = 3108.961, C = 159.091, dev = 1.0, range = 506 to 569.

2,2,3,3-Tetramethylbutane, CA 594-82-1: Flammable; soluble most organic solvents. Tm = 377.15; Tb = 379.15 to 380.15; Vm = 0.1761 at 383.15; Antoine I (sol) A = 6.85582, B = 1601.54, C = -48.39, *(continues)*

C_8H_{18} *(continued)*

dev = 0.1, range = 286 to 377; Antoine II (liq) A = 6.00155, B = 1329.93, C = -46.79, dev = 0.1, range = 377 to 390.

2,2,3-Trimethylpentane ±, CA 564-02-3: Flammable, flash point below 294; soluble most organic solvents. Tm = 160.85; Tb = 382.95; Vm = 0.1595 at 293.15; Antoine (liq) A = 5.94504, B = 1292.006, C = -55.019, dev = 0.02, range = 306 to 384.

2,2,4-Trimethylpentane, isooctane, CA 540-84-1: Flammable, flash point 261; soluble most organic solvents. Tf = 165.75; Tb = 372.35; Vm = 0.1651 at 293.15; Antoine I (liq) A = 5.93034, B = 1254.146, C = -52.831, dev = 0.02, range = 297 to 374; Antoine II (liq) A = 6.44016, B = 1650.17, C = 0, dev = 1.0, range = 423 to 523; Antoine III (liq) A = 6.33252, B = 1441.485, C = -36.695, dev = 1.0, range = 194 to 299; Antoine IV (liq) A = 5.97534, B = 1283.067, C = -49.166, dev = 0.1, range = 372 to 416; Antoine V (liq) A = 6.26002, B = 1501.036, C = -19.15, dev = 0.1, range = 413 to 494; Antoine VI (liq) A = 7.76427, B = 3268.783, C = 206.659, dev = 0.1 to 1.0, range = 490 to 544.

2,3,3-Trimethylpentane, CA 560-21-4: Flammable, flash point below 294; soluble most organic solvents. Tm = 172.45; Tb = 387.75; Vm = 0.1573 at 293.15; Antoine (liq) A = 5.96425, B = 1325.827, C = -52.988, dev = 0.02, range = 308 to 390.

2,3,4-Trimethylpentane, CA 565-75-3: Flammable, flash point 278; soluble most organic solvents. Tf = 163.95; Tb = 385.95; Vm = 0.1589 at 293.15; Antoine I (liq) A = 6.00347, B = 1330.047, C = -53.921, dev = 0.02, range = 288 to 400; Antoine II (liq) A = 6.37038, B = 1511.86, C = -38.054, dev = 0.1, range = 223 to 289.

$C_8H_{18}N_2$ MW 142.24

2,3,5,6-Tetramethylpiperazine, CA 6135-46-2: Exists as five stereoisomers; flammable; soluble water, ethanol. Antoine (liq) A = 5.61153, B = 1282.67, C = -100.89, dev = 1.0, range = 296 to 457.

$C_8H_{18}O$ MW 130.23

Butyl *tert*-butyl ether, CA 100-63-1: Flammable; forms explosive peroxides; soluble ethanol, ether. Antoine I (liq) A = 6.3471, B = 1518.2, C = -48.05, dev = 1.0, range = 356 to 397; Antoine II (liq) A = 5.81547, B = 1212.51, C = -77.985, dev = 1.0, range = 293 to 397.

Butyl isobutyl ether, CA 17071-47-5: Flammable; forms explosive peroxides; soluble ethanol, ether, acetone. Vm = 0.1632 at 295.15; Antoine (liq) A = 6.0182, B = 1354.5, C = -67.77, dev = 1.0, range = 328 to 406.

Dibutyl ether, CA 142-96-1: Toxic by inhalation; flammable, flash point 298; forms explosive peroxides; soluble ethanol, ether, acetone. Tm = 175.15; Tb = 415.15; Vm = 0.1690 at 293.15; Tc = 581, calculated; Pc = 2460, calculated; Vc = 0.487, calculated; Antoine I (liq) A = 6.4403, B = 1648.4, C = -43.15, dev = 0.1, range = 339 to 415; Antoine II (liq) A = 6.0537, B = 1398.8, C = -69.55, dev = 1.0, range = 336 to 415.

Di-*sec*-butyl ether, CA 6863-58-7: Flammable; forms explosive peroxides; soluble ethanol, ether, acetone; mixture of stereoisomers.   *(continues)*

$C_8H_{18}O$ *(continued)*

Tb = 395.15; Vm = 0.1723 at 298.15; Antoine (liq) A = 5.97706, B = 1289.804, C = -69.442, dev = 1.0, range = 284 to 464.

Di-*tert*-butyl ether, CA 6163-66-2: Flammable; forms explosive peroxides; soluble ethanol, ether. Tb = 381.15; Vm = 0.1709 at 293.15; Antoine I (liq) A = 5.71064, B = 1151.3, C = -71.15, dev = 1.0, range = 277 to 382; Antoine II (liq) A = 5.94934, B = 1269.314, C = -58.486, dev = 0.1, range = 289 to 382.

Diisobutyl ether, CA 628-55-7: Flammable; forms explosive peroxides; soluble ethanol, ether. Tm = 175.25; Tb = 396.15; Vm = 0.1710 at 288.15; Antoine (liq) A = 5.983, B = 1314.5, C = -65.27, dev = 1.0, range = 320 to 396.

2-Ethyl-1-hexanol, CA 104-76-7: Irritant; flammable, flash point 346; slightly soluble water; soluble ethanol, ether, acetone, benzene. Tm = 197.15; Tb = 457.15; Vm = 0.1564 at 293.15; Tc = 640.25; Pc = 2730, calculated; Vc = 0.485, calculated; Antoine (liq) A = 6.0138, B = 1325.88, C = -126.69, dev = 1.0, range = 347 to 457.

2-Ethyl-4-methyl-1-pentanol, CA 106-67-2: Flammable; slightly soluble water; soluble ethanol, ether. Antoine (liq) A = 6.0541, B = 1326.13, C = -122.63, dev = 1.0, range = 343 to 450.

2-Methyl-1-heptanol, CA 60435-70-3: Flammable; soluble ethanol, acetone, ether. Tm = 161.15; Antoine (liq) A = 5.28502, B = 989.2, C = -147.13, dev = 1.0, range = 350 to 449.

3-Methyl-1-heptanol, CA 1070-32-2: Flammable; soluble ethanol, ether, acetone. Tm = 183.15; Vm = 0.1580 at 297.15; Antoine (liq) A = 6.29982, B = 1570.3, C = -93.46, dev = 1.0, range = 360 to 459.

4-Methyl-1-heptanol, CA 817-91-4: Flammable; soluble ethanol, ether, acetone. Tb = 461.15 to 466.15; Vm = 0.1615 at 298.15; Antoine (liq) A = 4.96578, B = 841.5, C = -171.95, dev = 1.0 to 5.0, range = 357 to 456.

5-Methyl-1-heptanol ±, CA 57803-73-3: Flammable; soluble ethanol, acetone, ether. Tm = 169.15; Tb = 459.65; Vm = 0.1598 at 298.15; Antoine (liq) A = 5.59771, B = 1123.2, C = -147.05, dev = 1.0, range = 364 to 460.

6-Methyl-1-heptanol, CA 1653-40-3: Flammable; soluble ethanol, ether, acetone. Tm = 167.15; Tb = 460.75; Vm = 0.1593 at 298.15; Antoine (liq) A = 5.67384, B = 1119.4, C = -155.64, dev = 1.0, range = 368 to 461.

2-Methyl-2-heptanol, CA 625-25-2: Flammable; soluble ethanol, ether, acetone. Tm = 222.75; Tb = 434.35; Vm = 0.1599 at 293.15; Antoine (liq) A = 7.43008, B = 2059.1, C = -49.59, dev = 1.0, range = 343 to 430.

3-Methyl-2-heptanol, CA 31367-46-1: Flammable; soluble ethanol, ether, acetone. Tb = 445.15 to 446.15; Vm = 0.1593 at 298.15; Antoine (liq) A = 6.36406, B = 1604.9, C = -70.98, dev = 1.0, range = 341 to 440.

4-Methyl-2-heptanol: Flammable; soluble ethanol, ether, acetone. Tm = 171.15; Tb = 441.15; Antoine (liq) A = 5.88925, B = 1252.6, C = -122.19, dev = 1.0, range = 351 to 445.

5-Methyl-2-heptanol, CA 54630-50-1: Flammable; soluble ethanol, acetone, ether. Tm = 153.15; Tb = 440.15; Vm = 0.1593 at 294.15; Antoine (liq) A = 6.02559, B = 1366.9, C = -104.95, dev = 1.0, range = 348 to 445. *(continues)*

$C_8H_{18}O$ *(continued)*

6-Methyl-2-heptanol ±, CA 4730-22-7: Flammable; soluble ethanol, acetone, ether. Tm = 168.15; Tb = 449.15; Vm = 0.1585 at 293.15; Antoine (liq) A = 6.32719, B = 1467.1, C = -105.53, dev = 1.0, range = 354 to 445.

2-Methyl-3-heptanol ±, CA 18720-62-2: Flammable; soluble ethanol, acetone, ether. Tm = 188.15; Tb = 440.35; Vm = 0.1581 at 293.15; Antoine (liq) A = 5.87014, B = 1211.0, C = -126.89, dev = 1.0, range = 349 to 441.

3-Methyl-3-heptanol, CA 5582-82-1: Flammable; soluble ethanol, acetone, ether. Tm = 190.15; Tb = 436.15 to 438.15; Vm = 0.1572 at 293.15; Antoine (liq) A = 6.06206, B = 1280.8, C = -117.14, dev = 1.0, range = 344 to 433.

4-Methyl-3-heptanol, CA 14979-39-6: Flammable, flash point 327; soluble ethanol, ether, acetone. Tm = 150.15; Tb = 433.15 to 434.15; Vm = 0.1575 at 293.15; Antoine (liq) A = 7.32341, B = 2270.5, C = -1.66, dev = 1.0, range = 330 to 429.

5-Methyl-3-heptanol, CA 18720-65-5: Flammable; soluble ethanol, acetone, ether. Tm = 181.95; Tb = 440.15 to 441.15; Vm = 0.1546 at 298.15; Antoine (liq) A = 5.8554, B = 1278.8, C = -94.57, dev = 1.0, range = 330 to 427.

6-Methyl-3-heptanol ±, CA 18720-66-6: Flammable; soluble ethanol, acetone, ether. Tm = 214.65; Tb = 438.15 to 439.15; Antoine (liq) A = 5.51635, B = 1110.2, C = -115.27, dev = 1.0, range = 333 to 432.

2-Methyl-4-heptanol, CA 21570-35-4: Flammable; soluble ethanol, acetone, ether. Tm = 192.15; Antoine (liq) A = 6.01902, B = 1282.8, C = -119.71, dev = 1.0, range = 348 to 440.

3-Methyl-4-heptanol ±, CA 1838-73-9: Flammable; soluble ethanol, acetone, ether. Tb = 443.15; Vm = 0.1562 at 298.15; Antoine (liq) A = 6.37685, B = 1603.15, C = -71.06, dev = 1.0, range = 340 to 438.

4-Methyl-4-heptanol, CA 598-01-6: Flammable; soluble ethanol, ether, acetone. Tm = 191.15; Tb = 434.65; Vm = 0.1579 at 293.15; Antoine (liq) A = 5.72236, B = 1118.0, C = -133.57, dev = 1.0, range = 344 to 434.

1-Octanol, octyl alcohol, CA 111-87-5: Flammable, flash point 354; soluble ethanol, ether, chloroform. Tm = 356.45; Tb = 467.6; Vm = 0.1575 at 293.15; Antoine I (liq) A = 5.90052, B = 1273.291, C = -141.417, dev = 0.1, range = 386 to 480; Antoine II (liq) A = 9.342, B = 3343, C = 0, dev = 5.0, range = 267 to 282; Antoine III (liq) A = 18.014, B = 5507, C = 0, dev = 5.0, range = 238 to 251; Antoine IV (liq) A = 5.7934, B = 1208.201, C = -149.366, dev = 0.02, range = 430 to 474; Antoine V (liq) A = 6.39406, B = 1540.599, C = -115.618, dev = 0.1 to 1.0, range = 328 to 400; Antoine VI (liq) A = 5.90632, B = 1276.86, C = -140.996, dev = 0.1, range = 397 to 479; Antoine VII (liq) A = 6.00161, B = 1357.422, C = -128.524, dev = 1.0, range = 475 to 555.

2-Octanol ±, CA 4128-31-8: Flammable, flash point 344; soluble ethanol, ether, acetone. Tm = 241.55; Tb = 453.15; Vm = 0.1590 at 298.15; Antoine (liq) A = 6.2306, B = 1426.9, C = -114.25, dev = 1.0, range = 333 to 453.

3-Octanol ±, CA 20296-29-1: Flammable, flash point 338; soluble ethanol, ether. Tm = 228.15; Tb = 449.15 to 450.65; Vm = 0.1570 at 288.15; Antoine (liq) A = 5.9359, B = 1278.1, C = -124.95, dev = 1.0, range = 313 to 450.

(continues)

$C_8H_{18}O$ *(continued)*

4-Octanol ±, CA 589-62-8: Flammable; soluble ethanol, ether. Tm = 232.45; Tb = 449.45; Vm = 0.1591 at 293.15; Antoine (liq) A = 5.9879, B = 1305.7, C = -121.25, dev = 1.0, range = 343 to 450.

2,4,4-Trimethyl-1-pentanol ±, CA 16325-63-6: Flammable, flash point 333; soluble ethanol. Tb = 441.95 to 442.95; Vm = 0.1592 at 293.15; Antoine (liq) A = 6.16712, B = 1403.368, C = -108.316, dev = 1.0, range = 352 to 446.

$C_8H_{18}O_2$ MW 146.23

Di-*tert*-butyl peroxide, CA 110-05-4: Irritant; flammable, flash point 291; explosive; slightly soluble water; soluble acetone, light hydrocarbons. Tf = 233.15; Tb = 384.15; Vm = 0.1842 at 293.15; Antoine I (liq) A = 5.9209, B = 1233.2, C = -69.15, dev = 1.0, range = 273 to 384; Antoine II (liq) A = 6.21707, B = 1671.082, C = 0, dev = 1.0 to 5.0, range = 246 to 311.

Ethylene glycol, dipropyl ether, 1,2-dipropoxyethane, CA 18854-56-3: Flammable; forms explosive peroxides; soluble ether. Antoine (liq) A = 4.61779, B = 1095.06, C = -34.038, dev = 1.0, range = 234 to 453.

Ethylene glycol, mono(2-ethylbutyl) ether, CA 4468-93-3: Flammable, flash point 358; forms explosive peroxides; moderately soluble water; soluble ethanol, ether. Tm = 183.15; Tb = 469.85; Vm = 0.1630 at 293.15; Antoine (liq) A = 7.9653, B = 2792.3, C = 0, dev = 5.0, range = 357 to 470.

Ethylene glycol, monohexyl ether, CA 112-25-4: Flammable, flash point 364; forms explosive peroxides; moderately soluble water; soluble ethanol, ether. Tm = 223.05; Tb = 481.25; Vm = 0.1642 at 293.15; Antoine (liq) A = 7.9423, B = 2852.4, C = 0, dev = 1.0 to 5.0, range = 363 to 483.

3-Hydroxymethyl-4-heptanol: Flammable; soluble ethanol. Antoine (liq) A = 9.3887, B = 4246.3, C = 58.05, dev = 1.0 to 5.0, range = 375 to 518.

2,2,4-Trimethyl-1,3-pentanediol, CA 144-19-4: Flammable; soluble ethanol, water, ether. Tm = 325.15; Tb = 507.15; Antoine (liq) A = 8.0865, B = 3054.8, C = 0, dev = 1.0 to 5.0, range = 413 to 502.

$C_8H_{18}O_3$ MW 162.23

Diethylene glycol, diethyl ether, CA 112-36-7: Flammable, flash point 355; forms explosive peroxides; soluble water, ethanol, ether. Tm = 232.85; Tb = 462.15; Vm = 0.1790 at 293.15; Antoine (liq) A = 7.470, B = 2525, C = 0, dev = 1.0 to 5.0, range = 330 to 461.

Diethylene glycol, monobutyl ether, CA 112-34-5: Irritant; flammable, flash point 351; forms explosive peroxides; soluble water, ethanol, ether, acetone. Tm = 205.05; Tb = 503.75; Vm = 0.1698 at 293.15; Antoine (liq) A = 5.98752, B = 1542.8, C = -116.65, dev = 1.0, range = 415 to 505.

$C_8H_{18}O_4S_2$ MW 242.35

2,2-Butanediol, bis(ethylsulfonate), CA 76-20-0: Irritant; flammable; moderately soluble hot water; soluble ether, ethanol, benzene; *(continues)*

$C_8H_{18}O_4S_2$ *(continued)*

decomposes at boiling point. Tm = 349.15; Vm = 0.2021 at 358.15; Antoine
(liq) A = 8.8484, B = 3954.8, C = 0, dev = 1.0, range = 443 to 493.

$C_8H_{18}O_5$ MW 194.23

Tetraethylene glycol, CA 112-60-7: Irritant; flammable, flash point 449;
very hygroscopic; soluble water, ethanol, ether. Tm = 266.95; Tb = 601.15;
Vm = 0.1721 at 288.15; Antoine (liq) A = 9.4476, B = 4051.16, C = -36.521,
dev = 1.0, range = 426 to 581.

$C_8H_{18}S$ MW 146.29

Dibutyl sulfide, *alpha* form, CA 544-40-1: Toxic; stench; flammable, flash
point 349; soluble ethanol, ether, chloroform. Tm = 197.15; Tb = 458.15 to
458.65; Vm = 0.1744 at 293.15; Antoine (liq) A = 6.11001, B = 1573.284, C =
-78.769, dev = 1.0, range = 390 to 470.

Di-*tert*-butyl sulfide, CA 107-47-1: Toxic; stench; flammable, flash point
321; soluble ethanol, ether. Tb = 422.15; Vm = 0.1795 at 293.15; Antoine
(liq) A = 7.31554, B = 2216.64, C = 0, dev = 1.0 to 5.0, range = 324 to 420.

Diisobutyl sulfide, CA 592-65-4: Toxic; stench; flammable; soluble ether,
ethanol. Tm = 167.65; Tb = 445.15 to 446.15; Vm = 0.1749 at 283.15;
Antoine (liq) A = 7.55672, B = 2421.775, C = 0, dev = 1.0, range = 325
to 346.

1-Octanethiol, octyl mercaptan, CA 111-88-6: Tm = 223.95; Tb = 472.25;
Vm = 0.1735 at 293.15; Antoine (liq) A = 6.08472, B = 1580.8, C = -84.66,
dev = 1.0, range = 372 to 473.

2-Octanethiol ±, CA 10435-81-1: Toxic; stench; flammable; soluble ethanol,
ether, benzene. Tm = 194.15; Tb = 459.55; Vm = 0.1749 at 293.15; Antoine
(liq) A = 5.8915, B = 1426.9, C = -92.31, dev = 1.0, range = 361 to 460.

$C_8H_{18}S_2$ MW 178.35

Dibutyl disulfide, CA 629-45-8: Toxic; stench; flammable, flash point 366.
Tb = 499.15; Vm = 0.1901 at 293.15; Antoine (liq) A = 4.8846, B = 944.9,
C = -55.25, dev = 1.0 to 5.0, range = 283 to 353.

1,8-Octanedithiol, CA 1191-62-4: Toxic; stench; flammable. Tm = 272.25;
Antoine (liq) A = 6.2279, B = 1881.1, C = -96.93, dev = 1.0, range = 405
to 543.

$C_8H_{19}N$ MW 129.24

Butyl isobutyl amine, CA 20810-06-4: Irritant; corrosive; flammable;
moderately soluble water; soluble ethanol, ether. Antoine (liq) A = 7.115,
B = 2154, C = 0, dev = 5.0, range = 313 to 423.

Dibutyl amine, CA 111-92-2: Corrosive; toxic; irritant; flammable, flash
point 320; moderately soluble water; soluble ethanol, ether, acetone. Tf =
211.25; Tb = 432.75; Vm = 0.1685 at 293.15; Antoine (liq) A = 6.39093, B =
1616.4, C = -64.15, dev = 1.0, range = 343 to 479.

(continues)

$C_8H_{19}N$ *(continued)*

Diisobutyl amine, CA 110-96-3: Corrosive; irritant; flammable, flash point 302; moderately soluble water; soluble ethanol, ether, acetone, benzene. Tm = 203.15; Tb = 412.65; Vm = 0.1746 at 293.15; Antoine (liq) A = 6.63622, B = 1745.74, C = -35.62, dev = 1.0, range = 268 to 413.

2-Ethylhexyl amine, CA 104-75-6: Irritant; toxic; flammable, flash point 333. Tm = 197.15; Tb = 442.15; Vm = 0.1638 at 293.15; Antoine (liq) A = 7.3013, B = 2342.5, C = 0, dev = 5.0, range = 341 to 447.

Octyl amine, CA 111-86-4: Corrosive; flammable, flash point 333. Tm = 268.15 to 272.15; Tb = 458.15 to 460.15; Vm = 0.1651 at 293.15; Antoine (liq) A = 6.37446, B = 1694.81, C = -64.73, dev = 1.0, range = 308 to 453.

$C_8H_{19}O_2PS_3$ MW 274.39

O,O-Diethyl-*S*-[2-(ethylthio)ethyl] dithiophosphate, CA 298-04-4: Insecticide; highly toxic. Vm = 0.2399 at 293.15; Antoine (liq) A = 9.73892, B = 4420.5, C = 14.9, dev = 1.0, range = 283 to 401.

$C_8H_{19}O_3P$ MW 194.21

Dibutyl phosphite, CA 1809-19-4: Corrosive; moderately toxic; flammable, flash point 394; moisture-sensitive. Vm = 0.1952 at 293.15; Antoine (liq) A = 5.9249, B = 1973.2, C = 0, dev = 1.0, range = 298 to 438.

$C_8H_{19}O_3PS_2$ MW 258.33

O,O-Diethyl-*O*-[2-(ethylthio)ethyl] thiophosphate, CA 298-03-3: Insecticide; highly toxic; suspected carcinogen; decomposes above 403. Vm = 0.2309 at 294.15; Antoine (liq) A = 9.54584, B = 4110.9, C = 0, dev = 1.0, range = 283 to 411.

O,O-Diethyl-*S*-[2-(ethylthio)ethyl] thiophosphate, CA 126-75-0: Insecticide; highly toxic; suspected carcinogen; soluble ethanol, benzene. Vm = 0.2282 at 294.15; Antoine (liq) A = 9.1587, B = 3991, C = 0, dev = 1.0 to 5.0, range = 283 to 401.

$C_8H_{20}ClN$ MW 165.71

Dibutyl ammonium chloride, CA 6287-40-7: Tm = 554.15; Antoine (liq) A = 12.7639, B = 6096, C = 0, dev = 1.0, range = 553 to 563.

$C_8H_{20}N_2$ MW 144.26

Tetraethylhydrazine, CA 4267-00-9: Toxic; flammable. Antoine (liq) A = 6.04527, B = 1747, C = 0, dev = 1.0, range = 308 to 368.

N,N,N',*N'*-Tetramethyl-1,3-butanediamine, CA 97-84-7: Corrosive; flammable, flash point 319. Tm = below 173.15; Tb = 438.25; Antoine (liq) A = 7.092, B = 2229, C = 0, dev = 5.0, range = 335 to 439.

$C_8H_{20}N_2O_2S$ MW 208.32

Sulfamide, tetraethyl, CA 2832-49-7: Irritant; flammable, flash point
above 385. Tm = 311.65; Tb = 522.15 to 524.15; Antoine (liq) A = 7.84608,
B = 3087.089, C = 0, dev = 1.0 to 5.0, range = 407 to 528.

$C_8H_{20}O_5P_2S_2$ MW 322.31

Dithiopyrophosphoric acid, tetraethyl ester, CA 3689-24-5: Insecticide;
corrosive; highly toxic; slightly soluble water; soluble most organic solvents. Vm = 0.2695 at 298.15; Antoine (liq) A = 9.71747, B = 4211.7,
C = 0, dev = 1.0, range = 293 to 409.

$C_8H_{20}O_7P_2$ MW 290.19

Pyrophosphoric acid, tetraethyl ester, CA 107-49-3: Soluble water with decomposition; soluble most organic solvents; decomposes above 443. Vm =
0.2449 at 293.15; Antoine (liq) A = 9.9713, B = 4296, C = 0, dev = 5.0,
range = 283 to 411.

$C_8H_{23}N_5$ MW 189.30

Tetraethylene pentamine, CA 112-57-2: Irritant; corrosive; toxic; flammable, flash point 436; soluble water. Tm = below 233.15; Tb = 613.15 to
616.15; Antoine (liq) A = 8.1156, B = 3723.2, C = 0, dev = 5.0, range = 463
to 615.

$C_8H_{24}N_4O_3P_2$ MW 286.25

Pyrophosphoric acid, tetrakis(dimethylamide), CA 152-16-9: Insecticide;
highly toxic; soluble water, most organic solvents. Tm = 293.15; Vm =
0.2524 at 298.15; Antoine (liq) A = 8.6093, B = 4027, C = 24.45, dev = 5.0,
range = 273 to 415.

$C_9F_{17}NO_3S$ MW 525.14

1-Octanesulfonylisocyanate, heptadecafluoro, CA 34834-20-3: Irritant;
toxic; lachrymator. Tm = 295.15; Antoine (liq) A = 5.95921, B = 1342.956,
C = -129.793, dev = 1.0 to 5.0, range = 324 to 470.

$C_9F_{18}N_2$ MW 478.08

2,2-Propanediamine, 1,1,1,3,3,3-hexafluoro-N,N'-bis[2,2,2-trifluoro-1-
(trifluoromethyl)ethylidene], CA 34451-14-4: Antoine (liq) A = 6.885, B =
1857, C = 0, dev = 1.0, range = 314 to 381.

$C_9F_{18}O_3$ MW 498.07

Carbonic acid, bis[1,1,1,3,3,3-hexafluoro-2-(trifluoromethyl)-2-propyl]
ester, CA 40719-69-5: Tm = 316.15; Antoine (liq) A = 7.2874, B = 2074.8,
C = 0, dev = 1.0, range = 316 to 358.

$C_9F_{19}NO$ MW 499.07

Ethanimidic acid, 2,2,2-trifluoro-N-[1,2,2,2-tetrafluoro-1-(trifluoromethyl)ethyl]-1,2,2,2-tetrafluoro-(trifluoromethyl)ethyl ester, CA 54120-
06-8: Antoine (liq) A = 7.105, B = 1965, C = 0, dev = 1.0 to 5.0, range
not specified.

C_9F_{20} MW 488.07

Perfluorononane, CA 375-96-2: Tb = 396.15; Vm = 0.2712 at 298.15; Antoine
(liq) A = 7.41308, B = 2608.311, C = 94.031, dev = 1.0 to 5.0, range = 387
to 524.

$C_9F_{21}N$ MW 521.07

Tris(heptafluoropropyl) amine, CA 338-83-0: Tb = 403.15; Vm = 0.2860 at
277.15; Antoine (liq) A = 7.2679, B = 2121, C = 0, dev = 1.0, range = 333
to 403.

$C_9H_4O_5$ MW 192.13

Trimellitic acid, anhydride, CA 552-30-7: Irritant by inhalation; highly
toxic; moisture-sensitive; soluble acetone, ethyl acetate. Tm = 441.15;
Tb = 663.15; Antoine (liq) A = 7.381, B = 3429.2, C = 0, dev = above 5.0,
range = 558 to 595.

$C_9H_5Br_2NO$ MW 302.95

Quinoline, 5,7-dibromo-8-hydroxy, CA 521-74-4: Irritant; soluble ethanol,
acetone, benzene, chloroform. Tm = 469.15; Antoine (sol) A = 9.715, B =
4910, C = 0, dev = 5.0, range = 323 to 383.

C_9H_5ClINO MW 305.50

Quinoline, 7-chloro-5-iodo-8-hydroxy, CA 35048-13-6: Antoine (sol) A =
14.675, B = 6850, C = 0, dev = 5.0, range = 363 to 383.

$C_9H_5ClN_2O_2$ MW 208.60

Benzene, 5-chloro-2,4-diisocyanato-1-methyl, CA 15166-26-4: Toxic; lachrymator; flammable; reacts violently with water. Tm = 294.45; Antoine (liq)
A = 8.51121, B = 3484.305, C = 0, dev = 1.0 to 5.0, range = 373 to 433.

$C_9H_5Cl_2NO$ MW 214.05

Quinoline, 5,7-dichloro-8-hydroxy, CA 773-76-2: Irritant; soluble acetone,
benzene; bactericide used in shampoos. Tm = 453.15 to 454.15; Antoine
(sol) A = 10.405, B = 4860, C = 0, dev = 5.0, range = 363 to 383.

$C_9H_5Cl_3F_{12}$ MW 447.48

1,1,2,2,3,3,4,4,5,5,6,6-Dodecafluoro-1,7,7-trichlorononane, CA 16327-68-7:
Antoine (liq) A = 5.4407, B = 1730.7, C = 0, dev not specified, range not
specified.

294

$C_9H_5I_2NO$ MW 396.95

Quinoline, 5,7-diiodo-8-hydroxy, CA 83-73-8: Suspected carcinogen; soluble
hot pyridine, hot dioxane; used as fungicide. Tm = 483.15 to 488.15, decomposes; Antoine (sol) A = 10.995, B = 5790, C = 0, dev = 5.0, range = 323
to 383.

C_9H_6INO MW 271.06

Quinoline, 5-iodo-8-hydroxy, CA 13207-63-1: Tm = 408.15; Antoine (sol) A =
14.095, B = 6200, C = 0, dev = 5.0, range = 363 to 383.

$C_9H_6N_2O_2$ MW 174.16

2,4-Toluene diisocyanate, CA 584-84-9: Highly toxic by inhalation and skin
absorption; lachrymator; flammable, flash point 400; reacts violently with
water; soluble ethanol with decomposition; soluble ether, acetone, benzene;
raw material for polyurethanes. Tm = 295.15; Tb = 524.15; Vm = 0.1426 at
298.15; Antoine I (liq) A = 7.1493, B = 2453, C = -43.95, dev = 1.0 to 5.0,
range = 373 to 530; Antoine II (liq) A = 6.60938, B = 2071.438, C =
-74.713, dev = 1.0, range = 393 to 530.

2,6-Toluene diisocyanate, CA 91-08-7: Highly toxic by inhalation and skin
absorption; lachrymator; flammable, flash point above 385; reacts violently
with water; soluble ether, acetone, benzene; raw material for polyurethanes.
Tm = 291.45; Antoine (liq) A = 7.3601, B = 2636.6, C = -33.35, dev = 1.0
to 5.0, range = 373 to 463.

$C_9H_6O_2$ MW 146.15

Coumarin, 1,2-benzopyrone, CA 91-64-5: Toxic; suspected carcinogen; flammable; polymerizes; moderately soluble hot water; soluble ethanol, ether,
chloroform; used in perfumery. Tm = 343.75; Tb = 570.15 to 572.15; Antoine
I (sol) A = 11.209, B = 4530, C = 0, dev = 1.0 to 5.0, range = 293 to 326;
Antoine II (liq) A = 7.6995, B = 3163.95, C = -8.406, dev = 1.0 to 5.0,
range = 379 to 564.

$C_9H_6O_6$ MW 210.14

Trimesic acid, 1,3,5-benzenetricarboxylic acid, CA 554-95-0: Very soluble
hot water; soluble ethanol, ether. Tm = 653.15; Antoine (sol) A = 13.825,
B = 7492, C = -43.15; dev = 1.0 to 5.0, range = 553 to 593.

$C_9H_7Cl_3O_3$ MW 269.51

2,4,5-Trichlorophenoxyacetic acid, methyl ester, CA 1928-37-6: Irritant;
flammable. Antoine (liq) A = 5.26768, B = 1336.646, C = -193.999, dev =
1.0, range = 444 to 573.

$C_9H_7F_3O_2$ MW 204.15

Trifluoroacetic acid, 3-tolyl ester, CA 1736-09-0: Irritant; flammable.
Antoine (liq) A = 7.0079, B = 2017.3, C = -36.85, dev = 1.0, range = 363
to 439.
                                                         (continues)

$C_9H_7F_3O_2$ *(continued)*

Trifluoroacetic acid, 4-tolyl ester, CA 1813-29-2: Irritant; flammable. Antoine (liq) A = 6.8087, B = 1890, C = -49.15, dev = 1.0, range = 365 to 442.

C_9H_7N MW 129.16

Isoquinoline, CA 119-65-3: Toxic; irritant; flammable, flash point 380; strong base; slightly soluble water; soluble ethanol, ether, acetone, benzene. Tm = 297.75; Tb = 516.4; Vm = 0.1184 at 303.15; Tc = 803.15; Antoine (liq) A = 6.03204, B = 1719.5, C = -89.32, dev = 0.1, range = 439 to 517.

Quinoline, CA 91-22-5: Toxic; irritant; flammable, flash point 374; weak base; hygroscopic; moderately soluble hot water; soluble ethanol, acetone, ether, benzene. Tf = 257.55; Tb = 510.78; Vm = 0.1185 at 298.15; Tc = 782.15; Antoine I (liq) A = 5.92679, B = 1656.3, C = -88.37, dev = 0.1, range = 433 to 511; Antoine II (liq) A = 7.15102, B = 2846.253, C = 41.795, dev = 5.0, range = 463 to 794.

C_9H_7NO MW 145.16

Benzoyl acetonitrile, CA 614-16-4: Irritant; flammable; soluble ethanol, ether, benzene. Tm = 353.15 to 354.15; Antoine (sol) A = 16.2187, B = 5216, C = 0, dev = 1.0, range = 318 to 333.

8-Hydroxyquinoline, 8-quinolinol, CA 148-24-3: Irritant; suspected carcinogen; flammable; light-sensitive; soluble ethanol, acetone, chloroform, benzene. Tm = 348.15 to 349.15; Tb = 539.75; Antoine (sol) A = 14.615, B = 5690, C = 0, dev = 5.0, range = 308 to 333.

C_9H_8 MW 116.16

Indene, CA 95-13-6: Irritant; flammable, flash point 331; polymerizes on standing; easily oxidized; soluble ethanol, ether, acetone, benzene. Tm = 271.15; Tb = 455.35 to 455.55; Vm = 0.1166 at 298.15; Antoine I (liq) A = 7.28502, B = 2463.85, C = 12.046, dev = 1.0, range = 289 to 455; Antoine II (liq) A = 6.0443, B = 1541.955, C = -73.959, dev = 0.1 to 1.0, range = 369 to 457.

$C_9H_8Cl_2O_3$ MW 235.07

2,4-Dichlorophenoxyacetic acid, methyl ester, CA 1928-38-7: Irritant; suspected carcinogen; flammable; used as weed killer. Antoine (liq) A = 6.92535, B = 2566.432, C = -62.487, dev = 1.0 to 5.0, range = 403 to 548.

$C_9H_8N_2O_2$ MW 176.17

3-(Methylamino) phthalimide, CA 5972-09-8: Antoine (sol) A = 11.0289, B = 5480, C = 0, dev = 1.0 to 5.0, range = 402 to 450.

C_9H_8O MW 132.16

Cinnamaldehyde, 3-phenyl-2-propenal, CA 14371-10-9: Irritant, flammable, flash point 344; slightly soluble water; soluble ethanol,   *(continues)*

296

C_9H_8O *(continued)*

ether, chloroform; used in perfumes and flavorings. Tm = 265.65; Tb = 525.15; Vm = 0.1259 at 293.15; Antoine I (liq) A = 7.79733, B = 2989.85, C = -2.981, dev = 1.0 to 5.0, range = 349 to 519; Antoine II (liq) A = 9.8667, B = 3796.054, C = 0, dev = 1.0 to 5.0, range = 353 to 373.

1-Ethynyl-2-methoxybenzene, CA 767-91-9: Flammable; polymerizes; Antoine (liq) A = 3.3012, B = 1081.4, C = 0, dev = 5.0, range = 297 to 371.

$C_9H_8O_2$ MW 148.16

Cinnamic acid, *trans*, CA 140-10-3: Flammable; decomposes when distilled; slightly soluble water; soluble ethanol, ether, acetone, benzene. Tm = 406.15; Tb = 573.15; Antoine (liq) A = 8.18763, B = 3356.69, C = -30.188, dev = 1.0, range = 430 to 573.

$C_9H_8O_4$ MW 180.16

Terephthalic acid, monomethyl ester, CA 1679-64-7: Tm = 503.15; Antoine (sol) A = 8.880, B = 3765, C = 0, dev = 5.0, range = 433 to 493.

$C_9H_9F_6NO_5$ MW 325.16

N,O-Bis(trifluoroacetyl)-threonine, *levo*, methyl ester: Flammable. Tm = 323.85; Antoine (liq) A = 9.965, B = 3785, C = 0, dev = 5.0, range = 323 to 413.

C_9H_9N MW 131.18

3-Methylindole, skatole, CA 83-34-1: Stench; flammable; soluble hot water, ethanol, ether, acetone, benzene. Tm = 368.15; Tb = 538.15 to 539.15; Antoine I (sol) A = 11.27045, B = 4350.811, C = 0, dev = 1.0 to 5.0, range = 288 to 333; Antoine II (liq) A = 7.7123, B = 2932.12, C = -25.61, dev = 1.0, range = 368 to 540.

$C_9H_9NO_2$ MW 163.18

Benzaldehyde, 4-acetamido, CA 122-85-0: Soluble water, benzene. Tm = 429.15; Antoine (sol) A = 11.62765, B = 5173.328, C = 0, dev = 1.0, range = 328 to 346.

$C_9H_9NO_4$ MW 195.17

Benzoic acid, 3-nitro, ethyl ester, CA 618-98-4: Irritant; flammable. Tm = 320.15; Tb = 569.15 to 571.15; Antoine (liq) A = 7.17343, B = 2744.03, C = -40.103, dev = 1.0, range = 381 to 571.

$C_9H_9N_3O_6$ MW 255.19

Benzene, 1,3,5-trimethyl-2,4,6-trinitro, CA 602-96-0: Flammable; explodes above 688; soluble acetone, benzene. Tm = 506.15 to 508.15; Antoine (sol) A = 12.13, B = 5410, C = 0, dev not specified, range = 319 to 398.

C_9H_{10} MW 118.18

Indane, CA 496-11-7: Flammable, flash point 323; soluble ethanol, ether.
Tf = 221.75; Tb = 451.15; Vm = 0.1226 at 293.15; Antoine (liq) A = 6.11622,
B = 1580.375, C = -66.49, dev = 0.02, range = 374 to 466.

2-Methylstyrene, CA 611-15-4: Irritant; moderately toxic; lachrymator;
flammable; polymerizes on standing; soluble most organic solvents. Tf =
204.65; Tb = 442.15; Vm = 0.1302 at 298.15; Antoine (liq) A = 6.27762, B =
1628.405, C = -61.764, dev = 0.1, range = 305 to 385.

3-Methylstyrene, CA 100-80-1: Irritant; lachrymator; flammable, flash
point 324; polymerizes on standing; soluble most organic solvents. Tf =
186.85; Tb = 437.15; Vm = 0.1302 at 298.15; Antoine (liq) A = 6.07569, B =
1520.412, C = -71.183, dev = 1.0, range = 314 to 385.

4-Methylstyrene, CA 622-97-9: Irritant; lachrymator; flammable, flash
point 318; polymerizes on standing; soluble most organic solvents. Tf =
239.05; Tb = 443.15 to 448.15; Vm = 0.1288 at 298.15; Antoine (liq) A =
6.1732, B = 1594.747, C = -63.261, dev = 0.1, range = 304 to 370.

alpha-Methylstyrene, CA 98-83-9: Irritant; moderately toxic; lachrymator;
flammable, flash point 327; polymerizes; soluble most organic solvents.
Tm = 249.15; Tb = 440.15 to 443.15; Vm = 0.1298 at 293.15; Antoine I (liq)
A = 6.04856, B = 1486.88, C = -70.75, dev = 1.0 to 5.0, range = 343 to 493;
Antoine II (liq) A = 6.294, B = 1599.88, C = -63.72, dev = 1.0, range = 353
to 413.

beta-Methylstyrene, cis, CA 766-90-5: Irritant; lachrymator; flammable;
polymerizes; soluble most organic solvents. Tm = 211.55; Tb = 447.15 to
448.15; Vm = 0.1297 at 298.15; Antoine (liq) A = 6.04829, B = 1499.8, C =
-72.15, dev = above 5.0, range = 348 to 498.

beta-Methylstyrene, trans, CA 873-66-5: Irritant; lachrymator; flammable,
flash point 325; polymerizes; soluble most organic solvents. Tm = 243.85;
Tb = 448.15 to 449.15; Vm = 0.1310 at 298.15; Antoine (liq) A = 6.58873,
B = 1915.94, C = -33.996, dev = above 5.0, range = 291 to 452.

$C_9H_{10}N_2O$ MW 162.19

3-Pyrazolidinone, 1-phenyl, CA 92-43-3: Toxic; soluble hot water, hot
ethanol. Tm = 398.15 to 400.15; Antoine (sol) A = 9.17224, B = 4405.306,
C = 0, dev = 1.0, range = 327 to 348.

$C_9H_{10}O$ MW 134.18

Allyl phenyl ether, CA 1746-13-0: Irritant; lachrymator; flammable, flash
point 335; forms explosive peroxides; soluble ethanol, ether. Tb = 465.15
to 468.15; Vm = 0.1368 at 293.15; Antoine (liq) A = 5.98854, B = 1515.331,
C = -85.001, dev = 1.0 to 5.0, range = 349 to 456.

Cinnamyl alcohol, CA 4407-36-7: Flammable, flash point above 385; soluble
ethanol, ether; properties given for trans form which occurs naturally.
Tm = 306.15; Tb = 530.15; Vm = 0.1283 at 308.15; Antoine I (sol) A =
15.694, B = 5723, C = 0, dev = 5.0, range = 288 to 307; Antoine II (liq)
A = 10.683, B = 4171, C = 0, dev = 5.0, range = 310 to 328; Antoine III
(liq) A = 7.18006, B = 2580.55, C = -24.478, dev = 1.0 to 5.0, range = 373
to 523.

(continues)

298

$C_9H_{10}O$ *(continued)*

2,4-Dimethylbenzaldehyde, CA 15764-16-6: Flammable, flash point 361; soluble ethanol, ether, acetone, benzene. Tm = 264.15; Tb = 488.15 to 489.15; Vm = 0.1395 at 293.15; Antoine (liq) A = 7.57461, B = 2559.99, C = -28.425, dev = 1.0, range = 358 to 489.

5-Indanol, 5-hydroxyindane, CA 1470-94-6: Highly toxic; flammable, flash point above 385; soluble ethanol, ether. Tm = 329.15; Tb = 526.15; Antoine (liq) A = 8.21839, B = 3527.3, C = 42.28, dev = 1.0, range = 393 to 524.

4'-Methylacetophenone, CA 122-00-9: Flammable, flash point 365; soluble ethanol, ether, benzene, chloroform. Tm = 275.15; Tb = 498.15; Vm = 0.1335 at 293.15; Antoine (liq) A = 8.706, B = 3115, C = 0, dev = 1.0 to 5.0, range = 288 to 333.

3-Phenylpropionaldehyde, CA 104-53-0: Flammable, flash point 369; soluble ethanol, ether; used in flavorings and perfumery. Tm = 226.15; Tb = 494.15 to 497.15; Antoine (liq) A = 9.62933, B = 3525.299, C = 0, dev = 5.0, range = 330 to 363.

Propiophenone, ethyl phenyl ketone, CA 93-55-0: Flammable, flash point 358; slightly soluble water; soluble ethanol, ether, benzene. Tm = 291.35; Tb = 491.15; Vm = 0.1326 at 298.15; Antoine (liq) A = 6.33925, B = 1800.5, C = -75.15, dev = 1.0, range = 388 to 623.

2-Vinylanisole, CA 612-15-7: Irritant; lachrymator; flammable; polymerizes; soluble ethanol, ether, acetone, benzene. Tm = 302.15; Tb = 468.15 to 473.15; Antoine (liq) A = 7.00913, B = 2062.41, C = -54.884, dev = 1.0, range = 314 to 467.

3-Vinylanisole, CA 626-20-0: Irritant; lachrymator; flammable; polymerizes; soluble ethanol, ether, benzene. Vm = 0.1343 at 289.15; Antoine (liq) A = 7.1478, B = 2194.74, C = -43.885, dev = 1.0, range = 316 to 471.

4-Vinylanisole, CA 637-69-4: Irritant; lachrymator; flammable, flash point 349; polymerizes; soluble ethanol, ether, benzene. Tm = 275.15; Tb = 477.15 to 478.15; Vm = 0.1342 at 286.15; Antoine (liq) A = 7.00002, B = 2170.5, C = -43.179, dev = 1.0, range = 318 to 478.

$C_9H_{10}O_2$ MW 150.18

Acetic acid, 3-tolyl ester, CA 122-46-3: Irritant; flammable; soluble benzene, ethanol, ether, chloroform. Tm = 285.15; Tb = 485.15; Vm = 0.1433 at 299.15; Antoine (liq) A = 8.1614, B = 2967.5, C = 3.85, dev = 1.0 to 5.0, range = 385 to 480.

Acetic acid, 4-tolyl ester, CA 140-39-6: Irritant; flammable; soluble benzene, ethanol, ether, chloroform. Tb = 485.15 to 486.15; Vm = 0.1429 at 290.15; Antoine (liq) A = 8.1574, B = 2962.5, C = 2.95, dev = 1.0 to 5.0, range = 385 to 480.

4'-Acetylanisole, CA 100-06-1: Flammable, flash point above 385; light-sensitive; soluble ethanol, ether, acetone. Tm = 311.15 to 312.15; Tb = 531.15; Vm = 0.1388 at 314.15; Antoine I (sol) A = 13.342, B = 4892, C = 0, dev = 1.0, range = 286 to 311; Antoine II (liq) A = 8.818, B = 3474, C = 0, dev = 1.0, range = 311 to 334.

2H-1,5-Benzodioxepine: Antoine (liq) A = 7.92, B = 2918.3, C = 0, dev = 5.0, range not specified.   *(continues)*

$C_9H_{10}O_2$ *(continued)*

Benzyl acetate, CA 140-11-4: Toxic; irritant; flammable, flash point 385;
soluble ethanol, ether, acetone. Tm = 221.15; Tb = 488.15; Vm = 0.1430 at
298.15; Antoine (liq) A = 8.009, B = 2900, C = 0, dev = 1.0 to 5.0, range =
283 to 490.

Ethyl benzoate, CA 93-89-0: Moderately toxic; flammable, flash point 357;
soluble ethanol, ether, chloroform, benzene. Tf = 239.15; Tb = 486.05;
Vm = 0.1444 at 298.15; Antoine I (liq) A = 6.81152, B = 2174.3, C = -34.071,
dev = 1.0, range = 358 to 487; Antoine II (liq) A = 8.23958, B = 2922.167,
C = 0, dev = 1.0 to 5.0, range = 288 to 333.

2-Ethylbenzoic acid, CA 612-19-1: Irritant; flammable; soluble water, benzene, ethanol, ether. Tm = 341.15; Tb = 532.15; Vm = 0.1440 at 373.15;
Antoine (sol) A = 13.41862, B = 5249.283, C = 0, dev = 1.0, range = 298
to 313.

3-Ethylbenzoic acid, CA 619-20-5: Irritant; flammable; soluble water, benzene, ethanol, ether. Tm = 320.15; Vm = 0.1441 at 373.15; Antoine (sol)
A = 12.98469, B = 5177.326, C = 0, dev = 1.0, range = 290 to 318.

4-Ethylbenzoic acid, CA 619-64-7: Irritant; flammable; soluble water, benzene, ethanol, ether. Tm = 386.65; Antoine (sol) A = 12.32758, B =
5132.133, C = 0, dev = 1.0, range = 310 to 329.

Hydrocinnamic acid, 3-phenylpropionic acid, CA 501-52-0: Flammable, flash
point above 385; moderately soluble hot water; soluble most organic solvents. Tm = 321.65; Tb = 553.15; Vm = 0.1402 at 322.15; Antoine (liq) A =
7.51346, B = 2829.43, C = -39.208, dev = 1.0, range = 375 to 553.

3-Methylbenzoic acid, methyl ester, CA 99-36-5: Flammable, flash point
368; soluble ethanol, ether. Tb = 488.15; Vm = 0.1409 at 288.15; Antoine
(liq) A = 6.4049, B = 1840, C = -74.05, dev = 1.0 to 5.0, range = 359
to 500.

$C_9H_{10}O_3$ MW 166.18

3-Ethoxy-4-hydroxybenzaldehyde, ethyl vanillin, CA 121-32-4: Flammable;
slightly soluble water; soluble ethanol, ether, benzene; used in flavorings
and perfumery. Tm = 350.15 to 351.15; Antoine (sol) A = 13.095, B = 5302,
C = 0, dev = 1.0 to 5.0, range = 296 to 338.

Ethyl salicylate, CA 118-61-6: Flammable, flash point 380; light-sensitive;
slightly soluble water; soluble ethanol, ether. Tm = 274.45; Tb = 504.65;
Vm = 0.1467 at 293.15; Antoine I (liq) A = 7.21465, B = 2506.81, C =
-23.428, dev = 1.0, range = 334 to 505; Antoine II (liq) A = 8.33525, B =
3092.609, C = 0, dev = 1.0 to 5.0, range = 288 to 333.

2-Furanacrylic acid, ethyl ester, CA 623-20-1: Irritant; lachrymator;
flammable; polymerizes; soluble ethanol, ether. Tm = 297.65; Tb = 503.15
to 506.15; Vm = 0.1526 at 298.15; Antoine (liq) A = 7.852, B = 2969, C = 0,
dev = 5.0, range = 428 to 500.

$C_9H_{11}Br$ MW 199.09

1-Bromo-2-isopropylbenzene, CA 7073-94-1: Irritant; flammable; soluble
ether, chloroform, benzene. Tm = 213.91; Tb = 483.65; Vm =   *(continues)*

300

$C_9H_{11}Br$ *(continued)*

0.1531 at 293.15; Antoine (liq) A = 6.11844, B = 1666.7, C = -78.15, dev = 0.1, range = 378 to 528.

1-Bromo-4-isopropylbenzene, CA 586-61-8: Irritant; flammable; soluble benzene, ether, chloroform. Tm = 250.75; Tb = 492.15; Vm = 0.1558 at 298.15; Antoine (liq) A = 6.16897, B = 1732, C = -76.15, dev = 0.1, range = 388 to 528.

$C_9H_{11}Cl$ MW 154.64

1-Chloro-2-isopropylbenzene, CA 2077-13-6: Irritant; flammable; soluble ethanol, ether, benzene. Tm = 198.76; Antoine (liq) A = 6.11697, B = 1599.61, C = -75.15, dev = 0.1, range = 363 to 508.

1-Chloro-4-isopropylbenzene, CA 2621-46-7: Irritant; flammable; soluble ethanol, ether, benzene. Tm = 260.85; Tb = 463.15 to 468.15; Vm = 0.1515 at 293.15; Antoine (liq) A = 6.11274, B = 1623.51, C = -76.15, dev = 0.1, range = 368 to 513.

$C_9H_{11}ClN_2O$ MW 198.65

Urea, 3-(4'-chlorophenyl)-1,1-dimethyl, CA 150-68-5: Toxic; suspected carcinogen; slightly soluble water; soluble ethanol, acetone; used as weed killer. Tm = 443.65 to 444.65; Antoine (sol) A = 12.4301, B = 5988.39, C = 0, dev = 5.0, range = 303 to 379.

$C_9H_{11}ClO_2$ MW 186.64

Propylene glycol, mono(4-chlorophenyl) ether, CA 67146-43-4: Flammable; forms explosive peroxides. Antoine (liq) A = 7.4358, B = 2692.7, C = -47.01, dev = 1.0, range = 417 to 542.

$C_9H_{11}ClS$ MW 186.70

Benzyl (2-chloroethyl) sulfide, CA 4332-51-8: Toxic; stench; flammable. Antoine (liq) A = 6.50156, B = 2733.4, C = 0, dev = 1.0, range = 273 to 333.

$C_9H_{11}F_5O_2$ MW 246.18

Pentafluoropropionic acid, cyclohexyl ester, CA 24262-73-5: Antoine (liq) A = 7.3766, B = 2226.4, C = -14.35, dev = 1.0, range = 355 to 428.

$C_9H_{11}NO$ MW 149.19

Acetamide, *N*-(2-methylphenyl), CA 120-66-1: Soluble ethanol, ether, benzene. Tm = 383.15; Tb = 569.15; Antoine (sol) A = 11.88537, B = 5085.111, C = 0, dev = 1.0, range = 315 to 340.

Acetamide, *N*-(4-methylphenyl), CA 103-89-9: Soluble ethanol, ether, benzene. Tm = 421.65; Tb = 580.15; Antoine (sol) A = 11.58697, B = 5182.671, C = 0, dev = 1.0, range = 330 to 350.

(continues)

$C_9H_{11}NO$ *(continued)*

N-Methylacetanilide, exalgin, CA 579-10-2: Soluble hot water, ethanol, chloroform. Tm = 374.15 to 375.15; Tb = 526.15; Vm = 0.1487 at 378.15; Antoine (liq) A = 5.94087, B = 1639.42, C = -110.1, dev = 1.0, range = 383 to 519.

$C_9H_{11}NO_2$ MW 165.19

Anthranilic acid, ethyl ester, CA 87-25-2: Irritant; flammable, flash point above 385; soluble ethanol, ether. Tm = 286.15; Tb = 539.15 to 541.15; Vm = 0.1479 at 293.15; Antoine (liq) A = 6.04609, B = 1734.75, C = -113.48, dev = 1.0, range = 433 to 593.

Carbanilic acid, ethyl ester, CA 101-99-5: Flammable; soluble ethanol, ether, benzene. Tm = 326.15; Tb = 510.15, decomposes; Antoine (liq) A = 9.6173, B = 3595.2, C = -37.758, dev = 1.0, range = 380 to 510.

4'-Methoxyacetanilide, CA 51-66-1: Tm = 401.15; Antoine (liq) A = 5.80464, B = 1700.78, C = -177.08, dev = 1.0, range = 456 to 533.

Phenylalanine, *levo*, CA 63-91-2: Soluble water. Tm = 556.15 to 557.15, decomposes; Antoine (sol) A = 15.345, B = 8040, C = 0, dev = 5.0, range = 451 to 469.

C_9H_{12} MW 120.19

cis-Bicyclo(4,3,0)nona-3,7-diene, CA 38451-18-2: Flammable; soluble most organic solvents. Tb = 433.45; Antoine (liq) A = 5.95482, B = 1440.47, C = -69.639, dev = 0.1, range = 356 to 429.

5-Ethylidene-2-norbornene, CA 16219-75-3, mixed *cis* and *trans* isomers: Toxic; flammable, flash point 311; soluble most organic solvents. Vm = 0.1346 at 293.15; Antoine (liq) A = 5.17318, B = 966.032, C = -111.697, dev = 1.0, range = 337 to 374.

5-Ethylidene-2-norbornene, *cis*, CA 28304-66-7: Toxic; flammable; soluble most organic solvents. Antoine (liq) A = 6.07092, B = 1451.589, C = -63.423, dev = 0.02 to 0.1, range = 346 to 415.

5-Ethylidene-2-norbornene, *trans*, CA 28304-67-8: Toxic; flammable; soluble most organic solvents. Antoine (liq) A = 6.07437, B = 1456.254, C = -63.807, dev = 0.02, range = 346 to 416.

2-Ethyltoluene, CA 611-14-3: Flammable, flash point 312; soluble most organic solvents. Tm = 192.35; Tb = 437.95 to 438.15; Vm = 0.1365 at 293.15; Antoine (liq) A = 6.1229, B = 1532.449, C = -66.123, dev = 0.02, range = 353 to 443.

3-Ethyltoluene, CA 620-14-4: Flammable, flash point 310; soluble most organic solvents. Tm = 177.65; Tb = 430.15 to 432.15; Vm = 0.1390 at 293.15; Antoine (liq) A = 6.13801, B = 1527.983, C = -64.715, dev = 0.02, range = 348 to 438.

4-Ethyltoluene, CA 622-96-8: Flammable, flash point 309; soluble most organic solvents. Tm = 210.85; Tb = 433.15 to 435.15; Vm = 0.1395 at 293.15; Antoine (liq) A = 6.11098, B = 1519.486, C = -65.035, dev = 0.02, range = 349 to 442.

(continues)

C_9H_{12} *(continued)*

Isopropylbenzene, cumene, CA 98-82-8: Irritant; toxic; flammable, flash point 309; soluble most organic solvents; used for commercial production of phenol and acetone. Tm = 177.12; Tb = 425.54; Vm = 0.1395 at 293.15; Tc = 631.15; Pc = 3209; Vc = 0.428; Antoine (liq) A = 6.05949, B = 1459.975, C = -65.412, dev = 0.02, range = 339 to 433.

Propylbenzene, CA 103-65-1: Moderately toxic; flammable, flash point 303; soluble most organic solvents. Tm = 180.95; Tb = 432.6; Vm = 0.1394 at 293.15; Tc = 639.15; Pc = 3260; Vc = 0.440; Antoine (liq) A = 6.07438, B = 1490.61, C = -66.029, dev = 0.02, range = 348 to 433.

3a,4,7,7a-Tetrahydro-1-H-indene, CA 3048-65-5: Flammable. Tb = 433.45; Antoine (liq) A = 6.1083, B = 1547.413, C = -57.44, dev = 1.0, range = 338 to 440.

1,2,3-Trimethylbenzene, hemimellitene, CA 526-73-8: Irritant; moderately toxic; flammable, flash point 321; soluble most organic solvents. Tm = 247.77; Tb = 449.23; Vm = 0.1344 at 293.15; Antoine (liq) A = 6.16447, B = 1593.776, C = -66.032, dev = 0.02, range = 363 to 456.

1,2,4-Trimethylbenzene, pseudocumene, CA 95-63-6: Irritant; moderately toxic; flammable, flash point 317; soluble most organic solvents. Tm = 229.35; Tb = 442.5; Vm = 0.1372 at 293.15; Antoine (liq) A = 6.16695, B = 1572.687, C = -64.593, dev = 0.02, range = 357 to 450.

1,3,5-Trimethylbenzene, mesitylene, CA 108-67-8: Irritant; moderately toxic; flammable, flash point 323; soluble most organic solvents. Tm = 228.43, highest melting solid form; Tb = 437.87; Vm = 0.1389 at 293.15; Antoine I (liq) A = 6.19762, B = 1569.149, C = -63.565, dev = 0.02, range = 354 to 445; Antoine II (liq) A = 6.62312, B = 1810.653, C = -43.307, dev = 1.0, range = 249 to 356.

5-Vinyl-2-norbornene, CA 3048-64-4: Irritant; lachrymator; flammable, flash point 300; polymerizes; soluble most organic solvents. Tm = 193.15; Tb = 410.15; Vm = 0.1429 at 293.15; Antoine I (liq) A = 5.73107, B = 1267.834, C = -75.593, dev = 1.0, range = 301 to 410; Antoine II (liq) A = 5.71245, B = 1234.553, C = -81.213, dev = 0.1, range = 354 to 409.

$C_9H_{12}F_3N_3O_5$ MW 299.21

N-[N-(N-[Trifluoroacetyl]glycyl)glycyl] glycine, methyl ester, CA 651-18-3: Flammable. Antoine (sol) A = 14.875, B = 6969, C = 0, dev = 5.0, range = 343 to 433.

$C_9H_{12}NO_5PS$ MW 277.23

O,O-Dimethyl-O-(3-methyl-4-nitrophenyl) thiophosphate, CA 122-14-5: Toxic insecticide; flammable; soluble most organic solvents. Vm = 0.2095 at 298.15; Antoine (liq) A = 8.7652, B = 4075.1, C = 0, dev = 1.0, range = 293 to 382.

$C_9H_{12}N_2$ MW 148.21

Acetone, phenyl hydrazone, CA 103-02-6: Flammable; soluble ethanol, ether. Tm = 315.15; Antoine (liq) A = 9.7624, B = 3897, C = 0, dev = 5.0, range = 413 to 436.

$C_9H_{12}O$ MW 136.19

Benzyl ethyl ether, CA 539-30-0: Flammable; forms explosive peroxides; soluble ethanol, ether. Tb = 462.15; Vm = 0.1435 at 293.15; Antoine (liq) A = 6.92406, B = 2133.29, C = -24.38, dev = 1.0, range = 299 to 460.

2-Ethylanisole, CA 14804-32-1: Flammable; soluble ether, benzene. Tb = 459.15 to 461.15; Vm = 0.1413 at 292.15; Antoine (liq) A = 6.98521, B = 2147.69, C = -28.932, dev = 1.0, range = 302 to 460.

3-Ethylanisole, CA 10568-38-4: Flammable; soluble ether, benzene. Tb = 469.15 to 470.15; Vm = 0.1422 at 291.15; Antoine (liq) A = 6.96261, B = 2212.16, C = -23.384, dev = 1.0, range = 306 to 470.

4-Ethylanisole, CA 1515-95-3: Flammable; soluble ether, benzene. Tb = 468.15 to 469.15; Vm = 0.1415 at 288.15; Antoine (liq) A = 6.532, B = 1884.96, C = -53.185, dev = 1.0, range = 306 to 470.

5-Ethyl-3-methylphenol: Toxic; irritant; flammable; soluble ethanol, benzene, ether. Tm = 324.77; Antoine (liq) A = 6.16565, B = 1615.944, C = -120.476, dev = 0.02 to 0.1, range = 468 to 521.

2-Isopropylphenol, 2-cumenol, CA 88-69-7: Irritant; toxic; flammable, flash point 380; soluble ethanol, ether, benzene. Tm = 288.15 to 289.15; Tb = 485.15 to 487.15; Vm = 0.1346 at 293.15; Antoine (liq) A = 6.510, B = 1861.5, C = -75.35, dev = 1.0, range = 370 to 489.

3-Isopropylphenol, 3-cumenol, CA 618-45-1: Irritant; toxic; flammable, flash point 370; soluble ethanol, ether, benzene. Tm = 299.15; Tb = 501.15; Antoine (liq) A = 7.188, B = 2174, C = -76.57, dev = 1.0 to 5.0, range = 377 to 497.

4-Isopropylphenol, 4-cumenol, CA 99-89-8: Irritant; toxic; flammable; soluble ethanol, ether, benzene. Tm = 335.45; Tb = 496.15 to 498.15; Antoine (liq) A = 7.109, B = 2140, C = -76.52, dev = 1.0, range = 380 to 496.

Isopropyl phenyl ether, CA 2741-16-4: Flammable; forms explosive peroxides; soluble ether, benzene. Tm = 240.12; Tb = 451.15; Vm = 0.1393 at 293.15; Antoine (liq) A = 5.59029, B = 1208.29, C = -113.59, dev = 1.0, range = 345 to 448.

3-Phenyl-1-propanol, CA 122-97-4: Flammable, flash point 373; soluble ethanol, ether. Tm = 255.15; Tb = 508.15; Vm = 0.1351 at 293.15; Antoine (liq) A = 8.490, B = 3281, C = 0, dev = 1.0 to 5.0, range = 284 to 328.

2-Phenyl-2-propanol, CA 617-94-7: Flammable, flash point 360; soluble benzene, ethanol, ether. Tm = 308.15 to 310.15; Tb = 575.15; Antoine (liq) A = 7.84965, B = 2763.768, C = 0, dev = 0.1 to 1.0, range = 391 to 423.

Phenyl propyl ether, CA 622-85-5: Flammable; forms explosive peroxides; soluble ethanol, ether. Tm = 246.15; Tb = 463.05; Vm = 0.1438 at 293.15; Antoine (liq) A = 6.57955, B = 1920, C = -43.15, dev = 1.0, range = 374 to 463.

2-Propylphenol, CA 644-35-9: Irritant; flammable, flash point 366; soluble ethanol, ether, benzene. Tm = 280.15; Tb = 493.15 to 493.65; Vm = 0.1342 at 293.15; Antoine (liq) A = 8.2060, B = 3134.8, C = 10.53, dev = 1.0, range = 377 to 495.

3-Propylphenol, CA 621-27-2: Irritant; flammable; soluble ethanol, ether, benzene. Tm = 299.15; Tb = 502.15 to 504.15; Antoine (liq) A = *(continues)*

$C_9H_{12}O$ *(continued)*

6.96026, B = 2100.7, C = -77.15, dev = 1.0, range = 408 to 538.

4-Propylphenol, CA 645-56-7: Irritant; flammable, flash point 379; soluble ethanol, ether, benzene. Tm = 294.15 to 295.15; Tb = 503.15 to 505.15; Antoine (liq) A = 7.32632, B = 2550.1, C = -28.65, dev = 1.0, range = 383 to 508.

2,3,5-Trimethylphenol, CA 697-82-5: Corrosive; flammable; soluble ethanol, ether. Tm = 368.15 to 369.15; Tb = 506.15; Antoine (liq) A = 6.20676, B = 1687.869, C = -106.761, dev = 0.02 to 0.1, range = 459 to 521.

2,4,5-Trimethylphenol, pseudocumenol, CA 496-78-6: Corrosive; flammable; soluble ethanol, ether. Tm = 343.15; Tb = 505.15; Antoine (liq) A = 6.9232, B = 2230, C = -51.33, dev = 1.0 to 5.0, range = 379 to 505.

2,4,6-Trimethylphenol, mesitol, CA 527-60-6: Corrosive; flammable; soluble ethanol, ether. Tm = 342.15; Tb = 493.15; Antoine (liq) A = 6.82395, B = 2158.2, C = -45.2, dev = 1.0 to 5.0, range = 367 to 494.

3,4,5-Trimethylphenol, CA 527-54-8: Corrosive; flammable; soluble ethanol, ether. Tm = 381.15; Tb = 521.15 to 522.15; Antoine (liq) A = 7.33216, B = 2536.1, C = -44.56, dev = 1.0, range = 396 to 521.

$C_9H_{12}O_2$ MW 152.19

Cumene hydroperoxide, CA 80-15-9: Toxic; suspected carcinogen; flammable; strong oxidizer; potentially explosive; intermediate in manufacture of phenol and acetone. Antoine I (liq) A = 9.885, B = 3649.1, C = 0, dev = 5.0, range = 283 to 333; Antoine II (liq) A = 10.30852, B = 3865.795, C = 0, dev = 5.0, range = 347 to 390.

1,3-Dihydroxy-5-methyl-2-ethylbenzene, CA 27465-63-0: Flammable; soluble benzene. Tm = 405.15; Antoine (liq) A = 6.1247, B = 1751.1, C = -137.01, dev = 1.0, range = 388 to 453.

3,5-Dimethoxytoluene, CA 4179-19-5: Flammable; soluble ethanol, benzene, ether. Tb = 517.15; Antoine (liq) A = 5.220, B = 1156.8, C = -151.31, dev = 1.0 to 5.0, range = 374 to 520.

Ethylene glycol, monobenzylether, CA 622-08-2: Irritant; moderately toxic; flammable, flash point above 385; forms explosive peroxides; soluble water, ethanol, ether. Tm = below 198.15; Tb = 529.15; Vm = 0.1430 at 293.15; Antoine (liq) A = 7.8002, B = 3063.6, C = 0, dev = 1.0, range = 453 to 530.

Propylene glycol, 1-phenyl ether, CA 770-35-4: Irritant; flammable; forms explosive peroxides; soluble ethanol, ether. Tm = 291.15; Vm = 0.1433 at 293.15; Antoine (liq) A = 8.103, B = 3107, C = 0, dev = 5.0, range = 389 to 509.

$C_9H_{12}O_4$ MW 184.19

Ethylmalonic acid, diethyl ester, CA 133-13-1: Flammable; soluble ethanol, ether, acetone, chloroform. Tb = 480.15 to 482.15; Antoine (liq) A = 6.20215, B = 1521.974, C = =117.011, dev = 5.0, range = 365 to 479.

$C_9H_{12}S$ MW 152.25

Benzyl ethyl sulfide, CA 6263-62-3: Toxic; stench; flammable. Tb = 491.15 to 493.15; Antoine (liq) A = 7.88745, B = 2863.259, C = 0, dev = 1.0 to 5.0, range = 345 to 500.

$C_9H_{13}N$ MW 135.21

N,N-Dimethyl-2-toluidine, CA 609-72-3: Irritant; flammable; soluble ether, ethanol. Tm = 213.15; Tb = 457.95; Vm = 0.1456 at 293.15; Antoine (liq) A = 6.2307, B = 1632.23, C = -71.825, dev = 1.0, range = 301 to 458.

N,N-Dimethyl-4-toluidine, CA 99-97-8: Irritant; flammable, flash point 356; soluble ethanol, ether. Tb = 483.15 to 484.15; Vm = 0.1444 at 293.15; Antoine (liq) A = 6.41112, B = 1828.67, C = -67.786, dev = 1.0, range = 323 to 483.

4-Isopropylaniline, cumidine, CA 99-88-7: Irritant; flammable, flash point 365; soluble benzene. Tm = 210.15; Tb = 498.15; Vm = 0.142 at 293.15; Antoine (liq) A = 6.96537, B = 2250.16, C = -46.616, dev = 1.0, range = 333 to 500.

1-Phenyl-2-propylamine ±, amphetamine, CA 300-62-9: Highly toxic; flammable; slightly soluble water; soluble ethanol, ether, chloroform. Tm = 300.65; Tb = 476.15; Antoine (liq) A = 8.133, B = 2792, C = 0, dev = 5.0, range = 333 to 353.

2,4,6-Trimethylaniline, mesidine, CA 88-05-1: Irritant; toxic; flammable, flash point 369; soluble ethanol. Tb = 505.15 to 506.15; Antoine (liq) A = 7.08655, B = 2192.87, C = -67.789, dev = 1.0, to 5.0, range = 341 to 510.

C_9H_{14} MW 122.21

2-Methylbicyclo(2,2,2)oct-2-ene, CA 4893-13-4: Flammable; soluble most organic solvents. Antoine (liq) A = 6.02746, B = 1448.087, C = -63.897, dev = 0.02 to 0.1, range = 363 to 402.

2-Vinylbicyclo(2,2,1)heptane, CA 2146-39-6: Irritant; lachrymator; flammable; soluble most organic solvents. Antoine (liq) A = 4.11544, B = 546.801, C = -174.546, dev = 1.0, range = 350 to 385.

$C_9H_{14}F_3NO_3$ MW 241.21

N-Trifluoroacetyl-levo-leucine, methyl ester, CA 23635-25-8: Flammable. Antoine (liq) A = 7.845, B = 2922, C = 0, dev = 5.0, range = 273 to 463.

$C_9H_{14}N_2$ MW 150.22

Azelaic acid, dinitrile, CA 1675-69-0: Irritant; flammable, flash point above 385; soluble ethanol, ether, benzene. Vm = 0.1633 at 292.15; Antoine (liq) A = 9.6213, B = 4201, C = 0, dev = 1.0, range = 308 to 341.

$C_9H_{14}O$ MW 138.21

Isophorone, CA 78-59-1: Irritant; moderately toxic; lachrymator; flammable, flash point 357; soluble ethanol, ether, acetone. Tm = *(continues)*

$C_9H_{14}O$ *(continued)*

265.05; Tb = 488.45; Vm = 0.1498 at 293.15; Tc = 715.15; Pc = 3300; Antoine (liq) A = 6.51479, B = 2058.59, C = -32.274, dev = 1.0, range = 311 to 489.

Phorone, CA 504-20-1: Toxic; flammable, flash point 358. Tm = 301.15; Tb = 470.35; Vm = 0.1562 at 293.15; Antoine (liq) A = 7.22654, B = 2281.06, C = -33.504, dev = 1.0, range = 315 to 471.

$C_9H_{14}O_2$ MW 154.21

2-Octynoic acid, methyl ester, CA 111-12-6: Flammable; polymerizes. Vm = 0.1665 at 293.15; Antoine (liq) A = 9.452, B = 3370, C = 0, dev = 1.0 to 5.0, range = 283 to 312.

$C_9H_{14}O_4$ MW 186.21

Diethyl citraconate, CA 691-83-8: Flammable; soluble ethanol, ether, acetic acid. Tb = 504.15; Vm = 0.1775 at 293.15; Antoine (liq) A = 7.14462, B = 2456.12, C = -25.802, dev = 1.0 to 5.0, range = 332 to 504.

Diethyl itaconate, CA 2409-52-1: Flammable; soluble ethanol, ether, acetone, benzene. Tm = 331.15 to 332.15; Tb = 501.15; Antoine (liq) A = 7.04612, B = 2486.01, C = -8.222, dev = 1.0 to 5.0, range = 324 to 501.

Diethyl mesaconate, CA 2418-31-7: Flammable; soluble ethanol, ether, acetone, benzene. Tb = 502.15; Antoine (liq) A = 7.72566, B = 2874.53, C = 0.295, dev = 1.0, range = 335 to 502.

$C_9H_{14}O_5$ MW 202.21

Acetylmalonic acid, diethyl ester, CA 570-08-1: Flammable; soluble acetone. Tb = 505.15; Vm = 0.1866 at 299.15; Antoine (liq) A = 7.56609, B = 2820.288, C = 0, dev = 5.0, range = 363 to 510.

Ethyl[(1-allyloxycarbonyl)ethyl] carbonate: Toxic; flammable. Antoine (liq) A = 8.4554, B = 3201, C = 0, dev = 5.0, range = 342 to 496.

2-Lactyloxypropionic acid, allyl ester: Lachrymator; flammable; polymerizes. Antoine (liq) A = 7.01594, B = 2141.423, C = -90.15, dev = 5.0, range = 331 to 401.

$C_9H_{14}O_6$ MW 218.21

Triacetin, glycerol triacetate, CA 102-76-1: Flammable, flash point 411; very soluble water; soluble ethanol, ether, benzene. Tb = 531.15 to 532.15; Vm = 0.1887 at 298.15; Antoine (liq) A = 10.7922, B = 4285.8, C = 0, dev = 1.0, range = 284 to 319.

$C_9H_{14}O_7$ MW 234.21

Trimethyl citrate, CA 1587-20-8: Flammable; soluble ethanol, ether. Tm = 351.15 to 352.15; Tb = 556.15 to 560.15, decomposes; Antoine (liq) A = 8.77424, B = 3948.75, C = 23.121, dev = 1.0, range = 379 to 560.

$C_9H_{15}Cl_3O_2$ MW 261.58

Butyric acid, 3-chloro-2,2-bis(chloromethyl)propyl ester: Irritant; flammable. Antoine (liq) A = 7.06997, B = 2247.869, C = -103.762, dev = 1.0, range = 426 to 482.

$C_9H_{15}NOS$ MW 185.28

Carbamothioic acid, (1-methylethyl)-2-propynyl, S-ethyl ester, CA 59300-33-3: Irritant; flammable. Antoine (liq) A = 10.08035, B = 3801.808, C = 0, dev not specified, range = 298 to 313.

Carbamothioic acid, propyl-2-propynyl, S-ethyl ester, CA 59300-32-2: Irritant; flammable. Antoine (liq) A = 8.69454, B = 3375.554, C = 0, dev not specified, range = 298 to 313.

C_9H_{16} MW 124.23

Bicyclo(6,1,0)nonane, cis, CA 13757-43-2: Flammable; soluble most organic solvents. Antoine (liq) A = 8.290, B = 2631.1, C = 0, dev = 1.0, range = 297 to 360.

1,4-Dimethylbicyclo(2,2,1)heptane, CA 20454-81-3: Flammable; soluble most organic solvents. Antoine (liq) A = 5.89158, B = 1315.48, C = -59.299, dev = 0.02, range = 328 to 393.

2,3-Dimethylbicyclo(2,2,1)heptane, trans, CA 20558-16-1: Flammable; soluble most organic solvents. Antoine (liq) A = 5.97097, B = 1405.421, C = -61.977, dev = 0.02, range = 345 to 411.

2-Ethylbicyclo(2,2,1)heptane, CA 2146-41-0: Flammable; soluble most organic solvents. Antoine (liq) A = 8.45733, B = 3044.521, C = 53.181, dev = 1.0, range = 349 to 396.

Hexahydroindan, cis, CA 4551-51-3: Flammable; soluble most organic solvents. Tm = 236.49; Tb = 440.15; Vm = 0.1418 at 293.15; Antoine I (liq) A = 6.53231, B = 1793.955, C = -40.573, dev = 0.1, range = 263 to 293; Antoine II (liq) A = 6.16044, B = 1597.389, C = -55.799, dev = 0.1, range = 290 to 366; Antoine III (liq) A = 5.99176, B = 1497.436, C = -65.364, dev = 0.02, range = 363 to 463.

Hexahydroindan, trans, CA 3296-50-2: Flammable; soluble most organic solvents. Tm = 213.87; Antoine I (liq) A = 5.98306, B = 1474.284, C = -63.595, dev = 0.02, range = 356 to 457; Antoine II (liq) A = 6.80464, B = 1907.478, C = -29.475, dev = 0.1, range = 262 to 283; Antoine III (liq) A = 6.16906, B = 1583.804, C = -53.052, dev = 0.1, range = 281 to 362; Antoine IV (liq) A = 5.98723, B = 1477.062, C = -63.284, dev = 0.02, range = 358 to 479.

$C_9H_{16}ClN_5$ MW 229.71

2-Chloro-4,6-bis(isopropylamino)-1,3,5-triazine, propazine, CA 139-40-2: Toxic; flammable; used as herbicide. Tm = 486.15 to 487.15; Antoine (sol) A = 13.8789, B = 6533, C = 0, dev = 1.0 to 5.0, range = 323 to 403.

$C_9H_{16}Cl_4$ MW 266.04

1,1,1,9-Tetrachlorononane, CA 1561-48-4: Irritant; flammable. Antoine (liq) A = 11.605, B = 4650, C = 0, dev = 5.0, range = 298 to 338.

$C_9H_{16}NO_2$ MW 170.23

2,2,6,6-Tetramethyl-4-oxo-1-piperidinyloxy, CA 2896-70-0: Flammable. Tm = 309.15; Antoine (sol) A = 11.985, B = 4350, C = 0, dev = 5.0, range = 275 to 303.

$C_9H_{16}O$ MW 140.22

Cyclohexyl ethyl ketone, CA 1123-86-0: Flammable; soluble ether, acetone; used in air fresheners. Tb = 469.15; Vm = 0.1540 at 293.15; Antoine (liq) A = 7.3628, B = 2512.5, C = 0, dev = 1.0, range not specified.

Cyclononanone, CA 3350-30-9: Flammable, flash point 338; soluble ethanol, acetone. Tm = 307.15; Vm = 0.1462 at 293.15; Antoine (liq) A = 5.8682, B = 1611.52, C = -78.15, dev = 1.0 to 5.0, range = 333 to 413.

1-(1-Methyl-3-cyclohexen-3-yl)ethanol, CA 40213-09-0: Flammable. Antoine (liq) A = 8.0797, B = 2850.364, C = 0, dev = 1.0, range = 358 to 410.

Methyl (1-methylcyclohexyl) ketone, CA 2890-62-2: Flammable; soluble acetone. Antoine (liq) A = 7.24561, B = 2406.862, C = 0, dev = 1.0, range = 374 to 414.

2-Nonenal, *trans*, CA 18829-56-6: Irritant; flammable, flash point 357; polymerizes. Vm = 0.1657 at 293.15; Antoine (liq) A = 8.2419, B = 2933, C = 0, dev = 1.0 to 5.0, range = 363 to 398.

3,5,5-Trimethylcyclohexanone ±, CA 33496-91-2: Flammable. Tb = 462.15 to 462.65; Vm = 0.1572 at 292.15; Antoine (liq) A = 6.4472, B = 2052, C = 0, dev = 1.0 to 5.0, range = 423 to 463.

2,2,5-Trimethyl-4-hexene-1-al, CA 1000-30-2: Flammable; polymerizes. Antoine (liq) A = 9.05724, B = 2977.927, C = 0, dev = 5.0, range = 293 to 353.

$C_9H_{16}OS$ MW 172.28

4*H*-Thiopyran-4-one, tetrahydro-2,2,6,6-tetramethyl, CA 22842-41-7: Irritant; flammable. Antoine (liq) A = 5.615, B = 1813.8, C = 0, dev not specified, range = 300 to 360.

$C_9H_{16}O_2$ MW 156.22

Acetic acid, 2-methylcyclohexyl ester, mixed isomers, CA 5726-19-2: Flammable. Antoine (liq) A = 6.51389, B = 1809.459, C = -55.811, dev = 1.0, range = 337 to 457.

2-Butyl-4,7-dihydro-1,3-dioxepine, CA 61732-95-4: Flammable. Antoine (liq) A = 7.5469, B = 2658, C = 0, dev = 1.0 to 5.0, range = 318 to 453.

(continues)

$C_9H_{16}O_2$ *(continued)*

Hexyl acrylate, CA 2499-95-8: Lachrymator; flammable; polymerizes. Tm = 228.15; Vm = 0.1759 at 293.15; Antoine (liq) A = 7.4626, B = 2514.3, C = 0, dev = 1.0 to 5.0, range = 342 to 461.

Methacrylic acid, neopentyl ester: Lachrymator; flammable; polymerizes. Antoine (liq) A = 6.65446, B = 2113.654, C = 0, dev = 1.0, range = 313 to 338.

Oxo-2-cyclodecanone: Flammable. Antoine (liq) A = 8.80195, B = 3180.173, C = 0, dev = 1.0 to 5.0, range = 333 to 383.

Pentyl methacrylate, CA 2849-98-1: Lachrymator; flammable; polymerizes. Antoine (liq) A = 7.4685, B = 2489.1, C = 0, dev = 1.0, range = 339 to 456.

$C_9H_{16}O_3$ MW 172.22

Butyl levulinate, CA 2052-15-5: Flammable, flash point 364; soluble ethanol, ether, acetone, benzene. Tb = 510.15 to 511.15; Vm = 0.1769 at 293.15; Antoine (liq) A = 7.55112, B = 2793.4, C = -6.88, dev = 1.0, range = 358 to 511.

sec-Butyl levulinate: Flammable; soluble ethanol, ether, acetone. Tb = 498.95; Vm = 0.1781 at 293.15; Antoine (liq) A = 7.54776, B = 2834.5, C = 12.90, dev = 1.0, range = 393 to 499.

Isobutyl levulinate, CA 3757-32-2: Flammable; soluble ethanol, acetone, ether, benzene. Tb = 504.05; Vm = 0.1775 at 293.15; Antoine (liq) A = 6.44427, B = 1858.99, C = -84.359, dev = 1.0, range = 338 to 503.

$C_9H_{16}O_4$ MW 188.22

2-Acetoxypropionic acid, butyl ester: Flammable. Antoine (liq) A = 6.50779, B = 1780.07, C = -90.15, dev = 1.0 to 5.0, range = 325 to 485.

3-Acetoxypropionic acid, butyl ester, CA 40326-38-3: Flammable. Antoine (liq) A = 10.381, B = 3940, C = 0, dev = 5.0, range = 373 to 391.

Diethyl glutarate, CA 818-38-2: Flammable, flash point 369; soluble ether. Tm = 249.05; Tb = 509.15 to 510.15; Vm = 0.1842 at 293.15; Antoine (liq) A = 7.50461, B = 2751.52, C = -9.782, dev = 1.0, range = 338 to 510.

Ethylmalonic acid, diethyl ester, CA 133-13-1: Flammable, flash point 361; soluble ethanol, ether, acetone, chloroform. Tb = 480.15 to 482.15; Vm = 0.1873 at 293.15; Antoine (liq) A = 7.16371, B = 2318.55, C = -35.213, dev = 1.0 to 5.0, range = 323 to 485.

Nonanedioic acid, azelaic acid, CA 123-99-9: Irritant; flammable; moderately soluble hot water; soluble ethanol, ether. Tm = 379.65; Tb = 633.15, decomposes; Antoine (liq) A = 8.94128, B = 4287.4, C = -11.641, dev = 1.0, range = 451 to 630.

$C_9H_{16}O_5$ MW 204.22

Butyl[1-(methoxycarbonyl)ethyl] carbonate: Toxic; flammable. Antoine (liq) A = 8.3626, B = 3225, C = 0, dev = 5.0, range = 349 to 510.

(continues)

$C_9H_{16}O_5$ *(continued)*

Isobutyl[1-(methoxycarbonyl)ethyl] carbonate: Toxic; flammable. Antoine (liq) A = 8.1741, B = 3086.3, C = 0, dev = 1.0, range = 340 to 501.

2-Lactoylpropionic acid, propyl ester: Flammable. Antoine (liq) A = 6.89136, B = 2078.801, C = -90.15, dev = 1.0 to 5.0, range = 327 to 397.

Methyl[1-(butoxycarbonyl)ethyl] carbonate: Toxic; flammable. Antoine (liq) A = 8.220, B = 3147, C = 0, dev = 5.0, range = 311 to 503.

$C_9H_{17}N$ MW 139.24

Pelargononitrile, octyl cyanide, CA 2243-27-8: Highly toxic; flammable, flash point 354; soluble ethanol, ether. Tf = 238.95; Tb = 497.15; Vm = 0.1703 at 293.15; Antoine (liq) A = 6.36411, B = 1865.69, C = -70.74, dev = 1.0 to 5.0, range = 328 to 503.

$C_9H_{17}NO$ MW 155.24

2-Nonenoic acid, *trans*, amide, CA 14952-05-7: Tm = 404.65; Antoine (sol) A = 13.8618, B = 5848, C = 0, dev not specified, range = 383 to 393.

$C_9H_{17}NO_2$ MW 171.24

1-Hydroxy-2,2,6,6-tetramethyl-4-oxopiperidine, CA 3637-11-4: Tm = 368.15; Antoine (sol) A = 11.055, B = 4180, C = 0, dev = 5.0, range = 288 to 328.

$C_9H_{17}NO_3$ MW 187.24

N-Acetylvaline ±, ethyl ester, CA 56430-36-5: Flammable. Antoine (liq) A = 8.6562, B = 3535, C = 0, dev = 1.0, range = 382 to 466.

$C_9H_{17}NO_3S$ MW 219.30

N-Acetylmethionine ±, ethyl ester: Flammable. Antoine (liq) A = 9.2912, B = 4264, C = 0, dev = 1.0, range = 432 to 519.

$C_9H_{17}N_5O$ MW 221.27

2-Methoxy-4-ethylamino-6-isopropylamino-1,3,5-triazine, CA 1610-17-9: Irritant; used as herbicide. Antoine (sol) A = 10.428, B = 4933, C = 0, dev = 1.0 to 5.0, range = 323 to 403.

$C_9H_{17}N_5S$ MW 227.33

2-Methylthio-4-ethylamino-6-isopropylamino-1,3,5-triazine, CA 834-12-8: Irritant; used as herbicide. Tm = 361.15 to 362.15; Antoine (sol) A = 11.036, B = 5270, C = 0, dev = 1.0 to 5.0, range = 323 to 403.

C_9H_{18} MW 126.24

Butylcyclopentane, CA 2040-95-1: Flammable; soluble most organic solvents. Tf = 165.15; Tb = 429.15 to 430.15; Vm = 0.1605 at 293.15; Antoine (liq) A = 6.02425, B = 1457.08, C = -67.16, dev = 0.1, range = 413 to 432.

(continues)

C_9H_{18} *(continued)*

1-Ethyl-3-methylcyclohexane, *cis*, CA 19489-10-2: Flammable; soluble most organic solvents. Tb = 429.15; Vm = 0.1560 at 293.15; Antoine (liq) A = 6.01352, B = 1444.088, C = -61.297, dev = 1.0, range = 373 to 465.

Isopropylcyclohexane, CA 696-29-7: Flammable, flash point 308; soluble most organic solvents. Tf = 183.35; Tb = 427.85; Vm = 0.1573 at 293.15; Antoine (liq) A = 5.99603, B = 1452.423, C = -63.748, dev = 0.02, range = 295 to 431.

1-Nonene, CA 124-11-8: Flammable, flash point 299; soluble most organic solvents. Tf = 191.78; Tb = 420.02; Vm = 0.1741 at 298.15; Tc = 593.25; Pc = 2330; Vc = 0.528; Antoine (liq) A = 6.07341, B = 1432.453, C = -67.884, dev = 0.02, range = 339 to 423.

2-Nonene, *cis*, CA 6434-77-1: Flammable; soluble most organic solvents. Tb = 424.3; Antoine (liq) A = 6.06165, B = 1435.931, C = -69.92, dev = 0.1, range = 379 to 424.

2-Nonene, *trans*, CA 6434-78-2: Flammable, flash point 305; soluble most organic solvents. Tb = 422.15; Antoine (liq) A = 6.10962, B = 1456.475, C = -68.293, dev = 0.1, range = 379 to 422.

3-Nonene, *cis*, CA 20237-46-1: Flammable; soluble most organic solvents. Antoine (liq) A = 6.05689, B = 1427.818, C = -68.693, dev = 0.1, range = 376 to 422.

3-Nonene, *trans*, CA 20063-92-7: Flammable, flash point 305; soluble most organic solvents. Tb = 420.15 to 421.15; Vm = 0.1725 at 294.15; Antoine (liq) A = 6.05523, B = 1416.452, C = -71.532, dev = 0.1, range = 377 to 421.

4-Nonene, *cis*, CA 10405-84-2: Flammable, flash point 300; soluble most organic solvents. Antoine (liq) A = 6.04046, B = 1417.943, C = -69.116, dev = 0.1, range = 376 to 421.

4-Nonene, *trans*, CA 10405-85-3: Flammable, flash point 300; soluble most organic solvents. Tb = 416.15; Vm = 0.1725 at 293.15; Antoine (liq) A = 6.06953, B = 1428.685, C = -69.344, dev = 1.0, range = 376 to 420.

Propylcyclohexane, CA 1678-92-8: Flammable; soluble most organic solvents. Tf = 178.15; Tb = 427.65 to 428.65; Vm = 0.1591 at 293.15; Antoine (liq) A = 6.01136, B = 1460.8, C = -65.211, dev = 1.0, range = 346 to 431.

1,1,3-Trimethylcyclohexane, CA 3073-66-3: Flammable; soluble most organic solvents. Tf = 207.45; Tb = 410.15 to 411.15; Vm = 0.1629 at 298.15; Antoine (liq) A = 5.96449, B = 1395.396, C = -57.308, dev = 0.02, range = 348 to 411.

1,3,5-Trimethylcyclohexane +*cis*, CA 1795-27-3: Flammable; soluble most organic solvents. Tm = 223.45; Tb = 413.15 to 413.65; Vm = 0.1638 at 293.15; Antoine (liq) A = 7.95773, B = 2775.938, C = 59.095, dev = 1.0 to 5.0, range = 318 to 410.

$C_9H_{18}Br_2$ MW 286.05

1,1-Dibromononane: Flammable. Antoine (liq) A = 6.559, B = 2049, C = -83.15, dev = 5.0, range = 427 to 591.

$C_9H_{18}Cl_2$ MW 197.15

1,1-Dichlorononane, CA 821-88-5: Irritant; flammable. Antoine (liq) A = 6.407, B = 1866, C = -77.15, dev = 5.0, range = 398 to 556.

$C_9H_{18}F_2$ MW 164.24

1,1-Difluorononane: Flammable. Antoine (liq) A = 6.472, B = 1657, C = -65.15, dev = 5.0, range = 347 to 482.

$C_9H_{18}NO_2$ MW 172.25

4-Hydroxy-2,2,6,6-tetramethyl-1-piperidinyloxy, CA 2226-96-2: Flammable. Tm = 342.15 to 344.15; Antoine (sol) A = 13.585, B = 5300, C = 0, dev = 5.0, range = 293 to 318.

$C_9H_{18}N_2$ MW 154.25

Pentanenitrile, 2-(diethylamino), CA 19340-91-1: Irritant; flammable. Antoine (liq) A = 7.6835, B = 3070, C = 0, dev = 1.0, range = 283 to 326.

$C_9H_{18}N_2O_2S$ MW 218.31

2-Butanone, 3,3-dimethyl-1-(methylthio)-O-[(methylamino)carbonyl]oxime, CA 39196-18-4: Toxic insecticide; flammable. Tm = 329.65 to 330.65; Antoine (sol) A = 11.54046, B = 4905.485, C = 0, dev = 1.0, range = 298 to 328.

$C_9H_{18}O$ MW 142.24

1-Butylcyclopentanol, CA 1462-97-1: Flammable; soluble ethanol, ether. Antoine (liq) A = 5.71464, B = 1167.3, C = -151.81, dev = 1.0 to 5.0, range = 359 to 466.

2,6-Dimethyl-4-heptanone, diisobutyl ketone, CA 108-83-8: Flammable, flash point 322; slightly soluble water; soluble ethanol, ether. Tm = 231.65; Tb = 442.55; Vm = 0.1764 at 293.15; Tc = 617.15; Pc = 2500; Antoine (liq) A = 6.07029, B = 1476.4, C = -78.15, dev = 1.0, range = 336 to 451.

1-(1-Methylcyclohexyl)ethanol: Flammable. Antoine (liq) A = 8.23251, B = 2900.072, C = 0, dev = 1.0, range = 358 to 408.

Nonanal, pelargonaldehyde, CA 124-19-6: Flammable; oxidizes in air; soluble ether. Tb = 463.15 to 465.15; Vm = 0.1721 at 295.15; Antoine (liq) A = 7.30442, B = 2307.91, C = -22.653, dev = 1.0, range = 306 to 458.

2-Nonanone, heptyl methyl ketone, CA 821-55-6: Moderately toxic; flammable, flash point 337; soluble ethanol, ether, acetone, benzene. Tf = 258.15; Tb = 467.15 to 469.15; Vm = 0.1733 at 293.15; Antoine (liq) A = 6.33493, B = 1697.25, C = -74.71, dev = 1.0, range = 335 to 468.

5-Nonanone, dibutyl ketone, CA 502-56-7: Flammable, flash point 333; soluble ethanol, ether, acetone, chloroform. Tm = 268.35; Tb = 454.15 to 455.15; Vm = 0.1731 at 293.15; Antoine I (liq) A = 6.728, B = 2101, C = 0, dev = 5.0, range = 283 to 323; Antoine II (liq) A = 6.16176, *(continues)*

$C_9H_{18}O$ *(continued)*

B = 1578.543, C = -81.76, dev = 0.1, range = 357 to 468; Antoine III (liq)
A = 6.12976, B = 1556.651, C = -84.131, dev = 0.02, range = 443 to 486.

3,3,5-Trimethylcyclohexanol, mixed isomers, CA 116-02-9: Moderately toxic;
flammable; soluble ethanol, ether, chloroform. Antoine (liq) A = 6.0278,
B = 1401.5, C = -121.92, dev = 1.0 to 5.0, range = 343 to 473.

2,2,5-Trimethyl-4-hexene-l-ol, CA 53965-16-5: Flammable; soluble ethanol.
Antoine (liq) A = 9.22111, B = 3213.943, C = 0, dev = 1.0, range = 323
to 373.

$C_9H_{18}O_2$ MW 158.24

2-Butoxy-3-pentanone: Flammable. Antoine (liq) A = 6.7046, B = 1806.2,
C = -22.25, dev = 5.0, range = 333 to 398.

2-Butyl-1,3-dioxepane, CA 22432-66-2: Flammable. Antoine (liq) A =
8.74966, B = 2999.055, C = 0, dev = 5.0, range = 325 to 358.

2-Ethylheptanoic acid, CA 3274-29-1: Flammable. Antoine (liq) A = 8.405,
B = 3314, C = 0, dev = 5.0, range = 386 to 475.

2-Hexyl-1,3-dioxolane, CA 1708-34-5: Flammable. Antoine (liq) A = 8.26219,
B = 2873.317, C = 0, dev = 1.0 to 5.0, range = 325 to 353.

Isobutyl isovalerate, CA 589-59-3: Flammable; soluble ethanol, ether.
Tb = 443.15 to 445.15; Vm = 0.1855 at 293.15; Antoine (liq) A = 6.81152,
B = 1966.6, C = -32.699, dev = 1.0, range = 289 to 442.

Isopentyl butyrate, CA 106-27-4: Flammable; soluble ethanol, ether. Tb =
452.15; Vm = 0.1826 at 292.15; Antoine (liq) A = 6.76191, B = 1990.74, C =
-33.163, dev = 1.0, range = 294 to 452.

Isopentyl isobutyrate, CA 2050-01-3: Flammable; soluble ethanol, acetone,
ether. Tb = 442.15; Vm = 0.1834 at 293.15; Antoine (liq) A - 6.6372, B =
1861.35, C = -40.074, dev = 1.0, range = 287 to 442.

Isopropyl caproate, CA 2311-46-8: Flammable; soluble ethanol, ether. Vm =
0.1850 at 293.15; Antoine (liq) A = 6.8306, B = 1885.5, C = -52.55, dev =
1.0, range = 307 to 383.

Methyl caprylate, methyl octonoate, CA 111-11-5: Flammable, flash point
345; soluble ethanol, ether. Tm = 235.9; Tb = 466.15 to 467.15; Vm =
0.1803 at 293.15; Antoine (liq) A = 6.3832, B = 1711.9, C = -75.49, dev =
1.0, range = 347 to 470.

Nonanoic acid, pelargonic acid, CA 112-05-0: Irritant; flammable, flash
point 373; soluble ethanol, ether, chloroform. Tm = 288.15; Tb = 528.75;
Vm = 0.1745 at 293.15; Antoine (liq) A = 6.41498, B = 1728.01, C =
-136.009, dev = 1.0, range = 381 to 528.

Propyl caproate, CA 626-77-7: Flammable; soluble ethanol, ether. Tm =
204.45; Tb = 458.15 to 459.15; Vm = 0.1825 at 293.15; Antoine (liq) A =
7.14456, B = 2129.5, C = -40.08, dev = 1.0, range = 315 to 394.

$C_9H_{18}O_3$ MW 174.24

2-Butoxypropionic acid, ethyl ester: Flammable. Antoine (liq) A = 3.89114,
B = 353.32, C = -256.68, dev = 1.0, range = 348 to 438.

3-Ethoxypropionic acid, butyl ester, CA 14144-35-5: Flammable. Antoine
(liq) A = 7.6403, B = 2707.1, C = 0, dev = 1.0 to 5.0, range = 346 to 479.

3-Hydroxypropionic acid, hexyl ester: Flammable. Antoine (liq) A = 8.716,
B = 3636, C = 0, dev = 5.0, range = 408 to 432.

Lactic acid, hexyl ester, CA 20279-51-0: Flammable. Tm = 251.15; Antoine
(liq) A = 6.5112, B = 1822.386, C = -90.15, dev = 1.0 to 5.0, range = 307
to 494.

3-Methoxypropionic acid, pentyl ester, CA 10500-16-0: Flammable. Antoine
(liq) A = 7.7412, B = 2783, C = 0, dev = 1.0, range = 322 to 485.

3-Propoxypropionic acid, propyl ester, CA 14144-41-3: Flammable. Antoine
(liq) A = 7.5056, B = 2659, C = 0, dev = 5.0, range = 317 to 483.

$C_9H_{19}Br$ MW 207.15

1-Bromononane, CA 695-58-3: Irritant; flammable, flash point 324. Tm =
244.15; Tb = 474.15; Vm = 0.1918 at 293.15; Antoine (liq) A = 6.301, B =
1796.73, C = -76.25, dev = 1.0 to 5.0, range = 391 to 549.

$C_9H_{19}Cl$ MW 162.70

1-Chlorononane, CA 2473-01-0: Irritant; flammable, flash point 347; soluble ether, chloroform. Tm = 238.75; Tb = 476.15 to 477.15; Vm = 0.1866 at
293.15; Antoine (liq) A = 6.06553, B = 1586.937, C = -87.645, dev = 1.0,
range = 342 to 478.

$C_9H_{19}F$ MW 146.25

1-Fluorononane, CA 463-18-3: Irritant; flammable, flash point 321. Tm =
233.15; Antoine (liq) A = 6.3226, B = 1608.48, C = -65.55, dev = 1.0,
range = 333 to 473.

$C_9H_{19}I$ MW 254.15

1-Iodononane, CA 4282-42-2: Irritant; flammable, flash point 358; light-
sensitive. Tm = 253.15; Antoine (liq) A = 6.1894, B = 1830.37, C = -80.65,
dev = 1.0, range = 408 to 577.

$C_9H_{19}N$ MW 141.26

N,N-Diethyl-4-pentenylamine, CA 13173-21-2: Irritant; flammable. Antoine
(liq) A = 7.0541, B = 2170.1, C = 0, dev = 1.0, range = 338 to 430.

$C_9H_{19}NO$ MW 157.26

1-Cyclohexylamino-2-propanol, CA 103-00-4: Irritant; flammable. Tm =
318.74; Antoine (liq) A = 6.1930, B = 1696.9, C = -106.017, dev = 0.1,
range = 423 to 512.

(continues)

$C_9H_{19}NO$ *(continued)*

Nonamide, pelargonamide, CA 1120-07-6:   Tm = 372.15 to 373.15; Vm = 0.1873
at 383.15; Antoine (sol) A = 14.3739, B = 5997, C = 0, dev = 5.0, range =
353 to 370.

$C_9H_{19}NO_2$ MW 173.25

1,4-Dihydroxy-2,2,6,6-tetramethylpiperidine, CA 3637-10-3:   Tm = 431.15;
Antoine (sol) A = 12.155, B = 5240, C = 0, dev = 5.0, range = 313 to 348.

Heptylcarbamic acid, methyl ester, CA 35601-84-4:   Irritant; flammable.
Antoine (liq) A = 14.49105, B = 5737.885, C = 0, dev = 1.0 to 5.0, range =
368 to 408.

Lactic acid, *N*-hexylamide: Irritant; flammable.   Tm = 315.15; Antoine
(liq) A = 9.64317, B = 4301.052, C = 0, dev = 1.0, range = 383 to 452.

C_9H_{20} MW 128.26

3,3-Diethylpentane, CA 1067-20-5:   Flammable; soluble most organic solvents.
Tf = 240.05; Tb = 419.35; Vm = 0.1702 at 293.15; Antoine (liq) A = 6.0158,
B = 1450.634, C = -57.59, dev = 0.02, range = 335 to 426.

2-Methyloctane, CA 3221-61-2:   Flammable; soluble most organic solvents.
Tm = 193.05; Tb = 415.95; Vm = 0.1818 at 298.15; Antoine (liq) A = 6.1112,
B = 1449.77, C = -63.28, dev = 1.0, range = 305 to 417.

Nonane, CA 111-84-2:   Flammable, flash point 304; soluble most organic solvents.   Tm = 222.15; Tb = 423.95; Vm = 0.1787 at 293.15; Tc = 595.65; Pc =
2306; Vc = 0.548; Antoine I (liq) A = 6.0593, B = 1429.46, C = -71.33,
dev = 0.02, range = 344 to 426; Antoine II (liq) A = 8.17855, B = 2523.8,
C = 0, dev = 1.0 to 5.0, range = 219 to 308.

2,2,3,3-Tetramethylpentane, CA 7154-79-2:   Flammable; soluble most organic
solvents.   Tf = 264.15; Tb = 413.45; Vm = 0.1703 at 298.15; Antoine (liq)
A = 5.95517, B = 1398.942, C = -59.228, dev = 0.02, range = 328 to 415.

2,2,3,4-Tetramethylpentane, CA 1186-53-4:   Flammable; soluble most organic
solvents.   Tf = 152.15; Tb = 406.15; Vm = 0.1736 at 293.15; Antoine (liq)
A = 5.95157, B = 1371.297, C = -58.652, dev = 0.02, range = 325 to 413.

2,2,4,4-Tetramethylpentane, CA 1070-87-7:   Flammable; soluble most organic
solvents.   Tf = 206.65; Tb = 393.15 to 395.15; Vm = 0.1783 at 293.15;
Antoine (liq) A = 5.92882, B = 1329.726, C = -56.493, dev = 0.02, range =
313 to 397.

2,3,3,4-Tetramethylpentane, CA 16747-38-9:   Flammable; soluble most organic
solvents.   Tf = 171.05; Tb = 414.65; Vm = 0.1699 at 293.15; Antoine (liq)
A = 5.9947, B = 1424.544, C = -57.595, dev = 0.02, range = 331 to 416.

2,2,4-Trimethylhexane, CA 16747-26-5:   Flammable; soluble most organic solvents.   Tf = 153.15; Tb = 399.65; Vm = 0.1804 at 293.15; Antoine I (liq)
A = 6.11453, B = 1436.449, C = -50.097, dev = 0.1, range = 288 to 410;
Antoine II (liq) A = 6.26836, B = 1506.4, C = -44.643, dev = 0.1, range =
238 to 293.

2,2,5-Trimethylhexane, CA 3522-94-9:   Flammable; soluble most organic solvents.   Tf = 167.35; Tb = 397.25; Vm = 0.1814 at 293.15;   *(continues)*

C_9H_{20} *(continued)*

Antoine I (liq) A = 5.99253, B = 1343.85, C = -60.158, dev = 0.02, range = 288 to 399; Antoine II (liq) A = 6.25179, B = 1471.621, C = -48.981, dev = 0.02, range = 238 to 293.

2,3,3-Trimethylhexane, CA 16747-28-7: Flammable; soluble most organic solvents. Tf = 156.35; Tb = 410.85; Vm = 0.1746 at 298.15; Antoine I (liq) A = 6.19535, B = 1555.5, C = -39.6, dev = 1.0, range = 288 to 422; Antoine II (liq) A = 6.41432, B = 1592.346, C = -42.628, dev = 0.1, range = 238 to 303.

2,4,4-Trimethylhexane, CA 16747-30-1: Flammable; soluble most organic solvents. Tf = 159.75; Tb = 403.85; Vm = 0.1781 at 298.15; Antoine (liq) A = 5.97118, B = 1365.818, C = -59.382, dev = 0.02, range = 323 to 406.

$C_9H_{20}ClF_3N_2OS$ MW 296.78

Sulfur, chlorobis(*N*-ethylethanaminato)oxo(trifluoromethyl), CA 63265-74-7: Toxic; flammable. Antoine (liq) A = 6.755, B = 2309, C = 0, dev = 1.0 to 5.0, range not specified.

$C_9H_{20}ClF_3N_2S$ MW 280.78

Sulfur, chlorobis(*N*-ethylethanaminato)(trifluoromethyl), CA 63265-72-5: Toxic; flammable. Antoine (liq) A = 6.325, B = 2069, C = 0, dev = 1.0 to 5.0, range not specified.

$C_9H_{20}O$ MW 144.26

2,6-Dimethyl-4-heptanol, CA 108-82-7: Moderately toxic; flammable; soluble ethanol. Tb = 452.15; Vm = 0.1783 at 294.15; Antoine (liq) A = 5.66296, B = 1144.81, C = -138.15, dev = 1.0, range = 374 to 452.

1-Nonanol, nonyl alcohol, CA 143-08-8: Flammable, flash point 348; soluble ethanol, ether. Tf = 268.15; Tb = 488.15; Vm = 0.1744 at 293.15; Antoine I (liq) A = 5.9049, B = 1341.28, C = -142.64, dev = 0.1 to 1.0, range = 381 to 495; Antoine II (liq) A = 5.9454, B = 1366.566, C = -139.73, dev = 1.0, range = 368 to 500.

$C_9H_{20}O_2$ MW 160.26

4-*tert*-Butoxy-2-methyl-2-butanol, CA 22419-28-9: Flammable; soluble ethanol. Antoine (liq) A = 8.3499, B = 3211.4, C = 0, dev = 1.0 to 5.0, range = 367 to 483.

2,2,4-Trimethyl-1,6-hexanediol, CA 3089-24-5: Flammable; soluble water, ethanol. Tm = 311.15; Antoine (liq) A = 8.5771, B = 3550.5, C = 0, dev = 5.0, range = 419 to 541.

$C_9H_{20}O_3$ MW 176.26

Dipropylene glycol, isopropyl ether: Flammable; forms explosive peroxides. Antoine (liq) A = 6.96343, B = 2151.86, C = -44.89, dev = 1.0, range = 319 to 479.

$C_9H_{20}O_4$ MW 192.25

Tripropylene glycol, CA 1638-16-0: Flammable, flash point 414; hygroscopic; soluble water, ethanol. Vm = 0.1885 at 293.15; Antoine (liq) A =
8.04844, B = 3245.94, C = -3.443, dev = 1.0, range = 369 to 541.

$C_9H_{20}S$ MW 160.32

1-Nonanethiol, CA 1455-21-6: Toxic; stench; flammable, flash point 351.
Tm = 253.05; Tb = 493.35; Vm = 0.1904 at 293.15; Antoine (liq) A = 6.0066,
B = 1585.2, C = -97.18, dev = 1.0, range = 390 to 494.

2-Nonanethiol, CA 13281-11-3: Toxic; stench; flammable. Tm = 204.15;
Antoine (liq) A = 6.07231, B = 1605.6, C = -146.49, dev = 1.0, range = 379
to 482.

$C_9H_{20}S_2$ MW 192.38

1,9-Nonanedithiol, CA 3489-28-9: Toxic; stench; flammable, flash point
374. Tm = 255.65; Vm = 0.2021 at 293.15; Antoine (liq) A = 6.3728, B =
2016.5, C = -95.4, dev = 1.0, range = 418 to 557.

$C_9H_{21}N$ MW 143.27

N-Methyl octylamine, CA 2439-54-5: Irritant; flammable; soluble ethanol,
ether. Tm = 262.15; Antoine (liq) A = 6.41415, B = 1719.3, C = -69.15,
dev = 5.0, range = 365 to 508.

Nonyl amine, CA 112-20-9: Irritant; flammable, flash point 335; soluble
ethanol, ether. Tm = 272.15; Tb = 474.15; Vm = 0.1817 at 293.15; Antoine
(liq) A = 6.3659, B = 1756.1, C = -72.55, dev = 1.0, range = 377 to 478.

Tripropyl amine, CA 102-69-2: Highly toxic; flammable, flash point 314;
soluble ethanol, ether. Tm = 179.15; Tb = 429.15; Vm = 0.1896 at 293.15;
Antoine (liq) A = 6.38073, B = 1599.1, C = -64.15, dev = 1.0, range = 341
to 475.

$C_9H_{21}NO_3$ MW 191.27

Triisopropanolamine, CA 122-20-3: Corrosive; hygroscopic; flammable, flash
point 433; soluble water, ethanol. Tm = 331.15; Tb = 573.15; Vm = 0.1913
at 293.15; Antoine (liq) A = 9.60893, B = 4725.787, C = 47.771, dev = 1.0,
range = 428 to 573.

$C_9H_{21}O_4P$ MW 224.24

Tripropyl phosphate, CA 513-08-6: Irritant; flammable, flash point above
385; moderately soluble water; soluble ethanol, ether; used as fireproofing
agent. Tb = 525.15; Vm = 0.2216 at 293.15; Antoine (liq) A = 7.6437, B =
2960, C = 0, dev = 1.0, range = 394 to 525.

$C_9H_{22}ClN_2PS$ MW 256.77

Phosphorothioic diamide, *P*-(chloromethyl)-*N,N'*-bis(1-methylpropyl),
CA 58023-20-4: Toxic; flammable. Antoine (liq) A = 6.5986, B = 3490,
C = 0, dev not specified, range = 333 to 368.

$C_{10}F_8$ MW 272.10

Octafluoronaphthalene, CA 313-72-4: Tm = 360.15 to 361.15; Antoine (sol)
A = 11.95227, B = 4161.348, C = 0, dev = 1.0, range = 293 to 323.

$C_{10}F_{22}$ MW 538.07

Perfluorodecane, CA 307-45-9: Tm = 309.15; Antoine (liq) A = 7.9508, B =
3345.34, C = 156.09, dev = 1.0, range = 405 to 543.

$C_{10}F_{22}O$ MW 554.07

Bis(undecafluoropentyl) ether, CA 464-36-8: Antoine (liq) A = 9.4757, B =
2692, C = 0, dev = 1.0 to 5.0, range = 288 to 313.

$C_{10}HCl_5F_{14}O_2$ MW 596.36

2,2,3,4,4,5,6,6,7,8,8,9,10,10-Tetradecafluoro-3,5,7,9,10-pentachlorodeca-
noic acid: Irritant. Antoine (liq) A = 9.287, B = 4209, C = 0, dev = 5.0,
range = 373 to 578.

$C_{10}H_2O_6$ MW 218.12

Pyromellitic acid, dianhydride, CA 89-32-7: Irritant; moisture-sensitive.
Tm = 558.15; Tb = 663.15; Antoine (liq) A = 8.35788, B = 4171.119, C = 0,
dev = 1.0 to 5.0, range = 561 to 665.

$C_{10}H_6N_2O_4$ MW 218.17

1,5-Dinitronaphthalene, CA 605-71-0: Irritant; potentially explosive; soluble ether, pyridine, hot benzene. Tm = 492.15; Antoine (liq) A = 8.084,
B = 3902, C = 0, dev = 1.0 to 5.0, range = 506 to 642.

1,8-Dinitronaphthalene, CA 602-38-0: Irritant; potentially explosive; soluble acetone, pyridine. Tm = 446.15; Tb = 718.15, decomposes; Antoine
(liq) A = 7.767, B = 4102, C = 0, dev = 1.0 to 5.0, range = 553 to 715.

$C_{10}H_7Br$ MW 207.07

1-Bromonaphthalene, CA 90-11-9: Irritant; flammable, flash point above
385; moderately soluble water; soluble ethanol, ether, chloroform, benzene.
Tm = 279.35, highest melting solid form; Tb = 554.25; Vm = 0.1406 at
303.15; Antoine I (liq) A = 6.56365, B = 2303.73, C = -48.841, dev = 1.0,
range = 357 to 555; Antoine II (liq) A = 4.50697, B = 929.871, C =
-182.045, dev = 0.1, range = 469 to 559.

$C_{10}H_7Cl$ MW 162.62

1-Chloronaphthalene, CA 90-13-1: Irritant; flammable, flash point 394;
soluble ethanol, ether, benzene. Tm = 270.85; Tb = 533.15; Vm = 0.1368 at
298.15; Antoine (liq) A = 6.15143, B = 1861.65, C = -83.337, dev = 1.0,
range = 353 to 533.

(continues)

$C_{10}H_7Cl$ *(continued)*

2-Chloronaphthalene, CA 91-58-7: Irritant; flammable; soluble ethanol,
ether, benzene, chloroform. Tm = 332.65 to 333.15; Tb = 532.15; Vm =
0.1429 at 344.15; Antoine (liq) A = 7.8608, B = 3021.2, C = 0, dev = 1.0,
range = 400 to 435.

$C_{10}H_7Cl_7$ MW 375.34

Dihydroheptachlor, CA 2589-15-3: Antoine (sol) A = 9.22741, B = 4375.396,
C = 0, dev = 1.0 to 5.0, range = 333 to 353.

$C_{10}H_7F_5O_2$ MW 254.16

Pentafluoropropionic acid, 3-tolyl ester, CA 24271-51-0: Irritant; flammable. Antoine (liq) A = 7.1106, B = 2107.1, C = -34.05, dev = 1.0,
range = 371 to 446.

Pentafluoropropionic acid, 4-tolyl ester, CA 24271-52-1: Irritant; flammable. Antoine (liq) A = 6.7638, B = 1886.7, C = -52.25, dev = 1.0,
range = 371 to 448.

$C_{10}H_7NO_2$ MW 173.17

1-Nitronaphthalene, CA 86-57-7: Moderately irritant; flammable, flash
point 437; soluble ethanol, ether, benzene. Tm = 334.65; Tb = 577.15;
Antoine I (sol) A = 8.31261, B = 3579.698, C = 0, dev = 5.0, range = 309
to 326; Antoine II (sol) A = 14.223, B = 5584, C = 0, dev = 5.0, range =
325 to 332; Antoine III (liq) A = 7.8959, B = 3468.4, C = 0, dev = above
5.0, range = 332 to 580.

$C_{10}H_8$ MW 128.17

Azulene, CA 275-51-4: Flammable; soluble ethanol, ether, acetone; isomerizes to naphthalene above 543. Tm = 372.15; Tb = 515.15; Antoine I (sol)
A = 11.548, B = 4324.5, C = 0, dev = 1.0 to 5.0, range = 290 to 372;
Antoine II (liq) A = 7.3537, B = 2768.64, C = 0, dev = 1.0 to 5.0, range =
369 to 515.

Naphthalene, CA 91-20-3: Flammable, flash point 352; soluble ethanol,
ether, benzene, chloroform. Tm = 353.44; Tb = 491.14; Vm = 0.1321 at
363.15; Antoine I (sol) A = 8.70592, B = 2619.91, C = -52.499, dev = 1.0,
range = 310 to 353; Antoine II (sol) A = 9.45562, B = 3059.145, C =
-29.892, dev = 1.0, range = 263 to 353; Antoine III (sol) A = 11.9861, B =
4577.47, C = 30.394, dev = 1.0, range not specified; Antoine IV (liq) A =
6.19487, B = 1782.509, C = -65.637, dev = 0.1, range = 352 to 500; Antoine
V (liq) A = 6.14835, B = 1751.644, C = -68.319, dev = 0.1, range = 491 to
565; Antoine VI (liq) A = 6.53231, B = 2162.182, C = -12.108, dev = 0.1,
range = 563 to 665; Antoine VII (liq) A = 7.74783, B = 4042.567, C =
227.985, dev = 1.0, range = 661 to 750.

$C_{10}H_8N_2O_2$ MW 188.19

Benzene, 1,3-bis(isocyanatomethyl), CA 3634-83-1: Highly toxic; lachrymator; reacts violently with water. Antoine (liq) A = 5.35246, B =
2437.795, C = 0, dev = 1.0 to 5.0, range = 403 to 473.   *(continues)*

$C_{10}H_8N_2O_2$ *(continued)*

Benzene, 1,4-bis(isocyanatomethyl), CA 1014-98-8: Highly toxic; lachrymator; reacts violently with water. Tm = 318.95; Antoine (liq) A =
6.62597, B = 2974.207, C = 0, dev = 5.0, range = 403 to 473.

Benzene, ethyldiisocyanato, mixed isomers, CA 64711-83-7: Highly toxic;
lachrymator; reacts violently with water. Antoine (liq) A = 8.22041, B =
3244.819, C = 4.306, dev = 1.0 to 5.0, range = 363 to 473.

$C_{10}H_8N_2O_3$ MW 204.18

Phthalimide, 3-acetamido: Tm = 515.15; Antoine (sol) A = 11.020, B = 5667,
C = 0, dev = 1.0 to 5.0 range = 428 to 468.

$C_{10}H_8O$ MW 144.17

1-Naphthol, 1-hydroxynaphthalene, CA 90-15-3: Irritant; moderately toxic;
moderately soluble water; soluble ethanol, ether, benzene, chloroform.
Tm = 368.95 to 369.15; Tb = 551.15 to 553.15; Vm = 0.1312 at 372.15;
Antoine I (sol, *alpha* form) A = 12.20753, B = 4874.394, C = 0, dev = 1.0,
range = 298 to 312; Antoine II (sol, *beta* form) A = 10.70115, B = 4405.522,
C = 0, dev = 0.1 to 1.0, range = 314 to 324; Antoine III (liq) A = 7.53825,
B = 3083.8, C = 1.731, dev = 1.0, range = 399 to 556.

2-Naphthol, 2-hydroxynaphthalene, CA 135-19-3: Moderately toxic; flammable, flash point 426; light-sensitive; moderately soluble water; soluble
ethanol, ether, benzene. Tm = 395.15; Tb = 558.15 to 559.15; Vm = 0.1337
at 403.15; Antoine I (sol, *alpha* form) A = 12.48704, B = 5110.333, C = 0,
dev = 1.0, range = 298 to 312; Antoine II (sol, *beta* form) A = 10.80636,
B = 4586.029, C = 0, dev = 1.0, range = 314 to 332; Antoine III (sol) A =
9.273, B = 4112, C = 0, dev = 1.0 to 5.0, range = 283 to 323; Antoine IV
(liq) A = 7.22927, B = 2827.5, C = -19.868, dev = 1.0, range = 401 to 561.

$C_{10}H_9Cl_3O_3$ MW 283.54

(2,4,5-Trichlorophenoxy)acetic acid, ethyl ester, CA 1928-39-8: Irritant;
flammable. Antoine (liq) A = 5.7328, B = 1659.104, C = -162.882, dev =
1.0, range = 444 to 573.

$C_{10}H_9N$ MW 143.19

2-Methylquinoline, quinaldine, CA 91-63-4: Irritant; moderately toxic;
flammable, flash point 352; light-sensitive; slightly soluble water; soluble ethanol, ether, acetone, chloroform; anesthetic for fish. Tm = 271.15
to 272.15; Tb = 521.15; Vm = 0.1353 at 293.15; Antoine (liq) A = 6.30917,
B = 1862.84, C = -88.083, dev = 0.1, range = 443 to 521.

3-Methylquinoline, CA 612-58-8: Irritant; flammable; slightly soluble
water; soluble ethanol, ether, acetone. Tm = 289.15 to 290.15; Tb =
525.15; Vm = 0.1342 at 293.15; Antoine (liq) A = 6.08211, B = 1708.48, C =
-107.253, dev = 0.1, range = 443 to 528.

4-Methylquinoline, lepidine, CA 491-35-0: Irritant; flammable, flash point
above 385; light-sensitive; slightly soluble water; soluble ethanol, ether,
acetone, benzene. Tm = 282.15 to 283.15; Tb = 534.15 to   *(continues)*

$C_{10}H_9N$ (continued)

536.15; Vm = 0.1318 at 293.15; Antoine (liq) A = 6.39787, B = 1948.36; C = -95.244, dev = 0.1, range = 463 to 539.

6-Methylquinoline, CA 91-62-3: Irritant; flammable, flash point above 385; light-sensitive; slightly soluble water; soluble ethanol, ether, acetone. Tm = 305.55; Tb = 531.15; Vm = 0.1344 at 293.15; Antoine (liq) A = 6.04764, B = 1743.33, C = -106.971, dev = 1.0 to 5.0, range = 453 to 540.

7-Methylquinoline, CA 612-60-2: Irritant; flammable, flash point 383; slightly soluble water; soluble ethanol, ether, acetone. Tm = 308.15 to 311.15; Tb = 524.65 to 525.65; Vm = 0.1350 at 293.15; Antoine (liq) A = 5.9577, B = 1699.12, C = -123.243, dev = 0.1, range = 493 to 532.

8-Methylquinoline, CA 611-32-5: Irritant; flammable, flash point 378; slightly soluble water; soluble ethanol, ether, acetone. Tb = 521.15; Vm = 0.1336 at 293.15; Antoine (liq) A = 6.8115, B = 2310.92, C = -40.256, dev = 0.1, range = 493 to 523.

1-Naphthylamine, CA 134-32-7: Toxic by inhalation or skin absorption; suspected carcinogen; flammable, flash point 430; light-sensitive; slightly soluble water; very soluble ethanol, ether. Tm = 323.15; Tb = 574.15; Antoine (liq) A = 6.88407, B = 2570.55, C = -46.989, dev = 1.0, range = 377 to 574.

2-Naphthylamine, CA 91-59-8: Very toxic; potent carcinogen which produces bladder cancer; flammable; soluble hot water, ethanol, ether. Tm = 384.15 to 386.15; Tb = 579.15; Antoine I (sol) A = 8.4859, B = 3859, C = 0, dev = 5.0, range = 283 to 323; Antoine II (liq) A = 6.88978, B = 2604.31, C = -46.068, dev = 1.0, range = 388 to 579.

$C_{10}H_9NO$ MW 159.19

Propionitrile, 3-benzoyl, CA 5343-98-6: Irritant; flammable. Antoine (sol) A = 18.0639, B = 5669, C = 0, dev = 1.0 to 5.0, range = 318 to 333.

2-Methyl-8-hydroxyquinoline, CA 826-81-3: Irritant; flammable; soluble ether, benzene. Tm = 347.15; Tb = 539.15 to 540.15; Antoine (sol) A = 11.785, B = 4580, C = 0, dev = 5.0, range = 308 to 333.

$C_{10}H_{10}$ MW 130.19

1,3-Divinylbenzene, CA 108-57-6: Moderately toxic; lachrymator; flammable, flash point 349; soluble acetone, benzene. Tm = 220.85; Vm = 0.1401 at 293.15; Antoine (liq) A = 6.8295, B = 2167.64, C = -23.374, dev = 1.0, range = 305 to 453.

$C_{10}H_{10}Cl_2O_3$ MW 249.09

(2,4-Dichlorophenoxy)acetic acid, ethyl ester, CA 533-23-3: Irritant; flammable; soluble ethanol, ether, acetone; used as herbicide. Tm = 390.15 to 392.15; Antoine (liq) A = 5.64262, B = 1555.379, C = -164.852, dev = 1.0, range = 444 to 573.

$C_{10}H_{10}N_2O_2$ MW 190.20

3-(Dimethylamino)phthalimide: Antoine (sol) A = 9.8679, B = 4749, C = 0, dev = 1.0, range = 392 to 431.

$C_{10}H_{10}O$ MW 146.19

2-Methyl-3-phenyl-2-propenal, CA 101-39-3: Irritant; flammable. Antoine
(liq) A = 9.74851, B = 3736.644, C = 0, dev = 5.0, range = 343 to 393.

4-Phenyl-3-buten-2-one, benzal acetone, CA 122-57-6: Irritant; flammable,
flash point 338; light-sensitive; slightly soluble water; soluble ethanol,
ether, benzene, chloroform. Tm = 315.15; Tb = 533.15 to 535.15; Vm =
0.1448 at 318.15; Antoine (liq) A = 7.34421, B = 2751.17, C = -18.858,
dev = 1.0, range = 354 to 534.

1-Tetralone, CA 529-34-0: Flammable, flash point 403. Tm = 281.15; Tb =
531.15; Vm = 0.1330 at 293.15; Antoine (liq) A = 6.27045, B = 1883.562, C =
-94.142, dev = 1.0 to 5.0, range = 388 to 535.

$C_{10}H_{10}O_2$ MW 162.19

Cinnamic acid, methyl ester: CA 103-26-4: Flammable; soluble ether, benzene. Tm = 309.15; Tb = 536.15; Antoine I (liq) A = 8.28562, B = 3257.59,
C = 0, dev = 5.0, range = 288 to 333; Antoine II (liq) A = 6.67488, B =
2269.9, C = -49.866, dev = 1.0, range = 350 to 536.

1,3-Diacetylbenzene, CA 6781-42-6: Flammable, flash point 383; soluble
ethanol, benzene, chloroform. Tm = 305.15; Antoine (liq) A = 5.786, B =
2256.5, C = 0, dev = 1.0 to 5.0, range = 323 to 418.

1,4-Diacetylbenzene, CA 1009-61-6: Flammable; soluble ethanol, benzene,
chloroform. Tm = 384.15 to 386.15; Antoine (liq) A = 1.8178, B = 158.07,
C = -324.35, dev = 1.0 to 5.0, range = 388 to 431.

Isosafrole, CA 120-58-1: Flammable; soluble ethanol, ether, benzene. Tf =
279.85 to 279.95; Tb = 525.15; Vm = 0.1447 at 293.15; Antoine (liq) A =
5.9992, B = 1678.9, C = -107.75, dev = 1.0, range = 393 to 531.

alpha-Methylcinnamic acid, CA 1199-77-5: Flammable; soluble ethanol, benzene, ether. Tm = 354.15 to 355.15; Antoine (liq) A = 8.43038, B =
3352.08, C = -39.573, dev = 1.0, range = 398 to 561.

1-Phenyl-1,3-butanedione, CA 93-91-4: Flammable; slightly soluble water.
Tm = 331.15 to 333.15; Antoine (sol) A = 11.442, B = 4374.6, C = 0, dev =
1.0 to 5.0, range = 278 to 300.

Safrole, CA 94-59-7: Irritant; suspected carcinogen; flammable, flash
point 373; soluble ethanol, ether, chloroform. Tf = 284.35; Tb = 504.65
to 505.15; Vm = 0.148 at 293.15; Antoine (liq) A = 7.73042, B = 2924.73,
C = 4.488, dev = 1.0, range = 336 to 506.

$C_{10}H_{10}O_4$ MW 194.19

1,2-Benzenediol, diacetate, CA 635-67-6: Flammable; soluble ethanol,
ether, chloroform. Tm = 336.65; Antoine (liq) A = 7.83162, B = 3172.11,
C = -6.717, dev = 1.0, range = 371 to 551.

Dimethyl isophthalate, CA 1459-93-4: Irritant; flammable, flash point 411;
soluble ethanol, ether, benzene. Tm = 340.15; Tb = 555.15; Antoine (liq)
A = 7.023, B = 2525, C = -43.15, dev = 1.0 to 5.0, range = 393 to 550.

Dimethyl phthalate, CA 131-11-3: Irritant; toxic; suspected carcinogen;
flammable, flash point 419; soluble ethanol, ether, benzene. *(continues)*

$C_{10}H_{10}O_4$ *(continued)*

Tm = 273.15 to 275.15; Tb = 555.15; Vm = 0.1628 at 298.15; Antoine I (liq)
A = 10.185, B = 4113, C = 0, dev = 5.0, range = 304 to 371; Antoine II
(liq) A = 8.095, B = 3327, C = 0, dev = 5.0, range = 371 to 547.

Dimethyl terephthalate, CA 120-61-6: Suspected carcinogen; flammable,
flash point 426; soluble ether. Tm = 414.15; Tb = 561.15; Tc = 772, calculated; Pc = 2780, calculated; Vc = 0.529, calculated; Antoine I (sol) A =
10.582, B = 3893, C = -43.15, dev = 1.0 to 5.0, range = 373 to 413; Antoine
II (liq) A = 7.118, B = 2618, C = -43.15, dev = 1.0 to 5.0, range = 413
to 523.

$C_{10}H_{11}N_3O_2$ MW 205.22

3-Amino-6-(dimethylamino)phthalimide: Antoine (sol) A = 10.8809, B = 5684,
C = 0, dev = 1.0, range = 434 to 459.

$C_{10}H_{12}$ MW 132.20

Dicyclopentadiene, *endo*, CA 77-73-6: Toxic; flammable, flash point 299;
polymerizes; soluble ethanol, ether; commercial material is mostly this
isomer. Tm = 306.75; Tb = 443.15, decomposes; Vm = 0.1353 at 308.15;
Antoine (liq) A = 6.7121, B = 1984.79, C = -24.16, dev = 1.0, range = 350
to 446.

2,4-Dimethylstyrene, CA 2234-20-0: Irritant; lachrymator; flammable; polymerizes. Tm = 209.15; Vm = 0.1465 at 294.65; Antoine (liq) A = 6.62435,
B = 2014.28, C = -39.114, dev = 1.0, range = 307 to 453.

2,5-Dimethylstyrene, CA 2039-89-6: Irritant; lachrymator; flammable; polymerizes. Vm = 0.1457 at 290.65; Antoine (liq) A = 6.78661, B = 2098.69,
C = -27.153, dev = 1.0, range = 302 to 453.

1-Ethyl-2-vinylbenzene, CA 7564-63-8: Irritant; lachrymator; flammable;
polymerizes. Tm = 197.58; Antoine (liq) A = 6.55373, B = 1897.7, C =
-43.15, dev = 1.0, range = 363 to 413.

1-Ethyl-3-vinylbenzene, CA 7525-62-4: Irritant; lachrymator; flammable;
polymerizes. Tm = 171.85; Antoine (liq) A = 6.16418, B = 1614, C = -75.15,
dev = 1.0, range = 343 to 453.

1-Ethyl-4-vinylbenzene, CA 3454-07-7: Irritant; lachrymator; flammable;
polymerizes. Tm = 223.45; Antoine (liq) A = 6.02561, B = 1570.9, C =
-75.15, dev = 1.0, range = 341 to 448.

1,2,3,4-Tetrahydronaphthalene, tetralin, CA 119-64-2: Irritant; moderately toxic vapor; flammable, flash point 344; soluble most organic solvents. Tf = 237.4; Tb = 480.77; Vm = 0.1369 at 298.15; Antoine (liq) A =
6.35719, B = 1854.52, C = -54.257, dev = 1.0, range = 311 to 481.

$C_{10}H_{12}Cl_4NOPS$ MW 367.06

P-Chloromethyl-*N*-(1-methylethyl)amidothiophosphonic acid, *O*-(2,4,5-tri-
chlorophenyl) ester, CA 21844-03-1: Toxic; flammable. Antoine (liq) A =
8.492, B = 4091, C = 0, dev not specified, range = 323 to 363.

$C_{10}H_{12}N_2O_2$ MW 192.22

Acetylglycineanilide: Tm = 466.15; Antoine (sol) A = 12.9109, B = 6380.2,
C = 0, dev = 1.0, range = 362 to 365.

$C_{10}H_{12}O$ MW 148.20

Anethole, *cis*, CA 25679-28-1: Flammable; soluble most organic solvents.
Tm = 250.65; Vm = 0.1500 at 293.15; Antoine (liq) A = 9.8481, B = 3589.059,
C = 0, dev = 5.0, range = 333 to 363.

Anethole, *trans*, CA 4180-23-8: Flammable, flash point 363; soluble most
organic solvents. Tm = 294.15; Tb = 508.15; Vm = 0.1500 at 293.15; Antoine
(liq) A = 11.03042, B = 4090.18, C = 0, dev = 1.0 to 5.0, range = 333 to
363.

Anethole, mixed isomers, CA 104-46-1: Flammable; soluble most organic solvents. Tm = 295.65; Tb = 508.15; Vm = 0.1500 at 293.15; Antoine (liq) A =
6.92182, B = 2295.95, C = -41.509, dev = 1.0, range = 335 to 509.

Estragole, CA 140-67-0: Irritant; lachrymator; polymerizes; flammable;
soluble most organic solvents. Tb = 489.15; Vm = 0.1537 at 288.15; Antoine
(liq) A = 6.98682, B = 2205.51, C = -45.505, dev = 1.0, range = 325 to 488.

2'-Ethylacetophenone, CA 2142-64-5: Flammable. Tm = 254.15; Antoine (liq)
A = 7.63489, B = 2757.427, C = 0, dev = 1.0 to 5.0, range = 363 to 397.

3'-Ethylacetophenone, CA 22699-70-3: Flammable, flash point 356. Vm =
0.1594 at 293.15; Antoine (liq) A = 3.7742, B = 1290.5, C = 6.85, dev =
5.0, range = 292 to 416.

4'-Ethylacetophenone, CA 937-30-4: Flammable, flash point 363. Tm =
252.55; Vm = 0.1492 at 293.15; Antoine (liq) A = 4.6234, B = 1179.8, C =
-82.65, dev = 1.0, range = 294 to 368.

4-Isopropylbenzaldehyde, cumaldehyde, CA 122-03-2: Irritant; flammable,
flash point 366; soluble ethanol, ether. Tb = 508.15 to 509.15; Vm =
0.1519 at 293.15; Antoine (liq) A = 6.84988, B = 2250.78, C = -40.557,
dev = 1.0, range = 331 to 505.

2-Methyl-3-phenylpropanol, CA 5445-77-2: Flammable. Antoine (liq) A =
8.38585, B = 3089.186, C = 0, dev = 1.0 to 5.0, range = 333 to 373.

4'-Methylpropiophenone, CA 5337-93-9: Flammable, flash point 369; soluble
ethanol, ether, acetone, benzene. Tm = 280.35; Tb = 511.15 to 512.15; Vm =
0.1493 at 293.15; Antoine (liq) A = 6.9899, B = 2452.17, C = -19.195, dev =
1.0, range = 332 to 512.

4-Vinylphenetole, CA 5459-40-5: Irritant; lachrymator; flammable; polymerizes. Antoine (liq) A = 7.33669, B = 2451.23, C = -38.557, dev = 1.0,
range = 337 to 498.

$C_{10}H_{12}O_2$ MW 164.20

Acetic acid, phenethyl ester, CA 103-45-7: Irritant; flammable; soluble
ethanol, ether. Tm = 242.05; Tb = 505.75; Vm = 0.1509 at 293.15; Antoine I
(liq) A = 9.710, B = 3521, C = 0, dev = 1.0, range = 283 to 318; Antoine II
(liq) A = 6.79417, B = 2215, C = -43.15, dev = 0.1 to 1.0, range = 422
to 506.   *(continues)*

$C_{10}H_{12}O_2$ *(continued)*

4-Allyl-2-methoxyphenol, eugenol, CA 97-53-0: Irritant; lachrymator; flammable, flash point above 385; slightly soluble water; soluble most organic solvents; found in essential oils. Tm = 265.65; Tb = 528.15; Vm = 0.1540 at 293.15; Antoine I (liq) A = 8.851, B = 3455, C = 0, dev = 1.0, range = 285 to 333; Antoine II (liq) A = 7.4136, B = 2748.94, C = -18.303, dev = 1.0 to 5.0, range = 395 to 527.

Benzyl propionate, CA 122-63-4: Flammable. Tb = 495.15; Antoine (liq) A = 8.2841, B = 3084.5, C = 0, dev = 1.0, range = 298 to 378.

Chavibetol, 5-allyl-2-methoxyphenol, CA 501-19-9: Irritant; lachrymator; flammable; soluble ethanol, ether. Tm = 281.65; Tb = 527.15; Vm = 0.1547 at 298.15; Antoine (liq) A = 7.58647, B = 2809.12, C = -23.746, dev = 1.0 to 5.0, range = 356 to 527.

Isoeugenol, *cis*, CA 5912-86-7: Flammable; slightly soluble water; soluble ethanol, ether. Antoine (liq) A = 9.28912, B = 3641.687, C = 0, dev = 1.0, range = 373 to 403.

Isoeugenol, *trans*, 5932-68-3: Flammable; slightly soluble water; soluble ethanol, ether. Tm = 306.15; Antoine (liq) A = 8.9355, B = 3607.634, C = 0, dev = 1.0, range = 363 to 420.

Isoeugenol, mixed isomers, CA 97-54-1: Flammable, flash point above 385; slightly soluble water; soluble ethanol, ether. Tm = 263.15; Tb = 539.15; Vm = 0.1520 at 298.15; Antoine I (liq) A = 9.049, B = 3656, C = 0, dev = 1.0 to 5.0, range = 283 to 340; Antoine II (liq) A = 7.70052, B = 3124.91, C = 7.902, dev = 1.0 to 5.0, range = 405 to 541.

Phenylacetic acid, ethyl ester, CA 101-97-3: Irritant; flammable, flash point 350; soluble ethanol, ether. Tb = 500.15 to 500.75; Vm = 0.1589 at 293.15; Antoine (liq) A = 7.6588, B = 2818.7, C = 0, dev = 1.0, range = 393 to 500.

Propyl benzoate, CA 2315-68-6: Flammable; soluble ethanol, ether. Tm = 221.55; Tb = 503.15; Vm = 0.1605 at 293.15; Antoine (liq) A = 6.68614, B = 2165.28, C = -41.593, dev = 1.0, range = 327 to 504.

$C_{10}H_{12}O_3$ MW 180.20

Acetic acid, (2-phenoxyethyl) ester, CA 6192-44-5: Flammable, flash point 416; slightly soluble water. Tm = 270.45; Tb = 532.35; Vm = 0.1623 at 293.15; Antoine (liq) A = 8.1156, B = 3422.32, C = 27.331, dev = 1.0, range = 355 to 533.

$C_{10}H_{12}O_4$ MW 196.20

Maleic acid, diallyl ester, CA 999-21-3: Irritant; lachrymator; flammable, flash point 396; polymerizes; soluble most organic solvents. Vm = 0.1834 at 298.15; Antoine (liq) A = 10.1882, B = 4056.64, C = 0, dev = 1.0, range = 392 to 426.

$C_{10}H_{13}Br$ MW 213.12

2-Bromo-4-isopropyltoluene, CA 2437-76-5: Irritant; flammable. Tb = 506.15 to 508.15; Vm = 0.1679 at 288.15; Antoine (liq) A = 7.178, B = 2625, C = 0, dev = 1.0, range = 400 to 510. *(continues)*

$C_{10}H_{13}Br$ *(continued)*

3-Bromo-4-isopropyltoluene, CA 4478-10-8: Irritant; flammable. Tb = 505.15 to 506.15; Antoine (liq) A = 6.985, B = 2525, C = 0, dev = 5.0, range = 400 to 510.

$C_{10}H_{13}Cl$ MW 168.67

2-Chloro-4-isopropyltoluene, CA 4395-79-3: Irritant; flammable; soluble acetone, benzene. Tb = 487.15 to 489.15; Vm = 0.1669 at 298.15; Antoine (liq) A = 7.263, B = 2579, C = 0, dev = 1.0, range = 400 to 490.

3-Chloro-4-isopropyltoluene, CA 4395-80-6: Irritant; flammable; soluble acetone, benzene. Tb = 487.55 to 488.55; Vm = 0.1657 at 288.15; Antoine (liq) A = 7.031, B = 2463, C = 0, dev = 1.0, range = 400 to 490.

$C_{10}H_{13}ClO$ MW 184.67

2-Chloroethyl *alpha*-methylbenzyl ether, CA 4446-91-7: Irritant; flammable; forms explosive peroxides. Antoine (liq) A = 7.27989, B = 2591.37, C = -16.961, dev = 1.0, range = 335 to 508.

$C_{10}H_{13}ClO_3$ MW 216.66

Diethylene glycol, 4-chlorophenyl ether, CA 58498-77-4: Irritant; flammable; forms explosive peroxides. Antoine (liq) A = 4.8307, B = 1145.22, C = -214.76, dev = 1.0, range = 450 to 523.

$C_{10}H_{13}Cl_3NOPS$ MW 332.61

P-Chloromethyl-*N*-(1-methylethyl)amidothiophosphonic acid, *O*-(2,4-dichlorophenyl) ester, CA 18361-88-1: Toxic; flammable. Antoine (liq) A = 10.6648, B = 4864, C = 0, dev = 5.0, range = 323 to 368.

$C_{10}H_{13}NO$ MW 163.22

N,N-Dimethyl-3-toluamide, CA 6935-65-5: Irritant; flammable; soluble ethanol, ether. Antoine (sol) A = 4.0237, B = 1560.8, C = 0, dev = 1.0, range = 373 to 405.

$C_{10}H_{13}NO_2$ MW 179.22

4-Ethoxyacetanilide, phenacetin, CA 62-44-2: Suspected carcinogen; moderately soluble hot water; soluble ethanol, chloroform; decomposes at boiling point; important analgesic drug. Tm = 410.15 to 411.15; Antoine I (sol) A = 13.2101, B = 6036.09, C = 0, dev = 1.0, range = 312 to 388; Antoine II (liq) A = 7.99386, B = 3319.3, C = -58.63, dev = 1.0, range = 463 to 533.

2-Nitro-4-isopropyltoluene, CA 943-15-7: Irritant; flammable, flash point 383. Antoine (liq) A = 8.98356, B = 3538, C = 0, dev = 1.0, range = 370 to 415.

3-Nitro-4-isopropyltoluene, CA 35480-94-5: Irritant; flammable; decomposes on distillation at atmospheric pressure. Tm = 323.65; Antoine (liq) A = 7.365, B = 2820, C = 0, dev = 5.0, range = 330 to 430.

$C_{10}H_{14}$ MW 134.22

Butylbenzene, CA 104-51-8: Moderately toxic; flammable, flash point 344; soluble most organic solvents. Tf = 185.15; Tb = 455.15 to 457.15; Vm = 0.1561 at 293.15; Antoine (liq) A = 6.09809, B = 1571.648, C = -72.413, dev = 0.02, range = 369 to 463.

sec-Butylbenzene ±, CA 36383-15-0: Moderately toxic; flammable, flash point 325; soluble most organic solvents. Tf = 197.65; Tb = 446.15 to 447.15; Vm = 0.1557 at 293.15; Antoine (liq) A = 6.10298, B = 1559.452, C = -65.869, dev = 0.02, range = 384 to 448.

tert-Butylbenzene, CA 98-06-6: Moderately toxic; flammable, flash point 307; soluble most organic solvents. Tf = 215.35; Tb = 443.15 to 444.15; Vm = 0.1549 at 293.15; Antoine (liq) A = 6.06067, B = 1515.51, C = -68.551, dev = 0.02, range = 368 to 444.

1,2-Diethylbenzene, CA 135-01-3: Flammable, flash point 330; soluble most organic solvents. Tm = 241.95; Tb = 457.15; Vm = 0.1525 at 293.15; Antoine (liq) A = 6.11295, B = 1577.141, C = -72.619, dev = 0.02, range = 369 to 464.

1,3-Diethylbenzene, CA 141-93-5: Flammable, flash point 329; soluble most organic solvents. Tm = 189.25; Tb = 454.15 to 455.15; Vm = 0.1560 at 293.15; Antoine (liq) A = 6.12631, B = 1573.709, C = -72.374, dev = 0.02, range = 368 to 457.

1,4-Diethylbenzene, CA 105-05-5: Flammable, flash point 328; soluble most organic solvents. Tf = 230.35; Tb = 455.15 to 456.15; Vm = 0.1557 at 293.15; Antoine (liq) A = 6.12548, B = 1589.993, C = -70.995, dev = 0.02, range = 369 to 464.

1,2-Dimethyl-3-ethylbenzene, CA 933-98-2: Flammable; soluble most organic solvents. Tf = 223.65; Tb = 467.15; Vm = 0.1511 at 298.15; Antoine (liq) A = 6.1747, B = 1647.41, C = -71.95, dev = 1.0, range = 344 to 497.

1,2-Dimethyl-4-ethylbenzene, CA 934-80-5: Flammable; soluble most organic solvents. Tf = 206.25; Tb = 462.65; Vm = 0.1542 at 298.15; Antoine (liq) A = 6.1752, B = 1634.4, C = -70.94, dev = 1.0, range = 340 to 493.

1,3-Dimethyl-2-ethylbenzene, CA 2870-04-4: Flammable; soluble most organic solvents. Tf = 256.95; Tb = 463.15; Vm = 0.1514 at 298.15; Antoine (liq) A = 6.1699, B = 1633.4, C = -70.94, dev = 1.0, range = 341 to 493.

1,3-Dimethyl-4-ethylbenzene, CA 874-41-9: Flammable; soluble most organic solvents. Tf = 210.35; Tb = 461.35; Vm = 0.1532 at 298.15; Antoine (liq) A = 6.1686, B = 1630.4, C = -69.94, dev = 1.0, range = 339 to 492.

1,3-Dimethyl-5-ethylbenzene, CA 934-74-7: Flammable; soluble most organic solvents. Tf = 188.85; Tb = 456.75; Vm = 0.1552 at 293.15; Antoine (liq) A = 6.1718, B = 1616.39, C = -68.94, dev = 1.0, range = 336 to 487.

1,4-Dimethyl-2-ethylbenzene, CA 1758-88-9: Flammable; soluble most organic solvents. Tf = 219.45; Tb = 460.05; Vm = 0.1537 at 298.15; Antoine (liq) A = 6.156, B = 1623.39, C = -68.94, dev = 1.0, range = 338 to 490.

Isobutylbenzene, CA 538-93-2: Moderately toxic; flammable, flash point 328; soluble most organic solvents. Tf = 221.65; Tb = 443.65; Vm = 0.1548 at 293.15; Tc = 650.25; Pc = 3150; Antoine (liq) A = 6.06898, B = 1536.514, C = -67.788, dev = 0.02, range = 373 to 447.   *(continues)*

$C_{10}H_{14}$ *(continued)*

2-Isopropyltoluene, *ortho*-cymene, CA 527-84-4: Flammable; soluble most organic solvents. Tm = 201.61; Tb = 451.3; Vm = 0.1538 at 298.15; Antoine (liq) A = 6.54164, B = 1880.47, C = -36.878, dev = 0.1, range = 354 to 453.

3-Isopropyltoluene, *meta*-cymene, CA 535-77-3: Flammable; soluble most organic solvents. Tm = 209.4; Tb = 448.29; Vm = 0.1566 at 298.15; Antoine (liq) A = 6.44393, B = 1784.78, C = -45.917, dev = 0.1, range = 351 to 450.

4-Isopropyltoluene, *para*-cymene, CA 99-87-6: Moderately toxic; flammable, flash point 320; soluble most organic solvents. Tm = 205.21; Tb = 450.25; Vm = 0.1573 at 298.15; Antoine (liq) A = 6.16214, B = 1599.29, C = -65.491, dev = 0.1, range = 380 to 452.

2-Propyltoluene, CA 1074-17-5: Flammable; soluble most organic solvents. Tf = 212.85; Tb = 458.15; Vm = 0.1534 at 302.15; Antoine (liq) A = 6.1248, B = 1593.03, C = -71.24, dev = 1.0, range = 337 to 488.

3-Propyltoluene, CA 1074-43-7: Flammable; soluble most organic solvents. Tf = 190.65; Tb = 454.65 to 455.65; Vm = 0.1566 at 298.15; Antoine (liq) A = 6.1385, B = 1590.03, C = -70.25, dev = 1.0, range = 334 to 485.

4-Propyltoluene, CA 1074-55-1: Flammable; soluble most organic solvents. Tf = 209.55; Tb = 456.15 to 457.15; Vm = 0.1571 at 298.15; Antoine (liq) A = 6.1151, B = 1588.03, C = -70.05, dev = 1.0, range = 335 to 487.

1,2,3,4-Tetramethylbenzene, prehnitene, CA 488-23-3: Flammable; flash point 347; soluble most organic solvents. Tm = 266.75; Tb = 476.15 to 477.15; Vm = 0.1483 at 293.15; Antoine (liq) A = 6.1843, B = 1690.54, C = -73.67, dev = 1.0, range = 352 to 509.

1,2,3,5-Tetramethylbenzene, isodurene, CA 527-53-7: Flammable, flash point 344; soluble most organic solvents. Tf = 249.05; Tb = 471.15; Vm = 0.1508 at 293.15; Antoine (liq) A = 6.2028, B = 1675.43, C = -72.01, dev = 1.0, range = 348 to 502.

1,2,4,5-Tetramethylbenzene, durene, CA 95-93-2: Flammable, flash point 327; soluble most organic solvents. Tm = 352.39; Tb = 469.95; Vm = 0.1602 at 354.15; Antoine (liq) A = 6.2049, B = 1672.43, C = -71.72, dev = 1.0, range = 353 to 500.

$C_{10}H_{14}NO_5PS$ MW 291.26

Parathion, CA 56-38-2: Highly toxic liquid; absorbed through skin; cumulative poison; soluble ethanol, ether, acetone, chloroform. Tm = 279.15; Tb = 648.15; Vm = 0.2289 at 293.15; Antoine (liq) A = 8.91416, B = 3927, C = -31.55, dev = 1.0, range = 293 to 433.

Phosphorothioic acid, *O,O'*-diethyl-*S*-(4-nitrophenyl), CA 3270-86-8: Toxic; flammable. Antoine (liq) A = 7.4949, B = 3966, C = 0, dev = 1.0 to 5.0, range = 313 to 366.

Phosphorothioic acid, *O,S*-diethyl-*O'*-(4-nitrophenyl), CA 597-88-6: Toxic; flammable. Antoine (liq) A = 7.0499, B = 3924, C = 0, dev = 1.0 to 5.0, range = 332 to 364.

$C_{10}H_{14}NO_6P$ MW 275.20

O,O-Diethyl-*O*-(4-nitrophenyl) phosphate, CA 311-45-5: Highly toxic insecticide; slightly soluble water; soluble most organic solvents. Vm = 0.2160 at 293.15; Antoine (liq) A = 10.25163, B = 4815.5, C = 6.85, dev = 1.0, range = 273 to 422.

$C_{10}H_{14}N_2$ MW 162.23

Nicotine ±, CA 22083-74-5: Highly toxic; hygroscopic; flammable, flash point 374; soluble water, ethanol, ether, light hydrocarbons. Tb = 520.15, decomposes; Antoine (liq) A = 5.91387, B = 1650.347, C = -96.779, dev = 1.0, range = 406 to 520.

$C_{10}H_{14}O$ MW 150.22

2-Butylphenol, CA 3180-09-4: Corrosive; toxic; flammable; soluble ethanol, ether. Tm = 253.15; Tb = 507.15 to 510.15; Vm = 0.1541 at 293.15; Antoine (liq) A = 6.40976, B = 1889.3, C = -79.15, dev = 1.0, range = 403 to 733.

2-*sec*-Butylphenol, CA 89-72-5: Corrosive; toxic; flammable, flash point 385; soluble ethanol, ether. Tm = 289.15; Tb = 500.15 to 501.15; Vm = 0.1532 at 298.15; Antoine (liq) A = 6.07894, B = 1595.955, C = -109.068, dev = 0.02 to 0.1, range = 451 to 513.

2-*tert*-Butylphenol, CA 88-18-6: Corrosive; toxic; flammable, flash point above 385; soluble ethanol, ether. Tm = 267.53; Tb = 497.75; Vm = 0.1530 at 293.15; Antoine (liq) A = 6.47765, B = 1928.57, C = -65.966, dev = 0.1, range = 353 to 498.

3-Butylphenol, CA 4074-43-5: Corrosive; toxic; flammable; soluble ethanol, ether. Tb = 520.15 to 521.15; Vm = 0.1542 at 293.15; Antoine (liq) A = 6.4815, B = 1900.5, C = -97.15, dev = 1.0, range = 396 to 533.

3-*tert*-Butylphenol, CA 585-34-2: Corrosive; toxic; flammable, flash point 381; soluble ethanol, ether. Tm = 316.15; Tb = 513.15; Antoine I (sol) A = 9.45844, B = 3695.399, C = 0, dev = 1.0, range = 266 to 299; Antoine II (liq) A = 6.43496, B = 1818.9, C = -102.63, dev = 1.0, range = 391 to 524.

4-Butylphenol, CA 1638-22-8: Corrosive; toxic; flammable; soluble ethanol, ether. Tm = 295.15; Tb = 521.15; Vm = 0.1539 at 295.15; Antoine (liq) A = 6.7162, B = 2072.6, C = -81.15, dev = 1.0 to 5.0, range = 395 to 653.

4-*sec*-Butylphenol, CA 99-71-8: Corrosive; toxic; flammable, flash point 388; soluble ethanol, ether. Tm = 335.15; Tb = 513.15 to 515.15; Vm = 0.1520 at 293.15; Antoine (liq) A = 7.18941, B = 2480.29, C = -36.784, dev = 1.0, range = 344 to 516.

4-*tert*-Butylphenol, CA 98-54-4: Corrosive; toxic; flammable; soluble ethanol, ether. Tm = 372.15; Tb = 513.15; Antoine I (sol) A = 11.46945, B = 4405.873, C = 0, dev = 1.0, range = 280 to 304; Antoine II (liq) A = 6.13162, B = 1632.939, C = -117.258, dev = 0.02 to 0.1, range = 471 to 525.

Butyl phenyl ether, CA 1126-79-0: Irritant; flammable, flash point 355; forms explosive peroxides; soluble ether, acetone. Tm = 254.15; Tb = 483.15; Vm = 0.1606 at 293.15; Antoine (liq) A = 7.35311, B = 2446.14, C = -23.816, dev = 1.0, range = 391 to 483.

(continues)

$C_{10}H_{14}O$ *(continued)*

Carvacrol, 5-isopropyl-2-methylphenol, CA 499-75-2: Irritant; flammable; soluble ethanol, ether, acetone. Tm = 276.65; Tb = 510.15 to 511.15; Vm = 0.1536 at 298.15; Antoine (liq) A = 7.86519, B = 3011.79, C = 3.745, dev = 1.0, range = 343 to 510.

Carvone ±, CA 22327-39-5: Highly toxic; flammable, flash point 361; soluble ethanol, ether, chloroform. Tb = 504.15; Vm = 0.1573 at 293.15; Antoine (liq) A = 7.04816, B = 2364.37, C = -31.98, dev = 1.0, range = 330 to 501.

3,5-Diethylphenol, CA 1197-34-8: Corrosive; toxic; flammable; soluble ethanol, ether. Tm = 349.15; Antoine (liq) A = 7.4639, B = 2850.8, C = 0.91, dev = 1.0, range = 387 to 521.

4-Ethylphenetole, CA 1585-06-4: Flammable; soluble ethanol, acetone, benzene. Tb = 484.15; Vm = 0.1601 at 290.15; Antoine (liq) A = 7.35311, B = 2446.14, C = -23.816, dev = 1.0, range = 321 to 481.

2-(2-Ethylphenyl)ethanol: Irritant; flammable. Antoine (liq) A = 6.65889, B = 2055.2, C = -81.15, dev = 1.0, range = 420 to 653.

2-(4-Ethylphenyl)ethanol, CA 22545-13-7: Irritant; flammable. Tm = 281.02; Antoine (liq) A = 6.60742, B = 2029.3, C = -82.15, dev = 1.0, range = 420 to 653.

4-Isobutylphenol, CA 4167-74-2: Corrosive; toxic; flammable; soluble ethanol, ether, acetone. Tm = 326.15 to 327.15; Tb = 508.15 to 512.15; Antoine (liq) A = 7.97304, B = 3052.1, C = 1.102, dev = 5.0, range = 345 to 510.

4-Isopropylbenzyl alcohol, CA 536-60-7: Irritant; flammable, flash point above 385; Tm = 301.15; Tb = 519.15; Antoine (liq) A = 7.09745, B = 2430.77, C = -42.335, dev = 1.0, range = 347 to 520.

2-Methyl-3-phenyl-1-propanol, CA 7384-80-7: Irritant; flammable. Antoine (liq) A = 9.76157, B = 3755.659, C = 0, dev = 5.0, range = 343 to 393.

2,3,5,6-Tetramethylphenol, durenol, CA 527-35-5: Corrosive; toxic; flammable; soluble ethanol, ether. Tm = 392.15; Tb = 520.15; Antoine (liq) A = 7.1365, B = 2680.7, C = 0.55, dev = 1.0, range = 381 to 522.

Thymol, 2-isopropyl-5-methylphenol, CA 89-83-8: Corrosive; toxic; flammable, flash point 375; soluble ethanol, ether, chloroform. Tm = 324.65; Tb = 506.65; Vm = 0.1624 at 353.15; Antoine I (sol) A = 10.50206, B = 3925.658, C = 0, dev = 1.0, range = 273 to 295; Antoine II (sol) A = 13.365, B = 4780, C = 0, dev = 5.0, range = 283 to 323; Antoine III (liq) A = 6.68999, B = 2013.31, C = -74.3, dev = 1.0, range = 381 to 514.

$C_{10}H_{14}O_2$ MW 166.22

1,3-Dihydroxy-2-butylbenzene, CA 13331-20-9: Irritant; flammable; soluble ethanol, ether, benzene. Tm = 360.15; Antoine (liq) A = 4.531, B = 950.72, C = -217.15, dev = 1.0, range = 413 to 469.

2-Methoxy-4-propylphenol, CA 2785-87-7: Irritant; flammable; soluble ethanol, ether. Tm = 290.65; Antoine (liq) A = 10.456, B = 4077, C = 0, dev = 5.0, range = 373 to 413.

$C_{10}H_{14}O_5$ MW 214.22

Allyl[(1-allyloxycarbonyl)ethyl] carbonate: Toxic; lachrymator; flammable.
Antoine (liq) A = 8.2374, B = 3232, C = 0, dev = 1.0, range = 353 to 503.

$C_{10}H_{15}N$ MW 149.24

N-Butylaniline, CA 1126-78-9: Moderately toxic; flammable, flash point
380; soluble ethanol, ether. Tm = 258.75; Tb = 513.15; Vm = 0.1594 at
293.15; Antoine (liq) A = 6.41743, B = 1917.28, C = -80.15, dev = 0.1,
range = 413 to 643.

N,N-Diethylaniline, CA 91-66-7: Highly toxic by inhalation and skin absorption; cumulative poison; flammable, flash point 358; moderately soluble
water; soluble ethanol, ether, acetone. Tm = 234.35; Tb = 488.15 to
489.15; Vm = 0.1604 at 298.15; Antoine (liq) A = 6.78922, B = 2097.23, C =
-50.561, dev = 1.0, range = 343 to 493.

5-Isopropyl-2-methylaniline, carvacryl amine, CA 2051-53-8: Toxic; flammable; soluble ethanol, ether. Tf = 257.15; Tb = 515.15; Vm = 0.1501 at
293.15; Antoine (liq) A = 9.89034, B = 3763.5, C = 0, dev = 1.0, range =
360 to 386.

Phenethylamine, N,alpha-dimethyl, methamphetamine, CA 537-46-2: Toxic;
addictive drug; flammable; soluble ethanol, ether. Tb = 481.15 to 483.15;
Antoine (liq) A = 9.06327, B = 2756.214, C = 0, dev not specified, range =
270 to 304.

$C_{10}H_{15}NO$ MW 165.23

Phenol, 4-(butylamino), CA 103-62-8: Irritant; flammable. Tm = 343.15;
Antoine (liq) A = 8.48766, B = 3718.383, C = 0, dev = 1.0 to 5.0, range =
463 to 511.

$C_{10}H_{15}NO_2$ MW 181.23

N,N-Bis(2-hydroxyethyl)aniline, N-phenyl diethanolamine, CA 120-07-0:
Irritant; flammable; soluble ethanol, ether, acetone, benzene. Tm = 329.15
to 331.15; Antoine (liq) A = 7.38542, B = 2944.38, C = -63.771, dev = 1.0,
range = 418 to 611.

$C_{10}H_{15}O_3PS_2$ MW 278.32

O,O-Dimethyl-O-[3-methyl-4-(methylthio)phenyl] thiophosphate, CA 55-38-9:
Highly toxic insecticide; slightly soluble water; soluble ethanol, ether,
acetone, chloroform. Vm = 0.223 at 293.15; Antoine (liq) A = 8.0532, B =
3947.6, C = 0, dev = 1.0, range = 293 to 373.

$C_{10}H_{16}$ MW 136.24

Adamantane, CA 281-23-2: Tm = 541.15, sealed tube; Tb = 478.15 to 483.15,
sublimes; Antoine I (sol) A = 7.78373, B = 2585.114, C = -26.181, dev =
1.0, range = 278 to 368; Antoine II (sol) A = 7.6959, B = 2652, C = -14.39,
dev = 1.0, range = 328 to 373; Antoine III (sol) A = 7.912, B = 2838,
C = 0, dev = 1.0 to 5.0, range = 353 to 483.

(continues)

$C_{10}H_{16}$ *(continued)*

Camphene ±, CA 565-00-4: Flammable; soluble ethanol, ether, chloroform. Tm = 322.15; Tb = 431.15; Vm = 0.1618 at 327.15; Antoine (liq) A = 7.06988, B = 2135.16, C = -12.126, dev = 1.0, range = 320 to 434.

2-Carene, *dextro*, CA 4497-92-1: Flammable; soluble ethanol, acetone, benzene. Tb = 440.15; Vm = 0.1584 at 293.15; Antoine (liq) A = 6.3785, B = 1755.5, C = -43.15, dev = 1.0 to 5.0, range = 293 to 450.

3-Carene, *dextro*, CA 498-15-7: Flammable; oxidizes in air; soluble chloroform, oils. Tb = 443.15; Vm = 0.1580 at 303.15; Antoine (liq) A = 7.055, B = 2238, C = 0, dev = 1.0 to 5.0, range = 359 to 443.

Limonene, *dextro*, CA 5989-27-5: Irritant; flammable, flash point 321; soluble ethanol, ether. Tm = 198.85; Tb = 448.65 to 449.15; Vm = 0.1622 at 293.15; Antoine I (liq) A = 6.81591, B = 2075.62, C = -16.65, dev = 1.0, range = 287 to 448; Antoine II (liq) A = 7.67098, B = 2494.342, C = 0, dev = 1.0 to 5.0, range = 288 to 323.

Limonene, *levo*, CA 5989-54-8: Irritant; flammable, flash point 321; soluble ethanol, ether. Tb = 450.75 to 450.95; Vm = 0.1621 at 293.65; Antoine (liq) A = 6.3557, B = 1772.6, C = -43.15, dev = 1.0, range = 303 to 363.

Limonene ±, dipentene, CA 7705-14-8: Irritant; flammable; soluble ethanol, ether. Tm = 177.65; Tb = 448.65 to 449.65; Vm = 0.1622 at 294.15; Antoine (liq) A = 6.66342, B = 1946.23, C = -29.741, dev = 1.0, range = 287 to 448.

7-Methyl-3-methylene-1,6-octadiene, myrcene, CA 123-35-3: Flammable, flash point 312; polymerizes; soluble ethanol, ether, benzene, chloroform. Tb = 440.15; Vm = 0.1716 at 293.15; Antoine (liq) A = 6.73651, B = 1974.49, C = -27.299, dev = 1.0, range = 287 to 445.

alpha-Phellandrene, 5-isopropyl-2-methyl-1,3-cyclohexadiene, CA 99-83-2: Moderately irritant; flammable; soluble ethanol, ether. Tb = 448.15 to 449.15; Antoine (liq) A = 6.85536, B = 2025.9, C = -30.345, dev = 1.0, range = 293 to 448.

beta-Phellandrene, 3-isopropyl-6-methylenecyclohexene, CA 555-10-2: Flammable; soluble ethanol, ether. Tb = 444.15 to 445.15; Vm = 0.1599 at 293.15; Antoine (liq) A = 6.5617, B = 1858, C = -43.15, dev = 1.0, range = 303 to 363.

alpha-Pinene, *dextro*, CA 80-56-8: Irritant; flammable, flash point 305; soluble ethanol, ether. Tm = 223.15; Tb = 429.15; Vm = 0.1588 at 293.15; Antoine (liq) A = 5.92666, B = 1414.16, C = -68.64, dev = 1.0, range = 292 to 433.

beta-Pinene, *levo*, CA 127-91-3: Irritant; flammable, flash point 305; soluble ethanol, ether, benzene, chloroform. Tm = 223.15; Tb = 438.15; Vm = 0.1574 at 293.15; Antoine (liq) A = 6.04993, B = 1520.15, C = -62.75, dev = 1.0, range = 291 to 441.

Terpinolene, CA 586-62-9: Flammable; soluble ethanol, ether, benzene. Tb = 459.15; Vm = 0.1580 at 293.15; Antoine (liq) A = 7.36276, B = 2363.48, C = -17.069, dev = 1.0, range = 305 to 458.

Tetrahydrodicyclopentadiene, CA 54175-17-6: Flammable; soluble most organic solvents. Tm = 350.15; Tb = 466.15; Vm = 0.1493 at 352.15; Antoine (liq) A = 6.907, B = 2273.7, C = 0, dev not specified, range = 358 to 465.

$C_{10}H_{16}Cl_3NOS$ MW 304.66

Carbamothioic acid, bis(isopropyl), *S*-(2,3,3-trichloroallyl) ester,
CA 2303-17-5: Toxic; flammable; used as herbicide. Tm = 302.15 to 303.15;
Antoine (liq) A = 10.175, B = 4401, C = 0, dev not specified, range = 293
to 318.

$C_{10}H_{16}N_2$ MW 164.25

Sebaconitrile, CA 1871-96-1: Irritant; flammable. Vm = 0.1764 at 293.15;
Antoine (liq) A = 9.8409, B = 4370, C = 0, dev = 1.0 to 5.0, range = 303
to 343.

$C_{10}H_{16}O$ MW 152.24

3-Bornanone, *dextro*, CA 13854-85-8: Soluble ethanol, ether, light hydrocarbons. Tm = 455.15; Antoine I (sol) A = 7.995, B = 2830, C = 0, dev =
1.0 to 5.0, range = 273 to 408; Antoine II (sol) A = 8.10789, B = 2874.437,
C = 0, dev = 1.0, range = 323 to 339; Antoine III (sol) A = 6.3105, B =
1723.5, C = -78.55, dev = 1.0, range = 408 to 451; Antoine IV (liq) A =
6.8508, B = 2332, C = 0, dev = 1.0, range = 452 to 488.

Carvenone ±, 3-isopropyl-6-methyl-2-cyclohexen-1-one, CA 499-74-1: Flammable; soluble acetone. Tb = 506.15; Vm = 0.1644 at 293.15; Antoine (liq)
A = 5.76869, B = 1491.8, C = -110.3, dev = 1.0, range = 364 to 507.

Citral, geranial, CA 141-27-5: Flammable, flash point 364; soluble most
organic solvents. Tb = 501.15; Vm = 0.1713 at 293.15; Antoine I (liq) A =
8.568, B = 3184, C = 0, dev = 1.0, range = 283 to 333; Antoine II (liq) A =
7.44862, B = 2644.95, C = -15.487, dev = 1.0, range = 373 to 501.

2-Cyclohexen-1-one, 5-isopropyl-2-methyl, CA 43205-82-9: Flammable; soluble ethanol, ether, chloroform. Tb = 504.15; Vm = 0.1585 at 293.15;
Antoine (liq) A = 5.86662, B = 1521.1, C = -106.67, dev = 1.0, range = 361
to 503.

Dihydrocarvone ±, CA 4584-09-2: Flammable, flash point 354; soluble ether,
acetone. Tm = 262.15; Tb = 493.15 to 495.15; Vm = 0.1639 at 293.15;
Antoine (liq) A = 6.66941, B = 2154.24, C = -34.235, dev = 1.0, range =
319 to 496.

Fenchone, *dextro*, CA 4965-62-9: Flammable, flash point 325; soluble ethanol, ether, acetone. Tm = 278.15; Tb = 466.15; Vm = 0.1608 at 293.15;
Antoine (liq) A = 6.7476, B = 2041.24, C = -33.58, dev = 1.0, range = 301
to 464.

Pulegone, CA 89-82-7: Toxic; flammable, flash point 355; soluble ethanol,
ether, chloroform. Tb = 497.15; Vm = 0.1629 at 318.15; Antoine (liq) A =
4.03153, B = 524.14, C = -235.371, dev = 1.0 to 5.0, range = 331 to 494.

Thujone, *dextro*, CA 471-15-8: Toxic; ingestion may cause convulsions;
flammable, flash point 337; soluble ethanol, ether, acetone. Tb = 474.15
to 475.15; Vm = 0.1667 at 298.15; Antoine (liq) A = 6.84251, B = 2109.59,
C = -38.163, dev = 1.0 to 5.0, range = 311 to 474.

$C_{10}H_{16}O_2$ MW 168.24

Diosphenol, CA 490-03-9: Flammable; slightly soluble water; soluble ether,
chloroform, carbon disulfide. Tm = 356.15 to 357.15;   *(continues)*

$C_{10}H_{16}O_2$ *(continued)*

Tb = 496.15, decomposes; Vm = 0.1763 at 372.35; Antoine (liq) A = 7.7987, B = 2922.24, C = -0.93, dev = 1.0 to 5.0, range = 339 to 505.

(2,2,3-Trimethyl-3-cyclopentadienyl)acetic acid: Flammable. Antoine (liq) A = 7.83583, B = 2780.49, C = -52.153, dev = 1.0, range = 370 to 529.

$C_{10}H_{16}O_6$ MW 232.23

Lactic acid, *O*-ethoxycarbonyl, tetrahydrofurfuryl ester: Flammable. Antoine (liq) A = 8.7947, B = 3772.1, C = 0, dev = 1.0, range = 390 to 523.

$C_{10}H_{17}ClO_6$ MW 268.69

Lactic acid, *O*-ethoxycarbonyl, 2-(2-chloroethoxy)ethyl ester: Irritant; flammable. Antoine (liq) A = 9.8844, B = 4377.3, C = 0, dev = 1.0, range = 406 to 523.

$C_{10}H_{17}NOS$ MW 199.31

Carbamothioic acid, *N*-butyl-*N*-(2-propynyl), *S*-ethyl ester, CA 59300-35-5: Antoine (sol) A = 11.21434, B = 4290.519, C = 0, dev not specified, range = 298 to 313.

Carbamothioic acid, *N*,*N*-dipropyl, *S*-(2-propynyl) ester, CA 59300-36-6: Antoine (sol) A = 12.78856, B = 4828.192, C = 0, dev not specified, range = 298 to 313.

Carbamothioic acid, *N*-isopropyl-*N*-(2-propynyl), *S*-ethyl ester, CA 59300-34-4: Antoine (sol) A = 10.13027, B = 3865.681, C = 0, dev not specified, range = 298 to 313.

$C_{10}H_{17}NO_3$ MW 199.25

2-(2-Cyanoethoxy)propionic acid, butyl ester: Flammable. Antoine (liq) A = 7.57948, B = 3223.418, C = 0, dev = 5.0, range = 328 to 382.

$C_{10}H_{17}NO_5$ MW 231.25

N-Acetylaspartic acid, *levo*, diethyl ester, CA 1069-39-2: Flammable. Antoine (liq) A = 8.9065, B = 3970, C = 0, dev = 1.0, range = 418 to 508.

$C_{10}H_{18}$ MW 138.25

Bicyclo(5,3,0)decane, *cis*: Flammable; soluble most organic solvents. Antoine (liq) A = 7.7258, B = 2600.93, C = 0, dev = 1.0 to 5.0, range = 298 to 377.

Bicyclopentyl, CA 1636-39-1: Flammable; soluble most organic solvents. Tm = 237.81; Antoine (liq) A = 8.25156, B = 3384.294, C = 81.761, dev = 1.0, range = 350 to 393.

(continues)

$C_{10}H_{18}$ *(continued)*

Carane, *cis*, CA 18968-24-6: Flammable; soluble most organic solvents.
Tb = 442.15; Vm = 0.1644 at 293.15; Antoine (liq) A = 5.96254, B =
1466.669, C = -70.875, dev not specified, range = 362 to 445.

Carane, *trans*, CA 18968-23-5: Flammable; soluble most organic solvents.
Tb = 442.15; Vm = 0.1644 at 293.15; Antoine (liq) A = 5.95489, B =
1460.464, C = -72.285, dev not specified, range = 362 to 437.

Decahydronaphthalene, *cis*, decalin, CA 493-01-6: Flammable, flash point
331; soluble ethanol, ether, acetone, benzene. Tm = 230.17; Tb = 468.97;
Vm = 0.1548 at 298.15; Antoine (liq) A = 6.00019, B = 1595.176, C =
-69.622, dev = 0.02, range = 371 to 473.

Decahydronaphthalene, *trans*, CA 493-02-7: Flammable, flash point 327; soluble ethanol, ether, acetone, benzene. Tm = 242.77; Tb = 460.46; Vm =
0.1596 at 298.15; Antoine (liq) A = 5.99363, B = 1573.981, C = -65.77,
dev = 0.02, range = 363 to 461.

5-Decyne, CA 1942-46-7: Flammable; polymerizes; soluble most organic solvents. Tm = 200.15; Tb = 450.15; Vm = 0.1798 at 293.15; Antoine (liq) A =
7.2829, B = 2375.8, C = 0, dev = 1.0, range = 351 to 450.

$C_{10}H_{18}O$ MW 154.25

Borneol ±, CA 6627-72-1: Irritant; may cause nausea; flammable, flash
point 339; soluble ethanol, ether, benzene. Tm = 483.15; Antoine I (sol)
A = 5.87837, B = 1345.8, C = -142.12, dev = 1.0, range = 350 to 475;
Antoine II (liq) A = 7.48828, B = 2659.9, C = 0, dev = 1.0, range = 477
to 487.

1,4-Cineole, CA 470-67-7: Flammable; soluble ethanol, ether, benzene.
Tm = 274.15; Tb = 446.15 to 447.15; Vm = 0.1714 at 293.15; Antoine (liq)
A = 6.43164, B = 1809.58, C = -40.247, dev = 1.0, range = 288 to 449.

Citronellal, *dextro*, CA 2385-77-5: Flammable; slightly soluble water; soluble ethanol, ether. Tb = 477.15 to 478.15; Vm = 0.1799 at 293.15; Antoine
I (liq) A = 8.15104, B = 2869.05, C = 0, dev = 1.0 to 5.0, range = 288 to
333; Antoine II (liq) A = 6.97988, B = 2207.16, C = -36.182, dev = 1.0,
range = 317 to 480.

Cyclodecanone, CA 1502-06-3: Flammable, flash point 355; soluble ether,
benzene, chloroform. Tm = 302.15; Antoine (liq) A = 6.0171, B = 1725.93,
C = -83.15, dev = 1.0 to 5.0, range = 353 to 423.

Dihydrocarveol, *dextro,* CA 619-01-2: Flammable, flash point 364; soluble
ethanol, ether. Tb = 498.15; Antoine (liq) A = 7.4877, B = 2581.58, C =
-27.429, dev = 1.0, range = 336 to 498.

1,8-Epoxy-*para*-menthane, CA 470-82-6: Flammable; soluble ethanol, ether,
chloroform. Tm = 274.65; Tb = 449.15 to 450.15; Vm = 0.1665 at 293.15;
Antoine (liq) A = 6.37204, B = 1859.558, C = 0, dev = 1.0 to 5.0, range =
264 to 303.

Ethyl (1-methylcyclohexyl) ketone: Flammable. Antoine (liq) A = 6.91443,
B = 2362.228, C = 0, dev = 1.0, range = 388 to 431.

(continues)

$C_{10}H_{18}O$ *(continued)*

Fenchyl alcohol ±, CA 2217-01-8: Flammable, flash point 346; soluble ethanol, ether. Tm = 312.15; Tb = 473.15; Vm = 0.1650 at 313.15; Antoine (liq)
A = 4.54294, B = 682.62, C = -204.715, dev = 1.0 to 5.0, range = 318 to 474.

Geraniol, CA 106-24-1: Flammable, flash point above 373; soluble ethanol,
ether, acetone. Tm = below 258.15; Tb = 503.15; Vm = 0.1759 at 298.15;
Antoine I (liq) A = 8.64144, B = 3287.427, C = 0, dev = 1.0 to 5.0, range =
288 to 333; Antoine II (liq) A = 7.72316, B = 2769.9, C = -18.811, dev =
1.0, range = 342 to 503.

Isoborneol ±, CA 124-76-5: Flammable; soluble ethanol, ether, chloroform.
Tm = 485.15, sealed tube; Tb = 487.15; Antoine (sol) A = 6.3228, B = 2146,
C = 0, dev = 5.0, range = 373 to 457.

Isopulegol, *dextro*, CA 7786-67-6: Flammable, flash point 351; soluble
ethanol, ether. Tb = 485.15; Antoine (liq) A = 7.368, B = 2601, C = 0,
dev not specified, range = 335 to 485.

Linalool, *dextro*, CA 126-90-9: Flammable, flash point 344; soluble ether,
ethanol. Tb = 472.15; Vm = 0.1773 at 293.15; Antoine (liq) A = 7.11805,
B = 2253.97, C = -30.376, dev = 1.0, range = 313 to 471.

Menthone, CA 1074-95-9: Flammable, flash point 342; soluble ether, ethanol, acetone, benzene. Tm = 267.15; Tb = 480.15; Vm = 0.1723 at 293.15;
Antoine (liq) A = 5.97378, B = 1582.5, C = -84.16, dev = 1.0, range = 350
to 483.

1-(1-Methylcyclohexen-3-yl)-1-propanol: Flammable; soluble ethanol.
Antoine (liq) A = 7.76724, B = 2799.981, C = 0, dev = 1.0, range = 397 to
422.

Nerol, CA 106-25-2: Flammable, flash point 349; soluble ethanol. Tm =
below 258.15; Tb = 498.15; Vm = 0.1766 at 298.15; Antoine (liq) A = 7.79576,
B = 2889.27, C = -0.323, dev = 1.0, range = 334 to 499.

alpha-Terpineol ±, CA 98-55-5: Flammable, flash point 364; soluble ether,
ethanol, acetone, benzene. Tm = 308.15; Tb = 492.15; Antoine I (sol) A =
11.771, B = 4186, C = 0, dev = 5.0, range = 287 to 308; Antoine II (liq)
A = 7.26223, B = 2463.2, C = -22.201, dev = 1.0, range = 325 to 491.

$C_{10}H_{18}O_2$ MW 170.25

Citronellic acid, 3,7-dimethyl-6-octenoic acid, CA 502-47-6: Flammable;
polymerizes. Tb = 530.15; Antoine (liq) A = 8.49697, B = 3364.47, C =
-12.156, dev = 5.0, range = 372 to 530.

Heptyl acrylate, CA 2499-58-3: Lachrymator; flammable; polymerizes. Vm =
0.1925 at 293.15; Antoine (liq) A = 7.5653, B = 2672, C = 0, dev = 5.0,
range = 359 to 481.

Hexyl methacrylate, CA 142-09-6: Lachrymator; flammable, flash point 355;
polymerizes. Tb = 477.15 to 483.15; Vm = 0.1904 at 298.15; Antoine (liq)
A = 7.5511, B = 2636, C = 0, dev = 5.0, range = 354 to 475.

1-Methyl-3-isopropylcyclopentane carboxylic acid, fencholic acid, CA 512-
77-6: Flammable. Tm = 291.95 to 292.15; Tb = 528.15 to 532.15; Antoine
(liq) A = 5.43797, B = 1169.9, C = -196.204, dev = 1.0 to 5.0, range = 374
to 538.

$C_{10}H_{18}O_3$ MW 186.25

3-Hydroxy-2,3-dimethyl-4-hexenoic acid, ethyl ester: Flammable. Antoine
(liq) A = 8.29617, B = 2997.138, C = 0, dev = 1.0, range = 362 to 387.

Isopentyl levulinate: Flammable. Antoine (liq) A = 6.55274, B = 1992.21,
C = -82.926, dev = 1.0, range = 403 to 521.

Levulinic acid, 1-ethylpropyl ester: Flammable. Antoine (liq) A =
6.00777, B = 1587.5, C = -115.23, dev = 1.0, range = 397 to 513.

Levulinic acid, 1-methylbutyl ester: Flammable. Antoine (liq) A =
6.35136, B = 1845.6, C = -87.95, dev = 1.0, range = 397 to 513.

Levulinic acid, 2-methylbutyl ester: Flammable. Antoine (liq) A =
6.81756, B = 2270.7, C = -49.81, dev = 1.0, range = 391 to 473.

Levulinic acid, pentyl ester: Flammable. Antoine (liq) A = 6.27491, B =
1813.05, C = -101.824, dev = 1.0, range = 354 to 527.

$C_{10}H_{18}O_4$ MW 202.25

2-Acetoxypropionic acid, pentyl ester: Flammable. Antoine (liq) A =
6.55822, B = 1872.634, C = -90.15, dev = 1.0 to 5.0, range = 312 to 501.

Diethyl adipate, CA 141-28-6: Flammable, flash point above 383; soluble
ethanol, ether. Tm = 253.35; Tb = 513.15 to 518.15; Vm = 0.2007 at 293.15;
Antoine (liq) A = 9.9110, B = 4817.09, C = 96.256, dev = 1.0, range = 347
to 513.

Diisobutyl oxalate, CA 2050-61-5: Flammable; soluble ethanol, ether, acetone. Tb = 502.15; Vm = 0.2077 at 293.15; Antoine (liq) A = 7.57207, B =
2745.55, C = -9.423, dev = 1.0, range = 336 to 503.

Dipropyl succinate, CA 925-15-5: Flammable; soluble ether, acetone, benzene. Tm = 267.25; Tb = 523.95; Vm = 0.2018 at 293.15; Antoine (liq) A =
7.34584, B = 2651.28, C = -27.607, dev = 1.0, range = 350 to 524.

Ethylmethylmalonic acid, diethyl ester, CA 2049-70-9: Flammable. Antoine
(liq) A = 6.98804, B = 2218.47, C = -35.437, dev = 1.0, range = 317 to 481.

Sebacic acid, decanedioic acid, CA 111-20-6: Flammable; soluble ethanol,
acetone, ethyl acetate. Tm = 407.65; Antoine (sol) A = 18.036, B = 8395,
C = 0, dev = 1.0, range = 375 to 403.

$C_{10}H_{18}O_5$ MW 218.25

Ethyl[1-(butoxycarbonyl)ethyl] carbonate: Toxic; flammable. Antoine (liq)
A = 9.4169, B = 3666, C = 0, dev = 5.0, range = 324 to 473.

2-Lactoyloxypropionic acid, butyl ester: Flammable. Antoine (liq) A =
6.9147, B = 2155.356, C = -90.15, dev = 1.0 to 5.0, range = 336 to 407.

2-Lactoyloxypropionic acid, *sec*-butyl ester: Flammable. Antoine (liq) A =
6.95849, B = 2109.998, C = -90.15, dev = 1.0 to 5.0, range = 329 to 399.

Pentyl[(ethoxycarbonyl)methyl] carbonate: Toxic; flammable. Antoine (liq)
A = 8.7818, B = 3564, C = 0, dev = 5.0, range = 383 to 503.

(continues)

338

$C_{10}H_{18}O_5$ *(continued)*

Pentyl[1-(methoxycarbonyl)ethyl] carbonate: Toxic; flammable. Antoine (liq) A = 8.3709, B = 3327, C = 0, dev = 5.0, range = 360 to 524.

$C_{10}H_{18}O_6$ MW 234.25

Diisopropyltartrate, *dextro*, CA 62961-64-2: Flammable, flash point above 383; hygroscopic; soluble ethanol, ether, acetone. Tb = 548.15; Vm = 0.2073 at 293.15; Antoine (liq) A = 8.18817, B = 3366.68, C = -3.68, dev = 1.0, range = 376 to 548.

Dipropyltartrate, *dextro*, CA 2217-14-3: Flammable; soluble ethanol, ether, acetone. Tb = 576.15; Vm = 0.2057 at 293.15; Antoine (liq) A = 6.89695, B = 2434.74, C = -78.192, dev = 1.0, range = 388 to 576.

$C_{10}H_{19}ClNO_5P$ MW 299.69

Phosphamidon, CA 13171-21-6: Highly toxic insecticide; soluble water, most organic solvents. Vm = 0.2282 at 298.15; Antoine (liq) A = 10.60681, B = 4707.5, C = 0, dev = 1.0 to 5.0, range = 293 to 388.

$C_{10}H_{19}Cl_2N$ MW 224.17

N,*N*-Bis(2-chloroethyl)cyclohexylamine, CA 4261-59-0: Irritant; flammable. Tm = 270.15; Antoine (liq) A = 7.73387, B = 3258.8, C = 0, dev = 1.0, range = 273 to 376.

$C_{10}H_{19}N$ MW 153.27

Caprinitrile, decanitrile, CA 1975-78-6: Irritant; flammable; soluble acetone, ethanol, ether, chloroform. Tm = 258.7; Tb = 516.15; Vm = 0.1869 at 293.15; Antoine (liq) A = 6.22424, B = 1787.89, C = -91.555, dev = 0.1, range = 381 to 519.

$C_{10}H_{19}NO_3$ MW 201.26

N-Acetylisoleucine, *levo*, ethyl ester: Flammable. Antoine (liq) A = 8.6625, B = 3610, C = 0, dev = 1.0, range = 391 to 476.

N-Acetylleucine, *levo*, ethyl ester: Flammable. Antoine (liq) A = 9.2843, B = 3906, C = 0, dev = 1.0, range = 396 to 476.

$C_{10}H_{19}N_5O$ MW 225.29

2-Methoxy-4,6-bis(isopropylamino)-1,3,5-triazine, CA 1610-18-0: Toxic; nonselective herbicide; slightly soluble water. Tm = 364.15 to 365.15; Antoine (sol) A = 9.919, B = 4817, C = 0, dev = 5.0, range = 323 to 365.

$C_{10}H_{19}N_5S$ MW 241.35

2-Methylthio-4,6-bis(isopropylamino)-1,3,5-triazine, CA 7287-19-6: Toxic; selective herbicide; slightly soluble water; soluble most organic solvents. Tm = 391.15 to 393.15; Antoine (sol) A = 10.966, B = 5222, C = 0, dev = 5.0, range = 323 to 393.

$C_{10}H_{19}O_6PS_2$ MW 330.35

Malathion ±, CA 121-75-5: Moderately toxic insecticide; suspected carcinogen; easily hydrolyzed; slightly soluble water, decomposes; soluble most
organic solvents. Tm = 266.15; Vm = 0.2686 at 298.15; Antoine (liq) A =
7.8813, B = 3716.6, C = 0, dev = 1.0, range = 283 to 419.

$C_{10}H_{19}O_7PS$ MW 314.29

O,O-Dimethyl-*S*-[1,2-bis(ethoxycarbonyl)ethyl] thiophosphate: Toxic insecticide. Antoine (liq) A = 10.74313, B = 4880, C = 0, dev = 1.0, range =
283 to 406.

$C_{10}H_{20}$ MW 140.27

Butylcyclohexane, CA 1678-93-9: Flammable, flash point 314; soluble most
organic solvents. Tm = 198.45; Tb = 454.15; Vm = 0.1755 at 293.15; Antoine
(liq) A = 6.0352, B = 1538.518, C = -72.317, dev = 0.02, range = 367 to 457.

sec-Butylcyclohexane, CA 7058-01-7: Flammable; soluble most organic solvents. Tb = 452.45; Vm = 0.1725 at 293.15; Antoine (liq) A = 6.01102, B =
1527.963, C = -71.034, dev = 0.02, range = 369 to 455.

tert-Butylcyclohexane, CA 3178-22-1: Flammable, flash point 315; soluble
most organic solvents. Tf = 231.95; Tb = 444.65; Vm = 0.1726 at 293.15;
Antoine (liq) A = 5.98795, B = 1506.613, C = -66.434, dev = 0.02, range =
355 to 446.

Cyclodecane, CA 293-96-9: Flammable, flash point 338; soluble most organic
solvents. Tm = 283.15; Tb = 474.15; Vm = 0.1643 at 298.15; Antoine I (liq)
A = 6.08604, B = 1667.202, C = -66.627, dev = 0.02, range = 343 to 386;
Antoine II (liq) A = 6.00303, B = 1612.48, C = -72.122, dev = 0.02, range =
404 to 489.

1-Decene, CA 872-05-9: Flammable, flash point 320; polymerizes; soluble
most organic solvents. Tf = 206.85; Tb = 443.72; Vm = 0.1893 at 293.15;
Antoine (liq) A = 6.12458, B = 1528.811, C = -72.566, dev = 0.02, range =
383 to 445.

2-Decene, *cis*, CA 20348-51-0: Flammable; soluble most organic solvents.
Antoine (liq) A = 6.06755, B = 1499.584, C = -78.194, dev = 0.1, range =
401 to 447.

2-Decene, *trans*, CA 20063-97-2: Flammable; soluble most organic solvents.
Antoine (liq) A = 6.0844, B = 1499.327, C = -78.881, dev = 0.1, range = 401
to 447.

3-Decene, *cis*, CA 19398-86-8: Flammable; soluble most organic solvents.
Antoine (liq) A = 6.07151, B = 1496.965, C = -76.291, dev = 0.1, range =
398 to 444.

3-Decene, *trans*, CA 19150-21-1: Flammable; soluble most organic solvents.
Antoine (liq) A = 6.07107, B = 1490.592, C = -77.936, dev = 0.1, range =
398 to 445.

4-Decene, *cis*, CA 19398-88-0: Flammable; soluble most organic solvents.
Antoine (liq) A = 6.04539, B = 1481.351, C = -77.165, dev = 0.1, range =
397 to 444.

(continues)

$C_{10}H_{20}$ *(continued)*

4-Decene, *trans*, CA 19398-89-1: Flammable; soluble most organic solvents.
Antoine (liq) A = 6.05353, B = 1479.555, C = -78.457, dev = 0.1, range =
398 to 444.

5-Decene, *cis*, CA 7433-78-5: Flammable; soluble most organic solvents.
Tm = 161.15; Tb = 443.15; Vm = 0.1884 at 293.15; Antoine (liq) A = 6.09582,
B = 1515.224, C = -73.152, dev = 0.1, range = 397 to 443.

5-Decene, *trans*, CA 7433-56-9: Flammable, flash point 319; soluble most
organic solvents. Tm = 200.15; Tb = 443.15; Vm = 0.1895 at 293.15; Antoine
(liq) A = 6.09647, B = 1510.804, C = -75.144, dev = 0.1, range = 398 to 444.

Isobutylcyclohexane, CA 1678-98-4: Flammable; soluble most organic solvents. Tm = 178.35; Tb = 444.45; Vm = 0.1764 at 293.15; Antoine (liq) A =
5.99423, B = 1494.615, C = -69.77, dev = 0.02, range = 355 to 446.

1-Isopropyl-4-methylcyclohexane, *trans*, *para*-menthane, CA 1678-82-6: Flammable; soluble most organic solvents. Tb = 443.7; Vm = 0.1769 at 293.15;
Antoine (liq) A = 6.60447, B = 1927.68, C = -23.632, dev = 1.0, range = 282
to 443.

4-Propyl-3-heptene, CA 4485-13-6: Flammable; polymerizes; soluble most organic solvents. Tb = 433.65; Vm = 0.1866 at 290.15; Antoine (liq) A =
5.7073, B = 1320.3, C = -83.15, dev = 1.0 to 5.0, range = 333 to 371.

$C_{10}H_{20}Br_2$ MW 300.08

1,1-Dibromodecane, CA 59104-80-2: Irritant; flammable. Antoine (liq) A =
6.600, B = 2127, C = -87.15, dev = 5.0, range = 442 to 610.

1,2-Dibromodecane, CA 28467-71-2: Irritant; flammable. Antoine (liq) A =
9.04552, B = 3798.77, C = 16.041, dev = 1.0, range = 368 to 524.

$C_{10}H_{20}Cl_2$ MW 211.17

1,1-Dichlorodecane, CA 3162-62-7: Irritant; flammable. Antoine (liq) A =
6.459, B = 1955, C = -81.15, dev = 5.0, range = 415 to 577.

$C_{10}H_{20}F_2$ MW 178.26

1,1-Difluorodecane, CA 62127-43-9: Flammable. Antoine (liq) A = 6.530,
B = 1751, C = -69.15, dev = 5.0, range = 364 to 504.

$C_{10}H_{20}O$ MW 156.27

1-Butylcyclohexanol, CA 5445-30-7: Flammable; soluble ethanol. Antoine
(liq) A = 6.818, B = 2011, C = -63.55, dev = 5.0, range = 362 to 481.

Decanal, caprylaldehyde, CA 112-31-2: Irritant; flammable; soluble ethanol,
ether, acetone. Tm = about 268.15; Tb = 481.15 to 482.15; Vm = 0.1883 at
288.15; Antoine I (liq) A = 8.34238, B = 2994.55, C = 0, dev = 1.0, range =
288 to 333; Antoine II (liq) A = 8.351, B = 2995, C = 0, dev = 1.0 to 5.0,
range = 293 to 358; Antoine III (liq) A = 7.42091, B = 2446.93, C =
-29.764, dev = 1.0, range = 324 to 482.

(continues)

$C_{10}H_{20}O$ *(continued)*

2-Decanone, methyl octyl ketone, CA 693-54-9: Flammable, flash point 344; soluble ethanol, ether. Tm = 287.15; Tb = 483.15 to 484.15; Vm = 0.1899 at 295.15; Antoine I (liq) A = 6.92497, B = 2235.2, C = -29.808, dev = 1.0 to 5.0, range = 317 to 484; Antoine II (liq) A = 6.17067, B = 1657.808, C = -89.481, dev = 1.0, range = 357 to 560.

3,7-Dimethyl-6-octen-1-ol, citronellol, CA 106-22-9: Flammable; soluble ethanol, ether. Tb = 497.15; Vm = 0.1828 at 293.15; Antoine I (liq) A = 10.002, B = 3790, C = 0, dev = 5.0, range = 293 to 333; Antoine II (liq) A = 6.00131, B = 1443.2, C = -136.44, dev = 1.0, range = 373 to 500.

(2-Ethylhexyl) vinyl ether, CA 103-44-6: Irritant; lachrymator; flammable; forms explosive peroxides. Tm = 173.15; Vm = 0.1927 at 293.15; Antoine (liq) A = 7.1891, B = 2334, C = 0, dev = 5.0, range = 330 to 451.

Menthol, *levo*, CA 2216-51-5: Flammable, flash point 366; slightly soluble water; soluble most organic solvents. Tm = 315.65 to 316.15, highest melting solid form; Tb = 489.15; Antoine (liq) A = 5.2895, B = 1091.9, C = -156.45, dev = 1.0 to 5.0, range = 372 to 488.

1-(1-Methylcyclohexyl)-1-propanol: Flammable; soluble ethanol. Antoine (liq) A = 8.01859, B = 2890.71, C = 0, dev = 1.0, range = 396 to 420.

2-(1-Methylcyclohexyl)-2-propanol, CA 27331-02-8: Flammable; soluble ethanol. Antoine (liq) A = 7.75368, B = 2768.21, C = 0, dev = 1.0, range = 393 to 418.

$C_{10}H_{20}O_2$ MW 172.27

Acetic acid, 2-ethylhexyl ester, CA 103-09-3: Flammable, flash point 344; soluble ethanol, ether. Tm = 193.15; Tb = 472.15; Vm = 0.1969 at 293.15; Antoine (liq) A = 4.4339, B = 753.4, C = -160.85, dev = 5.0, range = 333 to 472.

Acetic acid, octyl ester, CA 112-14-1: Flammable, flash point 359; soluble ethanol, ether. Tm = 234.65; Tb = 483.15; Vm = 0.1979 at 293.15; Antoine I (liq) A = 6.39677, B = 1854.1, C = -49.63, dev = 1.0 to 5.0, range = 345 to 472; Antoine II (liq) A = 7.99933, B = 2867.714, C = 0, dev = 1.0, range = 334 to 417.

2-Butoxy-3-hexanone: Flammable. Antoine (liq) A = 6.907, B = 2062.6, C = 0, dev = 5.0, range = 333 to 418.

Caprylic acid, ethyl ester, CA 106-32-1: Flammable, flash point 352; soluble ethanol, ether. Tm = 230.05; Tb = 480.15 to 482.15; Vm = 0.1980 at 298.15; Antoine (liq) A = 7.83672, B = 2778.307, C = 0, dev = 1.0 to 5.0, range = 330 to 480.

Decanoic acid, capric acid, CA 334-48-5: Flammable, flash point above 383; soluble ethanol, ether, acetone, benzene. Tm = 304.75; Tb = 543.15; Vm = 0.1945 at 313.15; Antoine I (sol) A = 16.507, B = 6207, C = 0, dev = 1.0, range = 293 to 303; Antoine II (liq) A = 7.20526, B = 2303.92, C = -99.066, dev = 1.0, range = 398 to 543.

2-(1-Ethylpentyl)-1,3-dioxolane, CA 4359-47-1: Flammable. Antoine (liq) A = 8.1929, B = 2887, C = 0, dev = 1.0 to 5.0, range = 333 to 453.

(continues)

342

$C_{10}H_{20}O_2$ *(continued)*

2-Heptyl-1,3-dioxolane, CA 4359-57-3: Flammable. Antoine (liq) A = 8.9079, B = 3240, C = 0, dev = 1.0 to 5.0, range = 318 to 453.

4-Hexyl-1,3-dioxane, CA 2244-85-1: Flammable. Antoine (liq) A = 8.0449, B = 2973, C = 0, dev = 1.0 to 5.0, range = 318 to 453.

Hydroxycitronellal, CA 107-75-5: Flammable; soluble ethanol, acetone. Vm = 0.1868 at 293.15; Antoine (liq) A = 9.964, B = 3934, C = 0, dev = 1.0 to 5.0, range = 283 to 333.

Isopentyl isovalerate, CA 659-70-1: Flammable; soluble ethanol, ether. Tb = 464.15 to 467.15; Vm = 0.2007 at 292.15; Antoine (liq) A = 6.81759, B = 2149.04, C = -20.68, dev = 1.0, range = 341 to 479.

Nonanoic acid, methyl ester, CA 1731-84-6: Flammable, flash point 357; soluble ethanol, ether. Tb = 486.15 to 487.15; Vm = 0.1958 at 288.15; Antoine (liq) A = 6.14671, B = 1622.62, C = -95.42, dev = 1.0, range = 364 to 439.

3-Pentyl-4-hydroxytetrahydropyran, CA 61827-60-9: Flammable. Antoine (liq) A = 9.2089, B = 3795, C = 0, dev = 1.0 to 5.0, range = 383 to 453.

$C_{10}H_{20}O_3$ MW 188.27

3-Butoxypropionic acid, propyl ester: Flammable. Antoine (liq) A = 6.772, B = 2308, C = 0, dev = 5.0, range = 373 to 473.

3-Ethoxypropionic acid, pentyl ester, CA 14144-36-6: Flammable. Antoine (liq) A = 7.6810, B = 2827.5, C = 0, dev = 1.0, range = 374 to 498.

3-Hexyloxypropionic acid, methyl ester, CA 7419-97-8: Flammable. Antoine (liq) A = 7.7365, B = 2877.5, C = 0, dev = 1.0, range = 373 to 473.

$C_{10}H_{20}O_4$ MW 204.27

Diethylene glycol, monobutyl ether acetate, CA 124-17-4: Flammable, flash point 389; very soluble water; soluble ethanol, ether, acetone. Tm = 240.95; Tb = 519.85; Vm = 0.2074 at 293.15; Antoine (liq) A = 7.8107, B = 3012.9, C = 0, dev = 1.0, range = 393 to 520.

$C_{10}H_{21}Br$ MW 221.18

1-Bromodecane, decyl bromide, CA 112-29-8: Irritant; flammable, flash point 367; soluble ether, chloroform. Tm = 243.95; Tb = 513.75; Vm = 0.2067 at 293.15; Antoine (liq) A = 6.3585, B = 1888.67, C = -79.85, dev = 1.0, range = 383 to 570.

$C_{10}H_{21}Cl$ MW 176.73

1-Chlorodecane, decyl chloride, CA 1002-69-3: Irritant; flammable, flash point 357; soluble ether, chloroform. Tm = 241.85; Tb = 496.15 to 497.95; Vm = 0.2030 at 293.15; Antoine (liq) A = 6.11662, B = 1676.793, C = -91.133, dev = 1.0, range = 359 to 499.

$C_{10}H_{21}F$ MW 160.27

1-Fluorodecane, decyl fluoride, CA 334-56-5: Flammable; soluble ether. Tm = 238.15; Tb = 459.35; Vm = 0.1956 at 293.15; Antoine (liq) A = 6.3791, B = 1704.75, C = -69.55, dev = 1.0, range = 342 to 503.

$C_{10}H_{21}I$ MW 268.18

1-Iododecane, decyl iodide, CA 2050-77-3: Irritant; flammable, flash point above 383; light-sensitive; soluble ethanol, ether. Tm = 256.85; Vm = 0.2134 at 293.15; Antoine (liq) A = 6.2469, B = 1919.75; C = -84.25; dev = 1.0, range = 397 to 598.

$C_{10}H_{21}N$ MW 155.28

Cyclohexanethylamine, *N,alpha*-dimethyl, CA 101-40-6: Irritant; flammable. Antoine (liq) A = 8.72303, B = 2621.162, C = 0, dev not specified, range = 270 to 300.

$C_{10}H_{21}NO$ MW 171.28

Capramide, CA 2319-29-1: Irritant; flammable; soluble ethanol, ether, acetone. Tm = 381.15; Antoine (sol) A = 15.596, B = 6577, C = 0, dev = 5.0, range = 353 to 370.

Hexanamide, *N,N*-diethyl, CA 6282-97-9: Irritant; flammable. Antoine (liq) A = 2.96871, B = 499.05, C = -213.82, dev = 1.0, range = 373 to 443.

$C_{10}H_{22}$ MW 142.28

Decane, CA 124-18-5: Irritant; flammable, flash point 319; soluble most organic solvents. Tm = 243.15; Tb = 447.15; Vm = 0.1949 at 293.15; Tc = 618.45; Pc = 2123; Vc = 0.603; Antoine I (liq) A = 6.80914, B = 1900.343, C = -47.319, dev = 1.0, range = 252 to 383; Antoine II (liq) A = 6.09206, B = 1510.415, C = -77.646, dev = 0.02, range = 373 to 443; Antoine III (liq) A = 6.04899, B = 1482.502, C = -80.635, dev = 0.1, range = 447 to 526; Antoine IV (liq) A = 9.71412, B = 6858.314, C = 454.63, dev = 1.0, range = 524 to 617.

2,7-Dimethyloctane, CA 1072-16-8: Flammable; soluble most organic solvents. Tf = 218.25; Tb = 432.75; Vm = 0.1976 at 298.15; Antoine (liq) A = 6.2764, B = 1640.35, C = -48.73, dev = 1.0, range = 279 to 433.

2-Methylnonane, isodecane, CA 871-83-0: Flammable; soluble most organic solvents. Tm = 198.45; Tb = 439.95; Vm = 0.1954 at 293.15; Antoine (liq) A = 6.120, B = 1521.3, C = -70.4, dev not specified, range = 324 to 441.

4-Propylheptane, CA 3178-29-8: Flammable; soluble most organic solvents. Tb = 430.65; Vm = 0.1943 at 298.15; Antoine (liq) A = 7.393, B = 2305, C = 0, dev = above 5.0, range = 331 to 430.

$C_{10}H_{22}O$ MW 158.28

1-Decanol, decyl alcohol, CA 112-30-1: Irritant; moderately toxic; flammable, flash point 355; soluble ethanol, ether, acetone, benzene. Tf = 280.15; Tb = 506.05; Vm = 0.1908 at 293.15; Antoine I (sol) A = *(continues)*

$C_{10}H_{22}O$ *(continued)*

17.615, B = 6028, C = 0, dev = 1.0 to 5.0, range = 264 to 279; Antoine II (liq) A = 6.57397, B = 1761.708, C = -113.992, dev = 0.1, range = 349 to 410; Antoine III (liq) A = 5.8587, B = 1374.347, C = -147.547, dev = 0.1, range = 405 to 528; Antoine IV (liq) A = 5.85774, B = 1373.74, C = -147.614, dev = 0.1, range = 400 to 529; Antoine V (liq) A = 5.94825, B = 1438.46, C = -139.408, dev = 0.02 to 0.1, range = 474 to 529.

Diisopentyl ether, CA 544-01-4: Flammable, flash point 318; forms explosive peroxides; soluble ethanol, ether, acetone, chloroform. Tb = 446.15; Vm = 0.2035 at 293.15; Antoine (liq) A = 6.85022, B = 2014.89, C = -30.517, dev = 1.0, range = 291 to 447.

3,7-Dimethyl-1-octanol ±, CA 59204-02-3: Flammable; soluble ethanol, ether. Tb = 485.15 to 486.15; Vm = 0.1910 at 293.15; Antoine (liq) A = 8.00569, B = 2236.41, C = -93.855, dev = 5.0, range = 341 to 467.

Dipentyl ether, CA 693-65-2: Flammable, flash point 330; forms explosive peroxides; soluble ethanol, ether. Tm = 204.15; Tb = 457.15 to 459.15; Vm = 0.2031 at 298.15; Antoine (liq) A = 6.58087, B = 1906.7, C = -43.15, dev = 1.0, range = 373 to 460.

$C_{10}H_{22}O_2$ MW 174.28

Acetaldehyde, dibutyl acetal, CA 871-22-7: Flammable. Tb = 470.15 to 471.15; Vm = 0.2097 at 299.15; Antoine (liq) A = 7.357, B = 2470, C = 0, dev = 5.0, range = 303 to 464.

3,4-Diethyl-3,4-hexanediol, CA 6931-71-1: Flammable; soluble water, ether, ethanol. Tm = 301.15; Tb = 503.15; Vm = 0.1805 at 286.15; Antoine (liq) A = 6.34322, B = 1838.362, C = -82.995, dev = 1.0, range = 405 to 507.

Ethylene glycol, dibutyl ether, CA 112-48-1: Flammable, flash point 358; forms explosive peroxides; soluble ether. Tm = 204.05; Tb = 476.25; Vm = 0.2077 at 293.15; Antoine (liq) A = 5.5919, B = 1250.4, C = -127.91, dev = 1.0, range = 356 to 476.

Ethylene glycol, diisobutyl ether, CA 5669-09-0: Flammable; forms explosive peroxides; soluble ether. Tb = 444.45; Vm = 0.2125 at 293.15; Antoine (liq) A = 7.3078, B = 2411, C = 0, dev = 1.0 to 5.0, range = 336 to 456.

Ethylene glycol, mono(2-ethylhexyl) ether, CA 1559-35-9: Flammable, flash point 383; forms explosive peroxides; moderately soluble water; soluble ethanol, ether. Tm = 203.15; Tb = 502.15; Vm = 0.1966 at 293.15; Antoine (liq) A = 7.8943, B = 2949, C = 0, dev = 5.0, range = 381 to 502.

3-Ethyl-3-hydroxymethyl-2-heptanol: Flammable; soluble ethanol. Antoine (liq) A = 8.79376, B = 3733, C = 21.686, dev = 1.0, range = 338 to 500.

$C_{10}H_{22}O_3$ MW 190.28

Diethylene glycol, monohexyl ether, CA 112-59-4: Irritant; flammable, flash point above 383; forms explosive peroxides; moderately soluble water; soluble ethanol, ether. Tm = 232.95; Tb = 532.25; Vm = 0.2032 at 293.15; Antoine (liq) A = 8.1708, B = 3274.3, C = 0, dev = 5.0, range = 406 to 531.

Dipropylene glycol, monobutyl ether, CA 24083-03-2: Flammable, flash point 386; forms explosive peroxides; very soluble water; soluble   *(continues)*

$C_{10}H_{22}O_3$ *(continued)*

ethanol, ether. Tm = 200.15; Tb = 504.15; Vm = 0.2078 at 298.15; Antoine
(liq) A = 6.54668, B = 1881.79, C = -85.98, dev = 5.0, range = 337 to 500.

$C_{10}H_{22}O_4$ MW 206.28

Tripropylene glycol, monomethyl ether, CA 20324-33-8: Flammable, flash
point 394; forms explosive peroxides; soluble water, ethanol, ether.
Antoine (liq) A = 7.968, B = 3067, C = 0, dev = 5.0, range = 308 to 515.

$C_{10}H_{22}O_5$ MW 222.28

Tetraethylene glycol, dimethyl ether, CA 143-24-8: Flammable, flash point
414; forms explosive peroxides; soluble water, ethanol, ether. Tm =
243.45; Tb = 548.45; Vm = 0.2190 at 293.15; Antoine (liq) A = 13.497, B =
10684, C = 380.85, dev = 5.0, range = 419 to 553.

$C_{10}H_{22}S$ MW 174.34

1-Decanethiol, decyl mercaptan, CA 143-10-2: Toxic; stench; flammable,
flash point 371; soluble ethanol, ether. Tm = 247.15; Tb = 513.75; Vm =
0.2065 at 293.15; Antoine I (liq) A = 7.741, B = 3061.7, C = 0, dev = 1.0
to 5.0, range = 283 to 293; Antoine II (liq) A = 6.123, B = 1713.6, C =
-96.15, dev = 1.0, range = 413 to 534.

Diisopentyl sulfide, CA 544-02-5: Toxic; stench; flammable; soluble ethanol, ether. Tb = 489.15; Vm = 0.2118 at 293.15; Antoine I (liq) A =
-2.83491, B = 390.608, C = -492.484, dev = 5.0, range = 283 to 353; Antoine
II (liq) A = 8.47247, B = 3024.43, C = 0, dev = 1.0, range = 339 to 366.

Dipentyl sulfide, CA 872-10-6: Toxic; stench; flammable; soluble ethanol,
ether. Tm = 221.85; Tb = 503.15; Vm = 0.2073 at 293.15; Antoine (liq) A =
9.81574, B = 3589.481, C = 0, dev = 1.0 to 5.0, range = 346 to 366.

$C_{10}H_{22}S_2$ MW 206.40

1,10-Decanedithiol, CA 1191-67-9: Toxic; stench; flammable. Tm = 290.95;
Antoine (liq) A = 5.7228, B = 1505.7, C = -165.17, dev = 1.0, range = 434
to 571.

$C_{10}H_{23}N$ MW 157.30

Decylamine, CA 2016-57-1: Corrosive; highly toxic; flammable, flash point
358; soluble ethanol, ether, acetone, benzene. Tm = 288.15; Tb = 491.15 to
491.65; Vm = 0.1982 at 293.15; Antoine (liq) A = 6.4229, B = 1844.7, C =
-76.05, dev = 1.0, range = 410 to 506.

N,N-Dimethyl octylamine, CA 7378-99-6: Corrosive; toxic; flammable, flash
point 338. Tm = 198.15; Antoine (liq) A = 6.41512, B = 1746.1, C = -71.15,
dev = 5.0, range = 371 to 517.

Dipentyl amine, CA 2050-92-2: Corrosive; highly toxic; flammable, flash
point 325; soluble ethanol, ether, acetone. Tm = 240.15; Tb = 475.15 to
476.15; Vm = 0.2024 at 293.15; Antoine (liq) A = 6.42472, B = 1783.1, C =
-72.15, dev = 5.0, range = 379 to 527.

$C_{10}H_{23}N_3$ MW 185.31

2-Propanone, [2-(diethylamino)ethyl]methylhydrazone, CA 67752-90-3: Irritant; flammable. Antoine (liq) A = 8.4035, B = 3256, C = 0, dev = 1.0 to 5.0, range = 288 to 315.

$C_{10}H_{24}NO_3PS$ MW 269.34

O,O-Diethyl-S-[2-(diethylamino)ethyl] thiophosphate, CA 78-53-5: Toxic insecticide. Antoine (liq) A = 11.51993, B = 4934.9, C = 0, dev = 1.0, range = 358 to 407.

$C_{11}F_{24}O_2$ MW 620.08

Perfluoro(1,9-bis-methoxynonane): Antoine (liq) A = 7.1699, B = 2246, C = 0, dev = 5.0, range not specified.

$C_{11}H_8O_2$ MW 172.18

1-Benzoylfuran: Flammable; soluble ethanol, ether. Tm = 330.15; Antoine (liq) A = 5.9783, B = 1675, C = -136.85, dev = 1.0, range = 423 to 561.

1-Naphthoic acid, CA 86-55-5: Irritant; moderately toxic orally; slightly soluble hot water; soluble ethanol, ether, chloroform. Tm = 435.15; Tb = 573.15; Antoine I (sol) A = 12.585, B = 5764.4, C = 0, dev = 5.0, range = 340 to 361; Antoine II (liq) A = 9.3963, B = 3742.08, C = -66.807, dev = 1.0, range = 457 to 573.

2-Naphthoic acid, CA 93-09-4: Irritant; moderately toxic orally; slightly soluble hot water; soluble ethanol, ether, chloroform. Tm = 458.65; Tb = above 573.15; Antoine I (sol) A = 12.705, B = 5932.6, C = 0, dev = 5.0, range = 347 to 364; Antoine II (liq) A = 8.85947, B = 3349.48, C = -92.964, dev = 1.0, range = 463 to 582.

$C_{11}H_9Cl$ MW 176.65

1-(Chloromethyl)naphthalene, CA 86-52-2: Corrosive; lachrymator; flammable, flash point above 383; polymerizes on heating; soluble ethanol, benzene. Tm = 304.15 to 305.15; Tb = 564.15 to 565.15; Antoine (liq) A = 11.2429, B = 4713.7, C = 0, dev = 5.0, range = 407 to 447.

$C_{11}H_{10}$ MW 142.20

1-Methylnaphthalene, CA 90-12-0: Irritant; flammable, flash point 355; soluble ethanol, ether, benzene. Tm = 242.55; Tb = 517.75; Vm = 0.1394 at 293.15; Antoine I (liq) A = 7.03469, B = 3006.267, C = 0, dev = 1.0, range = 278 to 313; Antoine II (liq) A = 6.15928, B = 1826.402, C = -78.176, dev = 0.02, range = 415 to 526.

2-Methylnaphthalene, CA 91-57-6: Irritant; flammable, flash point 370; soluble ethanol, ether, benzene. Tm = 310.15 to 311.15; Tb = 514.25; Antoine (liq) A = 6.21475, B = 1858.19, C = -72.779, dev = 0.02, range = 423 to 515.

$C_{11}H_{11}Cl_3O_3$ MW 297.57

2,4,5-Trichlorophenoxyacetic acid, propyl ester, CA 1928-40-1: Toxic; flammable; used as weed killer. Antoine (liq) A = 5.53271, B = 1516.156, C = -187.697, dev = 1.0, range = 444 to 573.

$C_{11}H_{11}N$ MW 157.21

2,4-Dimethylquinoline, CA 1198-37-4: Irritant; flammable, flash point 383; slightly soluble water; soluble ethanol, ether. Tm = 288.0; Tb = 537.15 to 538.15; Vm = 0.1482 at 288.15; Antoine (liq) A = 6.04525, B = 1743.55, C = -108.778, dev = 1.0, range = 458 to 543.

2,6-Dimethylquinoline, CA 877-43-0: Irritant; flammable; slightly soluble hot water; soluble ethanol, ether, benzene. Tm = 333.15; Tb = 539.15 to 540.15; Antoine (liq) A = 6.05006, B = 1744.635, C = -107.227, dev = 0.1, range = 461 to 541.

$C_{11}H_{12}Cl_2O_3$ MW 263.12

2,4-Dichlorophenoxyacetic acid, isopropyl ester, CA 94-11-1: Toxic; flammable; used as weed killer. Tf = 278.15; Vm = 0.2078 at 298.15; Antoine (liq) A = 5.65483, B = 1591.17, C = -160.339, dev = 1.0, range = 460 to 573.

2,4-Dichlorophenoxyacetic acid, propyl ester, CA 1928-61-6: Toxic; flammable; used as weed killer. Antoine (liq) A = 5.26167, B = 1332.771, C = -194.934, dev = 1.0, range = 444 to 573.

$C_{11}H_{12}Cl_2O_4$ MW 279.12

2,4-Dichlorophenoxyacetic acid, 3-hydroxypropyl ester, CA 28191-20-0: Toxic; flammable; used as weed killer. Antoine (liq) A = 7.66717, B = 3765.71, C = 0, dev = 5.0, range = 463 to 483.

$C_{11}H_{12}O$ MW 160.22

2-Ethylidene-3-phenylpropanal: Flammable; polymerizes. Antoine (liq) A = 9.796, B = 3843.379, C = 0, dev = 1.0 to 5.0, range = 333 to 374.

$C_{11}H_{12}O_2$ MW 176.21

Benzyl methacrylate, CA 2495-37-6: Irritant; lachrymator; flammable; polymerizes. Vm = 0.1164 at 298.15; Antoine (liq) A = 9.65396, B = 3680.377, C = 0, dev = 1.0 to 5.0, range = 347 to 431.

Cinnamic acid, ethyl ester, CA 4192-77-2: Flammable, flash point above 383; soluble ethanol, ether, acetone, benzene. Tm = 285.15; Tb = 544.15; Vm = 0.1679 at 298.15; Antoine (liq) A = 7.44144, B = 2913.38, C = -8.226, dev = 1.0, range = 453 to 544.

1-Phenyl-1,3-pentanedione, CA 5331-64-6: Flammable; soluble acetone. Tm = 408.15; Antoine (liq) A = 7.42042, B = 2784.61, C = -35.341, dev = 1.0 to 5.0, range = 371 to 550.

$C_{11}H_{12}O_3$ MW 192.21

Benzoylacetic acid, ethyl ester, CA 94-02-0: Flammable, flash point 414;
light-sensitive; soluble ethanol, ether. Tb = 538.15 to 543.15, decomposes; Vm = 0.1713 at 288.15; Antoine (liq) A = 9.02469, B = 3784.81, C =
0.984, dev = 1.0, range = 380 to 538.

Myristicin, CA 607-91-0: Toxic alkaloid; flammable; soluble ether, benzene. Tm = 549.15 to 550.15; Antoine (liq) A = 7.38248, B = 2863.86, C =
-20.436, dev = 1.0, range = 368 to 553.

Propanal, 2-piperonyl: Irritant; flammable. Antoine (liq) A = 9.20069,
B = 3890.188, C = 0, dev = 1.0 to 5.0, range = 373 to 423.

$C_{11}H_{13}Cl_3$ MW 251.58

4-tert-Butyl-2,3,6-trichlorotoluene: Irritant; flammable. Tm = 322.15;
Antoine (liq) A = 6.2251, B = 2005.2, C = -95.05, dev = 1.0, range = 423
to 570.

$C_{11}H_{14}$ MW 146.23

1,1-Dimethylindane, CA 4912-92-9: Flammable; soluble ethanol, ether.
Tb = 464.15; Vm = 0.1591 at 293.15; Antoine I (liq) A = 6.10744, B =
1628.434, C = -70.186, dev = 1.0, range = 313 to 467; Antoine II (liq) A =
6.35302, B = 1772.347, C = -57.823, dev = 0.02, range = 313 to 348; Antoine
III (liq) A = 6.04901, B = 1586.63, C = -74.812, dev = 0.02, range = 387
to 467.

4,6-Dimethylindane, CA 1685-82-1: Flammable; soluble ethanol, ether.
Antoine I (liq) A = 6.30705, B = 1841.565, C = -69.582, dev = 1.0, range =
313 to 467; Antoine II (liq) A = 6.39954, B = 1902.886, C = -64.236, dev =
0.02 to 0.1, range = 313 to 363; Antoine III (liq) A = 6.20559, B =
1766.779, C = -77.206, dev = 0.02, range = 415 to 467.

4,7-Dimethylindane, CA 6682-71-9: Flammable; soluble ethanol, ether.
Vm = 0.1541 at 293.15; Antoine I (liq) A = 6.30346, B = 1851.143, C =
-69.949, dev = 1.0, range = 313 to 470; Antoine II (liq) A = 6.36076, B =
1891.219, C = -66.307, dev = 0.1, range = 313 to 363; Antoine III (liq) A =
6.1952, B = 1770.907, C = -78.143, dev = 0.02, range = 417 to 470.

4-Isopropylstyrene, CA 2055-40-5: Irritant; lachrymator; flammable; polymerizes; soluble most organic solvents. Tf = 228.45; Tb = 477.25; Vm =
0.1652 at 293.15; Antoine (liq) A = 6.22335, B = 1683.5, C = -78.15, dev =
0.1, range = 408 to 478.

5-Methyl-1,2,3,4-tetrahydronaphthalene, CA 2809-64-5: Flammable; soluble
benzene. Tm = 250.15; Tb = 507.35; Antoine (liq) A = 6.05435, B = 1653.42,
C = -99.11, dev = 0.1, range = 416 to 508.

6-Methyl-1,2,3,4-Tetrahydronaphthalene, CA 1680-51-9: Flammable; soluble
benzene. Tm = 233.42; Tb = 494.15; Antoine (liq) A = 5.41423, B = 1216.56,
C = -145.26, dev = 1.0, range = 411 to 502.

2,4,5-Trimethylstyrene, CA 3937-24-4: Irritant; lachrymator; flammable;
polymerizes; soluble most organic solvents. Tm = 275.65; Antoine (liq)
A = 6.01625, B = 1569.5, C = -98.95, dev = 1.0, range = 352 to 490.

(continues)

$C_{11}H_{14}$ *(continued)*

2,4,6-Trimethylstyrene, CA 769-25-5: Irritant; lachrymator; flammable, flash point 348; polymerizes; soluble most organic solvents. Tm = 236.15; Tb = 481.15 to 483.15; Vm = 0.1615 at 293.15; Antoine (liq) A = 6.34979, B = 1804.5, C = -66.45, dev = 1.0, range = 362 to 483.

$C_{11}H_{14}Cl_2$ MW 217.14

4-*tert*-Butyl-2,5-dichlorotoluene, CA 61468-35-7: Irritant; flammable. Tm = 258.65; Antoine (liq) A = 6.4254, B = 2064.5, C = -68.05, dev = 1.0, range = 395 to 538.

$C_{11}H_{14}O$ MW 162.23

tert-Butyl phenyl ketone, CA 938-16-9: Flammable. Tb = 492.15 to 494.15; Antoine (liq) A = 7.52284, B = 2633.52, C = -16.001, dev = 5.0, range = 330 to 493.

2-Ethyl-3-phenylpropanal: Irritant; flammable; polymerizes. Antoine (liq) A = 8.83619, B = 3374.397, C = 0, dev = 1.0 to 5.0, range = 343 to 388.

Isovalerophenone, isobutyl phenyl ketone, CA 582-62-7: Flammable; soluble ethanol, ether, acetone. Tb = 509.65; Vm = 0.1672 at 289.55; Antoine (liq) A = 7.00831, B = 2325.25, C = -36.61, dev = 1.0 to 5.0, range = 331 to 501.

2',3',5'-Trimethylacetophenone: Flammable; soluble ethanol, ether, benzene. Antoine (liq) A = 8.05732, B = 3222.74, C = 11.765, dev = 5.0, range = 352 to 557.

$C_{11}H_{14}O_2$ MW 178.23

3-Acetoxy-1-phenylpropane: Flammable. Antoine (liq) A = 10.466, B = 3879, C = 0, dev = 1.0 to 5.0, range = 293 to 333.

Butyl benzoate, CA 136-60-7: Irritant; flammable, flash point 380; soluble ethanol, ether, acetone. Tm = 251.15; Tb = 523.15; Vm = 0.1782 at 293.15; Antoine (liq) A = 8.04664, B = 3085.3, C = 0, dev = 1.0, range = 343 to 405.

3,5-Diethylbenzoic acid, CA 3854-90-8: Flammable. Antoine (sol) A = 12.5049, B = 5436.1, C = 0, dev = 1.0 to 5.0, range = 325 to 343.

1,2-Dimethoxy-4-(1-propenyl)benzene, CA 93-16-3: Flammable, flash point above 383; soluble acetone, benzene; present in essential oils. Tf = 269.15; Tb = 528.15; Vm = 0.1694 at 293.15; Antoine (liq) A = 8.66541, B = 3680.87, C = 31.379, dev = 1.0, range = 358 to 521.

Isobutyl benzoate, CA 120-50-3: Flammable; soluble ethanol, ether, acetone. Tb = 514.15; Vm = 0.1784 at 293.15; Antoine I (liq) A = 8.015, B = 3033, C = 0, dev = 1.0 to 5.0, range = 291 to 300; Antoine II (liq) A = 7.01319, B = 2360.72, C = -38.937, dev = 1.0, range = 338 to 510.

1-(4-Methoxyphenyl)-2-butanone, CA 53917-01-4: Flammable. Antoine (liq) A = 8.117, B = 3270, C = 0, dev = 5.0, range = 373 to 443.

$C_{11}H_{14}O_3$ MW 194.23

2-Piperonyl propanol: Flammable; soluble ethanol. Antoine (liq) A = 10.16174, B = 4431.129, C = 0, dev = 5.0, range = 373 to 443.

$C_{11}H_{15}Cl$ MW 182.69

4-*tert*-Butyl-2-chlorotoluene, CA 42597-10-4: Irritant; flammable. Antoine
(liq) A = 6.4081, B = 1917.6, C = -67.75, dev = 1.0, range = 372 to 503.

$C_{11}H_{15}N$ MW 161.25

Azetidine, 2-phenylethyl, CA 42525-65-5: Irritant; flammable. Antoine
(liq) A = 8.18829, B = 3249.837, C = 0, dev = 1.0, range = 302 to 333.

$C_{11}H_{15}NO$ MW 177.25

N,*N*-Diethylbenzamide, CA 1696-17-9: Irritant; flammable. Antoine (liq)
A = 7.1082, B = 2951, C = 0, dev = 5.0, range = 373 to 403.

$C_{11}H_{16}$ MW 148.25

4-*tert*-Butyltoluene, CA 98-51-1: Irritant; highly toxic by inhalation;
flammable, flash point 327; soluble most organic solvents. Tm = 219.15;
Tb = 462.15 to 465.15; Vm = 0.1721 at 293.15; Antoine (liq) A = 6.3227, B =
1731.2, C = -63.45, dev = 1.0, range = 342 to 465.

3,5-Diethyltoluene, CA 2050-24-0: Flammable; soluble most organic solvents.
Tf = 199.05; Tb = 473.85; Vm = 0.1718 at 293.15; Antoine (liq) A = 6.75459,
B = 2098.64, C = -31.998, dev = 1.0, range = 307 to 474.

1-Ethyl-3-isopropylbenzene, CA 4920-99-4: Flammable; soluble most organic
solvents. Tm = below 253.15; Tb = 465.15; Vm = 0.1726 at 293.15; Antoine
(liq) A = 6.62576, B = 1980.65, C = -37.348, dev = 1.0, range = 301 to 466.

1-Ethyl-4-isopropylbenzene, CA 4218-48-8: Flammable; soluble most organic
solvents. Tm = below 253.15; Tb = 469.75; Vm = 0.1727 at 293.15; Antoine
(liq) A = 6.71962, B = 2047.36, C = -34.705, dev = 1.0, range = 304 to 469.

2-Ethyl-1,3,5-trimethylbenzene, CA 3982-67-0: Flammable; soluble most
organic solvents. Tm = 260.95; Tb = 485.55; Vm = 0.1679 at 293.15; Antoine
(liq) A = 6.45181, B = 1894.19, C = -55.264, dev = 1.0, range = 312 to 481.

3-Ethyl-1,2,4-trimethylbenzene: Flammable; soluble most organic solvents.
Tb = 489.75; Vm = 0.1656 at 293.15; Antoine (liq) A = 4.8334, B = 896.0,
C = -170.05, dev = above 5.0, range = 347 to 488.

5-Ethyl-1,2,4-trimethylbenzene, CA 17851-27-3: Flammable; soluble most
organic solvents. Tf = 259.65; Tb = 486.15; Vm = 0.1679 at 293.15; Antoine
(liq) A = 6.62947, B = 1970.08, C = -55.34, dev = 1.0, range = 317 to 481.

Pentamethylbenzene, CA 700-12-9: Irritant; flammable, flash point 366;
soluble most organic solvents. Tm = 327.15; Tb = 505.15; Antoine (liq) A =
6.3509, B = 1867, C = -75.15, dev = 1.0, range = 338 to 503.

2-Phenylpentane ±, CA 2719-52-0: Moderately toxic; flammable; soluble most
organic solvents. Tb = 464.15 to 466.15; Vm = 0.1727 at 293.15; Antoine
(liq) A = 6.42129, B = 1825.06, C = -52.712, dev = 1.0, range = 302 to 466.

$C_{11}H_{16}O$ MW 164.25

2-*sec*-Butyl-4-methylphenol, CA 51528-17-7: Moderately toxic; flammable;
soluble ethanol, ether, benzene. Tm = 317.65; Tb = 510.15; Antoine (liq)
A = 6.70767, B = 2026.5, C = -79.15, dev = 1.0, range = 413 to 548.
(continues)

$C_{11}H_{16}O$ *(continued)*

2-*tert*-Butyl-4-methylphenol, CA 2409-55-4: Moderately toxic; flammable, flash point 373; soluble ethanol, ether, benzene. Tm = 328.15; Tb = 510.15; Vm = 0.1776 at 348.15; Antoine I (sol) A = 10.83523, B = 4043.245, C = 0, dev = 1.0, range = 274 to 294; Antoine II (liq) A = 7.50381, B = 2609.4, C = -31.56, dev = 1.0, range = 385 to 517.

2-*tert*-Butyl-5-methylphenol, CA 88-60-8: Moderately toxic; flammable, flash point 378; soluble ethanol, ether, benzene, acetone. Tm = 319.15 to 320.15; Tb = 497.15; Vm = 0.1781 at 353.15; Antoine I (liq) A = 6.4354, B = 1900.869, C = -86.195, dev = 1.0, range = 378 to 490; Antoine II (liq) A = 7.85768, B = 3198.8, C = 29.62, dev = 1.0 to 5.0, range = 383 to 518.

2-*tert*-Butyl-6-methylphenol, CA 2219-82-1: Moderately toxic; flammable, flash point 380; soluble ethanol, ether, benzene. Tm = 303.15 to 305.15; Tb = 503.15; Antoine (liq) A = 6.40166, B = 1890.8, C = -74.2, dev = 1.0, range = 375 to 505.

4-*tert*-Butyl-2-methylphenol, CA 98-27-1: Moderately toxic; flammable; soluble ethanol, ether, acetone, benzene. Tm = 300.15; Tb = 508.15 to 510.15; Antoine I (liq, subcooled) A = 10.33028, B = 3954.496, C = 0, dev = 1.0, range = 275 to 297; Antoine II (liq) A = 6.8323, B = 2213.87, C = -61.419, dev = 1.0, range = 347 to 520.

2-Ethyl-3-phenyl-1-propanol, CA 3968-87-4: Flammable; soluble ethanol, benzene. Antoine (liq) A = 9.29922, B = 3705.734, C = 0, dev = 1.0 to 5.0, range = 348 to 393.

4-Pentylphenol, CA 14938-35-3: Moderately irritant; flammable; soluble ethanol, ether, benzene. Tm = 296.15; Tb = 523.65; Antoine (liq) A = 6.76086, B = 2099.4, C = -82.15, dev = 1.0 to 5.0, range = 423 to 563.

4-*tert*-Pentylphenol, CA 80-46-6: Moderately irritant; flammable; soluble ethanol, ether, benzene, chloroform. Tm = 367.15 to 368.15; Tb = 535.65; Antoine (liq) A = 6.91623, B = 2428.5, C = -42.51, dev = 1.0, range = 385 to 548.

5-Phenyl-1-pentanol, CA 10521-91-2: Flammable, flash point above 383; soluble ethanol, ether. Vm = 0.1689 at 293.15; Antoine (liq) A = 9.20526, B = 3757.588, C = 0, dev = 1.0, range = 373 to 430.

1-(2,4,6-Trimethylphenyl)ethanol, CA 31108-34-6: Flammable; soluble ethanol, benzene. Tm = 314.15; Antoine (sol) A = 2.292, B = 297.2, C = 0, dev = 5.0, range = 282 to 313.

$C_{11}H_{16}O_2$ MW 180.25

2-*tert*-Butyl-4-methoxyphenol, CA 121-00-6: Moderately toxic; flammable; soluble ethanol, ether. Tm = 335.15 to 336.15; Antoine (liq) A = 7.038, B = 2843.5, C = 0, dev = 1.0, range = 403 to 463.

1,3-Dihydroxy-4-Pentylbenzene, CA 533-24-4: Irritant; flammable; soluble ethanol, ether, benzene. Tm = 347.05; Antoine (liq) A = 5.6781, B = 1578.9, C = -176.36, dev = 1.0, range = 423 to 488.

$C_{11}H_{16}O_5$ MW 228.24

Ethylcamphoric acid, anhydride: Flammable. Antoine (liq) A = 7.66126, B = 3001.83, C = -40.282, dev = 5.0, range = 391 to 571.

(continues)

$C_{11}H_{16}O_5$ *(continued)*

(1-Methylallyl)[1-(allyloxycarbonyl)ethyl] carbonate: Toxic; flammable.
Antoine (liq) A = 8.129, B = 3146, C = 0, dev = 5.0, range = 368 to 508.

$C_{11}H_{18}O_2$ MW 182.26

Borneol formate, CA 7492-41-3: Flammable. Vm = 0.1806 at 295.15; Antoine
(liq) A = 7.05418, B = 2323.75, C = -27.125, dev = 1.0, range = 320 to 487.

Formic acid, 3,7-dimethyl-*cis*-2,6-octadienyl ester, neryl formate, CA 2142-
94-1: Flammable; soluble ethanol, ether, acetone. Antoine (liq) A =
6.7414, B = 2073.24, C = -59.81, dev = 1.0, range = 330 to 498.

Formic acid, 3,7-dimethyl-*trans*-2,6-octadienyl ester, geranyl formate,
CA 105-86-2: Flammable; soluble ethanol, ether, acetone. Tb = 502.15, decomposes; Vm = 0.2006 at 298.15; Antoine (liq) A = 7.03828, B = 2327.5, C =
-40.727, dev = 1.0, range = 334 to 503.

Isoborneol formate, CA 1200-67-5: Flammable; soluble ethanol, acetone.
Vm = 0.1798 at 293.15; Antoine (liq) A = 7.727, B = 2794, C = 0, dev = 1.0,
range = 383 to 441.

$C_{11}H_{18}O_5$ MW 230.26

4-Oxononanedioic acid, dimethyl ester: Flammable. Antoine (liq) A =
8.8548, B = 3851.96, C = 3.038, dev = 5.0, range = 394 to 559.

$C_{11}H_{19}NO_5$ MW 245.27

N-Acetyl-*levo*-glutamic acid, diethyl ester, CA 1446-19-1: Flammable.
Antoine (liq) A = 8.0503, B = 3509, C = 0, dev = 5.0, range = 403 to 503.

$C_{11}H_{20}$ MW 152.28

Cyclopentylcyclohexane, CA 1606-08-2: Flammable; soluble most organic solvents. Tb = 488.25; Vm = 0.1739 at 293.15; Antoine (liq) A = 6.07478, B =
1706.7, C = -69.25, dev = 1.0, range = 383 to 488.

$C_{11}H_{20}Cl_4$ MW 294.09

1,1,1,11-Tetrachloroundecane, CA 3922-34-7: Irritant; flammable. Antoine
(liq) A = 11.345, B = 4830, C = 0, dev = 5.0, range = 303 to 353.

$C_{11}H_{20}O$ MW 168.28

Cycloundecanone, CA 878-13-7: Flammable, flash point 369. Tm = 289.15;
Antoine I (liq) A = 6.2686, B = 1850.72, C = -88.15, dev = 1.0, range = 363
to 433; Antoine II (liq) A = 6.58702, B = 2237.152, C = -41.845, dev = 0.1,
range = 448 to 501.

$C_{11}H_{20}O_2$ MW 184.28

2-Ethylhexyl acrylate ±, CA 103-11-7: Irritant; lachrymator; flammable,
flash point 353; polymerizes. Tm = 183.15; Tb = 487.15 to   *(continues)*

$C_{11}H_{20}O_2$ *(continued)*

493.15; Vm = 0.2094 at 298.15; Antoine (liq) A = 6.8095, B = 2117.56, C = -48.424, dev = 1.0, range = 323 to 489.

Formic acid, 3-*para*-menthol ester: Flammable. Tm = 282.15; Antoine (liq) A = 6.86816, B = 2246.69, C = -30.177, dev = 1.0, range = 320 to 492.

2-Hexyl-4,7-dihydro-1,3-dioxepin, CA 61732-96-5: Flammable. Antoine (liq) A = 9.0629, B = 3447, C = 0, dev = 1.0 to 5.0, range = 333 to 453.

Octyl acrylate, CA 2499-59-4: Irritant; lachrymator; flammable; polymerizes. Tb = 500.15 to 504.15; Vm = 0.2092 at 293.15; Antoine (liq) A = 7.29759, B = 2510.79, C = -25.921, dev = 1.0, range = 331 to 500.

Oxa-2-cyclododecanone, 11-hydroxyundecanoic acid, lactone, CA 1335-45-1: Flammable. Antoine (liq) A = 9.54924, B = 3682.712, C = 0, dev = 5.0, range = 353 to 413.

10-Undecenoic acid, CA 112-38-9: Flammable; soluble ethanol, ether, chloroform. Tm = 297.65; Tb = 548.15, decomposes; Vm = 0.2031 at 298.15; Antoine (liq) A = 10.0270, B = 5011.68, C = 76.646, dev = 1.0, range = 387 to 548.

$C_{11}H_{20}O_3$ MW 200.28

Hexyl levulinate, CA 24431-34-3: Flammable. Tb = 539.95; Vm = 0.2101 at 293.15; Antoine (liq) A = 6.60863, B = 2097.62, C = -84.288, dev = 1.0, range = 363 to 540.

$C_{11}H_{20}O_4$ MW 216.28

2-Acetoxypropionic acid, hexyl ester ±, CA 77008-66-3: Flammable. Antoine (liq) A = 6.6203, B = 1968.048, C = -90.15, dev = 1.0 to 5.0, range = 322 to 517.

Azelaic acid, dimethyl ester, CA 1732-10-1: Flammable; soluble ethanol, acetone, benzene. Vm = 0.2145 at 293.15; Antoine (liq) A = 8.1559, B = 3320, C = 0, dev = 1.0 to 5.0, range = 413 to 540.

Diethylmalonic acid, diethyl ester, CA 77-25-8: Flammable, flash point 367; soluble ethanol, ether. Tb = 503.15; Vm = 0.2243 at 303.15; Antoine (liq) A = 5.50554, B = 1100.293, C = -178.297, dev = 5.0, range = 386 to 491.

$C_{11}H_{20}O_5$ MW 232.28

Ethyl[1-methyl-1-(ethoxycarbonyl)ethyl] carbonate: Toxic; flammable; Antoine (liq) A = 8.6621, B = 3352.7, C = 0, dev = 1.0, range = 353 to 503.

Hexyl[1-(methoxycarbonyl)ethyl] carbonate: Toxic; flammable. Antoine (liq) A = 8.419, B = 3442, C = 0, dev = 5.0, range = 371 to 538.

Propyl[1-(butoxycarbonyl)ethyl] carbonate: Toxic; flammable. Antoine (liq) A = 8.600, B = 3471, C = 0, dev = 5.0, range = 330 to 463.

354

$C_{11}H_{21}N$ MW 167.29

Undecanonitrile, CA 2244-07-7: Irritant; flammable, flash point 383;
moisture-sensitive; soluble ethanol, ether. Tm = 267.35; Tb = 526.15 to
527.15; Vm = 0.2027 at 303.15; Antoine (liq) A = 6.44882, B = 2003.4, C =
-82.76, dev = 1.0 to 5.0, range = 355 to 534.

$C_{11}H_{21}NO$ MW 183.29

Piperidone, N-hexanoyl, CA 15770-38-4: Irritant; flammable. Antoine (liq)
A = 8.70332, B = 3461.321, C = 0, dev = 5.0, range = 383 to 433.

$C_{11}H_{22}$ MW 154.29

1-Undecene, CA 821-95-4: Flammable, flash point 335; soluble most organic
solvents. Tm = 223.96; Tb = 465.82; Vm = 0.2056 at 293.15; Antoine (liq)
A = 6.09158, B = 1563.163, C = -83.281, dev = 0.02, range = 378 to 473.

2-Undecene, cis, CA 821-96-5: Flammable; soluble most organic solvents.
Tf = 206.65; Tb = 469.15; Vm = 0.2037 at 293.15; Antoine (liq) A = 6.07376,
B = 1552.657, C = -87.762, dev = 1.0, range = 333 to 393.

2-Undecene, trans, CA 693-61-8: Flammable; soluble most organic solvents.
Tf = 224.85; Tb = 468.15; Vm = 0.2050 at 293.15; Antoine (liq) A = 5.97844,
B = 1497.271, C = -91.992, dev = 0.1 to 1.0, range = 333 to 393.

3-Undecene, cis, CA 821-97-6: Flammable; soluble most organic solvents.
Tf = 203.55; Tb = 466.65; Antoine (liq) A = 6.00835, B = 1512.206, C =
-88.987, dev = 1.0, range = 333 to 393.

3-Undecene, trans, CA 1002-68-2: Flammable; soluble most organic solvents.
Tf = 211.05; Tb = 466.65; Antoine (liq) A = 6.18984, B = 1618.345, C =
-79.197, dev = 1.0, range = 333 to 393.

4-Undecene, cis, CA 821-98-7: Flammable; soluble most organic solvents.
Tm = 176.15; Tb = 465.75; Vm = 0.2046 at 293.15; Antoine (liq) A = 6.14741,
B = 1597.23, C = -79.922, dev = 1.0, range = 333 to 393.

4-Undecene, trans, CA 693-62-9: Flammable; soluble most organic solvents.
Tf = 209.45; Tb = 466.15; Vm = 0.2055 at 293.15; Antoine (liq) A = 6.04444,
B = 1533.151, C = -86.617, dev = 0.1 to 1.0, range = 333 to 393.

5-Undecene, cis, CA 764-96-5: Flammable; soluble most organic solvents.
Tf = 166.65; Tb = 465.15; Vm = 0.2047 at 293.15; Antoine (liq) A = 6.15988,
B = 1604.431, C = -78.895, dev = 0.1 to 1.0, range = 333 to 393.

5-Undecene, trans, CA 764-97-6: Flammable; soluble most organic solvents.
Tf = 212.05; Tb = 466.15; Vm = 0.2058 at 293.15; Antoine (liq) A = 6.1124,
B = 1575.96, C = -82.274, dev = 0.1 to 1.0, range = 333 to 393.

$C_{11}H_{22}O$ MW 170.29

1-Hexylcyclopentanol, CA 36633-49-5: Flammable. Antoine (liq) A = 5.502,
B = 1121, C = -159.75, dev = 5.0, range = 387 to 509.

Undecanal, CA 112-44-7: Irritant; flammable, flash point 369; readily
polymerizes; soluble ethanol, ether. Tm = 269.15; Vm = 0.2064 *(continues)*

$C_{11}H_{22}O$ *(continued)*

at 296.15; Antoine (liq) A = 8.46246, B = 3144.732, C = 0, dev = 1.0 to 5.0, range = 288 to 400.

2-Undecanone, CA 112-12-9: Moderately toxic orally; flammable, flash point 361; soluble ethanol, ether, acetone, benzene. Tf = 288.15; Tb = 504.15 to 505.15; Vm = 0.2064 at 293.15; Antoine I (liq) A = 6.31779, B = 1804.95, C = -87.58, dev = 1.0, range = 335 to 433; Antoine II (liq) A = 6.17044, B = 1704.929, C = -96.92, dev = 0.1, range = 393 to 513; Antoine III (liq) A = 6.14696, B = 1687.68, C = -98.773, dev = 0.02, range = 461 to 538.

6-Undecanone, CA 927-49-1: Moderately toxic; flammable, flash point 361; soluble ethanol, ether, acetone. Tm = 287.15 to 288.15; Tb = 499.15; Vm = 0.2050 at 293.15; Antoine I (liq) A = 6.16734, B = 1689.401, C = -94.606, dev = 0.1, range = 388 to 532; Antoine II (liq) A = 6.14845, B = 1675.21, C = -96.183, dev = 0.02, range = 461 to 513.

$C_{11}H_{22}O_2$ MW 186.29

2-Butoxy-3-heptanone: Flammable. Antoine (liq) A = 2.665, B = 538.3, C = 0, dev = 5.0, range = 373 to 398.

Butyric acid, heptyl ester, CA 5870-93-9: Flammable; soluble ethanol. Tm = 215.65; Tb = 498.95; Vm = 0.2157 at 293.15; Antoine (liq) A = 8.1571, B = 3065.242, C = 0, dev = 5.0, range = 384 to 498.

4,5-Dimethyl-2-hexyl-1,3-dioxolane, CA 6454-22-4: Flammable. Antoine (liq) A = 9.5269, B = 3426, C = 0, dev = 1.0 to 5.0, range = 333 to 453.

4-Heptyl-1,3-dioxane, CA 2244-84-0: Flammable. Antoine (liq) A = 8.7689, B = 3364, C = 0, dev = 1.0 to 5.0, range = 353 to 453.

2-Hexyl-1,3-dioxepane, CA 4469-24-3: Flammable. Antoine (liq) A = 8.54472, B = 3204.508, C = 0, dev = 1.0 to 5.0, range = 328 to 368.

3-Hexyl-4-hydroxytetrahydro-2*H*-pyran, CA 41277-75-2: Flammable. Antoine (liq) A = 8.9909, B = 3847, C = 0, dev = 1.0 to 5.0, range = 383 to 453.

Isopropyl caprylate, CA 5458-59-3: Flammable. Vm = 0.2178 at 293.15; Antoine (liq) A = 7.49989, B = 2457.5, C = -33.68, dev = 1.0, range = 338 to 420.

Methyl caprate, CA 110-42-9: Flammable; soluble ethanol, ether, chloroform. Tm = 259.82; Tb = 497.15; Vm = 0.2134 at 293.15; Antoine (liq) A = 6.67657, B = 2038.05, C = -68.212, dev = 1.0 to 5.0, range = 379 to 500.

2-Octyl-1,3-dioxolane, CA 5432-30-4: Flammable. Antoine (liq) A = 8.2729, B = 3151, C = 0, dev = 1.0 to 5.0, range = 333 to 453.

Propyl caprylate, CA 624-13-5: Flammable; soluble ethanol, ether, acetone. Tm = 228.15; Tb = 498.15; Vm = 0.2151 at 293.15; Antoine (liq) A = 7.18638, B = 2283.5, C = -49.26, dev = 1.0 to 5.0, range = 343 to 500.

Undecanoic acid, CA 112-37-8: Flammable, flash point above 383; soluble ethanol, ether, acetone, benzene. Tm = 302.45; Tb = 557.15; Vm = 0.2092 at 303.15; Antoine (liq) A = 6.81379, B = 2093.94, C = -121.327, dev = 0.1, range = 393 to 557.

$C_{11}H_{22}O_3$ MW 202.29

2-Butoxypropionic acid, butyl ester, CA 38611-89-1: Flammable. Antoine
(liq) A = 5.938, B = 2131.5, C = 0, dev = 1.0 to 5.0, range = 373 to 398.

3-Butoxypropionic acid, butyl ester, CA 14144-48-0: Flammable. Antoine
(liq) A = 7.87141, B = 3006.6, C = 0, dev = 1.0, range = 343 to 493.

3-Ethoxypropionic acid, hexyl ester, CA 14144-37-7: Flammable. Antoine
(liq) A = 7.756, B = 2963, C = 0, dev = 5.0, range = 373 to 514.

Octyl lactate, CA 51191-33-4: Flammable. Tm = 269.15; Antoine (liq) A =
6.63976, B = 2027.035, C = -90.15, dev = 1.0 to 5.0, range = 328 to 528.

$C_{11}H_{23}Br$ MW 235.21

1-Bromoundecane, CA 693-67-4: Irritant; flammable, flash point above 383.
Tm = 263.45; Vm = 0.2232 at 293.15; Antoine (liq) A = 6.4131, B = 1977.14,
C = -83.35, dev = 1.0, range = 398 to 591.

$C_{11}H_{23}Cl$ MW 190.76

1-Chloroundecane, CA 2473-03-2: Irritant; flammable. Tm = 256.25; Antoine
(liq) A = 6.09739, B = 1713.225, C = -99.784, dev = 1.0, range = 374 to
519.

$C_{11}H_{23}F$ MW 174.30

1-Fluoroundecane, CA 506-05-8: Flammable. Tm = 257.15; Antoine (liq) A =
6.433, B = 1797.8, C = -73.15, dev = 5.0, range = 373 to 523.

$C_{11}H_{23}I$ MW 282.21

1-Iodoundecane, CA 4282-44-4: Irritant; light-sensitive; flammable, flash
point above 383. Tm = 275.15; Vm = 0.2313 at 293.15; Antoine (liq) A =
6.3021, B = 2006.28, C = -87.65, dev = 1.0, range = 412 to 618.

$C_{11}H_{23}NO$ MW 185.31

Decanamide, N-methyl, CA 23220-25-9: Irritant; flammable. Tm = 330.55;
Antoine (sol) A = 13.7189, B = 5370.6, C = 0, dev = 1.0, range = 303 to
325.

Nonamide, N,N-dimethyl, CA 6225-08-7: Irritant; flammable. Antoine (liq)
A = 5.54082, B = 1372.196, C = -163.45, dev = 1.0, range = 411 to 509.

$C_{11}H_{23}NO_2$ MW 201.31

Lactamide, N,N-dibutyl, CA 6288-16-0: Irritant; flammable. Antoine (liq)
A = 11.16067, B = 4613.506, C = 0, dev = 1.0 to 5.0, range = 393 to 418.

Lactamide, N-octyl: Irritant; flammable. Tm = 325.65; Antoine (liq) A =
10.84323, B = 5028.651, C = 0, dev = 1.0, range = 428 to 468.

$C_{11}H_{24}$ MW 156.31

2,3-Dimethylnonane, CA 2884-06-2: Flammable; soluble most organic solvents. Tm = 155.45; Antoine (liq) A = 7.32088, B = 2501.48, C = 10.65, dev = 1.0, range = 336 to 460.

2,4-Dimethylnonane, CA 17302-24-8: Flammable; soluble most organic solvents. Antoine (liq) A = 7.45691, B = 2468.67, C = 1.74, dev = 1.0, range = 334 to 452.

2-Methyldecane, CA 6975-98-0: Flammable; soluble most organic solvents. Tm = 224.25; Tb = 462.35; Vm = 0.2121 at 293.15; Antoine I (liq) A = 6.68586, B = 1877.01, C = -56.18, dev = 0.1, range = 273 to 353; Antoine II (liq) A = 6.07901, B = 1547.168, C = -82.537, dev = 0.02, range = 379 to 463.

3-Methyldecane, CA 13151-34-3: Flammable; soluble most organic solvents. Tm = 193.65; Tb = 461.25; Vm = 0.2106 at 293.15; Antoine (liq) A = 7.07709, B = 2307.88, C = -8.98, dev = 1.0, range = 340 to 464.

4-Methyldecane, CA 2847-72-5: Flammable; soluble most organic solvents. Tm = 195.65; Antoine (liq) A = 7.47287, B = 2568.91, C = 9.82, dev = 1.0, range = 339 to 460.

5-Methyldecane, CA 13151-35-4: Flammable; soluble most organic solvents. Tm = 183.15; Antoine (liq) A = 7.24332, B = 2412.15; C = 0.73, dev = 1.0, range = 334 to 452.

2,4,6-Trimethyloctane, CA 62016-37-9: Flammable; soluble most organic solvents. Antoine (liq) A = 7.36565, B = 2376.4, C = 2.17, dev = 1.0, range = 325 to 442.

Undecane, CA 1120-21-4: Flammable, flash point 338; soluble most organic solvents. Tf = 247.55; Tb = 469.15; Vm = 0.2112 at 293.15; Tc = 638.76; Pc = 1966; Vc = 0.657; Antoine (liq) A = 6.10154, B = 1572.477, C = -85.128, dev = 0.1, range = 278 to 470.

$C_{11}H_{24}O$ MW 172.31

Decyl methyl ether, CA 7289-52-3: Flammable; forms explosive peroxides. Antoine I (liq) A = 6.22006, B = 1694.153, C = -87.17, dev = 0.1 to 1.0, range = 341 to 465; Antoine II (liq) A = 6.22741, B = 1699.035, C = -86.7, dev = 0.1, range = 341 to 429.

1-Undecanol, CA 112-42-5: Flammable, flash point above 383; soluble ethanol, ether. Tm = 294.6; Tb = 516.15; Vm = 0.2077 at 293.15; Antoine I (liq) A = 5.87637, B = 1439.27, C = -149.42, dev = 0.1 to 1.0, range = 393 to 523; Antoine II (liq) A = 5.87911, B = 1443.695, C = -148.515, dev = 1.0, range = 393 to 534.

$C_{11}H_{24}O_4$ MW 220.31

Tripropylene glycol, monoethyl ether, CA 75899-69-3: Flammable; forms explosive peroxides. Antoine (liq) A = 8.036, B = 3132, C = 0, dev = 5.0, range = 317 to 521.

358

$C_{11}H_{24}S_2$ MW 220.43

1,11-Undecanedithiol, CA 63476-06-2: Toxic; stench; flammable. Tm =
267.75; Antoine (liq) A = 5.7503, B = 1538, C = -171.27, dev = 1.0, range =
444 to 582.

$C_{11}H_{25}N$ MW 171.33

Undecylamine, CA 7307-55-3: Irritant; flammable, flash point 365; soluble
ethanol. Tm = 290.15; Tb = 515.15; Vm = 0.2147 at 293.15; Antoine (liq)
A = 6.4489, B = 1931.6, C = -80.05, dev = 1.0, range = 428 to 527.

$C_{11}H_{26}NO_2PS$ MW 267.37

Methylthiophosphonic acid, O-ethyl-S-[2-(N,N-diisopropylamino)ethyl] ester,
CA 50782-69-9: Toxic insecticide; flammable. Antoine (liq) A = 13.16599,
B = 5275.129, C = 0, dev = 1.0 to 5.0, range = 280 to 315.

$C_{12}F_{10}$ MW 334.12

Decafluorobiphenyl, CA 434-90-2: Tm = 341.15; Tb = 479.15; Antoine (sol)
A = 12.87903, B = 4585.091, C = 0, dev = 1.0, range = 297 to 323.

$C_{12}F_{18}$ MW 486.10

Hexakis(trifluoromethyl)bicyclo(2,2,0)hexa-2,5-diene, CA 23174-55-2:
Antoine (liq) A = 7.9819, B = 2165, C = 0, dev = 1.0 to 5.0, range = 293
to 343.

Hexakis(trifluoromethyl)tetracyclo(2,2,0,0^{2,6},0^{3,5})hexane, CA 22736-20-5:
Tm = 307.15; Antoine I (sol) A = 9.2519, B = 2569, C = 0, dev = 1.0 to 5.0,
range = 293 to 306, Antoine II (liq) A = 6.5155, B = 1730, C = 0, dev =
1.0 to 5.0, range = 313 to 353.

Hexakis(trifluoromethyl)tricyclo(3,1,0,0^{2,6})hex-3-ene, CA 22186-64-7:
Antoine (liq) A = 6.9359, B = 2017, C = 0, dev = 1.0 to 5.0, range = 293
to 353.

$C_{12}F_{27}N$ MW 671.10

Perfluorotributylamine, CA 311-89-7: Corrosive; hygroscopic; nonflammable;
soluble acetone, benzotrifluoride, light hydrocarbons. Tb = 450.15 to
451.15; Antoine I (liq) A = 6.59116, B = 1716.35, C = -76.07, dev = 1.0,
range = 298 to 450; Antoine II (liq) A = 6.12403, B = 1441.359, C =
-101.863, dev = 0.02 to 0.1, range = 371 to 544.

$C_{12}H_4N_4$ MW 204.19

2,5-Cyclohexadiene-1,4-dimalononitrile, CA 1518-16-7: Irritant. Antoine
(sol) A = 10.2049, B = 5475, C = 0, dev = 1.0 to 5.0, range = 433 to 499.

$C_{12}H_7Cl_2NO_3$ MW 284.10

2,4-Dichlorophenyl para-nitrophenyl ether, CA 1836-75-5: Toxic; flammable;
forms explosive peroxides; slightly soluble water; selective   (continues)

$C_{12}H_7Cl_2NO_3$ *(continued)*

herbicide. Tm = 343.15 to 344.15; Antoine (liq) A = 9.55385, B = 4723.397, C = 0, dev = 1.0 to 5.0, range = 328 to 403.

$C_{12}H_8$ MW 152.20

Acenaphthylene, CA 208-96-8: Flammable. Tm = 365.15 to 366.15; Tb = 538.15 to 548.15, decomposes; Antoine (sol) A = 9.500, B = 3714, C = 0, dev = 1.0, range = 286 to 318.

Biphenylene, CA 259-79-0: Flammable. Tm = 383.15; Antoine (sol) A = 17.385, B = 6755.53, C = 0, dev = 1.0 to 5.0, range = 371 to 381.

$C_{12}H_8Cl_2$ MW 223.10

2,2'-Dichlorobiphenyl, CA 13029-08-8: Suspected carcinogen; flammable; soluble ethanol, acetone, benzene, chloroform. PCB, no longer produced. Tm = 333.65; Antoine (sol) A = 12.962, B = 5019, C = 0, dev = 1.0 to 5.0, range = 310 to 328.

4,4'-Dichlorobiphenyl, CA 2050-68-2: Suspected carcinogen; flammable; soluble ethanol, acetone, benzene, chloroform; PCB, no longer produced. Tm = 422.15 to 423.15; Tb = 588.15 to 592.15; Antoine (sol) A = 12.585, B = 5416, C = 0, dev = 1.0 to 5.0, range = 323 to 360.

$C_{12}H_8F_2$ MW 190.19

2,2'-Difluorobiphenyl, CA 388-82-9: Irritant; flammable; soluble ethanol, ether, acetone, benzene. Tm = 390.15 to 390.65; Antoine (sol) A = 13.005, B = 4967, C = 0, dev = 1.0 to 5.0, range = 301 to 319.

4,4'-Difluorobiphenyl, CA 398-23-2: Irritant; flammable; soluble ethanol, ether, chloroform, benzene. Tm = 367.15 to 368.15; Tb = 527.15 to 528.15; Antoine (sol) A = 12.709, B = 4771, C = 0, dev = 1.0 to 5.0, range = 294 to 319.

$C_{12}H_8N_2$ MW 180.21

Benzo*(c)*cinnoline, CA 230-17-1: Flammable; slightly soluble water; soluble ethanol, ether, benzene. Tm = 429.15; Tb = above 633.15; Antoine (sol) A = 11.4709, B = 5311.4, C = 0, dev = 1.0, range = 319 to 359.

Phenazine, CA 92-82-0: Highly toxic orally; soluble hot ethanol, ether, benzene. Tm = 449.15 to 450.15; Tb = above 633.15; Antoine I (sol) A = 6.11684, B = 2471.28, C = -83.621, dev = 1.0, range = 280 to 318; Antoine II (liq) A = 10.43658, B = 4721.636, C = 0, dev = 0.1 to 1.0, range = 303 to 328.

$C_{12}H_8N_2O_4$ MW 244.21

4,4'-Dinitrobiphenyl, CA 1528-74-1: Suspected carcinogen; soluble ethanol, ether, benzene, acetic acid. Tm = 512.15 to 512.65; Antoine (sol) A = 9.8689, B = 5458.2, C = 0, dev = 1.0 to 5.0, range = 411 to 429.

$C_{12}H_8O$ MW 168.19

Dibenzofuran, diphenylene oxide, CA 132-64-9: Flammable; slightly soluble water; soluble ethanol, ether, acetone. Tm = 359.15 to 360.15; Tb = 560.15; Vm = 0.1545 at 372.15; Antoine (liq) A = 5.8968, B = 1851.27, C = -82.64, dev = 1.0, range = 403 to 559.

$C_{12}H_8S$ MW 184.26

Dibenzothiophene, CA 132-65-0: Toxic; flammable; soluble ethanol, benzene. Tm = 372.15; Tb = 605.15 to 606.15; Antoine (liq) A = 7.18577, B = 3140.15, C = 0, dev not specified, range = 385 to 574.

$C_{12}H_9Br$ MW 233.11

4-Bromobiphenyl, CA 92-66-0: Irritant; flammable; soluble ethanol, ether, benzene. Tm = 364.65 to 365.15; Tb = 583.15; Antoine (liq) A = 6.24643, B = 2174.97, C = -70.067, dev = 1.0, range = 371 to 583.

$C_{12}H_9BrO$ MW 249.11

4-Bromodiphenyl ether, CA 101-55-3: Irritant; flammable, flash point above 383; forms explosive peroxides; soluble ether. Tm = 291.86; Tb = 578.15; Vm = 0.1548 at 293.15; Antoine (liq) A = 5.80633, B = 1683.84, C = -140.25, dev = 1.0, range = 463 to 673.

2-Bromo-4-phenylphenol, CA 92-03-5: Irritant; flammable; soluble ethanol, chloroform. Tm = 368.15 to 369.15; Antoine (liq) A = 6.95688, B = 2831.69, C = -12.124, dev = 1.0 to 5.0, range = 373 to 584.

$C_{12}H_9Cl$ MW 188.66

2-Chlorobiphenyl, CA 2051-60-7: Suspected carcinogen; flammable; soluble ethanol, ether, light hydrocarbons. Tm = 306.65; Tb = 546.15 to 547.15; Antoine I (liq) A = 9.5876, B = 5125.26, C = 135.232, dev = 1.0, range = 362 to 541; Antoine II (liq) A = 7.4309, B = 3018, C = 0, dev = 1.0 to 5.0, range = 409 to 540.

4-Chlorobiphenyl, CA 2051-62-9: Toxic; flammable; soluble ethanol, ether, light hydrocarbons. Tm = 350.15; Tb = 564.15; Antoine I (liq) A = 7.1989, B = 2868.98, C = -13.566, dev = 1.0, range = 369 to 566; Antoine II (liq) A = 8.1619, B = 3445, C = 0, dev = 1.0 to 5.0, range = 451 to 536.

$C_{12}H_9ClO$ MW 204.66

2-Chloro-3-phenylphenol: Irritant; toxic; flammable; soluble ethanol, acetone, ether, benzene. Tm = 279.15; Tb = 590.15 to 591.15; Vm = 0.165 at 298.15; Antoine (liq) A = 7.27476, B = 2976.26, C = -25.869, dev = 1.0, range = 391 to 591.

2-Chloro-6-phenylphenol, CA 85-97-2: Irritant; toxic; flammable; soluble ethanol, ether, acetone, benzene. Tm = 348.15 to 349.15; Antoine (liq) A = 7.11025, B = 2779, C = -45.968, dev = 1.0, range = 393 to 590.

$C_{12}H_9N$ MW 167.21

Carbazole, dibenzopyrrole, CA 86-74-8: Soluble hot ethanol, acetone, pyridine. Tm = 517.95; Tb = 627.85; Antoine I (sol) A = 10.1069, B = 4780, C = 0, dev = 5.0, range not specified; Antoine II (liq) A = 6.21123, B = 2179.424, C = -109.636, dev = 1.0, range = 525 to 631.

$C_{12}H_9NS$ MW 199.27

Phenothiazine, CA 92-84-2: Soluble ethanol, ether, acetone, benzene. Tm = 455.15; Tb = 644.15; Antoine (sol) A = 8.390, B = 4490, C = 0, dev = 5.0, range = 336 to 395.

$C_{12}H_9N_3O_3$ MW 243.22

4-Nitro-4'-hydroxyazobenzene, CA 1435-60-5: Flammable; potentially explosive. Tm = 492.15 to 492.65; Antoine (sol) A = 14.525, B = 7510, C = 0, dev = 5.0, range = 417 to 444.

$C_{12}H_9N_3O_4$ MW 259.22

2,4-Dinitrodiphenylamine, CA 961-68-2: Irritant; soluble ethanol, acetone, chloroform, pyridine. Tm = 430.15; Antoine (sol) A = 15.825, B = 7710, C = 0, dev = 5.0, range = 402 to 420.

$C_{12}H_9N_3O_5$ MW 275.22

2,4-Dinitro-4'-hydroxydiphenylamine, CA 119-15-3: Irritant; flammable. Antoine (sol) A = 14.925, B = 8160, C = 0, dev = 5.0, range = 440 to 470.

$C_{12}H_{10}$ MW 154.21

Acenaphthene, CA 83-32-9: Irritant; suspected carcinogen; flammable; soluble benzene, chloroform. Tm = 369.15; Tb = 552.15; Vm = 0.1506 at 372.15; Antoine I (sol) A = 10.883, B = 4290.5, C = 0, dev = 1.0, range = 290 to 311; Antoine II (sol) A = 9.4944, B = 3248.008, C = -48.055, dev = 0.1, range = 338 to 366; Antoine III (liq) A = 6.3539, B = 2082.356, C = -71.578, dev = 0.1, range = 368 to 413; Antoine IV (liq) A = 7.30401, B = 2975, C = 10.674, dev = 1.0, range = 388 to 552.

Biphenyl, CA 92-52-4: Irritant; toxic vapor; flammable, flash point 386; soluble ethanol, ether, benzene. Tm = 342.35; Tb = 528.35; Tc = 789.3; Pc = 3847; Vc = 0.502; Antoine I (sol) A = 11.71929, B = 4341.054, C = 0, dev = 1.0, range = 297 to 324; Antoine II (sol) A = 28.5175, B = 21141.5, C = 374.85, dev = 1.0, range = 283 to 342; Antoine III (liq) A = 6.37526, B = 1974.8, C = -75.85, dev = 1.0, range = 390 to 563.

$C_{12}H_{10}N_2$ MW 182.22

Azobenzene, *cis*, CA 1080-16-6: Suspected carcinogen; flammable; soluble ethanol, ether, benzene. Tm = 344.15; Antoine I (sol) A = 12.1899, B = 4853.5, C = 0, dev = 5.0, range = 273 to 323; Antoine II (sol) A = 8.7769, B = 3914, C = 0, dev = above 5.0, range = 303 to 333.

(continues)

$C_{12}H_{10}N_2$ *(continued)*

Azobenzene, *trans*, CA 17082-12-1: Suspected carcinogen; flammable; soluble ethanol, ether, benzene. Tm = 341.15; Tb = 566.15; Antoine I (sol) A = 12.4689, B = 4901.7, C = 0, dev = 5.0, range = 273 to 315; Antoine II (sol) A = 8.8459, B = 3911, C = 0, dev = above 5.0, range = 303 to 333; Antoine III (liq) A = 7.51528, B = 3075.9, C = -7.883, dev = 1.0 to 5.0, range = 376 to 566.

$C_{12}H_{10}N_2O_2$ MW 214.22

2-Nitrodiphenylamine, CA 119-75-5: Irritant; flammable; soluble ethanol. Tm = 348.65; Antoine (sol) A = 12.025, B = 5270, C = 0, dev = 5.0, range = 335 to 346.

4-Nitrodiphenylamine, CA 836-30-6: Irritant; flammable; soluble ethanol. Tm = 407.15; Antoine (sol) A = 14.225, B = 6820, C = 0, dev = 5.0, range = 382 to 403.

$C_{12}H_{10}N_4O_2$ MW 242.24

4'-Nitro-4-aminoazobenzene, disperse orange 3, CA 730-40-5: Irritant. Tm = 489.15; Antoine (sol) A = 13.350, B = 7181, C = 0, dev = 5.0, range = 404 to 424.

$C_{12}H_{10}N_4O_4$ MW 274.24

2,4-Dinitro-4'-aminodiphenylamine, disperse yellow 9, CA 6373-73-5: Irritant. Tm = 462.15; Antoine (sol) A = 15.025, B = 8180, C = 0, dev = 5.0, range = 437 to 460.

$C_{12}H_{10}O$ MW 170.21

1-Acetylnaphthalene, CA 941-98-0: Flammable; soluble ethanol, ether, acetone; used in perfumery. Tm = 307.15; Tb = 569.15 to 571.15; Antoine (liq) A = 8.32623, B = 3696.46, C = 16.23, dev = 1.0, range = 388 to 569.

2-Acetylnaphthalene, CA 93-08-3: Flammable; soluble ethanol, ether, acetone; used in perfumery. Tm = 329.15; Tb = 574.15; Antoine I (sol) A = 11.278, B = 4589.9, C = 0, dev = 1.0 to 5.0, range = 294 to 316; Antoine II (liq) A = 7.47556, B = 2810.2, C = -60.281, dev = 1.0, range = 393 to 574.

Diphenyl ether, phenyl ether, CA 101-84-8: Moderately toxic by inhalation; flammable, flash point 385; soluble ethanol, ether, benzene; heat transfer medium. Tm = 310.15 to 312.15; Tb = 532.15; Antoine I (liq) A = 8.7091, B = 3351.9, C = 0, dev = 1.0, range = 313 to 333; Antoine II (liq) A = 6.13647, B = 1800.743, C = -95.275, dev = 0.02 to 0.1, range = 477 to 544.

2-Phenylphenol, CA 90-43-7: Irritant; flammable, flash point 397; soluble ethanol, ether, acetone. Tm = 329.15; Tb = 548.15; Antoine I (liq) A = 10.8635, B = 4326.754, C = 0, dev = 1.0, range = 291 to 314; Antoine II (liq) A = 4.1553, B = 547.8, C = -298.55, dev = 5.0, range = 434 to 547.

4-Phenylphenol, CA 92-69-3: Irritant; flammable, flash point 438; soluble ethanol, ether, acetone. Tm = 437.15 to 438.15; Tb = 578.15 to 581.15; Antoine I (sol) A = 11.17513, B = 5066.004, C = 0, dev = 1.0,   *(continues)*

$C_{12}H_{10}O$ *(continued)*

range = 327 to 348; Antoine II (liq) A = 8.41978, B = 3684.9, C = -5.81, dev = 1.0, range = 450 to 581.

$C_{12}H_{10}O_2$ MW 186.21

2,2'-Biphenyldiol, CA 1806-29-7: Flammable; soluble ethanol, ether, acetone, benzene. Tm = 382.15; Tb = 598.15 to 599.15; Antoine (liq) A = 7.78079, B = 3601.2, C = 26.22, dev = 1.0, range = 444 to 598.

3-Phenoxyphenol, CA 713-68-8: Irritant; toxic; flammable, flash point 383. Tm = 315.15; Antoine (liq) A = 8.1349, B = 3630, C = 0, dev = 1.0 to 5.0, range = 416 to 494.

$C_{12}H_{10}O_4$ MW 218.21

Quinhydrone, CA 106-34-3: Irritant; toxic; soluble hot water, ethanol, ether. Tm = 444.15; Antoine (sol) A = 11.302, B = 4656.8, C = 0, dev = 1.0, range = 317 to 334.

$C_{12}H_{10}S$ MW 186.27

Diphenyl sulfide, CA 139-66-2: Toxic; stench; flammable, flash point above 383; soluble ether, benzene, carbon disulfide. Tm = 247.25; Tb = 569.15; Vm = 0.1673 at 293.15; Antoine (liq) A = 7.17935, B = 2869.01, C = -11.056, dev = 1.0, range = 369 to 566.

$C_{12}H_{10}S_2$ MW 218.33

Diphenyl disulfide, CA 882-33-7: Toxic; stench; flammable; soluble ethanol, ether, benzene. Tm = 335.15 to 336.15; Tb = 583.15; Antoine (liq) A = 8.42717, B = 3667.3, C = -11.913, dev = 1.0, range = 404 to 583.

$C_{12}H_{11}N$ MW 169.23

Diphenylamine, CA 122-39-4: Irritant; toxic; suspected carcinogen; flammable, flash point 426; soluble ethanol, ether, acetone, benzene. Tm = 327.15 to 328.15; Tb = 575.15; Antoine I (sol) A = 12.704, B = 5043.9, C = 0, dev = 5.0, range = 298 to 323; Antoine II (liq) A = 7.15045, B = 2778.28, C = -35.102, dev = 1.0, range = 381 to 575; Antoine III (liq) A = 6.5746, B = 2430.7, C = -43.15, dev = 1.0, range = 573 to 673.

$C_{12}H_{11}NO$ MW 185.23

1-Naphthylamine, *N*-acetyl, CA 575-36-0: Irritant; flammable; moderately soluble hot water; soluble ethanol. Tm = 433.15; Antoine (sol) A = 10.09143, B = 4914.677, C = 0, dev = 1.0, range = 337 to 360.

$C_{12}H_{11}N_3$ MW 197.24

4-Aminoazobenzene, CA 60-09-3: Highly toxic; suspected carcinogen; soluble ethanol, ether, chloroform, benzene. Tm = 399.15; Tb = above 633.15; Antoine (sol) A = 12.5339, B = 5802, C = 0, dev = 1.0 to 5.0, range = 356 to 373.

364

$C_{12}H_{12}$ MW 156.23

1,8-Dimethylnaphthalene, CA 569-41-5: Flammable; soluble ether, benzene.
Tm = 336.65 to 337.65; Tb = 543.15; Antoine I (sol) A = 10.56808, B =
4066.836, C = 0, dev = 1.0, range = 328 to 336; Antoine II (liq) A =
6.54778, B = 2197.524, C = -63.88, dev = 1.0, range = 338 to 547.

2,3-Dimethylnaphthalene, guaiene, CA 581-40-8: Flammable; soluble benzene,
ether. Tm = 377.15 to 377.65; Tb = 538.15 to 539.15; Antoine I (sol) A =
10.635, B = 4172.6, C = 0, dev = 1.0, range = 278 to 301; Antoine II (sol)
A = 8.97875, B = 2959.733, C = -59.936, dev = 0.1, range = 333 to 373;
Antoine III (liq) A = 5.57091, B = 1544.764, D = -116.821, dev = 1.0,
range = 378 to 408.

2,6-Dimethylnaphthalene, CA 581-42-0: Flammable; soluble ether, benzene.
Tm = 383.15 to 384.15; Tb = 534.15 to 535.15; Antoine I (sol) A = 11.290,
B = 4386.4, C = 0, dev = 1.0, range = 278 to 304; Antoine II (sol) A =
8.45107, B = 2512.509, C = -89.765, dev = 0.1, range = 348 to 383; Antoine
III (liq) A = 5.18084, B = 1320.21, C = -133.876, dev = 0.1, range 384
to 418.

2,7-Dimethylnaphthalene, CA 582-16-1: Flammable; soluble ether, benzene.
Tm = 369.15 to 370.15; Tb = 535.15; Antoine I (sol) A = 9.40197, B =
3047.828, C = -58.898, dev = 1.0, range = 333 to 368; Antoine II (liq) A =
6.4818, B = 2092.928, C = -66.181, dev = 1.0, range = 369 to 535.

1-Ethylnaphthalene, CA 1127-76-0: Flammable, flash point 384; soluble
ether, ethanol. Tm = 259.27; Tb = 531.82; Vm = 0.1550 at 293.15; Antoine
(liq) A = 6.15645, B = 1841.32, C = -87.87, dev = 0.1, range = 393 to 565.

2-Ethylnaphthalene, CA 939-27-5: Flammable, flash point 377; soluble
ethanol, ether. Tm = 265.65; Tb = 525.15; Vm = 0.1575 at 293.15; Antoine I
(liq) A = 7.46683, B = 3232.791, C = 0, dev = 1.0, range = 286 to 319;
Antoine II (liq) A = 6.20056, B = 1880.73, C = -82.74, dev = 1.0, range =
393 to 565.

$C_{12}H_{12}N_2$ MW 184.24

1,1-Diphenylhydrazine, CA 530-50-7: Toxic; flammable; soluble ether, ethanol, benzene, chloroform. Tm = 317.15; Antoine (liq) A = 7.2366, B =
2890.42, C = -42.726, dev = 1.0, range = 399 to 596.

$C_{12}H_{12}O_6$ MW 252.22

1,2,3-Benzenetricarboxylic acid, trimethyl ester, CA 2672-57-3: Flammable.
Tm = 375.15; Antoine (liq) A = 7.452, B = 3121, C = -43.15, dev = 5.0,
range = 453 to 513.

1,2,4-Benzenetricarboxylic acid, trimethyl ester, CA 2459-10-1: Flammable,
flash point above 383. Tf = 260.15; Antoine (liq) A = 6.437, B = 2620,
C = -43.15, dev = 5.0, range = 443 to 493.

1,3,5-Benzenetricarboxylic acid, trimethyl ester, CA 2672-58-4: Flammable.
Tm = 417.15; Antoine (liq) A = 7.750, B = 3230, C = -43.15, dev = 5.0,
range = 443 to 513.

$C_{12}H_{13}Cl_3O_3$ MW 311.59

2,4,5-Trichlorophenoxyacetic acid, butyl ester, CA 93-79-8: Toxic; flammable; used as weed killer. Antoine (liq) A = 5.33165, B = 1364.221, C = -214.921, dev = 1.0, range = 460 to 573.

$C_{12}H_{14}Cl_2O_3$ MW 277.15

2,4-Dichlorophenoxyacetic acid, butyl ester, CA 94-80-4: Toxic; flammable; used as weed killer. Antoine (liq) A = 6.00451, B = 1862.487, C = -145.062, dev = 1.0, range = 444 to 573.

2,4-Dichlorophenoxyacetic acid, sec-butyl ester, CA 94-79-1: Toxic; flammable; used as weed killer. Antoine (liq) A = 6.36811, B = 2120.36, C = -119.288, dev = 1.0, range = 444 to 573.

$C_{12}H_{14}Cl_2O_4$ MW 293.15

2,4-Dichlorophenoxyacetic acid, 2-ethoxyethyl ester, CA 74944-83-5: Toxic; flammable; used as weed killer. Antoine (liq) A = 6.97502, B = 3317.429, C = 0, dev = 5.0, range = 443 to 503.

2,4-Dichlorophenoxyacetic acid, 4-hydroxybutyl ester, CA 36227-43-7: Toxic; flammable; used as weed killer. Antoine (liq) A = 7.53016, B = 3764.686, C = 0, dev = 5.0, range = 443 to 503.

$C_{12}H_{14}N_2O_5$ MW 266.25

2-Cyclohexyl-4,6-dinitrophenol, CA 131-89-5: Toxic insecticide; fungicide; flammable; slightly soluble water; soluble benzene. Tm = 377.15; Antoine (liq) A = 7.57383, B = 2539.1, C = -108.675, dev = 1.0, range = 405 to 565.

$C_{12}H_{14}O_3$ MW 206.24

Eugenol acetate, CA 93-28-7: Flammable; soluble ethanol. Tm = 303.15 to 304.15; Tb = 553.15 to 554.15; Antoine (liq) A = 8.10829, B = 3500.29, C = 18.248, dev = 1.0, range = 374 to 555.

$C_{12}H_{14}O_4$ MW 222.24

Apiol, CA 523-80-8: Flammable; soluble most organic solvents. Tm = 303.15; Tb = 567.15; Antoine (liq) A = 8.34417, B = 3457.62, C = -12.739, dev = 1.0 to 5.0, range = 389 to 558.

Diethyl phthalate, CA 84-66-2: Irritant; flammable, flash point 434; soluble ethanol, ether, acetone, benzene. Tm = 272.15; Tb = 571.15; Vm = 0.1989 at 293.15; Antoine I (liq) A = 6.04308, B = 1866.05, C = -115.9, dev = 1.0, range = 345 to 453. Antoine II (liq) A = 10.6902, B = 6768.3, C = 209.45, dev = 5.0, range = 421 to 570.

$C_{12}H_{15}N$ MW 173.26

N,N-Diallyl aniline, CA 6247-00-3: Toxic; lachrymator; flammable. Antoine (liq) A = 7.5803, B = 2865, C = 0, dev = 5.0, range = 421 to 513.

$C_{12}H_{15}N_3O_2$ MW 233.27

3,6-Bis(dimethylamino) phthalimide, CA 5972-07-6: Antoine (sol) A = 10.8209, B = 5485, C = 0, dev not specified, range = 400 to 457.

$C_{12}H_{15}N_3O_6$ MW 297.27

2,4,6-Trinitro-1,3-dimethyl-5-*tert*-butylbenzene, musk xylene, CA 81-15-2: Flammable; soluble ethanol, ether; used in perfumery. Tm = 385.15 to 386.15; Antoine (sol) A = 11.7359, B = 5245, C = 0, dev = 1.0, range = 312 to 348.

$C_{12}H_{16}$ MW 160.26

Cyclohexylbenzene, CA 827-52-1: Flammable, flash point 372; soluble ethanol, ether. Tm = 280.15 to 281.15; Tb = 512.15 to 513.15; Vm = 0.1687 at 293.15; Antoine (liq) A = 5.94132, B = 1651.85, C = -93.571, dev = 0.1, range = 421 to 513.

Dicyclohexadiene: Flammable; soluble ether, acetone, benzene. Tb = 502.15 to 503.15; Antoine (liq) A = 3.6488, B = 383.37, C = -270.86, dev = 1.0, range = 377 to 505.

2,5-Diethylstyrene, CA 2715-29-9: Irritant; lachrymator; flammable; polymerizes. Antoine (liq) A = 6.82557, B = 2239.72, C = -31.595, dev = 1.0, range = 322 to 496.

1-Isopropenyl-4-isopropylbenzene, CA 2388-14-9: Flammable; polymerizes; soluble most organic solvents. Tm = 242.54; Antoine (liq) A = 6.73904, B = 2137.1, C = -43.15, dev = 0.1, range = 403 to 479.

$C_{12}H_{16}N_2O_5$ MW 268.27

2,6-Dinitro-1-methyl-3-methoxy-4-*tert*-butylbenzene, musk ambrette, CA 83-66-9: Flammable; can cause dermatitis. Tm = 360.15; Antoine (sol) A = 12.5649, B = 5372, C = 0, dev = 1.0, range = 303 to 346.

$C_{12}H_{16}N_3O_3PS_2$ MW 345.37

Azinphos-ethyl, CA 2642-71-9: Toxic insecticide. Tm = 326.15; Antoine (sol) A = 7.9308, B = 4532, C = 0, dev = 1.0, range = 326 to 420.

$C_{12}H_{16}O_2$ MW 192.26

Isopentyl benzoate, CA 94-46-2: Flammable; soluble ethanol. Tb = 533.15 to 535.15; Vm = 0.1915 at 293.15; Antoine (liq) A = 7.97383, B = 3487.08, C = 49.176, dev = 5.0, range = 345 to 535.

Pentyl benzoate, CA 2049-96-9: Flammable; soluble ethanol. Antoine (liq) A = 3.703, B = 313.2, C = -300.75, dev = 5.0, range = 395 to 492.

2-Phenylbutyric acid, ethyl ester, CA 119-43-7: Flammable. Antoine (liq) A = 6.85666, B = 2235.7, C = -52.58, dev = 1.0, range = 404 to 489.

$C_{12}H_{16}O_3$ MW 208.26

Isopentyl salicylate, CA 87-20-7: Flammable, flash point 405; soluble ethanol, chloroform. Tb = 549.15 to 550.15; Vm = 0.1977 at 293.15; Antoine (liq) A = 9.616, B = 3816, C = 0, dev = 1.0, range = 287 to 329.

Pentyl salicylate, CA 2050-08-0: Flammable; soluble ethanol, ether. Tb = 538.15; Vm = 0.1955 at 288.15; Antoine (liq) A = 8.4733, B = 3475, C = 0, dev = 5.0, range = 402 to 540.

$C_{12}H_{17}NO$ MW 191.27

N-Butylacetanilide, CA 91-49-6: Flammable; soluble chloroform. Tm = 297.65; Tb = 554.15; Antoine (liq) A = 6.45158, B = 2085.31, C = -85.07, dev = 0.1, range = 443 to 653.

N,N-Diethyl-2-phenyl acetamide, CA 2431-96-1: Irritant; flammable. Tm = 359.15; Tb = 570.15; Antoine (liq) A = 9.6932, B = 4325.3, C = 0, dev = 5.0, range = 404 to 570.

N,N-Diethyl-meta-toluamide, CA 134-62-3: Irritant; flammable, flash point above 383; soluble water, ethanol, ether, benzene. Vm = 0.1920 at 293.15; Antoine (liq) A = 4.38144, B = 1679.8, C = 0, dev = 5.0, range = 373 to 403.

$C_{12}H_{18}$ MW 162.27

Cyclododecatriene, 1-cis, 5-trans, 9-trans, CA 4904-61-4: Corrosive; flammable; soluble most organic solvents. Antoine I (liq) A = 6.2845, B = 2014.782, C = -43.15, dev = 1.0 to 5.0, range = 344 to 387; Antoine II (liq) A = 7.6403, B = 2512.659, C = -43.15, dev = 1.0 to 5.0, range = 400 to 423; Antoine III (liq) A = 6.4189, B = 2031.837, C = -43.15, dev = 1.0 to 5.0, range = 426 to 503.

Cyclododecatriene, 1-trans, 5-trans, 9-cis, CA 2765-29-9: Corrosive; flammable, flash point 360; soluble most organic solvents. Tm = 255.15; Tb = 504.15; Vm = 0.1711 at 293.15; Antoine (liq) A = 9.9859, B = 3552, C = 0, dev = 1.0, range = 286 to 373.

Cyclododecatriene, 1-trans, 5-trans, 9-trans, CA 676-22-2: Corrosive; flammable, flash point 354; soluble most organic solvents. Tm = 307.15; Tb = 510.15 to 511.15; Antoine (sol) A = 11.3309, B = 3930, C = 0, dev = 1.0, range = 273 to 307.

1,2-Diisopropylbenzene, CA 577-55-9: Flammable; soluble most organic solvents. Tm = 216.5; Tb = 477.15; Vm = 0.1865 at 293.15; Antoine (liq) A = 6.1049, B = 1619.5, C = -81.82, dev = 1.0, range = 388 to 476.

1,3-Diisopropylbenzene, CA 99-62-7: Flammable, flash point 349; soluble most organic solvents. Tm = 210.05; Tb = 476.35; Vm = 0.1896 at 293.15; Antoine (liq) A = 6.1054, B = 1616.6, C = -82.0, dev = 1.0, range = 387 to 477.

1,4-Diisopropylbenzene, CA 100-18-5: Flammable, flash point 349; soluble most organic solvents. Tm = 256.15; Tb = 483.45; Vm = 0.1894 at 293.15; Antoine (liq) A = 7.14678, B = 2485.33, C = 0, dev = 1.0, range = 393 to 485.

Hexamethylbenzene, mellitene, CA 87-85-4: Flammable; soluble most organic solvents. Tm = 437.15; Tb = 537.15; Antoine I (sol) A = *(continues)*

$C_{12}H_{18}$ *(continued)*

8.6223, B = 2965.633, C = -59.583, dev = 0.1, range = 303 to 343; Antoine II (liq) A = 5.89588, B = 1629.9, C = -118.46, dev = 1.0, range = 443 to 537.

2-Isopropenyl-1-methyl-1-vinyl-3-cyclohexene, CA 6902-73-4: Irritant; flammable; polymerizes; soluble most organic solvents. Antoine (liq) A = 7.3077, B = 2497.2, C = 0, dev = 1.0, range = 348 to 404.

1,2,4-Triethylbenzene, CA 877-44-1: Flammable, flash point 356; soluble most organic solvents. Tm = 228.25; Tb = 490.15 to 491.15; Vm = 0.1857 at 293.15; Antoine (liq) A = 6.83785, B = 2237.98, C = -28.303, dev = 1.0, range = 319 to 491.

$C_{12}H_{18}Cl_2NOPS$ MW 326.22

P-(Chloromethyl)-*N*-(1-methylpropyl)amidothiophosphonic acid, *O*-(2-chloro-4-methylphenyl) ester, CA 42585-08-0: Toxic; flammable. Antoine (liq) A = 6.1828, B = 3270, C = 0, dev = 1.0 to 5.0, range = 309 to 363.

$C_{12}H_{18}O$ MW 178.27

Benzyl pentyl ether, CA 6382-14-5: Flammable; forms explosive peroxides. Antoine (liq) A = 7.14347, B = 2571.7, C = -5.85, dev = 1.0, range = 363 to 513.

2,4-Diisopropylphenol, CA 2934-05-6: Toxic; flammable; soluble ethanol, ether, benzene. Antoine (liq) A = 6.74812, B = 2213.9, C = -60.73, dev = 1.0, range = 395 to 528.

2,3-Dimethyl-4-*tert*-butylphenol, CA 68189-19-5: Toxic; flammable; soluble ethanol, ether, benzene. Tb = 532.15; Antoine (liq) A = 7.0971, B = 2451, C = -50.75, dev = 5.0, range = 418 to 532.

2,3-Dimethyl-6-*tert*-butylphenol, CA 46170-85-8: Toxic; flammable; soluble ethanol, ether, benzene. Tm = 325.65; Antoine (liq) A = 6.6320, B = 2053.7, C = -81.25, dev = 5.0, range = 412 to 525.

2,4-Dimethyl-6-*tert*-butylphenol, CA 1879-09-0: Irritant; toxic; flammable, flash point 384; soluble ethanol, ether, benzene. Tm = 295.15; Tb = 523.15; Vm = 0.1944 at 353.15; Antoine (liq) A = 5.92906, B = 1620.9, C = -109.17, dev = 1.0 to 5.0, range = 388 to 522.

2,5-Dimethyl-4-*tert*-butylphenol, CA 17696-37-6: Toxic; flammable; soluble ethanol, ether, benzene. Tm = 344.15; Tb = 537.15; Vm = 0.1899 at 353.15; Antoine (liq) A = 7.69338, B = 2953, C = -17.95, dev = 1.0 to 5.0, range = 408 to 538.

2,6-Dimethyl-4-*tert*-butylphenol, CA 879-97-0: Toxic; flammable; soluble ethanol, ether, benzene. Tm = 356.15; Tb = 521.15; Antoine (liq) A = 6.63751, B = 2062, C = -75.81, dev = 1.0 to 5.0, range = 392 to 522.

3,4-Dimethyl-6-*tert*-butylphenol, CA 1445-23-4: Toxic; flammable; soluble ethanol, ether, benzene. Tm = 319.15; Vm = 0.1938 at 353.15; Antoine (liq) A = 6.02921, B = 1631.9, C = -125.59, dev = 1.0, range = 413 to 532.

(continues)

$C_{12}H_{18}O$ *(continued)*

2-Ethyl-4-*tert*-butylphenol, CA 63452-61-9: Toxic; flammable; soluble ethanol, ether, benzene. Tb = 530.15; Antoine (liq) A = 6.7495, B = 2120.5,
C = -83.15, dev = 5.0, range = 428 to 623.

2-Ethyl-6-*tert*-butylphenol, CA 63551-41-7: Toxic; flammable; soluble ethanol, ether, benzene. Antoine (liq) A = 7.9808, B = 3038.2, C = 0, dev =
5.0, range = 393 to 443.

3-Ethyl-6-*tert*-butylphenol, CA 4237-25-6: Toxic; flammable; soluble ethanol, ether, benzene. Antoine (liq) A = 6.8219, B = 2242.4, C = -64.55,
dev = 5.0, range = 415 to 530.

4-Ethyl-2-*tert*-butylphenol, CA 96-70-8: Toxic; flammable; soluble ethanol,
ether, benzene. Tb = 523.15; Antoine (liq) A = 6.9869, B = 2340.7, C =
-53.18, dev = 1.0, range = 394 to 523.

2-Methyl-4-*tert*-pentylphenol, CA 71745-63-6: Toxic; flammable; soluble
ethanol, ether, benzene. Antoine (liq) A = 6.9131, B = 2257.4, C = -86.15,
dev = 5.0, range = 443 to 653.

3-Methyl-4-*tert*-pentylphenol: Toxic; flammable; soluble ethanol, benzene,
ether. Antoine (liq) A = 6.87527, B = 2240, C = -86.15, dev = 5.0, range =
443 to 683.

4-Methyl-2-*tert*-pentylphenol, CA 34072-71-4: Toxic; flammable; soluble
ethanol, ether, benzene. Tm = 310.15; Antoine (liq) A = 6.77995, B = 2115,
C = -82.15, dev = 5.0, range = 423 to 653.

$C_{12}H_{18}O_2$ MW 194.27

1,3-Dihydroxy-2-hexylbenzene, 2-hexylresorcinol, CA 5673-09-6: Irritant;
flammable; soluble ethanol, acetone. Tm = 368.15 to 371.15; Antoine (liq)
A = 4.4547, B = 998.64, C = -224.15, dev = 1.0, range = 433 to 494.

1,3-Dihydroxy-4-hexylbenzene, 4-hexylresorcinal, CA 136-77-6: Irritant;
suspected carcinogen; flammable; soluble ethanol, acetone, ether, chloroform. Tm = 341.15 to 343.15; Tb = 606.15 to 608.15; Antoine (liq) A =
5.4052, B = 1445.9, C = -197.03, dev = 1.0, range = 434 to 494.

$C_{12}H_{18}O_4$ MW 226.27

3,4-Dihydro-2,2-dimethyl-4-oxo-2*H*-pyran-6-carboxylic acid, butyl ester,
CA 532-34-3: Flammable liquid; used as insect repellant. Tb = 529.15 to
543.15; Vm = 0.2143 at 293.15; Antoine (liq) A = 7.7983, B = 3378, C = 0,
dev = 1.0, range = 357 to 435.

$C_{12}H_{18}O_6$ MW 258.27

Aconitic acid, triethyl ester, CA 5349-99-5: Flammable; soluble ethanol,
ether. Tb = 548.15, decomposes; Vm = 0.2334 at 293.15; Antoine (liq) A =
9.69645, B = 4159.825, C = 0, dev = 1.0 to 5.0, range = 423 to 540.

$C_{12}H_{19}F_3N_2O_4$ MW 312.29

N[*N*-(Trifluoroacetyl)valyl]alanine, ethyl ester: Tm = 424.35; Antoine I
(sol) A = 13.395, B = 6036, C = 0, dev = 5.0, range = 323 to 424; Antoine
II (liq) A = 9.855, B = 4515, C = 0, dev = 5.0, range = 425 to 453.

$C_{12}H_{20}$ MW 164.29

1-Ethyladamantane, CA 770-69-4: Flammable. Tm = 215.75; Antoine (liq) A = 6.29135, B = 1854.18, C = -59.63, dev = 1.0, range = 383 to 492.

$C_{12}H_{20}O_2$ MW 196.29

Bornyl acetate, CA 76-49-3: Flammable, flash point 357; soluble ethanol, ether. Tm = 302.15; Tb = 496.15 to 497.15; Antoine (liq) A = 6.77225, B = 2234.78, C = -27.412, dev = 1.0, range = 319 to 496.

Geranyl acetate, CA 105-87-3: Flammable; soluble ethanol, ether. Antoine (liq) A = 7.63253, B = 2839.53, C = -11.931, dev = 1.0, range = 346 to 516.

Isobornyl acetate ±, CA 17283-45-3: Flammable; soluble ethanol, acetone. Vm = 0.1995 at 293.15; Antoine (liq) A = 4.63584, B = 813.29, C = -197.8, dev = 1.0, range = 404 to 450.

Linalyl acetate, CA 115-95-7: Flammable, flash point 363; soluble ethanol, ether. Tb = 493.15; Vm = 0.2193 at 293.15; Antoine (liq) A = 8.254, B = 3019, C = 0, dev = 5.0, range = 281 to 490.

Terpineol acetate, CA 80-26-2: Flammable; soluble ethanol, ether, benzene. Vm = 0.2032 at 293.15; Antoine (liq) A = 6.7143, B = 1983, C = -82.1, dev = 5.0, range = 310 to 424.

$C_{12}H_{20}O_4$ MW 228.29

Dibutyl maleate, CA 105-76-0: Flammable, flash point 383. Tb = 554.15; Vm = 0.2311 at 293.15; Antoine (liq) A = 13.04024, B = 9198.9, C = 288.06, dev = 5.0, range = 255 to 550.

$C_{12}H_{20}O_5$ MW 244.29

2-Ethoxycarbonylpropionic acid, cyclohexyl ester: Flammable. Antoine (liq) A = 8.4694, B = 3532.7, C = 0, dev = 1.0, range = 388 to 523.

$C_{12}H_{20}O_7$ MW 276.29

Triethyl citrate, ethyl citrate, CA 77-93-0: Flammable, flash point 383; soluble water, ethanol, ether. Tb = 567.15; Vm = 0.2430 at 293.15; Antoine (liq) A = 7.97481, B = 3522.76, C = 22.972, dev = 1.0, range = 380 to 567.

$C_{12}H_{21}N_2O_3PS$ MW 304.34

Diazinon, CA 333-41-5: Highly toxic insecticide; suspected teratogen; soluble most organic solvents. Vm = 0.2725 at 293.15; Antoine (liq) A = 10.6266, B = 4566, C = 0, dev = 5.0, range = 293 to 398.

$C_{12}H_{22}$ MW 166.31

Bicyclohexyl, *cis*, *cis*, CA 92-51-3: Flammable, flash point 374; slightly soluble water; soluble ethanol, ether. Tm = 277.15; Tb = 511.15; Vm = 0.1866 at 293.15; Antoine (liq) A = 6.6599, B = 2187.4, C = -40.53, dev = 1.0, range = 331 to 511.   *(continues)*

$C_{12}H_{22}$ *(continued)*

6-Dodecyne, CA 6975-99-1: Flammable; polymerizes; soluble ethanol, ether, acetone. Tb = 482.15; Vm = 0.2113 at 293.15; Antoine (liq) A = 8.7882, B = 3178.8, C = 0, dev = 1.0, range = 373 to 388.

$C_{12}H_{22}O$ MW 189.31

Cyclododecanone, CA 830-13-7: Flammable. Tm = 334.15; Vm = 0.2012 at 339.15; Antoine I (liq) A = 6.0938, B = 1837.26, C = -93.15, dev = 1.0, range = 373 to 443; Antoine II (liq) A = 6.28730, B = 2033.666, C = -75.98, dev = 0.1, range = 408 to 458; Antoine III (liq) A = 6.10384, B = 1891.116, C = -88.201, dev = 0.1, range = 450 to 556.

$C_{12}H_{22}O_2$ MW 198.30

Acetic acid, *para-tert*-butylcyclohexyl ester, CA 32210-23-4: Flammable; used in perfumery. Antoine (liq) A = 8.972, B = 3333, C = 0, dev = 1.0, range = 285 to 318.

Acetic acid, methol(+) ester, CA 16409-45-3: Flammable, flash point 365; slightly soluble water; soluble ethanol, ether. Tb = 500.15; Vm = 0.2158 at 293.15; Antoine (liq) A = 6.94767, B = 2283.94, C = -38.178, dev = 1.0, range = 330 to 500.

Citronellyl acetate, CA 150-84-5: Flammable. Antoine (liq) A = 8.12796, B = 2712.85, C = -47.079, dev = 1.0, range = 347 to 490.

2-(1-Ethylpentyl)-4,7-dihydro-1,3-dioxepin, CA 61732-97-6: Flammable. Antoine (liq) A = 9.1269, B = 3462, C = 0, dev = 1.0 to 5.0, range = 333 to 453.

Octyl methacrylate, CA 2157-01-9: Irritant; polymerizes; flammable. Antoine (liq) A = 7.676, B = 2907, C = 0, dev = above 5.0, range = 384 to 513.

10-Undecenoic acid, methyl ester, CA 111-81-9: Flammable; soluble ethanol, ether. Tf = 245.65; Tb = 521.15; Vm = 0.2231 at 288.15; Antoine (liq) A = 7.949, B = 3095, C = 0, dev = 5.0, range = 397 to 524.

$C_{12}H_{22}O_3$ MW 214.30

Heptyl levulinate: Flammable. Tb = 556.65; Vm = 0.2272 at 293.15; Antoine (liq) A = 7.03454, B = 2571.6, C = -46.21, dev = 1.0, range = 393 to 558.

3-Pentyl-4-acetoxytetrahydro-2*H*-pyran, CA 18871-14-2: Flammable. Antoine (liq) A = 8.5839, B = 3439, C = 0, dev = 1.0 to 5.0, range = 383 to 453.

$C_{12}H_{22}O_4$ MW 230.30

Adipic acid, dipropyl ester, CA 106-19-4: Flammable; soluble ethanol, ether, chloroform. Tm = 257.45; Vm = 0.2352 at 293.15; Antoine (liq) A = 8.1559, B = 3320, C = 0, dev = 1.0 to 5.0, range = 413 to 540.

Dodecanedioic acid, CA 693-23-2: Irritant; flammable. Tm = 402.15; Antoine (sol) A = 16.853, B = 8006, C = 0, dev = 5.0, range = 375 to 396.

(continues)

$C_{12}H_{22}O_4$ *(continued)*

Isopentylmalonic acid, diethyl ester, CA 5398-08-3: Flammable; soluble ethanol, ether, acetone. Tm = 375.15; Tb = 513.15 to 515.15; Antoine (liq) A = 8.309, B = 3348, C = 0, dev = 5.0, range = 377 to 420.

(1-Methylbutyl)malonic acid, diethyl ester, CA 22328-91-2: Flammable; soluble ethanol, ether, acetone. Antoine (liq) A = 5.89426, B = 1428.896, C = -148.501, dev = 1.0 to 5.0, range = 395 to 516.

Oxalic acid, diisopentyl ester, CA 2051-00-5: Flammable; soluble ethanol, ether. Tb = 540.15 to 541.15; Vm = 0.2378 at 284.15; Antoine (liq) A = 7.56278, B = 2953.91, C = -6.737, dev = 1.0, range = 358 to 538.

$C_{12}H_{22}O_4S$ MW 262.36

Thiodiglycolic acid, dibutyl ester, CA 4121-12-4: Irritant; flammable. Antoine (liq) A = 8.505, B = 3955, C = 0, dev = 5.0, range = 298 to 383.

$C_{12}H_{22}O_5$ MW 246.30

Butyl[1-(butoxycarbonyl)ethyl] carbonate: Toxic; flammable. Antoine (liq) A = 8.633, B = 3560, C = 0, dev = 5.0, range = 338 to 513.

Pentyl[1-ethoxycarbonyl)isopropyl] carbonate: Toxic; flammable. Antoine (liq) A = 8.402, B = 3332, C = 0, dev = 5.0, range = 368 to 513.

$C_{12}H_{22}O_6$ MW 262.30

Lactic acid, O-ethoxycarbonyl, 2-butoxyethyl ester: Flammable. Antoine (liq) A = 9.1457, B = 3896.9, C = 0, dev = 1.0, range = 383 to 521.

Tartaric acid, dibutyl ester, CA 87-92-3: Flammable, flash point above 383; soluble water, ethanol, acetone. Tm = 294.95; Tb = 593.15; Antoine (liq) A = 10.4202, B = 5665.5, C = 73.43, dev = 1.0 to 5.0, range = 428 to 511.

Tartaric acid(+), diisobutyl ester, CA 4054-82-4: Flammable; soluble ethanol. Tm = 346.65; Tb = 596.15 to 598.15; Vm = 0.2555 at 354.15; Antoine (liq) A = 6.87696, B = 2701.75, C = -42.501, dev = 1.0 to 5.0, range = 390 to 597.

$C_{12}H_{23}N$ MW 181.32

Dicyclohexylamine, CA 101-83-7: Corrosive; highly toxic; suspected carcinogen; skin irritant; flammable, flash point above 372; slightly soluble water; soluble most organic solvents. Tm = 293.15; Tb = 529.15, decomposes; Vm = 0.1988 at 293.15; Antoine (liq) A = 5.75276, B = 1587.7, C = -105.6, dev = 1.0, range = 408 to 529.

Lauronitrile, CA 2437-25-4: Irritant; flammable, flash point above 383; soluble ethanol, ether, acetone, benzene. Tm = 277.05; Tb = 550.15; Vm = 0.2200 at 293.15; Antoine (liq) A = 6.206196, B = 1859.981, C = -106.456, dev = 0.02, range = 440 to 556.

$C_{12}H_{24}$ MW 168.32

Cyclododecane, CA 294-62-2: Flammable. Tm = 334.15; Tb = 520.15; Vm = 0.2053 at 353.15; Antoine I (liq) A = 6.11502, B = 1817.481, C = -74.666, dev = 0.1, range = 386 to 441; Antoine II (liq) A = 5.98877, B = 1722.521, C = -84.722, dev = 0.02, range = 440 to 529.

1-Dodecene, CA 112-41-4: Flammable, flash point 350; soluble most organic solvents. Tm = 237.92; Tb = 487.01; Vm = 0.2219 at 293.15; Antoine (liq) A = 6.10139, B = 1621.45, C = -90.66, dev = 0.02, range = 396 to 493.

$C_{12}H_{24}N_2O_2$ MW 228.33

Dicyclohexyl ammonium nitrite, CA 3129-91-7: Used as corrosion inhibitor. Tm = 453.15; Antoine (sol) A = 12.815, B = 5178, C = 0, dev = 5.0, range = 290 to 298.

$C_{12}H_{24}O$ MW 184.32

Cyclododecanol, CA 1724-39-6: Flammable; soluble ethanol. Tm = 353.15; Antoine I (liq) A = 6.93787, B = 2289.071, C = -84.705, dev = 1.0, range = 405 to 468; Antoine II (liq) A = 7.50834, B = 3148.282, C = 13.153, dev = 1.0, range = 467 to 557.

Dodecanal, lauraldehyde, CA 112-54-9: Flammable, flash point 374; polymerizes; soluble ethanol, ether. Tm = 317.65; Antoine (liq) A = 7.73843, B = 3117.46, C = 13.541, dev = 1.0, range = 350 to 530.

2-Dodecanone, CA 6175-49-1: Flammable; soluble ethanol, ether, acetone. Tm = 294.15; Tb = 519.15 to 520.15; Antoine I (liq) A = 7.25097, B = 2510.02, C = -41.179, dev = 1.0 to 5.0, range = 350 to 420; Antoine II (liq) A = 6.19081, B = 1769.585, C = -101.407, dev = 1.0, range = 386 to 600.

Ethyl *para*-menthyl ether, CA 19321-39-2: Flammable; forms explosive peroxides. Antoine (liq) A = 7.5568, B = 2661.1, C = 0, dev = 1.0, range = 366 to 414.

1-Heptylcyclopentanol, CA 20999-39-7: Flammable; soluble ethanol. Antoine (liq) A = 7.3095, B = 2625.8, C = -30.15, dev = 5.0, range = 395 to 524.

1-Hexylcyclohexanol, CA 3964-63-4: Flammable; soluble ethanol. Antoine (liq) A = 8.2944, B = 2794.4, C = 0, dev = 5.0, range = 380 to 491.

$C_{12}H_{24}O_2$ MW 200.32

Decanoic acid, ethyl ester, CA 110-38-3: Flammable, flash point 375; soluble ethanol, ether, chloroform. Tm = 253.15; Tb = 516.15 to 518.15; Vm = 0.2324 at 293.15; Antoine (liq) A = 8.0869, B = 3112.732, C = 0, dev = 1.0 to 5.0, range = 359 to 515.

Decyl acetate, CA 112-17-4: Flammable; soluble ethanol, ether, benzene. Tm = 258.1; Tb = 517.15; Vm = 0.2310 at 293.15; Antoine (liq) A = 8.32159, B = 3231.008, C = 0, dev = 1.0 to 5.0, range = 363 to 515.

4,5-Dimethyl-2-(1-ethylpentyl)-1,3-dioxolane, CA 61732-91-0: Flammable. Antoine (liq) A = 9.4719, B = 3436, C = 0, dev = 1.0 to 5.0, range = 333 to 453.   *(continues)*

374

$C_{12}H_{24}O_2$ *(continued)*

4,5-Dimethyl-2-heptyl-1,3-dioxolane, CA 61732-90-9: Flammable. Antoine (liq) A = 9.8259, B = 3646, C = 0, dev = 1.0 to 5.0, range = 333 to 453.

Dodecanoic acid, lauric acid, CA 143-07-7: Irritant; flammable, flash point above 383; soluble ethanol, ether, benzene. Tm = 317.35; Tb = 572.05; Vm = 0.2308 at 323.15; Antoine I (sol) A = 16.710, B = 6683, C = 0, dev = 5.0, range = 293 to 303; Antoine II (liq) A = 6.39307, B = 1847.21, C = -150.29, dev = 1.0, range = 393 to 573.

2-(1-Ethylpentyl)-1,3-dioxepane, CA 61732-93-2: Flammable. Antoine (liq) A = 9.54221, B = 3558.038, C = 0, dev = 1.0 to 5.0, range = 333 to 373.

2-Heptyl-1,3-dioxepane, CA 61732-92-1: Flammable. Antoine (liq) A = 9.5568, B = 3674.958, C = 0, dev = 5.0, range = 328 to 373.

3-Heptyl-4-hydroxytetrahydro-2*H*-pyran, CA 62159-06-2: Flammable. Antoine (liq) A = 9.3379, B = 4055, C = 0, dev = 1.0 to 5.0, range = 383 to 453.

4-Octyl-1,3-dioxane, CA 23433-02-5: Flammable. Antoine (liq) A = 8.6599, B = 3420, C = 0, dev = 1.0 to 5.0, range = 353 to 453.

Undecanoic acid, methyl ester, CA 1731-86-8: Flammable. Vm = 0.2297 at 293.15; Antoine (liq) A = 6.31523, B = 1830.53, C = -98.299, dev = 1.0, range = 393 to 473.

$C_{12}H_{24}O_3$ MW 216.32

2-Butoxypropionic acid, pentyl ester: Flammable. Antoine (liq) A = 6.746, B = 2471.5, C = 0, dev = 5.0, range = 373 to 398.

3-Octyloxypropionic acid, methyl ester, CA 7419-98-9: Flammable. Antoine (liq) A = 7.8435, B = 3125.1, C = 0, dev = 1.0, range = 373 to 513.

$C_{12}H_{25}Br$ MW 249.23

1-Bromododecane, lauryl bromide, CA 143-15-7: Irritant; flammable, flash point 383; soluble ethanol, ether, acetone. Tm = 267.65; Tb = 549.15; Vm = 0.2401 at 293.15; Antoine (liq) A = 6.4639, B = 2061.93, C = -86.55, dev = 1.0, range = 411 to 610.

$C_{12}H_{25}Cl$ MW 204.78

1-Chlorododecane, lauryl chloride, CA 112-52-7: Irritant; flammable; soluble most organic solvents. Tf = 263.85; Tb = 533.15; Vm = 0.2357 at 293.15; Antoine (liq) A = 6.09466, B = 1754.079, C = -107.518, dev = 1.0, range = 389 to 519.

2-Chlorododecane ±, CA 51191-26-5: Irritant; flammable; soluble most organic solvents. Tm = 251.15; Antoine (liq) A = 8.861, B = 3410, C = 0, dev = 1.0, range = 283 to 328.

3-Chlorododecane ±, CA 51191-30-1: Irritant; flammable; soluble most organic solvents. Tm = 250.15; Antoine (liq) A = 8.986, B = 3445, C = 0, dev = 1.0, range = 283 to 328.

(continues)

$C_{12}H_{25}Cl$ *(continued)*

4-Chlorododecane ±, CA 51261-25-7: Irritant; flammable; soluble most organic solvents. Tm = 249.15; Antoine (liq) A = 8.690, B = 3350, C = 0, dev = 1.0, range = 283 to 328.

5-Chlorododecane ±, CA 51261-29-1: Irritant; flammable; soluble most organic solvents. Tm = 247.95; Antoine (liq) A = 9.022, B = 3442, C = 0, dev = 1.0, range = 283 to 328.

6-Chlorododecane, CA 26535-66-0: Irritant; flammable; soluble most organic solvents. Tm = 246.65; Antoine (liq) A = 8.969, B = 3421, C = 0, dev = 1.0, range = 283 to 328.

$C_{12}H_{25}F$ MW 188.33

1-Fluorododecane, lauryl fluoride, CA 334-68-9: Flammable. Tm = 260.15; Antoine (liq) A = 6.482, B = 1885.6, C = -77.15, dev = 5.0, range = 374 to 533.

$C_{12}H_{25}I$ MW 296.23

1-Iodododecane, lauryl iodide, CA 4292-19-7: Irritant; light-sensitive; flammable, flash point above 383; soluble ethanol, ether, acetone, chloroform. Tm = 273.45; Tb = 571.35; Vm = 0.2469 at 293.15; Antoine (liq) A = 6.3539, B = 2089.47, C = -90.85, dev = 1.0, range = 426 to 636.

$C_{12}H_{25}NO$ MW 199.34

Caprylamide, *N,N*-diethyl, CA 996-97-4: Irritant; flammable. Antoine (liq) A = 9.12786, B = 3718.9, C = 0, dev = 5.0, range = 373 to 450.

Lauramide, CA 1120-16-7: Irritant; soluble ethanol, acetone. Tm = 383.15; Antoine (sol) A = 18.2939, B = 7980, C = 0, dev = 1.0, range = 349 to 368.

$C_{12}H_{26}$ MW 170.34

2,3-Dimethyldecane, CA 17312-44-6: Flammable; soluble most organic solvents. Tm = 183.65; Antoine (liq) A = 6.88677, B = 2180.05, C = -33.26, dev = 1.0, range = 369 to 480.

2,4-Dimethyldecane, CA 2801-84-5: Flammable; soluble most organic solvents. Tm = 183.15; Antoine (liq) A = 7.45681, B = 2629.31, C = 10.68, dev = 1.0, range = 348 to 472.

Dodecane, CA 112-40-3: Hygroscopic; flammable, flash point 347; soluble most organic solvents. Tf = 261.15; Tb = 487.65; Vm = 0.2275 at 293.15; Tc = 658.2, Pc = 1824; Vc = 0.713; Antoine I (liq) A = 6.62064, B = 1942.122, C = -65.587, dev = 1.0, range = 278 to 400; Antoine II (liq) A = 6.12285, B = 1639.27, C = -91.315, dev = 0.02, range = 400 to 492.

2-Methylundecane, CA 7045-71-8: Flammable; soluble most organic solvents. Tm = 227.15; Antoine (liq) A = 7.17817, B = 2448.15, C = -10.04, dev = 1.0, range = 356 to 484.

(continues)

$C_{12}H_{26}$ *(continued)*

2-Methylundecane ±, CA 1002-43-3: Flammable; soluble most organic solvents. Tm = 215.15; Antoine (liq) A = 7.37737, B = 2633.55, C = 5.91, dev = 1.0, range = 357 to 485.

4-Methylundecane, CA 2980-69-0: Flammable; soluble most organic solvents. Tm = 204.15; Antoine (liq) A = 7.71287, B = 2769.11, C = 4.95, dev = 1.0, range = 359 to 481.

5-Methylundecane, CA 1632-70-8: Flammable; soluble most organic solvents. Tm = 203.15; Antoine (liq) A = 7.51083, B = 2651.23, C = 1.74, dev = 1.0, range = 357 to 480.

3,3,6,6-Tetramethyloctane, CA 62199-46-6: Flammable; soluble most organic solvents. Tm = 199.59; Antoine (liq) A = 5.29205, B = 1073.8, C = -136.1, dev = 1.0, range = 347 to 463.

2,4,6-Trimethylnonane, CA 62184-10-5: Flammable; soluble most organic solvents. Antoine (liq) A = 7.48819, B = 2577.66, C = 11.0, dev = 1.0, range = 339 to 459.

$C_{12}H_{26}O$ MW 186.34

Dihexyl ether, hexyl ether, CA 112-58-3: Irritant; flammable, flash point 350; forms explosive peroxides; soluble ether. Tb = 496.15; Vm = 0.2348 at 293.15; Antoine (liq) A = 7.53588, B = 2762, C = 0, dev = 1.0 to 5.0, range = 372 to 510.

1-Dodecanol, lauryl alcohol, CA 112-53-8: Irritant; flammable, flash point 400; soluble ethanol, ether. Tm = 299.15; Tb = 528.15 to 532.15; Vm = 0.2243 at 299.15; Antoine I (sol) A = 18.819, B = 6794, C = 0, dev = 5.0, range = 285 to 294; Antoine II (liq) A = 8.11108, B = 3297.391, C = 0, dev = 1.0 to 5.0, range = 288 to 333; Antoine III (liq) A = 6.20957, B = 1683.346, C = -134.83, dev = 0.1, range = 383 to 438; Antoine IV (liq) A = 5.88875, B = 1502.87, C = -150.744, dev = 0.1, range = 425 to 550; Antoine V (liq) A = 5.88546, B = 1500.675, C = -150.985, dev = 0.1, range = 434 to 550; Antoine VI (liq) A = 5.99065, B = 1579.439, C = -141.43, dev = 0.02, range = 505 to 550.

2-Dodecanol, CA 10203-28-8: Irritant; flammable; soluble ethanol, ether. Tm = 292.15; Tb = 525.15; Vm = 0.2249 at 293.15; Antoine (liq) A = 11.285, B = 4442, C = 0, dev = 5.0, range = 293 to 343.

3-Dodecanol ±, CA 10203-30-2: Irritant; flammable; soluble ethanol, ether. Tm = 298.15; Vm = 0.2266 at 305.15; Antoine (liq) A = 10.565, B = 4091, C = 0, dev = 5.0, range = 293 to 400.

4-Dodecanol, CA 10203-32-4: Irritant; flammable; soluble ethanol, ether. Tm = 285.15; Antoine (liq) A = 10.875, B = 4209, C = 0, dev = 5.0, range = 293 to 343.

5-Dodecanol, CA 10203-33-5: Irritant; flammable; soluble ethanol, ether. Tm = 292.15; Antoine (liq) A = 10.655, B = 4149, C = 0, dev = 5.0, range = 293 to 343.

6-Dodecanol, CA 6836-38-0: Irritant; flammable; soluble ethanol, ether. Tm = 303.15; Antoine (liq) A = 11.015, B = 4256, C = 0, dev = 5.0, range = 293 to 400.

$C_{12}H_{26}O_3$ MW 218.34

Diethylene glycol, dibutyl ether, CA 112-73-2: Irritant; flammable, flash point 320; forms explosive peroxides; soluble ethanol, ether. Tm = 213.15; Tb = 523.15 to 525.15; Vm = 0.2467 at 293.15; Antoine (liq) A = 7.6076, B = 2955, C = 0, dev = 1.0 to 5.0, range = 293 to 528.

$C_{12}H_{26}O_4$ MW 234.34

2,2-Bis(tert-butylperoxy)butane, CA 41407-59-4: Flammable; potentially explosive. Tm = 272.35; Antoine (liq) A = 11.891, B = 4028, C = 0, dev = 5.0, range = 299 to 323.

Tripropylene glycol, monoisopropyl ether: Flammable. Antoine (liq) A = 8.6723, B = 3924.88, C = 59.01, dev = 5.0, range = 355 to 530.

$C_{12}H_{26}S_2$ MW 234.46

1,12-Dodecanedithiol, CA 33528-63-1: Toxic; stench; flammable. Tm = 301.55; Antoine (liq) A = 5.8128, B = 1586.5, C = -175.72, dev = 1.0, range = 454 to 593.

$C_{12}H_{27}N$ MW 185.35

Dihexyl amine, CA 143-16-8: Corrosive; highly toxic by skin absorption; flammable, flash point 368; soluble ethanol, ether. Tm = 260.1; Tb = 506.15 to 516.15; Vm = 0.2331 at 293.15; Antoine (liq) A = 6.55843, B = 2013.2, C = -69.15, dev = 1.0, range = 408 to 569.

N,N-Dimethyl decylamine, CA 1120-24-7: Toxic; flammable; soluble ethanol, ether. Tm = 229.15; Antoine (liq) A = 6.44924, B = 1908.5, C = -78.15, dev = 5.0, range = 405 to 564.

Dodecylamine, lauryl amine, CA 124-22-1: Corrosive; toxic; flammable, flash point above 383; soluble ethanol, ether, benzene, chloroform. Tm = 300.15 to 301.15; Tb = 532.15; Antoine (liq) A = 6.4919, B = 2014, C = -83.35, dev = 1.0, range = 443 to 545.

Tributyl amine, CA 102-82-9: Corrosive; toxic; skin irritant; hygroscopic; flammable, flash point 336; slightly soluble water; soluble ethanol, ether, acetone, benzene. Tm = 203.15; Tb = 486.15; Vm = 0.2385 at 293.15; Antoine I (liq) A = 4.96696, B = 1088.83, C = -134.511, dev = 1.0 to 5.0, range = 298 to 337; Antoine II (liq) A = 7.169, B = 2515, C = 0, dev = 5.0, range = 333 to 487.

Triisobutyl amine, CA 1116-40-1: Corrosive; toxic; skin irritant; flammable; soluble ethanol, ether. Tf = 251.35; Tb = 464.65; Vm = 0.2412 at 293.15; Antoine (liq) A = 7.20344, B = 2127.51, C = -42.739, dev = 1.0 to 5.0, range = 305 to 452.

$C_{12}H_{27}O_4P$ MW 266.32

Tributyl phosphate, CA 126-73-8: Irritant; moderately toxic; flammable, flash point 419; moderately soluble water; soluble most organic solvents. Tm = below 193.15; Tb = 562.15, decomposes; Vm = 0.2738 at 298.15; Antoine (liq) A = 7.711, B = 3206.5, C = 0, dev = 5.0, range = 500 to 562.

(continues)

$C_{12}H_{27}O_4P$ *(continued)*

Triisobutyl phosphate, CA 126-71-6: Irritant; flammable; soluble water, most organic solvents. Tb = 537.15; Vm = 0.2751 at 293.15; Antoine (liq) A = 8.1194, B = 3283.1, C = 0, dev = 5.0, range = 411 to 537.

$C_{12}H_{28}N_2$ MW 200.37

1,12-Dodecanediamine, CA 2783-17-7: Irritant; flammable, flash point 428; soluble ethanol, ether. Tf = 339.15 to 340.15; Antoine (liq) A = 14.42331, B = 5750.884, C = 0, dev = 1.0 to 5.0, range = 313 to 353.

Tetrapropyl hydrazine, CA 60678-69-5: Toxic; flammable. Antoine (liq) A = 3.4359, B = 394.97, C = -247.86, dev = 1.0, range = 362 to 423.

$C_{13}H_9ClO_2$ MW 232.67

5-Chloro-2-hydroxybenzophenone, CA 85-19-8: Irritant; flammable. Tm = 369.15 to 370.15; Antoine I (sol) A = 10.215, B = 4800, C = 0, dev = 5.0, range = 293 to 367; Antoine II (liq) A = 7.575, B = 3830, C = 0, dev = 5.0, range = 367 to 493.

$C_{13}H_9N$ MW 179.22

Acridine, CA 260-94-6: Toxic; irritant; flammable; soluble ethanol, ether, benzene. Tm = 383.15, stable form; Tb = 619.15; Antoine I (sol) A = 8.30838, B = 3365.943, C = -48.723, dev = 1.0, range = 280 to 328; Antoine II (liq) A = 6.73664, B = 2699.39, C = -48.611, dev = 1.0, range = 402 to 619.

3,4-Benzoquinoline, CA 260-27-3: Flammable; soluble ethanol, ether, acetone, benzene. Tm = 382.15; Antoine (sol) A = 11.08139, B = 4956.705, C = 0, dev = 1.0, range = 288 to 323.

5,6-Benzoquinoline, CA 85-02-9: Flammable; soluble ethanol, ether, acetone, benzene. Tm = 367.15; Tb = 623.15; Antoine (sol) A = 9.37682, B = 4338.411, C = 0, dev = 1.0, range = 288 to 323.

7,8-Benzoquinoline, CA 230-27-3: Flammable, flash point above 383; soluble ethanol, ether, acetone, benzene. Tm = 325.15; Antoine (sol) A = 9.52199, B = 4188.8, C = 0, dev = 1.0, range = 293 to 323.

$C_{13}H_{10}$ MW 166.22

Fluorene, CA 86-73-7: Flammable; soluble ether, acetone, benzene. Tm = 389.15; Tb = 566.15 to 568.15; Antoine I (sol) A = 10.449, B = 4324, C = 0, dev = 1.0 to 5.0, range = 306 to 323; Antoine II (sol) A = 10.04542, B = 4122.908, C = 0, dev = 1.0, range = 348 to 388; Antoine III (liq) A = 8.31368, B = 4133.08, C = 86.582, dev = 1.0, range = 402 to 568.

$C_{13}H_{10}N_2$ MW 194.24

Carbodiimide, N,N'-diphenyl, CA 622-16-2: Flammable; soluble benzene. Tm = 441.15 to 443.15; Tb = 604.15; Antoine (liq) A = 7.715, B = 3425, C = 0, dev = 5.0, range = 500 to 599.

$C_{13}H_{10}N_4$ MW 222.25

1*H*-Tetrazole, 1,5-diphenyl, CA 7477-73-8: Flammable. Tm = 375.15; Antoine
(sol) A = 13.957, B = 6270.0, C = 0, dev not specified, range not specified.

$C_{13}H_{10}O$ MW 182.22

Benzophenone, diphenyl ketone, CA 119-61-9: Flammable, flash point above
383; soluble ethanol, ether, benzene, chloroform. Tm = 321.65 to 322.15;
Tb = 578.55; Vm = 0.1677 at 323.15; Antoine I (sol) A = 12.44989, B =
4924.329, C = 0, dev = 1.0, range = 293 to 318; Antoine II (sol) A =
11.736, B = 4698, C = 0, dev = 5.0, range = 298 to 318; Antoine III (liq)
A = 6.41427, B = 2144.6, C = -92.15, dev = 1.0, range = 433 to 673.

Xanthene, CA 92-83-1: Flammable; soluble ether, benzene, chloroform. Tm =
373.65; Tb = 583.15 to 585.15; Antoine (liq) A = 10.705, B = 4632.4, C = 0,
dev = 5.0, range = 413 to 433.

$C_{13}H_{10}O_2$ MW 198.22

Phenyl benzoate, CA 93-99-2: Flammable; soluble hot ethanol. Tm = 344.15;
Tb = 587.15; Antoine (liq) A = 6.6127, B = 2470.18, C = -50.986, dev = 1.0,
range = 379 to 587.

$C_{13}H_{10}O_3$ MW 214.22

2,4-Dihydroxybenzophenone, CA 131-56-6: Irritant; flammable; soluble
ethanol, ether, acetic acid. Tm = 418.15; Antoine I (sol) A = 14.905, B =
7000, C = 0, dev = over 5.0, range = 312 to 353; Antoine II (liq) A =
8.895, B = 4550, C = 0, dev = 5.0, range = 418 to 485.

Phenyl salicylate, salol, CA 118-55-8: Flammable, flash point above 383;
soluble ethanol, acetone, benzene, carbon tetrachloride; used as antiseptic.
Tm = 316.15; Antoine I (sol) A = 14.805, B = 5700, C = 0, dev = 5.0, range
= 279 to 315; Antoine II (liq) A = 6.09869, B = 1829.98, C = -139.08, dev =
1.0, range = 423 to 587.

$C_{13}H_{10}O_5$ MW 246.22

2,2',4,4'-Tetrahydroxybenzophenone, CA 131-55-5: Soluble water, ethanol,
ether, acetone. Tm = 469.15 to 471.15; Antoine (sol) A = 12.185, B = 7490,
C = 0, dev = above 5.0, range = 363 to 471.

$C_{13}H_{11}Cl$ MW 202.68

Chlorodiphenylmethane, CA 90-99-3: Irritant; corrosive; lachrymator; flammable, flash point above 383. Tm = 291.15 to 292.15; Vm = 0.1778 at
293.15; Antoine (liq) A = 8.756, B = 3680, C = 0, dev = 5.0, range = 381
to 450.

$C_{13}H_{11}ClO_2$ MW 234.68

Chlorodiphenoxymethane, CA 4431-86-1: Irritant; flammable. Antoine (liq)
A = 14.84855, B = 6605.798, C = 0, dev = 5.0, range = 385 to 453.

$C_{13}H_{11}N$ MW 181.24

Benzophenone, imine, CA 1013-88-3: Irritant; flammable, flash point above
383. Vm = 0.1636 at 293.15; Antoine (liq) A = 7.85058, B = 3255.985,
C = 0, dev = 5.0, range = 373 to 422.

$C_{13}H_{11}NO$ MW 197.24

Benzanilide, CA 93-98-1: Soluble hot ethanol. Tm = 436.15; Antoine (sol)
A = 10.38614, B = 5180.219, C = 0, dev = 1.0, range = 352 to 369.

4-Hydroxybenzal aniline: Soluble ethanol, ether. Tm = 467.15 to 468.15;
Antoine (sol) A = 13.155, B = 6679, C = 0, dev = 5.0, range = 348 to 408.

Salicylal aniline, CA 779-84-0: Flammable; soluble ethanol. Tm = 325.15;
Antoine (sol) A = 15.325, B = 6057, C = 0, dev = 5.0, range = 288 to 325.

$C_{13}H_{11}N_3O$ MW 225.25

2-(2'-Hydroxy-5'-methylphenyl)benzotriazole, CA 2440-22-4: Flammable; soluble acetone, ethyl acetate; used as ultraviolet screen. Tm = 404.15 to
405.15; Antoine I (sol) A = 14.625, B = 6540, C = 0, dev = 5.0, range = 293
to 333; Antoine II (liq) A = 7.655, B = 3690, C = 0, dev = 5.0, range = 404
to 435.

$C_{13}H_{12}$ MW 168.24

Diphenylmethane, CA 101-81-5: Flammable, flash point above 383; soluble
ethanol, ether, benzene, chloroform. Tm = 299.15 to 300.15; Tb = 534.15 to
535.15; Vm = 0.1681 at 300.15; Antoine I (liq) A = 5.8765, B = 1707.9, C =
-101.15, dev = 1.0, range = 295 to 383; Antoine II (liq) A = 6.28615, B =
1944.42, C = -83.15, dev = 0.1 to 1.0, range = 423 to 583.

$C_{13}H_{12}O$ MW 184.24

4-Benzylphenol, CA 101-53-1: Toxic; flammable; moderately soluble hot
water; soluble ethanol, ether, benzene, chloroform; used as fungicide.
Tm = 359.15; Tb = 593.15 to 595.15; Antoine (sol) A = 11.77562, B =
5088.747, C = 0, dev = 1.0, range = 313 to 335.

Benzyl phenyl ether, CA 946-80-5: Flammable; forms explosive peroxides;
soluble ether, benzene. Tm = 313.15; Tb = 559.15 to 560.15; Antoine (liq)
A = 7.53526, B = 3137.78, C = 7.281, dev = 1.0, range = 368 to 560.

Diphenylmethanol, benzhydrol, CA 91-01-0: Irritant; flammable; soluble
ether, ethanol, chloroform. Tm = 342.15; Tb = 570.15 to 571.15; Antoine
(liq) A = 8.09657, B = 3549.59, C = 8.63, dev = 1.0, range = 438 to 574.

Ethyl 1-naphthyl ketone, CA 2876-63-3: Flammable; soluble ethanol, ether,
carbon disulfide. Tb = 578.15 to 580.15; Vm = 0.1679 at 293.15; Antoine
(liq) A = 7.37924, B = 2771.83, C = -63.185, dev = 1.0, range = 397 to 579.

$C_{13}H_{13}N$ MW 183.25

N-Methyldiphenylamine, CA 552-82-9: Toxic; flammable; soluble ethanol,
ether. Tm = 265.55; Tb = 564.15; Vm = 0.1749 at 293.15;   *(continues)*

$C_{13}H_{13}N$ *(continued)*

Antoine (liq) A = 7.66324, B = 3008.43, C = -23.408, dev = 1.0 to 5.0, range = 376 to 555.

$C_{13}H_{14}$ MW 170.25

1-Isopropylnaphthalene, CA 6158-45-8: Flammable; soluble ethanol, ether, benzene. Tm = 257.15; Tb = 541.15; Vm = 0.1710 at 293.15; Antoine (liq) A = 6.01078, B = 1737.161, C = -107.31, dev = 1.0, range = 402 to 541.

2-Isopropylnaphthalene, CA 2027-17-0: Flammable; soluble ethanol, ether, benzene. Tb = 541.15; Vm = 0.1746 at 293.15; Antoine (liq) A = 5.80904, B = 1605.762, C = -119.152, dev = 1.0, range = 402 to 541.

$C_{13}H_{14}N_2$ MW 198.27

Diphenylmethane, 2,2'-diamino, CA 6582-52-1: Irritant; flammable; soluble ethanol, ether, benzene. Tm = 407.65 to 409.15; Antoine (sol) A = 13.71172, B = 6702.766, C = 26.351, dev = 0.1 to 1.0, range = 343 to 403.

Diphenylmethane, 2,4'-diamino, CA 1208-52-2: Irritant; flammable; soluble ethanol, ether, benzene. Tm = 361.15 to 362.15; Antoine (liq) A = 13.4189, B = 6560.013, C = 22.466, dev = 0.1 to 1.0, range = 353 to 403.

Diphenylmethane, 4,4'-diamino, CA 101-77-9: Carcinogenic; highly toxic by inhalation or skin absorption; flammable, flash point above 383; soluble ethanol, ether, benzene. Tm = 366.15; Tb = 671.15 to 672.15; Antoine I (liq) A = 12.7164, B = 6324.887, C = 18.779, dev = 1.0 to 5.0, range = 343 to 393; Antoine II (liq) A = 2.9429, B = 417.8, C = -357.15, dev = 5.0, range = 486 to 545.

$C_{13}H_{15}Cl_3O_3$ MW 325.62

2,4,5-Trichlorophenoxyacetic acid, pentyl ester, CA 120-39-8: Toxic; flammable; used as weed killer. Antoine (liq) A = 7.44504, B = 3039.036, C = -66.755, dev = 1.0, range = 460 to 573.

$C_{13}H_{16}Cl_2O_3$ MW 291.17

2,4-Dichlorophenoxyacetic acid, isopentyl ester, CA 67821-07-2: Toxic; flammable; used as weed killer. Antoine (liq) A = 6.44613, B = 2183.621, C = -122.067, dev = 1.0, range = 460 to 573.

2,4-Dichlorophenoxyacetic acid, pentyl ester, CA 1917-92-6: Toxic; flammable; used as weed killer. Antoine (liq) A = 7.04509, B = 2722.928, C = -72.562, dev = 1.0, range = 444 to 573.

$C_{13}H_{17}NO$ MW 203.28

Piperidine, 1-phenacyl, CA 3626-62-8: Irritant; flammable. Antoine (liq) A = 6.66293, B = 2685.9, C = 0, dev = 5.0, range = 381 to 446.

Piperidine, 1-(*meta*-tolyl), CA 13290-48-7: Irritant; flammable. Antoine (liq) A = 6.9983, B = 2808.3, C = 0, dev = 5.0, range = 373 to 403.

$C_{13}H_{17}NO_3$ MW 235.28

N-Acetylphenylalanine ±, ethyl ester, CA 4134-09-2: Flammable. Antoine
(liq) A = 9.2448, B = 4306, C = 0, dev = 1.0, range = 438 to 528.

Morpholine cinnamate: Flammable. Antoine (sol) A = 15.2254, B = 6205,
C = 0, dev = 5.0, range = 298 to 349.

$C_{13}H_{18}$ MW 174.29

1,1,4,6-Tetramethylindane, CA 941-60-6: Flammable. Antoine I (liq) A =
6.51331, B = 1979.945, C = -65.819, dev = 0.1 to 1.0, range = 313 to 383;
Antoine II (liq) A = 6.27742, B = 1834.819, C = -77.303, dev = 1.0, range =
313 to 469; Antoine III (liq) A = 6.17108, B = 1757.263, C = -85.365, dev =
0.02, range = 424 to 469.

1,1,4,7-Tetramethylindane, CA 1078-04-2: Flammable. Vm = 0.1866 at
298.15; Antoine I (liq) A = 6.46872, B = 2008.385, C = -64.406, dev = 0.1,
range = 313 to 388; Antoine II (liq) A = 6.28149, B = 1890.488, C =
-73.912, dev = 1.0, range = 313 to 469; Antoine III (liq) A = 6.13091, B =
1779.115, C = -85.127, dev = 0.02, range = 431 to 469.

$C_{13}H_{18}O$ MW 190.28

4,4-Dimethyl-1-phenyl-3-pentanone, CA 5195-24-4: Flammable. Antoine (liq)
A = 8.399, B = 3318, C = 0, dev = 5.0, range = 405 to 520.

para-Isopropyl-*alpha*-methylhydrocinnamaldehyde ±, CA 103-95-7: Irritant;
flammable. Antoine (liq) A = 9.674, B = 3794, C = 0, dev = 1.0, range =
283 to 400.

1-Phenyl-1-heptanone, CA 1671-75-6: Irritant; flammable; soluble ethanol,
ether, acetone. Tm = 290.15; Tb = 556.45; Vm = 0.2002 at 293.15; Antoine
(liq) A = 8.35809, B = 3506.45, C = 7.488, dev = 1.0 to 5.0, range = 373
to 550.

$C_{13}H_{19}NO$ MW 205.30

3-Phenylpropionic acid, *N*,*N*-diethylamide: Irritant; flammable. Antoine
(liq) A = 6.0397, B = 2426.7, C = 0, dev = 5.0, range = 353 to 439.

$C_{13}H_{19}NO_2$ MW 221.30

Cyclohexyl ammonium benzoate: Tm = 459.15; Antoine (sol) A = 13.335, B =
5384, C = 0, dev = 5.0, range = 289 to 298.

$C_{13}H_{20}$ MW 176.30

Heptylbenzene, CA 1078-71-3: Flammable, flash point 368; soluble most organic solvents. Tm = 225.15; Tb = 513.15; Vm = 0.2058 at 293.15; Antoine
(liq) A = 6.1255, B = 1761.2, C = -91.65, dev = 1.0, range = 423 to 527.

$C_{13}H_{20}O$ MW 192.30

alpha-Ionone, CA 24190-29-2: Flammable, flash point 377; soluble ethanol,
ether, acetone. Tb = 531.15; Vm = 0.2068 at 294.15; Antoine I *(continues)*

$C_{13}H_{20}O$ *(continued)*

(liq) A = 5.161, B = 3528, C = 0, dev = 1.0, range = 286 to 333; Antoine II
(liq) A = 7.24911, B = 2513.57, C = -43.634, dev = 1.0 to 5.0, range = 352
to 523.

beta-Ionone, CA 14901-07-6: Flammable, flash point above 383; soluble
ethanol, ether, benzene, chloroform. Tm = 238.15; Tb = 544.15; Vm = 0.2033
at 293.15; Antoine (liq) A = 9.210, B = 3605, C = 0, dev = 1.0, range = 291
to 334.

$C_{13}H_{20}O_2$ MW 208.30

1,3-Dihydroxy-5-heptylbenzene, CA 500-67-4: Flammable; soluble ethanol.
Tm = 322.45; Antoine (liq) A = 4.8115, B = 1240.6, C = -224.38, dev = 1.0,
range = 443 to 504.

1,3-Dihydroxy-5-methyl-2-hexylbenzene, CA 41395-27-1: Flammable; soluble
ethanol. Tm = 370.45; Antoine (liq) A = 4.4840, B = 1049, C = -226.93,
dev = 1.0, range = 433 to 493.

$C_{13}H_{21}NO_2$ MW 223.31

Butylamine, *N*-(3-phenoxy-2-hydroxypropyl), CA 3246-04-6: Irritant; flammable. Antoine (sol) A = 17.0202, B = 6997.4, C = 0, dev not specified,
range = 323 to 348.

$C_{13}H_{22}$ MW 178.32

2-Allyl-*cis*-decahydronaphthalene: Irritant; lachrymator; flammable; soluble ethanol, ether, benzene. Antoine (liq) A = 12.4752, B = 4695.9,
C = 0, dev not specified, range = 296 to 320.

2-Allyl-*trans*-decahydronaphthalene: Irritant; lachrymator; flammable; soluble ethanol, ether, benzene. Antoine (liq) A = 12.85154, B = 4788.85,
C = 0, dev not specified, range = 296 to 320.

Fluorene, dodecahydro, CA 5744-03-6: Flammable, flash point 337. Tb =
526.15; Vm = 0.1938 at 293.15; Antoine (liq) A = 7.58278, B = 2913.672,
C = 0, dev = 1.0, range = 332 to 525.

$C_{13}H_{22}Cl_2O_4$ MW 313.22

2,2-Bis(chloromethyl)-1,3-propanedioldibutyrate: Irritant; flammable.
Antoine (liq) A = 4.74178, B = 1951.42, C = -32.222, dev = 1.0, range = 454
to 572.

$C_{13}H_{22}O_2$ MW 210.32

Bornyl propionate, CA 20279-25-8: Flammable. Antoine (liq) A = 7.50723,
B = 2759.89, C = -6.653, dev = 1.0, range = 337 to 508.

$C_{13}H_{24}Cl_4$ MW 322.14

1,1,1,13-Tetrachlorotridecane, CA 3922-33-6: Irritant; flammable. Antoine
(liq) A = 11.375, B = 5088, C = 0, dev = 5.0, range = 320 to 370.

$C_{13}H_{24}O$ MW 196.33

4-Hexenal, 5-methyl-2-ethyl-2-butyl, CA 42023-59-6: Flammable; polymerizes. Antoine (liq) A = 9.52165, B = 3608.392, C = 0, dev = 5.0, range = 323 to 393.

$C_{13}H_{24}O_2$ MW 212.33

Decyl acrylate, CA 2156-96-9: Irritant; lachrymator; flammable. Vm = 0.2418 at 293.15; Antoine (liq) A = 7.812, B = 3112, C = 0, dev = 5.0, range = 404 to 536.

Oxa-2-cyclotetradecanone, CA 1725-04-8: Flammable. Antoine (liq) A = 8.48393, B = 3523.782, C = 0, dev = 5.0, range = 393 to 443.

10-Undecenoic acid, ethyl ester, CA 692-86-4: Flammable; soluble ethanol, ether. Tm = 235.15; Tb = 536.65 to 538.65; Vm = 0.2405 at 288.15; Antoine (liq) A = 6.41366, B = 1886, C = -104.27, dev = 1.0, range = 404 to 532.

$C_{13}H_{24}O_3$ MW 228.33

1,4-Dioxa-5-cyclopentadecanone, CA 1898-97-1: Flammable. Antoine (liq) A = 8.64228, B = 3633.749, C = 0, dev = 1.0, range = 403 to 443.

1,6-Dioxa-7-cyclopentadecanone, CA 36575-54-9: Flammable. Antoine (liq) A = 9.4021, B = 3956.233, C = 0, dev = 1.0, range = 403 to 443.

1,8-Dioxa-9-cyclopentadecanone, CA 36575-53-8: Flammable. Antoine (liq) A = 8.29326, B = 3475.49, C = 0, dev = 5.0, range = 403 to 443.

3-Hexyl-4-acetoxytetrahydro-2*H*-pyran, CA 18871-17-5: Flammable. Antoine (liq) A = 9.2209, B = 3768, C = 0, dev = 1.0 to 5.0, range = 383 to 453.

Octyl levulinate, CA 41780-57-8: Flammable. Antoine (liq) A = 7.50321, B = 2907.6, C = -35.78, dev = 1.0, range = 413 to 565.

$C_{13}H_{24}O_4$ MW 244.33

3-Acetoxypropionic acid, octyl ester: Flammable. Antoine (liq) A = 10.808, B = 4617, C = 0, dev = 5.0, range = 420 to 440

Ethylisopentylmalonic acid, ethylmethyl ester, CA 72030-39-8: Flammable. Antoine (liq) A = 9.60166, B = 3819.402, C = 0, dev = 5.0, range = 392 to 501.

$C_{13}H_{24}O_5$ MW 260.33

Octyl[1-(methoxycarbonyl)ethyl] carbonate: Toxic; flammable. Antoine (liq) A = 8.493, B = 3659, C = 0, dev = 5.0, range = 391 to 566.

Pentyl[1-(butoxycarbonyl)ethyl] carbonate: Toxic; flammable. Antoine (liq) A = 8.658, B = 3664, C = 0, dev = 5.0, range = 348 to 513.

$C_{13}H_{25}N$ MW 195.35

Tridecanonitrile, CA 629-60-7: Irritant; flammable; soluble ethanol, ether. Tm = 282.85; Tb = 566.15; Vm = 0.2366 at 293.15; Antoine (liq) A = 6.48331, B = 2116.33, C = -93.23, dev = 1.0 to 5.0, range = 380 to 566.

$C_{13}H_{25}NO$ MW 211.35

Piperidine, 1-octanoyl, CA 20299-83-6: Irritant; flammable. Antoine (liq)
A = 6.01872, B = 2612, C = 0, dev = 5.0, range = 373 to 443.

$C_{13}H_{26}$ MW 188.35

5-Butyl-4-nonene, CA 7367-38-6: Flammable; soluble most organic solvents.
Tb = 488.15 to 489.15; Vm = 0.2354 at 293.15; Antoine (liq) A = 6.5391, B =
1928.9, C = -60.42, dev = 1.0, range = 310 to 361.

1-Tridecene, CA 2437-56-1: Flammable, flash point 352; soluble most organic solvents. Tf = 260.15; Tb = 505.95; Vm = 0.2383 at 293.15; Antoine
(liq) A = 6.11633, B = 1679.71, C = -97.356, dev = 0.02, range = 413 to 509.

$C_{13}H_{26}O$ MW 198.35

1-Heptylcyclohexanol, CA 67639-25-2: Flammable; soluble ethanol. Antoine
(liq) A = 17.348, B = 6826, C = 0, dev = 5.0, range = 376 to 405.

5-Methyl-2-ethyl-2-butyl-4-hexene-1-ol, CA 53144-53-9: Flammable; soluble
ethanol. Antoine (liq) A = 10.18975, B = 4015.297, C = 0, dev = 1.0,
range = 333 to 393.

1-Octylcyclopentanol, CA 30089-09-9: Flammable; soluble ethanol. Tm =
255.65; Antoine (liq) A = 7.79734, B = 3092.2, C = -6.94, dev = 1.0, range
= 468 to 541.

2-Tridecanone, CA 593-08-8: Flammable, flash point 380; soluble ethanol,
ether, acetone, benzene. Tm = 302.15; Tb = 533.15 to 538.15; Vm = 0.2414
at 303.15; Antoine I (liq) A = 6.99847, B = 2313.06, C = -70.65, dev = 1.0,
range = 335 to 534; Antoine II (liq) A = 6.31566, B = 1901.101, C =
-99.759, dev = 0.1, range = 424 to 510.

7-Tridecanone, CA 462-18-0: Flammable, flash point above 383; soluble most
organic solvents. Tm = 306.15; Tb = 537.15; Antoine I (sol) A = 8.160, B =
1150, C = 0, dev = over 5.0, range = 284 to 296; Antoine II (liq) A =
6.20931, B = 1813.619, C = -104.614, dev = 1.0 to 5.0, range = 395 to 600.

$C_{13}H_{26}O_2$ MW 214.35

Decanoic acid, isopropyl ester, CA 2311-59-3: Flammable. Vm = 0.2509 at
293.15; Antoine (liq) A = 7.56879, B = 2730.8, C = -27.54, dev = 1.0,
range = 363 to 451.

Decanoic acid, propyl ester, CA 30673-60-0: Flammable. Vm = 0.2486 at
293.15; Antoine (liq) A = 7.81036, B = 2936.5, C = -19.63, dev = 1.0,
range = 369 to 459.

4,5-Dimethyl-2-octyl-1,3-dioxolane, CA 5452-11-9: Flammable. Antoine
(liq) A = 9.8659, B = 3803, C = 0, dev = 1.0 to 5.0, range = 333 to 453.

Lauric acid, methyl ester, CA 111-82-0: Flammable, flash point above 383;
soluble ethanol, ether, acetone, benzene. Tm = 278.15; Tb = 535.15; Vm =
0.2463 at 293.15; Antoine I (sol) A = 18.548, B = 6359, C = 0, dev = 5.0,
range = 262 to 273; Antoine II (liq) A = 6.04518, B = 1763.6, C = -109.67,
dev = 1.0, range = 287 to 333; Antoine III (liq) A = 6.27362, B = 1851, C =
-106.735, dev = 0.1 to 1.0, range = 407 to 540.   *(continues)*

$C_{13}H_{26}O_2$ *(continued)*

2-Octyl-1,3-dioxepane, CA 61732-94-3: Flammable. Antoine (liq) A = 10.75419, B = 4240.609, C = 0, dev = 1.0 to 5.0, range = 323 to 373.

Tridecanoic acid, CA 630-53-9: Flammable, flash point above 383; soluble ethanol, ether, acetone. Tm = 317.65 to 318.65; Tb = 585.55; Vm = 0.2534 at 353.15; Antoine (liq) A = 6.19777, B = 1768.57, C = -163.856, dev = 1.0, range = 409 to 585.

$C_{13}H_{26}O_3$ MW 230.35

Decyl lactate, CA 42175-34-8: Flammable. Tm = 281.15; Antoine (liq) A = 6.8651, B = 2261.504, C = -90.15, dev = 1.0 to 5.0, range = 349 to 556.

3-Ethoxypropionic acid, octyl ester: Flammable. Antoine (liq) A = 7.363, B = 2974, C = 0, dev = 5.0, range = 398 to 543.

3-Pentyloxypropionic acid, pentyl ester, CA 14144-56-0: Flammable. Antoine (liq) A = 8.035, B = 3257, C = 0, dev = 5.0, range = 378 to 498.

$C_{13}H_{27}Br$ MW 263.26

1-Bromotridecane, CA 765-09-3: Flammable, flash point above 383; soluble chloroform. Tm = 279.15; Tb = 569.15; Vm = 0.2566 at 293.15; Antoine (liq) A = 6.511, B = 2143, C = -89.15, dev = 5.0, range = 425 to 628.

$C_{13}H_{27}Cl$ MW 218.81

1-Chlorotridecane, CA 822-13-9: Irritant; flammable. Tm = 273.85; Antoine (liq) A = 6.516, B = 2087.9, C = -87.15, dev = 5.0, range = 414 to 611.

$C_{13}H_{27}F$ MW 202.35

1-Fluorotridecane: Flammable. Tm = 276.15; Antoine (liq) A = 6.531, B = 1969.1, C = -80.15, dev = 5.0, range = 387 to 558.

$C_{13}H_{27}I$ MW 310.26

1-Iodotridecane, CA 35599-77-0: Flammable. Tm = 285.45; Antoine (liq) A = 6.402, B = 2169.2, C = -94.15, dev = 5.0, range = 440 to 655.

$C_{13}H_{27}NO$ MW 213.36

Dodecanamide, *N*-methyl, CA 27563-67-3: Irritant; flammable; Tm = 341.75; Antoine (sol) A = 14.9269, B = 6092, C = 0, dev = 1.0 to 5.0, range = 323 to 337.

$C_{13}H_{27}NO_2$ MW 229.36

Lactamide, *N*-decyl: Irritant; flammable. Tm = 335.15; Antoine (liq) A = 10.70547, B = 5112.48, C = 0, dev = 1.0, range = 413 to 483.

Lactamide, *O*-decyl: Irritant; flammable. Antoine (liq) A = 10.385, B = 4963, C = 0, dev = 5.0, range = 413 to 483.

$C_{13}H_{28}$ MW 184.36

5-Butylnonane, CA 17312-63-9: Flammable; soluble most organic solvents.
Tb = 490.15 to 491.15; Vm = 0.2415 at 291.65; Antoine (liq) A = 9.92985,
B = 3613.2, C = 45.7, dev = 1.0, range = 298 to 365.

2,3-Dimethylundecane, CA 17312-77-5: Flammable; soluble most organic solvents. Tm = 205.65; Antoine (liq) A = 6.68275, B = 2088.3, C = -52.87,
dev = 1.0, range = 383 to 500.

2,4-Dimethylundecane, CA 17312-80-0: Flammable; soluble most organic solvents. Tm = 197.65; Antoine (liq) A = 7.55924, B = 2723, C = 0, dev = 1.0,
range = 365 to 490.

2-Methyldodecane, CA 1560-97-0: Flammable; soluble most organic solvents.
Tm = 247.15; Antoine (liq) A = 7.40299, B = 2693.3, C = -3.54, dev = 1.0,
range = 373 to 503.

3-Methyldodecane, CA 17312-57-1: Flammable; soluble most organic solvents.
Tm = 213.65; Antoine (liq) A = 7.49657, B = 2814.75, C = 9.27, dev = 1.0,
range = 372 to 504.

4-Methyldodecane, CA 6117-97-1: Flammable; soluble most organic solvents.
Tm = 220.35; Antoine (liq) A = 7.67473, B = 2920.8, C = 14.46, dev = 1.0,
range = 372 to 501.

5-Methyldodecane, CA 17453-93-9: Flammable; soluble most organic solvents.
Tm = 204.15; Antoine (liq) A = 7.36359, B = 2698.4, C = 3.98, dev = 1.0,
range = 368 to 500.

Tridecane, CA 629-50-5: Flammable, flash point 352; soluble most organic
solvents. Tm = 267.65; Tb = 507.15; Vm = 0.2437 at 293.15; Antoine (liq)
A = 6.13246, B = 1690.67, C = -98.93, dev = 0.1, range = 417 to 511.

2,4,6-Trimethyldecane: Flammable; soluble most organic solvents. Antoine
(liq) A = 7.29631, B = 2513.9, C = -2.28, dev = 1.0, range = 352 to 478.

$C_{13}H_{28}O$ MW 200.36

1-Tridecanol, CA 112-70-9: Toxic; flammable, flash point 394; soluble
ethanol, ether. Tm = 305.65 to 306.65; Vm = 0.2437 at 307.15; Antoine
(liq) A = 5.92508, B = 1576.5, C = -151.346, dev = 1.0, range = 431 to 568.

$C_{13}H_{28}O_4$ MW 248.36

Tripropylene glycol, monobutyl ether, CA 57499-93-1: Flammable. Antoine
(liq) A = 8.40594, B = 3455.57, C = -2.725, dev = 5.0, range = 374 to 543.

$C_{13}H_{29}N$ MW 199.38

Tridecylamine, CA 2869-34-3: Irritant; flammable, flash point above 383;
soluble ethanol, ether. Tm = 300.55; Tb = 549.15; Antoine (liq) A =
6.5349, B = 2094, C = -86.55, dev = 1.0, range = 458 to 562.

$C_{14}F_{30}O_2$ MW 770.10

Perfluoro(1,12-bis-methoxydodecane): Antoine (liq) A = 8.0039, B = 2819,
C = 0, dev = 1.0 to 5.0, range not specified.

388

$C_{14}H_6N_6O_{12}$ MW 450.23

1,2-Bis(2,4,6-trinitrophenyl)ethylene, CA 20062-22-0: Flammable; explosive.
Tm = 484.15, decomposes; Antoine (sol) A = 13.315, B = 9399, C = 0, dev =
5.0, range = 434 to 479.

$C_{14}H_7NO_4$ MW 253.21

1-Nitroanthraquinone, CA 82-34-8: Irritant; flammable; soluble ethanol,
ether. Tm = 506.15; Antoine (sol) A = 14.025, B = 7300, C = 0, dev = 5.0,
range = 407 to 440.

$C_{14}H_8BrNO_3$ MW 318.13

C. I. disperse yellow 64: Toxic. Antoine (sol) A = 11.25383, B = 6822.385,
C = 0, dev = 1.0, range = 483 to 523.

$C_{14}H_8O_2$ MW 208.22

9,10-Anthraquinone, CA 84-65-1: Irritant; mild allergen; may be carcinogenic; flammable, flash point 458; soluble hot ethanol, hot benzene; important dye intermediate. Tm = 559.15; Tb = 652.15 to 654.15; Antoine I (sol)
A = 13.435, B = 6604, C = 0, dev = 5.0, range = 343 to 403; Antoine II
(sol) A = 15.3836, B = 11078.9, C = 218.733, dev = 1.0, range = 461 to 558;
Antoine III (liq) A = 7.87399, B = 4237.5, C = 70.46, dev = 1.0, range =
559 to 660.

9,10-Phenanthraquinone, CA 84-11-7: Irritant; flammable; soluble hot ethanol, ether, benzene. Tm = 481.65 to 483.15; Tb = above 633.15; Antoine
(sol) A = 13.495, B = 6895, C = 0, dev = 5.0, range = 353 to 413.

$C_{14}H_8O_3$ MW 224.22

2,2'-Biphenyldicarboxylic acid, anhydride, CA 6050-13-1: Irritant; flammable, moisture-sensitive. Tm = 490.15; Antoine (sol) A = 9.3489, B =
4774.6, C = 0, dev = 5.0, range = 433 to 490.

1-Hydroxy-9,10-anthraquinone, CA 129-43-1: Irritant; flammable; soluble
ethanol, ether, benzene. Tm = 468.15 to 469.15; Antoine (sol) A = 12.965,
B = 6298, C = 0, dev = 5.0, range = 333 to 383.

2-Hydroxy-9,10-anthraquinone, CA 605-32-3: Irritant; flammable; soluble
ethanol, ether, benzene. Tm = 586.15 to 588.15; Antoine (sol) A = 14.355,
B = 7999, C = 0, dev = 5.0, range = 393 to 453.

$C_{14}H_8O_4$ MW 240.21

1,2-Dihydroxyanthraquinone, alizarin, CA 72-48-0: Irritant; slightly soluble water; soluble most organic solvents. Tm = 562.15 to 563.15; Tb =
703.15; Antoine (sol) A = 10.42713, B = 5543.9, C = -28.28, dev = 1.0,
range = 368 to 498.

1,4-Dihydroxyanthraquinone, quinizarin, CA 81-64-1: Irritant; slightly
soluble water; soluble most organic solvents. Tm = 473.15 to 475.15;
Antoine I (sol) A = 12.825, B = 6451, C = 0, dev = 0.1, range = *(continues)*

$C_{14}H_8O_4$ *(continued)*

338 to 463; Antoine II (liq) A = 6.98564, B = 3477.73, C = -24.872, dev = 1.0, range = 469 to 633.

1,5-Dihydroxyanthraquinone, anthrarufin, CA 117-12-4: Irritant; slightly soluble water; soluble most organic solvents. Tm = 553.15; Antoine (sol) A = 12.225, B = 6619, C = 0, dev = 5.0, range = 363 to 433.

1,8-Dihydroxyanthraquinone, chrysazin, CA 117-10-2: Irritant and cathartic; slightly soluble water; soluble most organic solvents. Tm = 466.15; Antoine (sol) A = 12.945, B = 6422, C = 0, dev = 5.0, range = 353 to 403.

2,6-Dihydroxyanthraquinone, anthraflavin, CA 84-60-6: Irritant; slightly soluble water; soluble most organic solvents. Tm = above 603.15, decomposes; Antoine (sol) A = 13.865, B = 9075, C = 0, dev = 5.0, range = 463 to 533.

$C_{14}H_8O_6$ MW 272.21

1,4,5,8-Tetrahydroxyanthraquinone, CA 81-60-7: Irritant; soluble ethanol, benzene. Tm = above 573.15, decomposes; Antoine (sol) A = 13.545, B = 7916, C = 0, dev = 5.0, range = 403 to 473.

$C_{14}H_9BrN_2O_4$ MW 349.14

C. I. disperse blue 56, CA 12217-79-7: Antoine (sol) A = 7.61165, B = 4874.2, C = 0, dev = 5.0, range = 483 to 533.

$C_{14}H_9Cl_5$ MW 354.49

2,2-Bis(4-chlorophenyl)-1,1,1-trichloroethane, DDT, CA 50-29-3: Irritant; toxic by skin absorption and oral ingestion; suspected carcinogen; soluble most organic solvents. Tm = 381.65 to 382.15; Tb = 533.15; Antoine (sol) A = 12.903, B = 6010, C = 0, dev = 1.0, range = 323 to 363.

$C_{14}H_9NO_2$ MW 223.23

1-Aminoanthraquinone, CA 82-45-1: Irritant; suspected carcinogen; soluble ethanol, ether, benzene, chloroform; dyestuff intermediate. Tm = 526.15 to 527.15; Antoine (sol) A = 12.725, B = 6610, C = 0, dev = 5.0, range = 413 to 443.

2-Aminoanthraquinone, CA 117-79-3: Irritant; suspected carcinogen; soluble acetone, benzene, chloroform; dyestuff intermediate. Tm = 576.15 to 579.15; Antoine (sol) A = 13.425, B = 7520, C = 0, dev = 5.0, range = 444 to 473.

$C_{14}H_9NO_3$ MW 239.23

1-Amino-4-hydroxyanthraquinone, CA 116-85-8: Irritant; soluble ethanol, acetone, benzene. Tm = 480.15 to 481.15; Antoine (sol) A = 13.125, B = 6860, C = 0, dev = 5.0, range = 418 to 438.

$C_{14}H_{10}$ MW 178.23

Anthracene, CA 120-12-7: Allergen; mild irritant; weak carcinogen; moderately soluble ethanol, benzene, carbon disulfide; dyestuff intermediate.
Tm = 489.25; Tb = 613.05; Antoine I (sol) A = 10.58991, B = 4903.3, C =
-1.58, dev = 0.1, range = 299 to 430; Antoine II (sol) A = 11.76139, B =
5315.532, C = 0, dev = 0.1 to 1.0, range = 313 to 363; Antoine III (sol)
A = 10.75544, B = 4947.951, C = 0, dev = 0.1 to 1.0, range = 363 to 393;
Antoine IV (liq) A = 7.47799, B = 3612.44, C = 44.906, dev = 1.0, range =
504 to 615.

Phenanthrene, CA 85-01-8: Irritant; suspected carcinogen; soluble ether,
benzene, carbon tetrachloride. Tm = 374.15; Tb = 613.15; Antoine I (sol)
A = 10.305, B = 4444, C = 0, dev = 5.0, range = 296 to 315; Antoine II
(sol) A = 10.70162, B = 4554.38, C = 0, dev = 1.0, range = 313 to 363;
Antoine III (liq) A = 6.64812, B = 2513.134, C = -65.345, dev = 1.0, range
= 373 to 423; Antoine IV (liq) A = 7.17186, B = 3235.19, C = 12.908, dev =
1.0, range = 391 to 613.

$C_{14}H_{10}N_2O_2$ MW 238.25

1,4-Diaminoanthraquinone, CA 128-95-0: Irritant; mutagenic; soluble ethanol, benzene, pyridine. Tm = 541.15; Antoine (sol) A = 14.225, B = 7900,
C = 0, dev = 5.0, range = 448 to 474.

$C_{14}H_{10}O_2$ MW 210.23

Benzil, diphenylglyoxal, CA 134-81-6: Irritant; flammable; soluble ethanol, ether, benzene, chloroform. Tm = 368.15; Tb = 619.15 to 621.15; Vm =
0.1939 at 375.15; Antoine I (sol) A = 11.833, B = 5140.8, C = 0, dev = 1.0,
range = 318 to 340; Antoine II (liq) A = 6.34651, B = 2339.54, C = -81.127,
dev = 1.0, range = 401 to 620.

$C_{14}H_{10}O_3$ MW 226.23

Benzoic acid, anhydride, CA 93-97-0: Mild irritant and allergen; moisture-
sensitive; flammable, flash point above 383; soluble most organic solvents.
Tm = 315.15; Tb = 633.15; Antoine (liq) A = 6.99609, B = 2956.76, C =
-40.787, dev = 1.0, range = 416 to 633.

$C_{14}H_{10}O_4$ MW 242.23

2,2'-Biphenyldicarboxylic acid, CA 482-05-3: Soluble most organic solvents. Tm = 505.15 to 506.15; Antoine (sol) A = 17.0229, B = 8674.9,
C = 0, dev = 5.0, range = 433 to 493.

$C_{14}H_{11}FO_3$ MW 246.24

2'-Fluoro-2-hydroxy-4-methoxybenzophenone, CA 3119-88-8: Antoine (sol) A =
12.972, B = 5712, C = 0, dev = 5.0, range = 307 to 318.

3'-Fluoro-2-hydroxy-4-methoxybenzophenone, CA 3506-35-2: Antoine (sol) A =
3.787, B = 902.3, C = 0, dev = 5.0, range = 322 to 343.

4'-Fluoro-2-hydroxy-4-methoxybenzophenone, CA 3602-47-9: Antoine (sol) A =
-0.591, B = 1967, C = 0, dev = 5.0, range = 322 to 343.

$C_{14}H_{12}$ MW 180.25

9,10-Dihydroanthracene, CA 613-31-0: Flammable; soluble ethanol, ether, benzene. Tm = 381.15 to 383.15; Tb = 578.15; Antoine I (sol) A = 11.855, B = 4878, C = 0, dev = 5.0, range = 279 to 328; Antoine II (sol) A = 11.73468, B = 4837.731, C = 0, dev = 1.0 to 5.0, range = 321 to 379.

9,10-Dihydrophenanthrene, CA 776-35-2: Flammable, flash point above 383; soluble ethanol, ether. Tm = 307.65 to 308.15; Vm = 0.1676 at 313.15; Antoine (liq) A = 7.86088, B = 3341.401, C = 0, dev = 5.0, range = 417 to 453.

1,1-Diphenylethylene, CA 530-48-3: Flammable, flash point above 383; forms an explosive peroxide; soluble ether, chloroform. Tm = 281.15; Tb = 550.15; Vm = 0.1762 at 293.15; Antoine (liq) A = 6.91564, B = 2522.93, C = -36.439, dev = 1.0, range = 360 to 550.

1,2-Diphenylethylene, cis, stilbene, CA 645-49-8: Flammable; reacts violently with oxygen; soluble ethanol, benzene. Tm = 278.15 to 279.15; Vm = 0.1777 at 293.15; Antoine (liq) A = 8.609, B = 3474, C = 0, dev not specified, range = 373 to 428.

1,2-Diphenylethylene, trans, stilbene, CA 103-30-0: Flammable; reacts violently with oxygen; soluble hot ethanol, benzene, ether. Tm = 397.15; Tb = 579.15 to 580.15; Antoine I (sol) A = 12.25604, B = 5201.358, C = 0, dev = 0.1 to 1.0, range = 298 to 343; Antoine II (liq) A = 6.97928, B = 2610.05, C = -54.759, dev = 1.0, range = 419 to 580.

$C_{14}H_{12}O$ MW 196.25

Benzyl phenyl ketone, CA 451-40-1: Flammable; soluble ethanol, ether, chloroform. Tm = 333.15; Tb = 593.15; Antoine (liq) A = 7.21197, B = 2879.33, C = -41.095, dev = 1.0, range = 396 to 594.

2-Methylbenzophenone, CA 131-58-8: Flammable, flash point above 383; soluble most organic solvents. Tm = below 255.15; Tb = 582.65; Vm = 0.1812 at 293.15; Antoine (liq) A = 7.93521, B = 3400.1, C = 0, dev = 1.0 to 5.0, range = 435 to 580.

3-Methylbenzophenone, CA 643-65-2: Flammable, flash point above 383; soluble most organic solvents. Tm = 275.15; Tb = 584.15 to 586.15; Vm = 0.1804 at 290.65; Antoine (liq) A = 8.14795, B = 3575.0, C = 0, dev = 1.0 to 5.0, range = 445 to 585.

4-Methylbenzophenone, CA 134-84-9: Flammable, flash point above 383; soluble most organic solvents. Tm = 332.15 to 333.15; Tb = 599.15; Antoine (liq) A = 8.46912, B = 3759.3, C = 0, dev = 1.0, range = 450 to 492.

$C_{14}H_{12}O_2$ MW 212.25

Benzoin ±, CA 579-44-2: Flammable; soluble hot ethanol, benzene, chloroform. Tm = 406.15 to 407.15; Tb = 617.15; Antoine (liq) A = 7.10513, B = 2930.4, C = -41.589, dev = 1.0, range = 408 to 616.

Benzyl benzoate, CA 120-51-4: Irritant; moderately toxic; flammable, flash point 421; soluble ethanol, ether, benzene, chloroform. Tm = 294.15; Tb = 596.15 to 597.15; Vm = 0.1898 at 298.15; Antoine I (liq) A = 9.240, B = 4057, C = 0, dev = 1.0, range = 297 to 353; Antoine II (liq) A = 5.63847, B = 1666.706, C = -137.564, dev = 1.0, range = 497 to 602.

$C_{14}H_{12}O_3$ MW 228.25

Benzyl salicylate, CA 118-58-1: Flammable; soluble ethanol, ether. Tm = 298.15; Tb = 593.15; Vm = 0.1934 at 303.15; Antoine (liq) A = 9.081, B = 4111, C = 0, dev = 1.0, range = 295 to 343.

2-Hydroxy-4-methoxybenzophenone, CA 131-57-7: Flammable. Tm = 338.15 to 339.15; Antoine I (sol) A = 14.905, B = 6210, C = 0, dev = 5.0, range = 281 to 337; Antoine II (liq) A = 8.075, B = 3900, C = 0, dev = 5.0, range = 337 to 413.

$C_{14}H_{12}O_4$ MW 244.25

2,2'-Dihydroxy-4-methoxybenzophenone, CA 131-53-3: Flammable; soluble ether, ethanol, benzene. Tm = 342.15; Antoine I (sol) A = 31.115, B = 11910, C = 0, dev = 5.0, range = 303 to 342; Antoine II (liq) A = 7.825, B = 3950, C = 0, dev = 5.0, range = 342 to 481.

$C_{14}H_{13}NO$ MW 211.26

N,N-Diphenylacetamide, CA 519-87-9: Irritant; flammable; soluble ethanol, ether. Tm = 376.15; Antoine (sol) A = 14.2809, B = 6409.2, C = 0, dev = 1.0, range = 343 to 376.

$C_{14}H_{14}$ MW 182.26

2,2'Dimethylbiphenyl, CA 605-39-0: Flammable; soluble ethanol, ether, acetone, benzene. Tm = 291.15; Tb = 531.15; Vm = 0.1840 at 293.15; Antoine (sol) A = 8.9539, B = 3428, C = 0, dev = 1.0, range = 283 to 288.

3,3'-Dimethylbiphenyl, CA 612-75-9: Flammable, flash point above 383; soluble ethanol, ether, acetone, benzene. Tm = 282.15 to 282.65; Vm = 0.1824 at 293.15; Antoine (liq) A = 9.210, B = 3754, C = 0, dev = 1.0, range = 288 to 308.

1,1-Diphenylethane, CA 612-00-0: Flammable, flash point above 373; soluble ethanol, ether, benzene. Tm = 255.45; Tb = 541.15 to 543.15; Vm = 0.1823 at 303.15; Antoine (liq) A = 8.18469, B = 3261.481, C = 0, dev = 1.0, range = 348 to 405.

1,2-Diphenylethane, CA 103-29-7: Flammable, flash point 402; soluble ether, chloroform, carbon disulfide. Tm = 325.15; Tb = 557.15; Vm = 0.1902 at 333.15; Antoine I (sol) A = 11.319, B = 4386, C = 0, dev = 5.0, range = 286 to 308; Antoine II (liq) A = 6.93271, B = 2636.21, C = -22.009, dev = 1.0, range = 359 to 557.

$C_{14}H_{14}NO_3$ MW 244.27

Bis(para-methoxyphenyl)nitrogen oxide, CA 2643-00-7: Tm = 423.15; Antoine (sol) A = 9.505, B = 5260, C = 0, dev = 5.0, range = 328 to 363.

$C_{14}H_{14}N_2O_3$ MW 258.28

4,4'-Dimethoxy azoxybenzene, CA 1562-94-3: Flammable; soluble ethanol, ether, benzene. Tm = 391.15 to 392.15; Antoine (liq) A = 8.679, B = 3852, C = 0, dev = 5.0, range = 395 to 418.

$C_{14}H_{14}N_4O_2$ MW 270.29

3-Nitro-4'-(N,N-dimethylamino)azobenzene, CA 3837-55-6: Flammable. Tm =
437.15; Antoine (sol) A = 13.2999, B = 6954, C = 0, dev = 1.0 to 5.0,
range = 392 to 410.

4-Nitro-4'-(N,N-dimethylamino)azobenzene, CA 2491-74-9: Flammable. Tm =
508.65; Antoine (sol) A = 12.7489, B = 7065, C = 0, dev = 1.0 to 5.0,
range = 412 to 428.

$C_{14}H_{14}O$ MW 198.26

Benzyl ether, dibenzyl ether, CA 103-50-4: Irritant; flammable, flash
point 408; forms explosive peroxides; soluble ethanol, ether, chloroform.
Tm = 276.75; Tb = 568.15 to 571.15, decomposes; Vm = 0.1988 at 298.15;
Antoine I (liq) A = 11.0848, B = 6741.3, C = 197.25, dev = 1.0, range = 275
to 417; Antoine II (liq) A = 6.84322, B = 2507.3, C = -43.15, dev = 5.0,
range = 413 to 561.

Isopropyl 2-naphthyl ketone, CA 59502-28-2: Flammable; soluble ethanol,
ether, benzene. Tb = 585.15 to 587.15; Vm = 0.1867 at 273.15; Antoine
(liq) A = 7.84681, B = 3160.74, C = -44.973, dev = 1.0, range = 406 to 586.

2-(1-Phenylethyl)phenol, CA 52857-29-1: Irritant; flammable. Antoine
(liq) A = 4.04635, B = 693.56, C = -274.12, dev = 1.0, range = 443 to 521.

4-(1-Phenylethyl)phenol, CA 52857-30-4: Irritant; flammable. Tm = 330.15;
Antoine (liq) A = 3.9583, B = 672.8, C = -287.43, dev = 1.0, range = 447
to 517.

$C_{14}H_{14}O_2$ MW 214.26

2-(2-Biphenyloxy)ethanol, CA 7501-02-2: Flammable. Tm = 346.15; Antoine
(liq) A = 7.46772, B = 3111.16, C = -38.098, dev = 0.1 to 1.0, range = 410
to 608.

$C_{14}H_{15}N$ MW 197.28

Dibenzylamine, CA 103-49-1: Irritant; causes burns; flammable, flash point
416; soluble ethanol, ether. Tm = 247.15; Tb = 573.15, decomposes; Vm =
0.1924 at 295.15; Antoine (liq) A = 7.53835, B = 2928.32, C = -43.782,
dev = 1.0, range = 391 to 573.

Diphenylamine, N-ethyl, CA 606-99-5: Irritant; flammable; soluble ethanol,
ether. Tb = 568.15 to 569.15; Vm = 0.1894 at 293.15; Antoine (liq) A =
7.11081, B = 2648.59, C = -40.213, dev = 1.0 to 5.0, range = 371 to 559.

$C_{14}H_{15}N_3$ MW 225.29

4-(Dimethylamino)azobenzene, methyl yellow, CA 60-11-7: Toxic; carcinogenic; soluble most organic solvents. Tm = 390.15, decomposes; Antoine
(sol) A = 12.9189, B = 6042, C = 0, dev = 1.0 to 5.0, range = 351 to 364.

$C_{14}H_{17}Cl_3O_3$ MW 339.65

2,4,5-Trichlorophenoxyacetic acid, hexyl ester, CA 2630-13-9: Toxic; flammable; used as weed killer. Antoine (liq) A = 6.2146, B = 2054.76, C =
-152.266, dev = 1.0, range = 460 to 573.

394

$C_{14}H_{18}$ MW 186.30

Octahydroanthracene, octhracene, CA 1079-71-6: Flammable; soluble ethanol,
benzene. Tm = 351.15; Tb = 566.15; Vm = 0.1920 at 353.15; Antoine (liq)
A = 5.911, B = 2383.9, C = 0, dev not specified, range = 437 to 498.

Octahydrophenanthrene, octathrene, CA 5325-97-3: Flammable; soluble acetone, benzene, acetic acid. Tm = 289.85; Tb = 568.15; Vm = 0.1816 at
293.15; Antoine (liq) A = 7.134, B = 2916.3, C = 0, dev = 5.0, range = 402
to 570.

$C_{14}H_{18}Cl_2O_3$ MW 305.20

2,4-Dochlorophenoxyacetic acid, hexyl ester, CA 1917-95-9: Toxic; flammable; used as weed killer. Antoine (liq) A = 6.36885, B = 2168.848, C =
-130.87, dev = 1.0, range = 444 to 573.

2,4-Dichlorophenoxyacetic acid, isohexyl ester: Toxic; flammable; used as
weed killer. Antoine (liq) A = 8.54639, B = 4313.152, C = 44.213, dev =
1.0, range = 460 to 573.

$C_{14}H_{18}N_2O_5$ MW 294.31

Acetophenone, 4'-*tert*-butyl-2',6'-dimethyl-3',5'-dinitro, musk ketone,
CA 81-14-1: Soluble chloroform; used in perfumery. Tm = 408.75 to 409.15;
Antoine (sol) A = 12.3409, B = 5633, C = 0, dev = 1.0, range = 323 to 354.

$C_{14}H_{18}O$ MW 202.30

alpha-Pentylcinnamaldehyde, jasmine aldehyde, CA 122-40-7: Irritant; flammable, flash point above 383; used in perfumery. Tb = 560.15 to 563.15;
Vm = 0.2086 at 293.15; Antoine (liq) A = 9.426, B = 3934, C = 0, dev = 1.0,
range = 282 to 333.

$C_{14}H_{18}O_4$ MW 250.29

Dipropyl phthalate, CA 131-16-8: Irritant; flammable, flash point above
383; soluble ethanol, ether. Tb = 577.15 to 578.15; Vm = 0.2322 at 293.15;
Antoine (liq) A = 8.625, B = 3824, C = 0, dev = 5.0, range = 403 to 578.

$C_{14}H_{19}NO$ MW 217.31

1*H*-Azepine, hexahydro-1-(phenylacetyl), CA 18494-61-6: Irritant; flammable.
Antoine (liq) A = 7.176, B = 2817, C = 0, dev = 5.0, range = 370 to 418.

$C_{14}H_{20}$ MW 188.31

1-Cyclohexyl-1-phenylethane, CA 4413-16-5: Flammable; soluble most organic
solvents. Antoine (liq) A = 9.39687, B = 3697.22, C = 0, dev = 1.0 to 5.0,
range = 359 to 400.

1-Cyclohexyl-2-phenylethane, CA 1603-61-8: Flammable; soluble most organic
solvents. Antoine (liq) A = 7.93156, B = 3170.788, C = 0, dev = 1.0,
range = 372 to 406.

(continues)

$C_{14}H_{20}$ *(continued)*

1-Cyclopentyl-3-phenylpropane, CA 2883-12-7: Flammable; soluble most organic solvents. Tb = 544.15; Vm = 0.2040 at 293.15; Antoine (liq) A = 8.00582, B = 3200.751, C = 0, dev = 1.0 to 5.0, range = 373 to 540.

1,8-Cyclotetradecadiyne, CA 1540-80-3: Tm = 370.15 to 371.15; Antoine (sol) A = 25.3199, B = 8758.3, C = 0, dev = 1.0, range = 316 to 333.

$C_{14}H_{20}Cl_2$ MW 259.22

1,2-Dichloro-3,4,5,6-tetraethylbenzene: Irritant; flammable. Antoine (liq) A = 6.59423, B = 2309.3, C = -71.762, dev = 1.0, range = 378 to 575.

1,4-Dichloro-2,3,5,6-tetraethylbenzene: Irritant; flammable. Antoine (liq) A = 6.53804, B = 2337.4, C = -53.748, dev = 1.0, range = 364 to 570.

$C_{14}H_{20}O_3$ MW 236.31

2-(4-*tert*-Butylphenoxy)ethyl acetate: Flammable. Antoine (liq) A = 6.2877, B = 1925.42, C = -128.063, dev = 1.0 to 5.0, range = 391 to 578.

$C_{14}H_{21}F_3N_2O_4$ MW 338.33

Proline, 1-[*N*-(trifluoroacetyl)-*levo*-leucyl]methyl ester: Flammable. Tm = 366.95; Antoine I (sol) A = 14.405, B = 6339, C = 0, dev = 5.0, range = 313 to 366; Antoine II (liq) A = 12.205, B = 5525, C = 0, dev = 5.0, range = 366 to 453.

$C_{14}H_{22}$ MW 190.33

1,3-Di-*tert*-butylbenzene, CA 1014-60-4: Flammable, flash point 356; soluble most organic solvents. Tm = 283.75; Vm = 0.2216 at 293.15; Antoine (liq) A = 8.2807, B = 3029.3, C = 0, dev = 5.0, range = 346 to 374.

1,4-Di-*tert*-butylbenzene, CA 1012-72-2: Flammable; soluble most organic solvents. Tm = 353.15 to 354.15; Tb = 510.15; Antoine (sol) A = 14.871, B = 4346, C = 0, dev = 1.0, range = 285 to 325.

Octylbenzene, CA 2189-60-8: Flammable, flash point 380; soluble most organic solvents. Tf = 266.15; Tb = 537.15 to 538.15; Vm = 0.2218 at 293.15; Antoine (liq) A = 8.35571, B = 3293.744, C = 0, dev = 1.0, range = 368 to 400.

2-Phenyloctane, CA 777-22-0: Flammable; soluble most organic solvents. Vm = 0.2210 at 293.15; Antoine (liq) A = 8.33643, B = 3215.816, C = 0, dev = 1.0 to 5.0, range = 361 to 392.

1,2,3,4-Tetraethylbenzene, CA 642-32-0: Flammable; soluble most organic solvents. Tm = 284.95; Tb = 525.75; Vm = 0.2145 at 293.15; Antoine (liq) A = 5.3269, B = 1161.1, C = -176.65, dev = 5.0, range = 423 to 525.

1,2,3,5-Tetraethylbenzene, CA 38842-05-6: Flammable; soluble most organic solvents. Tm = 252.15; Antoine (liq) A = 5.0319, B = 984.5, C = -196.75, dev = 5.0, range = 413 to 521.

1,2,4,5-Tetraethylbenzene, CA 635-81-4: Flammable; soluble most organic solvents. Tm = 284.75; Tb = 523.15; Vm = 0.2166 at 293.15; Antoine (liq) A = 6.86422, B = 2388.69, C = -29.573, dev = 1.0, range = 338 to 521.

$C_{14}H_{22}O$ MW 206.33

2,4-Di-*tert*-butylphenol, CA 96-76-4: Irritant; flammable, flash point 388; soluble most organic solvents. Tm = 329.65; Tb = 537.15; Antoine (liq) A = 6.86651, B = 2326.23, C = -58.06, dev = 1.0, range = 403 to 537.

2,6-Di-*tert*-butylphenol, CA 128-39-2: Irritant; flammable, flash point 391; soluble most organic solvents. Tm = 309.65; Tb = 526.15; Antoine (liq) A = 6.30104, B = 1887.772, C = -90.568, dev = 1.0 to 5.0, range = 386 to 530.

3,5-Di-*tert*-butylphenol, CA 1138-52-9: Irritant; flammable; soluble most organic solvents. Tm = 365.15 to 366.65; Antoine (sol) A = 7.90592, B = 3562.811, C = 0, dev = 5.0, range = 302 to 325.

4-(1,1-Diethylbutyl)phenol, CA 63264-81-3: Irritant; flammable; soluble most organic solvents. Antoine (liq) A = 7.8093, B = 2850.5, C = -47.65, dev = 5.0, range = 404 to 549.

2,4-Diisobutyl phenol: Irritant; flammable; soluble most organic solvents. Tm = 357.15; Antoine (liq) A = 6.81346, B = 2235.6, C = -87.15, dev = 5.0, range = 448 to 598.

4-[(1,2-Dimethyl-1-ethyl)butyl] phenol, CA 59048-99-6: Irritant; flammable; soluble most organic solvents. Antoine (liq) A = 8.0478, B = 3452.3, C = 4.55, dev = 5.0, range = 415 to 578.

4-[(1,3-Dimethyl-1-ethyl)butyl] phenol: Irritant; flammable; soluble most organic solvents. Antoine (liq) A = 9.1737, B = 4654.1, C = 88.55, dev = 5.0, range = 409 to 571.

4-[(2,2-Dimethyl-1-ethyl)butyl] phenol: Irritant; flammable; soluble most organic solvents. Antoine (liq) A = 7.643, B = 2807.1, C = -44.65, dev = 5.0, range = 413 to 553.

beta-Irone, CA 79-70-9: Flammable; soluble most organic solvents. Vm = 0.2187 at 294.15; Antoine (liq) A = 9.41568, B = 3768.811, C = 0, dev = 1.0 to 5.0, range = 288 to 333.

alpha-Isomethylionone, CA 127-51-5: Flammable; soluble most organic solvents. Antoine (liq) A = 9.31646, B = 3631.611, C = 0, dev = 1.0 to 5.0, range = 288 to 333.

4-[(1-Methyl-1-ethyl)pentyl] phenol, CA 1988-35-8: Irritant; flammable; soluble most organic solvents. Antoine (liq) A = 8.2721, B = 3729, C = 28.25, dev = 5.0, range = 413 to 578.

alpha-Methylionone, CA 127-42-4: Flammable; soluble most organic solvents. Antoine (liq) A = 9.35062, B = 3660.403, C = 0, dev = 1.0 to 5.0, range = 288 to 333.

beta-Methylionone, CA 127-43-5: Flammable; soluble most organic solvents. Antoine (liq) A = 9.30984, B = 3674.282, C = 0, dev = 1.0 to 5.0, range = 288 to 333.

4-(1,1,3,3-Tetramethylbutyl) phenol, CA 140-66-9: Irritant; flammable; soluble most organic solvents. Tm = 361.15; Tb = 553.15 to 575.15; Antoine (liq) A = 6.10803, B = 1812.48, C = -121.732, dev = 1.0, range = 381 to 563.

$C_{14}H_{22}O_{11}$ MW 366.32

Diethylene glycol, *O,O*-dicarboxylic acid, di[1-(methoxycarbonyl)ethyl] ester: Irritant; flammable. Antoine (liq) A = 10.0797, B = 5128, C = 0, dev = 5.0, range = 403 to 493.

$C_{14}H_{24}$ MW 192.34

Tetradecahydroanthracene, *trans-anti-trans*, CA 28071-99-0: Flammable. Tm = 320.15 to 322.15; Antoine (sol) A = 8.917, B = 3454, C = 0, dev = 5.0, range = 269 to 313.

Tetradecahydroanthracene, *trans-syn-trans*, CA 1755-19-7: Flammable. Tm = 359.15 to 361.65; Antoine (sol) A = 12.276, B = 4602, C = 0, dev = 5.0, range = 293 to 335.

$C_{14}H_{24}N_2$ MW 220.36

1,4-Phenylenediamine, *N,N'*-di-*sec*-butyl, CA 101-96-2: Irritant; flammable. Antoine (liq) A = 8.47097, B = 3673.71, C = 0, dev = 1.0 to 5.0, range = 370 to 507.

$C_{14}H_{24}O$ MW 208.34

4,8-Decadienal, 2,2,5,9-tetramethyl, CA 53131-20-7: Flammable; polymer-izes. Antoine (liq) A = 8.83723, B = 3470.931, C = 0, dev = 5.0, range = 353 to 416.

$C_{14}H_{24}O_2$ MW 224.34

Borneol butyrate: Flammable. Antoine (liq) A = 7.07503, B = 2419.38, C = -42.914, dev = 1.0, range = 347 to 520.

Borneol isobutyrate ±, CA 24717-86-0: Flammable. Antoine (liq) A = 7.02044, B = 2371.6, C = -43.276, dev = 1.0, range = 343 to 516.

Geraniol butyrate: Flammable. Antoine (liq) A = 8.03091, B = 2999.5, C = -32.65, dev = 1.0, range = 369 to 531.

Geraniol isobutyrate: Flammable. Antoine (liq) A = 7.86589, B = 2839.33, C = -39.524, dev = 1.0, range = 363 to 524.

$C_{14}H_{26}$ MW 194.36

1-Cyclohexyl-3-cyclopentylpropane, CA 2883-07-0: Flammable; soluble most organic solvents. Tb = 541.15 to 543.15; Vm = 0.2221 at 293.15; Antoine (liq) A = 8.49515, B = 3370.247, C = 0, dev = 1.0, range = 371 to 403.

1,1-Dicyclohexylethane, CA 2319-61-1: Flammable; soluble most organic sol-vents. Tm = 252.28; Vm = 0.2143 at 298.15; Antoine (liq) A = 8.19444, B = 3245.106, C = 0, dev = 1.0, range = 370 to 402.

1,2-Dicyclohexylethane, CA 3321-50-4: Flammable; soluble most organic sol-vents. Tm = 284.6; Antoine (liq) A = 8.61411, B = 3414.386, C = 0, dev = 1.0, range = 371 to 402.

$C_{14}H_{26}O$ MW 210.36

2,2,5,9-Tetramethyl-4,8-decadiene-1-ol, *cis*, CA 53965-17-6: Flammable; polymerizes. Antoine (liq) A = 12.34282, B = 4910.134, C = 0, dev = 1.0 to 5.0, range = 363 to 393.

2,2,5,9-Tetramethyl-4,8-decadiene-1-ol, *trans*, CA 53965-18-7: Flammable; polymerizes. Antoine (liq) A = 11.18036, B = 4507.423, C = 0, dev = 1.0 to 5.0, range = 363 to 393.

$C_{14}H_{26}O_2$ MW 226.36

Decyl methacrylate, CA 3179-47-3: Irritant; flammable; polymerizes. Antoine (liq) A = 8.061, B = 3276, C = 0, dev = 5.0, range = 350 to 541.

$C_{14}H_{26}O_3$ MW 242.36

1,7-Dioxa-8-cyclohexadecanone, CA 5963-13-3: Flammable. Antoine (liq) A = 8.83629, B = 3831.671, C = 0, dev = 5.0, range = 403 to 453.

3-Heptyl-4-acetoxytetrahydro-2*H*-pyran, CA 23144-23-2: Flammable. Antoine (liq) A = 9.1569, B = 3886, C = 0, dev = 1.0 to 5.0, range = 383 to 453.

Nonyl levulinate: Flammable. Antoine (liq) A = 8.04018, B = 3340.3, C = -17.53, dev = 1.0, range = 423 to 571.

$C_{14}H_{26}O_4$ MW 258.36

Adipic acid, dibutyl ester, CA 105-99-7: Flammable; soluble ethanol, ether. Tm = 240.75; Vm = 0.2688 at 293.15; Antoine (liq) A = 8.3809, B = 3590, C = 0, dev = 1.0 to 5.0, range = 435 to 563.

Isopentylmalonic acid, diethyl ester: Flammable. Antoine (liq) A = 5.5837, B = 1283.238, C = -172.451, dev = 5.0, range = 388 to 526.

2-Methylheptane-5,5-dicarboxylic acid, diethyl ester: Flammable. Antoine (liq) A = 9.1903, B = 3661.15, C = 0, dev = 1.0, range = 394 to 427.

Sebacic acid, diethyl ester, CA 110-40-7: Irritant; flammable, flash point above 383; moderately soluble hot water; soluble most organic solvents. Tm = 278.15; Tb = 580.15, decomposes; Vm = 0.2678 at 293.15; Antoine (liq) A = 7.56819, B = 2914.16, C = -54.683, dev = 1.0, range = 398 to 579.

$C_{14}H_{26}O_5$ MW 274.36

Ethyl[1-(octyloxycarbonyl)ethyl] carbonate: Toxic; flammable. Antoine (liq) A = 8.8905, B = 3866.6, C = 0, dev = 1.0, range = 413 to 513.

Hexyl[1-butoxycarbonyl)ethyl] carbonate: Toxic; flammable. Antoine (liq) A = 8.683, B = 3767, C = 0, dev = 5.0, range = 357 to 501.

$C_{14}H_{27}N$ MW 209.37

Myristonitrile, CA 629-63-0: Irritant; flammable; soluble most organic solvents. Tm = 292.15; Vm = 0.2528 at 292.15; Antoine (liq) A = 6.71082, B = 2328.3, C = -85.16, dev = 1.0 to 5.0, range = 391 to 580.

$C_{14}H_{28}$ MW 196.38

3-*tert*-Butyl-1-methyl-4-isopropylcyclohexane: Flammable; soluble most organic solvents. Antoine (liq) A = 6.771, B = 2226.7, C = -37.72, dev = 1.0, range = 329 to 505.

Cyclotetradecane, CA 295-17-0: Flammable; soluble most organic solvents. Tm = 327.45; Antoine (sol) A = 20.132, B = 6828, C = 0, dev = 5.0, range = 294 to 313.

2,2,3,5,5,6,6-Heptamethyl-3-heptene: Flammable; soluble most organic solvents. Antoine (liq) A = 10.2554, B = 4586.5, C = 98.05, dev = 1.0, range = 303 to 355.

(1-Methylheptyl)cyclohexane: Flammable; soluble most organic solvents. Antoine (liq) A = 8.07679, B = 3154.07, C = 0, dev = 1.0, range = 364 to 397.

Octylcyclohexane, CA 1795-15-9: Flammable; soluble most organic solvents. Tm = 253.15; Tb = 537.15; Vm = 0.2413 at 293.15; Antoine (liq) A = 8.33380, B = 3276.77, C = 0, dev = 1.0, range = 367 to 399.

1-Tetradecene, CA 1120-36-1: Flammable, flash point 388; soluble most organic solvents. Tm = 261.15; Tb = 524.15; Vm = 0.2546 at 293.15; Antoine (liq) A = 6.15309, B = 1752.085, C = -101.857, dev = 0.02, range = 430 to 527.

$C_{14}H_{28}O$ MW 212.37

1-Octylcyclohexanol, CA 5770-04-7: Flammable; soluble ethanol. Antoine (liq) A = 13.841, B = 5516, C = 0, dev = 5.0, range = 373 to 403.

Tetradecanal, myristaldehyde, CA 124-25-4: Flammable, flash point above 383; polymerizes; soluble ethanol, ether, acetone. Tm = 303.15; Antoine (liq) A = 6.46866, B = 2242.96, C = -68.31, dev = 1.0, range = 372 to 571.

2-Tetradecanone, CA 2345-27-9: Flammable; soluble ethanol, acetone. Tm = 306.15 to 307.15; Antoine I (liq) A = 7.51586, B = 2877.29, C = -29.098, dev = 1.0 to 5.0, range = 372 to 551; Antoine II (liq) A = 6.27079, B = 1909.569, C = -107.79, dev = 1.0 to 5.0, range = 411 to 560; Antoine III (liq) A = 6.43849, B = 2074.982, C = -87.434, dev = 1.0 to 5.0, range = 549 to 643.

7-Tetradecanone, CA 6137-34-4: Flammable; soluble ethanol, acetone. Tm = 304.15; Antoine (liq) A = 8.530, B = 3497, C = 0, dev = above 5.0, range = 438 to 462.

$C_{14}H_{28}O_2$ MW 228.37

Dodecyl acetate, lauryl acetate, CA 112-66-3: Flammable, flash point above 383. Tb = 533.15 to 543.15; Vm = 0.2640 at 293.15; Antoine (liq) A = 8.84841, B = 3683.683, C = 0, dev = 5.0, range = 398 to 540.

Lauric acid, ethyl ester, CA 106-33-2: Flammable; soluble ethanol, ether. Tf = 271.35; Tb = 542.15; Vm = 0.2650 at 293.15; Antoine (liq) A = 8.49839, B = 3509.089, C = 0, dev = 1.0, range = 386 to 435.

Methyl tridecanoate, CA 1731-88-0: Flammable, flash point above 383; soluble ethanol. Tm = 279.65; Vm = 0.2643 at 293.15; Antoine (liq) A = 6.03687, B = 1740.55, C = -126.272, dev not specified, range = 277 to 504.

(continues)

$C_{14}H_{28}O_2$ *(continued)*

Tetradecanoic acid, myristic acid, CA 544-63-8: Flammable, flash point above 383; soluble ethanol, ether, benzene, chloroform. Tm = 327.15; Tb = 599.35; Vm = 0.2662 at 333.15; Antoine I (liq) A = 6.78583, B = 2143.4, C = -143.27, dev = 1.0, range = 383 to 459; Antoine II (liq) A = 6.11582, B = 1751.35, C = -172.704, dev = 1.0, range = 423 to 599.

$C_{14}H_{28}O_3$ MW 244.37

3-Methoxypropionic acid, decyl ester: Flammable. Antoine (liq) A = 8.444, B = 3600, C = 0, dev = 5.0, range = 403 to 513.

$C_{14}H_{29}Br$ MW 277.29

1-Bromotetradecane, myristyl bromide, CA 112-71-0: Flammable, flash point above 383; soluble ethanol, acetone, benzene. Tm = 278.75; Tb = 580.15; Vm = 0.2724 at 298.15; Antoine (liq) A = 6.555, B = 2220.2, C = -92.15, dev = 5.0, range = 437 to 645.

$C_{14}H_{29}Cl$ MW 232.84

1-Chlorotetradecane, myristyl chloride, CA 2425-54-9: Irritant; flammable, flash point above 383; soluble most organic solvents. Tm = 278.05; Tb = 565.15; Vm = 0.2687 at 293.15; Antoine (liq) A = 6.01207, B = 1776.491, C = -126.97, dev = 1.0 to 5.0, range = 414 to 570.

$C_{14}H_{29}F$ MW 216.38

1-Fluorotetradecane, myristyl fluoride, CA 593-33-9: Flammable. Tm = 277.15; Antoine (liq) A = 6.574, B = 2048.3, C = -83.15, dev = 5.0, range = 400 to 593.

$C_{14}H_{29}I$ MW 324.29

1-Iodotetradecane, myristyl iodide, CA 19218-94-1: Flammable; light-sensitive. Tm = 286.75; Antoine (liq) A = 6.447, B = 2245.4, C = -97.15, dev = 5.0, range = 452 to 672.

$C_{14}H_{29}NO$ MW 227.39

Myristamide, CA 638-58-4: Irritant; flammable; soluble ethanol. Tm = 378.15 to 380.15; Antoine (sol) A = 20.0649, B = 8746, C = 0, dev = 1.0, range = 358 to 373.

$C_{14}H_{30}$ MW 198.39

2,3-Dimethyldodecane, CA 6117-98-2: Flammable; soluble most organic solvents. Tm = 222.55; Antoine (liq) A = 7.6138, B = 2989.35, C = 13.98, dev = 1.0, range = 385 to 519.

2,4-Dimethyldodecane, CA 6117-99-3: Flammable; soluble most organic solvents. Tm = 208.65; Antoine (liq) A = 7.4555, B = 2747.51, C = -5.04, dev = 1.0, range = 379 to 509.

(continues)

$C_{14}H_{30}$ *(continued)*

2,2,3,3,5,6,6-Heptamethylheptane, CA 7225-67-4: Flammable; soluble most organic solvents. Antoine (liq) A = 7.3122, B = 1890.69, C = -60.63, dev = 1.0, range = 313 to 366.

2-Methyltridecane, CA 1560-96-9: Flammable; soluble most organic solvents. Tm = 246.65; Antoine (liq) A = 7.30554, B = 2651.7, C = -20.25, dev = 1.0, range = 388 to 520.

3-Methyltridecane, CA 6418-41-3: Flammable; soluble most organic solvents. Tm = 235.65; Antoine (liq) A = 7.56334, B = 2910.8, C = 2.20, dev = 1.0, range = 389 to 521.

4-Methyltridecane, CA 26730-12-1: Flammable; soluble most organic solvents. Tm = 229.65; Antoine (liq) A = 7.3287, B = 2725.6, C = -7.65, dev = 5.0, range = 386 to 520.

5-Methyltridecane, CA 25117-31-1: Flammable; soluble most organic solvents. Tm = 224.65; Antoine (liq) A = 7.57655, B = 2934.2, C = 8.55, dev = 1.0, range = 385 to 518.

7-Methyltridecane, CA 26730-14-3: Flammable; soluble most organic solvents. Tm = 235.95; Vm = 0.2599 at 293.15; Antoine (liq) A = 8.05772, B = 3082.2, C = 0, dev = 1.0, range = 357 to 389.

Tetradecane, CA 629-59-4: Flammable, flash point 373; soluble most organic solvents. Tm = 279.05; Tb = 526.85; Vm = 0.2601 at 293.15; Antoine I (liq) A = 6.62828, B = 2063.84, C = -77.378, dev = 1.0, range = 313 to 433; Antoine II (liq) A = 6.1379, B = 1740.88, C = -105.43, dev = 0.02, range = 432 to 529.

2,4,6-Trimethylundecane: Flammable; soluble most organic solvents. Antoine (liq) A = 7.6595, B = 2768.67, C = -0.90, dev = 1.0, range = 368 to 491.

$C_{14}H_{30}O$ MW 214.39

Diheptyl ether, CA 629-64-1: Flammable; forms explosive peroxides; soluble ethanol, ether. Tb = 531.65; Vm = 0.2677 at 293.15; Antoine (liq) A = 7.0636, B = 2447.8, C = -51.85, dev = 1.0 to 5.0, range = 360 to 547.

1-Tetradecanol, myristyl alcohol, CA 112-72-1: Flammable, flash point 414; soluble ether, benzene. Tm = 312.15 to 312.65; Tb = 562.15; Vm = 0.2603 at 313.15; Antoine I (liq) A = 3.33507, B = 787.98, C = -207.71, dev = 1.0, range = 317 to 358; Antoine II (liq) A = 5.60897, B = 1412.907, C = -177.782, dev = 1.0 to 5.0, range = 424 to 569.

$C_{14}H_{30}O_2$ MW 230.39

2-(Dodecyloxy)ethanol, CA 4536-30-5: Flammable. Antoine (liq) A = 8.62969, B = 3737.075, C = 0, dev = 1.0 to 5.0, range = 414 to 467.

$C_{14}H_{31}N$ MW 213.41

Diheptyl amine, CA 2470-68-0: Irritant; flammable; soluble ethanol, ether. Tm = 274.15; Tb = 544.15; Antoine (liq) A = 6.46872, B = 2048.5, C = -86.15, dev = 5.0, range = 435 to 605.

(continues)

$C_{14}H_{31}N$ (continued)

Dodecyl amine, N,N-dimethyl, CA 112-18-5: Irritant; corrosive; flammable, flash point 383; soluble ethanol, ether. Tm = 253.15; Antoine (liq) A = 6.43576, B = 2015.7, C = -89.15, dev = 5.0, range = 380 to 604.

Tetradecyl amine, myristyl amine, CA 2016-42-4: Irritant; flammable, flash point above 383; soluble most organic solvents. Tm = 319.35; Tb = 564.15; Vm = 0.2642 at 293.15; Antoine (liq) A = 6.5759, B = 2170.6, C = -89.45, dev = 1.0, range = 471 to 577.

$C_{15}H_9N_3$ MW 231.26

Pyrido[2,3-f][1,7]phenanthroline: Tm = 498.15; Antoine (liq) A = 6.8119, B = 3400, C = 0, dev = 1.0, range = 648 to 707.

Pyrido[3,2-f][1,7]phenanthroline: Tm = 509.15; Antoine (liq) A = 6.9911, B = 3521, C = 0, dev = 1.0, range = 648 to 706.

$C_{15}H_{10}N_2O_2$ MW 250.26

Diphenylmethane, 2,2'-diisocyanate, CA 2536-05-2: Irritant; toxic; flammable; reacts violently with water. Tm = 320.15; Antoine (liq) A = 10.43882, B = 5356.802, C = 23.959, dev = 0.1 to 1.0, range = 343 to 413.

Diphenylmethane, 2,4'-diisocyanate, CA 5873-54-1: Irritant; toxic; flammable; reacts violently with water. Tm = 310.35; Antoine (liq) A = 10.19136, B = 5344.164, C = 25.226, dev = 0.1 to 1.0, range = 343 to 413.

Diphenylmethane, 4,4'-diisocyanate, CA 101-68-8: Irritant; toxic; causes dermatitis; flammable; reacts violently with water; used for polyurethanes. Tm = 313.75; Antoine I (liq) A = 10.1031, B = 4730, C = 0, dev = 1.0 to 5.0, range = 343 to 413; Antoine II (liq) A = 4.421, B = 970, C = -253.23, dev = 5.0, range = 442 to 530.

$C_{15}H_{10}O$ MW 206.24

2,3-diphenyl-2-cyclopropene-1-one, CA 886-38-4: Tm = 392.15 to 393.15; Antoine (sol) A = 17.14964, B = 7395.86, C = 0, dev = 1.0, range = 323 to 343.

$C_{15}H_{11}F_3O_3$ MW 296.25

2-Hydroxy-2'-trifluoromethyl-4-methoxybenzophenone, CA 3119-86-6: Antoine (sol) A = -4.740, B = 692.3, C = 0, dev = 5.0, range = 323 to 363.

2-Hydroxy-3'-trifluoromethyl-4-methoxybenzophenone, CA 7396-89-6: Antoine (sol) A = 11.849, B = 5422, C = 0, dev = 5.0, range = 313 to 323.

2-Hydroxy-4'-trifluoromethyl-4-methoxybenzophenone, CA 7396-90-9: Antoine (sol) A = 9.661, B = 4753, C = 0, dev = 5.0, range = 313 to 333.

$C_{15}H_{11}NO_2$ MW 237.26

Anthraquinone, 1-methylamino, CA 82-38-2: Irritant; soluble ethanol, benzene, chloroform. Tm = 443.15; Antoine I (sol) A = 12.914, B = 6477, C = 0, dev = 5.0, range = 384 to 405; Antoine II (liq) A = 10.19981, B = 5405.154, C = 0, dev = 5.0, range = 433 to 493.

$C_{15}H_{12}$ MW 192.26

9-Methylanthracene, CA 779-02-2: Irritant; flammable; soluble most organic solvents. Tm = 354.65; Vm = 0.1805 at 372.15; Antoine (liq) A = 11.683, B = 5168, C = 0, dev not specified, range = 354 to 402.

$C_{15}H_{12}N_2O_5$ MW 300.27

C. I. disperse blue 95, CA 1562-85-2: Antoine (sol) A = 7.86182, B = 4606.119, C = 0, dev = 5.0, range = 433 to 493.

$C_{15}H_{12}O_2$ MW 224.26

1,3-Diphenyl-1,3-propanedione, dibenzoyl methane, CA 120-46-7: Irritant; flammable; soluble ether, chloroform. Tm = 354.15; Antoine (liq) A = 7.5479, B = 3141.3, C = 0, dev = 1.0, range = 368 to 383.

$C_{15}H_{14}Cl_3O_2PS$ MW 395.67

(Chloromethyl)thiophosphonic acid, O,O'-bis(2-chloro-4-methylphenyl) ester, CA 57875-65-7: Toxic insecticide; flammable. Antoine (liq) A = 9.6606, B = 4867, C = 0, dev not specified, range = 343 to 365.

$C_{15}H_{14}O$ MW 210.28

Acetone, 1,3-diphenyl, CA 102-04-5: Flammable, flash point above 383; soluble ethanol, ether. Tm = 308.15 to 309.15; Tb = 604.15; Antoine (liq) A = 7.13943, B = 2942.76, C = -30.65, dev = 1.0, range = 398 to 604.

$C_{15}H_{14}O_2$ MW 226.27

1-Biphenyloxy-2,3-epoxypropane, 2-biphenylyl glycidyl ether, CA 7144-65-2: Irritant; flammable. Tm = 303.15 to 305.15; Antoine (liq) A = 6.15246, B = 2004.81, C = -129.732, dev = 1.0, range = 408 to 613.

$C_{15}H_{14}O_5$ MW 274.27

2,2'-Dihydroxy-4,4'dimethoxybenzophenone, CA 131-54-4: Irritant. Tm = 408.15; Antoine I (sol) A = 16.325, B = 7675, C = 0, dev = 5.0, range = 325 to 408; Antoine II (liq) A = 7.545, B = 4045, C = 0, dev = 5.0, range = 408 to 497.

$C_{15}H_{15}Cl$ MW 230.74

Chlorodi-4-tolylmethane, CA 13389-70-3: Irritant; flammable. Antoine (liq) A = 8.795, B = 3929, C = 0, dev = 5.0, range = 406 to 453.

$C_{15}H_{16}$ MW 196.29

1,3-Diphenylpropane, CA 1081-75-0: Irritant; flammable. Tm = 279.15; Tb = 571.15 to 572.15; Vm = 0.1949 at 293.15; Antoine (liq) A = 7.615, B = 3211, C = 0, dev = 5.0, range = 342 to 577.

$C_{15}H_{16}N_4O_2$ MW 284.32

3-Methyl-3'-nitro-4-N,N-dimethylamino azobenzene: Irritant; flammable.
Tm = 347.15; Antoine (liq) A = 9.6439, B = 5153, C = 0, dev = 5.0, range =
370 to 388.

3-Methyl-4'-nitro-4-N,N-dimethylamino azobenzene, CA 92114-99-3: Irritant;
flammable. Tm = 399.15; Antoine (sol) A = 13.0789, B = 6566, C = 0, dev =
5.0, range = 371 to 392.

$C_{15}H_{16}O$ MW 212.29

Di-(4-tolyl)methanol, CA 885-77-8: Flammable; soluble most organic solvents. Tm = 342.15; Antoine (liq) A = 9.389, B = 4267, C = 0, dev = 5.0,
range = 413 to 478.

1-Isovaleronaphthone: Flammable. Antoine (liq) A = 7.60262, B = 3009.91,
C = -55.246, dev = 1.0, range = 409 to 593.

$C_{15}H_{16}O_2$ MW 228.29

Bisphenol A, CA 80-05-7: Irritant; moderately toxic; soluble most organic
solvents; decomposes above 493; polymer intermediate. Tm = 429.15 to
430.15; Antoine (liq) A = 9.4293, B = 4439.51, C = -35.708, dev = 1.0,
range = 466 to 634.

$C_{15}H_{17}NO_2$ MW 243.30

Phenylamine, N-(2-hydroxy-3-phenoxypropyl): Irritant; flammable. Tm =
337.15; Antoine I (sol) A = 10.4205, B = 5220.8, C = 0, dev not specified,
range = 323 to 333; Antoine II (liq) A = 12.5775, B = 5948.8, C = 0, dev
not specified, range = 343 to 373.

$C_{15}H_{18}$ MW 198.31

1-Pentylnaphthalene, CA 86-89-5: Flammable. Tm = 251.15; Tb = 580.15;
Vm = 0.2054 at 293.15; Antoine (liq) A = 6.6412, B = 2281.24, C = -70.94,
dev = 1.0, range = 415 to 535.

$C_{15}H_{18}O$ MW 214.31

2,4,6-Triallylphenol, CA 20490-22-6: Irritant; toxic; flammable; soluble
most organic solvents. Tb = 567.15; Antoine (liq) A = 6.94835, B = 2596,
C = -42.69, dev = 1.0, range = 423 to 571.

$C_{15}H_{19}Cl_3O_3$ MW 353.67

2,4,5-Trichlorophenoxyacetic acid, heptyl ester: Toxic; flammable; used as
weed killer. Antoine (liq) A = 5.57535, B = 1619.175, C = -199.388, dev =
1.0, range = 460 to 573.

$C_{15}H_{20}Cl_2O_3$ MW 319.23

2,4-Dichlorophenoxyacetic acid, heptyl ester, CA 1917-96-0: Toxic; flammable; used as weed killer. Antoine (liq) A = 5.85825, B = 1750.755, C =
-182.114, dev = 1.0, range = 460 to 573.
                                                                            (continues)

$C_{15}H_{20}Cl_2O_3$ *(continued)*

2,4-Dichlorophenoxyacetic acid, 1-propylbutyl ester: Toxic; flammable; used as weed killer. Antoine (liq) A = 7.0787, B = 2719.532, C = -85.189, dev = 1.0, range = 460 to 573.

$C_{15}H_{20}Cl_2O_4$ MW 335.23

2,4-Dichlorophenoxyacetic acid, (1-methyl-2-butoxy)ethyl ester, CA 3966-11-8: Toxic; flammable; used as weed killer. Antoine (liq) A = 6.67838, B = 2372.605, C = -117.915, dev = 1.0, range = 443 to 573.

$C_{15}H_{20}O_2$ MW 232.32

Helenine, alantolactone, CA 1407-14-3: Flammable; soluble most organic solvents. Tm = 351.15 to 353.15; Tb = 548.15; Antoine (liq) A = 11.71, B = 4971.49, C = -36.022, dev = 5.0, range = 430 to 548.

$C_{15}H_{22}N_2O_2$ MW 262.35

Dicyclohexylmethane-4,4'-diisocyanate, CA 5124-30-1: Irritant; toxic; flammable; reacts violently with water. Tm = 333.15 to 344.15; Vm = 0.2550 at 343.15; Antoine (liq) A = 10.47001, B = 5657.416, C = 54.568, dev not specified, range = 326 to 404.

$C_{15}H_{22}O_2$ MW 234.34

3,5-Di-*tert*-butyl benzoic acid, CA 16225-26-6: Used in resins. Tm = 444.65 to 445.65; Antoine (sol) A = 12.4749, B = 5661.1, C = 0, dev = 1.0 to 5.0, range = 339 to 357.

$C_{15}H_{24}$ MW 204.35

Cadinene, probably mixed isomers: Flammable; soluble ether, light hydrocarbons. Tb = 547.15; Vm = 0.2214 at 293.15; Antoine (liq) A = 7.65354, B = 2910.63, C = -32.756, dev = 1.0, range = 374 to 548.

$C_{15}H_{24}O$ MW 220.35

2,4-Di-*tert*-butyl-5-methyl phenol, CA 497-39-2: Irritant; flammable; soluble most organic solvents. Tm = 335.25; Tb = 555.15; Vm = 0.2408 at 353.15; Antoine (liq) A = 7.5815, B = 2898.85, C = -35.199, dev = 1.0, range = 376 to 555.

2,4-Di-*tert*-butyl-6-methyl phenol, CA 616-55-7: Irritant; flammable; soluble most organic solvents. Tm = 324.15; Tb = 542.15; Vm = 0.2473 at 353.15; Antoine (liq) A = 7.11937, B = 2611.93, C = -31.852, dev = 1.0, range = 359 to 543.

2,6-Di-*tert*-butyl-4-methyl phenol, BHT, CA 128-37-0: Irritant; moderately toxic; suspected teratogen; soluble most organic solvents; used as antioxidant. Tm = 344.15; Tb = 538.15; Vm = 0.2466 at 348.15; Antoine I (sol) A = 11.9631, B = 4588.2, C = 0, dev = 5.0, range = 303 to 343; Antoine II (liq) A = 7.15913, B = 2553.55, C = -40.355, dev = 1.0, range = 358 to 536.

(continues)

$C_{15}H_{24}O$ *(continued)*

2-Methyl-4-(1,1,3,3-tetramethylbutyl) phenol, CA 2219-84-3: Irritant; flammable; soluble most organic solvents. Tm = 322.65; Antoine (liq) A = 7.02259, B = 2317.8, C = -86.15, dev = 5.0, range = 447 to 683.

3-Methyl-4-(1,1,3,3-tetramethylbutyl) phenol: Irritant; flammable; soluble most organic solvents. Tm = 322.95; Antoine (liq) A = 7.53114, B = 2794.4, C = -43.39, dev = 1.0, range = 436 to 549.

4-Methyl-2-(1,1,3,3-tetramethylbutyl) phenol, CA 4979-46-8: Irritant; flammable; soluble most organic solvents. Tm = 319.35; Antoine (liq) A = 7.33215, B = 2611.4, C = -52.87, dev = 1.0, range = 415 to 545.

4-Nonylphenol, CA 25154-52-3: Irritant; flammable, flash point 414; soluble most organic solvents. Tb = 568.15 to 593.15; Vm = 0.2311 at 293.15; Antoine (liq) A = 6.87147, B = 2547.289, C = -67.246, dev = 0.1, range = 487 to 595.

alpha-Santalol, CA 115-71-9: Flammable; soluble ethanol. Tb = 574.15 to 575.15; Vm = 0.2249 at 298.15; Antoine (liq) A = 7.154, B = 3044, C = 0, dev = 5.0, range = 293 to 450.

$C_{15}H_{24}O_2$ MW 236.35

2,5-Di-*tert*-butyl-4-methoxyphenol, CA 1991-52-2: Irritant; flammable; soluble most organic solvents. Tm = 373.15; Antoine (liq) A = 7.8651, B = 3362.26, C = 0, dev = 1.0, range = 423 to 453.

1,3-Dimethoxy-5-heptylbenzene, CA 6121-64-8: Flammable; soluble most organic solvents. Antoine (liq) A = 5.5760, B = 1600, C = -157.26, dev = 1.0, range = 419 to 488.

1,3-Dimethoxy-5-methyl-2-hexylbenzene, CA 41442-51-7: Flammable; soluble most organic solvents. Antoine (liq) A = 5.6090, B = 1544.8, C = -152.99, dev = 1.0, range = 410 to 475.

$C_{15}H_{24}O_6$ MW 300.35

Aconitic acid, tripropyl ester, CA 64617-28-3: Flammable. Vm = 0.2860 at 298.15; Antoine (liq) A = 8.93928, B = 4027.537, C = 0, dev = 1.0 to 5.0, range = 359 to 500.

$C_{15}H_{26}O$ MW 222.37

Guaiol, CA 489-86-1: Flammable; soluble ethanol, ether. Tm = 362.15 to 363.15; Tb = 561.15, decomposes; Vm = 0.2451 at 373.15; Antoine (liq) A = 7.34379, B = 2871.21, C = -23.248, dev = 1.0, range = 373 to 561.

$C_{15}H_{26}O_6$ MW 302.37

Camphorenic acid, triethyl ester: Flammable. Tm = 408.15; Antoine (liq) A = 7.71232, B = 3093.53, C = -32.108, dev = 1.0, range = 423 to 574.

1,2,3-Propanetricarboxylic acid, tripropyl ester, CA 5333-54-0: Flammable. Antoine (liq) A = 8.81199, B = 3994.924, C = 0, dev = 1.0 to 5.0, range = 360 to 460.

(continues)

$C_{15}H_{26}O_6$ *(continued)*

Tributyrin, glycerol tributyrate, CA 60-01-5: Flammable, flash point 446; soluble most organic solvents. Tm = 198.15; Tb = 578.15 to 583.15; Vm = 0.2921 at 293.15; Antoine (liq) A = 9.495, B = 4250, C = 0, dev = 5.0, range = 318 to 364.

$C_{15}H_{28}Cl_4$ MW 350.20

1,1,1,15-Tetrachloropentadecane: Irritant; flammable. Antoine (liq) A = 11.555, B = 5405, C = 0, dev = 5.0, range = 340 to 392.

$C_{15}H_{28}O_2$ MW 240.38

Dodecyl acrylate, CA 2156-97-0: Irritant; flammable. Vm = 0.2754 at 293.15; Antoine (liq) A = 7.9339, B = 3377, C = 0, dev = 5.0, range = 432 to 573.

Pentadecanolide, CA 32539-85-8: Flammable. Tm = 310.15; Antoine I (sol) A = 10.015, B = 4245, C = 0, dev = 5.0, range = 290 to 310; Antoine II (liq) A = 8.855, B = 3878, C = 0, dev = 1.0 to 5.0, range = 310 to 320; Antoine III (liq) A = 9.45607, B = 4083.652, C = 0, dev = 1.0 to 5.0, range = 363 to 443.

$C_{15}H_{28}O_3$ MW 256.38

Decyl levulinate, CA 37826-51-0: Flammable. Antoine (liq) A = 7.4701, B = 2771.7, C = -72.19, dev = 1.0, range = 423 to 580.

1,6-Dioxa-7-cycloheptadecanone, CA 6707-60-4: Flammable. Antoine (liq) A = 8.87269, B = 3964.672, C = 0, dev = 5.0, range = 403 to 463.

$C_{15}H_{28}O_5$ MW 288.38

Decyl[1-methoxycarbonyl)ethyl] carbonate: Toxic; flammable. Antoine (liq) A = 8.535, B = 3856, C = 0, dev = 5.0, range = 411 to 592.

$C_{15}H_{29}N$ MW 223.40

Pentadecane nitrile, CA 18300-91-9: Irritant; flammable. Tm = 296.15; Antoine (liq) A = 6.50002, B = 2201, C = -105.55, dev = 1.0 to 5.0, range = 403 to 596.

$C_{15}H_{29}NO_3$ MW 271.40

2-[2-Ethyl(hexanoyloxy)]propionic acid, butylamide: Irritant; flammable. Antoine (liq) A = 9.49672, B = 4231.363, C = 0, dev = 1.0 to 5.0, range = 378 to 433.

$C_{15}H_{30}$ MW 210.40

Decylcyclopentane, CA 1795-21-7: Flammable; soluble most organic solvents. Tf = 251.15; Tb = 552.15; Vm = 0.2594 at 293.15; Antoine I (liq) A = 6.37514, B = 1912.918, C = -105.104, dev = 1.0, range =   *(continues)*

408

C$_{15}$H$_{30}$ *(continued)*

358 to 411; Antoine II (liq) A = 6.12945, B = 1827.036, C = -109.54, dev = 0.02, range = 453 to 553.

1-Pentadecene, CA 13360-61-7: Flammable, flash point above 383; soluble most organic solvents. Tm = 270.35; Tb = 541.32; Vm = 0.2710 at 293.15; Antoine I (liq) A = 8.49811, B = 3405.146, C = 0, dev = 1.0, range = 375 to 407; Antoine II (liq) A = 6.14963, B = 1790.129, C = -109.62, dev = 0.02, range = 443 to 543.

C$_{15}$H$_{30}$O   MW 226.40

2-Pentadecanone, CA 2345-28-0: Flammable; soluble most organic solvents. Tm = 312.15; Tb = 567.15; Vm = 0.2767 at 312.15; Antoine I (liq) A = 6.32336, B = 1989.296, C = -109.303, dev = 1.0 to 5.0, range = 422 to 575; Antoine II (liq) A = 6.60672, B = 2268.449, C = -77.03, dev = 1.0 to 5.0, range = 559 to 658.

8-Pentadecanone, CA 818-23-5: Flammable, flash point 383; soluble most organic solvents. Tm = 316.15 to 317.15; Tb = 564.15; Antoine I (liq) A = 8.062, B = 3234, C = 0, dev = above 5.0, range = 438 to 462; Antoine II (liq) A = 6.33524, B = 1990.728, C = -107.896, dev = 0.1, range = 443 to 568; Antoine III (liq) A = 6.29330, B = 1959.015, C = -110.911, dev = 1.0, range = 443 to 589.

C$_{15}$H$_{30}$O$_2$ MW 242.40

Dodecanoic acid, isopropyl ester, CA 10233-13-3: Flammable; soluble ethanol, ether. Vm = 0.2840 at 293.15; Antoine (liq) A = 7.95633, B = 3207.2, C = -14.75, dev = 1.0, range = 390 to 469.

Dodecanoic acid, propyl ester, CA 3681-78-5: Flammable; soluble ethanol, ether. Vm = 0.2819 at 293.15; Antoine (liq) A = 7.15262, B = 2669.1, C = -51.61, dev = 1.0, range = 396 to 479.

Pentadecanoic acid, CA 1002-84-2: Flammable, flash point above 383; soluble most organic solvents. Tm = 326.15; Tb = 612.25; Vm = 0.2878 at 353.15; Antoine (liq) A = 5.92845, B = 1676.35, C = -185.165, dev = 1.0, range = 431 to 613.

Tetradecanoic acid, methyl ester, CA 124-10-7: Flammable, flash point above 383; soluble most organic solvents. Tm = 291.65; Tb = 568.15; Vm = 0.2835 at 293.15; Antoine (liq) A = 5.98041, B = 1744.95, C = -135.207, dev = 1.0 to 5.0, range = 389 to 519.

C$_{15}$H$_{30}$O$_3$ MW 258.40

Dodecyl lactate, CA 6283-92-7: Flammable. Tm = 298.15; Antoine (liq) A = 6.98081, B = 2451.555, C = -90.15, dev = 1.0 to 5.0, range = 367 to 583.

2-Ethoxypropionic acid, decyl ester, CA 70160-09-7: Flammable. Antoine (liq) A = 8.417, B = 3644, C = 0, dev = 5.0, range = 423 to 523.

C$_{15}$H$_{31}$Br   MW 291.31

1-Bromopentadecane, CA 629-72-1: Irritant; flammable, flash point above 383; soluble acetone, chloroform. Tm = 292.15; Tb = 595.15;   *(continues)*

$C_{15}H_{31}Br$ *(continued)*

Vm = 0.2729 at 293.15; Antoine (liq) A = 6.595, B = 2293.8, C = -95.15, dev = 5.0, range = 450 to 661.

$C_{15}H_{31}Cl$ MW 246.86

1-Chloropentadecane, CA 4862-03-7: Irritant; flammable. Tm = 287.85; Antoine (liq) A = 6.599, B = 2240.5, C = -93.15, dev = 5.0, range = 439 to 645.

$C_{15}H_{31}F$ MW 230.41

1-Fluoropentadecane, CA 1555-17-5: Flammable. Tm = 290.15; Antoine (liq) A = 6.613, B = 2123.4, C = -86.15, dev = 5.0, range = 413 to 593.

$C_{15}H_{31}I$ MW 338.31

1-Iodopentadecane, CA 35599-78-1: Flammable; light-sensitive. Tm = 297.15; Antoine (liq) A = 6.485 B = 2318.3, C = -99.15, dev = 5.0, range = 464 to 673.

$C_{15}H_{31}NO$ MW 241.42

Tetradecanamide, *N*-methyl, CA 7438-09-7: Irritant; flammable. Tm = 351.65; Antoine (sol) A = 16.1769, B = 6813, C = 0, dev = 1.0, range = 332 to 347.

$C_{15}H_{31}NO_2$ MW 257.42

Lactamide, *N,N*-dihexyl: Irritant; flammable. Antoine (liq) A = 9.16849, B = 4148.309, C = 0, dev = 1.0 to 5.0, range = 418 to 453.

Lactamide, *N*-dodecyl: Irritant; flammable. Tm = 342.15; Antoine (liq) A = 11.04135, B = 5427.059, C = 0, dev = 1.0, range = 408 to 476.

$C_{15}H_{32}$ MW 212.42

2,3-Dimethyltridecane, CA 18435-20-6: Flammable; soluble most organic solvents. Tm = 230.15; Antoine (liq) A = 7.54773, B = 2995.79, C = 3.72, dev = 1.0, range = 399 to 537.

2,4-Dimethyltridecane: Flammable; soluble most organic solvents. Tm = 229.15; Antoine (liq) A = 7.55878, B = 2829.79, C = -13.55, dev = 1.0, range = 393 to 523.

2-Methyltetradecane, CA 1560-95-8: Flammable; soluble most organic solvents. Tm = 264.25; Antoine (liq) A = 7.45871, B = 2841.74, C = -16.0, dev = 1.0, range = 402 to 537.

3-Methyltetradecane, CA 18435-22-8: Flammable; soluble most organic solvents. Tm = 237.15; Antoine (liq) A = 7.58216, B = 2970.07, C = -5.54, dev = 1.0, range = 403 to 538.

(continues)

410

$C_{15}H_{32}$ *(continued)*

4-Methyltetradecane, CA 25117-24-2: Flammable; soluble most organic solvents. Tm = 242.65; Antoine (liq) A = 7.52922, B = 2985.21, C = 4.60, dev = 1.0, range = 398 to 536.

5-Methyltetradecane, CA 25117-32-2: Flammable; soluble most organic solvents. Tm = 223.85; Antoine (liq) A = 7.57682, B = 3012.64, C = 5.81, dev = 1.0, range = 398 to 535.

Pentadecane, CA 629-62-9: Flammable, flash point 405; soluble most organic solvents. Tm = 283.15; Tb = 543.15; Vm = 0.2764 at 293.15; Antoine I (liq) A = 6.38149, B = 1945.479, C = -97.875, dev = 1.0, range = 366 to 409; Antoine II (liq) A = 6.14849, B = 1789.95, C = -111.77, dev = 0.02, range = 447 to 546.

2,4,6-Trimethyldodecane: Flammable; soluble most organic solvents. Tm = 161.15; Antoine (liq) A = 7.74816, B = 2919.04, C = 0.18, dev = 1.0, range = 382 to 508.

$C_{15}H_{32}O$ MW 228.42

1-Pentadecanol, CA 629-76-5: Flammable, flash point above 383; soluble most organic solvents. Tm = 318.15 to 319.15; Antoine I (liq) A = 6.07246, B = 1768.329, C = -148.541, dev = 1.0, range = 438 to 600; Antoine II (liq) A = 6.10359, B = 1794, C = -145.5, dev = 1.0, range = 453 to 584.

$C_{15}H_{32}O_5$ MW 292.41

Tetrapropylene glycol, monoisopropyl ether: Flammable. Antoine (liq) A = 7.88842, B = 3126.94, C = -34.192, dev = 5.0, range = 389 to 566.

$C_{15}H_{33}N$ MW 227.43

Pentadecylamine, 1-aminopentadecane, CA 2570-26-5: Irritant; flammable; soluble ethanol, ether. Tm = 313.15; Tb = 571.15 to 574.15; Antoine (liq) A = 6.6019, B = 2243.9, C = -92.55, dev = 1.0 to 5.0, range = 400 to 594.

$C_{16}F_{34}$ MW 838.12

Perfluorohexadecane, CA 355-49-7: Tm = 388.15; Antoine (sol) A = 13.735, B = 5464, C = 0, dev = 5.0, range = 288 to 303.

$C_{16}H_{10}$ MW 202.26

Fluoranthene, CA 206-44-0: Moderately toxic; flammable; soluble most organic solvents. Antoine I (sol) A = 11.96071, B = 5348.06, C = 0, dev = 1.0, range = 298 to 383; Antoine II (liq) A = 6.67549, B = 2957.01, C = -24.15, dev = 1.0, range = 503 to 658.

Pyrene, CA 129-00-0: Suspected carcinogen; flammable; soluble most organic solvents. Tm = 422.15 to 423.15; Tb = 666.15; Antoine I (sol) A = 10.75452, B = 5072.78, C = 0, dev = 1.0, range = 298 to 401; Antoine II (sol) A = 11.35052, B = 5286.784, C = 0, dev = 1.0, range = 360 to 419; Antoine III (liq) A = 5.5106, B = 1743.57, C = -170.83, dev = 1.0, range = 513 to 668.

$C_{16}H_{10}O$ MW 218.25

Benz[*b*]indeno[1,2-*e*]pyran, CA 243-24-3: Tm = 460.65; Antoine (sol) A =
13.81797, B = 6575.029, C = 0, dev = 1.0, range = 375 to 388.

$C_{16}H_{12}S_2$ MW 268.39

3,6-Diphenyl-1,2-dithiine, CA 16212-85-4: Irritant; soluble ethanol, benzene. Tm = 415.15; Antoine (sol) A = 20.645, B = 9113, C = 0, dev = 5.0,
range = 350 to 400.

$C_{16}H_{13}N$ MW 219.29

N-Phenyl-1-naphthylamine, CA 90-30-2: Irritant; suspected carcinogen;
flammable; moderately soluble most organic solvents. Tm = 335.15; Antoine
I (sol) A = 10.8227, B = 5043.4, C = 0, dev = 1.0, range = 313 to 333;
Antoine II (liq) A = 9.7442, B = 4679.2, C = 0, dev = 1.0, range = 338
to 368.

N-Phenyl-2-naphthylamine, CA 135-88-6: Irritant; suspected carcinogen;
flammable; moderately soluble most organic solvents. Tm = 381.15; Tb =
668.15 to 672.65; Antoine I (sol) A = 13.0031, B = 6047.1, C = 0, dev =
1.0, range = 333 to 363; Antoine II (liq) A = 9.2914, B = 4634.8, C = 0,
dev = 1.0, range = 383 to 520.

$C_{16}H_{13}NO$ MW 235.28

Acetamide, *N*-9-anthracenyl, CA 37170-96-0: Tm = 547.15; Antoine (sol) A =
12.9059, B = 7042, C = 0, dev = 5.0, range = 446 to 500.

$C_{16}H_{13}NO_2$ MW 251.28

9,10-Anthraquinone, 1-(dimethylamino), CA 5960-55-4: Irritant; flammable.
Tm = 413.65; Antoine I (sol) A = 2.55935, B = 188.105, C = 0, dev = 5.0,
range = 396 to 408; Antoine II (liq) A = 1.34355, B = 1792.083, C = 0,
dev = 5.0, range = 408 to 418.

$C_{16}H_{13}NO_3$ MW 267.28

Anthraquinone, 1-(2-hydroxyethylamino), CA 4465-58-1: Irritant; flammable.
Tm = 441.15; Antoine (sol) A = 14.9439, B = 7973, C = 0, dev = 1.0 to 5.0,
range = 423 to 438.

$C_{16}H_{14}$ MW 206.29

9,10-Dimethylanthracene, CA 781-43-1: Suspected carcinogen; soluble benzene, chloroform. Tm = 453.15 to 454.15; Antoine (sol) A = 11.266, B =
5391, C = 0, dev not specified, range = 381 to 434.

9,10-Dimethylphenanthrene, CA 604-83-1: Moderately toxic; soluble benzene,
chloroform, acetic acid. Tm = 417.15; Antoine (sol) A = 13.08391, B =
5984.001, C = 0, dev = 1.0, range = 372 to 382.

1,4-Diphenyl-1,3-butadiene, CA 886-65-7: Flammable; soluble most organic
solvents; three stereoisomers are known. Antoine (sol, mixed isomers) A =
9.662, B = 4545, C = 0, dev not specified, range = 362 to 420.

(continues)

$C_{16}H_{14}$ *(continued)*

1,2,3,10b-Tetrahydrofluoranthene, CA 20279-21-4: Flammable. Tm = 347.15; Antoine (liq) A = 7.70377, B = 3551.698, C = 0, dev = 1.0, range = 400 to 469.

$C_{16}H_{14}O_2$ MW 238.29

Benzyl cinnamate, CA 103-41-3: Flammable, flash point above 383; soluble ethanol, ether. Tm = 312.15; Tb = 623.15, decomposes; Antoine (liq) A = 8.60608, B = 3825.24, C = -43.737, dev = 1.0, range = 446 to 623.

$C_{16}H_{16}$ MW 208.30

[2,2]Metacyclophane, CA 2319-97-3: Flammable; soluble most organic solvents. Tm = 407.65 to 408.15; Tb = 563.15; Antoine (sol) A = 11.555, B = 4791, C = 0, dev not specified, range = 308 to 332.

Metaparacyclophane, CA 5385-36-4: Flammable; soluble most organic solvents. Tm = 354.15 to 354.65; Antoine (sol) A = 11.005, B = 4524, C = 0, dev not specified, range = 310 to 328.

[2,2]Paracyclophane, CA 1633-22-3: Flammable; soluble most organic solvents. Tm = 558.15 to 560.15; Antoine (sol) A = 9.968, B = 4849, C = 0, dev not specified, range = 343 to 383.

$C_{16}H_{18}$ MW 210.32

1,3-Diphenylbutane ±, CA 1520-44-1: Flammable; soluble most organic solvents. Tb = 568.15; Antoine (sol) A = 9.171, B = 3841, C = 0, dev = 1.0, range = 288 to 303.

Ethane, 1,1-di(4-tolyl), CA 530-45-0: Flammable; soluble most organic solvents. Tm = below 253.15; Tb = 569.15 to 571.15; Vm = 0.2159 at 293.15; Antoine (liq) A = 6.5538, B = 2200.3, C = -93.15, dev = 1.0 to 5.0, range = 298 to 575.

Ethane, 1-(2-tolyl)-1-(4-tolyl): Flammable; soluble most organic solvents. Antoine (liq) A = 6.6371, B = 2200.3, C = -93.15, dev = 0.1 to 1.0, range = 298 to 473.

$C_{16}H_{18}Cl_4O_4$ MW 416.13

Tetrachlorophthalic acid, dibutyl ester, CA 3015-66-5: Irritant; flammable. Antoine (liq) A = 10.325, B = 5210, C = 0, dev = 5.0, range = 368 to 421.

$C_{16}H_{18}NO_5$ MW 304.32

Nitrogen oxide, bis(2,4-dimethoxyphenyl), CA 3788-15-6: Irritant. Tm = 451.15; Antoine (sol) A = 15.205, B = 7530, C = 0, dev = 5.0, range = 333 to 363.

$C_{16}H_{18}N_4O_2$ MW 298.34

4-Nitro-4'-N,N-diethylamino azobenzene, CA 3025-52-3: Irritant; flammable. Tm = 425.15; Antoine (liq) A = 15.559, B = 7909, C = 0, dev = 5.0, range = 423 to 443.

$C_{16}H_{18}N_4O_3$ MW 314.34

4-Nitro-4'-[*N*-ethyl-*N*-(2-hydroxyethyl)amino] azobenzene, CA 2872-52-8: Irritant; flammable. Tm = 432.15; Antoine (sol) A = 17.433, B = 9230, C = 0, dev = 5.0, range = 420 to 433.

$C_{16}H_{18}O$ MW 226.32

Bis(*alpha*-methylbenzyl) ether, CA 93-96-9: Irritant; flammable. Tb = 553.35; Antoine (liq) A = 7.4809, B = 2929.82, C = -18.953, dev = 1.0, range = 369 to 554.

$C_{16}H_{21}Cl_3O_3$ MW 367.70

2,4,5-Trichlorophenoxyacetic acid, (2-ethylhexyl) ester, CA 1928-47-8: Toxic; flammable; used as weed killer. Antoine (liq) A = 7.7163, B = 3266.986, C = -68.477, dev = 1.0, range = 460 to 573.

2,4,5-Trichlorophenoxyacetic acid, octyl ester, CA 2630-15-1: Toxic; flammable; used as weed killer. Antoine (liq) A = 6.42482, B = 2209.578, C = -153.028, dev = 1.0, range = 460 to 573.

$C_{16}H_{22}Cl_2O_3$ MW 333.25

2,4-Dichlorophenoxyacetic acid, (2-ethylhexyl) ester, CA 1928-43-4: Toxic; flammable; used as weed killer. Antoine (liq) A = 6.57405, B = 2320.677, C = -127.489, dev = 1.0, range = 460 to 573

2,4-Dichlorophenoxyacetic acid, (1-methylheptyl) ester, CA 1917-97-1: Toxic; flammable; used as weed killer. Antoine (liq) A = 6.9617, B = 2596.904, C = -107.224, dev = 1.0, range = 460 to 573.

2,4-Dichlorophenoxyacetic acid, octyl ester, CA 1928-44-5: Toxic; flammable; used as weed killer. Antoine (liq) A = 6.25048, B = 2076.913, C = -155.322, dev = 1.0, range = 460 to 573.

$C_{16}H_{22}O_4$ MW 278.35

Dibutyl phthalate, CA 84-74-2: Toxic; flammable, flash point 430; soluble most organic solvents; used as plasticizer. Tm = 238.15; Tb = 613.15; Vm = 0.2671 at 298.15; Antoine I (liq) A = 6.8788, B = 2538.4, C = -92.25, dev = 1.0 to 5.0, range = 314 to 469; Antoine II (liq) A = 7.97157, B = 3385.9, C = -37.18, dev = 1.0 to 5.0, range = 468 to 605.

Di-*sec*-butyl phthalate, CA 4489-61-6: Toxic; flammable; soluble most organic solvents. Tm = 289.15; Antoine (liq) A = 11.275, B = 4899, C = 0, dev = 5.0, range = 313 to 373.

Dibutyl terephthalate, CA 1962-75-0: Flammable; soluble most organic solvents. Tm = 289.75 to 291.25; Vm = 0.2655 at 298.15; Antoine (liq) A = 6.0593, B = 2064.4, C = -131.45, dev = 1.0 to 5.0, range = 393 to 483.

$C_{16}H_{25}Cl$ MW 252.83

Chloropentaethylbenzene: Irritant; flammable. Antoine (liq) A = 7.12061, B = 2715.97, C = -27.027, dev = 1.0, range = 363 to 558.

414

$C_{16}H_{26}$ MW 219.38

Decylbenzene, CA 104-72-3: Flammable, flash point above 383; soluble most
organic solvents. Tf = 258.75; Tb = 566.15 to 567.15; Vm = 0.2553 at
293.15; Antoine I (liq) A = 4.03653, B = 876.208, C = -203.15, dev = 1.0,
range = 371 to 427; Antoine II (liq) A = 6.15658, B = 1900.918, C =
-113.16, dev = 1.0, range = 475 to 571.

Pentaethylbenzene, CA 605-01-6: Flammable; soluble most organic solvents.
Tm = below 253.15; Tb = 550.15; Vm = 0.2426 at 292.15; Antoine (liq) A =
7.5326, B = 3089.4, C = 8.795, dev = 1.0, range = 359 to 550.

$C_{16}H_{26}O$ MW 234.38

2,4-Di-*tert*-butyl-5,6-dimethyl phenol: Irritant; flammable; soluble most
organic solvents. Tm = 358.65; Tb = 557.15; Antoine (liq) A = 7.741, B =
2964.1, C = -41.85, dev = 5.0, range = 431 to 565.

2,4-Di-*tert*-butyl-5-ethyl phenol, CA 19245-41-1: Irritant; flammable; soluble most organic solvents. Tm = 353.65; Tb = 562.15; Antoine (liq) A =
7.56645, B = 2897.53, C = -42.052, dev = 5.0, range = 384 to 563.

2,4-Di-*tert*-butyl-6-ethyl phenol, CA 6287-47-4: Irritant; flammable; soluble most organic solvents. Tm = 303.15; Tb = 548.15; Antoine (liq) A =
6.9223, B = 2412.8, C = -58.15, dev = 5.0, range = 413 to 556.

2,6-Di-*tert*-butyl-4-ethyl phenol, CA 4130-42-1: Irritant; flammable; soluble most organic solvents. Tm = 317.15; Tb = 545.15; Antoine (liq) A =
7.79038, B = 3080.33, C = -9.201, dev = 5.0, range = 362 to 542.

2,4,5-Triisopropylbenzyl alcohol: Flammable; soluble ethanol. Antoine
(liq) A = 14.741, B = 5907, C = 0, dev = 1.0, range = 312 to 346.

$C_{16}H_{26}O_{11}$ MW 394.37

Diethylene glycol dicarboxylic acid, di[1-(ethoxycarbonyl)ethyl] ester:
Irritant; flammable. Antoine (liq) A = 10.0609, B = 5188.6, C = 0, dev =
1.0 to 5.0, range = 418 to 503.

$C_{16}H_{28}$ MW 220.40

Tricyclo(8,2,2,2^{4,7})hexadecane, CA 283-68-1: Flammable. Antoine (sol) A =
10.375, B = 4450, C = 0, dev not specified, range = 316 to 338.

Tricyclopentylmethane: Flammable; soluble most organic solvents. Antoine
I (liq) A = 9.950, B = 4065, C = 0, dev = 1.0, range = 273 to 351; Antoine
II (liq) A = 6.74316, B = 2282.073, C = -83.835, dev = 1.0, range = 371
to 429.

$C_{16}H_{30}O$ MW 238.41

3-Methylcyclopentadecanone, muscone, CA 541-91-3: Flammable; soluble ethanol, ether, acetone. Tb = 600.15 to 603.15; Vm = 0.2586 at 290.15; Antoine
(liq) A = 6.84557, B = 2726.58, C = -37.757, dev = 1.0, range = 391 to 601.

$C_{16}H_{30}O_2$ MW 254.41

Dodecyl methacrylate, CA 142-90-5: Irritant; flammable. Tm = 253.15; Vm = 0.2931 at 298.15; Antoine (liq) A = 7.8513, B = 3391, C = 0, dev = 5.0, range = 438 to 580.

Oxa-2-cycloheptadecanone, CA 109-29-5: Flammable. Tm = 306.15 to 307.15; Antoine (liq) A = 8.37258, B = 3738.338, C = 0, dev = 5.0, range = 403 to 463.

$C_{16}H_{30}O_3$ MW 270.41

1,7-Dioxa-8-cyclooctadecanone, CA 6720-22-5: Flammable. Antoine (liq) A = 8.28326, B = 3831.265, C = 0, dev = 1.0 to 5.0, range = 403 to 463.

1,9-Dioxa-2-cyclooctadecanone, CA 36575-58-3: Flammable. Antoine (liq) A = 8.44618, B = 3891.128, C = 0, dev = 5.0, range = 403 to 463.

$C_{16}H_{30}O_4$ MW 286.41

Adipic acid, dipentyl ester, CA 14027-78-2: Flammable. Antoine (liq) A = 8.7959, B = 3900, C = 0, dev = 1.0 to 5.0, range = 449 to 575.

Hexadecanedioic acid, thapsic acid, CA 505-54-4: Flammable; soluble ethanol, acetone. Tm = 399.15; Antoine (sol) A = 16.290, B = 7885, C = 0, dev = 1.0, range = 377 to 398.

$C_{16}H_{30}O_5$ MW 302.41

Octyl[1-(butoxycarbonyl)ethyl] carbonate: Toxic; flammable. Antoine (liq) A = 8.7662, B = 3979.5, C = 0, dev = 5.0, range = 374 to 503.

$C_{16}H_{31}N$ MW 237.43

Palmitonitrile, CA 629-79-8: Irritant; flammable; soluble most organic solvents. Tm = 306.15; Tb = 606.15; Antoine (liq) A = 6.45473, B = 2175.1, C = -118.62, dev = 1.0, range = 503 to 608.

$C_{16}H_{32}$ MW 224.43

Decylcyclohexane, CA 1795-16-0: Flammable; soluble most organic solvents. Tf = 271.45; Tb = 572.15; Vm = 0.2762 at 293.15; Antoine I (liq) A = 5.95662, B = 1792.111, C = -127.646, dev = 1.0, range = 371 to 425; Antoine II (liq) A = 6.1344, B = 1891.541, C = -112.666, dev = 0.02, range = 469 to 571.

1-Hexadecene, cetene, CA 629-73-2: Flammable, flash point 405; soluble most organic solvents. Tm = 277.25; Tb = 557.15; Vm = 0.2873 at 293.15; Antoine (liq) A = 6.16972, B = 1844.488, C = -115.133, dev = 0.02, range = 461 to 558.

$C_{16}H_{32}O$ MW 240.43

Hexadecanal, palmitic aldehyde, CA 629-80-1: Flammable; soluble most organic solvents; polymerizes on standing. Tm = 307.15; Antoine (liq) A = 6.97624, B = 2697, C = -51.566, dev = 1.0, range = 394 to 594. *(continues)*

416

$C_{16}H_{32}O$ (continued)

2-Hexadecanone, CA 18787-63-8: Flammable. Tm = 316.15 to 316.65; Antoine (liq) A = 7.02006, B = 2556.58, C = -70.216, dev = 1.0, range = 382 to 580.

$C_{16}H_{32}O_2$ MW 256.43

Dodecanoic acid, butyl ester, CA 106-18-3: Flammable. Tm = 266.31; Antoine (liq) A = 8.745, B = 3960, C = 0, dev = 5.0, range = 343 to 383.

Hexadecanoic acid, palmitic acid, CA 57-10-3: Irritant; flammable; soluble most organic solvents. Tm = 336.15 to 337.15; Tb = 624.65; Vm = 0.3007 at 337.15; Antoine I (sol) A = 19.342, B = 8069, C = 0, dev = 1.0, range = 319 to 334; Antoine II (liq) A = 5.85353, B = 1650.26, C = -195.73, dev = 5.0, range = 440 to 625.

Pentadecanoic acid, methyl ester, CA 7132-64-1: Flammable, flash point above 383; soluble ethanol, ether. Tm = 291.65; Vm = 0.2976 at 298.15; Antoine (liq) A = 5.93058, B = 1754.03, C = -143.093, dev not specified, range = 400 to 527.

Tetradecanoic acid, ethyl ester, CA 124-06-1: Flammable, flash point 383; soluble ethanol. Tm = 285.45; Tb = 568.15; Vm = 0.2991 at 298.15; Antoine (liq) A = 8.61137, B = 3748.727, C = 0, dev = 1.0, range = 407 to 568.

Tetradecyl acetate, CA 638-59-5: Flammable. Tm = 287.15; Antoine (liq) A = 8.6391, B = 3795.801, C = 0, dev = 1.0, range = 411 to 462.

$C_{16}H_{33}Br$ MW 305.34

1-Bromohexadecane, CA 112-82-3: Flammable, flash point above 383; soluble ether. Tm = 291.15; Tb = 609.15; Vm = 0.3056 at 293.15; Antoine (liq) A = 6.631, B = 2364, C = -98.15, dev = 5.0, range = 461 to 673.

$C_{16}H_{33}Cl$ MW 260.89

1-Chlorohexadecane, CA 4860-03-1: Irritant; flammable, flash point above 383; soluble ether. Tm = 291.05; Tb = 595.15; Vm = 0.3015 at 293.15; Antoine (liq) A = 6.12599, B = 1924.969, C = -132.809, dev = 1.0, range = 288 to 600.

$C_{16}H_{33}F$ MW 244.44

1-Fluorohexadecane, CA 408-38-8: Flammable; soluble ether, light hydrocarbons. Tm = 291.15; Tb = 562.15; Vm = 0.2938 at 293.15; Antoine (liq) A = 6.645, B = 2194.8, C = -89.15, dev = 5.0, range = 425 to 608.

$C_{16}H_{33}I$ MW 352.34

1-Iodohexadecane, CA 544-77-4: Irritant; light-sensitive; flammable, flash point above 383; soluble most organic solvents. Tm = 295.15; Tb = 630.15; Antoine (liq) A = 6.526, B = 2388, C = -102.15, dev = 5.0, range = 475 to 673.

$C_{16}H_{33}NO$ MW 255.44

Palmitamide, CA 629-54-9: Irritant; flammable; soluble most organic solvents. Tm = 379.15 to 380.15; Antoine (sol) A = 21.8149, B = 9489, C = 0, dev = 1.0, range = 364 to 378.

$C_{16}H_{34}$ MW 226.44

2,3-Dimethyltetradecane, CA 18435-23-9: Flammable; soluble most organic solvents. Tm = 245.25; Antoine (liq) A = 8.0636, B = 3598.3, C = 40.65, dev = 1.0, range = 412 to 554.

2,4-Dimethyltetradecane: Flammable; soluble most organic solvents. Tm = 236.05; Antoine (liq) A = 7.616, B = 2897.36, C = -17.98, dev = 1.0, range = 404 to 539.

Hexadecane, cetane, CA 544-76-3: Hygroscopic; flammable, flash point 408; soluble most organic solvents. Tm = 291.35; Tb = 560.15; Vm = 0.2928 at 293.15; Antoine I (liq) A = 6.77074, B = 2273.168, C = -80.252, dev = 1.0, range = 323 to 423; Antoine II (liq) A = 6.15357, B = 1830.51, C = -118.7, dev = 0.02, range = 467 to 563.

2-Methylpentadecane, CA 1560-93-6: Flammable; soluble most organic solvents. Tm = 263.75; Antoine (liq) A = 7.4925, B = 2919.29, C = -21.87, dev = 1.0, range = 417 to 554.

3-Methylpentadecane, CA 2882-96-4: Flammable; soluble most organic solvents. Tm = 249.35; Antoine (liq) A = 7.5231, B = 2991.99, C = -13.3, dev = 1.0, range = 417 to 555.

4-Methylpentadecane, CA 2801-87-8: Flammable; soluble most organic solvents. Tm = 249.65; Antoine (liq) A = 7.6045, B = 3143.27, C = 8.56, dev = 1.0, range = 411 to 553.

5-Methylpentadecane, CA 25117-33-3: Flammable; soluble most organic solvents. Tm = 242.15; Antoine (liq) A = 7.4702, B = 3018.8, C = 1.93, dev = 1.0, range = 408 to 551.

7-Propyltridecane: Flammable; soluble most organic solvents. Antoine (liq) A = 8.58072, B = 3464.5, C = 0, dev = 1.0, range = 355 to 410.

2,4,6-Trimethyltridecane: Flammable; soluble most organic solvents. Tm = 171.15; Antoine (liq) A = 7.9229, B = 3081.31, C = -0.41, dev = 1.0, range = 395 to 521.

$C_{16}H_{34}O$ MW 242.44

1-Hexadecanol, cetyl alcohol, CA 124-29-8: Moderately toxic; flammable, flash point above 383; soluble most organic solvents. Tm = 323.15; Tb = 607.15; Vm = 0.2965 at 323.15; Antoine I (sol) A = 22.773, B = 8736, C = 0, dev = 1.0, range = 308 to 320; Antoine II (sol) A = 13.376, B = 5717, C = 0, dev = 1.0, range = 323 to 335; Antoine III (liq) A = 6.1156, B = 1837.664, C = -150.085, dev = 0.1, range = 415 to 514; Antoine IV (liq) A = 6.12159, B = 1850.547, C = -147.978, dev = 0.1, range = 498 to 569; Antoine V (liq) A = 6.11647, B = 1846.628, C = -148.382, dev = 0.1, range = 509 to 569.

$C_{16}H_{34}O_3$ MW 274.44

2-[2-(Dodecyloxy)ethoxy] ethanol, CA 3055-93-4: Flammable; soluble ethanol. Antoine (liq) A = 9.2566, B = 4290.496, C = 0, dev = 1.0 to 5.0,
range = 448 to 489.

$C_{16}H_{35}N$ MW 241.46

N,N-Dimethyltetradecyl amine, CA 112-75-4: Irritant; flammable; soluble
most organic solvents. Tm = 270.15; Antoine (liq) A = 6.4933, B = 2172,
C = -91.15, dev = 5.0, range = 460 to 640.

Dioctyl amine, CA 1120-48-5: Irritant; flammable, flash point above 383;
soluble ethanol, ether. Tm = 287.15 to 288.15; Tb = 570.15 to 571.15; Vm =
0.3032 at 299.15; Antoine (liq) A = 6.4933, B = 2172, C = -91.15, dev =
5.0, range = 460 to 640.

Hexadecyl amine, cetyl amine, CA 143-27-1: Irritant; corrosive; flammable,
flash point 413; soluble most organic solvents. Tm = 319.35; Tb = 603.15;
Antoine (liq) A = 6.6319, B = 2313.9, C = -95.45, dev = 1.0, range = 498
to 609.

$C_{16}H_{36}N_2$ MW 255.47

Tetrabutyl hydrazine, CA 60678-70-8: Toxic; flammable. Antoine (liq) A =
6.9282, B = 2667, C = 0, dev = 5.0, range = 392 to 453.

$C_{17}H_{10}O$ MW 230.27

1,9-Benzanthr-10-one, CA 82-05-3: Irritant; soluble hot benzene. Tm =
447.15; Antoine I (sol) A = 12.557, B = 6030, C = 0, dev = 5.0, range not
specified; Antoine II (liq) A = 5.73375, B = 2358.04, C = -152.231, dev =
1.0 to 5.0, range = 498 to 673.

$C_{17}H_{13}N$ MW 231.30

5-Methyl-5H-indeno[2,1-b] quinoline, CA 6626-64-8: Flammable. Tm =
404.15; Antoine (sol) A = 12.64267, B = 6384.741, C = 0, dev = 1.0, range =
375 to 388.

$C_{17}H_{18}O_3$ MW 270.33

Salicylic acid, 4-($tert$-butylphenyl) ester, CA 87-18-3: Flammable; soluble
most organic solvents. Tm = 336.15; Antoine I (sol) A = 17.905, B = 7180,
C = 0, dev = 5.0, range = 293 to 336; Antoine II (liq) A = 10.645, B =
4725, C = 0, dev = 5.0, range = 336 to 438.

$C_{17}H_{24}O_2$ MW 260.38

Menthyl benzoate, CA 6284-35-1: Flammable. Tm = 327.65; Antoine (liq) A =
8.40485, B = 3717.7, C = 6.654, dev = 1.0 to 5.0, range = 396 to 574.

$C_{17}H_{28}O$ MW 248.41

4-Methyl-2,6-di-*tert*-pentyl phenol, CA 56103-67-4: Irritant; flammable;
soluble most organic solvents. Tb = 556.15; Antoine (liq) A = 6.77174,
B = 2230.5, C = -88.15, dev = 5.0, range = 438 to 556.

$C_{17}H_{28}O_2$ MW 264.41

1,3-Dimethoxy-2-nonylbenzene, CA 55095-35-7: Flammable. Antoine (liq)
A = 4.6303, B = 1058.4, C = -226.0, dev = 1.0, range = 443 to 509.

$C_{17}H_{32}Cl_4$ MW 378.25

1,1,1,17-Tetrachloroheptadecane: Irritant; flammable. Antoine (liq) A =
11.675, B = 5640, C = 0, dev = 5.0, range = 351 to 418.

$C_{17}H_{32}O_2$ MW 268.44

Oxa-2-cyclooctadecanone, CA 5637-97-8: Flammable. Antoine (liq) A =
8.3177, B = 3836.972, C = 0, dev = 5.0, range = 403 to 463.

Tetradecyl acrylate, CA 21643-42-5: Irritant; flammable. Vm = 0.3086 at
293.15; Antoine (liq) A = 8.0313, B = 3625, C = 0, dev = 5.0, range = 458
to 601.

$C_{17}H_{32}O_3$ MW 284.44

1,8-Dioxa-9-cyclononadecanone, CA 1725-00-4: Flammable. Antoine (liq) A =
8.48198, B = 4024.459, C = 0, dev = 5.0, range = 403 to 463.

$C_{17}H_{32}O_4$ MW 300.44

Azelaic acid, dibutyl ester, CA 2917-73-9: Flammable. Antoine (liq) A =
10.03563, B = 4615.672, C = 0, dev = 1.0, range = 313 to 450.

$C_{17}H_{32}O_5$ MW 316.44

Nonyl[1-(butoxycarbonyl)ethyl] carbonate: Toxic; flammable. Antoine (liq)
A = 8.316, B = 3855, C = 0, dev = 5.0, range = 420 to 534.

$C_{17}H_{33}N$ MW 251.45

Heptadecanonitrile, CA 5399-02-0: Irritant; flammable; soluble ether.
Tm = 307.15; Tb = 622.15; Antoine (liq) A = 6.6534, B = 2362.3, C = -111.4,
dev = 1.0 to 5.0, range = 425 to 620.

$C_{17}H_{34}$ MW 238.46

1-Heptadecene, CA 6765-39-5: Flammable, flash point above 383; soluble
most organic solvents. Tm = 284.35; Tb = 573.15; Vm = 0.3037 at 293.15;
Antoine (liq) A = 8.8608, B = 3776.5, C = 0, dev = 1.0, range = 376 to 432.

420

$C_{17}H_{34}O$ MW 254.46

2-Heptadecanone, CA 2922-51-2: Flammable; soluble most organic solvents.
Tm = 321.15; Tb = 592.15 to 593.15; Vm = 0.3100 at 321.15; Antoine (liq)
A = 6.76333, B = 2353.93, C = -97.76, dev = 1.0, range = 402 to 593.

9-Heptadecanone, CA 540-08-9: Flammable, flash point above 383; soluble
most organic solvents. Tm = 326.15; Tb = 523.15 to 526.15; Antoine (liq)
A = 9.166, B = 4092, C = 0, dev = above 5.0, range = 439 to 482.

$C_{17}H_{34}O_2$ MW 270.45

Heptadecanoic acid, margaric acid, CA 506-12-7: Flammable; soluble ether.
Tm = 335.15 to 336.15; Tb = 636.95, Antoine (liq) A = 5.72977, B = 1596.17,
C = -208.033, dev = 1.0, range = 449 to 637.

Hexadecanoic acid, methyl ester, CA 112-39-0: Flammable, flash point above
383; soluble most organic solvents. Tm = 303.65; Antoine I (sol) A =
21.462, B = 7993, C = 0, dev = 5.0, range = 291 to 301; Antoine II (liq)
A = 8.12224, B = 3637.502, C = 0, dev = 1.0, range = 287 to 322; Antoine
III (liq) A = 5.49492, B = 1528, C = -171.941, dev = 1.0, range = 411 to
543.

Tetradecanoic acid, isopropyl ester, CA 110-27-0: Flammable, flash point
above 383; soluble most organic solvents. Tf = 276.15; Vm = 0.3170 at
293.15; Antoine (liq) A = 8.0566, B = 3464.2, C = -11.87, dev = 1.0,
range = 413 to 466.

Tetradecanoic acid, propyl ester, CA 14303-70-9: Flammable; soluble most
organic solvents. Vm = 0.3148 at 293.15; Antoine (liq) A = 8.93411, B =
4302.7, C = 32.38, dev = 1.0, range = 420 to 474.

$C_{17}H_{34}O_3$ MW 286.45

Tetradecyl lactate, CA 1323-03-1: Flammable. Tm = 304.15; Antoine (liq)
A = 7.2519, B = 2717.473, C = -90.15, dev = 1.0 to 5.0, range = 388 to 608.

$C_{17}H_{35}Br$ MW 319.37

1-Bromoheptadecane, CA 3508-00-7: Irritant; flammable; soluble chloroform.
Tm = 300.15 to 302.15; Tb = 622.15; Antoine (liq) A = 6.665, B = 2430.9,
C = -94.15, dev = 5.0, range = 472 to 673.

$C_{17}H_{35}Cl$ MW 274.92

1-Chloroheptadecane, CA 62016-75-5: Irritant; flammable. Tm = 297.15;
Antoine (liq) A = 6.668, B = 2378.8, C = -98.15, dev = 5.0, range = 450
to 673.

$C_{17}H_{35}F$ MW 258.46

1-Fluoroheptadecane: Flammable. Tm = 302.15; Antoine (liq) A = 6.681,
B = 2262.5, C = -92.15, dev = 5.0, range = 437 to 623.

$C_{17}H_{35}I$ MW 366.37

1-Iodoheptadecane, CA 26825-83-2: Irritant; light-sensitive; flammable.
Tm = 307.95 to 308.45; Antoine (liq) A = 6.562, B = 2454.6, C = -105.15,
dev = 5.0, range = 517 to 673.

$C_{17}H_{35}NO$ MW 269.47

Hexadecanoamide, *N*-methyl, CA 7388-58-1: Irritant; flammable. Tm = 359.05;
Antoine (sol) A = 17.4339, B = 7530, C = 0, dev = 1.0, range = 345 to 355.

$C_{17}H_{35}NO_2$ MW 285.47

Lactamide, *N*-tetradecyl: Irritant; flammable. Tm = 348.15; Antoine (liq)
A = 11.07506, B = 5613.191, C = 0, dev = 1.0, range = 413 to 491.

$C_{17}H_{36}$ MW 240.47

2,3-Dimethylpentadecane, CA 2882-97-5: Flammable; soluble most organic
solvents. Tm = 249.35; Antoine (liq) A = 7.77899, B = 3364.82, C = 14.22,
dev = 1.0, range = 424 to 569.

2,4-Dimethylpentadecane: Flammable; soluble most organic solvents. Tm =
248.15; Antoine (liq) A = 8.23083, B = 3397.41, C = -0.41, dev = 1.0,
range = 419 to 546.

Heptadecane, CA 629-78-7: Flammable, flash point 421; soluble most organic
solvents. Tm = 295.15; Tb = 576.15; Antoine I (liq) A = 11.1197, B =
4757.087, C = 0, dev = 1.0 to 5.0, range = 289 to 320; Antoine II (liq) A =
6.1392, B = 1865.1, C = -123.95, dev = 0.1, range = 488 to 577.

2-Methylhexadecane, CA 1560-92-5: Flammable; soluble most organic solvents. Tm = 278.05; Antoine (liq) A = 7.58956, B = 3084.06, C = -15.82,
dev = 1.0, range = 428 to 569.

3-Methylhexadecane, CA 6418-43-5: Flammable; soluble most organic solvents. Tm = 258.45; Antoine (liq) A = 7.8156, B = 3283.08, C = -2.05,
dev = 1.0, range = 428 to 567.

4-Methylhexadecane, CA 25117-26-4: Flammable; soluble most organic solvents. Tm = 258.35; Antoine (liq) A = 7.33557, B = 2999.32, C = -4.7,
dev = 1.0, range = 420 to 567.

5-Methylhexadecane, CA 25117-34-7: Flammable; soluble most organic solvents. Tm = 241.25; Antoine (liq) A = 7.60293, B = 3198.92, C = 5.35,
dev = 1.0, range = 422 to 566.

2,4,6-Trimethyltetradecane: Flammable; soluble most organic solvents.
Tm = 181.15; Antoine (liq) A = 8.29037, B = 3348.45, C = -1.39, dev = 1.0,
range = 411 to 534.

$C_{17}H_{36}O$ MW 256.47

1-Heptadecanol, CA 1454-85-9: Flammable, flash point above 383; soluble
ethanol, ether. Tm = 327.15; Tb = 610.5; Antoine I (liq) A = 6.23865, B =
1969.73, C = -145.228, dev = 1.0, range = 460 to 620; Antoine II (liq) A =
6.33419, B = 2047.5, C = -137.25, dev = 1.0, range = 473 to 623.

422

$C_{17}H_{37}N$ MW 255.49

Heptadecyl amine, CA 4200-95-7: Irritant; flammable; soluble ethanol,
ether. Tm = 322.15; Tb = 608.15 to 609.15; Antoine (liq) A = 6.6639, B =
2380.7, C = -97.95, dev = 1.0, range = 522 to 636.

$C_{18}H_6N_8O_{16}$ MW 590.29

2,2',4,4',4'',6,6',6''-Octanitro-1,1':3,3''-terphenyl, CA 33491-88-2:
Flammable; potentially explosive. Antoine (liq) A = 9.0279, B = 8263.293,
C = 0, dev = above 5.0, range = 479 to 551.

$C_{18}H_{11}NO_3$ MW 289.29

C. I. disperse yellow 54, CA 12223-85-7: Flammable. Antoine (liq) A =
12.40543, B = 7328.236, C = 0, dev = 1.0, range = 483 to 513.

$C_{18}H_{12}$ MW 228.29

Benzanthracene, CA 56-55-3: Highly toxic; suspected carcinogen; soluble
ether, benzene, chloroform. Tm = 431.15 to 432.15; Tb = 710.75; Antoine I
(sol) A = 12.0507, B = 5925, C = 0, dev = 5.0, range = 330 to 390; Antoine
II (sol) A = 10.653, B = 5461, C = 0, dev = 5.0, range = 377 to 400.

Chrysene, CA 218-01-9: Highly toxic; suspected carcinogen; moderately soluble hot aromatics. Tm = 528.15 to 529.15; Tb = 721.15; Antoine (sol) A =
11.445, B = 6160, C = 0, dev = 5.0, range = 358 to 463.

Naphthacene, CA 92-24-0: Irritant; explosive; slightly soluble most organic solvents. Tm = 614.15; Antoine (sol) A = 11.505, B = 6540, C = 0,
dev = 5.0, range = 376 to 489.

Triphenylene, CA 217-59-4: Soluble ethanol, benzene, chloroform. Tm =
471.15; Tb = 711.15; Antoine I (sol) A = 9.435, B = 5620, C = 0, dev = 5.0,
range = 363 to 468; Antoine II (liq) A = 6.8974, B = 3527, C = 0, dev =
5.0, range = 600 to 720.

$C_{18}H_{12}N_2$ MW 256.31

2,2'-Biquinoline, CA 119-91-5: Soluble most organic solvents. Tm =
469.15; Antoine (liq) A = 7.125, B = 3720, C = 0, dev = 5.0, range not
specified.

$C_{18}H_{12}O$ MW 244.29

2-Phenylindeno[2,1-b]pyran, CA 10435-67-3: Tm = 471.15; Antoine (sol) A =
13.38872, B = 6935.884, C = 0, dev = 1.0, range = 394 to 424.

$C_{18}H_{14}$ MW 230.31

5,12-Dihydronaphthacene, CA 959-02-4: Soluble benzene. Tm = 481.15 to
482.15; Antoine (sol) A = 11.475, B = 6060, C = 0, dev = 5.0, range = 338
to 398.

(continues)

$C_{18}H_{14}$ *(continued)*

ortho-Terphenyl, CA 84-15-1: Irritant; toxic; flammable, flash point 444; soluble aromatic solvents. Tm = 331.15; Tb = 605.15; Vm = 0.2254 at 366.15; Tc = 891; Pc = 3903; Antoine (liq) A = 6.29309, B = 2160.24, C = -106.28, dev = 1.0, range = 462 to 650.

meta-Terphenyl, CA 92-06-8: Irritant; toxic; flammable, flash point 479; soluble aromatic solvents. Tm = 362.15; Tb = 638.15; Vm = 0.2217 at 366.15; Tc = 925; Pc = 3503; Antoine (liq) A = 6.48808, B = 2445.98, C = -102.76, dev = 1.0, range = 462 to 691.

para-Terphenyl, CA 92-94-4: Irritant; toxic; flammable, flash point 483; moderately soluble aromatic solvents. Tm = 486.15; Tb = 649.15; Vm = 0.2620 at 588.75; Tc = 926; Pc = 3330; Antoine I (sol) A = 12.515, B = 6210, C = 0, dev = 5.0, range = 338 to 431; Antoine II (liq) A = 6.16107, B = 2125.84, C = -145.29, dev = 1.0, range = 499 to 700.

$C_{18}H_{14}N_4O_2$ MW 318.33

1,4-Bis[(4-hydroxyphenyl)azo]benzene: Irritant; flammable. Antoine (liq) A = 4.81688, B = 3552.766, C = 0, dev = 1.0 to 5.0, range = 473 to 533.

$C_{18}H_{15}N$ MW 245.32

Triphenyl amine, CA 603-34-9: Irritant; flammable; soluble ether, benzene. Tm = 400.15; Tb = 638.15; Antoine I (sol) A = 10.085, B = 4590, C = 0, dev = 5.0, range = 322 to 373; Antoine II (liq) A = 9.0160, B = 5282.1, C = 119.51, dev = 1.0 to 5.0, range = 473 to 640.

$C_{18}H_{15}NO_2$ MW 277.32

N-9-Anthryldiacetamide, CA 3808-37-5: Tm = 432.15; Antoine (liq) A = 11.032, B = 5556, C = 0, dev = 5.0, range = 399 to 455.

$C_{18}H_{15}O_4P$ MW 326.29

Triphenyl phosphate, CA 115-86-6: Moderately toxic; flammable, flash point 493; soluble ether, acetone, benzene, chloroform. Tm = 322.15; Vm = 0.2707 at 323.15; Antoine (liq) A = 8.195, B = 4253, C = 0, dev = 5.0, range = 548 to 683.

$C_{18}H_{15}P$ MW 262.29

Triphenyl phosphine, CA 603-35-0: Irritant; moderately toxic; flammable; flash point 453; soluble ether, benzene, chloroform. Tm = 353.15; Tb = 650.15; Vm = 0.2440 at 353.15; Antoine (liq) A = 7.6639, B = 3717, C = 0, dev = 1.0 to 5.0, range = 483 to 660.

$C_{18}H_{16}O_2$ MW 264.32

2-*tert*-Butylanthraquinone, CA 84-47-9: Flammable. Tm = 377.65; Antoine (liq) A = 10.81482, B = 5299.04, C = 0, dev = 1.0, range = 483 to 523.

$C_{18}H_{18}$ MW 234.34

5-Butylanthracene: Flammable. Tm = 319.95: Antoine I (sol) A = 12.943, B = 5645, C = 0, dev = 1.0, range = 293 to 313; Antoine II (liq) A = 9.048, B = 4382, C = 0, dev = 1.0, range = 328 to 373; Antoine III (liq) A = 8.3155, B = 4030.342, C = 0, dev = 5.0, range = 422 to 492.

1-Methyl-7-isopropylphenanthrene, retene, CA 483-65-8: Flammable; soluble benzene, carbon disulfide. Tm = 373.65 to 374.15; Tb = 663.15; Antoine (liq) A = 14.292, B = 19156, C = 889.15, dev = 1.0, range = 539 to 678.

$C_{18}H_{18}N_2O_2$ MW 294.35

1,5-Bis(dimethylamino)-9,10-anthraquinone, CA 18084-37-2: Irritant. Tm = 435.15; Antoine (sol) A = 1.17798, B = 1872.977, C = 0, dev = 5.0, range = 411 to 433.

$C_{18}H_{20}$ MW 236.36

3,3-Paracyclophane, CA 2913-24-8: Flammable. Tm = 377.15 to 378.15; Antoine (sol) A = 11.875, B = 5109, C = 0, dev not specified, range = 321 to 343.

$C_{18}H_{20}ClN_3O_6$ MW 409.83

Hexamethylbenzene-1-chloro-2,4,6-trinitrobenzene, 1:1 complex: Tm = 424.15; Antoine (sol) A = 11.5309, B = 4893.4, C = 0, dev = 1.0, range = 325 to 343.

$C_{18}H_{22}$ MW 238.37

1,6-Diphenylhexane, CA 1087-49-6: Flammable; soluble most organic solvents. Tm = 261.15; Antoine (liq) A = 10.322, B = 4595, C = 0, dev not specified, range = 293 to 373.

2,2-Di-(4-tolyl)-butane: Flammable; soluble most organic solvents. Antoine (liq) A = 6.327, B = 2200.3, C = -93.15, dev = 1.0, range = 298 to 473.

1-(4-Tolyl)-1-(4-propylphenyl)-ethane: Flammable; soluble most organic solvents. Antoine (liq) A = 6.2626, B = 2200.3, C = -93.15, dev = 1.0, range = 298 to 473.

2-(2-Tolyl)-2-(4-tolyl)-butane: Flammable; soluble most organic solvents. Antoine (liq) A = 6.4068, B = 2200.3, C = -93.15, dev = 1.0, range = 298 to 473.

$C_{18}H_{24}$ MW 240.39

1,2,3,4,4a,7,8,9,10,12,12a-Dodecahydrochrysene, CA 1610-22-6: Flammable. Tm = 314.15; Antoine I (sol) A = 14.495, B = 6029, C = 0, dev not specified, range = 293 to 313; Antoine II (liq) A = 9.290, B = 4396, C = 0, dev not specified, range = 318 to 358; Antoine III (liq) A = 8.23714, B = 3972.168, C = 0, dev = 1.0 to 5.0, range not specified.

$C_{18}H_{24}O_4$ MW 304.39

Butylcyclohexylphthalate, CA 84-64-0: Flammable. Vm = 0.2829 at 298.15;
Antoine (liq) A = 13.319, B = 4924, C = 0, dev = above 5.0, range = 368
to 485.

$C_{18}H_{26}O_4$ MW 306.40

Diisopentyl phthalate, CA 605-50-5: Flammable; soluble ethanol. Tb =
603.15 to 611.15, decomposes; Vm = 0.2998 at 289.15; Antoine (liq) A =
9.0785, B = 4261, C = 0, dev = 5.0, range = 390 to 610.

Dipentyl phthalate, CA 131-18-0: Flammable; soluble ethanol. Tm = below
218.15; Tb = 615.15; Vm = 0.2998 at 289.15; Antoine (liq) A = 11.165, B =
5191, C = 0, dev = 5.0, range = 303 to 500.

$C_{18}H_{30}$ MW 246.44

Dodecylbenzene, CA 123-01-3: Flammable, flash point above 383; soluble
most organic solvents. Tm = 266.15; Tb = 604.15; Vm = 0.2882 at 293.15;
Tc = 774.26; Pc = 1579; Vc = 1.000; Antoine (liq) A = 6.1942, B = 1981.5,
C = -127.65, dev = 1.0, range = 496 to 609.

Hexaethylbenzene, CA 604-88-6: Flammable; soluble most organic solvents.
Tm = 402.15; Tb = 571.15; Vm = 0.2967 at 403.15; Antoine (liq) A = 6.4791,
B = 2226.89, C = -73.525, dev = 1.0, range = 407 to 572.

Perhydrochrysene: Flammable; soluble ethanol. Tm = 388.15; Tb = 626.15;
Antoine (sol) A = 9.736, B = 4298, C = 0, dev = 5.0, range = 273 to 353.

1,3,5-Tri-*tert*-butylbenzene, CA 1460-02-2: Flammable; soluble most organic
solvents. Tm = 345.15 to 347.15; Tb = 521.15; Antoine (sol) A = 10.6635,
B = 4167, C = 0, dev = 5.0, range = 273 to 315.

$C_{18}H_{30}O$ MW 262.43

2,4,6-Tri-*tert*-butylphenol, CA 732-26-3: Irritant; flammable; soluble
ethanol, acetone. Tm = 404.15; Tb = 551.15; Antoine I (sol) A = 10.757,
B = 4470.5, C = 0, dev = 1.0 to 5.0, range = 295 to 339; Antoine II (liq)
A = 6.57419, B = 2133.3, C = -84.35, dev = 1.0, range = 415 to 551.

$C_{18}H_{30}O_2$ MW 278.43

1,3-Dimethoxy-4-decylbenzene, CA 59968-12-6: Flammable; soluble ethanol,
ether. Antoine (liq) A = 5.7499, B = 1906.2, C = -141.66, dev = 1.0,
range = 443 to 493.

1,3-Dimethoxy-5-decylbenzene, CA 41442-52-8: Flammable; soluble ethanol,
ether. Antoine (liq) A = 5.9095, B = 1960.7, C = -145.8, dev = 1.0,
range = 459 to 519.

$C_{18}H_{30}O_4$ MW 310.43

1,4-Bis(1,1-diethoxyethyl)benzene, CA 47189-08-2: Flammable; soluble ethanol, ether. Tm = 326.61; Antoine I (sol) A = 14.880, B = 5865, *(continues)*

426

$C_{18}H_{30}O_4$ *(continued)*

C = 0, dev = 1.0, range = 306 to 327; Antoine II (liq) A = 11.100, B = 4625, C = 0, dev = 1.0, range = 329 to 347.

$C_{18}H_{30}O_6$ MW 342.43

Aconitic acid, *trans*, tributyl ester, CA 7568-58-3: Flammable. Antoine (liq) A = 9.58355, B = 4563.863, C = 0, dev = 1.0 to 5.0, range = 385 to 483.

$C_{18}H_{30}O_{11}$ MW 422.43

Diethylene glycol dicarboxylic acid, di[1-(isopropoxycarbonyl)ethyl] ester: Irritant; flammable. Antoine (liq) A = 9.8545, B = 5096.5, C = 0, dev = 1.0 to 5.0, range = 418 to 493.

Diethylene glycol dicarboxylic acid, di[1-(propoxycarbonyl)ethyl] ester: Irritant; flammable. Antoine (liq) A = 10.0697, B = 5302.3, C = 0, dev = 1.0 to 5.0, range = 418 to 514.

$C_{18}H_{32}$ MW 248.45

9-Butyltetrahydroanthracene: Flammable. Antoine (liq) A = 8.470, B = 3802.242, C = 0, dev = 1.0, range = 420 to 456.

1,2-Dicyclohexylcyclohexane: Flammable. Tm = 319.15; Antoine (liq) A = 6.518, B = 2370.9, C = -81.95, dev = 1.0, range = 375 to 563.

1,6-Dicyclohexylcyclohexane: Flammable. Antoine (liq) A = 10.167, B = 4474, C = 0, dev = 1.0, range = 288 to 373.

$C_{18}H_{32}O_6$ MW 344.45

1,2,3-Propanetricarboxylic acid, tributyl ester, CA 38094-11-0: Flammable. Antoine (liq) A = 9.64277, B = 4587.891, C = 0, dev = 1.0 to 5.0, range = 385 to 482.

$C_{18}H_{34}O_2$ MW 282.47

9-Octadecenoic acid, *cis*, oleic acid, CA 112-80-1: Flammable, flash point above 383; soluble most organic solvents; oxidizes easily. Tm = 289.15; Vm = 0.3147 at 298.15; Antoine (liq) A = 8.22018, B = 3711.59, C = -36.125, dev = 1.0, range = 441 to 633.

9-Octadecenoic acid, *trans*, elaidic acid, CA 112-79-8: Flammable, flash point above 383; soluble most organic solvents. Tm = 317.15 to 318.15; Vm = 0.3297 at 343.15; Antoine (liq) A = 8.69349, B = 4219.35, C = -4.415, dev = 1.0, range = 444 to 635.

Tetradecyl methacrylate, CA 2549-53-3: Irritant; lachrymator; flammable. Antoine (liq) A = 7.915, B = 3610.5, C = 0, dev = 5.0, range = 463 to 611.

$C_{18}H_{34}O_4$ MW 314.46

Adipic acid, dihexyl ester, CA 110-33-8: Flammable. Tm = 259.35; Vm =
0.3289 at 293.15; Antoine (liq) A = 9.0659, B = 4200, C = 0, dev = 1.0 to
5.0, range = 470 to 595.

Sebacic acid, dibutyl ester, CA 109-43-3: Flammable; soluble ether. Tm =
263.95; Tb = 617.15 to 618.15; Vm = 0.3344 at 288.15; Antoine (liq) A =
6.36069, B = 2139.9, C = -141.52, dev = 1.0, range = 401 to 520.

$C_{18}H_{34}O_5$ MW 330.46

Decyl[1-(butoxycarbonyl)ethyl] carbonate: Toxic; flammable. Antoine (liq)
A = 8.7373, B = 4143, C = 0, dev = 5.0, range = 391 to 503.

$C_{18}H_{34}O_6$ MW 346.46

Triethylene glycol, bis(2-ethylbutyrate), CA 95-08-9: Flammable. Vm =
0.3482 at 298.15; Antoine (liq) A = 8.3643, B = 3666.3, C = -40.95, dev =
1.0, range = 313 to 528.

$C_{18}H_{35}N$ MW 265.48

Stearonitrile, CA 638-65-3: Flammable, flash point above 383; soluble most
organic solvents. Tm = 314.15; Tb = 635.15; Antoine (liq) A = 6.733, B =
2454, C = -111.75, dev = 5.0, range = 478 to 631

$C_{18}H_{36}$ MW 252.49

Dodecylcyclohexane, CA 1795-17-1: Flammable; soluble most organic solvents. Tf = 285.65; Tb = 604.15; Vm = 0.3071 at 293.15; Antoine (liq) A =
11.5439, B = 4880, C = 0, dev = 5.0, range = 299 to 324.

1-Octadecene, CA 112-88-9: Flammable, flash point above 383; soluble most
organic solvents. Tm = 291.15; Tb = 587.35; Vm = 0.3200 at 293.15; Antoine
(liq) A = 6.2473, B = 1971.9, C = -122.83, dev = 1.0, range = 399 to 589.

$C_{18}H_{36}O$ MW 268.48

Octadecanal, stearaldehyde, CA 638-66-4: Irritant; flammable; polymerizes
readily. Tm = 328.15; Antoine (liq) A = 6.81286, B = 2557.29, C = -83.789,
dev = 1.0 to 5.0, range = 413 to 616.

$C_{18}H_{36}O_2$ MW 284.48

Acetic acid, hexadecyl ester, CA 629-70-9: Flammable; slightly soluble
ethanol. Tm = 297.35; Vm = 0.3318 at 298.15; Antoine (liq) A = 7.95649,
B = 3673.929, C = 0, dev = 1.0, range = 431 to 469.

Ethyl palmitate, CA 628-97-7: Flammable, flash point above 383; soluble
most organic solvents. Tm = 297.15; Vm = 0.3317 at 298.15; Antoine I (sol)
A = 20.875, B = 7879, C = 0, dev = 5.0, range = 286 to 294; Antoine II
(liq) A = 12.133, B = 5259, C = 0, dev = 5.0, range = 298 to 318; Antoine
III (liq) A = 8.41026, B = 3858.818, C = 0, dev = 1.0, range = 429 to 466.
(continues)

$C_{18}H_{36}O_2$ *(continued)*

Methyl heptadecanoate, CA 1731-92-6: Flammable, flash point above 383; soluble most organic solvents. Tm = 303.15; Antoine (liq) A = 5.43452, B = 1537.53, C = -178.227, dev not specified, range = 421 to 525.

Octadecanoic acid, stearic acid, CA 57-11-4: Flammable, flash point 469; soluble benzene, chloroform, carbon tetrachloride. Tm = 342.85; Tb = 649.15; Vm = 0.3359 at 343.15; Antoine I (liq) A = 6.17126, B = 2157.5, C = -153.78, dev = 1.0, range = 349 to 415; Antoine II (liq) A = 5.85188, B = 1717.93, C = -201.829, dev = 1.0, range = 457 to 649.

$C_{18}H_{37}Br$ MW 333.39

1-Bromooctadecane, CA 112-89-0: Flammable, flash point above 383; soluble most organic solvents. Tm = 301.65; Antoine (liq) A = 6.695, B = 2495.2, C = -103.15, dev = 5.0, range = 430 to 673.

$C_{18}H_{37}Cl$ MW 288.94

1-Chlorooctadecane, CA 3386-33-2: Irritant; flammable, flash point above 383. Tm = 301.75; Tb = 621.15; Antoine (liq) A = 6.698, B = 2443.5, C = -100.15, dev = 5.0, range = 472 to 673.

$C_{18}H_{37}F$ MW 272.49

1-Fluorooctadecane, CA 1649-73-6: Flammable. Tm = 302.15; Antoine (liq) A = 6.711, B = 2327.4, C = -94.15, dev = 5.0, range = 477 to 633.

$C_{18}H_{37}I$ MW 380.39

1-Iodooctadecane, CA 629-93-6: Irritant; light-sensitive; flammable, flash point 383. Tm = 307.65 to 308.15; Tb = 656.15; Antoine (liq) A = 6.594, B = 2518.7, C = -107.15, dev = 5.0, range = 496 to 673.

$C_{18}H_{37}NO$ MW 283.50

Stearamide, CA 124-26-5: Irritant; flammable; soluble ether, chloroform. Tm = 382.15; Antoine (sol) A = 23.5739, B = 10230, C = 0, dev = 1.0, range = 367 to 379.

$C_{18}H_{38}$ MW 254.50

4,9-Diisopropyldodecane: Flammable; soluble most organic solvents. Tm = 229.65; Antoine (liq) A = 8.74687, B = 3658.5, C = 0, dev = 1.0, range = 368 to 424.

2,3-Dimethylhexadecane: Flammable; soluble most organic solvents. Tm = 264.35; Antoine (liq) A = 6.86645, B = 2506.2, C = -67.45, dev = 1.0, range = 466 to 583.

2,4-Dimethylhexadecane: Flammable; soluble most organic solvents. Tm = 252.15; Antoine (liq) A = 8.4530, B = 3638.36, C = 2.18, dev = 1.0, range = 434 to 562.

(continues)

$C_{18}H_{38}$ *(continued)*

2-Methylheptadecane, CA 1560-89-0: Flammable; soluble most organic solvents. Tm = 278.85; Tb = 584.15, decomposes; Vm = 0.3247 at 288.15; Antoine (liq) A = 7.6839, B = 3148.4, C = -26.01, dev = 1.0, range = 442 to 581.

3-Methylheptadecane, CA 6418-44-6: Flammable; soluble most organic solvents. Tm = 266.95; Antoine (liq) A = 7.9389, B = 3479.27, C = 3.31, dev = 1.0, range = 441 to 583.

4-Methylheptadecane, CA 26429-11-8: Flammable; soluble most organic solvents. Tm = 264.85; Antoine (liq) A = 7.4854, B = 3244.71, C = 11.88, dev = 1.0, range = 429 to 580.

5-Methylheptadecane, CA 26730-95-0: Flammable; soluble most organic solvents. Tm = 253.25; Antoine (liq) A = 7.53373, B = 3224.0, C = 2.35, dev = 1.0, range = 432 to 581.

Octadecane, CA 593-45-3: Flammable, flash point 438; soluble most organic solvents. Tm = 301.15; Tb = 578.15 to 580.15; Vm = 0.3276 at 301.15; Antoine I (liq) A = 10.18833, B = 4404.095, C = 0, dev = 1.0 to 5.0, range = 318 to 361; Antoine II (liq) A = 6.127, B = 1894.3, C = -129.85, dev = 0.1, range = 501 to 548.

2,4,6-Trimethylpentadecane: Flammable; soluble most organic solvents. Tm = 198.15; Antoine (liq) A = 8.13923, B = 3380.15, C = 1.45, dev = 1.0, range = 420 to 550.

$C_{18}H_{38}O$ MW 270.50

1-Octadecanol, stearyl alcohol, CA 112-92-5: Flammable; soluble most organic solvents. Tm = 332.55 to 332.95; Vm = 0.3324 at 333.15; Antoine I (sol) A = 24.990, B = 9787, C = 0, dev = 5.0, range = 320 to 329; Antoine II (liq) A = 13.268, B = 5928, C = 0, dev = 5.0, range = 334 to 356; Antoine III (liq) A = 6.35729, B = 2076.321, C = -144.62, dev = 1.0, range = 435 to 504; Antoine IV (liq) A = 6.2459, B = 2014.81, C = -148.403, dev = 0.1, range = 494 to 575; Antoine V (liq) A = 6.21028, B = 1986.737, C = -151.16, dev = 0.1, range = 500 to 573.

$C_{18}H_{38}O_4$ MW 318.50

2-[2-(2-[Dodecyloxy]ethoxy)ethoxy] ethanol, CA 3055-94-5: Flammable; soluble ethanol. Antoine (liq) A = 10.88829, B = 5364.382, C = 0, dev = 1.0 to 5.0, range = 475 to 523.

$C_{18}H_{39}N$ MW 269.51

N,N-Dimethylhexadecyl amine, CA 112-69-6: Irritant; flammable; soluble ether, ethanol. Tm = 285.15; Antoine (liq) A = 6.5019, B = 2275.1, C = -97.15, dev = 5.0, range = 483 to 671.

Dinonyl amine, CA 2044-21-5: Irritant; flammable; soluble ethanol, ether. Tm = 298.15; Antoine (liq) A = 6.511, B = 2297.7, C = -97.15, dev = 5.0, range = 486 to 676.

(continues)

$C_{18}H_{39}N$ *(continued)*

N-Ethylhexadecyl amine, CA 5877-76-9: Irritant; flammable; soluble ethanol, ether. Antoine (liq) A = 7.30773, B = 3163.02, C = -18.837, dev = 1.0 to 5.0, range = 406 to 613.

Octadecyl amine, stearyl amine, CA 124-30-1: Irritant; corrosive; flammable, flash point above 383; soluble most organic solvents. Tm = 322.15 to 325.15; Antoine (liq) A = 6.6939, B = 2444.9, C = -100.45, dev = 1.0, range = 450 to 635.

$C_{19}H_{13}NO$ MW 271.32

2-(1-Naphthyl)-5-phenyloxazole, CA 846-63-9: Irritant; flammable. Antoine (liq) A = 9.025, B = 4660, C = 0, dev = 5.0, range = 510 to 595.

$C_{19}H_{15}N_3$ MW 285.35

Triphenylazidomethane, CA 14309-25-2: Flammable. Antoine (sol) A = 14.415, B = 6300, C = 0, dev = 1.0 to 5.0, range = 335 to 363.

$C_{19}H_{16}$ MW 244.34

Triphenylmethane, CA 519-73-3: Flammable; soluble most organic solvents. Tm = 367.15; Tb = 631.15 to 632.15; Vm = 0.2410 at 372.15; Antoine I (sol) A = 11.786, B = 5227.9, C = 0, dev = 1.0, range = 325 to 349; Antoine II (liq) A = 8.2921, B = 4833.92, C = 135.1, dev = 5.0, range = 512 to 643.

$C_{19}H_{16}O$ MW 260.33

Triphenylmethanol, tritanol, CA 76-84-6: Flammable; soluble ethanol, ether, benzene. Tm = 437.15 to 438.15; Tb = 633.15 to 653.15; Antoine (sol) A = 11.6949, B = 6370.6, C = 0, dev = 5.0, range = 353 to 373.

$C_{19}H_{17}NO_2$ MW 291.35

Anthraquinone, 1-piperidino, CA 4946-83-2: Irritant; flammable. Tm = 393.15; Antoine I (sol) A = -1.02478, B = 953.466, C = 0, dev = 5.0, range = 383 to 392; Antoine II (liq) A = 7.45841, B = 4285.499, C = 0, dev = 1.0 to 5.0, range = 395 to 404.

$C_{19}H_{20}O_4$ MW 312.36

Butylbenzylphthalate, CA 85-68-7: Flammable. Antoine (liq) A = 9.1472, B = 4647.5, C = 0, dev = 1.0, range = 416 to 516.

Ethylmalonic acid, dibenzyl ester, CA 74254-53-8: Flammable. Antoine (liq) A = 10.1361, B = 4914.3, C = 0, dev = 1.0, range = 403 to 483.

$C_{19}H_{24}$ MW 252.40

Dicumenylmethane, CA 25566-92-1: Flammable; soluble most organic solvents. Antoine (liq) A = 8.215, B = 3850, C = 0, dev = 5.0, range = 323 to 402.

$C_{19}H_{30}$ MW 258.45

7-Phenyl-6-tridecene: Flammable; soluble most organic solvents. Antoine
(liq) A = 9.09765, B = 4030.267, C = 0, dev = 1.0, range = 391 to 449.

$C_{19}H_{32}$ MW 260.46

7-Phenyltridecane, CA 2400-01-3: Flammable; soluble most organic solvents.
Vm = 0.2986 at 293.15; Antoine (liq) A = 9.04834, B = 3981.289, C = 0,
dev = 1.0, range = 413 to 470.

$C_{19}H_{32}O_2$ MW 292.46

Linolenic acid, methyl ester, CA 301-00-8: Flammable, flash point above
383. Tm = 220.75 to 221.35; Vm = 0.3264 at 293.15; Antoine (liq) A =
6.5641, B = 2298.8, C = -119.15, dev = 1.0, range = 394 to 459.

$C_{19}H_{34}$ MW 262.48

Tricyclohexylmethane, CA 1610-24-8: Flammable; soluble most organic solvents. Tm = 330.45; Tb = 595.15 to 602.15; Antoine I (sol) A = 15.221, B =
6133, C = 0, dev = 1.0, range = 301 to 321; Antoine II (liq) A = 9.491, B =
4250, C = 0, dev = 1.0, range = 333 to 365; Antoine III (liq) A = 8.37446,
B = 3828.804, C = 0, dev = 1.0, range = 428 to 605.

$C_{19}H_{34}O_2$ MW 294.48

Linoleic acid, methyl ester, CA 112-63-0: Flammable, flash point above
383; soluble most organic solvents. Tm = 238.15; Vm = 0.3314 at 283.15;
Antoine (liq) A = 6.9942, B = 2589.4, C = -98.15, dev = 1.0, range = 391
to 500.

$C_{19}H_{36}$ MW 264.49

1,1-Dicyclohexylheptane: Flammable; soluble most organic solvents.
Antoine I (liq) A = 10.541, B = 4585, C = 0, dev = 1.0, range = 293 to 368;
Antoine II (liq) A = 8.55293, B = 3857.525, C = 0, dev = 1.0, range = 422
to 458.

$C_{19}H_{36}O_2$ MW 296.49

Methyl oleate, CA 112-62-9: Flammable, flash point above 383; suspected
carcinogen; soluble ethanol, ether. Tm = 253.25; Vm = 0.3393 at 293.15;
Antoine (liq) A = 7.231702, B = 2728.049, C = -91.489, dev = 0.1, range =
428 to 486.

$C_{19}H_{36}O_5$ MW 344.49

Undecyl[1-(butoxycarbonyl)ethyl] carbonate: Toxic; flammable. Antoine
(liq) A = 8.338, B = 4024, C = 0, dev = 5.0, range = 438 to 637.

$C_{19}H_{37}NO$ MW 295.51

Octadecyl isocyanate, CA 112-96-9: Toxic; lachrymator; flammable, flash
point 457; reacts violently with water. Tm = 294.15; Vm =   *(continues)*

$C_{19}H_{37}NO$ *(continued)*

0.3433 at 298.15; Antoine (liq) A = 9.54025, B = 4830.117, C = 36.252, dev = 1.0 to 5.0, range = 388 to 494.

$C_{19}H_{37}NO_3$ MW 327.51

N,N-Dibutylpropionamide, 2-[2-ethyl-(hexanoyloxy)]: Irritant; flammable. Antoine (liq) A = 9.44014, B = 4337.108, C = 0, dev = 1.0 to 5.0, range = 403 to 448.

$C_{19}H_{38}$ MW 266.51

7-Cyclohexyltridecane: Flammable; soluble most organic solvents. Antoine (liq) A = 8.92029, B = 3950.208, C = 0, dev = 1.0, range = 391 to 449.

7-(Cyclopentylmethyl)tridecane: Flammable; soluble most organic solvents. Antoine (liq) A = 9.07607, B = 3993.563, C = 0, dev = 1.0, range = 389 to 446.

1-Nonadecene, CA 18435-45-5: Flammable, flash point above 383; soluble most organic solvents. Tm = 296.55; Tb = 601.15; Vm = 0.3365 at 293.15, undercooled liquid; Antoine (liq) A = 6.182, B = 1961.6, C = -131.95, dev = 1.0, range = 560 to 604.

$C_{19}H_{38}O_2$ MW 298.51

Isopropyl palmitate, CA 142-91-6: Irritant; flammable, flash point above 383; soluble most organic solvents. Tm = 287.15; Vm = 0.3552 at 311.15; Antoine (liq) A = 8.2972, B = 3843.6, C = 0, dev = 1.0, range = 433 to 471.

Methyl stearate, CA 112-61-8: Flammable, flash point above 383; soluble ethanol, ether, chloroform. Tm = 312.25; Tb = 715.15 to 716.15; Vm = 0.3513 at 313.15; Antoine I (sol) A = 21.070, B = 8250, C = 0, dev = 5.0, range = 301 to 310; Antoine II (liq) A = 7.869218, B = 3239.351, C = -60.283, dev = 0.1, range = 427 to 500.

Nonadecanoic acid, CA 646-30-0: Flammable; soluble most organic solvents. Tm = 341.85; Antoine (liq) A = 5.363, B = 1383, C = -247.15, dev = 5.0, range = 511 to 659.

Propyl palmitate, CA 2239-78-3: Flammable; soluble most organic solvents. Tm = 293.55; Vm = 0.3531 at 306.15; Antoine (liq) A = 8.2876, B = 3892.9, C = 0, dev = 1.0, range = 439 to 477.

$C_{19}H_{38}O_3$ MW 314.51

Hexadecyl lactate, CA 35274-05-6: Flammable. Tm = 314.15; Antoine (liq) A = 7.37766, B = 2914.312, C = -90.15, dev = 1.0 to 5.0, range = 405 to 556.

3-Octyloxypropionic acid, octyl ester: Flammable. Antoine (liq) A = 8.123, B = 3845, C = 0, dev = 5.0, range = 443 to 513.

$C_{19}H_{39}Br$ MW 347.42

1-Bromononadecane, CA 4434-66-6: Irritant; flammable. Tm = 311.65;
Antoine (liq) A = 6.722, B = 2557.2, C = -105.15, dev = 5.0, range = 493
to 673.

$C_{19}H_{39}Cl$ MW 302.97

1-Chlorononadecane: Irritant; flammable. Tm = 308.85; Antoine (liq) A =
6.725, B = 2505.7, C = -103.15, dev = 5.0, range = 483 to 673.

$C_{19}H_{39}F$ MW 286.52

1-Fluorononadecane: Flammable. Tm = 312.15; Antoine (liq) A = 6.737, B =
2389.7, C = -97.15, dev = 5.0, range = 458 to 648.

$C_{19}H_{39}I$ MW 394.42

1-Iodononadecane, CA 62127-51-9: Irritant; light-sensitive; flammable.
Tm = 315.65 to 316.65; Antoine (liq) A = 6.623, B = 2580.6, C = -109.15,
dev = 5.0, range = 506 to 673.

$C_{19}H_{39}NO_2$ MW 313.52

Lactamide, *N,N*-dioctyl: Irritant; flammable. Antoine (liq) A = 10.62187,
B = 5185.761, C = 0, dev = 1.0 to 5.0, range = 453 to 488.

Lactamide, *N*-hexadecyl: Irritant; flammable. Tm = 353.15; Antoine (liq)
A = 11.12176, B = 5801.523, C = 0, dev = 1.0, range = 423 to 508.

$C_{19}H_{40}$ MW 268.53

2,3-Dimethylheptadecane: Flammable; soluble most organic solvents. Tm =
263.65; Antoine (liq) A = 6.4007, B = 2137.3, C = -111.55, dev = 1.0,
range = 493 to 598.

2,4-Dimethylheptadecane: Flammable; soluble most organic solvents. Tm =
262.15; Antoine (liq) A = 8.5454, B = 3802.1, C = 7.15, dev = 1.0, range =
444 to 574.

7-Hexyltridecane, CA 7225-66-3: Flammable; soluble most organic solvents.
Tm = 244.85; Vm = 0.3409 at 293.15; Antoine (liq) A = 8.9786, B = 3926.6,
C = 0, dev = 1.0, range = 411 to 444.

2-Methyloctadecane, CA 1560-88-9: Flammable; soluble most organic solvents.
Tm = 284.35; Antoine (liq) A = 7.9119, B = 3506.4, C = -1.45, dev = 1.0,
range = 451 to 595.

3-Methyloctadecane, CA 6561-44-0: Flammable; soluble most organic solvents.
Tm = 275.35; Antoine (liq) A = 8.0749, B = 3624.3, C = 0.55, dev = 1.0,
range = 455 to 597.

4-Methyloctadecane, CA 10544-95-3: Flammable; soluble most organic solvents. Tm = 268.65; Antoine (liq) A = 7.6763, B = 3424.5, C = 8.25, dev =
1.0, range = 445 to 596.

(continues)

$C_{19}H_{40}$ *(continued)*

5-Methyloctadecane, CA 25117-35-5: Flammable; soluble most organic solvents. Tm = 257.35; Antoine (liq) A = 7.6219, B = 3351.9, C = 1.15, dev = 1.0, range = 445 to 595.

Nonadecane, CA 629-92-5: Flammable, flash point 441; soluble most organic solvents. Tm = 305.15; Tb = 603.15; Antoine (liq) A = 6.1402, B = 1932.8, C = -135.55, dev = 1.0, range = 456 to 606.

2,4,6-Trimethylhexadecane: Flammable; soluble most organic solvents. Tm = 208.15; Antoine (liq) A = 8.182, B = 3503.4, C = -0.95, dev = 1.0, range = 435 to 568.

$C_{19}H_{40}O$ MW 284.52

1-Nonadecanol, CA 1454-84-8: Flammable; soluble ether, acetone. Tm = 335.15 to 336.15; Antoine I (liq) A = 6.45643, B = 2207.247, C = -138.676, dev = 1.0, range = 479 to 640; Antoine II (liq) A = 6.63694, B = 2349.1, C = -127.17, dev = 1.0, range = 494 to 635.

$C_{19}H_{41}N$ MW 283.54

Nonadecyl amine, CA 14130-05-3: Irritant; flammable. Tm = 326.15; Antoine (liq) A = 6.7289, B = 2506.8, C = -102.65, dev = 1.0, range = 532 to 647.

$C_{20}H_{12}$ MW 252.31

Benzo[k]fluoranthene, CA 207-08-9: Carcinogenic; soluble ethanol, benzene, acetic acid. Tm = 490.15; Tb = 753.15; Antoine (sol) A = 12.8907, B = 6792, C = 0, dev = 5.0, range = 363 to 430.

Benzo[a]pyrene, CA 50-32-8: Potent carcinogen; soluble chloroform, aromatics. Tm = 450.15; Tb = 768.15; Antoine (sol) A = 11.6067, B = 6181, C = 0, dev = 5.0, range = 358 to 431.

Benzo[e]pyrene, CA 192-97-2: Weak carcinogen. Tm = 451.15 to 452.15; Tb = 583.15 to 585.15; Antoine (sol) A = 11.7417, B = 6220, C = 0, dev = 5.0, range = 358 to 423.

Perylene, *peri*-dinapththalene, CA 198-55-0: Soluble chloroform, carbon disulfide, benzene. Tm = 546.15 to 547.15; Antoine I (sol) A = 13.075, B = 7260, C = 0, dev = 5.0, range = 383 to 453; Antoine II (sol) A = 12.9379, B = 7210, C = 0, dev = 1.0 to 5.0, range = 383 to 518.

$C_{20}H_{13}NO_4$ MW 331.33

9,10-Anthraquinone, 1-amino-4-hydroxy-2-phenoxy, CA 17418-58-5: Irritant. Tm = 454.15; Antoine I (sol) A = 13.8743, B = 7968.482, C = 0, dev = 1.0, range = 359 to 366; Antoine II (sol) A = 1.03973, B = 2148.106, C = 0, dev = 5.0, range = 433 to 450.

$C_{20}H_{14}$ MW 254.33

9-Phenylanthracene, CA 602-55-1: Flammable; soluble most organic solvents. Tm = 425.15 to 427.15; Tb = 690.15; Antoine I (sol) A =   *(continues)*

$C_{20}H_{14}$ *(continued)*

12.4633, B = 6018.5, C = 0, dev = 1.0, range = 353 to 426; Antoine II (liq)
A = 8.7183, B = 4409.9, C = 0, dev = 1.0, range = 430 to 510.

$C_{20}H_{14}O_4$ MW 318.33

Dibenzoate resorcinol, CA 94-01-9: Flammable; soluble ethanol, ether.
Tm = 399.15; Antoine I (sol) A = 18.925, B = 8660, C = 0, dev = 5.0, range
= 323 to 399; Antoine II (liq) A = 7.155, B = 3970, C = 0, dev = 5.0,
range = 399 to 493.

$C_{20}H_{16}$ MW 256.35

7,12-Dimethylbenz[a]anthracene, CA 57-97-6: Potent carcinogen; flammable;
soluble acetone, benzene. Tm = 395.15 to 396.15; Antoine I (sol) A =
14.233, B = 7051, C = 0, dev = 5.0, range = 379 to 396; Antoine II (sol)
A = 10.70417, B = 5629.911, C = 0, dev = 1.0, range = 379 to 390; Antoine
III (liq) A = 11.357, B = 5897, C = 0, dev = 5.0, range = 396 to 408.

5,6-Dimethylchrysene, CA 3697-27-6: Suspected carcinogen; flammable; soluble ethanol, acetic acid. Tm = 400.15 to 401.15; Antoine (sol) A =
12.54913, B ≠ 6358.936, C = 0, dev = 1.0, range = 380 to 394.

$C_{20}H_{22}N_2O_4$ MW 354.40

Anthraquinone, 1,4-bis(propylamino): Irritant; flammable. Tm = 428.15;
Antoine (liq) A = 13.9089, B = 6180, C = 0, dev = 1.0 to 5.0, range = 409
to 463.

$C_{20}H_{26}O_4$ MW 330.42

Dicyclohexyl phthalate, CA 84-61-7: Flammable; soluble ethanol, ether.
Tm = 331.15 to 338.15; Antoine (liq) A = 10.065, B = 5069, C = 0, dev =
5.0, range = 391 to 475.

$C_{20}H_{30}O_4$ MW 334.45

Dihexyl phthalate, CA 84-75-3: Flammable; soluble ethanol, ether. Antoine
I (liq) A = -1.01167, B = 1483.636, C = 0, dev = 1.0 to 5.0, range = 288 to
303; Antoine II (liq) A = 11.105, B = 5381, C = 0, dev = 5.0, range = 343
to 387; Antoine III (liq) A = 9.785, B = 4805, C = 0, dev = 5.0, range =
453 to 533.

$C_{20}H_{34}$ MW 274.49

9-Cyclohexyltetradecahydroanthracene: Flammable. Tm = about 399.15;
Antoine (liq) A = 8.10735, B = 3892.235, C = 0, dev = 1.0, range = 419
to 488.

$C_{20}H_{34}O_2$ MW 306.49

Ethyl linolenate, CA 1191-41-9: Flammable, flash point above 383; light-
sensitive; soluble ethanol, ether. Vm = 0.3436 at 293.15; Antoine (liq)
A = 8.0238, B = 3800.2, C = 0, dev = 1.0, range = 447 to 491.

436

$C_{20}H_{34}O_{11}$ MW 450.48

Diethylene glycol dicarboxylic acid, di[1-(butoxycarbonyl)ethyl] ester:
Toxic; flammable. Antoine (liq) A = 10.0581, B = 5413, C = 0, dev = 5.0,
range = 433 to 525.

Diethylene glycol dicarboxylic acid, di[1-(sec-butoxycarbonyl)ethyl]
ester: Toxic; flammable. Antoine (liq) A = 10.2394, B = 5384, C = 0,
dev = 5.0, range = 418 to 513.

Diethylene glycol dicarboxylic acid, di[1-isobutoxycarbonyl)ethyl] ester:
Toxic; flammable. Antoine (liq) A = 10.1692, B = 5387, C = 0, dev = 5.0,
range = 415 to 513.

$C_{20}H_{36}O_{2}$ MW 308.50

Ethyl linoleate, CA 544-35-4: Flammable, flash point above 383; soluble
ethanol, ether, fat solvents. Vm = 0.3459 at 293.15; Antoine (liq) A =
8.005, B = 3791.9, C = 0, dev = 5.0, range = 448 to 497.

$C_{20}H_{38}O_{2}$ MW 310.52

Ethyl oleate, CA 111-62-6: Flammable, flash point above 383; light-
sensitive; soluble ethanol, ether. Tm = 241.15; Vm = 0.3561 at 293.15;
Antoine (liq) A = 6.5339, B = 2286, C = -124.15, dev = 5.0, range = 384
to 481.

Hexadecyl methacrylate, CA 2495-27-4: Irritant; flammable. Antoine (liq)
A = 7.971, B = 3820, C = 0, dev = 5.0, range = 431 to 541.

$C_{20}H_{38}O_{4}$ MW 342.52

Dioctyl succinate, CA 14491-66-8: Flammable. Antoine (liq) A = 10.245,
B = 4921, C = 0, dev = 5.0, range = 503 to 523.

Dipentyl sebacate, CA 6819-09-6: Flammable. Antoine (liq) A = 10.545,
B = 5180, C = 0, dev = 5.0, range = 353 to 408.

Eicosanedioic acid, CA 2424-92-2: Flammable; soluble ether. Tm = 398.15
to 399.65; Antoine (sol) A = 17.310, B = 8644, C = 0, dev = 5.0, range =
380 to 395.

$C_{20}H_{38}O_{5}$ MW 358.52

Dodecyl[1-(butoxycarbonyl)ethyl] carbonate: Toxic; flammable. Antoine
(liq) A = 8.784, B = 4326, C = 0, dev = 5.0, range = 408 to 498.

$C_{20}H_{40}$ MW 280.54

1-Eicosene, CA 3452-07-1: Flammable, flash point above 383; soluble most
organic solvents. Tf = 301.65; Tb = 614.35; Vm = 0.3559 at 303.15; Antoine
(liq) A = 6.1826, B = 1995.3, C = -137.15, dev = 1.0, range = 573 to 615.

$C_{20}H_{40}O_2$ MW 312.53

Acetic acid, octadecyl ester, CA 822-23-1: Flammable; soluble ethanol.
Tm = 306.15; Antoine (liq) A = 6.38278, B = 2403.291, C = -107.166, dev =
1.0, range = 341 to 500.

Butyl palmitate, CA 111-06-8: Flammable; soluble ethanol, ether. Tm =
290.05; Antoine (liq) A = 10.455, B = 4900, C = 0, dev = 5.0, range = 353
to 383.

Decanoic acid, decyl ester, CA 1654-86-0: Flammable; soluble ether. Tm =
282.85; Vm = 0.3640 at 293.15; Antoine (liq) A = 4.15535, B = 1372.922,
C = -170.952, dev = 1.0 to 5.0, range = 341 to 398.

Eicosanoic acid, arachidic acid, CA 506-30-9: Flammable; soluble ether,
benzene, chloroform. Tm = 350.15; Vm = 0.3793 at 373.15; Antoine I (sol)
A = 24.578, B = 10424, C = 0, dev = 5.0, range = 336 to 346; Antoine II
(liq) A = 5.296, B = 1359, C = -257.15, dev = 5.0, range = 477 to 670.

Ethyl stearate, CA 111-61-5: Flammable, flash point above 383; soluble
ether, ethanol, acetone. Tm = 306.55; Antoine I (sol) A = 21.595, B =
8430, C = 0, dev = 5.0, range = 297 to 306; Antoine II (liq) A = 12.257,
B = 5578, C = 0, dev = 1.0 to 5.0, range = 310 to 328; Antoine III (liq)
A = 12.29037, B = 5843.672, C = 0, dev not specified, range = 454 to 469.

Nonadecanoic acid, methyl ester, CA 1731-94-8: Flammable, flash point
above 383. Tm = 312.05; Antoine (liq) A = 4.80483, B = 1258.58, C =
-219.859, dev not specified, range = 441 to 529.

$C_{20}H_{41}Br$ MW 361.45

1-Bromoeicosane, CA 4276-49-7: Irritant; flammable. Tm = 310.05; Antoine
(liq) A = 6.746, B = 2617.5, C = -107.15, dev = 5.0, range = 502 to 673.

$C_{20}H_{41}Cl$ MW 317.00

1-Chloroeicosane, CA 42217-02-7: Irritant; flammable. Tm = 310.75;
Antoine (liq) A = 6.749, B = 2566.1, C = -105.15, dev = 5.0, range = 492
to 673.

$C_{20}H_{41}F$ MW 300.54

1-Fluoroeicosane: Flammable. Tm = 311.15; Antoine (liq) A = 6.761, B =
2450.1, C = -99.15, dev = 5.0, range = 468 to 663.

$C_{20}H_{41}I$ MW 408.45

1-Iodoeicosane, CA 34994-81-5: Irritant; light-sensitive; flammable. Tm =
315.15 to 316.15; Antoine (liq) A = 6.649, B = 2640.9, C = -111.15, dev =
5.0, range = 516 to 673.

$C_{20}H_{42}$ MW 282.55

5-Butylhexadecane, CA 6912-07-8: Flammable; soluble most organic solvents.
Antoine (liq) A = 8.96246, B = 4040.29, C = 0, dev = 0.1, range = 423 to
457. *(continues)*

438

$C_{20}H_{42}$ (continued)

2,3-Dimethyloctadecane: Flammable; soluble most organic solvents. Tm = 272.05; Antoine (liq) A = 7.7587, B = 3571.4, C = 9.35, dev = 1.0, range = 458 to 612.

2,4-Dimethyloctadecane, CA 61868-10-8: Flammable; soluble most organic solvents. Tm = 265.65; Antoine (liq) A = 8.8209, B = 3985.5, C = 1.55, dev = 1.0, range = 456 to 583.

Eicosane, CA 112-95-8: Flammable, flash point above 383; soluble most organic solvents. Tm = 309.85; Tb = 616.15; Antoine I (liq) A = 10.77373, B = 4872.63, C = 0, dev = 1.0, range = 344 to 380; Antoine II (liq) A = 6.2771, B = 2032.7, C = -141.05, dev = 0.1, range = 528 to 620.

2-Methylnonadecane, CA 1560-86-7: Flammable; soluble most organic solvents. Tm = 291.25; Antoine (liq) A = 8.1729, B = 3719.2, C = -3.90, dev = 1.0, range = 465 to 607.

3-Methylnonadecane, CA 6418-45-7: Flammable; soluble most organic solvents. Tm = 280.75; Antoine (liq) A = 7.7028, B = 3316.1, C = -27.05, dev = 1.0, range = 463 to 609.

4-Methylnonadecane, CA 25117-27-5: Flammable; soluble most organic solvents. Tm = 271.85; Antoine (liq) A = 7.7525, B = 3456.8, C = -7.65, dev = 1.0, range = 460 to 609.

5-Methylnonadecane, CA 57160-72-2: Flammable; soluble most organic solvents. Tm = 266.15; Antoine (liq) A = 7.9192, B = 3599.1, C = -0.55, dev = 1.0, range = 462 to 609.

4-Propylheptadecane: Flammable; soluble most organic solvents. Antoine (liq) A = 9.14612, B = 4136.9, C = 0, dev = 1.0, range = 425 to 459.

2,4,6-Trimethylheptadecane: Flammable; soluble most organic solvents. Tm = 225.15; Antoine (liq) A = 8.6123, B = 3875.2, C = 7.45, dev = 1.0, range = 449 to 579.

$C_{20}H_{42}O$ MW 298.55

1-Eicosanol, CA 629-96-9: Flammable; soluble most organic solvents. Tm = 345.65 to 346.15; Antoine I (sol) A = 28.860, B = 11393, C = 0, dev = 1.0 to 5.0, range = 327 to 337; Antoine II (liq) A = 13.378, B = 6213, C = 0, dev = 1.0 to 5.0, range = 339 to 358; Antoine III (liq) A = 6.52426, B = 2291.84, C = -138.263, dev = 1.0, range = 488 to 653; Antoine IV (liq) A = 6.8394, B = 2539.2, C = -120.03, dev = 1.0 to 5.0, range = 493 to 648.

$C_{20}H_{42}O_5$ MW 362.55

3,6,9,12-Tetraoxa-1-tetracosanol, CA 5274-68-0: Flammable. Antoine (liq) A = 13.74583, B = 7078.113, C = 0, dev = 1.0 to 5.0, range = 501 to 543.

$C_{20}H_{43}N$ MW 297.57

Didecyl amine, CA 1120-49-6: Irritant; flammable, flash point above 383. Tm = 307.15; Antoine (liq) A = 6.5209, B = 2393.1, C = -102.15, dev = 5.0, range = 506 to 705.

(continues)

$C_{20}H_{43}N$ *(continued)*

N,N-Diethylhexadecyl amine, CA 30951-88-3: Irritant; flammable. Tm = 281.15; Antoine (liq) A = 6.55036, B = 2500.3, C = -78.43, dev = 5.0, range = 412 to 628.

N,N-Dimethyloctadecyl amine, CA 124-28-7: Irritant; flammable. Tf = 296.05; Antoine (liq) A = 6.5144, B = 2376.1, C = -102.15, dev = 5.0, range = 504 to 701.

Eicosyl amine, CA 10525-37-8: Irritant; flammable. Tm = 331.15; Antoine (liq) A = 6.7539, B = 2567.1, C = -104.85, dev = 1.0, range = 543 to 659.

$C_{21}H_6N_{12}O_{18}$ MW 714.35

1,3,5-Triazine, 2,4,6-tris(2,4,6-trinitrophenyl), CA 49753-54-0: Irritant; flammable; potentially explosive. Antoine (sol) A = 9.64516, B = 8771.748, C = 0, dev = 5.0, range = 479 to 551.

$C_{21}H_8F_{28}O_8$ MW 920.25

Pentaerythritol, tetraperfluorobutyrate: Antoine (liq) A = 4.555, B = 1852, C = 0, dev = 5.0, range = 293 to 433.

$C_{21}H_{15}N_3$ MW 309.37

2,4,6-Triphenyl-1,3,5-triazine, CA 493-77-6: Irritant; flammable. Tm = 504.15; Antoine (liq) A = 8.2889, B = 4650, C = 0, dev = 1.0 to 5.0, range not specified.

$C_{21}H_{16}$ MW 268.36

3-Methylcholanthrene, CA 56-49-5: Potent carcinogen; soluble aromatic solvents. Tm = 452.15 to 453.15; Antoine (sol) A = 12.293, B = 6643, C = 0, dev = 1.0 to 5.0, range = 401 to 425.

$C_{21}H_{21}O_4P$ MW 368.37

Phosphoric acid, tri(2-tolyl) ester, CA 78-30-8: Highly toxic; flammable; soluble most organic solvents. Tm = 284.15; Tb = 683.15; Vm = 0.3081 at 293.15; Antoine (liq) A = 8.565, B = 4535, C = 0, dev = 5.0, range = 293 to 700.

Phosphoric acid, tri(3-tolyl) ester, CA 563-04-2: Highly toxic; flammable; soluble most organic solvents. Tm = 298.15 to 299.15; Antoine (liq) A = 11.8856, B = 6104.5, C = -10.81, dev = 1.0 to 5.0, range = 398 to 530.

Phosphoric acid, tri(4-tolyl) ester, CA 78-32-0: Highly toxic; flammable; soluble most organic solvents. Tm = 350.65 to 351.15; Antoine (liq) A = 10.245, B = 5480, C = 0, dev = 5.0, range = 388 to 530.

$C_{21}H_{26}$ MW 278.44

1,8-*para*-Cyclophane, CA 6169-94-4: Flammable. Antoine (sol) A = 11.935, B = 5480, C = 0, dev = 5.0, range = 354 to 376.

$C_{21}H_{30}$ MW 282.47

1-Undecylnaphthalene: Flammable. Tm = 296.05; Antoine (liq) A = 8.8934,
B = 4403.42, C = 0, dev = 1.0, range = 436 to 502.

$C_{21}H_{36}O_6$ MW 384.51

Aconitic acid, *trans*, triisopentyl ester: Flammable. Antoine (liq) A =
9.35287, B = 4611.549, C = 0, dev = 1.0 to 5.0, range = 396 to 499.

Aconitic acid, *trans*, tripentyl ester, CA 64617-29-4: Flammable. Antoine
(liq) A = 9.57596, B = 4776.861, C = 0, dev = 1.0 to 5.0, range = 403
to 505.

$C_{21}H_{38}O_6$ MW 386.53

Glycerol tricaproate, tricaproin, CA 621-70-5: Flammable; soluble most
organic solvents. Tm = 248.15; Vm = 0.3917 at 293.15; Antoine (liq) A =
9.945, B = 4920, C = 0, dev = 5.0, range = 356 to 410.

1,2,3-Propanetricarboxylic acid, triisopentyl ester: Flammable. Antoine
(liq) A = 9.34705, B = 4608.94, C = 0, dev = 1.0 to 5.0, range = 396 to 508.

1,2,3-Propanetricarboxylic acid, tripentyl ester, CA 5333-53-9: Flammable.
Antoine (liq) A = 9.39158, B = 4713.555, C = 0, dev = 1.0 to 5.0, range =
404 to 508.

$C_{21}H_{40}$ MW 292.55

1-Undecyldecahydronaphthalene: Flammable. Tm = 278.15; Antoine (liq) A =
9.04316, B = 4353.33, C = 0, dev = 1.0, range = 426 to 488.

$C_{21}H_{42}O_2$ MW 326.56

Eicosanoic acid, methyl ester, CA 1120-28-1: Flammable, flash point above
383; soluble most organic solvents. Tm = 318.95 to 319.45; Antoine I (sol)
A = 25.298, B = 9970, C = 0, dev = 5.0, range = 311 to 318; Antoine II
(liq) A = 4.67779, B = 1220.16, C = -230.822, dev = 1.0, range = 450 to 540.

Isopropyl stearate, CA 112-10-7: Flammable; soluble most organic solvents.
Tm = 301.15; Vm = 0.3886 at 311.15; Antoine (liq) A = 8.2371, B = 4004.0,
C = 0, dev = 1.0, range = 453 to 483.

Nonadecanoic acid, ethyl ester, CA 18281-04-4: Flammable. Tm = 309.25;
Antoine I (sol) A = 19.295, B = 7822, C = 0, dev = 5.0, range = 302 to 308;
Antoine II (liq) A = 12.684, B = 5798, C = 0, dev = 5.0, range = 312 to 328.

Propyl stearate, CA 3634-92-2: Flammable; soluble most organic solvents.
Tm = 301.75; Vm = 0.3864 at 311.15; Antoine (liq) A = 9.4145, B = 4593.6,
C = 0, dev = 1.0, range = 458 to 483.

$C_{21}H_{43}NO_2$ MW 341.58

Lactamide, *N*-octadecyl: Irritant; flammable. Tm = 357.15; Antoine (liq)
A = 10.99555, B = 5893.281, C = 0, dev = 1.0, range = 434 to 542.

$C_{21}H_{44}$ MW 296.58

2,3-Dimethylnonadecane, CA 75163-99-4: Flammable; soluble most organic solvents. Tm = 277.45; Antoine (liq) A = 7.1958, B = 3027.7, C = -41.55, dev = 1.0, range = 493 to 625.

2,4-Dimethylnonadecane: Flammable; soluble most organic solvents. Tm = 273.15; Antoine (liq) A = 9.0696, B = 4320.4, C = 17.45, dev = 1.0, range = 465 to 594.

8-Hexylpentadecane: Flammable; soluble most organic solvents. Antoine (liq) A = 8.92968, B = 4100.9, C = 0, dev = 1.0, range = 405 to 466.

2-Methyleicosane, CA 1560-84-5: Flammable; soluble most organic solvents. Tm = 295.75; Antoine (liq) A = 8.5265, B = 4320.6, C = 41.45, dev = 1.0, range = 473 to 621.

3-Methyleicosane, CA 6418-46-8: Flammable; soluble most organic solvents. Tm = 292.65; Antoine (liq) A = 8.4980, B = 4117.6, C = 14.05, dev = 1.0, range = 477 to 620.

4-Methyleicosane, CA 25117-28-6: Flammable; soluble most organic solvents. Tm = 279.35; Antoine (liq) A = 7.9809, B = 3738.6, C = 4.55, dev = 1.0, range = 471 to 621.

5-Methyleicosane, CA 25117-36-6: Flammable; soluble most organic solvents. Tm = 271.15; Antoine (liq) A = 7.3878, B = 3025, C = -58.95, dev = 1.0, range = 519 to 621.

2,4,6-Trimethyloctadecane: Flammable; soluble most organic solvents. Tm = 240.15; Antoine (liq) A = 8.5459, B = 3836.5, C = -4.55, dev = 1.0, range = 460 to 576.

Uneicosane, CA 629-94-7: Flammable, flash point above 383; soluble most organic solvents. Tm = 313.65; Tb = 629.65; Antoine (liq) A = 6.2019, B = 2022.5, C = -147.65, dev = 1.0, range = 422 to 630.

$C_{22}H_{10}O_2$ MW 306.32

Anthanthrone, dibenzochrysene-6,12-dione, CA 641-13-4: Tm = 573.15; Antoine (sol) A = 12.809, B = 7950, C = 0, dev = 5.0, range = 450 to 550.

$C_{22}H_{12}$ MW 276.34

Benzoperylene, CA 191-24-2: Carcinogenic. Tm = 545.15 to 546.15; Antoine I (sol) A = 11.5247, B = 6674, C = 0, dev = 5.0, range = 389 to 468; Antoine II (sol) A = 10.945, B = 6580, C = 0, dev = 5.0, range = 391 to 513.

Dibenzochrysene, anthanthrene, CA 191-26-4: Carcinogenic. Tm = 534.15, decomposes; Antoine (sol) A = 12.014, B = 7060, C = 0, dev = 1.0 to 5.0, range = 450 to 510.

$C_{22}H_{14}$ MW 278.35

Dibenz[a,h]anthracene, CA 53-70-3: Carcinogenic; soluble aromatics, light hydrocarbons. Tm = 539.15 to 540.15; Tb = 797.15; Antoine (sol) A = 12.515, B = 7420, C = 0, dev = 5.0, range = 403 to 513.

(continues)

442

$C_{22}H_{14}$ *(continued)*

1,2:6,7-Dibenzophenanthrene, CA 214-17-5: Suspected carcinogen; soluble
benzene, dioxane. Tm = 565.15 to 567.15; Antoine (sol) A = 12.075, B =
7150, C = 0, dev = 5.0, range = 398 to 513.

Pentacene, CA 135-48-8: Slightly soluble organic solvents. Tm = above
573.15; Antoine (sol) A = 12.725, B = 8260, C = 0, dev = 5.0, range = 444
to 566.

Picene, CA 213-46-7: Suspected carcinogen; slightly soluble hot benzene,
other organic solvents. Tm = 640.15 to 642.15; Tb = 791.15 to 793.15;
Antoine (sol) A = 12.075, B = 7350, C = 0, dev = 5.0, range = 409 to 527.

$C_{22}H_{16}O$ MW 296.37

3,8-Dimethylceroxene: Antoine (sol) A = 13.734, B = 7219, C = 0, dev =
5.0, range = 373 to 433.

$C_{22}H_{17}NO_3S$ MW 375.44

1*H*-Xantheno[2,1,9-*def*]isoquinoline-1,3(2*H*)-dione, 2-(3-methoxypropyl),
CA 36245-88-2: Antoine I (sol) A = 9.525, B = 5840, C = 0, dev = 5.0,
range = 605 to 647; Antoine II (sol) A = 12.725, B = 7880, C = 0, dev =
5.0, range = 647 to 685.

$C_{22}H_{18}O_4$ MW 346.38

Dibenzylphthalate, CA 523-31-9: Irritant; flammable; soluble ether, ethanol. Tm = 316.15; Antoine (liq) A = 3.7723, B = 1047.8, C = -272.53, dev =
1.0, range = 445 to 513.

$C_{22}H_{22}$ MW 286.42

Tribenzylmethane: Flammable; soluble most organic solvents. Antoine (liq)
A = 8.385, B = 4128, C = 0, dev = 1.0 to 5.0, range = 394 to 637.

$C_{22}H_{31}NO_4$ MW 373.49

Butyl amine, *N,N*-bis(3-phenoxy-2-hydroxypropyl), CA 23257-62-7: Antoine
(sol) A = 10.5984, B = 5972.9, C = 0, dev = 1.0, range = 363 to 411.

$C_{22}H_{38}$ MW 302.54

1,1-Bis(decahydro-1-naphthyl)ethane: Flammable. Antoine (liq) A = 8.14715,
B = 4036.692, C = 0, dev = 1.0, range = 432 to 503.

1,2-Bis(decahydro-1-naphthyl)ethane: Flammable. Antoine (liq) A = 9.32306,
B = 4664.953, C = 0, dev = 1.0, range = 440 to 507.

1,5-Dicyclopentyl-3-(2-cyclopentylethyl)-2-pentene: Flammable; soluble
most organic solvents. Antoine (liq) A = 8.75999, B = 4249.876, C = 0,
dev = 1.0, range = 427 to 492.

$C_{22}H_{40}$ MW 304.56

1,5-Dicyclopentyl-3-(2-cyclopentylethyl)pentane: Flammable; soluble most organic solvents. Antoine (liq) A = 8.96429, B = 4368.449, C = 0, dev = 1.0, range = 430 to 494.

$C_{22}H_{42}O_2$ MW 338.57

Butyl oleate, CA 142-77-8: Flammable; soluble ethanol. Tm = 246.75; Vm = 0.3914 at 298.15; Antoine (liq) A = 10.845, B = 5104, C = 0, dev = 5.0, range = 353 to 393.

13-Docosenoic acid, *cis*, erucic acid, CA 112-86-7: Flammable, flash point above 383; soluble ethanol, ether. Tm = 306.65; Vm = 0.3937 at 328.15; Antoine (liq) A = 9.6189, B = 4899.51, C = -11.174, dev = 1.0, range = 479 to 655.

13-Docosenoic acid, *trans*, brassidic acid, CA 506-33-2: Flammable; soluble ethanol, ether. Tm = 334.65; Antoine (liq) A = 8.98735, B = 4149.23, C = -61.317, dev = 1.0, range = 482 to 656.

$C_{22}H_{42}O_4$ MW 370.57

Adipic acid, dioctyl ester, CA 123-79-5: Flammable. Antoine (liq) A = 10.435, B = 5171, C = 0, dev = 5.0, range = 373 to 493.

$C_{22}H_{42}O_6$ MW 402.57

Sebacic acid, bis(2-butoxyethyl) ester, CA 141-19-5: Flammable. Antoine (liq) A = 12.465, B = 6287, C = 0, dev = 5.0, range = 368 to 423.

$C_{22}H_{44}O_2$ MW 340.59

Butyl stearate, CA 123-95-5: Flammable, flash point 433; soluble ethanol, ether. Tm = 300.15; Tb = 616.15; Antoine (liq) A = 11.005, B = 5220, C = 0, dev = 5.0, range = 352 to 399.

Docosanoic acid, behenic acid, CA 112-85-6: Flammable; slightly soluble most organic solvents. Tm = 354.15 to 355.15; Vm = 0.4142 at 363.15; Antoine (liq) A = 8.987, B = 4088, C = -77.55, dev = 5.0, range = 373 to 600.

Eicosanoic acid, ethyl ester, CA 18281-05-5: Flammable; soluble most organic solvents. Tm = 314.55 to 315.15; Antoine I (sol) A = 22.255, B = 8961, C = 0, dev = 5.0, range = 307 to 313; Antoine II (liq) A = 12.563, B = 5940, C = 0, dev = 5.0, range = 318 to 460.

Uneicosanoic acid, methyl ester, CA 6064-90-0: Flammable, flash point above 383. Tm = 322.15; Antoine (liq) A = 4.13393, B = 983.482, C = -263.217, dev not specified, range = 459 to 529.

$C_{22}H_{46}$ MW 310.61

2,4-Dimethyleicosane, CA 75163-98-3: Flammable; soluble most organic solvents. Tm = 275.65; Antoine (liq) A = 9.006, B = 4349.5, C = 17.8, dev = 1.0, range = 471 to 603.   *(continues)*

444

$C_{22}H_{46}$ *(continued)*

Docosane, CA 629-97-0: Flammable, flash point above 383; soluble most organic solvents. Tm = 320.15; Antoine (liq) A = 6.2091, B = 2054, C = -153.05, dev = 1.0, range = 431 to 642.

8-Heptylpentadecane, CA 71005-15-7: Flammable; soluble most organic solvents. Tm = 218.15; Antoine (liq) A = 12.897, B = 5626, C = 0, dev = 1.0, range = 298 to 313.

2-Methyluneicosane, CA 1560-82-3: Flammable; soluble most organic solvents. Antoine (liq) A = 6.535, B = 2390, C = -112.15, dev = 5.0, range = 485 to 640.

3-Methyluneicosane, CA 6418-47-9: Flammable; soluble most organic solvents. Tm = 292.95; Antoine (liq) A = 8.3624, B = 4097.8, C = 13.45, dev = 1.0, range = 484 to 631.

4-Methyluneicosane, CA 25117-29-7: Flammable; soluble most organic solvents. Tm = 285.65; Antoine (liq) A = 7.9287, B = 3770.2, C = 4.45, dev = 1.0, range = 479 to 632.

5-Methyluneicosane, CA 25117-37-7: Flammable; soluble most organic solvents. Tm = 278.55; Antoine (liq) A = 8.1877, B = 3939.8, C = 5.15, dev = 1.0, range = 483 to 632.

2,4,6-Trimethylnonadecane: Flammable; soluble most organic solvents. Tm = 261.15; Antoine (liq) A = 8.8226, B = 4155, C = 6.95, dev = 1.0, range = 470 to 587.

$C_{22}H_{46}O$ MW 326.60

1-Docosanol, CA 661-19-8: Flammable; soluble ethanol, chloroform. Tm = 342.75; Antoine I (sol) A = 26.191, B = 10793, C = 0, dev = 5.0, range = 335 to 341; Antoine II (liq) A = 12.2699, B = 6025, C = 0, dev = 5.0, range = 344 to 359.

$C_{23}H_{42}O_3$ MW 366.58

Tetrahydrofurfuryl oleate, CA 5420-17-7: Flammable. Antoine (liq) A = 10.045, B = 5156, C = 0, dev = 5.0, range = 353 to 398.

$C_{23}H_{45}NO_3$ MW 383.61

N,N-Dibutylpropionamide, 2-lauryloxy: Irritant; flammable. Antoine (liq) A = 9.34161, B = 4732.89, C = 0, dev = 1.0 to 5.0, range = 443 to 458.

$C_{23}H_{46}$ MW 322.62

9-Cyclohexylheptadecane: Flammable; soluble most organic solvents. Antoine (liq) A = 9.02504, B = 4380.367, C = 0, dev = 1.0, range = 456 to 492.

$C_{23}H_{46}O_2$ MW 354.62

Docosanoic acid, methyl ester, methyl behenate, CA 929-77-1: Flammable, flash point above 383. Tm = 327.15; Antoine (liq) A = 3.95414, B = 921.793, C = -277.222, dev not specified, range = 467 to 539.

$C_{23}H_{46}O_3$

MW 370.61

3-Decyloxypropionic acid, decyl ester: Flammable. Antoine (liq) A = 9.2744, B = 4712, C = 0, dev = 5.0, range = 453 to 523.

$C_{23}H_{47}NO_2$

MW 369.63

Lactamide, *N,N*-didecyl: Irritant; flammable. Antoine (liq) A = 31.77539, B = 14703.542, C = 0, dev = 5.0, range = 439 to 463.

$C_{23}H_{48}$

MW 324.63

9-Hexylheptadecane, CA 55124-79-3: Flammable; soluble most organic solvents. Tm = 253.75; Vm = 0.4070 at 293.15; Antoine (liq) A = 8.9993, B = 4314.12, C = 0, dev = 0.1, range = 450 to 486.

2-Methyldocosane, CA 1560-81-2: Flammable; soluble most organic solvents. Antoine (liq) A = 6.205, B = 2130, C = -145.15, dev = above 5.0, range = 495 to 652.

4-Methyldocosane, CA 25117-30-0: Flammable; soluble most organic solvents. Tm = 291.55; Antoine (liq) A = 8.1826, B = 3965.1, C = -1.25, dev = 1.0, range = 493 to 643.

5-Methyldocosane, CA 25163-52-4: Flammable; soluble most organic solvents. Tm = 284.95; Antoine (liq) A = 8.1515, B = 3955.9, C = 0.45, dev = 1.0, range = 492 to 644.

Tricosane, CA 638-67-5: Flammable, flash point above 383; soluble most organic solvents. Tm = 320.55; Tb = 653.35; Vm = 0.4170 at 321.15; Antoine I (liq) A = 11.94229, B = 5765.379, C = 0, dev = 1.0, range = 314 to 353; Antoine II (liq) A = 6.216, B = 2083.8, C = -158.35, dev = 1.0, range = 440 to 653.

$C_{24}H_{12}$

MW 300.36

Coronene, hexabenzobenzene, CA 191-07-1: Suspected carcinogen. Tm = 711.15 to 713.15; Tb = 798.15; Antoine I (sol) A = 11.1157, B = 7100, C = 0, dev = 5.0, range = 427 to 510; Antoine II (sol) A = 8.886, B = 5764, C = 0, dev = 5.0, range not specified; Antoine III (liq) A = 8.318, B = 5362, C = 0, dev = 5.0, range not specified.

$C_{24}H_{14}$

MW 302.37

Dibenzo[*fg,op*]naphthacene, CA 192-51-8: Tm = 624.15 to 625.15; Antoine (sol) A = 12.005, B = 7700, C = 0, dev = 5.0, range = 430 to 555.

Dibenzo[a,e]pyrene, CA 192-65-4: Suspected carcinogen. Tm = 506.15; Antoine (sol) A = 12.605, B = 7650, C = 0, dev = 5.0, range = 414 to 506.

$C_{24}H_{16}N_2O_2$

MW 364.40

Oxazole, 2,2'-(1,4-phenylene)bis(5-phenyl), CA 1806-34-4: Tm = 515.15 to 517.15; Antoine (sol) A = 8.225, B = 4930, C = 0, dev = 5.0, range = 600 to 680.

$C_{24}H_{18}$ MW 306.41

para-Quaterphenyl, CA 135-70-6: Tm = 593.15; Antoine (sol) A = 17.440,
B = 8131, C = 0, dev not specified, range not specified.

1,3,5-Triphenylbenzene, CA 612-71-5: Soluble ethanol, ether, benzene.
Tm = 445.15; Tb = 735.15; Antoine I (sol) A = 14.22393, B = 7420.508,
C = 0, dev = 1.0, range = 410 to 444; Antoine II (liq) A = 11.44116, B =
6164.197, C = 0, dev = 1.0, range = 454 to 500; Antoine III (liq) A =
7.501, B = 4048, C = 0, dev = 5.0, range = 500 to 735.

$C_{24}H_{20}O_6$ MW 404.42

Glycerol tribenzoate, tribenzoin, CA 614-33-5: Flammable; soluble most
organic solvents. Tm = 345.15; Antoine (liq) A = 11.425, B = 6450, C = 0,
dev = 5.0, range = 423 to 476.

$C_{24}H_{26}N_2O_5$ MW 374.48

Anthraquinone, 1,5-dipiperidyl, CA 14580-70-2: Tm = 479.15; Antoine (sol)
A = 16.255, B = 9053, C = 0, dev = 5.0, range = 408 to 458.

$C_{24}H_{27}NO_4$ MW 393.48

Phenylamine, bis-*N*,*N*-(2-hydroxy-3-phenoxypropyl), CA 3088-05-9: Irritant;
flammable. Tm = 363.15; Antoine (liq) A = 11.6052, B = 6842.4, C = 0,
dev = 1.0 to 5.0, range = 388 to 423.

$C_{24}H_{30}O_4$ MW 382.50

Dibenzyl sebacate, CA 140-24-9: Flammable. Tm = 301.45; Antoine (liq) A =
10.92812, B = 6027.8, C = 1.85, dev = 1.0, range = 368 to 550.

$C_{24}H_{32}$ MW 320.52

(6,6)-*para*-Cyclophane, CA 4384-23-0: Antoine (sol) A = 11.945, B = 5681,
C = 0, dev = 5.0, range = 352 to 371.

$C_{24}H_{38}O_4$ MW 390.56

Bis(2-ethylhexyl)phthalate, CA 117-81-7: Flammable, flash point 480. Tm =
227.15; Tb = 657.15; Vm = 0.3961 at 293.15; Tc = 806, calculated; Pc =
1180, calculated; Vc = 1.270, calculated; Antoine (liq) A = 11.8564, B =
6416.2, C = 36.74, dev = 1.0, range = 373 to 660.

Bis(1-methylheptyl)phthalate, CA 131-15-7: Flammable. Antoine (liq) A =
9.408, B = 4864, C = 0, dev = 5.0, range = 393 to 435.

Bis(6-methylheptyl)phthalate, CA 131-20-4: Flammable. Tm = below 223.15;
Antoine (liq) A = 9.387, B = 4829, C = 0, dev = 5.0, range = 383 to 490.

Dioctyl phthalate, CA 117-84-0: Flammable. Tm = 248.15; Antoine (liq) A =
9.897, B = 5197.4, C = 0, dev = 5.0, range = 423 to 523.

$C_{24}H_{42}$ MW 330.60

Hexapropylbenzene, CA 2456-68-0: Flammable; soluble most organic solvents.
Tm = 376.15; Antoine (liq) A = 7.8909, B = 3556.78, C = -0.98, dev = 1.0,
range = 458 to 606.

$C_{24}H_{42}O_6$ MW 426.59

Aconitic acid, *trans*, trihexyl ester, CA 64617-30-7: Flammable. Antoine
(liq) A = 9.8502, B = 5131.906, C = 0, dev = 1.0 to 5.0, range = 423 to 512.

$C_{24}H_{42}O_{11}$ MW 506.59

Diethylene glycol dicarboxylic acid, di[1-(2-ethylbutyl)oxycarbonyl]-ethyl
ester: Irritant; flammable. Antoine (liq) A = 10.4054, B = 5752, C = 0,
dev = 5.0, range = 448 to 538.

Diethylene glycol dicarboxylic acid, di[1-(hexyloxycarbonyl)ethyl] ester:
Irritant; flammable. Antoine (liq) A = 10.3901, B = 5799, C = 0, dev =
5.0, range = 443 to 548.

$C_{24}H_{44}$ MW 332.61

Tetradecahydroanthracene, 9-decyl: Flammable. Antoine (liq) A = 10.17709,
B = 5389.644, C = 0, dev = 1.0, range = 501 to 536.

Tetradecahydrophenanthrene, 9-decyl: Flammable. Antoine (liq) A =
8.99964, B = 4807.425, C = 0, dev = 1.0, range = 502 to 542.

$C_{24}H_{44}O_4$ MW 396.61

O-Acetylricinoleic acid, butyl ester, CA 140-04-5: Flammable. Antoine
(liq) A = 10.835, B = 5497, C = 0, dev = 5.0, range = 378 to 423.

$C_{24}H_{44}O_6$ MW 428.61

1,2,3-Propanetricarboxylic acid, trihexyl ester, CA 38094-13-2: Flammable.
Antoine (liq) A = 9.84847, B = 5122.975, C = 0, dev = 1.0 to 5.0, range =
422 to 526.

$C_{24}H_{46}O_4$ MW 398.62

Bis(3,5,5-trimethylhexyl)adipate, CA 20270-50-2: Flammable. Antoine (liq)
A = 11.525, B = 5622, C = 0, dev = 5.0, range = 353 to 413.

$C_{24}H_{48}O_2$ MW 368.64

Docosanoic acid, ethyl ester, CA 5908-87-2: Flammable; soluble ethanol,
ether. Tm = 323.15; Vm = 0.4180 at 324.15; Antoine I (sol) A = 25.455,
B = 10266, C = 0, dev = 5.0, range = 313 to 318; Antoine II (liq) A =
13.997, B = 6661, C = 0, dev = 5.0, range = 327 to 344.

Tricosanoic acid, methyl ester, CA 2433-97-8: Flammable. Tm = 326.55;
Antoine (liq) A = 4.74464, B = 1309.59, C = -243.086, dev not specified,
range = 473 to 528.

$C_{24}H_{49}Cl$ MW 373.10

1-Chlorotetracosane, CA 6422-18-0: Irritant; flammable. Tm = 325.15;
Antoine (liq) A = 6.18552, B = 2403.5, C = -113.15, dev = 5.0, range = 543
to 774.

$C_{24}H_{50}$ MW 338.66

2-Methyltricosane, CA 1928-30-9: Flammable; soluble light hydrocarbons,
most organic solvents. Tm = 315.15; Vm = 0.4492 at 363.15; Antoine (liq)
A = 6.245, B = 2190, C = -146.15, dev = above 5.0, range = 450 to 664.

5-Methyltricosane, CA 22331-09-5: Flammable; soluble light hydrocarbons,
most organic solvents. Tm = 290.85; Antoine (liq) A = 8.24425, B = 4021.0,
C = -8.69, dev = 1.0, range = 503 to 653.

Tetracosane, CA 646-31-1: Flammable, flash point above 383; soluble light
hydrocarbons, most organic solvents. Tm = 327.15; Tb = 664.45; Vm = 0.4418
at 343.15; Antoine (liq) A = 6.44054, B = 2289.02, C = -147.92, dev = 1.0,
range = 498 to 573.

$C_{24}H_{51}N$ MW 353.67

Trioctyl amine, CA 1116-76-3: Irritant; flammable, flash point above 383.
Tm = 238.55; Tb = 638.15; Vm = 0.4372 at 293.15; Antoine (liq) A = 6.5166,
B = 2381.7, C = -102.15, dev = 5.0, range = 505 to 702.

$C_{25}H_{20}$ MW 320.43

Tetraphenylmethane, CA 630-76-2: Tm = 558.15; Tb = 704.15; Antoine (sol)
A = 14.592, B = 7487, C = 0, dev = above 5.0, range = 404 to 466.

$C_{25}H_{28}$ MW 328.50

1-Pentadecylnaphthalene: Flammable. Tm = 314.75; Antoine (liq) A =
10.44905, B = 6057.3, C = 46.5, dev = 1.0, range = 474 to 540.

3-Phenethyl-1,5-diphenylpentane: Flammable; soluble most organic solvents.
Antoine (liq) A = 8.56305, B = 4558.016, C = 0, dev = 1.0, range = 498 to
542.

$C_{25}H_{36}$ MW 336.56

1-Phenyl-3-phenethylundecane, CA 7225-70-9: Flammable; soluble most organic solvents. Antoine (liq) A = 10.9489, B = 6442.2, C = 74.65, dev =
1.0, range = 456 to 521.

$C_{25}H_{38}$ MW 338.58

1-Pentadecylnaphthalene: Flammable. Tm = 314.75; Antoine (liq) A =
9.61056, B = 5123.915, C = 0, dev = 1.0, range = 474 to 524.

$C_{25}H_{40}$ MW 340.59

1-Cyclohexyl-6-cyclopentyl-3-phenethylhexane: Flammable; soluble most organic solvents. Antoine (liq) A = 8.84637, B = 4580, C = 0, dev = 1.0,
range = 486 to 525.

1,7-Dicyclopentyl-4-phenethylheptane: Flammable; soluble most organic solvents. Antoine (liq) A = 9.28699, B = 4807.197, C = 0, dev = 1.0, range =
487 to 525.

$C_{25}H_{42}$ MW 342.61

1-Hexadecylindane: Flammable; soluble most organic solvents. Antoine
(liq) A = 8.61245, B = 4546.139, C = 0, dev = 1.0, range = 495 to 536.

5-Pentadecyl-1,2,3,4-tetrahydronaphthalene: Flammable. Tm = 306.05;
Antoine (liq) A = 9.85078, B = 5195.118, C = 0, dev = 1.0, range = 471
to 534.

$C_{25}H_{44}$ MW 344.62

1,5-Dicyclohexyl-3-(2-cyclohexylethyl)-2-pentene: Flammable; soluble most
organic solvents. Antoine (liq) A = 8.89722, B = 4597.847, C = 0, dev =
1.0, range = 485 to 524.

1,7-Dicyclopentyl-4-(3-cyclopentylpropyl)-3-heptene: Flammable; soluble
most organic solvents. Antoine (liq) A = 9.03849, B = 4649.838, C = 0,
dev = 1.0, range = 483 to 522.

6-Octyl(hexahydrobenz[
9.09503, B = 4791.121, C = 0, dev = 1.0, range = 467 to 534.

3-Octyl-1-phenylundecane, CA 5637-96-7: Flammable; soluble most organic
solvents. Antoine (liq) A = 9.22376, B = 4668.973, C = 0, dev = 1.0,
range = 476 to 513.

9-(2-Phenethyl)-heptadecane, CA 5637-96-7: Flammable; soluble most organic
solvents. Tm = 246.45; Vm = 0.4026 at 293.15; Antoine (liq) A = 11.1852,
B = 5663.7, C = 49.85, dev = 1.0, range = 448 to 513.

9-(4-Tolyl)-octadecane: Flammable; soluble most organic solvents. Vm =
0.4031 at 293.15; Antoine (liq) A = 9.60467, B = 4807.853, C = 0, dev =
1.0, range = 472 to 507.

$C_{25}H_{46}$ MW 346.64

1-Cyclohexyl-3-(cyclohexylethyl)-6-cyclopentylhexane: Flammable; soluble
most organic solvents. Antoine (liq) A = 9.23029, B = 4772.238, C = 0,
dev = 1.0, range = 487 to 524.

4-(2-Cyclohexylethyl)-1,7-dicyclopentylheptane: Flammable; soluble most
organic solvents. Antoine (liq) A = 8.98008, B = 4641.026, C = 0, dev =
1.0, range = 471 to 524.

1,5-Dicyclohexyl-3-(2-cyclohexylethyl)pentane: Flammable; soluble most organic solvents. Antoine I (liq) A = 11.323, B = 5623, C = 0, dev = 1.0,
range = 318 to 418; Antoine II (liq) A = 8.65568, B = 4505.532, C = 0,
dev = 1.0, range = 488 to 528.

(continues)

$C_{25}H_{46}$ *(continued)*

1,7-Dicyclopentyl-4-(3-cyclopentylpropyl)heptane, CA 55429-35-1: Flammable; soluble most organic solvents. Tm = 249.45; Antoine I (liq) A = 11.5815, B = 7537.8, C = 133.75, dev = 1.0, range = 457 to 525; Antoine II (liq) A = 8.9760, B = 4643.363, C = 0, dev = 1.0, range = 486 to 525.

$C_{25}H_{48}$ MW 348.65

1-Cyclohexyl-3-(2-cyclohexylethyl)undecane, CA 7225-69-6: Flammable; soluble most organic solvents. Antoine (liq) A = 9.77245, B = 4974.939, C = 0, dev = 1.0, range = 480 to 516.

1-Cyclopentyl-4-(3-cyclopentylpropyl)dodecane, CA 7225-68-5: Flammable; soluble most organic solvents. Antoine (liq) A = 9.04687, B = 4622.779, C = 0, dev = 1.0, range = 480 to 518.

1-Hexadecylhexahydroindane: Flammable; soluble most organic solvents. Antoine (liq) A = 8.72425, B = 4576.143, C = 0, dev = 1.0, range = 492 to 532.

1-Pentadecyldecahydronaphthalene: Flammable. Antoine (liq) A = 9.35029, B = 4881.134, C = 0, dev = 1.0, range = 464 to 529.

$C_{25}H_{48}O_4$ MW 412.65

Nonanedioic acid, dioctyl ester, CA 2064-80-4: Flammable. Antoine (liq) A = 10.515, B = 5451, C = 0, dev = 5.0, range = 393 to 523.

$C_{25}H_{50}$ MW 350.67

9-(2-Cyclohexylethyl)heptadecane, CA 25446-35-9: Flammable; soluble most organic solvents. Antoine (liq) A = 9.14837, B = 4630.499, C = 0, dev = 1.0, range = 480 to 513.

9-(3-Cyclopentylpropyl)heptadecane: Flammable; soluble most organic solvents. Antoine (liq) A = 8.96482, B = 4540.496, C = 0, dev = 1.0, range = 476 to 514.

9-Octyl-8-heptadecene, CA 24306-18-1: Flammable; soluble most organic solvents. Tm = 233.15; Vm = 0.4337 at 293.15; Antoine (liq) A = 13.0975, B = 8365.9, C = 144.55, dev = 1.0, range = 441 to 500.

$C_{25}H_{50}O_2$ MW 382.67

Tetracosanoic acid, methyl ester, CA 2442-49-1: Flammable. Tm = 332.65 to 333.15; Antoine (liq) A = 4.62856, B = 1272.62, C = -252.836, dev not specified, range = 483 to 536.

Tricosanoic acid, ethyl ester, CA 18281-07-7: Flammable. Tm = 324.35; Antoine I (sol) A = 21.915, B = 9153, C = 0, dev = 5.0, range = 316 to 322; Antoine II (liq) A = 12.741, B = 6366, C = 0, dev = 5.0, range = 336 to 359.

$C_{25}H_{52}$ MW 352.69

9-Octylheptadecane, CA 7225-64-1: Flammable; soluble most organic solvents. Tm = 259.35; Vm = 0.4398 at 293.15; Antoine (liq) A = 9.7923, B = 4879.5, C = 0, dev = 1.0, range = 470 to 505.   *(continues)*

$C_{25}H_{52}$ *(continued)*

Pentacosane, CA 629-99-2: Flammable, flash point above 383; soluble light hydrocarbons, most organic solvents. Tm = 326.45; Tb = 675.05; Antoine (liq) A = 6.2287, B = 2138.8, C = -168.55, dev = 1.0, range = 457 to 675.

$C_{26}H_{16}$ MW 328.41

Dibenzo[*g*,*p*]chrysene, CA 191-68-4: Carcinogenic. Tm = 491.15 to 493.15; Antoine (sol) A = 12.395, B = 7430, C = 0, dev = 5.0, range = 408 to 493.

$C_{26}H_{18}$ MW 330.43

9,10-Diphenylanthracene, CA 1499-10-1: Tm = 521.15; Antoine (sol) A = 13.695, B = 7500, C = 0, dev = 5.0, range = 393 to 433.

$C_{26}H_{18}N_2O_4$ MW 422.44

C. I. disperse violet 31: Antoine (liq) A = 4.48059, B = 3127.07, C = 0, dev = above 5.0, range = 453 to 523.

$C_{26}H_{32}$ MW 344.54

6-Octyl-1,2,3,4-tetrahydronaphthacene: Flammable. Antoine (liq) A = 9.52557, B = 5393.22, C = 0, dev = 1.0, range = 503 to 574.

$C_{26}H_{34}$ MW 346.55

9-Dodecylanthracene, CA 2883-70-7: Tm = 322.65; Antoine I (liq) A = 12.817, B = 6655, C = 0, dev = 1.0, range = 353 to 403; Antoine II (liq) A = 9.30861, B = 5195.542, C = 0, dev = 1.0, range = 495 to 566.

9-Dodecylphenanthrene, CA 3788-61-2: Antoine (liq) A = 8.92385, B = 4998.815, C = 0, dev = 1.0, range = 495 to 568.

$C_{26}H_{38}$ MW 350.59

1,1-Diphenyltetradecane: Flammable; soluble most organic solvents. Tm = 291.05; Vm = 0.3816 at 293.15; Antoine (liq) A = 8.9780, B = 4360.9, C = -37.65, dev = 1.0, range = 467 to 530.

1,1-Di(4-tolyl)-dodecane: Flammable; soluble most organic solvents. Tm = 267.95; Vm = 0.3845 at 293.15; Antoine (liq) A = 11.1273, B = 6462.4, C = 58.65, dev = 1.0, range = 466 to 529.

$C_{26}H_{40}$ MW 352.60

5-Octyl-1,2,3,4,4*a*,5,7,8,9,10,12,12*a*-dodecahydronaphthacene: Flammable. Antoine (liq) A = 8.87208, B = 4803.133, C = 0, dev = 1.0, range = 479 to 549.

$C_{26}H_{42}$ MW 354.62

1,1-Bis(dodecahydroacenaphthylene-5-yl)ethane: Flammable. Antoine (liq)
A = 10.8289, B = 5792.055, C = 0, dev = 1.0, range = 482 to 541.

$C_{26}H_{42}O_4$ MW 418.62

Bis(3,5,5-trimethylhexyl)phthalate, CA 14103-61-8: Flammable. Antoine
(liq) A = 12.005, B = 5936, C = 0, dev = 5.0, range = 333 to 393.

Dinonyl phthalate, CA 84-76-4: Flammable. Antoine (liq) A = 10.535, B =
5690, C = 0, dev = 5.0, range = 333 to 393.

$C_{26}H_{46}$ MW 358.65

1,4-Didecylbenzene, CA 2655-95-0: Flammable; soluble most organic solvents.
Tm = 302.15; Antoine (liq) A = 9.4118, B = 4974.183, C = 0, dev = 1.0,
range = 468 to 536.

Octadecahydronaphthacene, 9-octyl: Flammable. Antoine (liq) A = 9.08464,
B = 4826.177, C = 0, dev = 1.0, range = 470 to 539.

1-Phenyleicosane, CA 2398-68-7: Flammable; soluble most organic solvents.
Tm = 315.45; Vm = 0.4355 at 333.15; Antoine (liq) A = 9.32803, B =
4948.107, C = 0, dev = 1.0, range = 499 to 538.

2-Phenyleicosane, CA 2398-66-5: Flammable; soluble most organic solvents.
Tm = 302.15; Antoine (liq) A = 9.02137, B = 4725.182, C = 0, dev = 1.0,
range = 492 to 531.

3-Phenyleicosane, CA 2400-02-4: Flammable; soluble most organic solvents.
Tm = 302.45; Antoine (liq) A = 9.26703, B = 4809.776, C = 0, dev = 1.0,
range = 489 to 526.

4-Phenyleicosane, CA 2400-03-5: Flammable; soluble most organic solvents.
Tm = 304.55; Antoine (liq) A = 8.87278, B = 4607.102, C = 0, dev = 1.0,
range = 487 to 527.

5-Phenyleicosane, CA 2400-04-6: Flammable; soluble most organic solvents.
Tm = 303.35; Antoine (liq) A = 9.58067, B = 4928.009, C = 0, dev = 1.0,
range = 485 to 521.

7-Phenyleicosane, CA 2398-64-3: Flammable; soluble most organic solvents.
Antoine (liq) A = 9.55193, B = 4898.892, C = 0, dev = 1.0, range = 483
to 520.

9-Phenyleicosane, CA 2398-65-4: Flammable; soluble most organic solvents.
Tm = 291.05; Vm = 0.4203 at 293.15; Antoine (liq) A = 9.35506, B =
4800.949, C = 0, dev = 1.0, range = 483 to 520.

8-(4-Tolyl)-nonadecane: Flammable; soluble most organic solvents. Antoine
(liq) A = 9.66739, B = 4934.587, C = 0, dev = 1.0, range = 482 to 517.

$C_{26}H_{50}$ MW 362.68

9-[alpha-(cis-Bicyclooctyl)methyl]heptadecane: Flammable; soluble most
organic solvents. Antoine (liq) A = 13.7608, B = 9782.8, C = 199.25, dev =
1.0, range = 455 to 518.

(continues)

$C_{26}H_{50}$ *(continued)*

1,1-Bis(4-methylcyclohexyl)dodecane: Flammable; soluble most organic solvents. Antoine (liq) A = 9.51284, B = 4882.586, C = 0, dev = 1.0, range = 484 to 520.

1,1-Dicyclohexyltetradecane: Flammable; soluble most organic solvents. Tm = 310.75; Tb = 679.15; Antoine (liq) A = 9.78017, B = 5104.058, C = 0, dev = 1.0, range = 493 to 529.

1,1-Dicyclopentylhexadecane: Flammable; soluble most organic solvents. Tm = 285.25; Antoine (liq) A = 11.37436, B = 5906.936, C = 0, dev = 1.0, range = 471 to 525.

2-Hexadecylbicyclopentyl: Flammable; soluble most organic solvents. Antoine (liq) A = 9.71015, B = 5101.139, C = 0, dev = 1.0, range = 495 to 532.

$C_{26}H_{50}O_4$ MW 426.68

Bis(2-ethylhexyl)sebacate ±, CA 122-62-3: Flammable, flash point above 383; soluble ethanol, acetone, benzene. Tm = 228.15; Vm = 0.4684 at 298.15; Antoine (liq) A = 11.625, B = 6000, C = 0, dev = 5.0, range = 308 to 453.

Dioctyl sebacate, CA 2432-87-3: Flammable; soluble ethanol, benzene. Tm = 291.15; Antoine (liq) A = 10.665, B = 5593, C = 0, dev = 5.0, range = 413 to 523.

$C_{26}H_{52}$ MW 364.70

1-Cyclohexyleicosane, eicosyl cyclohexane, CA 4443-55-4: Flammable; soluble most organic solvents. Tm = 321.65; Tb = 695.15; Antoine (liq) A = 9.26416, B = 4918.552, C = 0, dev = 1.0, range = 499 to 538.

2-Cyclohexyleicosane, CA 4443-56-5: Flammable; soluble most organic solvents. Antoine (liq) A = 9.82622, B = 5138.199, C = 0, dev = 1.0, range = 494 to 530.

3-Cyclohexyleicosane, CA 4443-57-6: Flammable; soluble most organic solvents. Antoine (liq) A = 9.38858, B = 4911.446, C = 0, dev = 1.0, range = 492 to 530.

4-Cyclohexyleicosane, CA 4443-58-7: Flammable; soluble most organic solvents. Antoine (liq) A = 9.93036, B = 5134.802, C = 0, dev = 1.0, range = 488 to 524.

5-Cyclohexyleicosane, CA 4443-59-8: Flammable; soluble most organic solvents. Antoine (liq) A = 9.93036, B = 5134.802, C = 0, dev = 1.0, range = 488 to 524.

7-Cyclohexyleicosane, CA 4443-60-1: Flammable; soluble most organic solvents. Antoine (liq) A = 9.48036; B = 4889.294, C = 0, dev = 1.0, range = 486 to 523.

9-Cyclohexyleicosane, CA 4443-61-2: Flammable; soluble most organic solvents. Antoine (liq) A = 9.48036, B = 4889.294, C = 0, dev = 1.0, range = 486 to 523.

(continues)

454

$C_{26}H_{52}$ *(continued)*

1-Cyclopentyluneicosane, CA 6703-82-8: Flammable; soluble most organic solvents. Tm = 315.15; Tb = 693.15; Antoine (liq) A = 9.2462, B = 4899.764, C = 0, dev = 1.0, range = 498 to 537.

11-Cyclopentyluneicosane: Flammable; soluble most organic solvents. Antoine (liq) A = 9.34381, B = 4827.574, C = 0, dev = 1.0, range = 486 to 524.

$C_{26}H_{54}$ MW 366.71

5-Butyldocosane, CA 55282-16-1: Flammable; soluble most organic solvents. Tm = 293.95; Vm = 0.4551 at 293.95; Antoine (liq) A = 9.6034, B = 4908.1, C = 0, dev = 1.0, range = 482 to 518.

7-Butyldocosane: Flammable; soluble most organic solvents. Tm = 276.35; Vm = 0.4558 at 293.15; Antoine (liq) A = 10.0046, B = 5078.5, C = 0, dev = 1.0, range = 480 to 514.

9-Butyldocosane: Flammable; soluble most organic solvents. Tm = 274.45; Vm = 0.4559 at 293.15; Antoine (liq) A = 9.433, B = 4800.3, C = 0, dev = 0.1 to 1.0, range = 479 to 516.

11-Butyldocosane, CA 13475-76-8: Flammable; soluble most organic solvents. Tm = 273.15; Vm = 0.4561 at 293.15; Antoine (liq) A = 9.5814, B = 4875.91, C = 0, dev = 0.1, range = 480 to 516.

5,14-Dibutyloctadecane: Flammable; soluble most organic solvents. Tm = 278.95; Antoine (liq) A = 9.30164, B = 4664.4, C = 0, dev = 1.0, range = 458 to 508.

6,11-Dipentylhexadecane, CA 15874-03-0: Flammable; soluble most organic solvents. Tm = 256.95; Vm = 0.4543 at 293.15; Antoine (liq) A = 9.3373, B = 4646.0, C = 0, dev = 1.0, range = 468 to 504.

3-Ethyl-5-(2-ethylbutyl)octadecane: Flammable; soluble most organic solvents. Tm = 233.15; Antoine (liq) A = 9.3077, B = 4617.3, C = 0, dev = 1.0, range = 467 to 503.

11-(1-Ethylpropyl)uneicosane: Flammable; soluble most organic solvents. Antoine (liq) A = 9.7326, B = 4886.8, C = 0, dev = 0.1, range = 474 to 509.

3-Ethyltetracosane: Flammable; soluble most organic solvents. Tm = 303.25; Vm = 0.4613 at 313.15; Antoine (liq) A = 9.0146, B = 4700.6, C = 0, dev = 1.0, range = 490 to 529.

Hexacosane, cerane, CA 630-01-3: Flammable; soluble light hydrocarbons; moderately soluble most organic solvents. Tm = 329.55; Tb = 685.35; Vm = 0.4712 at 333.15; Antoine I (liq) A = 6.2345, B = 2164.3, C = -173.55, dev = 1.0, range = 466 to 685; Antoine II (liq) A = 9.44384, B = 4935.909, C = 0, dev = 1.0, range = 478 to 530.

7-Hexyleicosane: Flammable; soluble most organic solvents. Tm = 283.35; Antoine (liq) A = 10.43982, B = 5282.87, C = 0, dev = 0.02, range = 479 to 512.

11-Neopentyluneicosane: Flammable; soluble most organic solvents. Tm = 252.15; Vm = 0.4566 at 293.15; Antoine (liq) A = 9.6276, B = 4858.0, C = 0, dev = 1.0, range = 476 to 511.

(continues)

$C_{26}H_{54}$ *(continued)*

11-Pentyluneicosane, CA 14739-72-1: Flammable; soluble most organic solvents. Tm = 264.05; Antoine (liq) A = 9.9467, B = 5031.1, C = 0, dev = 1.0, range = 478 to 512.

$C_{27}H_{19}NO$ MW 373.45

Oxazole, 2,5-bis(1,1'-biphenylyl), CA 2083-09-2: Tm = 511.15 to 513.15; Antoine (liq) A = 9.225, B = 5730, C = 0, dev = 5.0, range = 595 to 685.

$C_{27}H_{40}$ MW 364.61

5-Pentadecylacenaphthene: Flammable. Antoine (liq) A = 9.85292, B = 5522.199, C = 0, dev = 1.0, range = 500 to 568.

$C_{27}H_{46}O$ MW 386.66

Cholesterol, CA 57-88-5: Suspected carcinogen; flammable; soluble benzene, chloroform, ether, pyridine. Tm = 421.65; Tb = 633.15, decomposes; Antoine (liq) A = 11.056, B = 6000, C = 0, dev = 5.0, range = 411 to 447.

$C_{27}H_{48}$ MW 372.68

5-*alpha*-Cholestane, CA 481-21-0: Flammable; soluble ether, benzene, chloroform. Tm = 353.15 to 353.65; Vm = 0.4100 at 361.15; Antoine (liq) A = 11.35684, B = 6038.938, C = 0, dev = 1.0, range = 481 to 538.

11-Phenyluneicosane, CA 6703-80-6: Flammable; soluble most organic solvents. Tm = 293.95; Vm = 0.4463 at 293.95; Antoine (liq) A = 9.36087, B = 4882.904, C = 0, dev = 1.0, range = 491 to 529.

$C_{27}H_{50}$ MW 374.69

5-Pentadecyldodecahydroacenaphthylene: Flammable. Antoine (liq) A = 9.37032, B = 5123.286, C = 0, dev = 1.0, range = 486 to 554.

$C_{27}H_{50}O_6$ MW 470.69

Glycerol trioctanoate, tricaprylin, CA 538-23-8: Flammable; soluble most organic solvents. Tm = 283.15; Vm = 0.4934 at 293.15; Antoine (liq) A = 11.245, B = 6060, C = 0, dev = 5.0, range = 396 to 453.

$C_{27}H_{54}$ MW 378.72

11-Cyclohexyluneicosane, CA 6703-99-7: Flammable; soluble most organic solvents. Antoine (liq) A = 10.69644, B = 5588.912, C = 0, dev = 1.0, range = 485 to 529.

11-(Cyclopentylmethyl)uneicosane: Flammable; soluble most organic solvents. Antoine (liq) A = 9.44618, B = 4932.441, C = 0, dev = 1.0, range = 492 to 529.

456

$C_{27}H_{56}$ MW 380.74

Heptacosane, CA 593-49-7: Flammable; soluble light hydrocarbons; slightly
soluble ether. Tm = 332.65; Tb = 695.15; Vm = 0.4884 at 333.15; Antoine
(liq) A = 6.2401, B = 2188.5, C = -178.35, dev = 1.0, range = 473 to 695.

$C_{28}H_{14}$ MW 350.42

Phenanthro[1,10,9,8-*opqra*]perylene, CA 190-39-6: Antoine (sol) A = 12.914,
B = 9430, C = 0, dev = 5.0, range = 580 to 630.

$C_{28}H_{18}$ MW 354.45

9,9'-Bianthracene, CA 1055-23-8: Tm = above 633.15; Antoine (sol) A =
10.545, B = 6679, C = 0, dev = 5.0, range = 413 to 473.

$C_{28}H_{32}$ MW 368.56

1,7-Diphenyl-4-(3-phenylpropyl)-3-heptene: Irritant; flammable; soluble
most organic solvents. Antoine (liq) A = 9.32772, B = 5117.484, C = 0,
dev = 1.0, range = 488 to 556.

$C_{28}H_{34}$ MW 370.58

1,7-Diphenyl-4-(3-phenylpropyl)-heptane: Irritant; flammable; soluble most
organic solvents. Antoine (liq) A = 9.53336, B = 5241.497, C = 0, dev =
1.0, range = 490 to 557.

$C_{28}H_{44}O$ MW 396.66

Ergosterol, CA 57-87-4: Soluble benzene, chloroform; most important pro-
vitamin D. Tm = 441.15; Antoine (liq) A = 11.197, B = 6200, C = 0, dev =
5.0, range = 421 to 454.

$C_{28}H_{46}O_4$ MW 446.67

Diisodecyl phthalate, CA 26761-40-0: Flammable, flash point 505. Tm =
225.15; Tb = 723.0; Vm = 0.4648 at 298.15; Tc = 887, calculated; Pc = 1000,
calculated; Vc = 1.460, calculated; Antoine (liq) A = 14.0158, B = 10984,
C = 242.24, dev = 1.0, range = 371 to 496.

$C_{28}H_{50}$ MW 386.70

2-Decyl-1-phenyldodecane: Flammable; soluble most organic solvents.
Antoine (liq) A = 10.18666, B = 5354.289, C = 0, dev = 1.0, range = 497
to 532.

$C_{28}H_{50}O_{11}$ MW 562.70

Diethylene glycol dicarboxylic acid, di[1-(2-ethylhexyl)oxycarbonyl]ethyl
ester: Irritant; flammable. Antoine (liq) A = 10.7505, B = 6092, C = 0,
dev = 1.0 to 5.0, range = 463 to 553.

(continues)

$C_{28}H_{50}O_{11}$ *(continued)*

Diethylene glycol dicarboxylic acid, di[1-(octyloxycarbonyl)ethyl] ester:
Irritant; flammable. Antoine (liq) A = 10.143, B = 5878, C = 0, dev = 5.0,
range = 463 to 564.

$C_{28}H_{52}$ MW 388.72

1,7-Dicyclohexyl-4-(3-cyclohexylpropyl)heptane: Flammable; soluble most
organic solvents. Antoine (liq) A = 9.5094, B = 5153.514, C = 0, dev =
1.0, range = 482 to 549.

$C_{28}H_{54}$ MW 390.73

1-Cyclohexyl-2-(cyclohexylmethyl)pentadecane: Flammable; soluble most organic solvents. Antoine (liq) A = 10.40203, B = 5505.408, C = 0, dev =
1.0, range = 501 to 536.

$C_{28}H_{56}$ MW 392.75

11-(Cyclohexylmethyl)uneicosane: Flammable; soluble most organic solvents.
Antoine (liq) A = 9.26416, B = 4918.552, C = 0, dev = 1.0, range = 499
to 538.

2,2,4,10,12,12-Hexamethyl-7-(3,5,5-trimethylhexyl)-6-tridecene: Flammable;
soluble most organic solvents. Antoine (liq) A = 8.7818, B = 4102.9, C =
-14.15, dev = 1.0, range = 426 to 488.

$C_{28}H_{58}$ MW 394.77

2,2,4,10,12,12-Hexamethyl-7-(3,5,5-trimethylhexyl)tridecane: Flammable;
soluble most organic solvents. Antoine I (liq) A = 10.994, B = 5147, C = 0,
dev = 1.0, range = 308 to 393, Antoine II (liq) A = 9.2813, B = 4543.6, C =
5.45, dev = 1.0, range = 429 to 491.

7-Hexyldocosane: Flammable; soluble most organic solvents. Tm = 292.45;
Vm = 0.4886 at 293.15; Antoine (liq) A = 10.0277, B = 5260.5, C = 0, dev =
1.0, range = 506 to 531.

Octacosane, CA 630-02-4: Flammable; soluble light hydrocarbons, benzene,
chloroform. Tm = 334.35 to 334.55; Tb = 704.75; Vm = 0.5094 at 343.15;
Antoine I (liq) A = 13.99905, B = 6879.135, C = 0, dev = 1.0, range = 300
to 390; Antoine II (liq) A = 6.2454, B = 2211.6, C = -183.15, dev = 1.0,
range = 481 to 705.

9-Octyleicosane, CA 13475-77-9: Flammable; soluble most organic solvents.
Tm = 273.75; Vm = 0.4889 at 293.15; Antoine (liq) A = 10.6456, B = 5578.6,
C = 0, dev = 1.0, range = 460 to 530.

$C_{29}H_{50}$ MW 398.71

11-(2,5-Dimethylphenyl)-10-uneicosene: Flammable; soluble most organic
solvents. Antoine (liq) A = 9.82662, B = 5182.306, C = 0, dev = 1.0,
range = 471 to 534.

458

$C_{29}H_{52}$ MW 400.73

11-(2,5-Dimethylphenyl)uneicosane: Flammable; soluble most organic solvents. Antoine (liq) A = 9.96167, B = 5263.924, C = 0, dev = 1.0, range = 472 to 535.

$C_{29}H_{60}$ MW 408.79

Nonacosane, CA 630-03-5: Flammable; soluble light hydrocarbons, benzene, chloroform. Tm = 337.15; Tb = 713.95; Vm = 0.5358 at 373.15; Antoine (liq) A = 6.2505, B = 2233.6, C = -187.75, dev = 1.0, range = 488 to 714.

$C_{30}H_{14}O_2$ MW 406.44

Pyranthrone, 8,16-pyranthenedione, CA 128-70-1: Soluble aniline, hot nitrobenzene; orange vat dye. Antoine (sol) A = 16.605, B = 10330, C = 0, dev = 5.0, range = 503 to 543.

$C_{30}H_{16}$ MW 376.46

Pyranthrene, CA 191-13-9: Tm = 645.15 to 646.15; Antoine (sol) A = 13.950, B = 10150, C = 0, dev = 5.0, range = 550 to 640.

$C_{30}H_{30}$ MW 390.57

1,1,6,6-Tetraphenylhexane, CA 2819-41-2: Flammable; soluble most organic solvents. Tm = 397.15 to 398.15; Antoine (liq) A = 9.87214, B = 5644.86, C = 0, dev = 1.0, range = 511 to 579.

$C_{30}H_{34}$ MW 394.60

1,10-Di(1-naphthyl)decane, CA 40339-27-3: Flammable. Tm = 313.15; Antoine (liq) A = 9.3298, B = 5671.611, C = 0, dev = 1.0, range = 540 to 616.

$C_{30}H_{54}$ MW 414.76

1,10-Bis(decahydro-1-naphthyl)decane: Flammable. Antoine (liq) A = 10.8454, B = 6251.536, C = 0, dev = 1.0, range = 520 to 583.

1,1,6,6-Tetracyclohexylhexane: Flammable; soluble most organic solvents. Tm = 404.15; Antoine (liq) A = 9.56891, B = 5379.921, C = 0, dev = 1.0, range = 501 to 569.

$C_{30}H_{54}O_6$ MW 510.75

Aconitic acid, *trans*, tris(2-ethylhexyl) ester, CA 52193-50-7: Flammable. Antoine (liq) A = 9.32227, B = 5071.272, C = 0, dev = 1.0 to 5.0, range = 437 to 551.

$C_{30}H_{56}O_6$ MW 512.77

1,2,3-Propanetricarboxylic acid, tris(2-ethylhexyl) ester, CA 5400-99-7: Flammable. Antoine (liq) A = 9.3931, B = 5115.059, C = 0, dev = 1.0 to 5.0, range = 438 to 551.

$C_{30}H_{62}$ MW 422.82

2,6,10,15,19,23-Hexamethyltetracosane, squalane, CA 111-01-3: Flammable, flash point 490; soluble ether, benzene, chloroform, light hydrocarbons. Tm = 235.15; Vm = 0.5210 at 288.15; Antoine (liq) A = 7.190, B = 2974, C = -113.15, dev = 5.0, range = 363 to 513.

9-Octyldocosane, CA 55319-83-0: Flammable; soluble most organic solvents. Antoine (liq) A = 9.83312, B = 5710.266, C = 0, dev = 1.0, range = 518. to 588.

Triacontane, CA 638-68-6: Flammable; soluble benzene, light hydrocarbons. Tm = 339.05; Tb = 722.85; Vm = 0.5456 at 351.15; Antoine (liq) A = 6.2553, B = 2254.6, C = -192.25, dev = 1.0, range = 495 to 723.

$C_{30}H_{63}N$ MW 437.83

Tridecyl amine, CA 1070-01-5: Irritant; flammable. Tm = 272.15; Antoine (liq) A = 6.5411, B = 2576.1, C = -111.15, dev = 5.0, range = 545 to 759.

$C_{31}H_{34}$ MW 406.61

1,1-Di(1-naphthyl)-1-undecene: Flammable. Antoine (liq) A = 9.83312, B = 5710.266, C = 0, dev = 1.0, range = 518 to 588.

$C_{31}H_{48}$ MW 420.72

1-(1-Decylundec-1-enyl)naphthalene: Flammable. Tm = 276.25; Antoine (liq) A = 9.8127, B = 5489.771, C = 0, dev = 1.0, range = 499 to 567.

$C_{31}H_{56}$ MW 428.78

1,1-Bis(decahydro-1-naphthyl)undecane: Flammable. Antoine (liq) A = 10.42489, B = 5774.545, C = 0, dev = 1.0, range = 525 to 561.

13-Phenylpentacosane, CA 6006-90-2: Flammable; soluble most organic solvents. Tm = 304.85; Antoine (liq) A = 10.07687, B = 5574.536, C = 0, dev = 1.0, range = 495 to 560.

$C_{31}H_{60}$ MW 432.82

1-(1-Decylundecyl)decahydronaphthalene: Flammable. Antoine (liq) A = 10.10531, B = 5590.352, C = 0, dev = 1.0, range = 523 to 560.

$C_{31}H_{62}$ MW 434.83

13-Cyclohexylpentacosane: Flammable; soluble most organic solvents. Antoine (liq) A = 10.07687, B = 5574.536, C = 0, dev = 1.0, range = 495 to 560.

$C_{31}H_{64}$ MW 436.85

11-Decyluneicosane, CA 55320-06-4: Flammable; soluble most organic solvents. Tm = 283.15; Vm = 0.5383 at 293.15; Antoine (liq) A = 18.75, B = 7981, C = 0, dev = 1.0 to 5.0, range = 298 to 313.

(continues)

$C_{31}H_{64}$ *(continued)*

Untriacontane, CA 630-04-6: Flammable; soluble light hydrocarbons. Tm = 341.15; Tb = 731.15; Vm = 0.5593 at 341.15; Antoine (liq) A = 6.2605, B = 2276.9, C = -197.25, dev = 5.0, range = 503 to 732.

$C_{32}H_2Br_{16}N_8$ MW 1776.86

Phthalocyanine, hexadecabromo, CA 28746-04-5: Antoine (sol) A = 8.881, B = 5702, C = 0, dev = 5.0, range = 438 to 493.

$C_{32}H_2Cl_{16}N_8$ MW 1065.67

Phthalocyanine, hexadecachloro, CA 28888-81-5: Antoine (sol) A = 13.892, B = 7367, C = 0, dev = 5.0, range = 398 to 443.

$C_{32}H_{14}$ MW 398.46

Ovalene, CA 190-26-1: Tm = 746.15; Antoine (sol) A = 14.882, B = 11040, C = 0, dev = 5.0, range = 550 to 650.

$C_{32}H_{54}O_4$ MW 502.78

Phthalic acid, didodecyl ester, CA 2432-90-8: Flammable. Tm = 295.15 to 297.15; Antoine (liq) A = 0.15142, B = 1827.01, C = 0, dev = 5.0, range = 288 to 303.

$C_{32}H_{66}$ MW 450.87

11-Decyldocosane, CA 28261-97-4: Flammable; soluble most organic solvents. Tm = 274.15; Vm = 0.5545 at 293.15; Antoine (liq) A = 10.27909, B = 5678.1, C = 0, dev = 0.1, range = 523 to 559.

Dotriacontane, CA 544-85-4: Flammable; soluble benzene, light hydrocarbons. Tm = 342.85; Tb = 740.15; Antoine I (liq) A = 11.69026, B = 6818.289, C = 0, dev = 1.0, range = 361 to 395; Antoine II (liq) A = 6.2649, B = 2296.1, C = -201.55, dev = 5.0, range = 510 to 741.

9-Octyltetracosane: Flammable; soluble most organic solvents. Antoine (liq) A = 10.0382, B = 5257, C = -32.85, dev = 5.0, range = 501 to 563.

$C_{33}H_{62}O_6$ MW 554.85

Glycerol tricaprate, tricaprin, CA 621-71-6: Flammable; soluble most organic solvents. Tm = 305.15; Antoine (liq) A = 11.205, B = 6510, C = 0, dev = 5.0, range = 437 to 485.

$C_{33}H_{68}$ MW 464.90

Tritriacontane, CA 630-05-7: Flammable; soluble light hydrocarbons. Tm = 344.15 to 346.15; Antoine (liq) A = 6.2691, B = 2314.4, C = -205.85, dev = 5.0, range = 517 to 749.

$C_{34}H_{16}O_2$ MW 456.50

Dibenzanthrone, CA 116-71-2: Tm = 763.15 to 768.15, decomposes. Antoine
(sol) A = 17.045, B = 10910, C = 0, dev = 5.0, range = 513 to 548.

Isodibenzanthrone, CA 128-64-3: Soluble nitrobenzene; slightly soluble
other organic solvents; vat dye. Tm = 783.15 to 784.15; Antoine (sol) A =
18.015, B = 11550, C = 0, dev = 5.0, range = 523 to 553.

$C_{34}H_{18}$ MW 426.52

Benzo[rst]phenanthro[1,10,9-cde]pentaphene, CA 190-93-2: Tm = 603.15;
Antoine (sol) A = 10.885, B = 8050, C = 0, dev = 5.0, range = 478 to 603.

Tetrabenzopentacene, CA 191-79-7: Tm = 606.15 to 607.15; Antoine (sol) A =
11.92, B = 6190, C = 0, dev = 5.0, range = 348 to 448.

$C_{34}H_{20}$ MW 428.53

Isoviolanthrene, CA 4430-29-9: Tm = 783.15; Antoine (sol) A = 13.975, B =
1145, C = 0, dev = 5.0, range = 578 to 723.

Violanthrene, CA 81-31-2: Tm = 751.15; Antoine (sol) A = 16.657, B =
11680, C = 0, dev = 5.0, range = 560 to 620.

$C_{34}H_{70}$ MW 478.93

11-Decyltetracosane: Flammable; soluble most organic solvents. Tm =
283.95; Vm = 0.5869 at 293.15; Antoine (liq) A = 10.4097, B = 5906.3,
C = 0, dev = 1.0, range = 537 to 574.

9-Octylhexacosane: Flammable; soluble most organic solvents. Tm = 295.95;
Antoine (liq) A = 10.137, B = 5759.6, C = 0, dev = 1.0, range = 537 to 575.

Tetratriacontane, CA 14167-59-0: Flammable; soluble benzene, light hydrocarbons. Tm = 345.85; Vm = 0.6197 at 363.15; Antoine I (liq) A = 13.88969,
B = 7822.146, C = 0, dev = 1.0, range = 372 to 402; Antoine II (liq) A =
6.2731, B = 2331.9, C = -210.05, dev = 5.0, range = 523 to 756.

$C_{35}H_{64}$ MW 484.89

15-Phenylnonacosane: Flammable; soluble most organic solvents. Tm =
319.55; Antoine (liq) A = 11.35407, B = 6562.324, C = 0, dev = 1.0, range =
523 to 550.

$C_{35}H_{70}$ MW 490.94

15-Cyclohexylnonacosane: Flammable; soluble most organic solvents.
Antoine (liq) A = 11.72732, B = 6736.292, C = 0, dev = 1.0, range = 548
to 581.

$C_{35}H_{72}$ MW 492.95

Pentatriacontane, CA 630-07-9: Flammable; soluble light hydrocarbons.
Tm = 348.15; Tb = 763.15; Antoine (liq) A = 6.277, B = 2348.6, C = -214.05,
dev = 5.0, range = 529 to 764.

$C_{36}H_{62}O_4$ MW 558.88

Ditetradecylphthalate, CA 2915-60-8: Flammable. Antoine (liq) A = 12.095, B = 6580, C = 0, dev = 5.0, range = 416 to 465.

$C_{36}H_{74}$ MW 506.98

Hexatriacontane, CA 630-06-8: Flammable; soluble light hydrocarbons. Tm = 349.15; Vm = 0.6497 at 353.15; Antoine (liq) A = 6.2807, B = 2364.6, C = -218.05, dev = 5.0, range = 535 to 571.

13-Undecylpentacosane: Flammable; soluble most organic solvents. Tm = 282.85; Vm = 0.6207 at 293.15; Antoine (liq) A = 12.0876, B = 6941.2, C = 0, dev = 1.0, range = 548 to 580.

$C_{36}H_{75}N$ MW 522.00

Triodecyl amine, trilauryl amine, CA 102-87-4: Irritant; flammable. Tm = 289.15 to 291.15; Antoine (liq) A = 6.5574, B = 2740.1, C = -119.15, dev = 5.0, range = 579 to 807.

$C_{37}H_{70}O_6$ MW 610.96

Glycerol, 1-caprylic-2-lauryl-3-myristic, CA 30283-10-4: Flammable; soluble most organic solvents. Tm = 310.15; Antoine (liq) A = 11.025, B = 6880, C = 0, dev = 5.0, range = 464 to 526.

$C_{37}H_{76}$ MW 521.01

Heptatriacontane, CA 7194-84-5: Flammable; soluble light hydrocarbons. Tm = 350.85; Antoine (liq) A = 6.2842, B = 2380, C = -221.95, dev = 5.0, range = 541 to 778.

$C_{38}H_{30}$ MW 486.66

1-Diphenylmethylene-4-triphenylmethyl-2,5-cyclohexadiene, CA 18909-18-7: Flammable; soluble most organic solvents. Tm = 418.15 to 420.15, decomposes; Antoine (sol) A = 11.975, B = 5987, C = 0, dev = above 5.0, range = 348 to 394.

$C_{38}H_{30}O_2$ MW 518.65

Bis(triphenylmethyl)peroxide, CA 596-30-5: Flammable. Tm = 459.15; Antoine (sol) A = 16.105, B = 8259, C = 0, dev = 5.0, range = 392 to 434.

$C_{38}H_{74}O_4$ MW 595.00

Ditetradecyl sebacate, CA 26719-47-1: Flammable. Antoine (liq) A = 12.545, B = 7080, C = 0, dev = 5.0, range = 431 to 483.

$C_{38}H_{78}$ MW 535.03

Octatriacontane, CA 7194-85-6: Flammable; soluble light hydrocarbons. Tm = 352.15; Antoine (liq) A = 6.2876, B = 2394.7, C = -225.75, dev = 5.0, range = 546 to 785.

$C_{39}H_{30}N_6$ MW 582.71

Hexaphenyl melamine, CA 18343-40-3: Tm = 573.15; Antoine (liq) A = 10.315, B = 6860, C = 0, dev = 5.0, range not specified.

$C_{39}H_{72}$ MW 541.00

17-Phenyltritriacontane: Flammable; soluble most organic solvents. Antoine (liq) A = 12.89548, B = 7683.075, C = 0, dev = 1.0, range = 544 to 571.

$C_{39}H_{74}O_6$ MW 639.01

Glycerol trilaurate, trilaurin, CA 538-24-9: Flammable; soluble most organic solvents. Tm = 319.65; Vm = 0.7111 at 328.15; Antoine (liq) A = 11.705, B = 7190, C = 0, dev = 5.0, range = 458 to 520.

$C_{39}H_{78}$ MW 547.05

17-Cyclohexyltritriacontane: Flammable; soluble most organic solvents. Antoine (liq) A = 11.49012, B = 6890.323, C = 0, dev = 1.0, range = 570 to 602.

$C_{39}H_{80}$ MW 549.06

Nonatriacontane, CA 7194-86-7: Flammable; soluble light hydrocarbons. Tm = 353.15 to 353.35; Antoine (liq) A = 6.2909, B = 2408.8, C = -229.45, dev = 5.0, range = 552 to 791.

$C_{40}H_{82}$ MW 563.09

Tetracontane, CA 4181-95-7: Flammable; soluble hydrocarbons. Tm = 354.65; Antoine (liq) A = 6.2940, B = 2422.3, C = -233.05, dev = 5.0, range = 557 to 798.

$C_{41}H_{84}$ MW 577.11

Untetracontane, CA 7194-87-8: Flammable; soluble hydrocarbons. Antoine (liq) A = 6.2970, B = 2435.2, C = -236.65, dev = 5.0, range = 562 to 804.

$C_{42}H_{28}$ MW 532.68

Rubrene, 5,6,11,12-tetraphenylnaphthacene, CA 517-51-1: Soluble benzene. Tm = 604.15; Antoine (sol) A = 12.835, B = 8397, C = 0, dev = 5.0, range = 453 to 523.

$C_{42}H_{30}$ MW 534.70

Hexaphenylbenzene, CA 992-04-1: Tm = 727.15 to 729.15; Antoine I (sol) A = 13.181, B = 8550, C = 0, dev = 5.0, range not specified; Antoine II (liq) A = 7.569, B = 4444, C = 0, dev = 5.0, range not specified.

464

$C_{42}H_{86}$ MW 591.14

Dotetracontane, CA 7098-20-6: Flammable; soluble hydrocarbons. Tm =
356.15; Antoine (liq) A = 6.2999, B = 2447.7, C = -240.15, dev = 5.0,
range = 567 to 810.

2,2,4,15,17,17-Hexamethyl-7,12-bis(3,5,5-trimethylhexyl)octadecane: Flammable; soluble most organic solvents. Antoine (liq) A = 9.6072, B = 4912,
C = -57.05, dev = 5.0, range = 512 to 575.

$C_{42}H_{87}N$ MW 606.16

Tritetradecyl amine, CA 27911-72-4: Irritant; flammable. Tm = 306.15;
Antoine (liq) A = 6.5690, B = 2879.4, C = -126.15, dev = 5.0, range = 609
to 848.

$C_{43}H_{88}$ MW 605.17

Tritetracontane, CA 7098-21-7: Flammable; soluble hydrocarbons. Tm =
358.65; Vm = 0.7747 at 363.15; Antoine (liq) A = 6.3026, B = 2459.7, C =
-243.55, dev = 5.0, range = 572 to 820.

$C_{44}H_{30}N_4$ MW 614.75

5,10,15,20-Tetraphenylporphine, CA 917-23-7: Antoine (sol) A = 6.130, B =
5787, C = 0, dev = 5.0, range = 588 to 678.

$C_{44}H_{90}$ MW 619.20

Tetratetracontane, CA 7098-22-8: Flammable; soluble hydrocarbons. Tm =
359.65; Antoine (liq) A = 6.3053, B = 2471.2, C = -246.85, dev = 5.0,
range = 577 to 821.

$C_{45}H_{86}O_6$ MW 723.17

Glycerol, 1-lauryl-2-myristic-3-palmitic ±, CA 60138-25-2: Flammable; soluble most organic solvents. Tm = 322.15; Antoine (liq) A = 11.925, B =
7720, C = 0, dev = 5.0, range = 491 to 551.

Glycerol, 1-myristic-2-capric-3-stearic ±: Flammable; soluble most organic
solvents. Tm = 325.65; Antoine (liq) A = 12.005, B = 7750, C = 0, dev =
5.0, range = 490 to 551.

Glycerol trimyristate, trimyristin, CA 555-43-1: Flammable; soluble most
organic solvents. Tm = 331.65; Vm = 0.8173 at 333.15; Antoine (liq) A =
11.905, B = 7720, C = 0, dev = 5.0, range = 488 to 551.

$C_{45}H_{92}$ MW 633.22

Pentatetracontane, CA 7098-23-9: Flammable; soluble hydrocarbons. Antoine
(liq) A = 6.3078, B = 2482.3, C = -250.15, dev = 5.0, range = 582 to 827.

$C_{46}H_{94}$ MW 647.25

Hexatetracontane, CA 7098-24-0: Flammable; soluble hydrocarbons. Tm = 362.15; Antoine (liq) A = 6.3103, B = 2493.0, C = -253.35, dev = 5.0, range = 586 to 832.

$C_{47}H_{90}O_6$ MW 751.22

Glycerol, 1-myristic-2-lauryl-3-stearic ±: Flammable; soluble most organic solvents. Tm = 328.15; Antoine (liq) A = 11.965, B = 7860, C = 0, dev = 5.0, range = 493 to 558.

Glycerol, 1-palmitic-2-capric-3-stearic ±: Flammable; soluble most organic solvents. Tm = 328.15; Antoine (liq) A = 12.425, B = 8090, C = 0, dev = 5.0, range = 507 to 559.

$C_{47}H_{96}$ MW 661.28

Heptatetracontane, CA 7098-25-1: Flammable; soluble hydrocarbons. Antoine (liq) A = 6.3127, B = 2503.3, C = -256.45, dev = 5.0, range = 591 to 837.

$C_{48}H_{98}$ MW 675.30

Octatetracontane, CA 7098-26-2: Flammable; soluble hydrocarbons. Tm = 372.15; Antoine (liq) A = 6.3149, B = 2513.1, C = -259.45, dev = 5.0, range = 595 to 843.

$C_{49}H_{94}O_6$ MW 779.28

Glycerol, 1-palmitic-2-lauryl-3-stearic ±: Flammable; soluble most organic solvents. Tm = 330.65; Antoine (liq) A = 12.675, B = 8360, C = 0, dev = 5.0, range = 506 to 567.

$C_{49}H_{100}$ MW 689.33

Nonatetracontane, CA 7098-27-3: Flammable; soluble hydrocarbons. Antoine (liq) A = 6.3171, B = 2522.7, C = -262.45, dev = 5.0, range = 599 to 847.

$C_{50}H_{102}$ MW 703.36

Pentacontane, CA 6596-40-3: Flammable; soluble hydrocarbons. Tm = 365.15; Antoine (liq) A = 6.3193, B = 2531.8, C = -265.35, dev = 5.0, range = 603 to 852.

$C_{51}H_{98}O_6$ MW 807.33

Glycerol, 1-myristic-2-palmitic-3-stearic, CA 60138-20-7: Flammable; soluble most organic solvents. Tm = 329.15; Antoine (liq) A = 12.305, B = 8250, C = 0, dev = 5.0, range = 508 to 572.

Glycerol tripalmitate, tripalmitin, CA 555-44-2: Flammable; soluble most organic solvents. Tm = 339.55; Vm = 0.9248 at 343.15; Antoine (liq) A = 12.525, B = 8400, C = 0, dev = 5.0, range = 506 to 572.

$C_{51}H_{104}$ MW 717.38

Unpentacontane, CA 7667-76-7: Flammable; soluble hydrocarbons. Antoine
(liq) A = 6.3213, B = 2540.7, C = -268.25, dev = 5.0, range = 607 to 857.

$C_{52}H_{106}$ MW 731.41

Dopentacontane, CA 7719-79-1: Flammable; soluble hydrocarbons. Tm =
363.45; Antoine (liq) A = 6.3233, B = 2549.2, C = -271.05, dev = 5.0,
range = 611 to 861.

$C_{53}H_{108}$ MW 745.44

Tripentacontane, CA 7719-80-4: Flammable; soluble hydrocarbons. Antoine
(liq) A = 6.3252, B = 2557.3, C = -273.85, dev = 5.0, range = 615 to 866.

$C_{54}H_{110}$ MW 759.46

Tetrapentacontane, CA 5856-66-6: Flammable; soluble hydrocarbons. Antoine
(liq) A = 6.3270, B = 2565.5, C = -276.45, dev = 5.0, range = 618 to 870.

$C_{55}H_{112}$ MW 773.49

Pentapentacontane, CA 5846-40-2: Flammable; soluble hydrocarbons. Antoine
(liq) A = 6.3288, B = 2573.1, C = -279.15, dev = 5.0, range = 622 to 874.

$C_{56}H_{114}$ MW 787.52

Hexapentacontane, CA 7719-82-6: Flammable; soluble hydrocarbons. Tm =
363.15; Antoine (liq) A = 6.3305, B = 2580.6, C = -281.65, dev = 5.0,
range = 625 to 878.

$C_{57}H_{108}O_6$ MW 889.48

Glycerol, 1,3-distearic-2-oleic, CA 2846-04-0: Flammable; soluble most organic solvents. Tm = 313.65 to 315.65; Antoine (liq) A = 12.545, B = 8660,
C = 0, dev = 5.0, range = 523 to 593.

$C_{57}H_{110}O_6$ MW 891.49

Glycerol tristearate, tristearin, CA 555-43-1: Flammable; soluble most organic solvents. Tm = 346.65; Vm = 1.0342 at 353.15; Antoine (liq) A =
12.725, B = 8750, C = 0, dev = 5.0, range = 521 to 588.

$C_{57}H_{116}$ MW 801.54

Heptapentacontane, CA 5856-67-7: Flammable; soluble hydrocarbons. Antoine
(liq) A = 6.3322, B = 2587.7, C = -284.25, dev = 5.0, range = 629 to 882.

$C_{58}H_{118}$ MW 815.57

Octapentacontane, CA 7667-78-9: Flammable; soluble hydrocarbons. Tm =
372.15; Antoine (liq) A = 6.3338, B = 2594.7, C = -286.65, dev = 5.0,
range = 632 to 886.

$C_{59}H_{120}$ MW 829.60

Nonapentacontane, CA 7667-79-0: Flammable; soluble hydrocarbons. Antoine
(liq) A = 6.3353, B = 2601.4, C = -289.15, dev = 5.0, range = 635 to 890.

$C_{60}H_{122}$ MW 843.62

Hexacontane, CA 7667-80-3: Flammable; soluble hydrocarbons. Tm = 371.65
to 372.45; Antoine (liq) A = 6.3368, B = 2607.9, C = -291.45, dev = 5.0,
range = 638 to 893.

$C_{61}H_{124}$ MW 857.65

Unhexacontane, CA 7667-81-4: Flammable; soluble hydrocarbons. Antoine
(liq) A = 6.3383, B = 2614.2, C = -293.75, dev = 5.0, range = 642 to 897.

$C_{62}H_{126}$ MW 871.68

Dohexacontane, CA 7719-83-7: Flammable; soluble hydrocarbons. Antoine
(liq) A = 6.3397, B = 2620.2, C = -296.05, dev = 5.0, range = 645 to 901.

$C_{63}H_{128}$ MW 885.70

Trihexacontane, CA 7719-84-8: Flammable; soluble hydrocarbons. Antoine
(liq) A = 6.3410, B = 2626.1, C = -298.25, dev = 5.0, range = 647 to 904.

$C_{64}H_{130}$ MW 899.73

Tetrahexacontane, CA 7719-87-1: Flammable; soluble hydrocarbons. Tm =
375.15; Antoine (liq) A = 6.3423, B = 2631.8, C = -300.95, dev = 5.0,
range = 650 to 907.

$C_{65}H_{132}$ MW 913.76

Pentahexacontane, CA 7719-88-2: Flammable; soluble hydrocarbons. Antoine
(liq) A = 6.3436, B = 2637.4, C = -302.65, dev = 5.0, range = 653 to 910.

$C_{66}H_{134}$ MW 927.78

Hexahexacontane, CA 7719-89-3: Flammable; soluble hydrocarbons. Tm =
376.75; Antoine (liq) A = 6.3448, B = 2642.7, C = -304.75, dev = 5.0,
range = 656 to 914.

$C_{67}H_{136}$ MW 941.81

Heptahexacontane, CA 7719-90-6: Flammable; soluble hydrocarbons. Antoine
(liq) A = 6.3460, B = 2647.9, C = -306.75, dev = 5.0, range = 659 to 937.

$C_{68}H_{138}$ MW 955.84

Octahexacontane, CA 7719-91-7: Flammable; soluble hydrocarbons. Antoine
(liq) A = 6.3472, B = 2652.9, C = -308.75, dev = 5.0, range = 661 to 920.

$C_{69}H_{140}$ MW 969.86

Nonahexacontane, CA 7719-92-8: Flammable; soluble hydrocarbons. Antoine
(liq) A = 6.3483, B = 2657.8, C = -310.75, dev = 5.0, range = 664 to 923.

$C_{70}H_{142}$ MW 983.89

Heptacontane, CA 7719-93-9: Flammable; soluble hydrocarbons. Tm = 378.15;
Antoine (liq) A = 6.3494, B = 2662.5, C = -312.65, dev = 5.0, range = 666
to 926.

$C_{71}H_{144}$ MW 997.92

Unheptacontane, CA 7667-82-5: Flammable; soluble hydrocarbons. Antoine
(liq) A = 6.3505, B = 2667.1, C = -314.55, dev = 5.0, range = 669 to 928.

$C_{72}H_{146}$ MW 1011.95

Doheptacontane, CA 7667-83-6: Flammable; soluble hydrocarbons. Antoine
(liq) A = 6.3515, B = 2671.6, C = -316.45, dev = 5.0, range = 671 to 931.

$C_{73}H_{148}$ MW 1025.97

Triheptacontane, CA 7667-84-7: Flammable; soluble hydrocarbons. Antoine
(liq) A = 6.3525, B = 2675.9, C = -318.25, dev = 5.0, range = 674 to 934.

$C_{74}H_{150}$ MW 1040.00

Tetraheptacontane, CA 7667-85-8: Flammable; soluble hydrocarbons. Antoine
(liq) A = 6.3535, B = 2680.1, C = -320.05, dev = 5.0, range = 676 to 936.

$C_{75}H_{152}$ MW 1054.03

Pentaheptacontane, CA 7667-86-9: Flammable; soluble hydrocarbons. Antoine
(liq) A = 6.3544, B = 2684.2, C = -321.85, dev = 5.0, range = 678 to 939.

$C_{76}H_{154}$ MW 1068.05

Hexaheptacontane, CA 7667-87-0: Flammable; soluble hydrocarbons. Antoine
(liq) A = 6.3553, B = 2688.2, C = -323.55, dev = 5.0, range = 680 to 941.

$C_{77}H_{156}$ MW 1082.08

Heptaheptacontane, CA 7719-94-0: Flammable; soluble hydrocarbons. Antoine
(liq) A = 6.3562, B = 2692.0, C = -325.25, dev = 5.0, range = 682 to 944.

$C_{78}H_{158}$ MW 1096.11

Octaheptacontane, CA 7719-85-9: Flammable; soluble hydrocarbons. Antoine
(liq) A = 6.3571, B = 2695.8, C = -326.85, dev = 5.0, range = 685 to 946.

$C_{79}H_{160}$ MW 1110.13

Nonaheptacontane, CA 7719-86-0: Flammable; soluble hydrocarbons. Antoine
(liq) A = 6.3579, B = 2699.4, C = -328.45, dev = 5.0, range = 687 to 949.

$C_{80}H_{162}$ MW 1124.16

Octacontane, CA 7667-88-1: Flammable; soluble hydrocarbons. Antoine (liq)
A = 6.3587, B = 2702.9, C = -330.05, dev = 5.0, range = 689 to 951.

$C_{81}H_{164}$ MW 1138.19

Unoctacontane, CA 7667-89-2: Flammable; soluble hydrocarbons. Antoine
(liq) A = 6.3595, B = 2706.4, C = -331.65, dev = 5.0, range = 691 to 953.

$C_{82}H_{166}$ MW 1152.21

Dooctacontane, CA 7719-95-1: Flammable; soluble hydrocarbons. Tm = 383.45
to 383.55; Antoine (liq) A = 6.3603, B = 2709.7, C = -333.15, dev = 5.0,
range = 693 to 955.

$C_{83}H_{168}$ MW 1166.24

Trioctacontane, CA 7667-90-5: Flammable; soluble hydrocarbons. Antoine
(liq) A = 6.3610, B = 2713.0, C = -334.65, dev = 5.0, range = 694 to 957.

$C_{84}H_{170}$ MW 1180.27

Tetraoctacontane, CA 7667-91-6: Flammable; soluble hydrocarbons. Antoine
(liq) A = 6.3618, B = 2716.2, C = -336.15, dev = 5.0, range = 696 to 960.

$C_{85}H_{172}$ MW 1194.29

Pentaoctacontane, CA 7719-96-2: Flammable; soluble hydrocarbons. Antoine
(liq) A = 6.3625, B = 2719.2, C = -337.55, dev = 5.0, range = 698 to 962.

$C_{86}H_{174}$ MW 1208.32

Hexaoctacontane, CA 7667-92-7: Flammable; soluble hydrocarbons. Antoine
(liq) A = 6.3632, B = 2722.2, C = -338.95, dev = 5.0, range = 700 to 964.

$C_{87}H_{176}$ MW 1222.35

Heptaoctacontane, CA 7667-93-8: Flammable; soluble hydrocarbons. Antoine
(liq) A = 6.3638, B = 2725.2, C = -340.35, dev = 5.0, range = 702 to 966.

$C_{88}H_{178}$ MW 1236.37

Octaoctacontane, CA 7667-94-9: Flammable; soluble hydrocarbons. Antoine
(liq) A = 6.3645, B = 2728.0, C = -341.75, dev = 5.0, range = 703 to 967.

$C_{89}H_{180}$ MW 1250.40

Nonaoctacontane, CA 7719-76-8: Flammable; soluble hydrocarbons. Antoine
(liq) A = 6.3651, B = 2730.8, C = -343.05, dev = 5.0, range = 705 to 969.

$C_{90}H_{182}$ MW 1264.43

Nonacontane, CA 7719-76-8: Flammable; soluble hydrocarbons. Antoine (liq)
A = 6.3657, B = 2733.5, C = -344.45, dev = 5.0, range = 707 to 971.

$C_{91}H_{184}$ MW 1278.45

Unnonacontane, CA 7719-97-3: Flammable; soluble hydrocarbons. Antoine
(liq) A = 6.3664, B = 2736.1, C = -345.65, dev = 5.0, range = 708 to 973.

$C_{92}H_{186}$ MW 1292.48

Dononacontane, CA 7667-95-0: Flammable; soluble hydrocarbons. Antoine
(liq) A = 6.3669, B = 2738.7, C = -346.95, dev = 5.0, range = 710 to 975.

$C_{93}H_{188}$ MW 1306.51

Trinonacontane, CA 7667-96-1: Flammable; soluble hydrocarbons. Antoine
(liq) A = 6.3675, B = 2741.2, C = -348.25, dev = 5.0, range = 711 to 977.

$C_{94}H_{190}$ MW 1320.54

Tetranonacontane, CA 1574-32-9: Flammable; soluble hydrocarbons. Tm =
387.22 to 387.62; Antoine (liq) A = 6.3681, B = 2743.6, C = -349.45, dev =
5.0, range = 713 to 978.

$C_{95}H_{192}$ MW 1334.56

Pentanonacontane, CA 7667-97-2: Flammable; soluble hydrocarbons. Antoine
(liq) A = 6.3686, B = 2746.0, C = -350.65, dev = 5.0, range = 714 to 980.

$C_{96}H_{194}$ MW 1348.59

Hexanonacontane, CA 7763-13-5: Flammable; soluble hydrocarbons. Antoine
(liq) A = 6.3692, B = 2748.3, C = -351.85, dev = 5.0, range = 716 to 982.

$C_{97}H_{196}$ MW 1362.62

Heptanonacontane, CA 7670-25-9: Flammable; soluble hydrocarbons. Antoine
(liq) A = 6.3697, B = 2750.6, C = -352.95, dev = 5.0, range = 717 to 983.

$C_{98}H_{198}$ MW 1376.64

Octanonacontane, CA 7670-26-0: Flammable; soluble hydrocarbons. Antoine
(liq) A = 6.3702, B = 2752.8, C = -354.05, dev = 5.0, range = 719 to 985.

$C_{99}H_{200}$ MW 1390.67

Nonanonacontane, CA 7670-27-1: Flammable; soluble hydrocarbons. Antoine
(liq) A = 6.3707, B = 2755.0, C = -355.25, dev = 5.0, range = 720 to 986.

$C_{100}H_{202}$ MW 1404.70

Hectane, CA 6703-98-6: Flammable; soluble hydrocarbons. Tm = 388.25;
Antoine (liq) A = 6.3712, B = 2757.1, C = -356.25, dev = 5.0, range = 721
to 988.

Properties of Organometallic Compounds

CAsCl$_2$F$_3$ MW 214.83

  Dichloro(trifluoromethyl) arsine, CA 421-32-9: Soluble acetone, ethanol;
  irritant; toxic.   Tb = 344.15; Vm = 0.1502 at 293.15; Antoine (liq) A =
  5.56689, B = 1078.175, C = -50.543, dev = 1.0, range = 225 to 295.

CCl$_3$NSSi   MW 192.52

  Silane, trichloroisothiocyanate, CA 18157-00-1: Irritant. Tm = 198.15;
  Tb = 402.65; Vm = 0.1318 at 297.15; Antoine (liq) A = 7.1234, B = 2060,
  C = 0, dev = 1.0, range = 340 to 403.

CCl$_3$NSi   MW 160.46

  Trichlorosilane carbonitrile, CA 18157-01-2: Irritant. Antoine (liq) A =
  6.8759, B = 1687, C = 0, dev = 1.0, range = 227 to 293.

CHCl$_5$Ge   MW 262.87

  Trichloro(dichloromethyl) germane, CA 21572-22-5: Irritant. Antoine (liq)
  A = 6.14902, B = 1568.571, C = -64.127, dev = 1.0, range = 313 to 433.

CH$_2$Cl$_4$Ge   MW 228.43

  Trichloro(chloromethyl) germane, CA 21572-18-9: Sensitive to air;
  irritant.   Antoine (liq) A = 5.95029, B = 1381.551, C = -75.824, dev = 1.0,
  range = 303 to 423.

CH$_2$Cl$_4$OSi   MW 199.92

  Trichloro(chloromethoxy) silane: Irritant. Antoine (liq) A = 7.5199,
  B = 2159, C = 0, dev = 1.0, range = 273 to 323.

CH$_2$Cl$_6$Si$_2$ MW 282.92

  Bis(trichlorosilyl) methane, CA 4142-85-2: Soluble ether, ethanol;
  irritant.   Vm = 0.1817 at 293.15; Antoine (liq) A = 7.2646, B = 2380,
  C = 0, dev = 1.0 to 5.0, range = 328 to 453.

CH$_3$AsCl$_2$ MW 160.86

  Dichloromethyl arsine, CA 593-89-5: Toxic; irritant; vesicant; flash point
  above 378.   Tm = 230.65; Tb = 407.65; Vm = 0.0876 at 293.15; Antoine I
  (liq) A = 7.3230, B = 2142.6, C = 0, dev = 1.0, range = 271 to 338; Antoine
  II (liq) A = 5.9880, B = 1339.0, C = -70.41, dev = 1.0, range = 338 to 408.

CH$_3$AsF$_2$ MW 127.95

  Difluoromethyl arsine, CA 420-24-6: Irritant; toxic. Tm = 243.45; Tb =
  349.65; Vm = 0.0665 at 291.15; Antoine (liq) A = 6.4048, B = 1285.6, C =
  -57.35, dev = 1.0, range = 282 to 350.

CH$_3$BCl$_2$O   MW 112.75

Dichloromethoxy borane, CA 867-46-9: Irritant. Tm 258.15; Antoine (liq)
A = 5.9275, B = 1067.7, C = -58.88, dev = 1.0, range = 267 to 331.

CH$_3$BO   MW 41.84

Borane carbonyl, CA 13205-44-2: Irritant. Tm = 136.15; Antoine (liq) A =
6.34176, B = 827.76, C = -19.06, dev = 1.0, range = 136 to 194.

CH$_3$BrHg   MW 295.53

Methylmercuric bromide, CA 506-83-2: Soluble ethanol; toxic. Tm = 445.15;
Antoine (sol) A = 9.098, B = 3530, C = 0, dev = 5.0, range = 258 to 297.

CH$_3$ClHg   MW 251.08

Methylmercuric chloride, CA 115-09-03: Soluble ethanol; toxic. Tm =
443.15; Antoine (sol) A = 8.5537, B = 3371.4, C = 0, dev = 1.0, range =
267 to 285.

CH$_3$Cl$_2$FSi   MW 133.02

Fluorodichloromethyl silane, CA 420-58-6: Irritant. Tm = 174.45; Antoine
(liq) A = 6.8702, B = 1472.3, C = 0, dev = 1.0, range = 243 to 303.

CH$_3$Cl$_3$Ge   MW 193.98

Trichloromethyl germane, CA 993-10-2: Tb = 384.15; Vm = 0.1138 at 293.15;
Antoine (liq) A = 6.16795, B = 1412.518, C = -46.05, dev = 1.0, range =
293 to 385.

CH$_3$Cl$_3$Si   MW 149.48

Methyltrichloro silane, CA 75-79-6: Flammable, irritant, moderately toxic;
reacts violently with water. Tm = 195.35; Tb = 339.15; Vm = 0.1174 at
293.15; Antoine I (liq) A = 5.9469, B = 1157, C = -45.75, dev = 1.0,
range = 307 to 340; Antoine II (liq) A = 6.29909, B = 1368.411, C =
-20.599, dev = 1.0, range = 339 to 516.

CH$_3$Cl$_5$Si$_2$ MW 248.47

Pentachloromethyl disilane, CA 26980-40-5: Irritant. Tm = 279.15;
Antoine (liq) A = 7.04337, B = 2186.113, C = 8.138, dev = 1.0, range = 308
to 379.

CH$_3$F$_3$Si   MW 100.12

Trifluoromethyl silane, CA 373-74-0: Irritant. Tm = 200.35; Tb = 242.95;
Antoine (liq) A = 7.1001, B = 1238.1, C = 0, dev = 1.0, range = 213 to 243.

$CH_3F_6NSi_2$ MW 199.20

Hexafluoro-*N*-methyl disilazane, CA 7270-39-5: Irritant. Tm = 223.15;
Antoine (liq) A = 7.4789, B = 1640, C = 0, dev = 1.0 to 5.0, range = 223
to 274.

CH_3HgI MW 342.53

Methylmercuric iodide, CA 143-36-2: Toxic; soluble ethanol, methanol;
Tm = 418.15; Antoine (sol) A = 8.6281, B = 3393.7, C = 0, dev = 1.0,
range = 263 to 290.

CH_3NSi MW 57.13

Isocyano silane, CA 18081-38-4: Irritant. Tm = 305.55; Antoine I (sol)
A = 10.0759, B = 2550, C = 0, dev = 1.0, range = 253 to 305; Antoine II
(liq) A = 6.8599, B = 1567, C = 0, dev = 1.0, range = 305 to 319.

CH_4ClFSi MW 98.58

Chlorofluoromethyl silane: Irritant. Tm = 152.95; Antoine (liq) A =
6.7315, B = 1296.6, C= 0, dev = 1.0, range = 198 to 273.

CH_4Cl_2Ge MW 159.54

Dichloromethyl germane, CA 1111-82-6: Tm = 210.05; Antoine (liq) A =
5.99598, B = 1383.993, C = -39.976, dev = 1.0 to 5.0, range = 280 to 297.

CH_4Cl_2OSi MW 131.03

Dichloromethoxy silane, CA 6485-89-8: Irritant. Tm = 168.15; Vm = 0.1088
at 293.15; Antoine (liq) A = 6.96490, B = 1600.0, C = 0, dev = 1.0, range =
273 to 311.

CH_4Cl_2Si MW 115.03

(Dichloromethyl) silane, CA 42430-97-7: Irritant, flammable. Antoine
(liq) A = 5.83160, B = 1073.158, C = -61.201, dev = 1.0 to 5.0, range =
283 to 319.

Dichloromethyl silane, CA 75-54-7: Highly irritant, highly flammable,
flash point 241. Tm = 180.15; Tb = 314.15; Vm = 0.1041 at 293.15; Antoine
(liq) A = 6.7317, B = 1484.94, C = 0, dev = 1.0, range = 208 to 314.

CH_4F_2Si MW 82.12

Difluoromethyl silane, CA 420-34-8: Irritant. Tm = 163.05; Antoine (liq)
A = 6.8904, B = 1160.4, C = 0, dev = 1.0, range = 198 to 237.

CH_4Se MW 95.00

Selenomethane, CA 6486-05-1: Tb = 298.65; Antoine (liq) A = 6.0749,
B = 1060, C = -38.15, dev = 1.0, range = 228 to 288.

476

CH$_5$As   MW 91.97

Methyl arsine, CA 593-52-2: Highly toxic; ignites spontaneously in air;
soluble acetone, ether, ethanol. Tm = 130.15; Tb = 275.15; Antoine (liq)
A = 6.1273, B = 1109.2, C = -6.05, dev = 1.0, range = 199 to 276.

CH$_5$BO$_2$ MW 59.86

Dihydroxymethyl borane, CA 13061-96-6: Readily dehydrates. Tm = 368.15
to 373.15 (decomposes); Antoine (sol) A = 10.7868, B = 3347.5, C = 0,
dev = 1.0, range = 293 to 362.

CH$_5$ClSi   MW 80.59

(Chloromethyl) silane, CA 10112-09-1: Irritant, flammable. Tm = 183.55;
Tb = 304.85; Antoine (liq) A = 7.21792, B = 1719.28, C = 25.449, dev = 1.0,
range = 246 to 297.

Chloromethyl silane, CA 993-00-0: Irritant, flammable. Tm = 138.15;
Antoine (liq) A = 5.92766, B = 963.08, C = -36.23, dev = 1.0, range = 178
to 282.

CH$_5$FGe   MW 108.64

Fluoromethyl germane, CA 30123-02-5: Tm = 226.15; Antoine (liq) A =
7.5379, B = 1728, C = 0, dev = 1.0 to 5.0, range = 226 to 273.

CH$_6$Ge   MW 90.65

Methylgermane, CA 1449-65-6: Tm = 118.65; Tb = 239.05; Antoine (liq) A =
6.41379, B = 1013.5, C = -9.06, dev = 1.0, range = 154 to 228.

CH$_6$GeS   MW 122.71

(Methylthio) germane, CA 16643-16-6: Tm = 176.15; Antoine (liq) A =
6.35225, B = 1556.15, C = 0, dev = 1.0, range = 223 to 291.

CH$_6$OSi   MW 62.14

Methoxy silane, CA 2171-96-2: Irritant, flammable. Tm = 174.65; Antoine
(liq) A = 5.57614, B = 722.556, C = -53.375, dev = 1.0, range = 183 to 216.

CH$_6$Si   MW 46.14

Methyl silane, CA 992-94-9: Irritant, flammable. Tm = 116.34; Tb =
215.65; Antoine (liq) A = 6.1839, B = 868.94, C = -8.0, dev = 1.0, range =
142 to 321.

CH$_6$Sn   MW 136.75

Methyl stannane, CA 1631-78-3. Toxic. Tb = 274.55; Antoine (liq) A =
6.5999, B = 1255, C = 0, dev = 1.0 to 5.0, range = 193 to 273.

CH_7ClSi_2 MW 110.69

(Chloromethyl) disilane, CA 54713-74-5: Irritant, flammable. Tm = 165.35;
Antoine (liq) A = 7.21772, B = 1875.073, C = 0, dev = 1.0, range = 250
to 299.

Disilane, 1-chloro-2-methyl, CA 35483-45-5: Irritant, flammable. Tm =
153.15; Antoine (liq) A = 6.96331, B = 1674.13, C = 0, dev = 1.0, range =
273 to 301.

$CH_8B_3N_3$ MW 94.52

N-Methyl borazine, CA 21127-94-6: Irritant. Antoine (liq) A = 6.7939, B =
1713, C = -0.05, dev = 1.0, range = 288 to 325.

B-Methyl borazine, CA 21127-95-7: Irritant. Tm = 214.15; Antoine (liq)
A = 7.0049, B = 1800, C = -0.05, dev = 1.0, range = 273 to 299.

CH_8Ge_2 MW 165.25

Methyldigermane, CA 20420-08-0: Antoine (liq) A = 7.0019, B = 1637.7,
C = 0, dev = 1.0, range = 209 to 297.

CH_8Si_2 MW 76.25

1,3-Disilapropane, CA 1759-88-2: Irritant, flammable. Tb = 287.85;
Antoine (liq) A = 6.5439, B = 1370, C = -0.05, dev = 5.0, range = 247
to 299.

Methyldisilane, CA 13498-43-6: Irritant; ignites in air. Tm = 138.25;
Tb = 289.75; Antoine (liq) A = 6.41854, B = 1205.778, C = -17.162, dev =
1.0, range = 189 to 265.

CH_9NSi_2 MW 91.26

2-Methyl disilazane, CA 4459-06-7: Irritant. Tm = 149.05; Antoine (liq)
A = 6.37714, B = 1226.07, C = -26.7, dev = 1.0, range = 197 to 307.

$CH_{10}Si_3$ MW 106.35

2-Silyl-1,3-disilapropane, CA 4335-85-7: Irritant. Antoine (liq) A =
6.2374, B = 1290.4, C = -28.6, dev = 1.0, range = 212 to 283.

COSe   MW 106.97

Carbonyl selenide, CA 1603-84-5: Toxic. Tm = 148.78; Antoine (liq) A =
6.60240, B = 1159.07, C = 0.65, dev = 0.1, range = 220 to 251.

CSSe   MW 123.03

Carbon sulfide selenide, CA 5951-19-9: Tm = 197.95; Antoine (liq) A =
6.15882, B = 1358.54, C = -31.71, dev = 1.0, range = 225 to 359.

CSe$_2$ MW 169.93

   Carbon diselenide, CA 506-80-9: Tm = 229.45; Tb = 399.15; Vm = 0.06335 at
   293.15; Antoine I (sol) A = 8.914, B = 2418, C = 0, dev = 1.0, range = 218
   to 229; Antoine II (liq) A = 5.79722, B = 1378.3, C = -43.71, dev = 1.0,
   range = 230 to 337.

C$_2$AsClF$_6$ MW 248.39

   Chlorobis(trifluoromethyl) arsine, CA 359-53-5: Toxic. Antoine (liq) A =
   7.33345, B = 1705.613, C = 1.905, dev = 1.0, range = 208 to 295.

C$_2$BrF$_5$Se   MW 277.88

   (Pentafluoroethane) selenyl bromide: Tm = 178.15; Antoine (liq) A = 7.385,
   B = 1800, C = 0, dev = 5.0, range = 242 to ?93.

C$_2$ClF$_5$Se   MW 233.43

   (Pentafluoroethane) selenyl chloride: Tm = 155.15; Antoine (liq) A =
   6.905, B = 1580, C = 0, dev = 5.0, range = 215 to 289.

C$_2$Cl$_4$Si   MW 193.92

   Trichloro(chloroethynyl) silane, CA 29442-46-4: Irritant. Antoine (liq)
   A = 3.03838, B = 374.672, C = -159.857, dev = 1.0, range = 253 to 295.

C$_2$F$_3$NSSe   MW 206.04

   Trifluoromethanesulfenyl selenocyanate, CA 21438-06-2: Tm = 206.15;
   Antoine (liq) A = 6.4469, B = 1741, C = 0, dev = 1.0 to 5.0, range = 263
   to 313.

C$_2$F$_6$HgS$_2$ MW 402.72

   Bis(trifluoromethylthio) mercury: Toxic. Tm = 312.15 to 313.15; Vm =
   0.1351 at 318.55; Antoine (liq) A = 7.95591, B = 2760.586, C = 10.7, dev =
   1.0, range = 353 to 423.

C$_2$FeN$_2$O$_4$ MW 171.88

   Dicarbonyldinitrosyl iron, CA 13682-74-1: Insoluble water, soluble organic
   solvents; decomposes in air; toxic. Tm = 291.65; Antoine I (sol) A =
   8.8419, B = 2467, C = 0, dev = 1.0, range = 272 to 291; Antoine II (liq)
   A = 7.3139, B = 2021, C = 0, dev = 1.0, range = 297 to 356.

C$_2$HAsF$_6$ MW 213.94

   Bis(trifluoromethyl) arsine, CA 371-74-4: Toxic. Antoine (liq) A =
   5.57793, B = 852.737, C = -55.268, dev = 1.0, range = 207 to 273.

$C_2H_2AsCl_3$ MW 207.32

Dichloro-*cis*-(2-chlorovinyl) arsine, CA 34461-56-8: Powerful vesicant;
carcinogen; highly toxic by skin absorption. Tm = 226.55; Tb = 442.95;
Vm = 0.1110 at 293.15; Antoine (liq) A = 4.61278, B = 785.094, C =
-157.538, dev = 1.0, range = 345 to 383.

Dichloro-*trans*-(2-chlorovinyl) arsine, CA 50361-05-2: Reacts with air and
light; powerful vesicant; carcinogen; highly toxic by skin absorption; used
in chemical warfare. Tm = 271.95; Tb = 469.75; Vm = 0.1103 at 298.15;
Antoine (liq) A = 7.7779, B = 2648, C = 0, dev = 1.0 to 5.0, range = 273
to 373.

C_2H_3BrSi MW 135.04

(Bromoethynyl) silane, CA 68196-78-1: Irritant. Tm = 234.15; Antoine
(liq) A = 7.3189, B = 1763, C = 0, dev = 1.0 to 5.0, range not specified.

C_2H_3ClGe MW 135.09

(Chloroethynyl) germane, CA 68196-76-9: Irritant. Tm = 117.15; Antoine
(liq) A = 7.3569, B = 1761, C = 0, dev = 1.0 to 5.0, range not specified.

C_2H_3ClSi MW 90.58

(Chloroethynyl) silane, CA 58468-21-6: Irritant, flammable. Tm = 187.15;
Antoine (liq) A = 7.3879, B = 1645, C = 0, dev = 1.0 to 5.0, range not
specified.

$C_2H_3Cl_3Si$ MW 161.49

Trichlorovinyl silane, CA 75-94-5: Irritant; may ignite in air; flash
point below 283. Tm = 178.15; Tb = 363.15 to 365.15; Vm = 0.1272 at
293.15; Antoine (liq) A = 6.3257, B = 1440.2, C = -30.55, dev = 1.0,
range = 299 to 364.

$C_2H_3Cl_5Si$ MW 232.40

Trichloro(1,2-dichloroethyl) silane, CA 684-00-4: Irritant. Antoine
(liq) A = 7.06049, B = 2227, C = -13.15, dev = 1.0, range = 334 to 454.

$C_2H_3F_3O_2Si$ MW 144.13

Silyl (trifluoroacetate), CA 6876-44-4: Irritant. Tm = 387.95; Antoine
(sol) A = 7.2929, B = 1604, C = 0, dev = 1.0 to 5.0, range = 273 to 293.

$C_2H_3F_3Se$ MW 163.00

Methyl (trifluoromethyl) selenide, CA 1544-45-4: Antoine (liq) A = 6.785,
B = 1448, C = 0, dev = 1.0 to 5.0, range = 209 to 294.

480

$C_2H_4Cl_6Si_2$ MW 296.94

1,1,1,4,4,4-Hexachloro-1,4-disilabutane, CA 2504-64-5: Irritant. Tm =
297.65 to 298.65; Tb = 475.15; Antoine (liq) A = 6.94521, B = 2242, C =
-22.15, dev = 1.0, range = 364 to 476.

1,1,1,3,3,3-Hexachloro-2-methyl-1,3-disilapropane, CA 18076-92-1: Irritant.
Antoine (liq) A = 7.2338, B = 2425, C = 0, dev = 5.0, range = 328 to 453.

C_2H_4Si MW 56.14

Ethynyl silane, CA 1066-27-9: Irritant, flammable. Tm = 182.45; Tb =
250.65; Antoine (liq) A = 6.5899, B = 1150, C = 0, dev = 5.0, range = 215
to 251.

$C_2H_5AsBr_2$ MW 263.79

Dibromoethyl arsine, CA 683-43-2: Toxic. Tb = 465.15; Vm = 0.1099 at
295.15; Antoine (liq) A = 7.62048, B = 2608.7, C = 0, dev = 5.0, range =
273 to 333.

$C_2H_5AsCl_2$ MW 174.89

Dichloroethyl arsine, CA 598-14-1: Highly toxic by inhalation; very
irritant; soluble water, ethanol, benzene. Tb = 428.45; Vm = 0.1054 at
293.15; Antoine (liq) A = 5.33629, B = 1035.2, C = -117.65, dev = 1.0,
range = 343 to 429.

$C_2H_5AsF_2$ MW 141.98

Difluoroethyl arsine: Toxic. Tm = 234.45; Tb = 376.45; Vm = 0.0831 at
290.15; Antoine (liq) A = 6.78565, B = 1608.3, C = -30.17, dev = 1.0,
range = 248 to 368.

$C_2H_5BCl_2O$ MW 126.78

Dichloroethoxy borane, CA 16339-28-9: Irritant; sensitive to moisture.
Tb = 352.15 to 353.15; Antoine (liq) A = 5.8879, B = 1101.1, C = -67.4,
dev = 1.0, range = 229 to 351.

C_2H_5BrHg —   MW 309.56

Ethylmercuric bromide, CA 107-26-6: Toxic; insoluble water; slightly
soluble ethanol, chloroform. Tm = 471.15; Antoine (sol) A = 9.5709, B =
3994, C = 0, dev = 1.0, range = 285 to 303.

$C_2H_5ClF_2Si$ MW 130.60

Difluorochloroethyl silane: Irritant. Antoine (liq) A = 6.9029, B =
1470.7, C = 0, dev = 1.0 to 5.0, range = 233 to 298.

C_2H_5ClHg 　　　　　　　　　　　　　　　　　　　　　　　MW 265.10

Ethylmercuric chloride, CA 107-27-7: Highly toxic; causes skin burns;
insoluble water; slightly soluble ethanol, ether; soluble chloroform. Tm =
469.15 to 471.15; Antoine (sol) A = 9.6040, B = 3980, C = 0, dev = 1.0,
range = 283 to 303.

$C_2H_5Cl_2FSi$ 　　　　　　　　　　　　　　　　　　　　　　　MW 147.05

Fluorodichloroethyl silane: Irritant, flammable. Antoine (liq) A = 6.806,
B = 1609.5, C = 0, dev = 1.0 to 5.0, range = 243 to 333.

$C_2H_5Cl_3OSi$ 　　　　　　　　　　　　　　　　　　　　　　　MW 179.51

Trichloroethoxy silane, CA 1825-82-7: Irritant. Tm = 138.15; Tb = 375.05;
Vm = 0.1463 at 293.15; Antoine (liq) A = 6.29033, B = 1438.83, C = -39.825,
dev = 1.0, range = 241 to 376.

$C_2H_5Cl_3Si$ 　　　　　　　　　　　　　　　　　　　　　　　MW 163.51

Dichloro(chloromethyl)methyl silane, CA 1558-33-4: Irritant, flammable.
Tb = 393.65 to 394.65; Vm = 0.1272 at 293.15; Antoine (liq) A = 6.055, B =
1265, C = -79.15, dev = 5.0, range = 335 to 398.

Ethyltrichloro silane, CA 115-21-9: Irritant; highly flammable; reacts
violently with water; flash point 287. Tm = 167.55; Tb = 370.15 to 372.15;
Vm = 0.1321 at 293.15; Antoine I (liq) A = 6.05229, B = 1305.166, C =
-49.72, dev = 0.1 to 1.0, range = 303 to 372; Antoine II (liq) A = 5.36406,
B = 833.443, C = -123.806, dev = 1.0, range = 372 to 473; Antoine III (liq)
A = 8.96746, B = 5223.549, C = 399.153, dev = 1.0, range = 469 to 559.

$C_2H_5F_3OSi$ 　　　　　　　　　　　　　　　　　　　　　　　MW 130.14

Trifluoroethoxy silane, CA 460-55-9: Irritant. Tm = 151.15; Tb = 266.15;
Antoine (liq) A = 4.329, B = 358.5, C = -115.05, dev = 5.0, range = 206
to 248.

$C_2H_5F_3Si$ 　　　　　　　　　　　　　　　　　　　　　　　MW 114.14

Trifluoroethyl silane, CA 353-89-9: Irritant. Tm = 168.15; Tb = 268.75;
Vm = 0.0930 at 197.15; Antoine (liq) A = 6.61516, B = 1165.14, C = -16.33,
dev = 1.0, range = 201 to 269.

C_2H_5HgI 　　　　　　　　　　　　　　　　　　　　　　　MW 356.56

Ethylmercuric iodide, CA 2440-42-8: Highly toxic; insoluble water;
slightly soluble ethanol, chloroform. Tm = 458.15 to 459.15; Antoine (sol)
A = 10.277, B = 4163, C = 0, dev = 5.0, range = 286 to 303.

C_2H_5Li 　　　　　　　　　　　　　　　　　　　　　　　MW 36.00

Ethyl lithium, CA 811-49-4: Spontaneously flammable; sensitive to air and
water; soluble benzene. Tm = 368.15; Antoine (sol) A = 15.405, B = 6090,
C = 0, dev = 5.0, range = 298 to 333.

$C_2H_6BClO_2$ MW 108.33

Chlorodimethoxy borane, CA 868-81-5: Irritant; unstable; reacts with water. Tm = 185.65; Antoine (liq) A = 6.47934, B = 1394.8, C = -36.07, dev = 1.0, range = 238 to 348.

$C_2H_6BCl_2N$ MW 125.79

Dichloro(dimethylamino) borane, CA 1113-31-1: Irritant. Tm = 227.05; Antoine (liq) A = 6.10345, B = 1392.6, C = -47.82, dev = 1.0, range = 283 to 343.

$C_2H_6BF_3O$ MW 113.87

Dimethyl ether, boron trifluoride complex, CA 353-42-4: Corrosive, flammable. Tm = 259.15; Tb = 399.15 to 400.15 (decomposes); Vm = 0.0918 at 293.15; Antoine (liq) A = 8.9309, B = 2775, C = 0, dev = 1.0, range = 311 to 346.

$C_2H_6BF_3S$ MW 129.93

Dimethyl sulfide, boron trifluoride complex, CA 353-43-5: Irritant. Tm = 194.65; Antoine (liq) A = 8.9309, B = 2227, C = 0, dev = 1.0, range = 273 to 298.

$C_2H_6B_4$ MW 73.31

1,6-Dicarbahexaborane, CA 20693-67-8: Irritant. Tm = 243.65; Tb = 295.85; Antoine I (sol) A = 7.0786, B = 1385.0, C = -15.5, dev = 1.0, range = 190 to 209; Antoine II (liq) A = 5.8219, B = 958.36, C = -44.7, dev = 1.0, range = 243 to 288.

C_2H_6Cd MW 142.48

Dimethyl cadmium, CA 506-82-1: Toxic; pyrophoric; may explode at temperatures over 373. Tm = 270.75; Tb = 378.85; Vm = 0.07178 at 291.05; Antoine (liq) A = 5.84232, B = 1165.0, C = -75.15, dev = 1.0, range = 274 to 379.

$C_2H_6Cl_2Ge$ MW 173.57

Dichlorodimethyl germane, CA 1529-48-2: Irritant. Tm = 251.15; Tb = 397.15; Vm = 0.1165 at 299.15; Antoine (liq) A = 6.22804, B = 1454.929, C = -51.93, dev = 1.0, range = 251 to 396.

$C_2H_6Cl_2OSi$ MW 145.06

Dichloroethoxy silane, CA 6485-90-1: Irritant; flammable. Tm = 159.15; Vm = 0.1230 at 273.15; Antoine (liq) A = 7.0253, B = 1744.5, C = 0, dev = 1.0, range = 235 to 335.

$C_2H_6Cl_2Si$ MW 129.06

Dichlorodimethyl silane, CA 75-78-5: Irritant; highly flammable; flash
point 264. Tm = 197.15; Tb = 343.15; Vm = 0.1206 at 293.15; Antoine I
(liq) A = 6.26844, B = 1328, C = -32.15, dev = 1.0, range = 263 to 346;
Antoine II (liq) A = 5.75189, B = 987.116, C = -80.082, dev = 1.0, range =
343 to 443; Antoine III (liq) A = 7.52582, B = 2716.240, C = 161.348, dev =
1.0, range = 433 to 520.

Dichloroethyl silane, CA 1789-58-8: Irritant; flammable; flash point below
296. Tm = 166.15; Tb = 347.15 to 348.15; Antoine (liq) A = 6.73916, B =
1664, C = 2.85, dev = 1.0, range = 279 to 347.

$C_2H_6Cl_4Si_2$ MW 228.05

Disilane, 1,1,1,2-tetrachloro-2,2-dimethyl, CA 26980-43-8: Irritant.
Antoine (liq) A = 6.26977, B = 1680.792, C = -36.306, dev = 1.0, range =
300 to 376.

$C_2H_6Ge_2$ MW 175.25

Digermanyl acetylene, CA 68196-77-0: Flammable. Tm = 255.15; Antoine
(liq) A = 7.2029, B = 1824, C = 0, dev = 1.0 to 5.0, range not specified.

C_2H_6Hg MW 230.66

Dimethyl mercury, CA 593-74-8: Highly toxic. Tb = 365.15; Vm = 0.0779 at
293.15; Antoine (liq) A = 6.14178, B = 1342.2, C = -41.15, dev = 1.0,
range = 261 to 359.

$C_2H_6O_2Si$ MW 90.15

Silyl acetate, CA 6876-41-1: Irritant; flammable. Tm = 210.75; Antoine
(liq) A = 7.1219, B = 1685, C = 0, dev = 1.0 to 5.0, range = 273 to 293.

C_2H_6Se MW 109.03

Dimethyl selenide, CA 593-79-3: Flammable. Tm = 185.95; Tb = 330.85; Vm =
0.0762 at 298.15; Antoine I (liq) A = 6.6389, B = 1544.5, C = 0, dev = 5.0,
range = 303 to 337; Antoine II (liq) A = 6.8719, B = 1610, C = 0, dev = 1.0
to 5.0, range = 280 to 303.

C_2H_6Si MW 58.15

Vinyl silane, CA 7291-09-0: Irritant; flammable. Tm = 94.07; Tb = 250.35;
Antoine (liq) A = 5.39664, B = 675.43, C = -51.41, dev = 1.0, range = 183
to 253.

$C_2H_6Si_2$ MW 86.24

Disilyl acetylene, CA 70277-88-2: Irritant; flammable. Tm = 214.15;
Antoine (liq) A = 6.5419, B = 1434, C = 0, dev = 5.0, range not specified.

C_2H_6Te MW 157.67

Dimethyl telluride, CA 593-80-6: Foul-smelling oil; irritant. Tm = 263.15; Tb = 364.15 to 365.15; Antoine (liq) A = 6.6519, B = 1720.5, C = 0, dev = 1.0 to 5.0, range = 306 to 353.

C_2H_6Zn MW 95.45

Dimethyl zinc, CA 544-97-8: Pyrophoric. Tm = 230.65; Tb = 319.15; Vm = 0.0689 at 283.65; Antoine (liq) A = 5.88761, B = 1009.3, C = -57.15, dev = 1.0, range = 238 to 313.

C_2H_7As MW 106.00

Dimethyl arsine (Cacodyl hydride), CA 593-57-7: Highly toxic; ignites spontaneously in air. Tm = 137.05; Tb = 309.15; Vm = 0.0876 at 297.15; Antoine (liq) A = 6.6569, B = 1443, C = 0, dev = 1.0, range = 194 to 257.

$C_2H_7BO_2$ MW 73.89

Dimethoxy borane, CA 4542-61-4: Irritant; hydrolyzes in air. Tm = 142.55; Tb = 247.25; Antoine (liq) A = 6.15234, B = 1071.4, C = -39.82, dev = 1.0, range = 177 to 287.

$C_2H_7B_5$ MW 85.13

Dicarbaheptaborane, CA 20693-69-0: Tm = 265.15; Antoine (liq) A = 5.56337, B = 932.8, C = -70.37, dev = 1.0, range = 295 to 331.

$C_2H_7ClO_2Si$ MW 126.61

Chlorodimethoxy silane, CA 4861-14-7: Irritant; flammable. Tm = 154.15; Tb = 343.65; Vm = 0.1174 at 273.15; Antoine (liq) A = 7.1865, B = 1774.0, C = 0, dev = 1.0, range = 234 to 331.

C_2H_7ClSi MW 94.62

Chlorodimethyl silane, CA 1066-35-9: Irritant; flammable; flash point 245. Tm = 162.15; Tb = 307.85; Vm = 0.1040 at 273.15; Antoine (liq) A = 6.7410, B = 1458.0, C = 0, dev = 1.0, range = 204 to 296.

C_2H_7FGe MW 122.67

Fluorodimethyl germane, CA 34117-35-6: Flammable. Tm = 240.65; Antoine (liq) A = 7.5349, B = 1822, C = 0, dev = 1.0 to 5.0, range = 242 to 278.

C_2H_8Ge MW 104.68

Dimethyl germane, CA 1449-64-5: Flammable. Tm = 128.85; Tb = 276.15; Antoine (liq) A = 6.00035, B = 978.0, C = -30.91, dev = 1.0, range = 178 to 258.

C_2H_8GeO MW 120.67

(Methoxymethyl) germane, CA 16284-75-6: Flammable. Tm = 151.55; Antoine (liq) A = 6.45745, B = 1304.805, C = -27.837, dev = 1.0, range = 209 to 273.

$C_2H_8O_2Si$ MW 92.17

Dimethoxy silane, CA 5314-52-3: Irritant; flammable. Tm = 175.15; Vm = 0.1075 at 273.15; Antoine I (liq) A = 7.3679, B = 1645, C = 0, dev = 1.0, range = 218 to 259; Antoine II (liq) A = 7.0146, B = 1558.6, C = 0, dev = 1.0, range = 213 to 301.

C_2H_8Si MW 60.17

Dimethyl silane, CA 1111-74-6: Irritant; flammable. Tm = 122.93; Tb = 253.55; Vm = 0.0885 at 193.15; Antoine (liq) A = 5.40874, B = 692.99, C = -50.06, dev = 0.1 to 1.0, range = 183 to 253.

Ethyl silane, CA 2814-79-1: Irritant; flammable. Tm = 93.45; Tb = 259.45; Antoine (liq) A = 5.53935, B = 758.83, C = -45.29, dev = 1.0, range = 193 to 268.

C_2H_9BS MW 75.96

Borane dimethyl sulfide, CA 13292-87-0: Flammable; sensitive to moisture. Tm = 235.15 to 246.15; Antoine (liq) A = 8.3449, B = 2346, C = 0, dev = 1.0, range = 273 to 314.

$C_2H_9FOSi_2$ MW 124.26

Disiloxane, 1-fluoro-1,3-dimethyl, CA 35192-38-2: Irritant; flammable. Antoine (liq) A = 5.44743, B = 1162.742, C = 1.319, dev = 1.0, range = 195 to 250.

C_2H_9NSi MW 75.19

(Dimethylamino) silane, CA 2875-98-1: Irritant; flammable. Tm = 276.55; Antoine I (sol) A = 12.851, B = 3070, C = 0, dev = 5.0, range = 228 to 264; Antoine II (liq) A = 6.75337, B = 1387.073, C = 0, dev = 1.0, range = 276 to 289.

$C_2H_{10}B_2$ MW 55.72

1,1-Dimethyl diborane, CA 16924-32-6: Highly flammable; flash point below 263. Tm = 122.95; Tb = 270.55; Antoine (liq) A = 6.4879, B = 1212, C = 0, dev = 1.0, range = 205 to 253.

1,2-Dimethyl diborane, CA 17156-88-6: Highly flammable; flash point below 218. Tm = 148.25; Tb = 224.15; Antoine (liq) A = 4.6699, B = 571.4, C = -72.85, dev = 1.0, range = 193 to 233.

$C_2H_{10}Ge_2$ MW 179.28

Ethyl digermane, CA 20549-65-9: Flammable. Antoine (liq) A = 6.4549. B =
1609.7, C = 0, dev = 1.0, range = 248 to 296.

$C_2H_{10}Si_2$ MW 90.27

1,2-Disilylethane, CA 4364-07-2: Irritant; flammable. Tm = 258.15 to
259.15; Tb = 340.15 to 341.65; Vm = 0.129 at 293.15; Antoine (liq) A =
5.97801, B = 1112.886, C = -39.373, dev = 1.0, range = 258 to 311.

$C_2H_{11}B_2N$ MW 70.74

(Dimethylamino) diborane, CA 23273-02-1: Tm = 218.35; Antoine (liq) A =
7.46495, B = 1917.352, C = 29.58, dev = 0.1, range = 234 to 287.

$C_2H_{12}B_{10}$ MW 144.22

1,2-Dicarbadodecaborane, CA 16872-09-6: Soluble aromatic solvents. Tm =
593.15; Antoine I (sol) A = 6.39571, B = 2361.004, C = -15.411, dev not
specified, range 283 to 333; Antoine II (sol) A = 6.21790, B = 2247.741,
C = -23.248, dev not specified, range = 333 to 423.

1,7-Dicarbadodecaborane, CA 16986-24-6: Soluble aromatic solvents. Tm =
546.15; Antoine I (sol) A = 7.70016, B = 2397.280, C = -52.199, dev not
specified, range = 283 to 333; Antoine II (sol) A = 6.88827, B = 1951.956,
C = -80.388, dev not specified, range = 333 to 423.

C_3AsF_9 MW 281.94

Tris(trifluoromethyl) arsine, CA 432-02-0: Highly toxic; soluble ether.
Tb = 306.15; Antoine (liq) A = 6.9909, B = 1528.0, C = 0, dev = 1.0 to 5.0,
range = 250 to 301.

$C_3BF_9S_3$ MW 314.01

Thioboric acid, tris(trifluoromethyl) ester, CA 36884-78-3: Tf below
195.15; Antoine (liq) A = 6.44805, B = 1772.539, C = 0, dev = 5.0, range =
242 to 298.

C_3BrF_7Se MW 327.89

Heptafluoro-1-propane selenyl bromide: Antoine (liq) A = 6.9769, B = 1829,
C = 0, dev = 5.0, range = 251 to 298.

C_3ClF_7Se MW 283.43

Heptafluoro-1-propane selenyl chloride: Antoine (liq) A = 7.005, B = 1742,
C = 0, dev = 5.0, range = 223 to 289.

C_3CoNO_4 MW 172.97

Nitrosyl cobalt carbonyl, CA 14096-82-3: Highly toxic. Tm = 262.15; Tb = 351.75; Antoine (liq) A = 7.097, B = 1787, C = 0, dev = 5.0, range = 278 to 338.

$C_3H_2F_6Se_2$ MW 309.96

Bis[(trifluoromethyl)seleno] methane: Antoine (liq) A = 6.8709, B = 1850, C = 0, dev = 1.0 to 5.0, range = 278 to 359.

$C_3H_4Cl_3NSi$ MW 188.52

Trichloro-(2-cyanoethyl) silane, CA 1071-22-3: Irritant. Tm = 309.15; Tb = 493.15; Vm = 0.1386 at 293.15; Antoine (liq) A = 5.29199, B = 1172.734, C = -133.607, dev = 5.0, range = 342 to 442.

$C_3H_5Cl_3Si$ MW 175.52

Allyltrichloro silane, CA 107-37-9: Irritant; toxic; corrosive; flammable; flash point 308. Tb = 389.15; Vm = 0.1436 at 293.15; Antoine (liq) A = 5.83743, B = 1179.265, C = -82.351, dev = 1.0, range = 318 to 389.

C_3H_5FOSe MW 155.03

Selenoacetic acid, methyl ester, CA 367-52-2: Tb = 403.15 to 405.15; Vm = 0.0986 at 293.15; Antoine (liq) A = 8.1619, B = 2420.0, C = 0, dev = 1.0 to 5.0, range = 273 to 333.

$C_3H_5F_3Se$ MW 177.03

Ethyl(trifluoromethyl) selenide: Tm = 148.15; Antoine (liq) A = 7.015, B = 1650, C = 0, dev = 1.0 to 5.0, range = 223 to 292.

$C_3H_6Cl_2Si$ MW 141.07

Dichloromethylvinyl silane, CA 124-70-9: Irritant; flammable. Tb = 365.15; Vm = 0.1306 at 293.15; Antoine (liq) A = 6.2192, B = 1383.0, C = -38.95, dev = 1.0, range = 294 to 366.

$C_3H_6Cl_4Si$ MW 211.98

Trichloro-(2-chloropropyl) silane, CA 7787-89-5: Irritant. Antoine (liq) A = 6.01084, B = 1461.48, C = -74.67, dev = 0.1 to 1.0, range = 313 to 443.

Trichloro-(3-chloropropyl) silane, CA 2550-06-3: Irritant. Tb = 454.65; Vm = 0.1560 at 293.15; Antoine (liq) A = 6.23358, B = 1646.28, C = -66.6, dev = 0.1 to 1.0, range = 313 to 443.

$C_3H_6F_3NSe$ MW 192.04

N,N-Dimethyl(trifluoromethyl) selenide: Antoine (liq) A = 6.275, B = 1470, C = 0, dev = 5.0, range = 228 to 321.

$C_3H_7AsCl_2$ MW 188.92

Dichloropropyl arsine, CA 926-53-4: Irritant; toxic. Tm = 301.35; Vm =
0.1228 at 293.15; Antoine (liq) A = 7.9765, B = 2572.0, C = 0, dev = 5.0,
range = 293 to 333.

Dichloroisopropyl arsine, CA 683-67-0: Irritant; toxic. Antoine (liq) A =
5.1736, B = 971.49, C = -135.49, dev = 1.0, range = 340 to 442.

$C_3H_7BO_2$ MW 85.90

1,3,2-Dioxaborinane, CA 6253-16-3: Antoine (liq) A = 4.66273, B = 734.744,
C = -112.743, dev = 1.0 to 5.0, range = 250 to 292.

$C_3H_7Cl_3Si$ MW 177.53

Trichloroisopropyl silane, CA 4170-46-1: Irritant; flammable. Tm =
185.45; Tb = 392.55; Vm = 0.1517 at 293.15; Antoine (liq) A = 5.8732, B =
1293.5, C = -57.97, dev = 1.0, range = 283 to 393.

$C_3H_8BF_3O$ MW 127.90

Methylethyl ether boron trifluoride complex: Tm = 371.15; Antoine (liq)
A = 9.175, B = 2860, C = 0, dev = 5.0, range not specified.

$C_3H_8Cl_2OSi$ MW 159.09

Dichloroethoxymethyl silane, CA 1825-75-8: Irritant; flammable. Tm =
181.35; Tb = 373.25; Vm = 0.1405 at 273.15; Antoine (liq) A = 6.34285, B =
1452.98, C = -38.71, dev = 1.0, range = 239 to 374.

$C_3H_8Cl_2Si$ MW 143.09

Chlorodimethyl(chloromethyl) silane, CA 1719-57-9: Irritant; flammable;
flash point 294. Tb = 386.85 to 388.35; Vm = 0.1325 at 293.15; Antoine
(liq) A = 8.1599, B = 2635, C = 40.85, dev = 1.0 to 5.0, range = 318
to 391.

C_3H_9Al MW 72.09

Trimethyl aluminum, CA 75-24-1: Pyrophoric; explodes with water; dimerizes
at 343.15. For dimer, Tm = 288.55; Tb = 403.15; Vm = 0.0959 at 293.15;
Antoine (liq) A = 6.67901, B = 1724.24, C = -31.346, dev = 0.1, range = 335
to 401.

C_3H_9As MW 120.03

Trimethyl arsine, CA 593-88-4: Highly toxic; spontaneously flammable in
air; soluble benzene, ether, ethanol. Tm = 185.85; Tb = 325.15; Vm =
0.1049 at 288.15; Antoine (liq) A = 6.8368, B = 1563, C = 0, dev = 1.0,
range = 248 to 288.

$C_3H_9AsO_3$ MW 168.02

Trimethyl arsenite, CA 6596-95-8: Toxic. Antoine (liq) A = 7.4829, B = 2200, C = 0, dev = 1.0 to 5.0, range = 300 to 335.

C_3H_9B MW 55.91

Trimethyl borane, CA 593-90-8: Highly toxic; spontaneously flammable; soluble ether, ethanol. Tm = 111.65; Tb = 252.95; Antoine (liq) A = 6.12799, B = 951.13, C = -22.21, dev = 1.0, range = 189 to 253.

$C_3H_9BF_3N$ MW 126.92

Trimethylamine boron trifluoride complex, CA 420-20-2: Tm = 413.15 to 419.15; Tb = 511.15; Antoine I (sol) A = 9.365, B = 3600, C = 0, dev = 5.0, range = 373 to 413; Antoine II (liq) A = 7.8599, B = 2963, C = 0, dev = 1.0, range = 418 to 503.

$C_3H_9BN_4$ MW 111.93

4,5-Dihydro-1,4,5-trimethyl-1H-tetrazaboral, CA 20546-18-3: Antoine (liq) A = 5.795, B = 1730, C = 0, dev = 5.0, range = 289 to 346.

$C_3H_9BO_3$ MW 103.91

Trimethoxy borane, CA 121-43-7: Moderately toxic; flammable; soluble benzene, ether, ethanol. Tm = 243.85; Tb = 341.15 to 342.15; Vm = 0.1136 at 293.15; Antoine I (liq) A = 4.5354, B = 468.84, C = -156.06, dev = 1.0, range = 294 to 342; Antoine II (liq) A = 6.19042, B = 1164.0, C = -60.0, dev = 1.0 to 5.0, range = 453 to 501.

C_3H_9BS MW 87.97

Dimethyl(methylthio) borane, CA 19163-05-4: Moderately toxic; air- and moisture-sensitive; gradually decomposes at room temperature; soluble acetone, ether. Tm = 189.15; Tb = 344.15; Antoine (liq) A = 6.8019, B = 1651, C = 0, dev = 1.0, range = 227 to 304.

$C_3H_9BS_2$ MW 120.03

Methyl bis(methylthio) borane, CA 19163-08-7: Moderately toxic; air- and moisture-sensitive; soluble acetone, ether. Tm = 214.15; Tb = 428.15; Antoine (liq) A = 6.06150, B = 1468.5, C - -65.05, dev = 1.0, range = 300 to 373.

$C_3H_9BS_3$ MW 152.09

Tris(methylthio) borane, CA 997-49-9: Soluble acetone, ether. Tm = 278.05; Tb = 491.35; Vm = 0.1351 at 293.15; Antoine (liq) A = 6.7740, B = 2345.1, C = 0, dev = 1.0, range = 325 to 462.

$C_3H_9B_3Cl_3N_3$ MW 225.91

2,4,6-Trichloro-1,3,5-trimethylborazine, CA 703-86-6: Moisture-sensitive;
soluble benzene, ether. Tm = 435.15 to 436.15; Antoine (sol) A = 7.95964,
B = 3024.76, C = 0, dev = 1.0, range = 363 to 404.

C_3H_9Bi MW 254.08

Trimethyl bismuth, CA 593-91-9: Highly toxic; ignites in air. Tm =
187.15; Tb = 383.15; Vm = 0.1105 at 291.15; Antoine (liq) A = 6.7529, B =
1816, C = 0, dev = 1.0, range = 248 to 288.

C_3H_9ClSi MW 108.64

Chlorotrimethyl silane, CA 75-77-4: Irritant; highly flammable; flash
point 245; suspected carcinogen. Tm = 233.15; Tb = 330.15; Vm = 0.1263 at
293.15; Antoine I (liq) A = 6.04257, B = 1161.6, C = -43.15, dev = 1.0,
range = 241 to 337; Antoine II (liq) A = 6.29681, B = 1344.762, C = -17.57,
dev = 1.0, range = 332 to 419; Antoine III (liq) A = 6.93745, B = 1931.109,
C = 64.84, dev = 1.0, range = 409 to 497.

$C_3H_9Cl_3Si_2$ MW 207.63

1,1,1-Trichloro-2,2,2-trimethyl disilane, CA 18026-87-4: Irritant;
flammable. Tm = 296.65 to 297.65; Antoine (liq) A = 6.29144, B = 1604.432,
C = -49.244, dev = 1.0, range = 297 to 371.

C_3H_9FGe MW 136.69

Fluorotrimethyl germane, CA 661-37-0: Flammable. Tm = 275.05; Antoine I
(sol) A = 6.70576, B = 1390.418, C = -48.244, dev = 1.0, range = 250 to
275; Antoine II (liq) A = 6.31061, B = 1403.141, C = -29.056, dev = 1.0,
range = 280 to 345.

C_3H_9FSi MW 92.19

Fluorotrimethyl silane, CA 420-56-4: Irritant; flammable. Tm = 198.85;
Tb = 289.15 to 290.15; Antoine (liq) A = 6.8575, B = 1405, C = 0, dev = 1.0
to 5.0, range = 232 to 290.

C_3H_9Ga MW 114.82

Trimethyl gallium, CA 1445-79-0: Toxic; ignites in air; explodes with
water; very reactive to oxygen. Tm = 257.25 to 257.45; Tb = 328.85;
Antoine I (sol) A = 9.6913, B = 2362.2, C = 0, dev = 1.0 to 5.0, range =
247 to 257; Antoine II (liq) A = 6.74839, B = 1438.0, C = -25.63, dev =
1.0, range = 258 to 329.

C_3H_9In MW 159.92

Trimethyl indium, CA 3385-78-2: Toxic; may ignite spontaneously in air;
decomposes with water or alcohols. Tm = 362.15 to 362.95; Tb = 408.95;
Antoine I (sol) A = 9.645, B = 3014, C = 0, dev = 5.0, range = 323 to 362;
Antoine II (liq) A = 7.3629, B = 2190, C = 0, dev = 1.0, range = 363
to 408.

C_3H_9Sb MW 166.85

Trimethyl antimony (Trimethyl stibine), CA 594-10-5: Highly toxic;
pyrophoric; soluble ether, ethanol. Tm = 211.15; Tb = 355.15; Vm = 0.1096
at 288.15; Antoine (liq) A = 6.8529, B = 1709, C = 0, dev = 1.0, range =
248 to 288.

C_3H_9Tl MW 249.47

Trimethyl thallium, CA 3003-15-4: Toxic; ignites in air; explodes above
368. Tm = 311.65; Antoine I (sol) A = 9.5773, B = 2827.4, C = -7.05, dev =
1.0 to 5.0, range = 258 to 304; Antoine II (liq) A = 4.6607, B = 799.5, C =
-128.69, dev = 1.0, range = 312 to 360.

$C_3H_{10}BN$ MW 70.93

Methyl(dimethylamino) borane, CA 18494-94-5: Moderately toxic; flammable.
Tm = 136.85; Antoine (liq) A = 5.52102, B = 897.842, C = -57.798, dev =1.0,
range = 214 to 250.

$C_3H_{10}OSi$ MW 90.20

Dimethylmethoxy silane, CA 18033-75-5: Irritant; flammable. Tm = 139.15;
Tb = 309.15; Antoine (liq) A = 6.76926, B = 1415.649, C = -10.572, dev =
1.0, range = 209 to 253.

(2-Methoxyethyl) silane, CA 5624-62-4: Irritant; flammable. Tm = 113.15;
Antoine (liq) A = 6.71765, B = 1590.34, C = 0, dev = 1.0 to 5.0, range =
227 to 283.

Trimethyl silanol, CA 1066-40-6: Irritant; flammable; unstable; may
hydrolyze and polymerize. Antoine (liq) A = 7.25149, B = 1657.645, C =
-53.957, dev = 1.0, range = 291 to 358.

$C_3H_{10}O_2Si$ MW 106.20

Dimethoxymethyl silane, CA 16881-77-9: Irritant; flammable. Tm = 137.15;
Antoine (liq) A = 7.1288, B = 1712.5, C = 0, dev = 1.0, range = 227 to 322.

$C_3H_{10}O_3Si$ MW 122.20

Trimethoxy silane, CA 2487-90-3: Irritant; flammable. Tm = 159.65; Tb =
357.15; Vm = 0.1311 at 273.15; Antoine I (liq) A = 7.4219, B = 1919, C = 0,
dev = 1.0, range = 273 to 317; Antoine II (liq) A = 7.4349, B = 1929,
C = 0, dev = 1.0, range = 273 to 344.

$C_3H_{10}Si$ MW 74.20

Propyl silane, CA 13154-66-0: Irritant; flammable. Tb = 296.15; Antoine
(liq) A = 6.7069, B = 1390, C = 0, dev = 5.0, range = 211 to 296.

Trimethyl silane, CA 993-07-7: Irritant; flammable. Tm = 137.26; Tb =
279.85; Antoine (liq) A = 6.17799, B = 1064.88, C = -24.29, dev = 0.1 to
1.0, range = 203 to 273.

$C_3H_{12}BN$ MW 72.94

Trimethylamine borane, CA 75-22-9: Tm = 365.15 to 367.15; Tb = 445.15;
Antoine I (sol) A = 9.2993, B = 3128.6, C = 7.75, dev = 1.0, range = 296 to
367; Antoine II (liq) A = 2.8879, B = 189.51, C = -269.23, dev = 1.0 to
5.0, range = 367 to 393.

$C_3H_{12}B_3N_3$ MW 122.58

1,3,5-Trimethyl borazine, CA 1004-35-9: Tm = 264.95 to 265.65; Tb = 404.15
to 406.15; Antoine (liq) A = 7.37911, B = 2304.5, C = 22.96, dev = 1.0,
range = 323 to 406.

$C_4F_{10}Se$ MW 316.99

Bis(pentafluoroethyl) selenide: Moderately toxic. Tm = 151.15; Antoine
(liq) A = 7.185, B = 1650, C = 0, dev = 5.0, range = 232 to 295.

$C_4F_{10}Se_2$ MW 395.95

Bis(pentafluoroethyl) diselenide, CA 6123-49-5: Moderately toxic. Tm =
173.15; Antoine (liq) A = 7.775, B = 2090, C = 0, dev = 5.0, range = 272
to 318.

$C_4GeN_4O_4$ MW 240.66

Germane, tetraisocyanato, CA 4756-66-5: Moderately toxic. Tm = 265.15;
Antoine (liq) A = 7.7827, B = 2757, C = 0, dev = 5.0, range not specified.

$C_4HF_{10}NSe_2$ MW 410.96

Bis[(pentafluoroethyl)seleno] amine: Moderately toxic. Antoine (liq) A =
6.825, B = 2000, C = 0, dev = 5.0, range = 270 to 322.

$C_4H_3F_7Se$ MW 263.02

Methyl(heptafluoropropyl) selenide: Moderately toxic. Antoine (liq) A =
7.075, B = 1608, dev = 1.0 to 5.0, range = 232 to 324.

$C_4H_4AsCl_3$ MW 233.36

Chlorobis(2-chlorovinyl) arsine, CA 40334-69-8: Irritant; toxic; possible
carcinogen. Antoine (liq) A = 8.0982, B = 3065.3, C = 0, dev = 1.0 to 5.0,
range = 385 to 503.

C_4H_4Se MW 131.04

Selenothene, CA 288-05-1: Moderately toxic; flammable. Tm = 235.15; Tb =
383.05 to 383.25; Vm = 0.08565 at 288.15; Antoine I (sol) A = 6.42990, B =
1435.6, C = -51.29, dev = 1.0, range = 208 to 235; Antoine II (liq) A =
8.01630, B = 2312.0, C = 11.62, dev = 1.0, range = 235 to 301.

$C_4H_5F_6Se$ MW 246.03

Ethyl(pentafluoroethyl) selenide: Moderately toxic. Tm = 151.15; Antoine
(liq) A = 7.335, B = 1815, C = 0, dev = 1.0 to 5.0, range = 241 to 311.

$C_4H_6F_5NSe$ MW 242.05

1,1,2,2,2-Pentafluoro-*N*,*N*-dimethylethane selenamide: Moderately toxic.
Tm = 198.15; Antoine (liq) A = 7.195, B = 1820, C = 0, dev = 5.0, range =
256 to 320.

C_4H_8OSe MW 151.07

1,4-Oxaselenane, CA 5368-46-7: Moderately toxic. Tm = 251.65; Antoine
(liq) A = 5.55419, B = 1198.177, C = -105.324, dev = 1.0, range = 352
to 430.

$C_4H_9ClF_2Si$ MW 158.65

Butylchlorodifluoro silane: Irritant; flammable. Antoine (liq) A =
6.91788, B = 1754.4, C = 0, dev = 1.0, range = 283 to 358.

$C_4H_9Cl_2FSi$ MW 175.10

Butyldichlorofluoro silane: Irritant; flammable. Antoine (liq) A =
6.81803, B = 1873.3, C = 0, dev = 1.0, range = 313 to 388.

$C_4H_9Cl_3Si$ MW 191.56

Butyltrichloro silane, CA 7521-80-4: Irritant; corrosive; moderately
toxic; flammable; soluble benzene, ether. Tb = 421.15 to 422.15; Vm =
0.1651 at 293.15; Antoine (liq) A = 6.92272, B = 2075.1, C = 0, dev = 1.0,
range = 343 to 423.

$C_4H_9F_3Si$ MW 142.20

Butyltrifluoro silane, CA 371-93-7: Irritant; flammable. Tm = 176.25;
Antoine (liq) A = 6.99227, B = 1623.4, C = 0, dev = 1.0, range = 263
to 323.

C_4H_9Li MW 64.06

Butyl lithium, CA 109-72-8: Moderately toxic; highly flammable; soluble
ethers, hydrocarbons. Antoine (sol) A = 12.9949, B = 5732, C = 0, dev =
1.0, range = 333 to 368.

$C_4H_{10}AlCl$ MW 120.56

Chlorodiethyl aluminum, CA 51466-48-9: Irritant; ignites spontaneously in
air; exists as dimer in liquid phase. Tm = 199.15; Vm = 0.1255 at 298.15;
Antoine (liq) A = 7.35460, B = 2484.531, C = -17.703, dev = 1.0, range =
317 to 398.

$C_4H_{10}BF_3O$ MW 141.93

Diethyl ether boron trifluoride complex, CA 109-63-7: Corrosive; moisture-sensitive; flammable; flash point 338. Tm = 215.45; Tb = 399.15; Vm = 0.1262 at 298.15; Antoine (liq) A = 9.2069, B = 2879, C = 0, dev = 1.0, range = 283 to 363.

$C_4H_{10}BF_3S$ MW 157.99

Diethyl sulfide boron trifluoride complex: Irritant; flammable. Tm = below 195.15: Antoine (liq) A = 7.6599, B = 1893, C = 0, dev = 1.0, range = 273 to 293.

$C_4H_{10}Be$ MW 67.14

Diethyl beryllium, CA 542-63-2: Highly toxic; pyrophoric; decomposes above 358; reacts violently with water and alcohols. Tm = 260.15 to 262.15; Antoine (liq) A = 6.715, B = 2200, C = 0, dev = 5.0, range = 260 to 358.

$C_4H_{10}Cl_2Si$ MW 157.11

Dichlorodiethyl silane, CA 1719-53-5: Corrosive; toxic; flammable; flash point 297. Tm = 176.65; Tb = 402.15 to 403.55; Vm = 0.1421 at 288.15; Antoine I (liq) A = 6.20202, B = 1483, C = -50.15, dev = 1.0, range = 321 to 401; Antoine II (liq) A = 5.01039, B = 681.169, C = -176.509, dev = 1.0, range = 403 to 497; Antoine III (liq) A = 8.72799, B = 5052.916, C = 367.871, dev = 1.0, range = 487 to 596.

$C_4H_{10}F_2Si$ MW 124.21

Diethyldifluoro silane, CA 358-06-5: Irritant; flammable. Tm = 194.45; Tb = 334.65; Vm = 0.1329 at 293.15; Antoine (liq) A = 5.90167, B = 1067.94, C = -59.87, dev = 1.0, range = 244 to 334.

$C_4H_{10}Hg$ MW 258.71

Diethyl mercury, CA 627-44-1: Highly toxic; flammable; soluble ether, slightly soluble ethanol. Tb = 432.15; Vm = 0.1065 at 293.15; Antoine (liq) A = 7.4320, B = 2344.7, C = 0, dev = 1.0, range = 321 to 403.

$C_4H_{10}Se$ MW 137.08

Diethyl selenide, CA 627-53-2: Highly toxic; flammable. Tb = 381.15; Vm = 0.1114 at 293.15; Antoine (liq) A = 6.56487, B = 1585.87, C = -33.364, dev = 1.0, range = 247 to 381.

$C_4H_{10}Te$ MW 185.72

Diethyl telluride, CA 627-54-3: Irritant; foul-smelling; flammable. Tb = 410.65; Vm = 0.1161 at 288.15; Antoine (liq) A = 7.285, B = 2168, C = 0, dev = 1.0 to 5.0, range = 298 to 410.

$C_4H_{10}Zn$ MW 123.50

Diethyl zinc, CA 557-20-0: Toxic; pyrophoric; explodes with water. Tm =
245.15; Tb = 391.15; Vm = 0.1040 at 291.15; Antoine (liq) A = 7.4049, B =
2109, C = 0, dev = 1.0, range = 273 to 363.

$C_4H_{11}ClSi$ MW 122.67

Chlorodiethyl silane, CA 1609-19-4: Irritant; flammable. Tm = 130.15;
Tb = 372.85; Vm = 0.1378 at 293.15; Antoine (liq) A = 6.5444, B = 1689.5,
C = 0, dev = 1.0, range = 273 to 367.

$C_4H_{12}BN$ MW 84.96

Dimethyl(dimethylamino) borane, CA 1113-30-0: Irritant; air- and moisture-
sensitive. Tm = 180.95; Tb = 336.15 to 338.15; Antoine (liq) A = 6.25681,
B = 1279.603, C = -36.762, dev = 1.0, range = 222 to 332.

$C_4H_{12}Cl_2OSi_2$ MW 203.22

Disiloxane, 1,3-dichloro-1,1,3,3-tetramethyl, CA 2401-73-2: Irritant;
flammable. Tm = 236.15; Tb = 411.15; Vm = 0.1958 at 293.15; Antoine (liq)
A = 6.10241, B = 1410.62, C = -65.82, dev = 1.0, range = 303 to 403.

$C_4H_{12}Ge$ MW 132.73

Tetramethyl germane, CA 865-52-1: Moderately toxic; flammable. Tm =
185.15; Tb = 316.55; Vm = 0.1319 at 273.15; Antoine (liq) A = 5.95287, B =
1091.47, C = -41.0, dev = 0.1 to 1.0, range = 189 to 297.

$C_4H_{12}GeO$ MW 148.73

(3-Methoxypropyl) germane, CA 54832-77-8: Moderately toxic; flammable.
Tm = 175.45; Antoine (liq) A = 7.2975, B = 1583.7, C = 0, dev = 1.0 to 5.0,
range = 257 to 299.

Trimethylmethoxy germane, CA 6163-67-3: Moderately toxic; flammable. Tm =
170.95; Tb = 360.15 to 361.15; Vm = 0.1384 at 298.15; Antoine (liq) A =
4.97011, B = 796.825, C = -98.719, dev = 1.0, range = 273 to 335.

$C_4H_{12}GeO_4$ MW 196.73

Tetramethoxy germane: Moderately toxic; flammable. Tm = 255.15; Antoine
(liq) A = 6.9909, B = 2098.8, C = 0, dev = 1.0 to 5.0, range = 293 to 313.

$C_4H_{12}O_2Si$ MW 120.22

Diethoxy silane, CA 18165-68-9: Irritant; flammable. Tm = 150.15; Vm =
0.1391 at 273.15; Antoine (liq) A = 7.1255, B = 1810, C = 0, dev = 1.0 to
5.0, range = 241 to 341.

$C_4H_{12}O_4Si$ MW 152.22

Tetramethoxy silane, CA 681-84-5: Irritant; lachrymator; highly toxic;
flammable. Tm = 278.5; Tb = 393.15 to 394.15; Vm = 0.1478 at 295.15;
Antoine (liq) A = 7.4019, B = 2128.2, C = 0, dev = 1.0, range = 278 to 381.

$C_4H_{12}Pb$ MW 267.34

Tetramethyl lead, CA 75-74-1: Highly toxic; explosive; flash point 311.
Tm = 245.65; Tb = 383.15; Vm = 0.1340 at 293.15; Antoine (liq) A = 6.12969,
B = 1369.97, C = -50.762, dev = 0.1, range = 273 to 333.

$C_4H_{12}Sb_2$ MW 303.64

Tetramethyl distibine, CA 41422-43-9: Toxic; ignites in air. Tm = 289.15
to 290.15; Antoine (liq) A = 7.36977, B = 2573.42, C = -4.35, dev = 1.0,
range = 317 to 484.

$C_4H_{12}Si$ MW 88.22

Butyl silane, CA 1600-29-9: Extremely flammable; toxic. Tm = 134.95; Tb =
329.55; Vm = 0.1297 at 293.15; Antoine (liq) A = 5.4342, B = 887.09, C =
-72.06, dev = 0.1 to 1.0, range = 240 to 293.

Diethyl silane, CA 542-91-6: Irritant; extremely flammable; toxic. Tf =
138.76; Tb = 329.14; Vm = 0.1292 at 293.15; Antoine (liq) A = 5.65694, B =
980.57, C = -60.58, dev = 0.1 to 1.0, range = 233 to 313.

Isobutyl silane, CA 18165-87-2: Irritant; extremely flammable; toxic.
Tb = 322.65; Antoine (liq) A = 5.76143, B = 1009.68, C = -53.55, dev = 0.1
to 1.0, range = 233 to 293.

Tetramethyl silane, CA 75-76-3: Irritant; extremely flammable; flash point
below 246; toxic; moisture-sensitive. Tm = 174.13; Tb = 299.75; Vm =
0.1361 at 292.15; Antoine I (liq) A = 5.94729, B = 1033.724, C = -37.527,
dev = 0.1, range = 209 to 294; Antoine II (liq) A = 5.96432, B = 1009.86,
C = -45.08, dev = 1.0, range = 300 to 451; Antoine III (liq) A = 5.95719,
B = 1037.715, C = -37.164, dev = 0.1, range = 209 to 294; Antoine IV (liq)
A = 5.94009, B = 1035.448, C = -36.642, dev = 1.0, range = 290 to 375;
Antoine V (liq) A = 7.05239, B = 1941.853, C = 90.423, dev = 1.0, range =
373 to 449.

$C_4H_{12}Sn$ MW 178.83

Tetramethyl tin, CA 594-27-4: Highly toxic; highly flammable; flash point
below 261. Tm = 218.15; Tb = 351.15; Vm = 0.1386 at 298.65; Antoine I
(liq) A = 6.45361, B = 1483.83, C = -17.635, dev = 1.0, range = 222 to 351;
Antoine II (liq) A = 5.8628, B = 1137.076, C = -56.415, dev = 1.0, range =
273 to 353; Antoine III (liq) A = 6.16027, B = 1313.538, C = -34.958, dev =
1.0, range = 350 to 430; Antoine IV (liq) A = 6.88103, B = 1991.912, C =
62.792, dev = 1.0, range = 425 to 522.

$C_4H_{13}NSi$ MW 103.24

(Diethylamino) silane, CA 14660-24-3: Irritant; toxic; flammable. Tm =
124.15; Antoine (liq) A = 8.99164, B = 3061.595, C = 95.933, dev = 1.0,
range = 225 to 294.

$C_4H_{14}B_2$ MW 83.77

1,1-Diethyl diborane, CA 62133-33-9: Antoine (liq) A = 7.1799, B = 1760,
C = 0, dev = 1.0, range = 245 to 284.

1,1,2,2-Tetramethyl diborane, CA 21482-59-7: Tm = 200.65; Antoine (liq)
A = 6.63711, B = 1542.81, C = -8.671, dev = 1.0, range = 213 to 342.

$C_4H_{14}N_2Si$ MW 118.25

N,N,N',N'-Tetramethylsilane diamine, CA 4693-04-3: Irritant; toxic;
flammable. Tm = 169.15; Antoine (liq) A = 6.63801, B = 1704.914, C = 1.84,
dev = 1.0 to 5.0, range = 288 to 344.

$C_4H_{16}N_2Si_2$ MW 148.35

1,1-Bis(dimethylamino) disilane: Irritant; toxic; flammable. Tm = 139.15;
Antoine (liq) A = 5.88223, B = 1295.827, C = -70.626, dev = 1.0, range =
310 to 354.

C_4NiO_4 MW 170.74

Nickel tetracarbonyl, CA 13463-39-3: Extremely toxic; carcinogen; causes
dermatitis; may ignite in air; flash point below 269; explodes at about
333; soluble organic solvents. Tm = 248.15; Tb = 316.15; Tc = 473; Pc =
3040; Antoine I (sol) A = 9.3146, B = 1555.4, C = 0, dev = 1.0, range not
specified; Antoine II (liq) A = 7.0092, B = 1578, C = 0, dev = 1.0, range
not specified; Antoine III (liq) A = 6.9372, B = 1555.4, C = 0, dev = 5.0,
range = 277 to 315.

C_5FeO_5 MW 195.90

Iron pentacarbonyl, CA 13463-40-6: Highly toxic; pyrophoric in air; flash
point 258; insoluble water, soluble organic solvents. Tf = 253.15 to
253.65; Tb = 375.95; Vm = 0.1333 at 291.15; Tc = 558.15 to 561.15; Pc =
3000; Antoine I (liq) A = 5.9500, B = 1183.6, C = -78.15, dev = 1.0,
range = 313 to 364; Antoine II (liq) A = 7.3487, B = 2010.7, C = 0, dev =
1.0 to 5.0, range = 266 to 353; Antoine III (liq) A = 7.6208, B = 2096.7,
C = 0, dev = 1.0 to 5.0, range = 254 to 304.

$C_5H_3F_{10}NSe_2$ MW 424.99

N,N-Bis[(pentafluoroethyl)seleno] methylamine: Moderately toxic. Tm =
228.15; Antoine (liq) A = 6.965, B = 2000, C = 0, dev = 1.0 to 5.0, range =
282 to 324.

$C_5H_5F_7Se$ MW 277.04

Ethyl(heptafluoropropyl) selenide: Moderately toxic. Antoine (liq) A =
7.1319, B = 1880, C = 0, dev = 1.0 to 5.0, range = 243 to 333.

$C_5H_6F_7NSe$ MW 292.06

1,1,2,2,3,3,3-Heptafluoro-*N*,*N*-dimethyl-1-propaneselenamide: Moderately toxic. Tm = 179.15; Antoine (liq) A = 6.285, B = 1610, C = 0, dev = 1.0 to 5.0, range = 228 to 321.

C_5H_8Ge MW 140.71

2,4-Cyclopentadien-1-yl germane, CA 35682-28-1: Moderately toxic; flammable. Antoine (liq) A = 6.94039, B = 1853.199, C = 0, dev not specified, range = 283 to 305.

$C_5H_{10}Cl_2Si$ MW 169.13

Allyldichloroethyl silane: Irritant; flammable. Antoine (liq) A = 6.23104, B = 1593.69, C = -46.288, dev = 1.0, range = 270 to 424.

$C_5H_{10}Cl_3Ti$ MW 224.39

Cyclopentadienyl titanium chloride. Antoine (sol) A = 10.596, B = 4693, C = 0, dev = 5.0, range = 354 to 404.

$C_5H_{11}NSi$ MW 113.23

(Trimethylsilyl) acetonitrile, CA 18293-53-3: Irritant; flammable. Vm = 0.1364 at 293.15; Antoine (liq) A = 6.1306, B = 1536.3, C = -67.45, dev = 1.0, range = 286 to 358.

$C_5H_{12}Si$ MW 100.24

1,1-Dimethyl-1-silacyclobutane, CA 2295-12-7: Irritant; flammable. Tm = 155.15; Tb = 354.15 to 355.15; Vm = 0.130 at 293.15; Antoine (liq) A = 6.24475, B = 1368.951, C = -33.114, dev = 1.0, range = 281 to 344.

Trimethylvinyl silane, CA 754-05-2: Irritant; highly flammable; flash point 239. Tm = 141.65; Tb = 328.15; Vm = 0.154 at 293.15; Antoine (liq) A = 6.8749, B = 1580, C = 0, dev = 1.0, range = 289 to 315.

$C_5H_{13}NO_2Si$ MW 147.25

N-Methyl(trimethylsilyl) carbamate, CA 18147-09-6: Irritant; flammable. Antoine (liq) A = 7.9864, B = 2207.101, C = -43.852, dev = 1.0, range = 339 to 411.

$C_5H_{14}OSi$ MW 118.25

Ethoxytrimethyl silane, CA 1825-62-3: Irritant; flammable; flash point 255; soluble acetone, ether, ethanol. Tb = 348.15; Vm = 0.1562 at 293.15; Antoine (liq) A = 6.34183, B = 1375.44, C = -31.553, dev = 1.0, range = 222 to 349.

$C_5H_{14}O_2Si$ MW 134.25

Diethoxymethyl silane, CA 2031-62-1: Irritant; flammable. Tb = 372.55;
Vm = 0.1619 at 298.15; Antoine (liq) A = 7.1877, B = 1922.8, C = 0, dev =
5.0, range = 253 to 358.

$C_5H_{14}O_3Si$ MW 150.25

Ethyltrimethoxy silane, CA 5314-55-6: Irritant; flammable. Tb = 397.45;
Vm = 0.1584 at 293.15; Antoine (liq) A = 7.0009, B = 1968.1, C = 0, dev =
1.0, range = 286 to 394.

$C_5H_{14}Pb$ MW 281.37

Ethyltrimethyl lead, CA 1762-26-1: Highly toxic; flammable. Antoine (liq)
A = 6.4009, B = 1602.5, C = -43.15, dev = 1.0, range not specified.

$C_5H_{14}Si$ MW 102.25

Ethyltrimethyl silane, CA 3439-38-1: Irritant; flammable. Tb = 335.15;
Vm = 0.1504 at 293.15; Antoine (liq) A = 6.04566, B = 1191.18, C = -40.25,
dev = 1.0, range = 212 to 335.

$C_5H_{14}Sn$ MW 192.86

Ethyltrimethyl tin, CA 3531-44-0: Toxic; flammable. Tb = 377.15 to
378.65; Antoine I (liq) A = 7.1186, B = 1934.3, C = 0, dev = 1.0, range =
273 to 336; Antoine II (liq) A = 6.7879, B = 1822.7, C = 0, dev = 1.0,
range = 336 to 384.

$C_5H_{15}AsN_2$ MW 178.11

Methylbis(dimethylamino) arsine, CA 41813-33-6: Highly toxic; flammable.
Tm = 211.15; Antoine (liq) A = 6.8399, B = 2067.651, C = 0, dev = 1.0 to
5.0, range = 273 to 334.

$C_5H_{15}NbO_5$ MW 248.08

Pentamethylniobate: Moderately toxic; flammable. Antoine (liq) A = 6.265,
B = 3214, C = 0, dev = 5.0, range = 360 to 412.

$C_5H_{15}O_5Ta$ MW 336.12

Pentamethyltantalate: Moderately toxic; flammable. Antoine (liq) A =
8.065, B = 3634, C = 0, dev = 5.0, range = 390 to 422.

C_6CrO_6 MW 220.06

Chromium hexacarbonyl, CA 13007-92-6: Highly toxic; flammable; explodes
at 483; insoluble water, ethanol; soluble ether, carbon tetrachloride.
Tm = 425.15 to 428.15, decomposes; Antoine (sol) A = 10.4739, B = 3575.9,
C = 0, dev = 1.0 to 5.0, range = 288 to 423.

$C_6F_{14}Se$ MW 417.00

Bis(heptafluoropropyl) selenide, CA 755-81-7: Moderately toxic. Tm =
165.15; Antoine (liq) A = 6.835, B = 1803, C = 0, dev = 1.0 to 5.0, range =
228 to 343.

$C_6F_{14}Se_2$ MW 495.96

Bis(heptafluoropropyl) diselenide, CA 755-51-1: Moderately toxic. Tm =
222.15; Antoine (liq) A = 6.6609, B = 1967, C = 0, dev = 1.0 to 5.0,
range = 260 to 348.

C_6HO_6Re MW 355.27

Rhenium hydridehexacarbonyl complex: Toxic; flammable. Antoine (sol) A =
7.723, B = 2353.6, C = 0, dev = 5.0, range = 279 to 369.

$C_6H_3O_5Re$ MW 341.29

Rhenium methylpentacarbonyl complex, CA 14524-92-6: Toxic; flammable.
Tm = 393.15; Antoine (sol) A = 10.0606, B = 3406, C = 0, dev = 1.0, range =
315 to 380.

$C_6H_4Cl_4Si$ MW 246.00

Trichloro(2-chlorophenyl) silane, CA 2003-90-9: Highly irritant. Antoine
(liq) A = 7.06054, B = 2434.935, C = -28.903, dev = 1.0 to 5.0, range =
406 to 472.

Trichloro(3-chlorophenyl) silane, CA 2003-89-6: Highly irritant. Antoine
(liq) A = 5.9607, B = 1572.274, C = -105.968, dev = 1.0, range = 398 to 463.

$C_6H_5AsCl_2$ MW 222.93

Dichlorophenyl arsine, CA 696-28-6: Highly toxic; powerful vesicant;
flammable; soluble benzene, ether, ethanol. Tf = 257.55; Tb = 525.15 to
528.15; Vm = 0.1350 at 293.15; Antoine (liq) A = 7.88526, B = 3048.3,
C = 0, dev = 5.0, range = 273 to 333.

$C_6H_5Cl_3Si$ MW 211.55

Trichlorophenyl silane, CA 98-13-5: Corrosive; toxic; flammable; flash
point 364; moisture-sensitive; soluble chloroform. Tb = 474.15; Vm =
0.1601 at 293.15; Antoine (liq) A = 6.01734, B = 1584.31, C = -80.02, dev =
0.1 to 1.0, range = 333 to 453.

$C_6H_5F_3Si$ MW 162.19

Trifluorophenyl silane, CA 368-47-8: Irritant; flammable; soluble benzene,
ethanol. Tb = 374.15 to 375.15; Vm = 0.1350 at 299.65; Antoine (liq) A =
7.57187, B = 2224.2, C = 25.14, dev = 1.0, range = 263 to 375.

$C_6H_6AsCl_3$ MW 259.39

Tris(2-chlorovinyl) arsine, CA 40334-70-1: Highly toxic; flammable. Tm = 296.15; Vm = 0.165 at 293.15; Antoine (liq) A = 8.1844, B = 3295.0, C = 0, dev = 5.0, range = 409 to 533.

$C_6H_6Cl_2Si$ MW 177.10

Dichlorophenyl silane, CA 1631-84-1: Irritant; flammable. Vm = 0.1462 at 293.15; Antoine (liq) A = 7.06251, B = 2397.556, C = 0, dev = 1.0, range = 378 to 474.

C_6H_6Se MW 157.07

Benzene selenol, CA 645-96-5: Moderately toxic; stench; flammable; oxidizes in air; soluble ether, ethanol. Tb = 456.75; Vm = 0.1057 at 288.15; Antoine (liq) A = 7.185, B = 2370, C = 0, dev not specified, range = 331 to 458.

C_6H_8Si MW 108.21

Phenyl silane, CA 694-53-1: Irritant; flammable. Tb = 391.15 to 393.15; Vm = 0.124 at 293.15; Antoine (liq) A = 6.6209, B = 1814, C = 0, dev = 5.0, range = 279 to 393.

$C_6H_{10}Cl_2Si$ MW 181.14

Dichlorodiallyl silane, CA 3651-23-8: Irritant; flammable. Antoine (liq) A = 6.44025, B = 1759.15, C = -41.806, dev = 1.0, range = 282 to 439.

$C_6H_{11}NSi_2$ MW 153.33

2-Phenyl disilazane, CA 4459-07-8: Irritant; flammable. Tm = 204.15; Antoine (liq) A = 5.79189, B = 1763.681, C = 0, dev = 5.0, range = 297 to 356.

$C_6H_{12}BCl_3O_3$ MW 249.33

Tris(2-chloroethyl) borate, CA 22238-19-3: Irritant; flammable. Antoine (liq) A = 7.5393, B = 3012, C = 0, dev = 5.0, range = 390 to 448.

$C_6H_{12}CuN_2S_4$ MW 303.96

Bis(dimethyldithiocarbamato) copper complex, CA 137-29-1: Tm = 599.15; Antoine (sol) A = 13.1839, B = 7698, C = 0, dev = 1.0 to 5.0, range = 443 to 473.

$C_6H_{12}N_2NiS_4$ MW 299.11

Bis(dimethyldithiocarbamato) nickel complex, CA 15521-65-0: Tm = 595.15; Antoine (sol) A = 11.5979, B = 7307, C = 0, dev = 5.0, range = 448 to 478.

$C_6H_{14}Hg$ MW 286.77

Diisopropyl mercury, CA 1071-39-2: Highly toxic; flammable; sensitive to
oxygen; decomposes in light. Vm = 0.1434 at 293.15; Antoine (liq) A =
8.2722, B = 2748.4, C = 0, dev = 1.0, range = 298 to 333.

Dipropyl mercury, CA 628-85-3: Highly toxic; flammable; sensitive to oxygen; soluble ether, ethanol. Tb = 462.15 to 464.15; Vm = 0.1420 at 293.15;
Antoine (liq) A = 8.2855, B = 2803.5, C = 0, dev = 1.0, range = 308 to 345.

$C_6H_{14}Zn$ MW 151.56

Diisopropyl zinc, CA 625-81-0: Moderately toxic; pyrophoric; reacts with
water. Tb = 412.15; Antoine (liq) A = 7.1119, B = 1858, C = -43.15, dev =
1.0, range = 310 to 338.

Dipropyl zinc, CA 628-91-1: Moderately toxic; pyrophoric; reacts with
water. Tm = 192.15; Tb = 430.15; Vm = 0.1403 at 293.15; Antoine (liq) A =
6.832, B = 2067, C = 0, dev = 5.0, range = 313 to 376.

$C_6H_{15}Al$ MW 114.17

Triethyl aluminum, CA 97-93-8: Highly toxic; pyrophoric; flash point 210;
reacts violently with water or alcohols; dimerizes in liquid. Tm = 227.15;
Tb = 467.15; Vm = 0.1372 at 298.15; Antoine (liq) A = 9.2769, B = 3382,
C = 0, dev = 5.0, range = 330 to 399.

$C_6H_{15}AlO_3$ MW 162.16

Triethoxy aluminum, CA 555-75-9: Highly irritant; flammable; strongly
affected by traces of moisture; slightly soluble high-boiling solvents.
Tm = 413.15; Antoine (liq) A = 9.365, B = 4410, C = 0, dev = 5.0, range =
430 to 523.

$C_6H_{15}As$ MW 162.11

Triethyl arsine, CA 617-75-4: Highly toxic; flammable; fumes in air; suspected carcinogen; soluble acetone, ether, ethanol. Tb = 413.15; Vm =
0.1410 at 293.15; Antoine I (liq) A = 7.0009, B = 2070, C = 0, dev = 1.0
to 5.0, range = 265 to 423; Antoine II (liq) A = 6.2272, B = 1740, C = 0,
dev = 1.0, range = 265 to 423.

$C_6H_{15}B$ MW 97.99

Triethyl borane, CA 97-94-9: Toxic; pyrophoric; soluble ether, ethanol.
Tm = 180.65; Tb = 368.15; Vm = 0.140 at 296.15; Antoine I (liq) A = 4.77563,
B = 694.266, C = -121.276, dev = 1.0, range = 322 to 351; Antoine II (liq)
A = 6.49083, B = 1500.75, C = -33.126, dev = 0.1, range = 273 to 372.

$C_6H_{15}BO_3$ MW 145.99

Boric acid, triethyl ester: Moderately toxic; flammable; flash point 284;
moisture-sensitive; soluble ether, ethanol. Tm = 188.35; Tb = 391.75; Vm =
0.169 at 299.15; Antoine (liq) A = 6.63603, B = 1641.651, C = -36.867,
dev = 1.0, range = 302 to 382.

$C_6H_{15}Bi$ MW 296.16

Triethyl bismuth, CA 617-77-6: Toxic; foul odor; pyrophoric; explodes at
423. Vm = 0.1627 at 293.13; Antoine (liq) A = 7.077, B = 2293, C = 0,
dev = 5.0, range = 301 to 343.

$C_6H_{15}ClSi$ MW 150.72

Chlorotriethyl silane, CA 994-30-9: Irritant; corrosive; flammable; flash
point 302. Tb = 418.15 to 420.15; Vm = 0.1678 at 293.15; Antoine (liq) A =
6.27431, B = 1596.93, C = -45.413, dev = 1.0, range = 269 to 420.

$C_6H_{15}ClSn$ MW 241.33

Chlorotriethyl stannane, CA 994-31-0: Highly toxic; flammable. Tm =
288.65; Tb = 483.15; Vm = 0.1676 at 293.15; Antoine (liq) A = 5.93335, B =
1490.44, C = -105.71, dev = 0.1 to 1.0, range = 333 to 473.

$C_6H_{15}FO_3Si$ MW 182.27

Fluorotriethoxy silane, CA 358-60-1: Irritant; flammable. Tb = 407.75;
Antoine (liq) A = 7.1649, B = 2104, C = 0, dev = 1.0, range = 290 to 374.

$C_6H_{15}FSi$ MW 134.27

Fluorotriethyl silane, CA 358-43-0: Irritant; flammable; soluble petroleum
ether. Tb = 383.15 to 384.15; Vm = 0.1607 at 298.15; Antoine (liq) A =
6.10465, B = 1352.104, C = -53.415, dev = 0.1 to 1.0, range = 303 to 384.

$C_6H_{15}Ga$ MW 156.90

Triethyl gallium, CA 1115-99-7: Moderately toxic; ignites in air; soluble
organic solvents. Tm = 190.85; Tb = 415.75; Vm = 0.1483 at 303.15; Antoine
(liq) A = 6.8234, B = 2003.3, C = 0, dev = 5.0, range = 296 to 408.

$C_6H_{15}In$ MW 202.00

Triethyl indium, CA 923-34-2: Moderately toxic; ignites in air. Tm =
241.15; Tb = 457.15; Vm = 0.1603 at 293.15; Antoine (liq) A = 7.0049, B =
2340, C = 0, dev = over 5.0, range = 241 to 403.

$C_6H_{15}NO_2Si$ MW 161.28

N,N-Dimethyl(trimethylsilyl) carbamate, CA 32115-55-2: Irritant; flammable.
Antoine (liq) A = 9.62965, B = 4367.226, C = 149.718, dev = 1.0 to 5.0,
range = 292 to 411.

$C_6H_{15}O_3Sb$ MW 256.93

Triethoxy stibine, CA 2155-74-0: Moderately toxic; flammable. Antoine
(liq) A = 13.29862, B = 7161.17, C = 178.14, dev = 5.0, range = 346 to 457.

$C_6H_{15}O_4V$ MW 202.12

Vanadic acid, triethyl ester, CA 1686-22-2: Flammable. Antoine (liq) A = 3.28273, B = 399.732, C = -240.463, dev = 1.0, range = 336 to 367.

$C_6H_{16}O_2Si$ MW 148.28

Diethoxydimethyl silane, CA 78-62-6: Irritant; flammable; flash point 284; moisture-sensitive. Tm = 186.15; Tb = 387.15; Vm = 0.1724 at 293.15; Antoine (liq) A = 6.19317, B = 1349.05, C = -64.416, dev = 1.0, range = 254 to 387.

Diethyldimethoxy silane, CA 15164-57-5: Irritant; flammable. Antoine (liq) A = 7.0369, B = 2018.5, C = 0, dev = 1.0, range = 291 to 402.

$C_6H_{16}Si$ MW 116.28

Dipropyl silane, CA 871-77-2: Irritant; flammable. Tb = 383.65; Antoine (liq) A = 6.8629, B = 1867, C = 0, dev = 5.0, range = 276 to 385.

Triethyl silane, CA 617-86-7: Irritant; flammable; flash point 270. Tf = 116.25; Tb = 381.92; Vm = 0.1593 at 293.15; Antoine (liq) A = 5.50165, B = 1038.197, C = -86.104, dev = 0.1, range = 300 to 373.

Trimethylpropyl silane, CA 3510-70-1: Irritant; flammable. Tb = 363.15; Vm = 0.166 at 293.15; Antoine (liq) A = 6.1507, B = 1383.79, C = -29.375, dev = 1.0, range = 227 to 363.

$C_6H_{16}Si_2$ MW 144.36

1,1,3,3-Tetramethyl-1,3-disilacyclobutane, CA 1627-98-1: Irritant; flammable. Tm = 266.15; Tb = 396.15; Antoine (liq) A = 7.13105, B = 2035.669, C = 5.127, dev = 1.0, range = 290 to 364.

$C_6H_{16}Sn$ MW 206.88

Triethyltin hydride, CA 997-50-2: Toxic; flammable; easily oxidized. Tb = 421.15 to 423.15; Antoine (liq) A = 7.515, B = 2273, C = 0, dev = 5.0, range = 308 to 415.

Trimethylpropyl tin, CA 3531-45-1: Toxic; flammable. Antoine (liq) A = 6.02512, B = 1387.9, C = -59.624, dev = 1.0, range = 265 to 405.

$C_6H_{18}Cl_2O_2Si_3$ MW 277.37

1,5-Dichloro-1,1,3,3,5,5-hexamethyl trisiloxane, CA 3582-71-6: Irritant; flammable. Tm = 220.15; Tb = 457.15; Vm = 0.2725 at 293.15; Antoine (liq) A = 6.61061, B = 1884.61, C = -47.766, dev = 1.0, range = 299 to 457.

$C_6H_{18}Ge_2O$ MW 251.39

Hexamethyl digermane, CA 2237-93-6: Moderately toxic; flammable. Tm = 212.05; Tb = 410.15; Antoine (liq) A = 5.34962, B = 1003.897, C = -109.614, dev = 1.0, range = 291 to 346.

$C_6H_{18}OSi_2$ MW 162.38

Hexamethyl disiloxane, CA 107-46-0: Irritant; flammable; flash point 271. Tm = 214.15; Tb = 374.15; Vm = 0.2137 at 293.15; Antoine I (liq) A = 5.89514, B = 1200.259, C = -65.073, dev = 0.02, range = 309 to 412; Antoine II (liq) A = 6.47357, B = 1654.111, C = 0, dev = 1.0, range = 491 to 519.

$C_6H_{18}O_3Si_3$ MW 222.46

Hexamethyl cyclotrisiloxane, CA 541-05-9: Irritant; flammable; flash point 308; moisture-sensitive. Tm = 337.15; Tb = 407.15; Antoine I (sol) A = 9.4596, B = 2883.6, C = 0, dev = 1.0 to 5.0, range = 297 to 336; Antoine II (liq) A = 7.6773, B = 2519.91, C = 37.45, dev = 1.0, range = 353 to 403.

2,4,6-Triethyl cyclotrisiloxane, CA 18442-02-9: Irritant; flammable. Antoine (liq) A = 5.93192, B = 1486.596, C = -109.756, dev = 1.0, range = 333 to 453.

$C_6H_{18}SSi_2$ MW 178.44

Hexamethyl disilathiane, CA 3385-94-2: Irritant; very toxic; flammable. Tb = 436.15; Vm = 0.2097 at 293.15; Antoine (liq) A = 6.32455, B = 1975.437, C = 10.132, dev = 1.0 to 5.0, range = 302 to 375.

$C_6H_{18}Si_2$ MW 146.38

Hexamethyl disilane, CA 1450-14-2: Irritant; toxic; flammable; flash point 272. Tm = 260.15; Tb = 385.15 to 386.15; Vm = 0.2016 at 293.15; Antoine I (liq) A = 7.6752, B = 2105.37, C = 0, dev = 1.0, range = 274 to 287; Antoine II (liq) A = 7.1354, B = 1953.12, C = 0, dev = 1.0, range = 288 to 323.

$C_6H_{19}B_5Si_2$ MW 201.44

2,4-Bis(dimethylsilyl)-2,4-dicarbaheptaborane, CA 59351-11-0: Irritant. Tm = 193.15; Antoine (liq) A = 6.38922, B = 1915.763, C = -24.006, dev = 1.0 to 5.0, range = 343 to 438.

$C_6H_{24}O_6Si_6$ MW 360.76

2,4,6,8,10,12-Hexamethyl cyclohexasiloxane, CA 6166-87-6: Irritant; flammable. Tm = 194.15; Vm = 0.359 at 293.15; Antoine (liq) A = 7.3842, B = 2550, C = 0, dev not specified, range = 320 to 378.

C_6MoO_6 MW 264.00

Molybdenum hexacarbonyl, CA 13939-06-5: Highly toxic; flammable; soluble organic solvents. Tm = 423.15, decomposes; Antoine (sol) A = 10.4188, B = 3607.93, C = 0, dev = 1.0, range = 316 to 423.

C_6O_6W MW 351.91

Tungsten hexacarbonyl, CA 14040-11-0: Highly toxic; flammable; slightly soluble hexane. Tm = 423.15, decomposes; Antoine (sol) A = 10.6629, B = 3886.394, C = 0, dev = 1.0, range = 333 to 433.

506

$C_7H_5MnO_5$ MW 224.05

Ethylpentacarbonyl manganese complex, CA 15694-83-4: Toxic; flammable.
Tm = 331.15; Antoine (liq) A = 15.5175, B = 5017.368, C = 0, dev = 1.0 to
5.0, range = 332 to 351.

$C_7H_7IrO_4$ MW 347.35

Dicarbonyl-2,4-pentanedionato iridium complex, CA 14023-80-4: Toxic; flammable. Antoine (sol) A = 11.48672, B = 4873.609, C = 0, dev = 1.0 to 5.0,
range = 286 to 325.

$C_7H_7O_4Rh$ MW 258.04

Dicarbonyl-2,4-pentanedionato rhodium complex, CA 14874-82-9: Toxic; flammable. Tm = 428.15; Antoine (sol) A = 11.64023, B = 4543.849, C = 0, dev =
1.0 to 5.0, range = 276 to 301.

$C_7H_8Cl_2Si$ MW 191.13

Dichlorobenzyl silane, CA 18173-99-4: Irritant; flammable. Antoine (liq)
A = 7.01253, B = 2033.07, C = -61.405, dev = 1.0, range = 318 to 468.

Dichloromethylphenyl silane, CA 149-74-6: Corrosive; flammable; flash
point 355; moisture-sensitive. Tb = 478.15; Vm = 0.162 at 293.15; Antoine
I (liq) A = 6.35814, B = 1844.84, C = -54.83, dev = 1.0, range = 308 to
479; Antoine II (liq) A = 6.31956, B = 1781.798, C = -64.071, dev = 1.0,
range = 323 to 463.

Dichloro-4-tolyl silane, CA 13272-80-5: Irritant; flammable. Antoine (liq)
A = 7.07273, B = 2094.47, C = -56.211, dev = 1.0, range = 319 to 470.

$C_7H_8F_2Si$ MW 158.22

Difluoromethylphenyl silane, CA 328-57-4: Irritant; flammable. Tb =
415.15; Vm = 0.1432 at 293.15; Antoine (liq) A = 6.26082, B = 1507.411, C =
-62.055, dev = 1.0, range = 303 to 413.

C_7H_8Se MW 171.10

Methylphenyl selenide, CA 4346-64-9: Flammable. Tb = 475.15 to 476.15;
Vm = 0.1227 at 298.15; Antoine (liq) A = 7.875, B = 2740, C = 0, dev = 1.0
to 5.0, range = 273 to 291.

$C_7H_9F_9N_2OSSi$ MW 368.29

Methanesulfonimidamide, 1,1,1-trifluoro-N-[2,2,2-trifluoro-1-
(trifluoromethyl)ethylidene]-N'-(trimethylsilyl), CA 62609-67-0: Antoine
(liq) A = 6.825, B = 2058, C = 0, dev = 1.0 to 5.0, range not specified.

$C_7H_{13}Cl_2IrO_2$ MW 392.30

Bis(chloroethylene)-2,4-pentanedionato iridium complex: Antoine (sol) A =
10.56835, B = 4739.252, C = 0, dev = 1.0 to 5.0, range = 281 to 298.

$C_7H_{13}Cl_2O_2Rh$ MW 302.99

Bis(chloroethylene)-2,4-pentanedionato rhodium complex: Antoine (sol) A = 16.31407, B = 6242.719, C = 0, dev = 1.0 to 5.0, range = 274 to 288.

$C_7H_{15}B_3F_3N_3$ MW 230.64

1,2,3,4,5-Pentamethyl-6-(trifluorovinyl) borazine, CA 20453-68-3: Irritant; flammable. Tm = 280.15; Antoine (liq) A = 3.14324, B = 958.261, C = 0, dev = 5.0, range = 280 to 324.

$C_7H_{17}ClSi$ MW 164.75

Methyldiethyl(1-chloroethyl) silane: Irritant; flammable. Antoine (liq) A = 6.9121, B = 2184.2, C = 0, dev = 1.0, range = 353 to 445.

$C_7H_{17}NOSi_2$ MW 187.39

(Pentamethyldisiloxanyl)methyl cyanide: Irritant; flammable. Antoine (liq) A = 7.6694, B = 2641.4, C = 0, dev = 1.0, range = 348 to 401.

$C_7H_{18}OSi$ MW 146.30

Trimethylbutoxy silane, CA 1825-65-6: Irritant; flammable. Tb = 398.15; Vm = 0.188 at 293.15; Antoine (liq) A = 7.0148, B = 1977.6, C = -3.05, dev = 1.0, range = 344 to 398.

$C_7H_{18}O_3Si$ MW 178.30

Methyltriethoxy silane, CA 2031-67-6: Irritant; flammable; flash point 296; moisture-sensitive. Tb = 417.65; Vm = 0.1992 at 293.15; Antoine (liq) A = 6.3406, B = 1560.95, C = -56.574, dev = 1.0, range = 271 to 417.

$C_7H_{18}Si_2$ MW 130.30

Butyltrimethyl silane, CA 1000-49-3: Irritant; flammable. Tb = 388.15; Vm = 0.181 at 293.15; Antoine (liq) A = 6.2960, B = 1476.78, C = -43.872, dev = 1.0, range = 249 to 388.

Methyltriethyl silane, CA 757-21-1: Irritant; flammable. Tb = 400.15; Vm = 0.176 at 293.15; Antoine (liq) A = 6.39578, B = 1615.72, C = -32.189, dev = 1.0, range = 255 to 400.

$C_7H_{20}Si_2$ MW 160.41

Methylenebis(trimethylsilane), CA 2117-28-4: Irritant; flammable. Tm = 202.15; Tb = 405.15; Vm = 0.213 at 293.15; Antoine (liq) A = 7.1771, B = 2102.6, C = 0, dev = 1.0, range = 323 to 407.

$C_8Co_2O_8$ MW 341.95

Octacarbonyl dicobalt, CA 10210-68-1: Highly toxic; air-sensitive; insoluble water; soluble organic solvents. Tm = 324.15 to 325.15; *(continues)*

$C_8Co_2O_8$ *(continued)*

Antoine I (sol) A = 16.7252, B = 5420.6, C = 0, dev = 1.0, range = 288 to 315; Antoine II (liq) A = 9.474, B = 3416, C = 0, dev = 1.0 to 5.0, range = 287 to 298.

$C_8H_6N_2Se$ MW 209.11

4-Phenyl-1,2,3-selenadiazole, CA 25660-64-4: Antoine (sol) A = 11.4698, B = 4735.6, C = 0, dev not specified, range = 327 to 345.

$C_8H_{10}Cl_2OSi$ MW 221.16

Dichloroethoxyphenyl silane, CA 18236-80-1: Irritant; flammable. Antoine (liq) A = 6.48117, B = 1927.38, C = -64.751, dev = 1.0, range = 325 to 496.

$C_8H_{10}Cl_2Si$ MW 205.16

Dichloroethylphenyl silane, CA 1125-27-5: Irritant; flammable. Antoine (liq) A = 6.41670, B = 2032.21, C = -42.712, dev = 1.0, range = 321 to 503.

$C_8H_{10}Pd$ MW 212.57

Cyclopentadienylallyl palladium, CA 1271-03-0: Soluble petroleum ether. Tm = 336.15 to 336.65; Antoine (sol) A = 7.43808, B = 2604.719, C = 0, dev = 5.0, range = 291 to 333.

$C_8H_{11}B$ MW 117.98

Dimethylphenyl borane, CA 54098-94-1: Antoine (liq) A = 7.3749, B = 2270, C = 0, dev = 1.0 to 5.0, range = 293 to 323.

$C_8H_{11}BN_4$ MW 174.01

4,5-Dihydro-1,4-dimethyl-5-phenyl-1*H*-tetrazaborole: Antoine (liq) A = 6.615, B = 2480, C = 0, dev = 1.0 to 5.0, range = 339 to 383.

$C_6H_{11}ClSi$ MW 170.71

Chlorodimethylphenyl silane, CA 768-33-2: Corrosive; moisture-sensitive; flammable; flash point 334. Tb = 467.15 to 470.15; Vm = 0.1657 at 293.15; Antoine (liq) A = 6.26422, B = 1703.69, C = -66.476, dev = 1.0, range = 302 to 467.

$C_8H_{11}FSi$ MW 154.26

Fluorodimethylphenyl silane, CA 454-57-9: Irritant; flammable. Tb = 435.15 to 436.15; Antoine (liq) A = 5.99628, B = 1402.788, C = -84.006, dev = 1.0, range = 303 to 423.

$C_8H_{12}Si$ MW 136.27

Dimethylphenyl silane, CA 766-77-8: Irritant; flammable; flash point 308.
Tb = 430.15; Vm = 0.1533 at 293.15; Antoine (liq) A = 6.55091, B = 1819.27,
C = -32.278, dev = 1.0, range = 278 to 433.

$C_8H_{13}NSi$ MW 151.28

1,1-Dimethyl-1-phenyl silylamine, CA 60755-66-0: Irritant; flammable.
Antoine (liq) A = 16.220, B = 14465.43, C = 461.83, dev = 1.0, range = 384
to 503.

$C_8H_{14}Pt$ MW 305.29

Trimethyl-cyclopentadienyl platinum complex, CA 1271-07-4: Soluble
methanol. Tm = 338.15; Antoine (liq) A = 6.5049, B = 2540, C = 0, dev =
5.0, range not specified.

$C_8H_{16}Cl_4O_4Si$ MW 346.11

Tetrakis(2-chloroethoxy) silane, CA 18290-84-1: Irritant; flammable.
Antoine (liq) A = 9.2883, B = 4237, C = 0, dev = 1.0, range = 447 to 500.

$C_8H_{18}Zn$ MW 179.61

Dibutyl zinc, CA 1119-90-0: Flammable; easily oxidized and hydrolyzed.
Tm = 215.45; Antoine (liq) A = 7.0819, B = 2366, C = 0, dev = 5.0, range =
377 to 444.

$C_8H_{19}NO_2Si$ MW 189.33

N,N-Diethyl(trimethylsilyl) carbamate, CA 18279-61-3: Irritant; flammable.
Antoine (liq) A = 7.03076, B = 2001.338, C = -46.53, dev = 1.0 to 5.0,
range = 326 to 443.

$C_8H_{20}Cl_2OSi_2$ MW 259.32

1,3-Dichloro-1,1,3,3-tetraethyl disiloxane, CA 18825-03-1: Irritant; flam=
mable. Antoine (liq) A = 6.22612, B = 1842.18, C = -67.63, dev = 0.1 to
1.0, range = 343 to 463.

$C_8H_{20}Ge$ MW 188.84

Tetraethyl germane, CA 597-63-7: Moderately toxic; flammable; decomposed
by water. Tm = 183.15; Tb = 438.65; Vm = 0.1575 at 293.15; Antoine (liq)
A = 5.69014, B = 1253.9, C = -97.34, dev = 1.0, range = 337 to 438.

Trimethylpentyl germane, CA 57596-76-6: Moderately toxic; flammable.
Antoine (liq) A = 6.28789, B = 1587.463, C = - 54.214, dev = 1.0, range =
303 to 423.

$C_8H_{20}GeO_2$ MW 220.83

Dimethyldiisopropoxy germane, CA 5314-29-4: Moderately toxic; flammable. Antoine (liq) A = 5.6278, B = 1212.723, C = -98.897, dev = 1.0, range = 303 to 423.

$C_8H_{20}GeO_4$ MW 252.83

Tetraethoxy germane: Moderately toxic; flammable. Tm = 201.15; Antoine (liq) A = 6.8874, B = 2250.0, C = 0, dev = 5.0, range = 293 to 413.

$C_8H_{20}O_4Si$ MW 208.33

Tetraethoxy silane, CA 78-10-4: Moderately toxic; irritant; flammable; flash point 319; insoluble water; soluble ethanol. Tm = 199.15; Tb = 439.15; Vm = 0.223 at 293.15; Antoine (liq) A = 6.005, B = 1770, C = 0, dev = 5.0, range = 275 to 442.

$C_8H_{20}Pb$ MW 323.45

Tetraethyl lead, CA 78-00-2: Highly toxic; flammable; flash point 367; insoluble water; soluble benzene, petroleum ether. Tb = 471.15 to 475.15; Vm = 0.1957 at 293.15; Antoine (liq) A = 6.2586, B = 1721.7, C = -69.75, dev = 5.0, range = 273 to 413.

$C_8H_{20}Si$ MW 144.33

Tetraethyl silane, CA 631-36-7: Irritant; flammable. Tm = 190.15; Tb = 425.95 to 426.95; Vm = 0.1884 at 293.15; Antoine (liq) A = 6.59221, B = 1848.66, C = -23.202, dev = 1.0, range = 272 to 426.

$C_8H_{20}Sn$ MW 234.94

Tetraethyl tin, CA 597-64-8: Highly toxic; flammable; flash point 326. Tm = 137.15 to 148.15; Tb = 448.15; Vm = 0.196 at 298.15; Antoine (liq) A = 7.369, B = 2448, C = 0, dev = 5.0, range = 329 to 453.

$C_8H_{22}B_2$ MW 139.88

1,1,2,2-Tetraethyl diborane, CA 12081-54-8: Moderately toxic. Tm = 217.15; Antoine (liq) A = 12.93234, B = 3872.961, C = 0, dev = 5.0, range = 263 to 283.

$C_8H_{24}Cl_2O_3Si_4$ MW 351.52

1,7-Dichloro-1,1,3,3,5,5,7,7-octamethyl tetrasiloxane, CA 2474-02-4: Irritant; flammable. Tm = 211.15; Antoine (liq) A = 7.10018, B = 2387.91, C = -26.469, dev = 1.0, range = 326 to 495.

$C_8H_{24}N_4Si$ MW 204.39

Octamethylsilane tetramine, CA 1624-01-7: Irritant; flammable. Tm = 271.15; Tb = 453.15; Antoine (liq) A = 6.22888, B = 1996.563, C = 0, dev = 5.0, range = 364 to 415.

$C_8H_{24}O_2Si_3$ MW 236.53

Octamethyl trisiloxane, CA 107-51-7: Irritant; moisture-sensitive; flammable; flash point 302; soluble benzene, hydrocarbons. Tf = 193.15; Tb = 426.15 to 427.15; Vm = 0.2885 at 293.15; Antoine (liq) A = 5.5579, B = 1134.3, C = -106.8, dev = 1.0, range = 345 to 428.

$C_8H_{24}O_4Si_4$ MW 296.62

Octamethyl cyclotetrasiloxane, CA 556-67-2: Irritant; moisture-sensitive; flammable; flash point 333. Tm = 290.65; Tb = 448.15; Vm = 0.309 at 293.15; Antoine (liq) A = 5.2866, B = 1002.1, C = -143.95, dev = 1.0 to 5.0, range = 378 to 446.

$C_8H_{24}O_{12}Si_8$ MW 536.95

Octamethyldodecaoxoocta silicon, CA 57348-79-5: Tm = approximately 623; Antoine (sol) A = 11.115, B = 5770, C = 0, dev = 5.0, range = 463 to 563.

$C_9Fe_2O_9$ MW 363.79

Iron nonacarbonyl, CA 15321-51-4: Highly toxic; flammable; flash point 308; insoluble organic solvents. Tm = 373.15 to 393.15, decomposes; Antoine (sol) A = 17.825, B = 7070, C = 0, dev = 5.0, range = 296 to 314.

$C_9H_{15}IrO_2$ MW 347.44

Bis(ethylene)-2,4-pentanedionato iridium complex, CA 52654-27-0: Antoine (sol) A = 9.93924, B = 4379.618, C = 0, dev = 5.0, range = 283 to 311.

$C_9H_{15}O_2Rh$ MW 258.12

Bis(ethylene)-2,4-pentanedionato rhodium complex, CA 12082-47-2: Soluble ether. Tm = 417.15 to 419.15, decomposes; Antoine (sol) A = 12.49893, B = 4998.383, C = 0, dev = 1.0 to 5.0, range = 282 to 301.

$C_9H_{20}O_3Si$ MW 204.34

Allyltriethoxy silane, CA 2550-04-1: Irritant; moisture-sensitive; flammable; flash point 294. Vm = 0.2263 at 293.15; Antoine (liq) A = 7.46665, B = 2434.569, C = -1.534, dev = 1.0, range = 333 to 447.

$C_9H_{21}AlO_3$ MW 204.24

Triisopropoxy aluminum, CA 555-31-7: Moisture-sensitive; flammable; decomposed by water; soluble organic solvents. Antoine I (liq) A = 4.645, B = 2030, C = 0, dev = 5.0, range = 353 to 373; Antoine II (liq) A = 6.695, B = 2820, C = 0, dev = 5.0, range = 378 to 399; Antoine III (liq) A = 11.155, B = 4690, C = 0, dev = 1.0 to 5.0, range not specified.

Tripropoxy aluminum, CA 4073-85-2: Moisture-sensitive; flammable. Antoine (liq) A = 9.665, B = 4880, C = 0, dev = 5.0, range = 475 to 540.

$C_9H_{21}BO_3$ MW 188.07

Triisopropoxy boron, CA 5419-55-6: Moisture-sensitive; flammable; flash
point 290; soluble organic solvents. Tb = 412.15 to 413.15; Vm = 0.231 at
301.15; Antoine (liq) A = 7.19475, B = 2120.204, C = -4.025, dev = 1.0,
range = 358 to 412.

Tripropoxy boron, CA 688-71-1: Moisture-sensitive; flammable; flash point
305; soluble organic solvents. Tb = 450.15; Vm = 0.2192 at 293.15; Antoine
(liq) A = 6.52473, B = 1741.1, C = -66.738, dev = 1.0, range = 373 to 452.

$C_9H_{21}BS_3$ MW 236.25

Trithioboric acid, tripropyl ester, CA 998-38-9: Antoine (liq) A = 9.445,
B = 3980, C = 0, dev = 5.0, range = 423 to 483.

$C_9H_{21}ClO_3Si$ MW 240.80

Triethoxy(3-chloropropyl) silane, CA 5089-70-3: Irritant; flammable. Vm =
0.241 at 298.15; Antoine (liq) A = 6.45366, B = 183.12; C = -83.31, dev =
1.0, range = 363 to 463.

$C_9H_{21}In$ MW 244.08

Triisopropyl indium, CA 17144-80-8: Highly flammable. Antoine (liq) A =
7.5779, B = 2665.8, C = 0, dev = 1.0, range = 394 to 478.

Tripropyl indium, CA 3015-98-3: Highly flammable; burns spontaneously in
air; soluble organic solvents. Tb = 451.15; Vm = 0.206 at 293.15; Antoine
(liq) A = 7.6069, B = 2716.8, C = 0, dev = 1.0 to 5.0, range = 400 to 483.

$C_9H_{21}O_4V$ MW 244.20

Vanadic acid, triisopropyl ester, CA 5588-84-1: Flammable. Antoine (liq)
A = 4.25412, B = 718.277, C = -202.9, dev = 1.0, range = 342 to 377.

Vanadic acid, tripropyl ester, CA 1686-23-3: Flammable. Antoine (liq) A =
1.9848, B = 173.22, C = -308.475, dev = 1.0 to 5.0, range = 367 to 402.

$C_9H_{24}Ge_3$ MW 350.06

1,1,3,3,5,5-Hexamethyl-1,3,5-trigermacyclohexane, CA 1077-32-3: Moderately
toxic; flammable. Antoine (liq) A = 5.89344, B = 1610.867, C = -97.98,
dev = 1.0, range = 334 to 453.

$C_{10}H_2F_{12}O_6U$ MW 684.13

Bis(1,1,1,5,5,5-hexafluoro-2,4-pentanedionato) uranium dioxide complex,
CA 67316-66-9: Tm = 473.15; Antoine (sol) A = 16.5, B = 7680, C = 0,
dev = over 5.0, range = 370 to 425.

$C_{10}H_{10}Cl_2Hf$ MW 379.58

Bis(cyclopentadienyl) hafnium dichloride, CA 12116-66-4: Antoine (sol) A =
10.556, B = 5241, C = 0, dev = 1.0 to 5.0, range = 394 to 447.

$C_{10}H_{10}Cl_2Ti$ MW 249.00

Bis(cyclopentadienyl) titanium dichloride, CA 1271-19-8: Soluble aromatics.
Tm = 562.15; Antoine (sol) A = 12.345, B = 6500, C = 0, dev = 5.0, range =
418 to 533.

$C_{10}H_{10}Cl_2Zr$ MW 292.32

Bis(cyclopentadienyl) zirconium chloride, CA 1291-32-3: Tm = 521.15;
Antoine (sol) A = 10.386, B = 5241, C = 0, dev = 5.0, range = 393 to 457.

$C_{10}H_{10}Fe$ MW 186.04

Ferrocene, CA 102-54-5: Toxic; insoluble water; soluble ethanol, benzene.
Tm = 445.65 to 446.15; Tb = 522.15; Antoine I (sol) A = 9.395, B = 3680,
C = 0, dev = 5.0, range not specified; Antoine II (sol) A = 8.7119, B =
3377, C = 0, dev = 1.0 to 5.0, range = 348 to 446; Antoine III (liq) A =
6.740, B = 2470, C = 0, dev = 5.0, range = 456 to 523; Antoine IV (liq)
A = 7.0149, B = 2601, C = 0, dev = 1,0 to 5.0, range = 451 to 523.

$C_{10}H_{10}Mn$ MW 185.13

Bis(cyclopentadienyl) manganese complex, CA 1271-27-8: Flammable; can ex-
plode in air; soluble benzene.   Tm = 446.15; Antoine I (sol) A = 9.705,
B = 3780, C = 0, dev = 5.0, range = 298 to 445; Antoine II (liq) A = 7.055,
B = 2615, C = 0, dev = 1.0 to 5.0, range = 446 to 525.

$C_{10}H_{10}V$ MW 181.13

Bis(cyclopentadienyl) vanadium complex, CA 1277-47-0: Soluble benzene.
Tm = 440.15 to 441.15; Antoine (sol) A = 5.925, B = 3000, C = 0, dev = 1.0,
range = 323 to 338.

$C_{10}H_{11}NSi_4$ MW 257.55

1,1,3,3-Tetramethyl-1,3-bis(trimethylsilyl) disilazane: Irritant; flammable. Antoine (liq) A = 7.65988, B = 2649.302, C = -25.47, dev = 1.0,
range = 378 to 435.

$C_{10}H_{14}O_5V$ MW 265.16

Bis(2,4-pentanedionato) vanadium oxide complex, CA 3153-26-2: Antoine
(sol) A = 9.13, B = 4782, C = 0, dev = 5.0, range = 418 to 443.

$C_{10}H_{16}AsNO_3$ MW 273.16

Diethyl arsanilate: Toxic; flammable. Antoine (liq) A = 7.95011, B =
2628.85, C = -11.878, dev = 1.0, range = 311 to 454.

$C_{10}H_{20}CdN_2S_4$ MW 408.93

Bis(diethyldithiocarbamate) cadmium complex, CA 14239-68-0: Tm = 526.15;
Antoine (sol) A = 11.8179, B = 6960, C = 0, dev = 1.0 to 5.0, range = 433
to 469.

$C_{10}H_{20}CoN_2S_4$ MW 355.45

Bis(diethyldithiocarbamate) cobalt complex, CA 15974-34-2: Tm = 552.15; Antoine (sol) A = 15.6029, B = 9273, C = 0, dev = 5.0, range = 458 to 482.

$C_{10}H_{20}CuN_2S_4$ MW 360.07

Bis(diethyldithiocarbamate) copper complex, CA 13681-87-3: Tm = 475.15; Antoine (sol) A = 14.3529, B = 7789, C = 0, dev = 1.0 to 5.0, range = 420 to 465.

$C_{10}H_{20}HgN_2S_4$ MW 497.11

Bis(diethyldithiocarbamate) mercury complex, CA 14239-51-1: Tm = 403.15; Antoine (sol) A = 1.9909, B = 2488, C = 0, dev = 5.0, range = 378 to 403.

$C_{10}H_{20}N_2NiS_4$ MW 355.22

Bis(diethyldithiocarbamate) nickel complex, CA 14267-17-5: Tm = 509.15; Antoine (sol) A = 14.1029, B = 7940, C = 0, dev = 5.0, range = 440 to 478.

$C_{10}H_{20}N_2PbS_4$ MW 503.72

Bis(diethyldithiocarbamate) lead complex, CA 17549-30-3: Tm = 483.15; Antoine (sol) A = 11.5549, B = 6785, C = 0, dev = 5.0, range = 444 to 482.

$C_{10}H_{20}N_2S_4Zn$ MW 361.90

Bis(diethyldithiocarbamate) zinc complex, CA 14324-55-1: Tm = 421.15; Antoine (sol) A = 14.3899, B = 7477, C = 0, dev = 1.0 to 5.0, range = 401 to 444.

$C_{10}H_{20}O_2Si$ MW 200.35

Diallyldiethoxy silane, CA 13081-67-9: Irritant; flammable. Antoine (liq) A = 7.55434, B = 2563.286, C = 2.817, dev = 1.0, range = 342 to 459.

$C_{10}H_{25}NO_2Si_3$ MW 275.57

1,1,1,3,5,5,5-Heptamethyl-3-(2-cyanoethyl) trisiloxane: Irritant; flammable. Antoine (liq) A = 5.78279, B = 1512.128, C = -115.356, dev = 5.0, range = 367 to 511.

$C_{10}H_{25}NO_4Si_4$ MW 335.65

2,2,4,4,6,6,8-Heptamethyl-7-(2-cyanoethyl) cyclotetrasiloxane, CA 6506-66-7: Irritant; flammable. Antoine (liq) A = 6.16169, B = 1827.766, C = -93.185, dev = 1.0 to 5.0, range = 396 to 518.

$C_{10}H_{25}NbO_5$ MW 318.21

Pentaethylniobate: Antoine (liq) A = 12.305, B = 5619, C = 0, dev = 5.0, range = 376 to 414.

$C_{10}H_{25}O_5Ta$ MW 406.25

Pentaethyltantalate: Antoine (liq) A = 8.065, B = 3794, C = 0, dev = 1.0
to 5.0, range = 388 to 424.

$C_{10}H_{28}O_4Si_3$ MW 296.59

1,5-Diethoxy-1,1,3,3,5,5-hexamethyl trisiloxane, CA 17928-13-1: Irritant;
flammable. Tm = 147.15; Antoine (liq) A = 6.7901, B = 1960.15, C =
-60.014, dev = 1.0, range = 314 to 470.

$C_{10}H_{30}O_3Si_4$ MW 310.69

Decamethyl tetrasiloxane, CA 141-62-8: Irritant; moisture-sensitive; flam-
mable, flash point 335; soluble benzene, light hydrocarbons. Tf = 203.15;
Tb = 467.15; Vm = 0.364 at 293.15; Antoine (liq) A = 7.3972, B = 2514.9,
C = 0, dev = 1.0, range = 343 to 454.

$C_{10}H_{30}O_5Si_5$ MW 370.77

Decamethyl cyclopentasiloxane, CA 541-02-6: Irritant; flammable. Tm =
235.15; Tb = 483.15; Vm = 0.3865 at 293.15; Antoine (liq) A = 7.3116, B =
2561, C = 0, dev = 1.0, range = 364 to 472.

$C_{10}O_{10}Re_2$ MW 652.50

Dirhenium decacarbonyl, CA 14285-68-8: Toxic; flammable; insoluble water;
soluble organic solvents. Tm = 450.15; Antoine (sol) A = 9.6559, B =
4054.6, C = 0, dev = 1.0, range = 363 to 450.

$C_{11}H_{19}IrO_2$ MW 375.49

Bis(propylene)-2,4-pentanedionato iridium complex, CA 66467-05-8: Antoine
(sol) A = 12.17591, B = 4762.405, C = 0, dev = 1.0 to 5.0, range = 269 to 304.

$C_{11}H_{19}O_2Rh$ MW 286.18

Bis(propylene)-2,4-pentanedionato rhodium complex, CA 12282-38-1: Antoine
(sol) A = 11.47447, B = 4451.175, C = 0, dev = 1.0 to 5.0, range = 270
to 296.

$C_{11}H_{20}OSi$ MW 196.36

Triallylethoxy silane, CA 17962-20-8: Irritant; flammable. Antoine (liq)
A = 7.31439, B = 2504.853, C = -1.603, dev = 1.0, range = 349 to 473.

$C_{11}H_{20}OSi_2$ MW 224.45

Pentamethylphenyl disiloxane, CA 14920-92-4: Irritant; flammable. Antoine
(liq) A = 6.01859, B = 1576.981, C = -89.459, dev = 1.0, range = 347 to 474.

$C_{11}H_{28}O_4Si_4$ MW 336.68

8,8,10,10,12,12-Hexamethyl-7,9,11,13-tetraoxa-6,8,10,12-tetrasilaspiro
[5,7] tridecane, CA 35331-58-9: Antoine (liq) A = 7.87488, B = 3400.656,
C = 68.943, dev = 1.0 to 5.0, range = 393 to 504.

516

$C_{12}H_9Cl_3Si$ MW 287.65

2-(Trichlorosilyl) biphenyl, CA 18030-62-1: Irritant; flammable. Tm = 338.04; Antoine (liq) A = 8.1349, B = 3505, C = 0, dev = 1.0 to 5.0, range = 461 to 552.

4-(Trichlorosilyl) biphenyl, CA 18030-61-0: Irritant; flammable. Tm = 371.9; Antoine (liq) A = 8.6619, B = 3956, C = 0, dev = 1.0 to 5.0, range = 479 to 573.

$C_{12}H_{10}Cl_2Si$ MW 253.20

Dichlorodiphenyl silane, CA 80-10-4: Corrosive; highly irritant; flammable; flash point 430; soluble acetone, benzene, ether, ethanol. Tm = 251.15; Tb = 578.15; Vm = 0.211 at 293.15; Antoine (liq) A = 6.0797, B = 1884, C = -115.15, dev = 1.0, range = 465 to 555.

$C_{12}H_{10}F_2Si$ MW 220.29

Difluorodiphenyl silane, CA 312-40-3: Irritant; flammable. Tb = 525.15; Vm = 0.1902 at 293.15; Antoine (liq) A = 8.33689, B = 3702.5, C = 74.08, dev = 1.0, range = 392 to 516.

$C_{12}H_{10}Se$ MW 233.17

Diphenyl selenide, CA 1132-39-4: Irritant; flammable; soluble benzene, ether, ethanol. Tm = 275.65; Tb = 574.15 to 575.15; Vm = 0.1762 at 298.15; Antoine (liq) A = 7.13563, B = 2811.79, C = -26.545, dev = 1.0, range = 378 to 575.

$C_{12}H_{12}Cr$ MW 208.22

Dibenzene chromium, CA 1271-54-1: Air-sensitive; soluble benzene; slightly soluble ether. Tm = 557.15 to 558.15; Antoine (sol) A = 6.4155, B = 2443.89, C = -93.297, dev = 0.1, range = 323 to 363.

$C_{12}H_{17}O_4V$ MW 276.21

Vanadic acid, tributyl ester: Flammable. Antoine (liq) A = 3.90871, B = 679.918, C = -253.675, dev = 1.0, range = 395 to 435.

Vanadic acid, triisobutyl ester: Flammable. Antoine (liq) A = 4.55962, B = 905.95, C = -214.75, dev = 1.0, range = 383 to 418.

Vanadic acid, tri-*sec*-butyl ester: Flammable. Antoine (liq) A = 3.44618, B = 520.381, C = -255.602, dev = 1.0, range = 378 to 413.

Vanadic acid, tri-*tert*-butyl ester: Flammable. Antoine (liq) A = 6.89726, B = 1910.567, C = -102.915, dev = 1.0, range = 348 to 385.

$C_{12}H_{18}Be_4O_{13}$ MW 406.32

Beryllium, hexakis(aceto)-oxotetra, CA 19049-40-2: Carcinogenic; insoluble water; soluble chloroform, light hydrocarbons. Tm = 559.15; Tb = 603.15 to 604.15, decomposes; Antoine (sol) A = 11.9019, B = 6025.2, C = 0, dev = 1.0 to 5.0, range = 390 to 451.

$C_{12}H_{21}B$ MW 176.11

Dodecahydro-9-boraphenalene, CA 16664-33-8: Irritant; sensitive to air.
Tm = 304.15; Vm = 0.189 at 308.15; Antoine (liq) A = 7.36912, B =
2776.316, C = 0, dev = 5.0, range = 304 to 404.

$C_{12}H_{26}F_6O_4Si_4$ MW 460.67

2,2,4,4,6,8-Hexamethyl-6,8-bis(3,3,3-trifluoropropyl) cyclotetrasiloxane,
CA 15445-52-0: Irritant; flammable. Antoine (liq) A = 5.34443, B =
1175.225, C = -156.968, dev = 1.0, range = 381 to 455.

$C_{12}H_{26}N_2O_4Si_4$ MW 374.69

2,2,4,4,6,8-Hexamethyl-6,8-bis(2-cyanoethyl) cyclotetrasiloxane,
CA 6500-74-9: Irritant; flammable. Antoine (liq) A = 4.38245, B =
940.96, C = -247.146, dev = 5.0, range = 454 to 581.

$C_{12}H_{27}AlO_3$ MW 246.32

Tributoxy aluminum, CA 3085-30-1: Flammable. Antoine (liq) A = 10.035,
B = 5440, C = 0, dev = 5.0, range = 503 to 533.

Triisobutoxy aluminum, CA 3453-79-0: Flammable. Antoine (liq) A = 14.115,
B = 7280, C = 0, dev = 5.0, range = 500 to 550.

Tri-*sec*-butoxy aluminum, CA 2269-22-9: Corrosive; flammable. Antoine
(liq) A = 9.495, B = 4260, C = 0, dev = 5.0, range = 425 to 469.

$C_{12}H_{27}Ga$ MW 241.07

Tributyl gallium, CA 15677-44-8: Flammable; sensitive to air, light, and
water; soluble organic solvents. Antoine (liq) A = 7.7959, B = 2934,
C = 0, dev = 1.0 to 5.0, range = 426 to 507.

Triisobutyl gallium, CA 17150-84-4: Flammable; sensitive to air, light,
and water; soluble organic solvents. Vm = 0.256 at 293.15; Antoine (liq)
A = 7.625, B = 2724, C = 0, dev = 1.0 to 5.0, range not specified.

$C_{12}H_{27}In$ MW 286.17

Tributyl indium, CA 15676-66-1: Flammable. Antoine (liq) A = 7.6939, B =
3055.1, C = 0, dev = 1.0 to 5.0, range = 444 to 539.

Triisobutyl indium, CA 6731-23-3: Flammable. Antoine (liq) A = 7.935,
B = 3009, C = 0, dev = 1.0 to 5.0, range not specified.

$C_{12}H_{28}Ge$ MW 244.94

Tetrapropyl germane, CA 994-65-0: Flammable. Tm = 200.15; Tb = 498.15;
Antoine (liq) A = 7.6974, B = 2831.8, C = 0, dev = 1,0 to 5.0, range = 353
to 493.

518

$C_{12}H_{28}GeO_4$ MW 308.94

Tetraisopropoxy germane: Flammable. Antoine (liq) A = 5.74866, B = 1395.909, C = -109.61, dev = 5.0, range = 313 to 453.

Tetrapropoxy germane: Flammable. Antoine (liq) A = 7.22292, B = 2410.485, C = -52.311, dev = 5.0, range = 343 to 453.

$C_{12}H_{28}O_4Si$ MW 264.44

Tetrapropoxy silane, CA 682-01-9: Irritant; moisture-sensitive; flammable; flash point 368. Tb = 498.15 to 500.15; Vm = 0.289 at 293.15; Antoine (liq) A = 6.27592, B = 1738.5, C = -94.58, dev = 1.0, range = 307 to 563.

$C_{12}H_{28}O_4Ti$ MW 284.25

Tetraisopropyl titanate: Flammable; decomposes with water; soluble ether, benzene, chloroform. Tm = approximately 293.15; Tb = 493.15; Vm = 0.293 at 293.15; Antoine (liq) A = 6.13278, B = 1650.086, C = -100.8, dev = 1.0, range = 336 to 459.

Tetrapropyl titanate: Flammable; flash point below 295; decomposes with water. Antoine (liq) A = 5.4116, B = 972.759, C = -251.729, dev = 1.0, range = 411 to 479.

$C_{12}H_{28}Sn$ MW 291.04

Tetrapropyl tin, CA 2176-98-9: Toxic; flammable. Tm = 164.15; Tb = 501.15; Antoine (liq) A = 7.330, B = 2745, C = 0, dev = 5.0, range = 361 to 470.

$C_{12}H_{30}O_3Si_3$ MW 306.62

Hexaethyl cyclotrisiloxane, CA 2031-79-0: Irritant; flammable. Tm = 283.05; Tb = 521.15 to 523.15; Vm = 0.321 at 293.15; Antoine I (liq) A = 5.50807, B = 1276, C = -159.15, dev = 1.0, range = 434 to 516; Antoine II (liq) A = 6.21013, B = 1844.509, C = -87.529, dev = 5.0, range = 385 to 524.

$C_{12}H_{36}O_4Si_5$ MW 384.84

Dodecamethyl pentasiloxane, CA 141-63-9: Irritant; flammable; soluble benzene, light hydrocarbons. Tf = 193.15; Vm = 0.4396 at 293.15; Antoine (liq) A = 7.212, B = 2627.8, C = 0, dev = 1.0, range = 389 to 498.

$C_{12}H_{36}O_6Si_6$ MW 444.93

Dodecamethyl cyclohexasiloxane, CA 540-97-6: Irritant; flammable. Tm = 270.15; Tb = 518.15; Vm = 0.460 at 293.15; Antoine (liq) A = 6.50199, B = 1914.72, C = -83.153, dev = 1.0 to 5.0, range = 340 to 509.

$C_{13}H_{10}AsN$ MW 255.15

Diphenylarsine carbonitrile, CA 23525-22-6: Highly irritant; suspected carcinogen; flammable. Tm = 304.65; Antoine (liq) A = 9.8489, B = 4420, C = 0, dev = 5.0, range = 296 to 326.

$C_{13}H_{26}O_2Si_3$ MW 298.60

1,1,1,3,5,5,5-Heptamethyl-3-phenyl trisiloxane, CA 546-44-1: Irritant;
flammable. Antoine (liq) A = 5.97179, B = 1579.244, C = -111.05, dev = 1.0
to 5.0, range = 367 to 492.

$C_{13}H_{26}O_4Si_4$ MW 358.69

2,4,4,6,6,8,8-Heptamethyl-2-phenyl cyclotetrasiloxane, CA 10448-09-6:
Irritant; flammable. Antoine (liq) A = 5.29123, B = 1182.093, C =
-169.714, dev = 1.0 to 5.0, range = 397 to 514.

$C_{14}H_{28}CuN_2S_4$ MW 416.17

Bis(diisopropyldithiocarbamate) copper complex, CA 14354-07-5: Tm =
588.15; Antoine (sol) A = 11.4039, B = 6767, C = 0, dev = 5.0, range = 440
to 465.

Bis(dipropyldithiocarbamate) copper complex, CA 14354-08-6: Tm = 375.15;
Antoine (liq) A = 10.6069, B = 6187, C = 0, dev = 5.0, range = 422 to 453.

$C_{14}H_{28}N_2NiS_4$ MW 411.33

Bis(diisopropyldithiocarbamate) nickel complex, CA 15694-55-0: Tm =
602.15; Antoine (sol) A = 12.2279, B = 7492, C = 0, dev = 5.0, range = 442
to 477.

Bis(dipropyldithiocarbamate) nickel complex, CA 14516-30-4: Tm = 406.15;
Antoine (liq) A = 11.2159, B = 6586, C = 0, dev = 1.0 to 5.0, range = 433
to 462.

$C_{14}H_{42}O_5Si_6$ MW 459.00

Tetradecamethyl hexasiloxane, CA 107-52-8: Irritant; flammable; flash
point 391.5; soluble benzene, light hydrocarbons. Tm = 214.15; Tb =
518.65; Vm = 0.515 at 293.15; Antoine (liq) A = 7.5871, B = 2959.3, C = 0,
dev = 5.0, range = 397 to 522.

$C_{14}H_{42}O_7Si_7$ MW 519.08

Tetradecamethyl cycloheptasiloxane, CA 107-50-6: Irritant; flammable.
Tm = 241.15; Vm = 0.535 at 293.15; Antoine (liq) A = 7.64235, B = 3010.3,
C = -3.245, dev = 1.0, range = 359 to 537.

$C_{16}H_{12}Ge$ MW 276.86

Diethynyldiphenyl germane, CA 1675-59-8: Flammable. Tm = 319.94; Antoine
(liq) A = 14.51091, B = 5785.913, C = 0, dev = 1.0, range = 305 to 337.

$C_{16}H_{26}O_3Si_3$ MW 350.64

2,2,4,4-Tetramethyl-6,6-diphenyl cyclotrisiloxane: Irritant; flammable.
Antoine (liq) A = 7.3363, B = 3181.6, C = 0, dev not specified, range not
specified.

$C_{16}H_{32}O_4Si_4$ MW 400.77

6,12,18,24-Tetraoxa-5,7,13,19-tetrasilatetraspiro-[4,1,4,1,4,1,4,1]
tetracosane, CA 177-49-1: Irritant; flammable. Antoine (liq) A = 8.41055,
B = 4314.464, C = 49.017, dev = 1.0 to 5.0, range = 452 to 583.

$C_{16}H_{36}O_4Si$ MW 320.54

Tetrabutoxy silane, CA 4766-57-8: Irritant; flammable; flash point 351.
Tb = 541.15; Vm = 0.356 at 293.15; Antoine (liq) A = 6.43084, B = 1957.63,
C = -108.91, dev = 1.0, range = 333 to 479.

$C_{16}H_{36}O_4Ti$ MW 340.36

Tetrabutoxy titanium: Irritant; moisture-sensitive; flammable; flash point
349. Antoine I (liq) A = 6.6083, B = 1750.9, C = -185.11, dev = 5.0,
range = 462 to 564; Antoine II (liq) A = 9.8649, B = 4440, C = 0, dev = 1.0
to 5.0, range = 443 to 493.

Tetraisobutoxy titanium: Irritant; moisture-sensitive; flammable. Tm =
approximately 313.15; Antoine (liq) A = 7.42719, B = 2379.08, C = -105.01,
dev = 1.0, range = 436 to 529.

Tetra-sec-butoxy titanium: Irritant; moisture-sensitive; flammable.
Antoine I (liq) A = 5.06119, B = 1142.1, C = -182.86, dev = 1.0, range =
378 to 414; Antoine II (liq) A = 6.4182, B = 2034.354, C = -91.605, dev =
1.0, range = 370 to 476.

Tetra-tert-butoxy titanium: Irritant; moisture-sensitive; flammable.
Antoine (liq) A = 8.6636, B = 2918, C = 0, dev = 5.0, range = 386 to 486.

$C_{16}H_{36}O_4Zr$ MW 383.68

Tetra-tert-butoxy zirconium: Flammable. Antoine (liq) A = 5.93715, B =
1583.239, C = -104.141, dev = 1.0, range = 374 to 587.

$C_{16}H_{40}O_4Si_4$ MW 408.83

Octaethyl cyclotetrasiloxane, CA 1451-99-6: Irritant; flammable. Tm =
209.15; Vm = 0.424 at 293.15; Antoine (liq) A = 5.80189, B = 1651.292, C =
-140.786, dev = 1.0 to 5.0, range = 420 to 574.

$C_{16}H_{46}O_7Si_6$ MW 519.05

1,11-Diethoxy-1,1,3,3,5,5,7,7,9,9,11,11-dodecamethyl hexasiloxane,
CA 18143-15-2: Irritant; flammable. Antoine (liq) A = 7.85002, B =
3043.51, C = -25.985, dev = 1.0, range = 376 to 547.

$C_{16}H_{48}O_6Si_7$ MW 533.15

Hexadecamethyl heptasiloxane, CA 541-01-5: Irritant; flammable; soluble
benzene, light hydrocarbons. Tm = 195.15; Tb = 543.15; Vm = 0.592 at
293.15; Antoine (liq) A = 7.719, B = 3176.9, C = 0, dev = 1.0 to 5.0,
range = 443 to 551.

$C_{16}H_{48}O_8Si_8$ MW 593.23

Hexadecamethyl cyclooctasiloxane, CA 556-68-3: Irritant; flammable. Tm = 241.65; Tb = 563.15; Vm = 0.504 at 293.15; Antoine (liq) A = 6.86729, B = 2424.8, C = -64.386, dev = 1.0 to 5.0, range = 376 to 563.

$C_{18}H_{15}As$ MW 306.24

Triphenyl arsine, CA 603-32-7: Highly toxic; suspected carcinogen; flammable; soluble ethanol, benzene, chloroform. Tm = 333.65; Tb = 633.15; Vm = 0.242 at 291.15; Antoine (liq) A = 8.0789, B = 3953, C = 0, dev = 1.0, range = 493 to 563.

$C_{18}H_{15}B$ MW 242.13

Triphenyl borane, CA 960-71-4: Flammable; soluble ethanol, benzene. Tm = 415.15; Antoine (liq) A = 7.4049, B = 3360, C = 0, dev = 5.0, range = 423 to 568.

$C_{18}H_{15}Sb$ MW 353.07

Triphenyl antimony, CA 603-36-1: Irritant; toxic; flammable; flash point over 385; soluble benzene, light hydrocarbons. Tm = 326.15; Tb = 650.15; Vm = 0.2462 at 298.15; Antoine (liq) A = 8.7029, B = 4350, C = 0, dev = 1.0 to 5.0, range = 503 to 553.

$C_{18}H_{28}O_4Si_4$ MW 420.76

2,2,4,4,6,6-Hexamethyl-8,8-diphenyl cyclotetrasiloxane, CA 1693-44-3: Irritant; flammable. Antoine (liq) A = 7.4218, B = 3260, C = 0, dev not specified, range not specified.

2,2,4,4,6,8-Hexamethyl-6,8-diphenyl cyclotetrasiloxane, CA 18604-02-9: Irritant; flammable. Vm = 0.395 at 298.15; Antoine (liq) A = 6.27344, B = 2040.292, C = -121.151, dev = 1.0, range = 459 to 576.

$C_{18}H_{36}CuN_2S_4$ MW 472.28

Bis(dibutyldithiocarbamate) copper complex, CA 13927-71-4: Flammable. Tm = 347.15; Antoine (liq) A = 10.3949, B = 6360, C = 0, dev = 1.0 to 5.0, range = 423 to 468.

Bis(diisobutyldithiocarbamate) copper complex, CA 51205-55-1: Flammable. Tm = 420.15; Antoine (liq) A = 8.6429, B = 5317, C = 0, dev = 5.0, range = 425 to 445.

$C_{18}H_{36}N_2NiS_4$ MW 467.44

Bis(dibutyldithiocarbamate) nickel complex, CA 13927-77-0: Tm = 360.15; Antoine (liq) A = 11.8599, B = 7134, C = 0, dev = 1.0 to 5.0, range = 438 to 562.

Bis(diisobutyldithiocarbamate) nickel complex, CA 28371-07-5: Tm = 449.15; Antoine I (sol) A = 14.3429, B = 7945, C = 0, dev = 1.0 to 5.0, range = 423 to 443; Antoine II (liq) A = 11.0659, B = 6476, C = 0, dev = 1.0 to 5.0, range = 453 to 473.

$C_{18}H_{54}O_7Si_8$ MW 607.30

Octadecamethyl octasiloxane, CA 556-69-4: Irritant; flammable; soluble
benzene, light hydrocarbons. Tm = 210.15; Vm = 0.665 at 293.15; Antoine
(liq) A = 6.91793, B = 2440.2, C = -66.436, dev = 1.0 to 5.0, range = 378
to 563.

$C_{18}H_{54}O_9Si_9$ MW 667.39

Octadecamethyl cyclononasiloxane, CA 556-71-8: Irritant; flammable;
soluble benzene, light hydrocarbons. Antoine (liq) A = 8.0786, B = 3547.8,
C = 0, dev not specified, range = 463 to 584.

$C_{20}H_{34}O_5Si_5$ MW 494.91

Cyclopentasiloxane, 2,2,4,4,6,6,8,8-octamethyl-10,10-diphenyl,
CA 13438-48-7: Irritant; flammable. Antoine (liq) A = 6.7315, B = 3023.5,
C = 0, dev not specified, range not specified.

$C_{20}H_{44}HfO_4$ MW 527.06

Tetra-*tert*-pentoxy hafnium: Antoine (liq) A = 5.88343, B = 1732.9, C =
-123.05, dev = 1.0, range = 346 to 478.

$C_{20}H_{44}O_4Si$ MW 376.65

Tetra(1-ethylpropoxy) silane: Irritant; flammable. Antoine (liq) A =
3.26406, B = 543.11, C = -253.5, dev = 1.0, range = 371 to 427.

$C_{20}H_{44}O_4Ti$ MW 396.47

Tetra(1,1-dimethylpropoxy) titanium: Flammable. Antoine (liq) A = 8.585,
B = 3640, C = 0, dev = 5.0, range = 397 to 430.

Tetra(1-ethylpropoxy) titanium: Flammable. Antoine (liq) A = 12.4028, B =
5410, C = 0, dev = 5.0, range = 385 to 445.

Tetra(3-methylbutoxy) titanium: Flammable. Antoine (liq) A = 6.06459, B =
1564.925, C = -210.52, dev = 1.0, range = 407 to 493.

Tetrapentoxy titanium: Flammable. Antoine I (liq) A = 11.3882, B = 5401,
C = 0, dev = 5.0, range = 484 to 558; Antoine II (liq) A = 5.58487, B =
1396.177, C = -229.9, dev = 1.0, range = 416 to 506.

Tetra-*tert*-pentoxy titanium: Flammable. Antoine (liq) A = 5.93017, B =
1870.498, C = -109.006, dev = 1.0, range = 361 to 423.

$C_{20}H_{44}O_4Zr$ MW 439.79

Tetra(1,1-dimethylpropoxy) zirconium: Flammable. Antoine (liq) A = 8.465,
B = 3550, C = 0, dev = 5.0, range = 392 to 426.

Tetrapentoxy zirconium: Flammable. Antoine (liq) A = 24.58, B = 14529,
C = 0, dev = 5.0, range = 529 to 549.   *(continues)*

$C_{20}H_{44}O_4Zr$ *(continued)*

Tetra-*tert*-pentoxy zirconium: Flammable. Antoine (liq) A = 5.41361, B = 1501.6, C = -141.44, dev = 1.0, range = 361 to 435.

$C_{20}H_{58}O_9Si_8$ MW 667.36

1,15-Diethoxy-1,1,3,3,5,5,7,7,9,9,11,11,13,13,15,15-hexadecamethyl octasiloxane: Irritant; flammable. Tm = 163.15; Antoine (liq) A = 7.19885, B = 2566.35, C = -90.368, dev = 1.0 to 5.0, range = 406 to 585.

$C_{20}H_{60}O_8Si_9$ MW 681.46

Eicosamethyl nonasiloxane, CA 2652-13-3: Irritant; flammable; soluble benzene, light hydrocarbons. Tb = 580.65; Vm = 0.743 at 293.15; Antoine (liq) A = 8.37952, B = 3314.78, C = -60.684, dev = 1.0 to 5.0, range = 417 to 581.

$C_{20}H_{60}O_{10}Si_{10}$ MW 741.54

Eicosamethyl cyclodecasiloxane, CA 18772-36-6: Irritant; flammable. Antoine (liq) A = 8.1809, B = 3725.51, C = 0, dev not specified, range = 480 to 603.

$C_{21}H_{22}Si$ MW 302.49

Tribenzyl silane, CA 1747-92-8: Irritant; flammable; soluble light hydrocarbons. Tm = 364.65 to 365.15; Antoine (liq) A = 8.7349, B = 4280, C = 0, dev = 5.0, range = 460 to 637.

$C_{21}H_{24}OSi_2$ MW 348.59

1,1,3-Trimethyl-1,3,3-triphenyl disiloxane, CA 14920-93-5: Irritant; flammable. Antoine (liq) A = 5.71714, B = 1701.088, C = -184.053, dev = 1.0 to 5.0, range = 494 to 624.

$C_{22}H_{44}CuN_2S_4$ MW 528.39

Bis[bis(3-methylbutyl)dithiocarbamate] copper complex, CA 69090-74-0: Tm = 337.15; Antoine (liq) A = 13.2009, B = 7785, C = 0, dev = 1.0 to 5.0, range = 427 to 458.

$C_{22}H_{44}N_2NiS_4$ MW 523.54

Bis[bis(3-methylbutyl)dithiocarbamate] nickel complex, CA 55935-69-8: Tm = 361.15; Antoine (liq) A = 14.9709, B = 8590, C = 0, dev = 5.0, range = 429 to 468.

$C_{22}H_{66}O_{11}Si_{11}$ MW 815.70

Docosamethyl cycloundecasiloxane, CA 18766-38-6: Irritant; flammable. Antoine (liq) A = 8.2805, B = 3893.57, C = 0, dev not specified, range = 496 to 620.

$C_{23}H_{30}O_2Si_3$ MW 422.75

1,1,1,3,5-Pentamethyl-3,5,5-triphenyl trisiloxane, CA 67102-99-2: Irritant; flammable. Antoine (liq) A = 6.41728, B = 2679.275, C = -76.484, dev = 1.0 to 5.0, range = 521 to 678.

$C_{24}H_{20}Ge$ MW 381.01

Tetraphenyl germane, CA 1048-05-1: Soluble aromatics. Tm = 510.15 to 511.15; Tb = over 673.15; Antoine (sol) A = 15.09238, B = 7763.713, C = 0, dev = 5.0, range = 402 to 480.

$C_{24}H_{20}Pb$ MW 515.62

Tetraphenyl lead, CA 595-89-1: Highly toxic; soluble aromatics. Antoine (sol) A = 14.71603, B = 7886.298, C = 0, dev = 5.0, range = 412 to 480.

$C_{24}H_{20}Si$ MW 336.51

Tetraphenyl silane, CA 1048-08-4: Flammable; soluble ethanol, carbon disulfide. Tm = 510.65; Tb = 803.15; Antoine (sol) A = 4.28904, B = 2675.3, C = 0, dev = 1.0, range = 428 to 484.

$C_{24}H_{20}Sn$ MW 427.11

Tetraphenyl tin, CA 595-90-4: Highly toxic; soluble chloroform; slightly soluble ethanol, ether. Tm = 498.85; Tb = over 693.15; Antoine (sol) A = 14.8217, B = 7927, C = 0, dev = 1.0 to 5.0, range = 393 to 461.

$C_{24}H_{22}N_2Si$ MW 366.54

N,N,N',N'-Tetraphenyl silane diamine, CA 22519-45-5: Tm = 360.65; Antoine (liq) A = 6.98427, B = 3087.391, C = 0, dev = over 5.0, range = 410 to 473.

$C_{24}H_{52}O_4Si$ MW 432.76

Tetrahexyloxy silane, CA 7425-86-7: Flammable. Antoine (liq) A = 9.455, B = 4545, C = 0, dev = 5.0, range = 454 to 573.

$C_{24}H_{52}O_4Ti$ MW 452.57

Tetra(1,1-dimethylbutoxy) titanium: Antoine (liq) A = 11.04645, B = 4941.6, C = 0, dev = 1.0 to 5.0, range = 414 to 454.

Tetrahexyloxy titanium: Antoine (liq) A = 9.7784, B = 4954, C = 0, dev = 5.0, range = 520 to 581.

Tetra(1-methyl-1-ethylpropoxy) titanium: Antoine (liq) A = 9.97942, B = 4505.5, C = 0, dev = 1.0, range = 412 to 460.

$C_{24}H_{52}O_4Zr$ MW 495.89

Tetra(1,1-dimethylbutoxy) zirconium: Antoine (liq) A = 11.0228, B = 4875.4, C = 0, dev = 1.0 to 5.0, range = 406 to 449.   *(continues)*

$C_{24}H_{52}O_4Zr$ *(continued)*

Tetra(1-methyl-1-ethylpropoxy) zirconium: Antoine (liq) A = 10.53891, B = 4776.6, C = 0, dev = 1.0 to 5.0, range = 423 to 460.

$C_{24}H_{72}O_{12}Si_{12}$ MW 889.85

Tetracosamethyl cyclododecasiloxane, CA 18919-94-3: Antoine (liq) A = 8.2976, B = 4003.58, C = 0, dev not specified, range = 508 to 636.

$C_{26}H_{26}OSi_2$ MW 410.66

1,3-Dimethyl-1,1,3,3-tetraphenyl disiloxane, CA 807-28-3: Antoine (liq) A = 4.98416, B = 1377.074, C = -249.408, dev = 5.0, range = 518 to 616.

$C_{26}H_{26}O_3Si_3$ MW 470.75

2,2-Dimethyl-4,4,6,6-tetraphenyl cyclotrisiloxane, CA 1438-86-4: Antoine (liq) A = 4.7749, B = 2276.8, C = 0, dev not specified, range not specified.

$C_{28}H_{32}O_2Si_3$ MW 484.82

1,1,1,3-Tetramethyl-3,5,5,5-tetraphenyl trisiloxane, CA 67103-00-8: Antoine (liq) A = 6.9099, B = 2969.22, C = -95.958, dev = 1.0 to 5.0, range = 549 to 678.

1,1,3,5-Tetramethyl-1,3,5,5-tetraphenyl trisiloxane, CA 67142-05-6: Antoine (liq) A = 5.95795, B = 2004.977, C = -203.332, dev = 1.0 to 5.0, range = 566 to 666.

1,3,3,5-Tetramethyl-1,1,5,5-tetraphenyl trisiloxane, CA 3982-82-9: Antoine (liq) A = 5.86949, B = 2053.327, C = -185.935, dev = 1.0, range = 544 to 686.

$C_{28}H_{32}O_4Si_4$ MW 544.90

2,2,4,4-Tetramethyl-6,6,8,8-tetraphenyl cyclotetrasiloxane, CA 1693-47-6: Antoine (liq) A = 6.9321, B = 3633.4, C = 0, dev not specified, range not specified.

2,2,6,6-Tetramethyl-4,4,8,8-tetraphenyl cyclotetrasiloxane, CA 1693-48-7: Antoine (liq) A = 7.0043, B = 3628.1, C = 0, dev not specified, range not specified.

2,4,6,8-Tetramethyl-2,4,6,8-tetraphenyl cyclotetrasiloxane, CA 77-63-4: Soluble methanol. Four isomers, Tm varies from 328.15 to 373.15; Antoine (liq) A = 7.125, B = 3756.5, C = 0, dev not specified, range not specified.

$C_{30}H_{28}CuN_2S_4$ MW 608.35

Bis(dibenzyldithiocarbamate) copper complex, CA 34409-33-1: Tm = 477.15; Antoine (sol) A = 6.6129, B = 5582, C = 0, dev = 1.0 to 5.0, range = 453 to 478.

$C_{33}H_{34}O_2Si_3$ `   MW 546.89

1,1,3-Trimethyl-1,3,5,5,5-pentaphenyl trisiloxane, CA 67103-01-9: Antoine (liq) A = 7.87036, B = 4158.207, C = -40.983, dev = 5.0, range = 603 to 711.

1,3,5-Trimethyl-1,1,3,5,5-pentaphenyl trisiloxane, CA 3390-61-2: Antoine (liq) A = 7.54372, B = 3954.601, C = -40.429, dev = 1.0, range = 575 to 625.

Vapor-Liquid Critical Constants of Fluids

The critical temperature Tc, critical pressure Pc, and critical molar volume Vc are the values of the temperature, pressure, and molar volume at which the densities of the liquid and gaseous phases become identical. The critical density ρc = M/Vc, where M is the molar mass. At temperature above the critical temperature a gas cannot be liquified. With the appropriate conditions of temperature, pressure, and volume, the critical point may be identified by visual observation of the disappearance and reappearance of the meniscus between liquid and vapor.

Compilations of critical constants have been published: Kobe and Lynn;[1] Kudchadker, Alani, and Zwolinski;[2] and for inorganic compounds, Matthews[4] (these papers include reviews of the methods of measurement). At temperature greater than 750°K the determination becomes progressively more difficult because indirect methods have to be used to identify the critical point; no values are given here for substances with critical temperatures greater than 2000°K. Ohse and Tippelskirsch have reviewed the results obtained in this range by estimation and special experimental methods.[5] Correlations useful for the estimation of the critical properties of more volatile, primarily organic, compounds are to be found in the book by Reid, Prausnitz, and Sherwood.[6]

This table is based on a survey of the literature to the end of 1986, including the reviews just quoted, and lists those elements and compounds for which there are observed values of the critical properties. Where applicable, standard refrigerant numbers are given (in parentheses) with the name. In some instances the critical pressure and critical volume (or critical density) given by authors are estimates, or were calculated from the experimental observation of other properties, and in a few instances the critical temperature and critical pressure have been adjusted as a result of comparative studies of a series of compounds. The number of digits given is an indication of the accuracy attributed to the values of the normal, i.e., 101.325kPa, boiling point Tb, also included, and Tc and pc. However, at temperatures greater than 750°K, irrespective of the number of figures given, possible errors in 10° K or more in the critical temperature must be expected, with corresponding errors in the critical pressure. Where temperatures are given to only 0.1°K and have been obtained from Celsius values the difference 273.15°K has been rounded to 273.2°K. Where appropriate, temperatures have been converted to IPTS-68. Critical densities are given uniformly with three significant figures; the majority of these values may be assumed to have uncertainties of several percent because the quantity is experimentally ill-defined.

An important use of critical properties is for the calculation of other thermodynamic properties by use of the principle of corresponding states. As now applied, the principle of corresponding states requires an additional parameter, usually the critical compression factor $Zc = pcVc/RTc$, or the acentric factor ω, both of which are given in the table. The critical compression factor has been calculated in accordance with the expression just given; values of the acentric factor have been calculated from the normal boiling point (or, for carbon dioxide, uranium hexafluoride, sulfur hexafluride, and perfluorocyclohexane, from a calculated value for the supercooled liquid) by use of the equations given by Lee and Kesler,[3] and the values may differ slightly from those found elsewhere that have been calculated in some other way; the acentric factor is not an exact property and differences of at least 0.005 may arise from different methods of calculation. Correlations dependent on the acentric factor do not apply to quantum fluids or associated compounds; values of ω are nevertheless included for these elements and compounds, but they are omitted for most compounds with triple-point pressures greater than 101.325 kPa. Errors in the critical properties may be suspected if the critical compression factor or the acentric factor lies outside the range of values found for related substances.

REFERENCES

1. Kobe, K. A. and Lynn, Jr, R. E., *Chem. Rev.* 52: 117 (1953).
2. Kudchadker, A. P., Alani, G. H., and Zwolinski, B. J., *Chem. Rev.* 68: 659 (1968).
3. Lee, B. I. and Kesler, M. B., *Am. Inst. Chem. Engr. J.* 21: 510 (1975).
4. Matthews, J. F., *Chem. Rev.* 72: 71 (1972).
5. Ohse, R. W. and von Tippelskirch, H., *High Temperatures High Pressures* 9: 367 (1977).
6. Reid, R. C., Prausnitz, J. M. and Sherwood, T. K., *The Properties of Gases and Liquids, Their Estimation and Correlation*, 3rd Ed., McGraw-Hill, New York, 1977.

Formula	Name	M	$\dfrac{T_b}{K}$	$\dfrac{T_c}{K}$	$\dfrac{p_c}{MPa}$	$\dfrac{\rho_c}{\text{g cm}^{-3}}$	$\dfrac{V_c}{\text{cm}^3\ \text{mol}^{-1}}$	Z_c	ω
elements									
H2	hydrogen (normal)	2.016	20.38	33.20	1.297	0.031	65	0.306	-0.220
H2	hydrogen (equilibrium)	2.016	20.28	32.98	1.293	0.031	65	0.307	-0.218
HD	hydrogen deuteride	3.024	22.1	36.	1.48	0.048	63	0.311	-0.182
D2	deuterium (normal)	4.032	23.45	38.40	1.66				-0.163
D2	deuterium (equilibrium)	4.032	23.64	38.2	1.65	0.067	60	0.313	-0.140
T2	tritium	6.040	25.04	40.0		0.109	55		
4He	helium-4 (704)	4.003	4.25	5.19	0.227	0.070	57	0.301	-0.371
3He	helium-3	3.000	3.20	3.31	0.114	0.041	73	0.303	-0.490
Ne	neon (720)	20.18	27.07	44.40	2.76	0.484	42	0.311	-0.032
Ar	argon (740)	39.98	87.28	150.86	4.898	0.536	75	0.291	-0.005
Kr	krypton	83.80	119.92	209.4	5.50	0.919	91	0.288	0.000
Xe	xenon	131.30	165.1	289.73	5.840	1.110	118	0.287	0.002
Rn	radon	222.00	211.4	377.	6.28				-0.014
Hg	mercury	200.59	629.88	1750.	172.00	4.700	43	0.505	-0.155
N2	nitrogen (728)	28.01	77.35	126.2	3.39	0.313	89	0.289	0.035
P	phosphorus	30.97	553.	994.					
As	arsenic	74.92	888.	1673.	22.3	2.150	35	0.056	0.114
O2	oxygen (732)	32.00	90.18	154.58	5.043	0.436	73	0.288	0.020
O3	ozone	48.00	161.8	261.1	5.57	0.540	89	0.228	0.202
S	sulphur	32.06	717.82	1314.	20.7				0.164
Se	selenium	78.96	1009.	1766.	27.2				0.340
F2	fluorine	38.00	85.02	144.30	5.215	0.574	66	0.288	0.050
Cl2	chlorine	70.91	239.2	416.9	7.977	0.573	124	0.285	0.085
Br2	bromine	159.81	331.9	588.	10.34	1.260	127	0.268	0.102
I2	iodine	253.81	457.5	819.		1.640	155		

Formula	Name	M	$\dfrac{Tb}{K}$	$\dfrac{Tc}{K}$	$\dfrac{pc}{MPa}$	$\dfrac{lc}{g\ cm^{-3}}$	$\dfrac{Vc}{cm^3\ mol^{-1}}$	Zc	ω
	deuterides								
D3N	nitrogen trideuteride	20.05	242.1	405.6					
D3P	phosphorus trideuteride	37.02	187.	323.6					
D3As	arsenic trideuteride	80.97	220.	372.1					
	nitrogen compounds								
H3N	ammonia (717)	17.03	239.82	405.5	11.35	0.235	72	0.244	0.244
H4N2	hydrazine	32.05	386.7	653.	14.7				0.310
HCN	hydrogen cyanide	27.03	299.	456.7	5.39	0.195	139	0.197	0.385
C2N2	cyanogen	52.04	251.9	400.	5.98				0.274
	oxides, sulphides and selenides								
H2O	water	18.02	373.15	647.14	22.06	0.322	56	0.229	0.321
D2O	heavy water	20.03	374.58	643.89	21.67	0.356	56	0.228	0.345
T2O	tritium oxide	22.04	374.66	641.7	21.41	0.393	56	0.225	0.354
CO	carbon monoxide	28.01	81.65	132.91	3.499	0.301	93	0.295	0.049
CO2	carbon dioxide (744)	44.01	s	304.14	7.375	0.468	94	0.274	0.234
N2O	nitrous oxide	44.01	184.67	309.57	7.255	0.452	97	0.274	0.160
NO	nitric oxide	30.01	121.41	180.	6.48	0.520	58	0.250	0.586
N2O4	dinitrogen tetroxide	92.01	294.3	431.	10.1	0.550	167	0.472	0.831
O2S	sulphur dioxide	64.06	263.1	430.8	7.884	0.525	122	0.269	0.251
O3S	sulphur trioxide	80.06	318.	491.0	8.2	0.630	127	0.255	0.477
O7Re2	rhenium oxide	484.40	723.	942.		1.450	334		
O4Os	osmium tetroxide	254.20	408.	678.					
H2S	hydrogen sulphide	34.08	213.6	373.2	8.94	0.346	98	0.284	0.104
D2S	deuterium sulphide	36.09	215.	372.3					
CS2	carbon disulphide	76.13	319.	552.	7.90	0.440	173	0.298	0.103
COS	carbonyl sulphide	60.07	223.	375.	5.88	0.440	137	0.257	0.101

Formula	Name								
H2Se	hydrogen selenide	80.98	231.9	411.	8.92				0.070
D2Se	deuterium selenide	82.99	233.	412.4					
	fluorides								
HF	hydrogen fluoride	20.01	293.	461.	6.48	0.290	69	0.117	0.325
F2Xe	xenon difluoride	169.30	387.5	631.	9.32	1.140	149	0.264	0.312
F4Xe	xenon tetrafluoride	207.29	388.9	612.	7.04	1.100	188	0.261	0.353
BF3	boron trifluoride	67.81	172.	260.8	4.98	0.590	115	0.264	0.391
F6U	uranium hexafluoride	352.02	s	505.8	4.66	1.410	250	0.277	0.268
F4Si	silicon tetrafluoride	104.08	187.	259.0	3.72				0.755
N2F4	tetrafluorohydrazine	104.01	199.	309.	3.75				0.203
N2F2	cis-difluorodiazine	66.01	167.4	272.	7.09				0.247
N2F2	trans-difluorodiazine	66.01	161.7	260.	5.57				0.213
NF3	nitrogen trifluoride	71.00	144.4	234.0	4.53				0.131
HNF2	difluoramine	53.01	250.	403.					
F3P	phosphorus trifluoride	87.97	178.	271.2	4.33				0.323
F5Nb	niobium pentafluoride	187.90	502.	737.	6.28	1.210	155	0.159	0.632
OF2	oxygen fluoride	54.00	128.4	215.					
F4S	sulphur tetrafluoride	108.05	232.7	364.	3.77	0.735	199	0.283	0.198
F6S	sulphur hexafluoride	146.05	s	318.69					
F6Se	selenium hexafluoride	192.95	227.	345.5	4.75	0.930	226	0.273	0.309
F6Mo	molybdenum hexafluoride	209.93	307.	473.					
F6Te	tellurium hexafluoride	241.59	234.	356.	4.34	1.280	233	0.273	0.309
F6W	tungsten hexafluoride	297.84	290.	444.					
	chlorides								
HCl	hydrogen chloride	36.46	188.	324.7	8.31	0.450	81	0.249	0.116
DCl	deuterium chloride	37.47	191.6	323.5					
Cl2Hg	mercuric chloride	271.50	577.	973.		1.560	174	0.245	
BCl3	boron trichloride	117.17	285.8	455.	3.87	0.490	239	0.245	0.137
AlCl3	aluminium trichloride	133.34	s	620.	2.63	0.518	257	0.132	

Formula	Name	M	$\dfrac{Tb}{K}$	$\dfrac{Tc}{K}$	$\dfrac{pc}{MPa}$	$\dfrac{lc}{\mathrm{g\ cm^{-3}}}$	$\dfrac{Vc}{\mathrm{cm^3\ mol^{-1}}}$	Zc	ω
Cl3Ga	gallium trichloride	176.08	474.	694.		0.670	263		
SiCl4	silicon tetrachloride	169.90	330.8	508.1	3.593	0.521	326	0.277	0.230
Cl4Ti	titanium tetrachloride	189.71	409.6	638.	4.66	0.560	339	0.298	0.264
Cl4Ge	germanium tetrachloride	214.40	359.7	553.2	3.861	0.650	330	0.277	0.249
Cl4Zr	zirconium tetrachloride	233.03	s	778.	5.77	0.730	319	0.284	
Cl4Sn	tin tetrachloride	260.50	387.3	591.9	3.75	0.742	351	0.267	0.262
Cl4Hf	hafnium tetrachloride	320.30	s	725.7	5.42	1.020	314	0.282	
H4NCl	ammonium chloride	53.49	s	882.	1.635				
PCl3	phosphorus trichloride	137.33	349.1	563.		0.520	264		
PCl5	phosphorus pentachloride	208.24	s	646.					
Cl3As	arsenic trichloride	181.28	403.	654.		0.720	252		
Cl5Nb	niobium pentachloride	270.17	527.2	803.5	4.88	0.680	397	0.290	0.362
Cl3Sb	antimony trichloride	228.11	493.4	794.		0.840	272		
Cl5Ta	tantalum pentachloride	358.21	512.5	767.		0.890	402		
Cl3Bi	bismuth chloride	315.34	720.	1179.	12.0	1.210	261	0.319	0.362
Cl5Mo	molybdenum pentachloride	273.21	541.	850.		0.740	369		
Cl4Te	tellurium tetrachloride	269.41	597.	1002.	8.56	0.870	310	0.318	0.198
Cl6W	tungsten hexachloride	396.57	619.9	923.		0.940	422		
bromides									
HBr	hydrogen bromide	80.91	206.77	363.2	8.55				0.082
DBr	deuterium bromide	81.92	208.	362.0					
Br2Hg	mercuric bromide	360.40	595.	1012.					
BBr3	boron tribromide	250.52	364.	581.		0.920	272		
AlBr3	aluminium tribromide	266.69	528.	763.	2.89	0.860	310	0.141	0.398
GaBr3	gallium tribromide	309.43	552.	806.7		1.020	303		
SiBr4	silicon tetrabromide	347.70	427.	663.		0.910	382		
TiBr4	titanium tetrabromide	367.52	503.	795.7		0.940	391		

Formula	Name								
GeBr4	germanium tetrabromide	392.21	459.5	718.		1.000	392		
Br4Zr	zirconium tetrabromide	410.84	s	805.		0.970	424		
Br4Sn	tin tetrabromide	438.31	478.	744.		1.050	417		
Br4Hf	hafnium tetrabromide	498.11	s	746.		1.200	415		
PBr3	phosphorus tribromide	270.69	446.1	711.		0.902	300		
Br5Nb	niobium pentabromide	492.43	634.8	1010.		1.050	469		
Br3Sb	antimony tribromide	361.46	553.	904.		1.205	300		
Br5Ta	tantalum pentabromide	580.47	622.0	974.		1.260	461		
Br3Bi	bismuth bromide	448.69	726.	1220.		1.490	301		
iodides									
HI	hydrogen iodide	127.91	237.6	424.0	8.31			0.043	
DI	deuterium iodide	128.92	237.0	421.8					
I2Hg	mercuric iodide	454.40	627.	1072.					
BI3	boron triiodide	391.52	483.	773.		1.100	356		
AlI3	aluminium triiodide	407.70	655.	983.		1.000	408		
GaI3	gallium triiodide	450.43	613.	951.		1.140	395		
SiI4	silicon tetraiodide	535.70	560.5	944.		0.960	558		
TiI4	titanium tetraiodide	555.52	650.	1040.		1.100	505		
GeI4	germanium tetraiodide	580.21	650.	973.		1.160	500		
ZrI4	zirconium tetraiodide	598.84	704.	960.		1.130	530		
SnI4	tin tetraiodide	626.31	637.5	968.		1.180	531		
I4Hf	hafnium tetraiodide	686.11	s	916.		1.300	528		
SbI3	antimony triiodide	502.46	674.	1102.					
mixed halides									
F5Cl	chlorine pentafluoride	130.45	260.0	416.	5.27	0.560	233	0.355	0.213
F3SiCl	chlorotrifluorosilane	120.53	203.1	307.7	3.46				0.270
BrI	iodine bromide	206.81	s	719.		1.490	139		
F5SCl	sulphur chloride pentafluoride	162.51	254.1	390.9					
F2SiCl2	dichlorodifluorosilane	136.99	240.9	369.0	3.50				0.232
FSiCl3	trichlorofluorosilane	153.44	285.4	438.6	3.58				0.226
NF2Cl	nitrogen chloride difluoride	87.46	207.	337.5	5.15				0.149

Formula	Name	M	$\dfrac{T_b}{K}$	$\dfrac{T_c}{K}$	$\dfrac{p_c}{MPa}$	$\dfrac{\rho_c}{g\ cm^{-3}}$	$\dfrac{V_c}{cm^3\ mol^{-1}}$	Z_c	ω
F2PCl	phosphorus(III)chloride difluoride								
		104.42	225.9	362.4	4.52				0.160
FPCl2	phosphorus(III)dichloride fluoride								
		120.88	287.0	463.0	4.96				0.170
	oxyhalides								
NO2F	nitryl fluoride	65.00	213.2	349.5		0.593	147	0.375	0.209
NOF3	trifluoramine oxide	87.00	185.6	303.	6.43	0.520	190	0.285	0.201
COCl2	phosgene	98.92	281.	455.	5.67				
NOCl	nitrosyl chloride	65.46	267.6	440.					
OCl2Se	selenium oxychloride	165.87	390.4	730.	7.09	0.705	235	0.275	-0.074
OClV	vanadium oxychloride	102.39	400.	636.		0.600	171		
OCl4W	tungsten oxide tetrachloride	341.66	500.7	782.		1.010	338		
OCl4Re	rhenium oxide tetrachloride	344.01	496.	781.		0.950	362		
O3FCl	perchloryl fluoride	102.45	226.4	368.4	5.37	0.637	161	0.282	0.166
	miscellaneous inorganic compounds								
H6B2	diborane	27.67	185.6	289.8	4.05				0.214
H3P	phosphine	34.00	185.4	324.5	6.54				0.033
H4Ge	germane	76.62	183.	312.2	4.95	0.522	147	0.280	0.026
H3As	arsine	77.95	218.	373.1					
H4PCl	phosphonium chloride	70.46	s	322.3	7.37				
N3F6P3	phosphonitrilic fluoride trimer	248.93	324.	461.0					
N4F8P4	phosphonitrilic fluoride tetramer	331.91	363.	496.4					
N5F10P5	phosphonitrilic fluoride pentamer	414.89	392.5	524.0					
F3PS	thiophosphoryl trifluoride	120.03	220.9	346.0	3.82				0.185

Code	Name								
F2PSCl	thiophosphoryl chloride difluoride	136.48	279.	439.2	4.14				0.198
HS1Cl3	trichlorosilane	135.45	306.	479.		0.505	268		
HS1Br3	tribromosilane	268.81	382.	610.0		0.880	305		
	alkanes								
H4C	methane (50)	16.04	111.63	190.53	4.604	0.162	99	0.288	0.008
CD4	deuteromethane	20.07	111.68	189.2	4.66	0.205	98	0.290	0.027
H6C2	ethane (170)	30.07	184.55	305.40	4.884	0.203	148	0.285	0.095
H8C3	propane (290)	44.10	231.05	369.82	4.250	0.217	203	0.281	0.149
H10C4	butane (600)	58.12	272.65	425.14	3.784	0.228	255	0.273	0.196
H10C4	isobutane (600a)	58.12	261.42	407.85	3.630	0.226	257	0.275	0.181
H12C5	pentane	72.15	309.21	469.69	3.364	0.232	311	0.268	0.248
H12C5	isopentane	72.15	301.03	460.43	3.381	0.236	306	0.270	0.225
H12C5	2,2-dimethylpropane	72.15	282.63	433.8	3.197	0.235	307	0.272	0.195
H14C6	hexane	86.18	341.88	507.70	3.010	0.233	370	0.264	0.295
H14C6	2-methylpentane	86.18	333.41	497.7	3.031	0.235	367	0.269	0.278
H14C6	3-methylpentane	86.18	336.42	504.5	3.126	0.235	367	0.273	0.271
H14C6	2,2-dimethylbutane	86.18	322.88	488.8	3.090	0.240	359	0.273	0.231
H14C6	2,3-dimethylbutane	86.18	331.13	500.0	3.131	0.241	358	0.269	0.245
H16C7	heptane	100.21	371.58	540.3	2.756	0.234	428	0.263	0.351
H16C7	2-methylhexane	100.21	363.19	530.4	2.734	0.238	421	0.261	0.328
H16C7	3-methylhexane	100.21	364.99	535.3	2.814	0.248	404	0.255	0.321
H16C7	3-ethylpentane	100.21	366.64	540.7	2.891	0.241	416	0.267	0.309
H16C7	2,2-dimethylpentane	100.21	352.35	520.5	2.773	0.241	416	0.266	0.286
H16C7	2,3-dimethylpentane	100.21	362.93	537.4	2.908	0.255	393	0.256	0.295
H16C7	2,4-dimethylpentane	100.21	353.64	519.8	2.737	0.240	418	0.264	0.301
H16C7	3,3-dimethylpentane	100.21	359.21	536.4	2.946	0.242	414	0.274	0.265
H16C7	2,2,3-trimethylbutane	100.21	354.01	531.2	2.954	0.252	398	0.266	0.248
H18C8	octane	114.23	398.82	568.9	2.493	0.232	492	0.260	0.399
H18C8	2-methylheptane	114.23	390.81	559.7	2.484	0.234	488	0.261	0.378
H18C8	3-methylheptane	114.23	392.09	563.7	2.546	0.246	464	0.252	0.370

Formula	Name	M	$\dfrac{Tb}{K}$	$\dfrac{Tc}{K}$	$\dfrac{pc}{MPa}$	$\dfrac{lc}{g\ cm^{-3}}$	$\dfrac{Vc}{cm^{3}\ mol^{-1}}$	Zc	ω
H18C8	4-methylheptane	114.23	390.87	561.8	2.542	0.240	476	0.259	0.371
H18C8	3-ethylhexane	114.23	391.70	565.5	2.608	0.251	455	0.252	0.361
H18C8	2,2-dimethylhexane	114.23	380.01	549.9	2.529	0.239	478	0.264	0.337
H18C8	2,3-dimethylhexane	114.23	388.77	563.5	2.628	0.244	468	0.263	0.346
H18C8	2,4-dimethylhexane	114.23	382.59	553.6	2.556	0.242	472	0.262	0.343
H18C8	2,5-dimethylhexane	114.23	382.27	550.1	2.487	0.237	482	0.262	0.356
H18C8	3,3-dimethylhexane	114.23	385.12	562.1	2.654	0.258	443	0.251	0.320
H18C8	3,4-dimethylhexane	114.23	390.88	568.9	2.692	0.245	466	0.265	0.337
H18C8	2-methyl-3-ethylpentane	114.23	388.81	567.1	2.700	0.258	443	0.254	0.329
H18C8	3-methyl-3-ethylpentane	114.23	391.42	576.6	2.808	0.251	455	0.267	0.302
H18C8	2,2,3-trimethylpentane	114.23	383.00	563.5	2.730	0.262	436	0.254	0.296
H18C8	2,2,4-trimethylpentane	114.23	372.37	544.0	2.568	0.244	468	0.266	0.302
H18C8	2,3,3-trimethylpentane	114.23	387.90	573.6	2.820	0.251	455	0.269	0.288
H18C8	2,3,4-trimethylpentane	114.23	386.62	566.5	2.730	0.248	461	0.267	0.314
H18C8	2,2,3,3-tetramethylbutane	114.23	379.6	567.8	2.87	0.248	461	0.280	0.249
H20C9	nonane	128.26	423.97	594.6	2.288				0.446
H20C9	2-methyloctane	128.26	416.43	587.0	2.310				0.424
H20C9	2,2-dimethylheptane	128.26	405.85	576.8	2.350				0.391
H20C9	2,2,5-trimethylhexane	128.26	397.24	568.					
H20C9	2,2,3,3-tetramethylpentane	128.26	413.44	607.7	2.741				0.302
H20C9	2,2,3,4-tetramethylpentane	128.26	406.18	592.7	2.602				0.312
H20C9	2,2,4,4-tetramethylpentane	128.26	395.44	574.7	2.485				0.311
H20C9	2,3,3,4-tetramethylpentane	128.26	414.72	607.7	2.716				0.312
H22C10	decane	142.29	447.30	617.7	2.104				0.491
H22C10	3,3,5-trimethylheptane	142.29	428.85	609.7	2.317				0.383
H22C10	2,2,3,3-tetramethylhexane	142.29	433.48	623.2	2.510				0.364
H22C10	2,2,5,5-tetramethylhexane	142.29	410.63	581.6	2.186				0.375
H24C11	undecane	156.31	469.08	638.8	1.966				0.539

H26C12	dodecane	170.34	489.47	658.2	1.824				0.579
H28C13	tridecane	184.37	508.62	676.	1.72				0.624
H30C14	tetradecane	198.39	526.73	693.	1.62				0.660
H34C16	hexadecane	226.45	560.01	722.					
H38C18	octadecane	254.50	589.5	748.					
	cycloalkanes								
H6C3	cyclopropane	42.08	240.34	398.30	5.579	0.259	162	0.274	0.125
H8C4	cyclobutane	56.11	285.7	460.0	4.98	0.267	210	0.274	0.177
H10C5	cyclopentane	70.13	322.40	511.7	4.508	0.270	260	0.275	0.192
H12C6	methylcyclopentane	84.16	344.95	532.73	3.784	0.264	319	0.272	0.228
H14C7	ethylcyclopentane	98.19	376.62	569.5	3.397	0.262	375	0.269	0.269
H12C6	cyclohexane	84.16	353.88	553.5	4.07	0.273	308	0.273	0.209
H14C7	methylcyclohexane	98.19	374.08	572.20	3.471	0.267	368	0.268	0.234
H16C8	1,trans-4-dimethylcyclohexane	112.22	392.50	587.7					
H18C9	1,trans-3,5-trimethylcyclohexane	126.24	413.7	602.2					
H14C7	cycloheptane	98.19	391.63	604.2	3.81	0.278	353	0.268	0.234
H16C8	cyclooctane	112.22	422.	647.2	3.56	0.274	410	0.271	0.234
H12C10	tetrahydronaphthalene	132.21	480.	716.5	3.35				0.317
H18C10	cis-decahydronaphthalene	138.25	468.96	702.3	3.20				0.285
H18C10	trans-decahydronaphthalene	138.25	460.46	687.1					
	alkenes								
H4C2	ethylene (1150)	28.05	169.38	282.30	5.030	0.215	130	0.280	0.085
H6C3	propene (1270)	42.08	225.46	364.85	4.601	0.233	181	0.274	0.140
H8C4	but-1-ene	56.11	266.89	419.57	4.023	0.234	240	0.277	0.188
H8C4	cis-but-2-ene	56.11	276.86	435.58	4.197	0.240	234	0.271	0.199
H8C4	trans-but-2-ene	56.11	274.03	428.63	3.985	0.236	238	0.266	0.202
H8C4	2-methylpropene	56.11	266.2	417.90	4.000	0.235	239	0.275	0.191
H10C5	pent-1-ene	70.13	303.11	464.78	3.527				0.230
H10C5	cis-pent-2-ene	70.13	310.08	475.0	3.69				0.249

Formula	Name	M	$\dfrac{T_b}{K}$	$\dfrac{T_c}{K}$	$\dfrac{\rho c}{MPa}$	$\dfrac{\ell c}{g\ cm^{-3}}$	$\dfrac{V_c}{cm^3\ mol^{-1}}$	Z_c	ω
H10C5	trans-pent-2-ene	70.13	309.49	471.	3.52				0.257
H10C5	2-methylbut-1-ene	70.13	304.30	470.	3.8				0.229
H10C5	2-methylbut-2-ene	70.13	311.71	481.	3.91				0.242
H12C6	hex-1-ene	84.16	336.63	504.1					
H14C7	hept-1-ene	98.19	366.79	537.3					
H16C8	oct-1-ene	112.22	394.44	566.7					
	alkadienes								
H4C3	propadiene	40.07	238.7	393.					
H6C4	1,3-butadiene	54.09	268.74	425.	4.33	0.245	221	0.270	0.191
H10C6	1,5-hexadiene	82.15	332.60	507.					
H18C10	1,3-decadiene	138.25	442.	615.					
	cycloalkenes								
H8C5	cyclopentene	68.12	317.39	506.0					
H10C6	cyclohexene	82.15	356.13	560.48					
	alkynes								
H2C2	acetylene	26.04	188.43	308.33	6.139	0.231	113	0.270	0.186
H4C3	propyne	40.07	249.95	402.38	5.628	0.245	164	0.275	0.211
H6C4	but-1-yne	54.09	281.23	463.7					
H6C4	but-2-yne	54.09	300.10	488.7					
H8C5	pent-1-yne	68.12	313.33	493.5					
	aromatic hydrocarbons								
H6C6	benzene	78.11	353.24	562.16	4.898	0.302	259	0.271	0.208
H8C7	toluene	92.14	383.78	591.79	4.104	0.292	316	0.263	0.260
H10C8	ethylbenzene	106.17	409.34	617.20	3.600	0.284	374	0.262	0.300
H10C8	ortho-xylene	106.17	417.58	630.3	3.730	0.288	369	0.262	0.308

Code	Name								
H10C8	meta-xylene	106.17	412.27	617.05	3.535	0.282	376	0.259	0.323
H10C8	para-xylene	106.17	411.52	616.2	3.511	0.280	379	0.260	0.318
H12C9	propylbenzene	120.20	432.39	638.32	3.200	0.273	440	0.265	0.343
H12C9	isopropylbenzene	120.20	425.56	631.1	3.209				0.325
H12C9	1,2,3-trimethylbenzene	120.20	449.27	664.47	3.454				0.364
H12C9	1,2,4-trimethylbenzene	120.20	442.53	649.17	3.232				0.375
H12C9	1,3,5-trimethylbenzene	120.20	437.89	637.25	3.127				0.398
H14C10	butylbenzene	134.22	456.46	660.5	2.887	0.270	497	0.261	0.392
H14C10	isobutylbenzene	134.22	445.94	650.	3.05				0.380
H14C10	4-isopropyl-1-methylbenzene	134.22	450.28	651.	2.73				0.373
H14C10	1,4-diethylbenzene	134.22	456.94	657.88	2.803				0.403
H14C10	1,2,4,5-tetramethylbenzene	134.22	469.99	675.	2.9				0.429
H18C12	hexamethylbenzene	162.28	536.6	758.					
H8C10	naphthalene	128.17	491.14	748.4	4.051	0.310	413	0.269	0.300
H10C11	1-methylnaphthalene	142.20	517.89	772.					
H10C11	2-methylnaphthalene	142.20	514.26	761.					
H10C12	biphenyl	154.21	529.25	789.	3.85	0.307	502	0.295	0.370
H10C14	phenanthrene	178.23	613.	873.		0.322	554		
H10C14	anthracene	178.23	613.1	869.3			554		
H14C18	o-terphenyl	230.31	605.	891.0	3.90	0.306	753	0.396	0.430
H14C18	m-terphenyl	230.31	636.	924.9	3.51	0.300	768	0.350	0.448
H14C18	p-terphenyl	230.31	649.	926.0	3.32	0.302	763	0.329	0.523
H12C13	diphenylmethane	168.24	538.2	770.	2.86				0.442
H10C9	indane	118.18	451.12	684.9	3.95				0.305

halogenated alkanes

Code	Name								
CF3Br	bromotrifluoromethane (13B1)	148.91	215.26	340.2	3.97	0.760	196	0.275	0.168
CF3Cl	chlorotrifluoromethane (13)	104.46	193.2	302.0	3.870	0.579	180	0.278	0.195
CF2Cl2	dichlorodifluoromethane (12)	120.91	245.2	384.95	4.136	0.558	217	0.280	0.201
CF2ClBr	bromochlorodifluoromethane (12B1)	165.36	269.43	426.88	4.254	0.673	246	0.294	0.180
CF2Br2	dibromodifluoromethane (12B2)	209.82	298.	471.30					
CFCl3	trichlorofluoromethane (11)	137.37	296.9	471.2	4.41	0.554	248	0.279	0.186

Formula	Name	M	$\dfrac{T_b}{K}$	$\dfrac{T_c}{K}$	$\dfrac{p_c}{\text{MPa}}$	$\dfrac{I_c}{\text{g cm}^{-3}}$	$\dfrac{V_c}{\text{cm}^3\ \text{mol}^{-1}}$	Z_c	ω
CCl4	carbon tetrachloride (10)	153.82	349.9	556.6	4.516	0.558	276	0.269	0.186
CF4	carbon tetrafluoride (14)	88.00	145.13	227.6	3.74	0.630	140	0.276	0.174
HCF2Cl	chlorodifluoromethane (22)	86.47	232.4	369.30	4.99	0.513	169	0.274	0.218
HCFCl2	dichlorofluoromethane (21)	102.92	282.1	451.58	5.18	0.524	196	0.271	0.206
HCCl3	chloroform (20)	119.38	334.32	536.4	5.47	0.500	239	0.293	0.214
HCF3	trifluoromethane (23)	70.01	191.0	299.30	4.858	0.528	133	0.259	0.257
H2CCl2	dichloromethane (30)	84.93	313.0	510.	6.10	0.430	121	0.241	0.196
H2CF2	difluoromethane (32)	52.02	221.46	351.6	5.830				0.267
H3CCl	chloromethane (40)	50.49	249.06	416.25	6.679	0.363	139	0.268	0.148
H3CF	fluoromethane (41)	34.03	194.74	317.8	5.88	0.300	113	0.252	0.183
H3CI	iodomethane	141.94	315.7	528.					
C2F5Cl	chloropentafluoroethane (115)	154.47	235.2	353.2	3.229	0.613	252	0.277	0.277
C2F4Cl2	1,2-dichlorotetrafluoroethane (114)	170.92	276.2	418.78	3.252	0.576	297	0.277	0.243
C2F4Cl2	1,1-dichlorotetrafluoroethane (114)	170.92	277.0	418.6	3.30	0.582	294	0.279	0.261
C2F4Br2	1,2-dibromotetrafluoroethane (114B2)	259.82	320.5	487.8	3.393	0.762	341	0.285	0.243
C2F3Cl3	1,2,2-trichlorotrifluoroethane (113)	187.38	320.8	487.3	3.42	0.576	325	0.274	0.253
C2F3ClBr2	1,2-dibromo-1-chloro-1,2,2-trifluoroethane (113B2)	276.28	366.	560.7	3.61	0.750	368	0.285	0.246
C2F2Cl4	tetrachloro-1,2-difluoroethane (112)	203.83	366.0	551.					
C2F6	perfluoroethane (116)	138.01	194.9	293.0		0.622	222		
HC2F4Cl	chloro-1,1,2,2-tetrafluoroethane (124a)	136.48	263.0	399.9	3.72	0.559	244	0.273	0.279
H3C2F2Cl	1-chloro-1,1-difluoroethane (142)	100.50	263.4	409.60	4.332	0.435	231	0.294	0.247
H3C2Cl3	1,1,1-trichloroethane (140a)	133.40	347.24	545.	4.30				0.214

Formula	Name								
H3C2F3	1,1,1-trifluoroethane (143a)	84.04	225.6	346.3	3.76	0.434	194	0.253	0.248
H4C2Br2	1,2-dibromoethane	187.86	404.7	583.0	7.2				0.798
H4C2Cl2	1,1-dichloroethane (150a)	98.96	330.5	523.	5.07	0.420	236	0.275	0.236
H4C2Cl2	1,2-dichloroethane	98.96	356.7	561.	5.4	0.440	225	0.260	0.276
H4C2F2	1,1-difluoroethane (152a)	66.05	248.2	386.7	4.50	0.365	181	0.253	0.253
H5C2Br	bromoethane	108.97	311.49	503.9	6.23	0.507	215	0.320	0.225
H5C2Cl	chloroethane (160)	64.51	285.45	460.4	5.3				0.189
H5C2F	fluoroethane	48.06	235.5	375.31	5.028				0.211
C3F8	perfluoropropane (218)	188.02	236.5	345.1	2.680	0.628	299	0.280	0.325
H3C3F5	1,1,1,2,2-pentafluoropropane	134.05	255.71	380.11	3.137	0.491	273	0.271	0.307
H7C3Cl	chloropropane	78.54	320.4	503.	4.58				0.231
C4F10	perfluorobutane	238.03	271.2	386.4	2.323	0.629	378	0.274	0.374
C4F8Br2	1,4-dibromooctafluorobutane	359.84	370.	532.5	2.39				0.340
C4F10	perfluoro-2-methylpropane	238.03	273.	395.4					
C5F12	perfluoropentane	288.04	302.4	420.59	2.045	0.609	473	0.277	0.437
HC5F11	1H-undecafluoropentane	270.05	318.2	444.0					
C6F14	perfluorohexane	338.04	329.8	448.77	1.868	0.558	606	0.303	0.517
HC6F13	1H-tridecafluorohexane	320.05	344.7	471.8					
C6F14	perfluoro-2-methylpentane	338.04	330.81	455.3	1.923	0.635	532	0.270	0.465
C6F14	perfluoro-3-methylpentane	338.04	331.52	450.	1.69				0.479
C6F14	perfluoro-2,3-dimethylbutane	338.04	332.93	463.	1.87	0.644	525	0.256	0.395
C7F16	perfluoroheptane	388.05	355.66	474.8	1.62	0.584	664	0.273	0.560
HC7F15	1H-pentadecafluoroheptane	370.06	369.2	495.8					
C8F18	perfluoro-octane	438.06	376.7	502.	1.66				0.586
C9F20	perfluorononane	488.07	398.5	524.	1.56				0.643
C10F22	perfluorodecane	538.07	417.4	542.	1.45				0.691
	halogenated cycloalkanes								
C4F8	perfluorocyclobutane (C318)	200.03	267.16	388.46	2.784	0.617	324	0.279	0.355
C6F12	perfluorocyclohexane	300.05	323.76	457.2	2.43				0.438
HC6F11	undecafluorocyclohexane	282.06	335.2	477.7					
C7F14	perfluoromethylcyclohexane	350.05	349.5	485.91	2.019	0.614	570	0.285	0.433

Formula	Name	M	T_b/K	T_c/K	p_c/MPa	l_c/g cm^{-3}	V_c/cm^3 mol^{-1}	Z_c	ω
	halogenated alkenes and cycloalkenes								
C2Cl4	perchloroethylene	165.83	394.4	620.2					
C2F3Cl	chlorotrifluoroethylene (1113)	116.47	245.3	379.	4.05	0.550	212	0.272	0.249
C2F4	perfluoroethylene (1114)	100.02	197.2	306.5	3.94	0.580	172	0.267	0.220
HC2Cl3	trichloroethylene	131.39	360.36	544.2	5.02				0.410
HC2F2Cl	1-chloro-2,2-difluoro-ethylene	98.48	254.6	400.6	4.46	0.499	197	0.264	0.217
H2C2F2	1,1-difluoroethylene (1132a)	64.03	187.45	302.9	4.46	0.416	154	0.273	0.136
H2C2Cl2	cis-1,2-dichloroethylene	96.94	333.34	544.2					
H2C2Cl2	trans-1,2-dichloroethylene	96.94	321.88	516.5	5.51				0.216
H3C2F	vinyl fluoride (1141)	46.04	201.0	327.9	5.24	0.320	144	0.277	0.153
H3C3F3	trifluoropropene	96.05	244.	376.2	3.80	0.455	211	0.256	0.235
H5C3Cl	allyl chloride	76.53	318.30	514.					
C6F12	perfluorohex-1-ene	300.05	330.2	454.4					
C7F14	perfluorohept-1-ene	350.05	354.2	478.2					
C6F10	perfluorocyclohexene	262.05	325.2	461.8					
	halogenated aromatics								
C6F5Br	bromopentafluorobenzene	246.96	410.0	601.	3.0				0.349
C6F3Cl3	1,3,5-trichlorotrifluorobenzene								
C6F4Cl2	dichlorotetrafluorobenzene	235.42	471.52	684.8	3.27	0.526	448	0.257	0.426
C6F5Cl	chloropentafluorobenzene	218.97	430.9	626.	5.32	0.539	376	0.256	0.621
C6F6	perfluorobenzene	202.51	391.11	570.81	3.238	0.555	335	0.255	0.399
HC6F5	pentafluorobenzene	186.06	353.41	516.73	3.273	0.518	324	0.260	0.395
H2C6F4	1,2,3,4-tetrafluorobenzene	168.07	358.89	530.97	3.531	0.480	313	0.259	0.371
H2C6F4	1,2,3,5-tetrafluorobenzene	150.08	367.51	550.83	3.791				0.342
H2C6F4	1,2,5-tetrafluorobenzene	150.08	357.61	535.25	3.747				0.344
H2C6F4	1,2,4,5-tetrafluorobenzene	150.08	363.41	543.35	3.801				0.353

Code	Name								
H4C6Cl2	dichlorobenzene	147.00	452.	729.	4.40				0.296
C6H4F2	1,4-difluorobenzene	114.09	362.0	556.	4.52	0.485	324	0.263	0.248
H5C6Br	bromobenzene	157.01	429.21	670.					
H5C6Cl	chlorobenzene	112.56	404.87	632.4	4.52	0.365	308	0.265	0.246
H5C6F	fluorobenzene	96.10	357.88	560.09	4.551	0.357	269	0.263	0.241
H5C6I	iodobenzene	204.01	461.60	721.	4.52	0.581	351	0.265	0.245
C7F8	perfluorotoluene	236.06	377.73	534.47	2.705	0.552	428	0.260	0.476
H3C7F5	2,3,4,5,6-pentafluorotoluene	182.09	390.65	566.52	3.126	0.474	384	0.255	0.414
C10F8	perfluoronaphthalene	272.10	482.	673.1					
C10F18	perfluorodecalin	462.08	415.	566.	1.52				0.394
	alcohols								
H4CO	methanol	32.04	337.70	512.64	8.092	0.272	118	0.224	0.553
H6C2O	ethanol	46.07	351.44	513.92	6.137	0.276	167	0.240	0.643
H8C3O	propan-1-ol	60.10	370.30	536.78	5.170	0.275	219	0.253	0.622
H8C3O	propan-2-ol	60.10	355.39	508.30	4.762	0.273	220	0.248	0.665
H10C4O	butan-1-ol	74.12	390.88	563.05	4.423	0.270	275	0.259	0.593
H10C4O	butan-2-ol	74.12	372.66	536.05	4.179	0.276	269	0.252	0.577
H10C4O	2-methylpropan-1-ol	74.12	381.04	547.78	4.300	0.272	273	0.257	0.592
H10C4O	2-methylpropan-2-ol	74.12	355.50	506.21	3.973	0.270	275	0.259	0.612
H12C5O	pentan-1-ol	88.15	411.13	587.7	3.909	0.270	326	0.261	0.583
H12C5O	pentan-2-ol	88.15	392.4	560.4					
H12C5O	pentan-3-ol	88.15	389.4	559.6					
H12C5O	3-methylbutan-1-ol	88.15	405.2	579.40					
H12C5O	2-methylbutan-2-ol	88.15	375.50	545.	0.268	381			
H14C6O	hexan-1-ol	102.18	430.7	611.					
H14C6O	hexan-2-ol	102.18	411.	586.2					
H14C6O	2-methylpentan-2-ol	102.18	394.2	559.5					
H14C6O	4-methylpentan-1-ol	102.18	424.8	603.5					
H14C6O	4-methylpentan-2-ol	102.18	404.8	574.4					
H16C7O	heptan-1-ol	116.20	449.6	632.3	0.267	435			
H16C7O	heptan-2-ol	116.20	432.	611.4					

Formula	Name	M	$\dfrac{T_b}{K}$	$\dfrac{T_c}{K}$	$\dfrac{p_c}{MPa}$	$\dfrac{l_c}{g\ cm^{-3}}$	$\dfrac{V_c}{cm^3\ mol^{-1}}$	Z_c	ω
H16C7O	heptan-3-ol	116.20	430.	605.4					
H16C7O	heptan-4-ol	116.20	429.0	602.60					
H18C8O	octan-1-ol	130.23	468.31	652.5	2.86	0.266	490	0.258	
H18C8O	octan-2-ol	130.23	452.	637.					0.589
H18C8O	4-methylheptan-3-ol	130.23	443.	623.5					
H18C8O	5-methylheptan-3-ol	130.23	445.	621.2					
H18C8O	2-ethylhexan-1-ol	130.23	457.8	640.2					
H6C2O2	ethylene glycol	62.07	470.49	790.					
H10C5O	cyclopentanol	86.13	413.57	619.5	4.90				0.437
H12C6O	cyclohexanol	100.16	433.99	650.0	4.26				0.387
	ethers								
H6C2O	dimethyl ether	46.07	248.3	400.0	5.37	0.242	190	0.307	0.196
H8C3O	ethyl methyl ether	60.10	280.6	437.8	4.40	0.272	221	0.267	0.241
H10C4O	diethyl ether	74.12	307.58	466.74	3.638	0.265	280	0.262	0.279
H10C4O	methyl propyl ether	74.12	311.72	476.25	3.801				0.268
H10C4O	isopropyl methyl ether	74.12	303.92	464.48	3.762				0.264
H12C5O	butyl methyl ether	88.15	343.31	512.78	3.371	0.268	329	0.260	0.314
H12C5O	tert-butyl methyl ether	88.15	328.30	497.1	3.430				0.267
H12C5O	ethyl propyl ether	88.15	336.36	500.23	3.370	0.260	339	0.275	0.331
H14C6O	methyl pentyl ether	102.18	372.	546.53	3.042	0.261	392	0.262	0.346
H14C6O	dipropyl ether	102.18	363.23	530.60	3.028				0.368
H14C6O	diisopropyl ether	102.18	341.66	500.32	2.832	0.265	386	0.263	0.330
H18C8O	dibutyl ether	130.23	413.43	584.1	3.01				0.532
H18C8O	di-tert-butyl ether	130.23	380.38	550.					0.283
H8C3O2	dimethoxymethane	76.10	315.	480.6	3.95	0.357	213	0.211	
H10C4O2	1,2-dimethoxyethane	90.12	358.	536.	3.87	0.333	271	0.235	0.356
H14C6O2	1,1-diethoxyethane	118.18	375.4	527.					

H8C4O	ethyl vinyl ether	72.11	308.7	475.	4.07				0.265
H10C5O	allyl ethyl ether	86.13	340.8	518.					

aldehydes and ketones

H4C2O	acetaldehyde	44.05	294.	461.					
H6C3O	acetone	58.08	329.2	508.10	4.700	0.278	209	0.232	0.301
H8C4O	butan-2-one	72.11	352.74	536.78	4.207	0.270	267	0.252	0.318
H10C5O	pentan-2-one	86.13	375.41	561.08	3.694	0.286	301	0.238	0.345
H10C5O	pentan-3-one	86.13	375.11	561.46	3.729	0.256	336	0.269	0.342
H10C5O	3-methylbutan-2-one	86.13	367.48	553.4	3.85	0.278	310	0.259	0.329
H12C6O	hexan-2-one	100.16	400.7	587.0	3.32				0.391
H12C6O	hexan-3-one	100.16	396.6	582.82	3.320				0.377
H12C6O	4-methylpentan-2-one	100.16	389.6	571.	3.27				0.384
H14C7O	heptan-2-one	114.19	424.2	611.5	3.436				0.483
H18C9O	nonan-5-one	142.24	461.6	640.					
H8C5O	cyclopentanone	84.12	403.72	624.5	4.60				0.285
H10C6O	cyclohexanone	98.15	428.58	653.0	4.00				0.296

anhydrides, acids and esters

H6C4O3	acetic anhydride	102.09	412.7	606.	4.0				0.453
H4C2O2	acetic acid	60.05	391.05	592.71	5.786	0.351	171	0.201	0.444
H6C3O2	propanoic acid	74.08	414.3	604.	4.53	0.334	222	0.200	0.538
H8C4O2	butanoic acid	88.11	436.9	624.	4.03	0.304	290	0.225	0.601
H8C4O2	2-methylpropanoic acid	88.11	427.6	605.	3.70	0.302	292	0.215	0.667
H10C5O2	pentanoic (valeric) acid	102.13	459.3	643.	3.58				0.648
H10C5O2	3-methylbutanoic (isovaleric) acid	102.13	449.7	629.	3.40				
H12C6O2	hexanoic (caproic) acid	116.16	478.4	662.	3.20				0.687
H14C7O2	heptanoic (oenanthic) acid	130.19	495.4	679.	2.90				0.701
H16C8O2	octanoic (caprylic) acid	144.21	512.0	695.	2.64				0.718
H18C9O2	nonanoic (pelargonic) acid	158.24	527.7	711.	2.40				0.719
H20C10O2	decanoic (capric) acid	172.27	541.9	726.	2.23				0.718
H4C2O2	methyl formate	60.05	304.85	487.2	5.998	0.349	172	0.255	0.253

Formula	Name	M	T_b / K	T_c / K	p_c / MPa	l_c / g cm⁻³	V_c / cm³ mol⁻¹	Z_c	ω
H6C3O2	ethyl formate	74.08	327.5	508.5	4.74	0.323	229	0.257	0.282
H8C4O2	propyl formate	88.11	354.05	538.0	4.06	0.309	285	0.259	0.311
H10C5O2	isobutyl formate	102.13	371.4	551.	3.88	0.290	352	0.298	0.395
H12C6O2	pentyl formate	116.16	403.6	576.	3.46				0.538
H12C6O2	isopentyl formate	116.16	396.7	578.					
H6C3O2	methyl acetate	74.08	330.4	506.8	4.69	0.325	228	0.254	0.324
H8C4O2	ethyl acetate	88.11	350.30	523.2	3.83	0.308	286	0.252	0.361
H10C5O2	propyl acetate	102.13	374.70	549.4	3.33	0.296	345	0.252	0.389
H12C6O2	n-butyl acetate	116.16	399.25	579.					
H12C6O2	isobutyl acetate	116.16	389.7	561.	3.16				0.455
H14C7O2	isopentyl acetate	130.19	415.7	599.					
H8C4O2	methyl propanoate	88.11	352.80	530.6	4.004	0.312	282	0.256	0.348
H10C5O2	ethyl propanoate	102.13	372.2	546.0	3.362	0.296	345	0.256	0.390
H12C6O2	propyl propanoate	116.16	395.8	578.					
H14C7O2	isobutyl propanoate	130.19	410.0	592.					
H16C8O2	isopentyl propanoate	144.21	433.4	611.					
H10C5O2	methyl butanoate	102.13	375.9	554.4	3.47	0.300	340	0.257	0.379
H10C5O2	methyl 2-methylpropanoate	102.13	365.5	540.8	3.43	0.301	339	0.259	0.361
H12C6O2	ethyl butanoate	116.16	394.7	566.	3.06	0.276	421	0.274	0.461
H12C6O2	ethyl 2-methylpropanoate	116.16	383.2	553.	3.07	0.276	421	0.281	0.430
H14C7O2	propyl butanoate	130.19	416.2	600.					
H14C7O2	propyl 2-methylpropanoate	130.19	408.6	589.					
H16C8O2	isobutyl butanoate	144.21	430.1	611.					
H16C8O2	isobutyl 2-methylpropanoate	144.21	421.80	602.					
H18C9O2	isopentyl butanoate	158.24	452.	619.					
H12C6O2	methyl pentanoate	116.16	400.5	567.	3.19				0.547
H14C7O2	ethyl pentanoate	130.19	419.2	570.					
H14C7O2	ethyl 3-methylbutanoate	130.19	408.2	588.					

Code	Compound	MW					
H16C802	propyl 3-methylbutanoate	144.21	429.1	609.			
H18C902	isobutyl 3-methylbutanoate	158.24	441.7	621.			
H20C1002	ethyl octanoate	172.27	481.7	659.			
H22C1102	ethyl nonanoate	186.30	500.2	674.			
H26C1302	methyl laurate	214.35	540.	712.			
H10C602	ethyl but-2-enoate (crotonate)	114.14	411.	599.			
H6C404	dimethyl oxalate	118.09	436.5	628.	3.98		0.555
H14C804	ethyl succinate	174.20	490.90	663.			
H8C803	methyl salicylate	152.15	496.1	709.			

aromatic oxygen compounds

Code	Compound	MW					
H6C60	phenol	94.11	455.02	694.2	6.13		0.435
H8C70	o-cresol	108.14	464.19	697.6	5.01	309	0.431
H8C70	m-cresol	108.14	475.42	705.8	4.56	0.240	0.452
H8C70	p-cresol	108.14	475.13	704.6	5.15	0.350	0.503
H10C80	o-ethylphenol	122.17	477.67	703.0			
H10C80	m-ethylphenol	122.17	491.57	718.8			
H10C80	p-ethylphenol	122.17	491.13	716.4			
H10C80	2.3-xylenol	122.17	490.07	722.8			
H10C80	2.4-xylenol	122.17	484.13	707.6			
H10C80	2.5-xylenol	122.17	484.33	706.9			
H10C80	2.6-xylenol	122.17	474.22	701.0			
H10C80	3.4-xylenol	122.17	500.15	729.8			
H10C80	3.5-xylenol	122.17	494.89	715.6			
H14C100	thymol (p-cymen-3-ol)	150.22	505.7	698.			
H20C100	menthol	156.27	489.5	694.			
H8C70	methyl phenyl ether	108.14	426.8	645.6	4.25		0.345
H10C80	ethyl phenyl ether	122.17	442.96	647.	3.42		0.417
H10C120	diphenyl ether	170.21	531.2	766.8			
H6C70	benzaldehyde	106.12	452.2	695.	4.65		0.313

nitrogen compounds

Formula	Name	M	$\dfrac{Tb}{K}$	$\dfrac{Tc}{K}$	$\dfrac{pc}{MPa}$	$\dfrac{lc}{g\ cm^{-3}}$	$\dfrac{Vc}{cm^3\ mol^{-1}}$	Zc	ω
H5CN	methylamine	31.06	266.83	430.7	7.614				0.287
H7C2N	ethylamine	45.08	289.70	456.	5.62	0.248	182	0.270	0.286
H9C3N	propylamine	59.11	321.7	497.0	4.74				0.300
H9C3N	isopropylamine	59.11	305.6	471.8	4.54	0.268	221	0.255	0.288
H11C4N	butylamine	73.14	349.46	531.9	4.20				0.316
H11C4N	s-butylamine	73.14	336.15	514.3	5.0				0.355
H11C4N	tert-butylamine	73.14	316.8	483.9	3.84	0.250	293	0.279	0.272
H7C2N	dimethylamine	45.08	280.03	437.22	5.340				0.299
H11C4N	diethylamine	73.14	328.60	499.99	3.758				0.280
H15C6N	dipropylamine	101.19	382.45	555.8	3.63				0.465
H15C6N	di-isopropylamine	101.19	357.05	523.1	3.02				0.354
H19C8N	dibutylamine	129.25	432.75	607.5	3.11				0.584
H19C8N	di-isobutylamine	129.25	412.84	584.4	3.20	0.233	254	0.288	0.549
H9C3N	trimethylamine	59.11	276.02	432.79	4.087	0.260	389	0.265	0.202
H15C6N	triethylamine	101.19	362.5	535.60	3.032	0.170	271	0.474	0.319
H6CN2	methylhydrazine	46.07	362.	567.	8.24				0.421
H3C2N	ethanenitrile	41.05	354.8	545.5	4.85	0.237	173	0.185	0.326
H5C3N	propanenitrile	55.08	370.29	561.3	4.26	0.240	230	0.209	0.337
H7C4N	butanenitrile	69.11	391.1	585.4	3.88				0.356
H9C5N	pentanenitrile	83.13	414.	610.3	3.58				0.394
H11C6N	hexanenitrile	97.16	436.8	633.8	3.30				0.434
H15C8N	octanenitrile	125.21	478.4	674.4	2.85				0.519
H7C6N	aniline	93.13	457.6	699.	5.31	0.324	287	0.263	0.381
H9C7N	o-toluidine	107.16	473.5	694.	3.75				0.437
H9C7N	m-toluidine	107.16	476.6	709.	4.15				0.408
H9C7N	p-toluidine	107.16	473.7	667.	2.38				0.443

H9C7N	N-methylaniline	107.16	469.40	701.	5.20				0.473
H11C8N	N,N-dimethylaniline	121.18	467.3	687.	3.63				0.410
H11C8N	N-ethylaniline	121.18	476.2	698.					0.484
H13C9N	N,N-dimethyl-o-toluidine	135.21	467.30	668.	3.12				0.360
H5C7N	benzonitrile	103.12	464.3	699.4	4.21				
H7C8N	p-tolunitrile	117.15	490.2	723.					
	heterocyclic oxygen compounds								
H4C2O	epoxyethane (ethylene oxide)	44.05	283.7	469.	7.19	0.314	140	0.259	0.197
H6C3O	1,2-epoxypropane (propylene oxide)	58.08	308.	482.2	4.92	0.312	186	0.229	0.266
H8C4O	tetrahydrofuran	72.11	338.	540.1	5.19	0.322	224	0.259	0.213
H10C5O	2-methyltetrahydrofuran	86.13	351.	537.	3.76	0.322	267	0.225	0.261
H4C4O	furan	68.08	304.6	490.2	5.50	0.312	218	0.295	0.205
H6C5O	2-methylfuran (sylvane)	82.10	338.	527.	4.72	0.333	247	0.266	0.267
H4C5O2	furfural	96.09	434.9	670.	5.89				0.380
H8C5O	dihydropyran	84.12	359.	561.7	4.56	0.314	268	0.262	0.244
H10C5O	tetrahydropyran	86.13	361.	572.2	4.77	0.328	263	0.263	0.214
H8C4O2	dioxan	88.11	374.6	587.	5.21	0.370	238	0.254	0.278
H12C6O3	paraldehyde (2,4,6-trimethyl-s-trioxane)	132.16	397.50	563.					
	heterocyclic nitrogen compounds								
H5C4N	pyrrole	67.09	403.0	639.8	5.61	0.286	249	0.295	0.270
H9C4N	pyrrolidine	71.12	359.64	568.6	5.01	0.333	283	0.268	0.312
H6C5N2	2-methylpyrazine	94.12	410.	634.3					0.248
H11C5N	piperidine	85.15	379.55	594.0	4.65				
H5C5N	pyridine	79.10	388.41	620.0	5.63	0.312	254	0.277	0.239
H7C6N	2-methylpyridine	93.13	402.56	621.	4.60				0.296
H7C6N	3-methylpyridine	93.13	417.29	645.					
H7C6N	4-methylpyridine	93.13	418.51	646.	4.66				0.298

Formula	Name	M	$\dfrac{Tb}{K}$	$\dfrac{Tc}{K}$	$\dfrac{pc}{MPa}$	$\dfrac{lc}{g\ cm^{-3}}$	$\dfrac{Vc}{cm^3\ mol^{-1}}$	Zc	ω
H9C7N	2,3-dimethylpyridine)	107.16	434.41	655.4					
H9C7N	2,4-dimethylpyridine	107.16	431.55	647.					
H9C7N	2,5-dimethylpyridine	107.16	430.16	644.2					
H9C7N	2,6-dimethylpyridine	107.16	417.20	623.8					
H9C7N	3,4-dimethylpyridine	107.16	452.28	683.8					
H9C7N	3,5-dimethylpyridine	107.16	445.1	667.2					
H7C9N	quinoline	129.16	510.78	782.					
H7C9N	isoquinoline	129.16	516.39	803.					
sulphur compounds									
H6C2S	dimethylsulphide	62.13	310.48	503.0	5.53	0.309	201	0.266	0.187
H8C3S	ethylmethylsulphide	76.16	339.80	533.	4.26				0.213
H10C4S	diethylsulphide	90.18	365.25	557.	3.96	0.284	318	0.272	0.289
H22C10S	diisopentylsulphide	174.35	484.15	664.					
H10C6S	diallylsulphide	114.21	411.75	653.					
H10C4S2	diethyldisulphide	122.24	427.13	642.					
H4CS	methanethiol	48.10	279.11	470.0	7.23	0.332	145	0.268	0.148
H6C2S	ethanethiol	62.13	308.20	499.	5.49	0.300	207	0.274	0.187
H12C5S	3-methylbutane-1-thiol	104.21	393.	604.					
H8C4S	tetrahydrothiophene	88.17	394.15	632.0					
H4C4S	thiophene	84.14	357.15	579.4	5.69	0.385	219	0.258	0.191
miscellaneous organic compounds									
C3OF6	perfluoroacetone	166.02	245.7	357.14	2.84	0.505	329	0.314	0.364
HC2O2F3	trifluoroacetic acid	114.02	346.	491.3	3.258	0.559	204	0.163	0.541
C3OF5Cl	chloropentafluoroacetone	182.48	281.0	410.6	2.878				0.346
H2C5O2F6	hexafluoroacetylacetone	208.06	327.3	485.1	2.767				0.272

Formula	Name								
C8OF16	perfluoro-2-butyltetrahydrofuran	416.06	375.8	500.2	1.607	0.707	588	0.227	0.574
HC6OF5	pentafluorophenol	184.07	418.79	609.	4.00	0.529	348	0.275	0.501
H3CNO2	nitromethane	61.04	374.34	588.	5.87	0.352	173	0.208	0.306
H3C3NO	isoxazole	69.06	368.	552.0					
C2NF3	trifluoroacetonitrile	95.02	204.33	311.11	3.618	0.470	202	0.283	0.264
H9BC3O3	methyl borate	103.91	342.	501.7	3.59				0.414
alkylsilanes and stannanes									
H4Si	silane	32.12	161.	269.6	4.84				
H6CSi	methylsilane	46.15	215.6	352.5					0.063
H12C4Si	tetramethylsilane	88.23	299.8	448.64	2.821	0.244	362	0.273	0.241
H20C8Si	tetraethylsilane	144.33	427.9	603.7	2.602				0.473
H20C7Si2	hexamethyldisilmethylene	160.41	407.	573.9	1.99	0.249	644	0.269	0.354
H12C4Sn	tetramethylstannane	178.83	351.	521.8	2.981				0.289
halogenated alkylsilanes									
H5CSiCl	chloromethylsilane	80.59	280.	517.8					
H6C2SiCl2	dichlorodimethylsilane	129.06	343.45	520.4	3.49	0.369	350	0.282	0.270
H9C3SiCl	chlorotrimethylsilane	108.64	333.15	497.8	3.20	0.297	366	0.283	0.294
H3CSiCl3	trichloromethylsilane	149.48	338.8	517.	3.28	0.430	348	0.265	0.222
H5C2SiCl3	trichloroethylsilane	163.51	371.55	560.0	3.33	0.406	403	0.288	0.274
H10C4SiCl2	dichlorodiethylsilane	157.12	403.6	595.8	3.06	0.345	455	0.281	0.326
siloxanes									
H18C6OSi2	hexamethyldisiloxane	162.38	372.7	516.6	1.910	0.258	629	0.280	0.421
H30C12OSi2	hexaethyldisiloxane	246.54	504.	693.					
H24C8O2Si3	octamethyltrisiloxane	236.54	425.	562.9	1.420	0.261	906	0.275	0.536
H30C10O3Si4	decamethyltetrasiloxane	310.69	465.	599.4	1.15	0.257	1209	0.279	0.595
H50C20O3Si4	decaethyltetrasiloxane	450.96	598.	788.					
H36C12O4Si5	dodecamethylpentasiloxane	384.85	502.	629.0	0.95	0.255	1509	0.273	0.686

Formula	Name	M	T_b / K	T_c / K	p_c / MPa	l_c / g cm^{-3}	V_c / cm^3 mol^{-1}	Z_c	ω
$H_{60}C_{24}O_4Si_5$	dodecaethylpentasiloxane	553.17	633.	823.					
$H_{42}C_{14}O_5Si_6$	tetradecamethylhexasiloxane	459.00	536.	653.2	0.80	0.254	1807	0.268	0.819
$H_{48}C_{16}O_6Si_7$	hexadecamethylheptasiloxane	533.16	561.	671.8	0.68	0.250	2133	0.258	0.851
$H_{54}C_{18}O_7Si_8$	octadecamethyloctasiloxane	607.31	584.	688.9	0.62	0.246	2469	0.269	0.957
$H_{30}C_{12}O_3Si_3$	hexaethylcyclotrisiloxane	306.63	520.	673.					
$H_{24}C_8O_4Si_4$	octamethylcyclotetrasiloxane	296.62	448.	586.5	1.34	0.301	985	0.271	0.579
$H_{40}C_{16}O_4Si_4$	octaethylcyclotetrasiloxane	408.84	569.	698.					
$H_{30}C_{10}O_5Si_5$	decamethylcyclopentasiloxane	370.78	483.	617.4	1.035	0.288	1287	0.260	0.584
$H_{12}C_4O_4Si$	tetramethoxysilane	152.22	394.	562.8	2.873	0.342	445	0.273	0.455
$H_{20}C_8O_4Si$	tetraethoxysilane	208.33	442.	592.2	2.045	0.320	651	0.270	0.668
$H_{28}C_{12}O_4Si$	tetrapropoxysilane	264.44	502.	647.3	1.696	0.310	853	0.269	0.857
Mixtures									
	air (729)	28.96	78.78	132.55	3.796	0.313	93	0.319	-0.007
	dichlorodifluoromethane/difluoroethane azeotrope (500)	99.10	240.	378.7	4.426	0.497	199	0.280	0.205
	chlorodifluoromethane/chloropentafluoroethane azeotrope (502)	111.60	228.	363.	4.27	0.559	200	0.282	0.167
	chlorotrifluoromethane/trifluoromethane azeotrope (503)	87.24	185.3	292.6	4.357	0.564	155	0.277	0.198
	difluoromethane/pentafluoroethane azeotrope (504)	73.64	216.	340.	4.76	0.494	149	0.251	0.235
	chlorofluoromethane/1,2-dichlorotetrafluoroethane azeotrope	93.69	260.71	415.2	5.16	0.539	174	0.260	0.221
	perfluorocyclobutane/chlorotetrafluoroethane azeotrope	151.09	260.	380.	2.94	0.540	280	0.260	0.354
	chloropentafluoroethane/propane azeotrope	86.20	226.6	353.9	3.58				0.173
	hydrogen sulphide/ethane azeotrope	30.87	182.	309.5	5.34				0.051
	hydrogen sulphide/propane azeotrope	34.70	210.	368.5	8.26				0.077